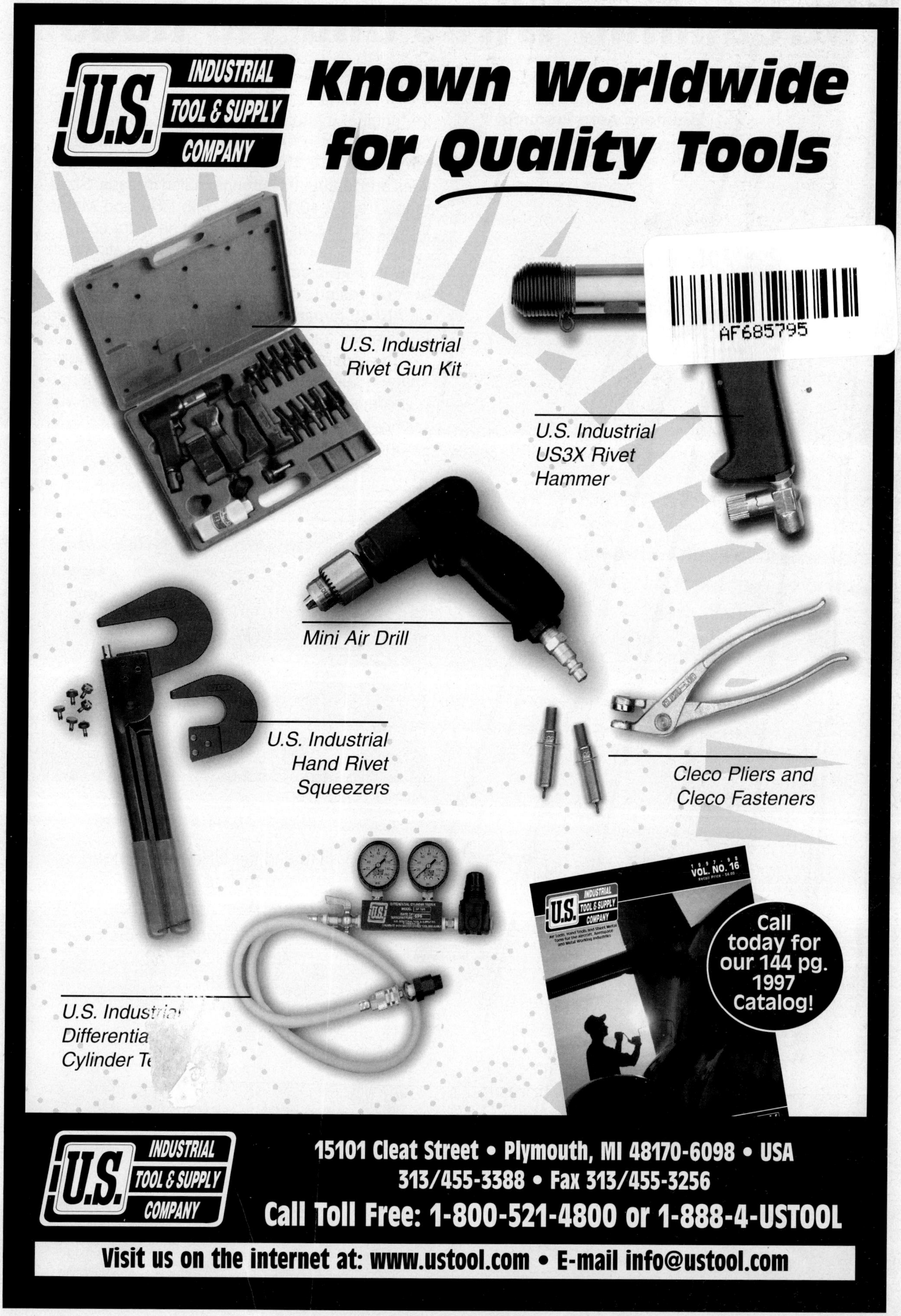
U.S. INDUSTRIAL TOOL & SUPPLY COMPANY
Known Worldwide for Quality Tools
U.S. Industrial Rivet Gun Kit
U.S. Industrial US3X Rivet Hammer
Mini Air Drill
U.S. Industrial Hand Rivet Squeezers
Cleco Pliers and Cleco Fasteners
U.S. Industri
Differentia
Cylinder Te
VOL. NO. 16
Call today for our 144 pg. 1997 Catalog!
U.S. INDUSTRIAL TOOL & SUPPLY COMPANY
15101 Cleat Street • Plymouth, MI 48170-6098 • USA
313/455-3388 • Fax 313/455-3256
Call Toll Free: 1-800-521-4800 or 1-888-4-USTOOL
Visit us on the internet at: www.ustool.com • E-mail info@ustool.com

Streamline Wires and Tie Rods

Manufactured By
Bruntons Aero Products
Musselburgh, Scotland

Streamline and Round aircraft tie rods from the original manufacturer with over 75 years experience. These high quality wires are manufactured from 316 stainless steel to AN specifications and feature the stronger rolled threads. Sizes range from 6-40 to 3/4-16 with BSF and Metric thread profiles also available at no extra cost.

Finish is normally high polish with satin finish available at lower cost for absolute realism on antique restorations.

AN665 terminal assemblies are also supplied in stainless for total corrosion resistance.

Bruntons Aero Products maintains strict aerospace quality standards and all wires and terminals are proof loaded. Shipments are supplied with quality certifications with traceability to material lot and date of manufacture.

Amateur Built: SKYBOLT

Available in the USA and Canada
exclusively from
Steen Aero Lab, Inc.
1210 Airport Road
Marion, NC 28752
704-652-7382
(Phone/Fax)

Amateur Built: PITTS S1C

Replica: WEDELL WILLIAMS

Certified: WACO YMF-5

AeroCrafter™
HOMEBUILT AIRCRAFT SOURCEBOOK

CONTENTS

AeroCrafter Fifth Edition ISBN 0-9636409-4-1 is published by BAI Communications Inc. • PO Box 1516 • Benicia CA 94510 (707) 747-1509 • Fax (707) 747-1534 • email: aerocrafter@baicorp.com

Web Site: www.baicorp.com/aerocrafter

INQUIRIES/CONTRIBUTIONS: Address all editorial communications to Editor, ***AeroCrafter*, PO Box 1516, Benicia, CA 94510.** In Europe contact: **Lucio Turchet, Via Galilei 15/12, 31044 Montebelluna, TV, Italia.** We are happy to consider contributions in the form of manuscripts, drawings, and photographs. All material must be identified with the sender's name and address. Include sufficient postage for return of materials if they are unsuited to our use. ***AeroCrafter*** is not responsible for loss or damage to unsolicited material.

The Spirit is Alive

I was pleasantly surprised at the number of new aircraft kits, plans, and related suppliers and services that appear in this Fifth Edition of AeroCrafter. Not only does the number reflect the growth of the industry, it also, and perhaps more importantly, confirms the well-being of the "spirit" of experimental aviation.

That spirit, born of self reliance, determination, and ingenuity, which inspires engineers and designers to push the envelope and prompts ordinary folks to build real airplanes in basements and garages, is responsible for many, if not most, of the advances and innovations in general aviation today.

Another aspect of the number of aircraft available is the incredible array of choices presented to the potential builder. While somewhat daunting, making that choice can be an educational and enjoyable endeavor.

One of the time-tested designs may be just the ticket for you. And, no matter what type of aircraft best fills your needs, there are plenty of proven designs available. But, if your tastes and abilities run toward more cutting edge technology, there is no shortage of exciting developments.

Check out the "In Development" Section in this Edition. To be sure, some of these aircraft may not make it into production for one reason or another. But, a number will be successful and the dynamic evolution of Experimental Aviation will continue.

An open cockpit ultralight might be the most appropriate aircraft for your time, budget, etc., but look at something like the *Viperfan*. It's got to stir your spirit a little, and that's a good thing. It's what keeps this business so interesting.

Cover Aircraft

Avid Aircraft
MAGNUM

Cosmos
PHASE II 582 TOP

Stewart 51, Inc.
STEWART S-51D
Photo by Xavier Meal

Rotary Air Force
RAF 2000

Glassic Composites
SQ 2000

Hy-Tek Hurricane
CLIP WING HP

Using the AeroCrafter Sourcebook

- Kits and Plans are listed alphabetically by Company Name within the major categories. **An Aircraft Name Index is located on page 418. A Company Index is located on page 422.**
- All information has been provided by the manufacturers and other sources. Listing of a company in AeroCrafter does not constitute the publisher's endorsement of the quality of the aircraft, plans, product or service offered, nor guarantee the accuracy of any claims or specifications.
- Aircraft Performance Data is dependent upon manufacturer's testing and can vary with engine installation, weight, environmental factors, etc. **Contact the manufacturer for actual specifications**.
- N/A in the Kit and Plans listings indicates that the information was not available at the time of publication.
- Kit, Plan, Supplier, and Service information is current at the time of publication but, as with all directories, addresses, phone numbers, etc. can change. We appreciate your feedback regarding such changes.
- When contacting a company, please be sure to say you found them in AeroCrafter. If you know of a company that belongs in AeroCrafter, drop us a note at the address below.
- Your comments on AeroCrafter and experiences with the companies listed are welcome. Send us email at AEROCFTR@BAICORP.COM or contact us at PO Box 1516 Benicia CA 94510 • 707-747-1509.

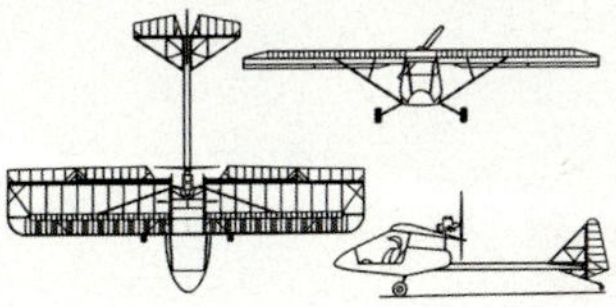

Lowell Metz's Kolb Firestar

My son Eric and I had a dream to be able to fly side by side one day, so after the FP 202 Koala was in the air we purchased a Kolb Firestar. After 602 hours of construction and painting we realized our dream and flew formation to our first fly-in, October 5th, 1996.

The first test flight was quite an event as my son unexpectedly lifted off during a taxi practice while the test pilot and I watched in awe. His experience paid off as he immediately gained some altitude, did some stall airspeed checks and brought the Kolb in for a picture perfect three point landing.

Everything about the Kolb is top notch, from company support, detailed plans, ease of construction, and delightful to fly. The paint scheme (all Poly Fiber) was the greatest challenge as the pattern was airbrushed free hand.

Lowell S. Metz
Adrian, MI

GILBERT GEISZ'S GIZMO

The 'Bullet' is my own original design and was intended to look like a 1930's type racer. I began the design drawing in the spring of 1991, started cutting metal in April 1993 and our first flight was on April 1, 1996. This may seem like a fast build, but I designed the aircraft around an existing set of wings and an existing engine mount. I have built wings and engine mounts before and decided I've had enough of that.

The wings are from a Navy U8-D and are very close to the outboard panels of a Twin Bonanza. The engine mount is Beech as well from a C-45. All the rest was fabricated from my plans. The engine is a Pratt & Whitney R-985-14B and it spins a 22D30 Hamilton Standard prop.

Flying characteristics of 'Gizmo' are excellent. It is light on the controls (nimble) but has a very strong feel. Stalls (both power on and power off) are straight forward and it doesn't seem to have any bad flight characteristics. It is however difficult to land because of visibility problems and twitchiness on the ground

Stall: 85 knots clean
78 knots full flaps
Initial rate of climb: 3300 fpm
Best cruise: 170 KIAS @ 18gph
Top speed (so far): 212 Kts @ 90%

Gilbert A. Geisz
721 Glenhaven Ct
Boulder, CO 80303

Built from Scratch

The Plans-Only Option

by Jamie Beckett

If you spend much time in the world of experimental aircraft you'll soon become familiar with the famous 51% rule. In a nutshell, the rule says that to be accepted as an amateur-built aircraft, where the owner/builder is awarded the repairman's certificate, the builder has to perform 51% of the actual building of the aircraft. This assures the FAA that the builder has enough time and effort invested in the project to have gained a significant understanding of the aircraft from the ground up.

Many manufacturers have had their kits approved by the FAA as conforming with the requirements of the rule. By going that extra step, they can assure their buyers that there will be no unexpected problems at the end of the project concerning whether or not the airplane was at least 51% away from completion when it came out of the delivery crate.

But some builders have no interest in percentages and insist on doing it all. A case in point is Jerry

Cornwell, of central Florida. He built his aircraft from scratch, starting with nothing but a set of plans and a strong desire to turn out one beautiful machine.

Over the course of the project Jerry enlisted the help of a few friends. Some with special skills, others just folks with a bit of experience and free time. In either case Jerry tapped into one of the unique resources of experimental aviation, a very talented and social group of people.

Jerry built a Boredom Fighter, a relatively modern creation reminiscent of, but not actually modeled after, any particular World War I aircraft. It is a single seat biplane, traditionally built of wood and fabric, and usually powered by one of the "modern" aircraft engines. Jerry did things a little differently.

Gigi (named for the builder's wife) is constructed of steel tubing, not wood. And her powerplant, is a Subaru engine that Jerry and his buddies converted to aviation use themselves.

Every aspect of the aircraft is meticulously finished, from the instrument panel, which is basic but tidy; to the replica machine gun based on drawings obtained from the Smithsonian Institution; to the paint and artwork on the fuselage of the aircraft. The finished product is a winner and *Gigi* gets noticed wherever she goes. Heads are poked into her cockpit and cowling and under her belly looking for hidden details or technical innovations. Grown men get down on their hands and knees to inspect Jerry's handiwork.

The project took some three and a half years. But no record was kept on the actual number of hours required to build *Gigi*. And no record was kept of how much money was spent aside from the price of the plans themselves ($125). Jerry says he doesn't know, and that he doesn't want to know. His quest was to build the airplane, to make it absolutely exquisite, and to fly it here and there at whim. And so he did.

William Postula's Eagle

Construction of my Eagle N106JP began on September 10, 1991. First flight was August 23, 1996. It has met or exceeded all performance criteria. Maneuvers performed have included loops to flat spins pulling 6G's and pushing to -5. King radio and transponder, internal antennas in the upper wing, and a high performance package by Dick Demars Aero AEIO-360AID were some of the modifications made. Empty weight came in at 1089 lbs. with a gross weight of 1600 lbs. Cruise speed is 165 mph at 25" manifold pressure and 2500 rpm. Factory support and quality of the kit were frustrating at times, but the factory is now under new management. The Eagle won Best Biplane at Sun 'N Fun 1997.

William J. Postula
9825 Woodlawn Drive
Portage MI 49002
EAA 356619

Rules and Regulations

Going by the Book

by Ron Alexander

What are the rules pertaining to building an airplane? Are there controls concerning what can be built and how to build it? Do I need a pilot's license to fly an experimental airplane? These questions and many others are common to the person interested in building an airplane. Often the person does not know where to obtain answers to these questions. With this in mind, what follows is a discussion of the regulations involved in aircraft building.

As you might imagine, the Federal government does regulate this industry. However, the regulations are not too restrictive. To understand this, let's first take a look at the requirements to manufacture an airplane that apply to companies such as Piper and Cessna. In order to build an aircraft that will be manufactured in quantity and sold to the general public, a company or individual is subject to Federal Aviation Regulation Part 23. This FAR outlines in detail the restrictions of design as they apply to weight, speed, etc. The manufacturer must comply with FAR Part 23 in order to obtain authorization to construct and sell the aircraft. A person desiring to build an amateur-built aircraft does not have to meet any restrictions in FAR part 23. They simply do not apply because as an amateur builder we are constructing one airplane for our own personal use that will not be sold to the general public. Experimental aircraft must have an airworthiness certificate however, just like production aircraft, before they can legally fly. The airworthiness certificate is issued by the Federal Aviation Administration. There are two categories of airworthiness certificates as defined by FAR 21.175. These categories are termed *standard and special.* We are interested in the *special* category which is further broken down into several additional categories, one of which is *experimental.* Experimental airworthiness certificates are issued for different purposes. One of the purposes is for operation of an amateur-built airplane. This is the primary purpose that will be discussed.

The majority of custom built airplanes fall under the experimental category for amateur-built purposes. FAR 21.191(g) defines this purpose as "operating an aircraft the major portion of which has been fabricated and assembled by persons who undertook the project solely for their own education or recreation." This regulation is the essence of custom aircraft building. The intent of the classification is very clear. One or more persons may build an airplane and obtain an airworthiness certifi-

CONTINUED

cate under the experimental category. However, this aircraft must be built solely for their education or recreation; absolutely no commercial purposes whatsoever. So, you cannot have someone build your airplane and then obtain an experimental airworthiness certificate. This can only be accomplished by a company that has FAA authorization to mass produce a particular airplane.

Another class of airplanes that fall under a different set of rules is called "Ultra-light" aircraft. If a completed airplane is smaller in design and meets certain restrictions as defined under FAR 103.1, that airplane does not require an airworthiness certificate. In order to qualify as an ultralight the airplane must not weigh more than 254 pounds, not carry more than 5 gallons of fuel, and not be capable of more than 55 knots of speed in level flight. In addition, the aircraft must not be built to carry more than one person. The operator of an ultralight is not required to be the holder of a pilot's license. Even though a license is not necessary we certainly recommend that individuals building ultralight aircraft obtain flight instruction prior to their first flight. Meeting the requirements for solo in a certified aircraft should be the minimum training received before flying an ultralight.

A pilot flying an experimental aircraft must be the holder of a pilot's certificate. Of course, if you are carrying passengers you must have at least a Private Pilot certificate. To continue our discussion of FAR 21.191(g) as it applies to building an airplane, it is clear that we must build at least 51% of the aircraft. This is often referred to as the major portion rule. Recall in the definition that it is stated, "an aircraft the *major portion* of which has been fabricated and assembled by persons who undertook the project solely for their own education or recreation." Major portion means that if you purchase an aircraft kit you must build and assemble a minimum of 51% of

the completed aircraft. You cannot hire someone else to build it for you. You are building only for your own education and recreation. When you complete the aircraft and apply for the airworthiness certificate, you will be required to sign a legal document stating that you did build 51% or more of the airplane. The FAA inspector who will inspect your airplane and issue the certificate of airworthiness will want to see documentation in the form of a builder's log to prove you have built 51% or more of the aircraft. This log is nothing more than a history of the building process showing pictures of construction, work that was accomplished during each time period, receipts for materials, inspections that you or other knowledgeable people performed, samples of work, etc. It is important that this log be maintained properly to insure compliance with the major portion rule. The builder's log also will provide you with a history of construction that you will find interesting in years to come.

Notice that it is acceptable to have more than one person build an airplane as long as it is for education and recreation. There is no limit to the number of people who can participate as long as no one receives compensation. You are allowed however, to hire people to perform certain tasks. Examples of this are painting of the airplane, engine installation, upholstery work, avionics installation, etc. A listing of these items is contained in FAA Advisory Circular 20-139. This circular explains in detail what can be legally hired and what you must do. The 51% rule is based upon the parts of the airplane and not time. The FAA maintains a listing of aircraft kits they have evaluated for the major portion rule. If a kit appears on this listing the FAA has determined that no more than 49% of the aircraft is being fabricated by the kit manufacturer. As a word of caution, if a kit appears on this listing the FAA is in no way endorsing the kit as being safe or of high quality.

After making the decision to build an airplane you should contact your local FAA Regional Office and request their package of information regarding amateur-built aircraft. If you attend Oshkosh or Sun 'N Fun, the FAA passes out all Advisory Circulars bound together in a handy book form. You can locate these at the FAA display. It is important that you obtain Advisory Circular 20 27D. This circular lists most of the information needed to begin your project. From the very beginning you should establish a relationship with the FAA inspector you will use to inspect your project. Begin your builder's log and find someone to

CONTINUED ON PAGE 20

George and Becki Orndorff's 'Yellar' RV-6A

This 'yellar' RV-6A is the second RV we have completed. With both of our RV-6A's, we have been very pleased with the quality of the kits and the factory support. We found the kits straight forward in assembly and the experience and support from other RV builders always easily offered. The airplane performance matches the numbers advertised by Van's Aircraft.

In building RV's we have found ourselves members of a large and friendly family of other RV builders. We have also found other RV builders to be a very caring and sharing group. Help has always been willingly offered, whether it be for airplane building, flying, or other personal needs.

The yellow RV-6A in this picture is probably one of the better known RVs. It is the plane constructed during the creation of our series of videos detailing RV construction available through Vans Aircraft. The first flight is documented on the "Aircraft Systems" video.

George and Becki Orndorff

FK 12 Comet

B&F Technik Launches A New Biplane

Peter Funk of B&F Technik developed the very successful FK9 as a derivative of the FK 1 through 8 series of aircraft which had been designed by his father, Otto Funk. About one hundred FK9's have been built to date.

Having always been inspired by the classic biplanes, Peter Funk recently turned his talents to the idea of a new sport biplane. But since the realities of the marketplace don't always coincide with a designer's particular interests, B&F undertook a market research program to test the feasibility of the concept from a business standpoint.

A widespread market of potential buyers was presented with the essential criteria for such a plane along with the disadvantages of a biplane design. From this research it was found that the following aspects of the FK12 project needed to be addressed:

In spite of the configuration, it should provide good performance with low fuel consumption.

The design should allow for the plane to be flown with either an open cockpit or a closed canopy to permit year 'round flying.

Stowing a biplane in a hangar often necessitates the removal of nearby high or low wing aircraft thus requiring a larger (more expensive) bay. The aircraft should have folding wings which would accommodate a small hangar space or even trailer stowage.

At the end of 1994 the project started on the drawing board. The demand for good flight performance and relatively compact design gave it some similarity to a Christen Eagle. The wing concept was based on the FK9 with a Wortmann Laminar pro-

file. The upper and lower wings have flaperons, positioned negative during cruise flight. For landing, the flaps can be in a locked position, giving a 10 degree gradient.

Drag was reduced as much as possible. The aircraft presents a very clean appearance with a single flying wire per wing, no exposed torsion cables on the empennage and low profile carbon spring gear.

To increase in-flight comfort, an easily installed, lockable canopy was developed in which the front windshields remain attached to the fuselage. The cockpit section of the fuselage can be flipped open to get into the plane. This allows for only small head openings to remain, and the windshields can then go far back to protect the pilots.

The Rotax 912 was selected as the powerplant. Dimensions and tolerances have been calculated to accommodate the future installation of a turbo engine (and the higher speed).

At the beginning of 1995 the initial production drawings were completed and at the AERO 95 airshow a 1:1 scale model by B&F was shown. The large interest that it evoked led to almost immediate construction of the first FK12 prototype, which had the code name, *Comet*.

The anticipated completion date of December '95 soon became unrealistic, due to the large demand for the new FK9 model and the constraints it imposed on production capacity. In 1996, the production force was greatly enlarged and a team solely responsible for the *Comet* was organized. By the end of the year the brickwork was finished and all of the load/resistance tests had been completed.

After the maiden flight of the first prototype at Speyer in May, 1997, the first series of five Comets went under construction to be delivered as kit planes.

The fuselage is similar to the FK9 in the frontal area with a steel tube frame. The rudder is made from aluminum rods. One reason for the clean aerodynamic quality of the fuselage is that a large proportion of it is covered with fiberglass sandwich parts. The wings have a CFK leading edge and are covered from the main spar to the aluminum rear spar. The flaps are also made out of aluminum and covered. The stabilizer sits within a box spar of aluminum which is able to cope with the torsion placed on it. The nose is aluminum and, as in the wings, there are insulating ribs which give a greater stability and strength. The main landing gear is a carbon spring, and the wheels are faired. The tail skid is in its own box compartment, flexibly positioned and tucked away in the fuselage. The range of the tailwheel when unlocked is 360 degrees.

When the prototype, D-MPLI, had completed a basic program of 70 flight hours, the plane had shown itself to be extremely nimble in all aspects. A roll change of 45 degrees to 45 degrees could be achieved in just over a second. The *Comet* is truly unique and the view over the long nose is relatively good which is a surprise for the experienced pilot. The plane can easily show about 185 km/h in level flight. Certification in the European ultralight category should be finished this year.

assist you with inspections. The FAA only requires one inspection for certification. That inspection is accomplished when you have completed the project and it is ready for flight. It is advisable to find another builder, an EAA Technical Counselor, or someone with building experience to regularly inspect your airplane. These inspections are termed "in-process" inspections and should be documented in the builder's log.

The FAA or a designated repre-

sentative of the FAA called a Designated Airworthiness Representative (DAR) should be contacted when your airplane is ready for final inspection. All necessary placards, markings, N numbers, etc. must be in place. The FAA then inspects the airplane and if it is found to be airworthy they will issue an airworthiness certificate. In addition, they will also issue you operating limitations which are rules you must follow when flying the airplane. These operating limitations are issued for two phases of operation, Phase One which pertains to the test flying portion of flight, and Phase Two limitations which go into effect upon successful completion of the test flying phase. Operating limitations define where you can fly, number of flight test hours required, limit commercial activity, and in general spell out what you can and cannot do with the airplane. Of course, all of the general operating rules for aircraft found in FAR part 91 apply to amateur-built airplanes. FAR 91.319 is directed only toward experimental airplanes. A copy of the operating limitations must be carried on board the aircraft during flight. Experimental aircraft are also required to carry a copy of the weight and balance data on board the airplane.

A few of the regulations regarding normal operation of amateur-built aircraft follow. First of all, you may not carry passengers or property for hire. Secondly, any major changes that are made to the airplane as defined by FAR 21.193 require inspection by the FAA prior to further flight. Thirdly, you may not operate your aircraft unless it has received a condition inspection annually. You must advise any passengers that you carry that the airplane is experimental. These are just a few regulations that apply to amateur-built aircraft. You will find them in FAR 91.319 and also in the specific operating limitations for the airplane.

Rules that relate to maintaining your amateur-built airplane are fairly simple. You can maintain your own airplane including the engine. Each year the airplane must be inspected by a FAA licensed mechanic or you may do the inspection if you built the airplane. In order to do the condition inspection, you must obtain from the FAA a repairman's certificate. AC 65-23A defines the requirements for this certificate. Basically, you must be the primary builder of the airplane and be able to demonstrate to the FAA that you can adequately inspect the airplane. It is noteworthy that the primary builder must be one person. If a group of people build the airplane one of the

group must be designated as the primary builder. This person may then apply for the repairman's certificate. Normal maintenance may be performed by anyone. Common sense should rule. Do not overhaul your engine if you are not qualified or perform maintenance on items with inadequate knowledge.

A listing of Federal Aviation Regulations pertaining to amateur-built aircraft follows:

FAR 21.191 - Basic definition of Amateur-built
21.175 - Airworthiness classifications
21.193 - Licensing information
45. 22 - N number information
91.305 - Flight testing area
91.319 - Operating rules

There are a number of other regulations that apply, but these are the major ones. Advisory Circular 20-27D provides more information on this subject. If you are contemplating building an airplane obtain as much information as possible prior to beginning. The more prepared you are from the start the better your chances of successfully completing your project. The FAA is not overly restrictive regarding amateur-built airplanes. They have left it up to the industry to police itself. We must be cognizant of the rules and abide by them to insure the FAA does not have to impart additional regulations.

Ron Alexander conducts the SPORTAIR/EAA Workshops. For more information contact Alexander SportAir Center at 800-967-5746 or visit www.sportair.com

Larry Maynard's Titan Tornado

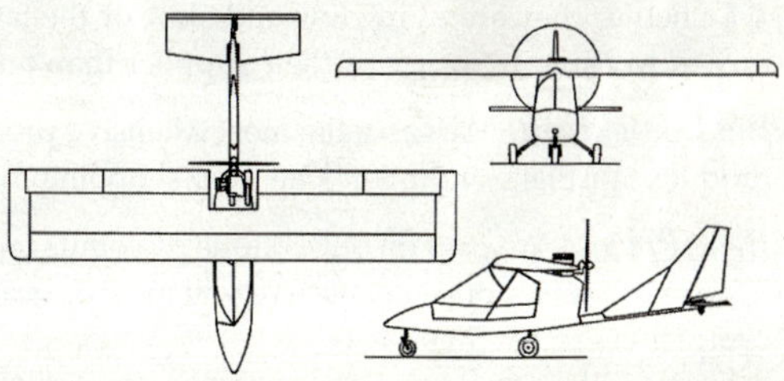

It took about 450 hours to complete my Titan Tornado II. Because of its design, and fighter-like cockpit, it draws attention wherever I go. It's powered by an 80 hp Rotax 912 engine. It cruises at 110 at 75% power and climbs at 1500 fpm, single pilot. With two, it cruises at 100 and climbs at 1000 fpm. It stalls at 45 clean and 40 with flaps. With the 15 gallon tank, I can go 400 miles with a thirty minute reserve burning mogas or avgas at 3.5 gallons per hour.

This airplane is fairly easy to build, the factory gives great support, and it's very well designed. The builder's manual is very detailed and one of the best on the market. The airplane has several options and has improved over the years.

For an enclosed design, it has the best visibility on the market. It's a blast to fly and I enjoy the panoramic view from the front seat.

Larry D. Maynard
Alpharetta GA

PROVEN PERFORMANCE
The GLASS GOOSE
The two-seater amphibian that lives up to its claims!
NOW IN FULL PRODUCTION
Quikkit Corp.
9002 Summer Glen
Dallas, TX 75243
Phone and Fax
(214) 349-0462
e-mail: quikkit@airmail.net
Info & VHS video $25.00
Foreign $27.50, PAL $35.00 Visa/MC
Web Site: http://webwrks.com/glassgoose/
Fast, comfortable, easy to build... no better value in a homebuilt.

PAY FUEL HERE
HOURS:
7:00AM
8:00PM

WELCOME TO THE FUTURE

...meet the SKY ARROW!

Discover today the features of tomorrow's sportplanes. The Italian designed and manufactured Sky Arrow incorporates modern materials and technology, to achieve unsurpassed performance and functionality. Beautiful, yet practical... the all-new Sky Arrow!

HI-TECH CONSTRUCTION

"3 I - Meteor S.p.A.", an Italian aircraft manufacturer established in 1947, proudly presents an all-new sportplane. The Sky Arrow utilizes carbon fiber sandwich throughout, resulting in a extremely strong yet lightweight aircraft, with outstanding take-off and flight performance and a high useful load. Stress tested to +7.6 and -3.8 g's, the Sky Arrow is a responsive, strong, safe and economical sportplane.

ERGONOMIC CABIN

Tandem comfortable seats with plenty of headroom, dual side-stick controls, a spacious cabin and wrap-around canopy provide the ultimate in comfort and visibility from both positions.
Enjoy a spacious instrument panel, electric flaps and elevator trim, low cockpit noise levels and even plenty of storage for those long trips.

MODERN POWERPLANTS

Choose the proven and economical four-stroke, 80 hp Rotax 912 for 95 kt. cruise on 4 gph, or demand the high performance of the new turbo-charged Rotax 914 for 100 hp up to 15,000 feet.

PRODUCTION OR KIT

As a production aircraft in the Primary Category or JAR-VLA, the Sky Arrow is ideal for flight training, observation, or as a fun personal sportplane. Also available as an easy-build kit, the Sky Arrow can be easily assembled in as little as 500 hours from pre-fabricated parts.

OPTIONS

All composite amphibious floats, variable pitch propeller, glider towing package, wheel pants... the Sky Arrow will meet and exceed your requirements.

The SKY ARROW:
Tomorrow's airplane... Today!

PACIFIC AEROSYSTEM, INC. (PAI)
5760 Chesapeake Ct., San Diego, CA. 92123
Telephone: (619) 571-1441, Fax: (619) 571-0803

Buying A Used Homebuilt

Stalking The Great Deal

by Scott "Sky" Smith

There are over 700 kit or plans-only aircraft available for pilots to build. But if you don't want to build there are only two options. Buy an aircraft already completed or hire a professional builder. If you have been following the FAA and their quest to eliminate the professional builders...the first option may soon be your only choice.

There are almost as many options as there are aircraft designs. Many people feel that you "can't trust someone else's work." They also feel that the advantages of amateur-built aircraft are lost if you don't hold the repairman certificate for that aircraft. Then there is another contingent that says you shouldn't fly an experimental, if you don't build it! Aircraft mechanics are notorious for not wanting to trust anyone else's work. But...evidence shows the demand for completed kits has increased. An indicator of this is in the number of companies that are offering quick build kits. More and more people want an aircraft that is almost complete or ready to fly. New affordable production aircraft are still not readily available.

If you are the type of person who hates to have anything that needs work, needs adjustments, or doesn't have readily available parts, maybe this isn't for you. Some people will never be comfortable in an amateur-built aircraft. Those individuals should stay with a certified aircraft and skip this area completely.

But if you want something special, unique, individualized, and designed with a specific purpose in mind, yet you don't want to wait years for a completed aircraft, check out the used experimental aircraft market.

Where do you start? First analyze your needs and the type of aircraft that fits them. This is the time to be realistic. How many people fly with you regularly? Where do you fly? What airport will you be flying out of? How big are you? Are you willing to take advanced training? What is your financial ability to buy and then maintain an aircraft? The initial cost is the cheap part of the deal. Paying the regular fixed costs is what most people forget about.

> "The initial cost is the cheap part of the deal."

Personal example: I have no patience to build. I wanted an aircraft that is fun and capable of aerobatics. I'm very conscious of the fixed costs that are required. Things like hangar rent and insurance are important factors. Maintenance was an issue. I want to do as much as possible, myself. I also wanted to have fun with the aircraft. Fixing up cosmetics was okay, major repairs would be out. My family was not interested in all of this, so I wasn't worried about hauling passengers. Plus I decided that aerobatics didn't need passengers (I rent an aircraft when we want to go somewhere).

I wanted something that would get me into the basement of the aerobatics market. Low entry cost. Learn, fly, sell, and move up. I wanted to at least get my money back or take as little loss as possible.

I started my search for a single place aerobatic capable aircraft. I located a hangar space that was really cheap. But it meant the aircraft had to have a maximum wing span of 19 feet and a length of less than 18 feet. Still, that gave me a lot of models to chose from. I also did research on handling, specifications, abilities, and construction. Did I mention that I didn't want to spend much? Researching Trade-A-Plane and all the kit magazines, I decided on a few airplanes that met the requirements and were readily available. Most were biplanes — Pitts S1C, EAA Biplane, Acro 1, Smith Miniplane, and Baby Lakes.

I found a Smith Miniplane that had 300 hours on it and a bad engine. It was reasonably priced, and I had an engine available. Additionally, the aircraft already had hours on it. It wasn't a hangar queen. Empty weight was within the recommendations on the plans, fabric was still good, and it had been improved with all of the modifications that were recommended.

Once the specs and the pictures were studied, research completed, and engine located, I bought the aircraft sight unseen with an inspection planned. Big risk if it's not what you want! But I did my research. I new where to look for cracks and what tubes failed. I also had an idea how big the cockpit was before I went. I was at the maximum height to be comfortable. It was open cockpit which provided lots of clearance for my head and headsets. Make sure that you "try on" a comparable aircraft for size. Weight and balance of the Smith would be fine with full fuel and it would fit in my cheap hangar.

How did I do the research? I joined the type club. They are an excellent way of getting good tips on a particular aircraft. Read reviews of the

BUYING USED CONTINUED

aircraft. Call the EAA and ask for safety information or accident reports. They can also get you a list of information that they have published through the years. Talk to owners. Ask the seller lots of questions. I try to find out as much as possible from each seller and why they are selling the aircraft. Some people have actually told me the aircraft was too much for their abilities. Those people should be commended for making a safe decision.

How can you locate an aircraft? Read Trade-A-Plane, magazines, and trade advertisements. You can always send direct mailers to aircraft owners (usually you can buy the mailing list for less then $50). Place "want to buy" ads and visit the local airports, putting up a "want to buy" sign.

Make sure you get all the details before you bother to go look. If possible, get pictures. Ask for pictures of fittings, landing gear, cockpit layout, etc. You might get copies of the logbook inspection entries. Don't expect do get this stuff for free. Pictures and copies cost money. Be considerate. If you are going to request all of these things, offer to pay for the expenses.

Do a title search. The FAA will put any lien on the aircraft and transfer the ownership without a worry. Make sure the aircraft your buying doesn't have any liens on it.

Remember, you can never ask enough questions. Basic information that you need includes the year, make, and model of the aircraft. FAA Registration Number and Serial Number will be necessary for a title search.

Make sure that the aircraft registration is in the name of the person you are buying it from or that you have the correct paperwork allowing that person to sell the aircraft.

Any aircraft, certified or experimental, has to have the airworthiness certificate included. Also make sure you have the original and correct airframe, engine, and propeller (if applicable) log books. The log books should have the entries for the last annual safety inspection. Look for the date to see if it is current and compare the engine and the airframe logs to see if they correspond.

Make sure that either the pilot manual, builder's log, or the log books have the airspeeds — stall, cruise, maximum, etc.

As to the engine make and model, data plate information is important. If it is a certified engine, it needs to have the applicable AD's complied with. Check the engine total time. Has it had a major overhaul? Has there been a top overhaul? Are the cylinders chrome or steel? Check the compression. Use your mechanic's advice for a minimum. Generally I like Lycomings to have compression in the 70 lbs. per square inch or better range. Continentals can drop into the 60's. The key is to have an even compression in all the cylinders. If you have one that has a large discrepancy, that could indicate a problem.

Filters or screens should be checked for metal particles. In fact, that is probably the first thing to look for. Don't worry about minor oil leaks, they are usually nothing more than a nuisance.

Check the propeller for nicks and delaminating. Establish a total time for the propeller. The prop hub needs to be inspected for wear and "crushing."

Inspection of the airframe should include going through the basic numbers such as: Airframe total time; Recommended empty weight; Actual empty weight.

The inspections should also include a fabric test, if applicable. If you don't have a tester, you really need a knowledgeable fabric person to explain what good fabric is. Better yet, take that person along. Check for delamination or moisture in composite or wood.

Look at the firewall material and check for cracks and or delamination. Look for a list of materials used in construction. Are they aircraft quality or scrap metal/wood? A lot can be discovered in the appearance of the welds or joints. High quality, penetrating welds look good and neat. This is an additional indication as to the care and quality of the construction.

Same goes for the sheet metal work. Wrinkles, extensions, and lack of doublers are examples of other possible concerns. Even screw heads that are stripped and damaged, or rivets that are poorly installed can all be indicative of the overall quality. But be prepared to expect reasonable wear and tear.

Composites are a little different. Always get professional help. Most composites should be boroscoped (looking inside with a light tube). I always look for fibers, wrinkles, lots of fillers and twists. Surprisingly, there are a few aircraft that have flaps and ailerons that are warped and separating. Any time a problem, no matter how small, is found it should remind the buyer that the inspection is very important.

Other things to check include: avionics certifications (if applicable), instruments, fuel system and capacity, wheels and brakes (condition, size, brand).

These are just the basics. You need to like the looks, be comfortable with the construction, and be able to picture yourself handling the airplane. The seller will never tell you how bad it flies (nobody told me the

Smith would be so much fun to fly). You have to ask a lot of questions and get a feel for the true story. But don't waste the seller's time if your not interested. Move on.

After the inspection and the deal are done, how do you get it home? In my case there was no engine so I trailered it back. If it is flyable, get the right training and the correct insurance coverage or get a qualified pilot to fly it. Don't take a chance of breaking it before you get a chance to have fun. Before you leave make sure you have all the correct paper work; Aircraft log book, Airworthiness papers, and builder's log. Anything that is there needs to be with you.

You also need a signed bill of sale. Use an FAA Bill of Sale if possible. If not, make sure you get a good description written down and signed by the registered owner. The key is to get a printed name just as on the registration and a dated signature. Take a look at the FAA form. In the area that says value, put "$1+OVC" (this is for "other valuable considerations") don't put the actual price. Make sure that each page has an original signature.

If you are flying home you need a FAA registration form in your possession. Complete and keep the pink copy with the aircraft.

If the aircraft is out of license get a ferry permit and have a mechanic complete the required entries for the ferry permit (a mechanic has to tell you if the aircraft is ferryable!). Check with insurance and see if they will cover it while on a ferry permit. Many won't.

If you do your research and make a careful inspection you might be the next owner of a well built, fun flying, affordable aircraft.

Scott "Sky" Smith owns an aviation insurance agency and conducts forums at Oshkosh and other fly-ins. For more information contact: SkySmith International, Ltd. at 800-743-1439 or visit www.skysmith.com

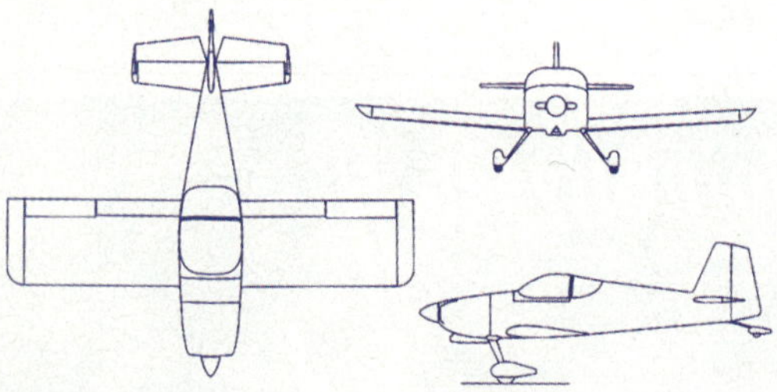

James Daron Sr.'s RV6

I built my RV6 in 18 months and 2700 hours with very little extra help. My wife has about 10 to 12 hours of bucking bar experience now that the project is finished. Other outside help was about 4 hours for final assembly after transporting to the airport.

Kit quality and plans are the very best as is factory support. I think I called the factory twice during construction and one of the questions I had was right there in the plans and I had missed it. Factory support is great. My only other building experience was an Avid Flyer a few years earlier which was of tube and fabric construction. This was my first experience with any type of metal work and I thoroughly enjoyed it. I am completely satisfied with my RV6 and it does perform as the factory says. I find it hard to believe I built an airplane that flies and handles absolutely great.

About the only tips and pitfalls I could pass on would be: If you are going to build it in your spare time don't count on building it in 18 months. I am retired and tried to work on it every day, some days as little as an hour. Don't work on it after you get tired; about eight hours a day is enough. You're not going to finish it today anyway.

James R. Daron Sr.
3970 Miller Lane
Valkaria, FL, 32950

Aircraft Fabrics

Wrapping It Up

by Jon Goldenbaum

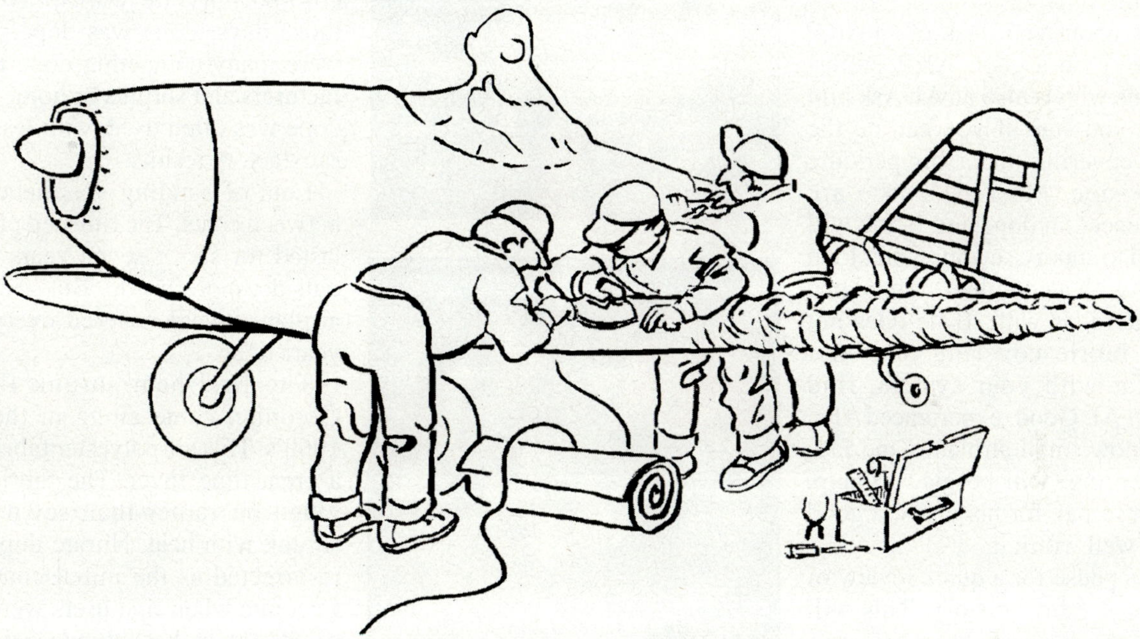

You know Leroy. He's your local expert on fabric covering. Been around the airport as long as anyone can remember. Probably helped dope on the Wright Flyer. Always ready to give advice or pass on tidbits of knowledge to the uninitiated. Good old Leroy.

Leroy loves to tell you how difficult it is to cover an airplane. He says it took him years of apprenticeship before he got any good at it. The hardest part is the rib stitching, he explains. According to Leroy, it takes forever to learn the knots. Almost beyond mortal man.

He regales you with tales of shrinking cotton with distilled water and sanding 40 coats of dope to completely hide the weave of the fabric and the tapes. Often, he continues, blush ruined the whole job and he had to rip the whole thing off and start all over again. Six months down the drain! But, that's fabric covering, according to Leroy. Happens all the time.

His other thing is to tell you about all the substitutions he has come up with over the years for those expensive covering systems the salesmen are always trying to foist on unwary. Heck, he has techniques they haven't even thought of. If you are lucky he might show you sometime. But you will need a lot of extra time, could take months.

We have found that any FAA approved system works fine if you follow the manufacture's instructions and don't improvise. Each system has its differences, but they all work. More on how to pick a system in a minute.

Let's also put the cost question to rest right off. All the systems today basically cost the same. You will spend about $2000.00 in raw materials to cover an airplane no matter what system you choose. Although different quantities of products are required for different systems, when the smoke clears they all cost about the same. Beware of quotes that are less than $2000.00. You are being low-balled.

How about your lack of an A&P license? Find a good A&P, preferably one who is also an AI. Ask him before you start if you can do the fabric covering under his periodic supervision. Most AI's who are experienced in dope and fabric will be glad to supervise your work. Plan to stay with a complete system, one he is familiar with. If he does not know fabric covering or is not familiar with your system, find another AI. Good, experienced AI's, who know small airplanes and fabric covering, will be glad to help. Expect to pay for his visits; a good one is well worth it.

Let's pause for a quick review of dope and fabric history. This will help you sound like an old hand when you next see Leroy.

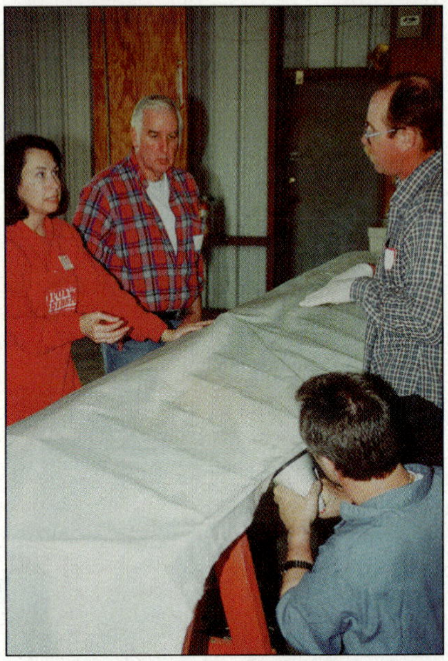

Prior to the mid 1960's, all covering was done with either cotton or linen fabrics. Grade A cotton was the fabric of choice. Highly flammable nitrate dope was primarily used in the years from WWI to WWII to seal and tighten the fabric. Nitrate dope was generally replaced in WWII with butyrate dope which burned less readily. Covering in those years involved sewing grade A cotton to the airframe, shrinking first with distilled water, then adding butyrate dope. The first coat of dope contained a fungicide to protect the cotton from rot. Additional coats of clear butyrate were sprayed on to build the coating. Silver butyrate was added to protect from UV radiation. Finally, colored butyrate was sprayed. In those days, dope was dope. There were many competing dope manufacturers and surplus or home made dope was often used, with less than satisfactory results.

Lots of sanding was necessary between coats. The finished product lasted for six or seven years if left outside, more inside. But generally airplanes were covered every five years or so.

The first heat shrunk fabric, Ceconite, came along in the mid 1960's. This all-polyester fabric was a great time saver. The fabric was glued on rather than sewn, then shrunk with heat. Nitrate dope was resurrected as the initial coat over Ceconite when first users were horrified to learn that butyrate would not stick directly to slick polyester fabric. Peeled off in sheets.

A new kind of "non-tautening" dope was developed for polyester fabric. Since the fabric was shrunk partially with heat, full shrinking dope was not necessary.

The big advantage with polyester fabrics was longevity. Ceconite jobs easily last twenty years when properly done. Unlike cotton, polyester did not rot and, if protected with silver dope, lasted far longer.

Ray Stits developed the Poly-Fiber system over thirty years ago. Like Ceconite, it used heat shrunk polyester fabric but used no dope at all; Poly-Fiber coatings are all flexible vinyl. By the Eighties, the Poly-Fiber and Ceconite systems dominated the market.

Complete, 100% polyurethane systems were introduced about ten years ago. Both Cooper Superflite II and Air Tech systems utilized high build polyurethane coatings over heat shrunk polyester fabrics to give wet-look, high-gloss finishes. These polyurethanes were based on automotive technology with additives to make them flex.

Finally, the water based Ceconite 7600 system was introduced about the same time as the polyurethanes. Water-based systems have had their ups and downs but show promise for the future.

A couple of systems have fallen by the wayside. Grade A cotton is almost impossible to find today. It seems that in our "I'll sue you" society, nobody wants to make cotton for airplanes anymore.

Now let's look at the specific systems on the market today.

CECONITE AND DOPE. Ceconite fabric comes in three styles for different speeds and wing loadings. The fabric is cemented to the airframe with nitrate Seam Sealer cement and heat shrunk. Three coats of nitrate dope are initially applied to the fabric followed by about three coats of clear, non-tautening butyrate for fill. Three coats of silver butyrate follow for UV protection. Two to three coats of colored butyrate dope are sprayed for topcoat. All coats except the first nitrate are sprayed.

ADVANTAGES: Any brand dope may be used over Ceconite. Randolph, and Classic Aero coatings (by Poly-Fiber) are the predominant names. Classic Aero offers a 42 page manual for step-by-step instructions. Ceconite has a small STC manual. Randolph also offers a pamphlet. Dope may be rejuvenated to restore flexibility over years and repairs are easy.

DISADVANTAGES: Dope is susceptible to temperature and humidity changes when it is being applied. Funny things happen when the weather changes. Dope is also pretty flammable and gets brittle over the years. Note that lots of coats and sanding are required for a good finish. Solvent exposure requires precautions.

2. STITS POLY-FIBER Poly-Fiber offers five fabric styles. Fabric is cemented to the airframe with Poly-Tak fabric cement. One coat of vinyl Poly-Brush is brushed on followed by one more spray coat. Three coats of silver Poly-Spray are sprayed on next for fill and UV protection. Two coats of Poly-Tone colored paint are then applied. As an option, Aero-Thane polyurethane may be used for the color coat to give a wet look, high gloss finish.

ADVANTAGES: Only seven coats. Does not burn. Little sanding and finish. Smoothing is done with an iron. Vinyl remains flexible over the years. All coats are the same generic chemical, so all bond together. May be rejuvenated to extend service life. Repairs are easy. Complete manual and toll free customer support phones available. Learn-to-cover seminars offered throughout U.S. and Canada. Free workshops at Oshkosh and Lakeland.

DISADVANTAGES: Solvent exposure requires precautions. Fresh air breather mask required for Aerothane.

POLYURETHANE SYSTEMS: The major brands here are Cooper Superflite II and Air Tech. One fabric is offered. Fabric is cemented to the airframe and heat shrunk. Three coats of primer coat are sprayed. Two coats of polyurethane follow for the color coat. Polyurethane systems rely on flex agent additives.

ADVANTAGES: Extremely high gloss, wet look finish. Only five coats. Little sanding required.

DISADVANTAGES: Although fewer coats are required, coats are generally thicker. System can be heavy. Cannot be easily repaired. Cannot be rejuvenated. Since all coats are urethane, a fresh air breather mask system is required for all spraying.

CECONITE 7600 (WATER-BASED): Ceconite 7600 was formerly known as the Blue River system. Two styles of fabric are offered. Fabric is cemented to the airframe with a water soluble cement, then heat shrunk. Two coats of filler coat are applied with a pad. No color coats are offered by butyrate dope. Poly-Tone, or polyurethanes may be used.

ADVANTAGES: No solvent exposure problems. Only five coats. Filler coat is easily applied with the pad. No HAZMAT shipping charges, no solvent fumes. Subcoats not flammable.

DISADVANTAGES: No color coats are offered. Past problems with color coats and sub-surface corrosion have made some skeptical of water-based systems. All previous problems with the systems appear to be fixed.

So how to choose?

Ask reputable covering veterans who have recently completed a project using a whole system. If your advisor says he used mixed systems or substituted something, pass his opinions right on by. Talk to the guys in your EAA chapter who followed the manual to the letter. Go to fly-ins and talk to the owners.

Choose a system as if you were going to have to do the job alone on a desert island. In other words, how complete are the instructions? Is there a manual? Can you reach the manufacturer by phone when you need to? Is there a video available? Can you get products easily?

What kind of equipment do you have available or are you willing to purchase? Do you have a spray rig? Can you borrow a good one? If you are considering a polyurethane system, you need a fresh air breather mask. The catalyst in urethanes are polyisocyanates, as in CYANIDE. You cannot use a standard charcoal mask. If you don't use a fresh air breathing mask you risk serious health problems.

Do you have a good place to work? Can you spray solvent based products without risking divorce or arrest?

Most importantly, avoid Leroy. If you have any questions or just want to talk about dope and fabric, call me at: 1-800-362-3490

John Oglesby's
Avid MK IV

The Hobbs meter read 200 one year after the first flight of my Avid MK IV. The required forty hours was flown off in three weeks and there have been very few days I did not get at least a short flight in. I guess you could say I like this airplane.

I installed a 100 HP Stratus Subaru with a three blade Warp Drive prop. The performance is really exciting and the visibility is terrific. After ground adjusting the prop for cruise we took off on an 1100 miles cross country. Average speed was 115 mph burning 4.8 gph. The two fuel stops were necessitated mostly for bladder relief. Range is over 600 miles although my 6'1", 240 pound body needed to straighten out a little more often than that. Actually, with the LP Aero Plastics full bubble doors there is more than enough elbow room for two. The doors are so clear they don't photograph but they are there. Although a capable traveler, the Avid really shines just playing around. Takeoffs, landings, flying low over the sea and sand dunes, it's all great fun. Building my Avid was enjoyable. I believe it to be the most rewarding endeavor I have involved myself with. Factory support from Avid, Stratus and Warp Drive continues to be outstanding. Truth be told, I don't just like my Avid, I love it!

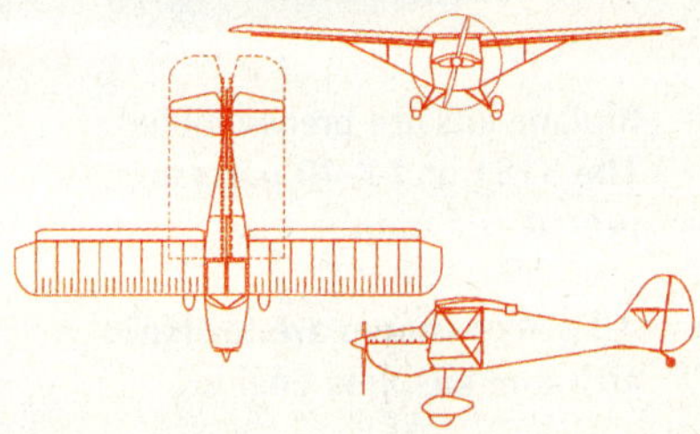

John M. Oglesby
285 Lakeview Beach Dr
Destin, Fl 32541-4192
(805) 654-1051

Randy Brooks'

"Bratz"

The Bratz (Brooks - Hatz) was completed in November 1996 after three years and eleven months of construction. The FAA certification was on December 5, 1996. Due to weather, the first flight did not take place until February 22, 1997. Thirty hours were flown, the requirements of Operations Specifications Phase 1 were met, and the Bratz was released on May 16, 1997.

Several modifications were made such as aluminum ribs in wings and ailerons, modified wing-tips, friese ailerons on lower wings only and 100% static balance. The spars are ½ inch deeper, center section raised 2" with Stearman type cutout, lower wing dihedral four degrees, adjustable stagger, adjustable stabilizer, 15 gallon auxiliary fuel tank in fuselage, toe brakes in both cockpits, landing gear lengthened 3", and aluminum side and lower panels were installed from firewall to the rear of the aft cockpit.

The gross weight is 1659 lbs. which includes 198 lbs. of fuel and 40 lbs. of baggage and crew of two. Some performance figures are: takeoff distance- 300 ft., landing distance- 400 ft., stall speed- 40 mph, approach speed- 65 mph, rate of climb- 600-700 fpm at 60-80 mph, cruise speeds- 88 mph @ 2300 rpm, 95 mph @ 2400 rpm, and 100 mph @ 2500 rpm.

The Bratz is powered by a Lycoming O-320A at 150 HP, turning a Sensenich W74-56 propeller and is equipped with Cleveland wheels and brakes with chrome discs and Maule tailwheel.

The Bratz NX24B is a pleasure to fly and has no bad habits. Controls are well-balanced with elevators and ailerons responsive but not overly sensitive.

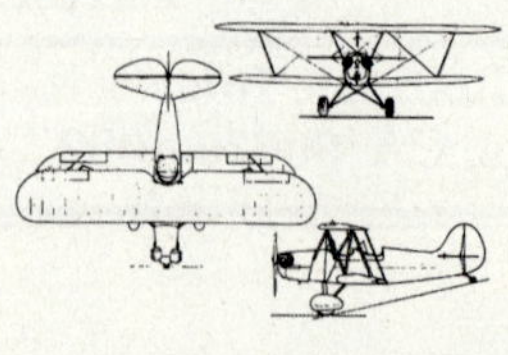

Randy Brooks
Yukon, OK 73099

Homebuilts on the Internet

Experimental Surfing

by Gail Allinson
&
Jim Weir

My mission: to write a guide to homebuilt aircraft and kit planes on the Internet.

I started my research the way I usually start any Internet search: I fired up my browser, went to the Alta Vista search engine at "http://altavista.digital.com" and typed in the word "kitplanes." I was instantly rewarded with a long list of Web sites to visit. Those Web sites had links to other related sites. I had expected it to be real work, but it was easy.

To be truthful, any A&P, student pilot, or airline captain with a working computer and a paid up phone bill can find the information he or she needs on the Internet. If you have never done Web research, here's how to start. Both MS Internet Explorer and Netscape Navigator (the two most popular browsers) have ways to access search engines built right in. Look for "search" or something similar on the toolbar or menu of your browser. If that doesn't work for you, try using your browser to go to Search.Com at "http://ww.search.com."

Once you have gotten to a search engine, type in the keywords of your choice (for example, the names of companies that sell aircraft plans, aircraft kits, or the kinds of supplies you are looking for). Also try the words "kitplane," "homebuilt," or "homebuilder." By the way, don't bother capitalizing your keywords — it's not necessary and will limit your search in some cases.

If your first search tactic doesn't return the results you hoped for, try your chosen keywords one at a time or grouped in a different way than for the first try. If you really want to get specific with your search, you'll have to read the help page for each particular search engine you use. No two search engines operate exactly the same way nor do they necessarily attempt to catalog the whole Internet. A search with no results from one engine may return 20 or more hits for the same word in the next search engine. Besides that, the first result on one search engine may very well be way down the list on the next. Patience plus practice equals success in finding specific information via the search engines.

During my homebuilt search, one link led to another and I quickly had a mountain of printed Web pages (my old analog brain still likes to organize certain things by shuffling paper). Indeed, I had not even scratched the proverbial surface of the most obvious sites. The good and the bad news is that there is simply too much information out there to compile into the "complete guide" I had envisioned. Besides, with new Web sites coming on-line everyday, anything I write today may be different tomorrow.

So I changed tactics. Since anyone can type a keyword into a search engine and be rewarded with a "click trip" to a Web site, I will share a representative selection of the Web pages I found. Many are jump pages with long lists of aircraft manufacturers and homebuilding suppliers. My favorites include the personal homebuilding photo essays and stories recounting successes and learning experiences. With all that said and done, I present the results of some serious mouse clicking, keyboard tapping, and Web surfing. All URL's are preceded by "http://" and typed without quotes.

AeroCrafter Homebuilt Aircraft Sourcebook "www.baicorp.com/aerocrafter/" Kit of the month, Homebuilt Supplier Directory, ads, jump page, and more.

Aircraft for the Homebuilder "www.computeractive.com/cgi-bin/WebObjects/Aircraft/" ComputerActive's aircraft search. Try it. It's fun.

AirNav "www.airnav.com" Airport information including where to get mogas and 80.

Building an Airplane "/ic.net/~patk/rv6a.htm" Personal story of building an airplane in an apartment.

Canard Aviator's Page "www.canard.com/" Post a free listing about your canard aircraft, resources for owners and builders, canard accident stats.

Compressible Aerodynamics Calculator "www.aoe.vt.edu/aoe3114/calcbody.html" JavaScript, on-line calculator for the aeronautical engineer in all of us.

Dave, Carey and Ed's Lancair ES Airplane Construction Progress Page "members.aol.com/lancaired/index.htm" Large personal Web site with lots of info and photos.

Europages Europa Homebuilt Aircraft "www.kaon.co.nz/europa/europa.html" Personal effort serves Europa builders — from New Zealand.

Experimental Aircraft Association "www.eaa.org" A professional quality, full service Web site

Experimental Aircraft Association Club Home Page "acro.harvard.edu/EAA/eaa_homepg.html" Developed by EAA members, this is a good guide to Chapters with Web sites.

Experimental Balloon and AirshipAssociation "www.hotairship.com/ebaa/index.html" This site will get you started on that balloon or airship you've been dreaming of building.

Homebuilt Aviation "www.pec.co.nz/~frankv/homebilt.htm" Personal jump page from New Zealand with RV links and more.

Homebuilt Homepage "www.azstarnet.com/~cmddata/homebuilt/" A volunteer effort, one of the finer homebuilt jump pages around.

James Redmon's Proto Page "http: //web2.airmail.net/redmonj/" A personal Berkut kit plane project.

Jim's Plane Links "home.earthlink.net/~jaroot/htmls/plnlnk.html" Yet another well crafted personal jump page.

Jon Ruthven's BD-5 Homebuilt Aircraft Project "home.regent.edu/ruthven/bd-5.html" Personal BD-5 project photo essay.

Kitfox Jim's Page "therockies.com/av8r" Personal jump page, Kitfox construction photos, and flying adventures from Canada.

Landings "www.landings.com" Huge aviation resource Web site with special sections for homebuilts and ultralights.

Marc J. Zeitlin's Unofficial Cozy Web Page "www.ultranet.com/~marcz/cozy_mkIV/" Super personal effort — professional quality Cozy resource with personal Cozy building story.

Marcus van Lerssel's Homebuilt Aircraft "www.eecg.toronto.edu/

CONTINUED ON PAGE 43

David Flurry's
Aventura II

I completed N106D during a four month period, Feb. 1996 - May 1996, investing 200 hours in construction or more accurately, assembly. The Aventura II kit from Arnet Pereya included everything but gasoline. The manufacturer support has been excellent and my bird was the first builder completed 2 seater.

The Standard BRS recovery system and the amphibian's ability to land on a dime, wet or dry, takes the "fear of 2 stroke," off the pilot's mind. The Rotax 582 gets you off the concrete or water in 200' - 300'.

With a cruise of 70 mph, a stall of 35 mph, and visibility approaching that of a breezy, she's a great sight seeing platform. She was completed in May 1996. My first passenger was my 17 year old daughter, Sara. My 15 year old son, Justin was my second passenger. They both love the plane, especially water operations, and refer to the plane as my third child. The building process was a very positive experience and the flying can only be described as FUN.

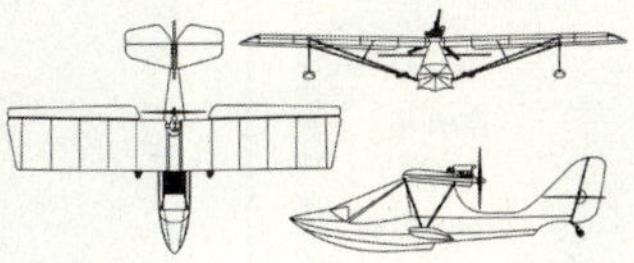

David Flurry
5905 Elizabeth Dr.
Trussville, AL 35173

strong impression that an increasing number of the people now active in the general aviation and sport flying scene tend to take propellers as much for granted as they do the tires on their cars.

Just as a champion race car crew can tell us things about tires that we never knew before, a real expert on aircraft propellers can scrutinize a prop for several minutes, say "Change this a little, whittle that down a mite, and twitch the gizmo a little more." And when we do these things, low and behold, we often find that we have changed a dog of a plane into a tiger!

Unfortunately, old-timers with a real and profound "feel" for propellers are few and far between. However, we should point out that it's increasingly common for present-day makers of custom propellers for homebuilt and other specialized aircraft to use sophisticated computer programs in their work. These print out specifications for props that are able to coax the best possible results from any particular aircraft and engine combination.

A leading reason why it can be hard to ferret out a grassroots propeller man who has an uncanny feel for these things is this. In the first thirty years of aviation's history, aircraft were comparatively simple and slow, yet progress was rapid, and the aviation technical press was full of technical and engineering articles that spread understandable and useable information to grassroots users. But today, aviation technology is so sophisticated that less and less that is understandable to, and usable by, homebuilt airplane enthusiasts, gets into print.

With its graceful curves and hardwood grain showing through the varnish, an old wooden propeller hanging on a lucky owner's wall is an attractive and impressive example of woodworking skill. But anyone who is making a homebuilt airplane or restoring an antique one has to make a strong effort to overlook this beauty enough to realize that above and beyond the matter of workmanship, any successful propeller is the result of a vast amount of knowledge and experience in a very specialized branch of the field of aeronautical engineering. Ferret out a pile of propeller literature in a good aeronautical library and you will be deeply impressed with amount and sophistication of mathematical and laboratory investigation that has gone into developing the propellers that we tend to take for granted.

When one stops to think of it, it's math that gives us answers to such questions as, "How much diameter? How much pitch? How much strength? How much weight?" and so on. It's interesting to learn that some airplanes in the past have been fitted with propellers made of very heavy or very light wood to solve center of gravity problems. It's a vast, complex subject.

We have been around small airplanes long enough to have encountered several instances of small planes flying either poorly or not at all because something was not quite right about the propellers initially fitted to them. The stock propellers that pilots take for granted on the noses of popular airplanes have dimensions arrived at not by guess work but as the outcome of knowledgeable calculations and much test flying. We have seen planes get into serious trouble as a result of uninformed owners replacing damaged propellers with others found in hangar corners that "looked about right."

It's both interesting and educational to review propeller history. Aviation pioneers were very much on their own when it came to designing and constructing propellers for their primitive aircraft. When they looked into literature on marine propellers, they found little of use because water is so much denser than air.

For one example of pioneer resourcefulness, the Wright brothers had constructed a small wind tunnel with which to obtain basic data on the behavior of wings in moving air. After months of debate, they finally made a pair of propellers for their first succesful airplane of 1903 that looked like nothing ever seen on a boat. They had realized that a propeller blade is basically a small wing that travels in a circle instead of straight ahead, and basic laws of wing aerodynamics therefore apply to propeller blades.

While early airplane builders were for the most part practical men and cut-and-try experimenters, scientists working in universities were also intrigued

Mike Leger's *RV-4*

My father and I began construction of our RV-4 on March 11, 1991. For four and a half years we built this wonderful aircraft to the absolute best of our ability. The extra time and patience exerted were greatly rewarded when we hear our workmanship is award winning quality.

Our RV-4 is an absolute joy to fly. It is very nimble, both in the air and on the ground. The controls are light and very effective which make the aircraft most predictable. We are very impressed with the stall performance. We operate out of a 1200 ft. private strip with very poor approaches an have no problem at all.

Our powerplant is a Lycoming O-320 E2A 150 hp. We used a Warnke 70 x 72 prop. This combination works very well. Our plane is equipped with

basic VFR/TKM MX-11 radio and electric flaps that were added this spring. We used automotive Imron by Dupont to finish the outside as well as interior cockpit. The registration and tapered striping was computer cut out of vinyl material; a procedure I would highly recommend to anyone. It is very easy to apply and makes for nice clean edges. Van's Aircraft is a very professional and efficient enterprise which produces a very nice kit. We chose the RV because we felt it is an excellent value for the dollar.

Van's staff is first class for ordering and delivery of parts and kits. The kit itself is quite complete. The workmanship is good, although we found that the fiberglass fairings and cowlings needed some alterations to achieve the proper fit that we were looking for. All in all both my father and I were very pleased with our kit and with Van's Aircraft (an RV-8 may be in the future).

Mike Leger
P.O. box 1493
Alexandria, Ont.
KOC 1AO Canada

with the challenge of powered flight. Over a hundred years ago they were engaged in mathematical analysis of the problems posed by air propeller design. Both practical and theoretical people came to an early realization that propellers are just as important to successful flight as are light and reliable engines and good wing curves.

By the outbreak of World War I in 1914, the principles of air propeller design were well understood and in general use. Watching the Sopwith Pup replica fly drove home the truth of this.

Concerned as they are with operating their aircraft safely and effectively, few pilots give thought to the forces acting on the propellers that pull or push them along. But it's very much worth appreciating that there are many of them and most are of surprising magnitude.

Centrifugal force pulls outward on the blade. A pull of well over a ton acts on each blade of a typical ultralight prop, and there can be 10 to 15 tons on larger props. To provide a generous margin of safety, root areas of prop blades for the big and powerful military and airline planes that appeared late in the piston engine era, were stressed to handle pulls on the order of 150 tons. Hubs had to be designed and manufactured so as to be able to stand such stresses dependably over years of service life. Quite a number of engineering graduates made careers in the field of propeller design.

While centrifugal force is pulling outward on blades, thrust developed by them is simultaneously bending them forward. This puts the material on the blade faces under tension and

KOLB Aircraft
RD 3, Box 38
Phoenixville, PA
19460
Tel (610) 948-4136
Slingshot
1 or 2 seat
FireFly
Legal ultralight w/
Rotax 447
INFORMATION PACK
Domestic $5.00 Foreign $10.00
20 minute Video
Domestic $12.00 Foreign $15.00
FireStar
1 or 2 seat
All Models Feature
Open or closed cockpit
Factory welded steel frame
Stits aircraft covering
10 minute wing folding
MARK III
2 Seat

• 1,000,000 Hours Continuos Life.
• No Radio Interference or Audible Noise.
• Will Fit Original O.E.M. Nav. Light Fixtures.
• Operates 5 - 64 Volts @ VERY Low Current.
• Cool Bright Light, Reduces Danger of Fire.
• Available in Red, Green, White, Rotating & Strobe Beacons! Call For Catalog.
The FIRST High Intensity SOLID STATE
TERFLIGHT
Nav Lights & Strobes For Aircraft!
1(810) 229 - 2082
Phone / Fax 8am-11pm EST
$325.00 Visa, MC, Check
No templates Included.
RUTAN AIRCRAFT FACTORY
CD-ROM ENCYCLOPEDIA
Long-EZ, Defiant, Solitaire,
VariEze & VariViggen
All Canard Pusher Newsletters #1 through #83!
In Computer Searchable Format!
Plans, Texts, Brochures, Operators Manuals & Color Photos.
A 3 CD Set with Over 1.8 GB of Original RAF Material!
More than 2100 Documents, the ONLY RAF Authorized Source!
Windows™ 3.1x, 95, NT, Macintosh™, DOS™ & UNIX Formats.
www.terf.com
By TERF Inc.
11411 Culver
Brighton, MI 48114

THE ULTRALIGHT EXPERTS!
Ultralight Flying! Magazine
Each monthly issue of Ultralight Flying! is action packed with flight reports on new ultralights, the latest ultralight news, new products and flying accessories, do-it-yourself projects, flying techniques, humor articles, exciting flying stories and much, much more!
• 4 month trial subscription $9.95 • 1 year subscription $29.95 (in U.S.)
• Ultralight Flying Calendar $10.95 (plus shipping and handling)
• 17 Ultralight Aircraft Reviews $14.95 (plus shipping and handling)
• 1 Year U.S. Ultralight Assoc. membership $39.95 (includes 1 year UF! Magazine subscription), call USUA at: (301) 695-9100
Cards
Call (423) 629-5375 today to Subscribe or Order
Ultralight Flying! Magazine • PO Box 6009 • Dept. AC • Chattanooga, TN 37401
ULTRALIGHT FLYING!
THE NEW DRIFTER
Visit our Website at: www.ulflyingmag.com
or USUA's Website at: www.usua.com

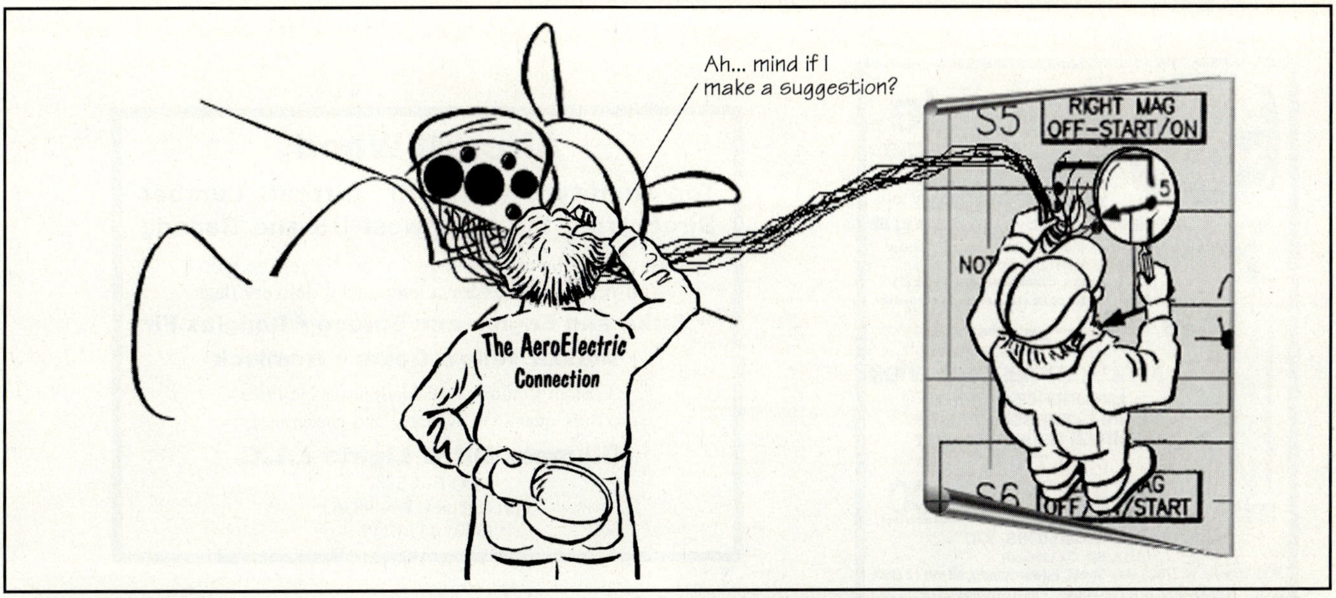

PROPS CONTINUED

that on their backs under compression. When doing preflight checks the backs of blades are usually easier to see and get the closest scrutiny. But it's scratches and nicks on the faces that are more likely to result in blade failures, and they are thus what we should look for.

The engine power that keeps a propeller's blades whirling through the air also puts a backward bend on them. Depending on the number of cylinders, a variable number of power impulses is also sent into the blades in any given length of time. Just as the center of pressure moves forward and backward on a wing in response to varying flight attitudes, so also it does on the smaller kind of wing called a propeller blade. This applies complex twisting forces to it. Weight distribution of the material in a blade adds to the complexity of the situation.

Propellers designed without respect for this mixture of stresses can bend, twist, flutter, and fail in weird and usually disastrous ways.

The fact that we can't really see a propeller when it is doing what it is made to do complicates the matter. When pulled forward on the ground, a thin prop blade may bend alarmingly. But in the air, the strong pull of centrifugal force may very well keep it from bending nearly as much.

We usually think of a gyroscope as being a rapidly spinning wheel. Although of rather different appearance, a whirling propeller is in fact a gyroscope. If you have ever played with a toy gyroscope, you will have had some experience with the strong and weird ways in which it reacts to being moved out of a fixed plane of rotation. When an airplane pitches or yaws quickly in response to rough air or sudden control movements, the much greater weight of a propeller produces powerful gyroscopic reactions which are most certainly felt by the blades and hub. When we walk down the line of planes at an airport, we should make the effort to remember that each and every propeller to be seen is a miracle of skilled

Peter Ashley's Thorp S-18

My Thorp Sunderland S-18 won Concours D'Elegance at our Sport Aircraft Association Australian National Championships at Mangalore in 1996. It first flew in October 1992 following four years and 4,000 hours of construction time. It is powered by a Lycoming O-360, 180 hp engine and has a Hartzell constant speed propeller. It was built with solid flush rivets. Its true airspeed is about 180 knots at 8,000 feet and 75% power. I usually cruise at 65% power where at the same height I true at about 170 knots. I have an auxiliary fuel tank supported under the flat deck of the luggage compartment and this gives me a total capacity of 188 litres or 4.5 hours range plus 0.75 hours reserve. It is an ideal aircraft for traversing Australia's vast distances. My supplier of plans and materials was Sport Aircraft Inc. of Lancaster, California run by Phil Tucker. He provided first class service throughout the construction period.

Peter Ashley
68Bay St.
Cleveland
Queensland 4163
Australia

Ed note: S-18 now available from Classic Sport Aircraft

PROPS CONTINUED

engineering and manufacturing methods. Also, when talking or reading about propellers, we should be wary of pat statements.

Propellers are designed and made in so many ways, of such a variety of materials, driven by so many kinds and sizes of engines, and used for such a variety of purposes, that a statement that is perfectly true in one case can be very incorrect in another.

For example, you may hear an airplane owner proclaim that metal propellers are the best. The outer portions of a metal prop's blades can be made thinner than a wooden one's and still be adequately strong, stiff and split-resistant. Aerodynamic efficiency and therefore airplane performance are improved. The propeller attached to an airplane engine serves as its flywheel. Therefore a heavier metal propeller can make an engine run smoother than a lighter wooden one. But a metal blade is subject to fatigue cracking from long exposure to vibrations produced by piston engine power impulses. Bells and gongs are made of metal because this material tends to keep on vibrating after being struck or otherwise excited. Wood on the other hand tends to snub out vibrations. In engineer's language, it has good fatigue resistance.

What it comes down to is that each airplane, engine, and propeller combination is a separate and distinct engineering problem and must be evaluated as such. Metal propellers are to be seen on many factory-built airplanes, so a person unaware of technical complexities might ask the maker of homebuilt or ultralight airplane props why he too does not use metal. His reply will make perfect sense; "I don't make propellers of metal because I don't have the very expensive machinery needed to process it."

In recent years blades made of composite materials have come into wide use in the ultralight field. They are formed in molds that cost much less than the forging, milling, and heat treating equipment needed to manufacture dependable metal blades. Less material is wasted than is the case when a lot of the fine and therefore expensive wood in a prop blank is reduced to shavings and wood dust. Composite blades are thinner and smoother than wooden ones. So for many in the ultralight field, composite blades are obviously the best choice.

Blades of props for smaller aircraft tend to be slim. A large number of laminations is thus desirable. By comparing glue lines on the two blades as the carving and finish sanding progresses, symmetry of the blades is facilitated.

Blades have been made in a bewildering variety of shapes, for reasons ranging from fanciful to scientific. Some props have had blades so curved as to suggest the letter "S" and are popularly called "scimitar" props. Air pressure on their outer areas causes them to bend as speed increases, changing pitch. It requires much knowledge and testing to make such a prop so that it works as desired. Straighter blades are more common today because they are better adapted to the characteristics of production machinery. Tip shapes can affect both efficiency and noise making properties.

The trend in recent years is to manufacture blades and hubs separately. Because blades blanks can be made by laminating together shorter pieces of wood, there is less waste of raw stock. Fitted into metal hubs, the feature of adjustable pitch is obtained. A basic assortment of blade sizes and hub designs thus enables makers of these (and also composite) props to put together propellers suited to any of the considerable number of makes and models of ultralight aircraft now on the market. Owners of planes equipped with these props can try various pitch settings to get the kind of takeoff, climb and cruising performance they desire.

The reason why wooden propellers are made from laminated blanks is primarily to achieve better resistance to warping with changes in the weather. Laminating also allows closer inspection of all the wood going into a blank. Smaller propellers sometimes have as few as three

CONTINUED ON PAGE 57

OSHKOSH
EXPRESS
INTERNATIONAL

George Stearns'

Spacewalker

It took me two and one half years and 1500 hours of work to complete my Spacewalker. I purchased a welded fuselage (including tail feathers) and built the wings from scratch. Although this was my first full size airplane building project (I have built many model airplanes) I had no trouble constructing the aircraft to the plans. There were no instructions and Spacewalker was out of business at this time.

My biggest problem was figuring out the best (I hope) way to do things. I spent several weeks measuring and setting up a jig so I could build the wings straight and with the proper wash out etc.

I purchased a newly overhauled 65 hp Continental (75 hp now) and a 69" x 48" McCauley propeller for the airplane. The engine installation was straight forward except for getting the cowling to fit.

I did fuel flow, weight and balance, cooling, taxi and other tests before I flew the airplane for the first time on July 24, 1996.

I took off without looking at the gauges and was surprised to see a 90 mph indicated air speed as I went over the fence. The first flight was uneventful but I did have a cooling problem and a significant amount of back pressure was required to keep the plane level. I stalled the aircraft several times and found it stalled at 75 mph indicated instead of a normal 45 mph. Thus the 90 mph over the fence speed was really 60 mph. I also experienced a lot of buffeting on my crash helmet.

I flew the aircraft twice more before fixing the problems. I made a new larger windshield, relocated the airspeed static port, put a trim tab on the elevator to eliminate the back pressure problem and built "eyebrows" for the engine to give it more cooling air. The next time I flew the plane the airspeed was right on the button. The engine ran cool and I had to apply forward stick to keep it level. I then found that the larger windshield trimmed out the airplane perfectly in pitch. I later completely removed the trim tab and the airplane now flies hands off. The Spacewalker is a safe and easy airplane to fly. It does, however, slow down rapidly due to its light weight and you need to have it a foot or so off the ground while you still have some speed. It will drop out from under you if you aren't careful.

George Stearns
9249 SW 62nd Court
Trenton FL 32693

PROPS CONTINUED FROM PAGE 54

laminations. Where high warp and split resistance is very important, a large number of laminations is better, each lamination being sometimes as thin as one-sixteenth of an inch. Talk to any experienced propeller maker and he will give you sound and sometimes surprising reasons why he makes things the way he does.

The first metal propeller to go into general use was the Reed type, which appeared in the early 1920's. It was developed by Dr. S.A. Reed of New York and manufactured under license in both the U.S. and Europe. Billets of aluminum alloy were put through rolling mills, to bring the metal down to about one inch thickness and strengthen it. Blanks were then worked down to airfoil shape and twisted to the desired degree of pitch. Reed's propeller is historically important because it represented a major step forward in propeller manufacture.

In the mid 1920's a firm called Standard Steel Propeller Company was at work developing ground adjustable propellers having forged and machined steel hubs and forged, heat treated aluminum alloy blades. They soon merged with the Hamilton airplane firm to form the Hamilton Standard Company. Handsomely reconditioned propellers bearing this name are frequently seen at present day antique airplane fly-ins.

CONTINUED ON PAGE 417

EMC Solutions for the Homebuilder

What's All This Noise About?

by Bob Nuckolls

Electromagnetic Compatibility. How's that for a mouthful? The guys at Raytheon and Boeing call it "EMC" for short. Originally the title made reference to "noise" but the influences we're going to discuss run much further afield; we'll talk about some interference problems that some people would not classify as "noise."

I've been putting this discussion off for some time even though questions about noise and other compatibility problems have been plentiful. My reluctance was rooted in uncertainty about how to approach the subject. For some manufacturers, noise control in electrical systems and avionics is a whole separate department employing dozens of engineers. Of course, their systems integration task may be much larger than ours. Nonetheless, noise and interference control are nebulous topics compared with Ohm's law or voltage regulation! So, after ten years and much discussion with many of you over the phone, I'll tackle the task.

What It Is

For our purposes, electromagnetic incompatibility is any phenomenon which degrades a system's accuracy or ability to measure, display or communicate. For example: a common noise complaint is that an alternator seems to cause a Loran receiver to lose lock on the transmitting chain. In this instance, the victim receiver is degraded by some energy (other than DC power) from an antagonist alternator. In another example, you might open the squelch on a communications receiver and suffer a cacophony of hiss sprinkled with crackles from the ignition system. All of this may go away when someone transmits on the monitored frequency and the voice is clearly heard. Here is a case where noise is present but its amplitude is small compared to the desired signal. Except in a simple system, it is impossible to create a noise-free environment. Therefore, discussions on noise must balance our ability to tolerate noise with our ability to limit noise. In engineering parlance, the trade-off is often expressed numerically as a "signal-to-noise ratio." For our purposes, we're not so concerned about putting numbers on noise levels; we're more likely to express our perceptions with terms like, "I can live with it" or "Good grief! That's terrible!"

When some device exhibits sensitivity to another device the antagonistic phenomenon may not be something we'd necessarily refer to as "noise." A good example comes to mind: I recently had discussions with a builder about his autopilot. It seems that every time he spoke on his VHF comm transmitter the autopilot tried to do aileron rolls! The issue was an electromagnetic compatibility problem due to poor design of an autopilot servo. A few years ago, a reader called to describe how his engine gauges all wiggled when he cycled the landing light or turned the alternator on and off. This problem was traced to an improper grounding technique which induced errors into each instrument depending on the alternator's load. The problem showed up in a canard-pusher with a very long, undersized ground lead combined with engine instruments "grounded" to the battery at the front of the fuselage.

Most interference problems experienced by both amateur and professional airplane builders are INDUCED by poor systems integration techniques... how the equipment is installed. However, there are cases (like the aerobatic autopilot) where the design of an accessory may make it vulnerable to outside interference in otherwise perfectly ordinary situations. There's an old adage that asks, "If you haven't got time to do it right the first time, where are you going to find the time to do it over?" Of course, to "do it right the first time" presupposes that the designer/builder knows what the right way is. In both the experimental and certified worlds, compatibility problems don't show up until operation of an installed system AFTER it's built. Some interference scenarios may be mitigated by normal variability in operation, components, assembly technique and time in service. In the latter case, factories deliver product with a potential for problems that may not show up for years!

Readers of these words will likely fall into one of three categories: (1) still planning your installation and in a position to "do it right the first time," (2) you're already flying and have no problems that you are aware of (or are willing to fix) or (3) some piece of equipment is beating up on another piece of equipment in some manner that cannot be tolerated. If you're in the first group, consider what you read here as prophylactic advice for reducing your chances of future problems. If you're a member of the third, you will find advice for analysis of the problem and techniques for reducing if not eliminating the effects. A majority of builders occupy the second category, but some of those airplanes might be more pleasant to fly and perhaps perform better if a tolerance for "nuisance" levels of interference was not quite so high.

Irrespective of your situation, what follows is not an "eye of newt and toe of frog" potion for warding off evil spirits. Every compatibility problem has causes and solutions based on the physics of the matter. Design rules I'm going to suggest have evolved over years of collective experience within the aircraft industry.

The Nature of Electrical "Noise"

Noise is the electronic version of environmental pollution. It comes in lots of forms, has many ways to propagate, and affects victims in different ways. And like environmental pollution, some level of most antagonistic phenomenon is considered "safe," or in our case studies, "acceptable." Any periodic electrical event that includes rapidly changing levels of voltage or current is going to generate a "noise." The key word here is "rapid." In the non-aviation world we can identify a number of common noise sources. Can you recall driving under a high tension power line and hearing a raucous buzz in the radio for the short time you were traversing the ground beneath the lines? If a high-tension line's insulators are in perfect condition, there is no leakage of power and no noise. If an insulator becomes damaged or filmed with dust and/or pollution, little streamers of electrical conduction may form around it, each one behaving much like a subminiature electrical storm. Mini-strokes of lightning form up and die 120 times per second in synchronization with the AC voltage waveform. Unlike the smooth transitions of the current in the wires, these little sparks are erratic. In addition to their fundamental periodic rate of 120 Hz, their erratic nature causes new laws of physics to apply. The most important law says that when a voltage or current waveform departs from a smooth, mathematically defined, sinusoidal waveform, harmonic energy is generated which may extend many decades higher in the electromagnetic spectrum. If we had a receiver capable of tuning in to as low as 120 Hz and we stood under the offending power line, we would detect the strongest components of the noise at 120 Hz. Tuning further up the spectrum we would find prominent signals at 240 Hz, 360 Hz, and still more harmonics until we were tuned into the AM broadcast band.

If we tune in a radio station at 1200 KHz, the 10,000th harmonic of the 120 Hz noise would set right on top of the station's center frequency. The total bandwidth assigned to a single AM radio broadcast signal is about 15 KHz. This means that with 120Hz spacing of harmonic energy packets, over 100 separate little raspy noises will blanket the RF spectrum occupied by an AM radio station! Indeed if we sat under a noisy power line and tuned over the entire broadcast band I dare say every frequency would be awash with the same noise. If we changed to FM broadcast a distinct difference would be noted. First, power line noise is amplitude modulated: FM receivers are rather immune to AM noise. Further, the energy in each successive harmonic of the original noise gets weaker as the harmonic multiplier goes up. Harmonic multipliers on the order of 10,000 are required for 120 Hz noise to be heard on the AM broadcast band. A multiplier greater than 800,000 is needed to make the pesky varmint audible at 100 MHz, and it will be a much weaker signal at this point in the spectrum. This illustrates the basic physics which raises the quality of FM broadcast transmissions. In addition to the wider bandwidth occupied by an FM signal (better audio bandwidth), the FM receiver is inherently resistant to AM noise. Further, due to the high operating frequencies, AM noise from common sources is much weaker.

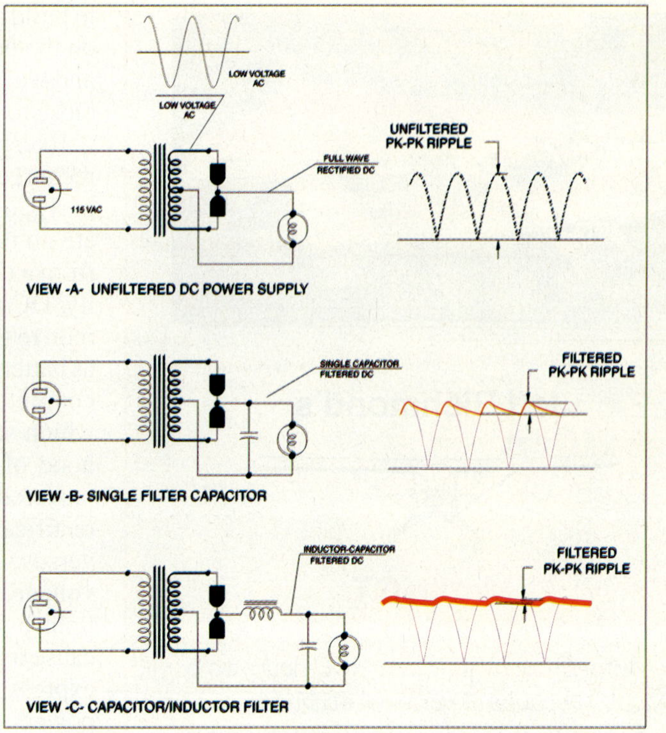

Figure 1

So far, we've discussed noise forms which occur on or near the frequency occupied by a radio signal of interest. Interference in other forms may invade avionics components through other routes. Noises may propagate around the DC power distribution system. One very common antagonist is the ship's alternator: a 3-phase AC device having an array of diodes at the rear which converts AC into DC for purposes of powering the bus. Ideally, we'd like

Jeff Richmond's

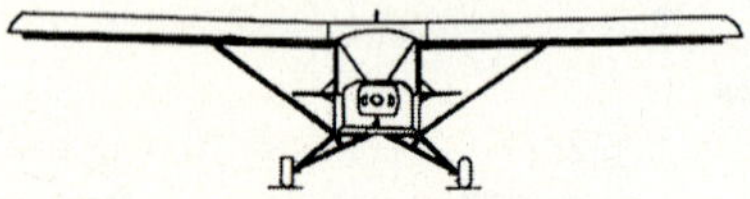

Merlin GT

My Merlin GT took just over 400 hours to complete. Mostly because of personal stubbornness, I did every bit of work on the aircraft myself, except for a little help with final assembly and painting. The aircraft is powered by a Rotax 912, which lets me take off in less than 200 feet and climb at 800 to 1000 fpm. The large tundra tires are just what is needed on our sod, often soggy, runway. Some reshaping of the lower cowl was necessary to get proper cooling and air flow to the carburetors. This was my first experience putting fabric on an aircraft, but the Poly Fabric process produced excellent results. I added navigation and anti-collision lights so I could fly in the evening after sunset. The aircraft cruises comfortably at 85 to 90 mph. Stall is around 28 to 32 mph. This would make an awesome float plane. FAA sign-off was accomplished on October 13, 1996, and first flight was the next day. Flight characteristics are as expected. It handles much like an Aeronca Champ. I was pleased with the completeness of the kit and the detail in the instructions.

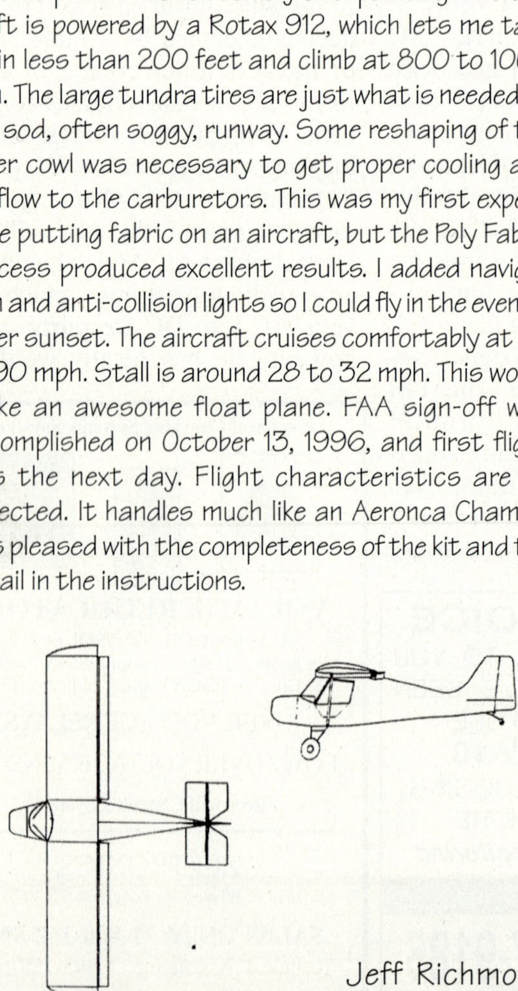

Jeff Richmond
3544 Ridgeway Trail
Maryville, TN 37801

to build a power distribution system which has zero noise on the bus. If our source of electrical power is a battery and we have no alternator or strobe lights, we'll have as close to a zero-noise system as we can get.

Except for the disadvantage of having to replace or recharge them, batteries are probably the most "neighborly" of electrical system components. Batteries generate no noise. When you rectify AC into DC, the output from a diode array is anything but smooth, battery quality, DC power. *Figure 1* shows some schematics common to the wall-plug power supplies we've encountered as battery eliminators for radios, recorders, video games, computers, etc. These devices contain a transformer which steps wall socket voltages down to a neighborhood of 3 to 24 volts, depending on the device to be powered. The circuit in *View -A-* produces "full wave" rectification of the AC power. I've illustrated the time versus voltage waveform of the unfiltered, rectified AC voltage. As you can see, it's pretty lumpy. However, a light bulb would illuminate and operate normally because its output qualities (light and heat) have no way to express the "roughness" or "noise" in the energy that powers it. However, if instead of light bulbs, you tried to power a pocket radio or tape recorder, you'd be lucky to hear any desired signals through an overpowering "hum."

Obviously, the wall plug power supply must contain something besides a transformer and rectifier. In most cases the extra component is a capacitor.

Capacitor, the Basic Filter

Capacitors have been around since the very earliest experiments in electrical phenomena. In fact, the Leyden jar, which was discovered to have electrical storage capabilities, is probably the earliest example of a capacitor. The archaic term for capacitor is "condenser," a word widely used in the mechanic's vernacular. The "condenser" connected across the points of a magneto or automotive ignition distributor is in fact a capacitor not unlike those used in millions of electronic devices.

The capacitor is an electronic component having two "plates" consisting of many square inches of conductive foil, rolled up like a jelly-roll with an appropriate insulating medium in between. The schematic symbol for a capacitor suggests this construction. Seen in *Figure 1, View -B-*, the device is depicted as two "plates" or conductors with nothing in the middle. Quite often, one of the plates is curved. For polarity sensitive electrolytic capacitors, this will represent the negative (-) terminal.

Capacitors can store electrical energy. Capacitors are used to store the electrical energy used to flash your anticollision strobes. Further, many of you have heard of "capacitor-discharge" ignition systems. In addition to their service as energy storage devices, they also contribute strongly in noise filters. If you connect a capacitor across a battery, electrons will rush out onto the negative plate with every intention of jumping across the insulation to the positive plate. Insulators (very poor conductors) inhibit electron migration, therefore excess electrons pile up in the insulation on the surface of the negative plate. A law of physics says, "current flow in every part of a series circuit is the same." It follows that for every electron that left the battery's negative terminal another electron had to enter the positive terminal. If electrons cannot jump the insulating gap, where did the electrons leaving the capacitor come from? Simple, from the insulation!

It is a property of all insulators that when confined between two conductive surfaces, individual molecules of insulation will allow electrons to leave (giving the molecule a net positive charge) or take on an extra electron or two (giving the molecule a negative charge). We're all familiar with the static charge which can accumulate on our bodies when moving across a carpet in dry weather. The charge accumulates because of a difference in affinities for holding or shedding electrons by different insulators. (If you wore shoes soled with materials used to weave carpet, you could not develop a static charge, since both shoes and carpet have the same affinity for electrons).

When static charges build between shoe and carpet, your body has become one plate of a capacitor with electrons piling up all over the surface of your skin, using the air as a dielectric storage medium. When your finger approaches another plate (a door handle), the insulating quality of the dielectric (air) is limited. Depending on the voltage you've acquired and at some distance from the knob, insulation breaks down; a miniature lightning stroke occurs!

The ability of any dielectric to store energy is a function of its insulating (or dielectric) material. Air has a very low capability, its dielectric constant is 1.0. Oils and plastics have constants ranging from 4 to 10, meaning they can store 4 to 10 times more energy in same volume. More exotic materials have been developed with dielectric constants of 100 or more. The catalog description or name for a capacitor usually offers a clue as to the type of dielectric used in its construction. Words like paper, Mylar, mica, polystyrene, electrolytic, oil, ceramic, and many others speak to how the capacitor is built, and will give some clue as to its physical size and suitability to the task.

Referring to *Figure 2, View -A-* at time of switch closure, there are zero volts on the capacitor and a large current will flow limited only by internal resistance of the battery and interconnecting wires. As the charge potential approaches battery potential, the current falls. After some time, charge on the capacitor equals the battery voltage. The capacitor has become fully charged and charging current falls to zero. It's a capacitor's major role in life to RESIST any changes in voltage across its terminals. So, if we place a suitably sized capacitor across the terminals of our rectifier as shown in *Figure 1, View -B-*, the capacitor charges during the peak wave-

George Reynolds'
Corbin Baby Ace

Wanting an open cockpit, easy to fly, safe sport plane and preferring the old tube, wood and fabric construction; I chose the Corbin Baby Ace and I've had no regrets! With an honest 90 mph cruise (Continental A-65, 72 x 44 prop), it's a fun airplane to fly and usually draws a crowd. It has no bad characteristics except it is nose heavy on the ground. No problem if you taxi slowly with the stick back and leave the brakes alone.

The plans were purchased in 1982 and are excellent. The entire aircraft was scratch built and everything fits! The only design problem is the aileron pully/landing gear attaching brackets at the lower longerons. It can be modified to improve the overall strength, however.

I would suggest installing a sump drain at the aft edge of the fuel tank to allow proper sumping in the 3 point attitude. I reduced my tank to 12 gallons from the 17 gallon plans tank to give a little more knee room by taking a 3" section sliced out the middle.

The seat back can be hinged to fold forward. This allows a look into the fuselage interior on preflight. I also installed a zippered canvas "baggage bin" on the rear cockpit cross member to carry 15 lbs. of whatever.

J-3 cub exhaust systems are expensive! A muffler shop bent mine from automotive pipe for $25. Volkswagen engine exhaust flanges with minor rework fit an A-65 fine. Total cost for the complete system was $31.

I didn't like the plans windshield so I hammered my PT-19 type from a single piece of .063 soft aluminum.

The Corbin is an attention getter and wheel pants really dress it up!

It takes time, money, effort, and patience to build an airplane. I was also lucky to have the understanding and support of my good wife, Flora. Thanks, honey. Also some immense help from my good friend, Lowell Huffman—an ingenious craftsman and fellow aviator.

The enjoyment and satisfaction of building and flying your own airplane is well worth the effort and expense!

George Reynolds

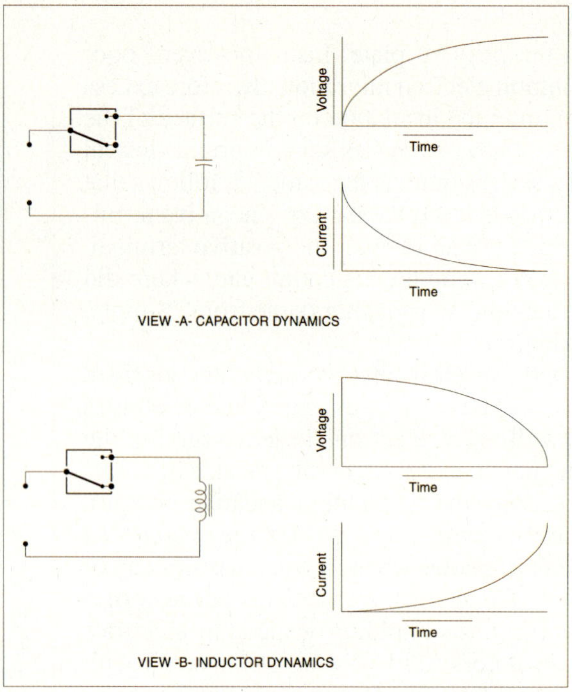

Figure 2

form cycles and provides a source of energy to "fill in" the valleys between peaks. The voltage excursions experienced by the load are now much less. We have "filtered" some of the "noise" from the rectifier's output energy.

INDUCTORS- Make Filters Work Even Better

Unlike the capacitor whose major working component is an insulator, the inductor's most prominent feature is a conductor, a length of wire wound into a coil. When a wire carries a flow of electrons, a magnetic field proportional to current flow is created around the wire. In most cases, the magnetic field was is used to do WORK, to drive a pointer upscale on an indicator or spin the armature of a motor to raise or lower flaps. In the case of noise filters, we'll take advantage of the inductor's ability to store energy in its magnetic field.

In *Figure 2, View -B-* shows a series circuit consisting of an inductor (a coil of wire with a core of anything from air to very specialized magnetic materials), a series switch, and a battery. The instant that the switch is closed, electrons begin to flow throughout the inductor and a magnetic field begins to build. The increasing magnetic field generates

an opposing electromotive force (counter EMF) that tends to oppose the rising current in the inductor. Therefore, in the first few microseconds after the switch is closed, the current through the inductor climbs slowly at first, then increases in rate as the magnetic field becomes established. Compare the two curves in *Views -A-* and *-B-*. In the case for the capacitor series circuit, current is initially high and the rate of change (slope of line) for voltage across the capacitor is initially steep. As the capacitor becomes "charged" the current approaches zero and the rate of change for voltage also approaches zero. For the inductor, the current is initially zero and begins to increase in rate of rise some time after switch closure. The initial voltage impressed across the inductor is maximum at time of switch closure and decreases toward zero as the current in the circuit climbs and finally stabilizes at some value determined by inductance, circuit resistance, and applied voltage. The charging curves for capacitors and inductors are exactly the same shape and are mirror images of each other. The characteristics of inductors and capacitors as depicted by their voltage/ current curves suggest their ability to help us control noise in electrical and electronic systems.

In *Figure 1, View -C-* I show a capacitor and inductor used in combination along with the relative effects upon ripple voltage being delivered to the load. Note that the resulting ripple voltage is still lower than that depicted in *View -B-*.

A physical analogy to the capacitor-inductor filter may be realized as follows: Suppose you have an inner tube floating on a pond and you are able to move it by means of a long pole tied to the inner tube. Your task is to push and pull on the pole such that the inner tube moves back and forth, say two feet peak-to-peak on the surface of the water once per second. Now, suppose a kid gets on board the inner tube for a ride: the forces required to achieve the two foot shake will obviously go up. If you are force-limited, then for the same amount of effort, the inner tube excursions will become damped as a result of having added mass to the equation.

Inductors are analogous to the pole's stretch. Added mass and loss of rigidity have sharply diminished your ability to displace the innertube. Suppose I said, "slow it down. You can take ten seconds to make the two foot, peak-to-peak requirement." Hmmmm . . . that just might be possible in spite of the changes. For any given set of filtering conditions, lowering the frequency of the wiggle increases its chances of getting past any given filter. The opposite is also true, "Now move the inner tube over the two foot interval 10 times per second." Even if your end of the pole could be moved over that distance in that period of time, I dare say motions at the other end would hardly be noticed. As the frequency of interest goes up any combination of filter components become more effective.

Filters made up from combinations of capacitors and inductor-capacitor combinations attenuate the electri-

---

Fritz Bachem's

Avid Mark IV

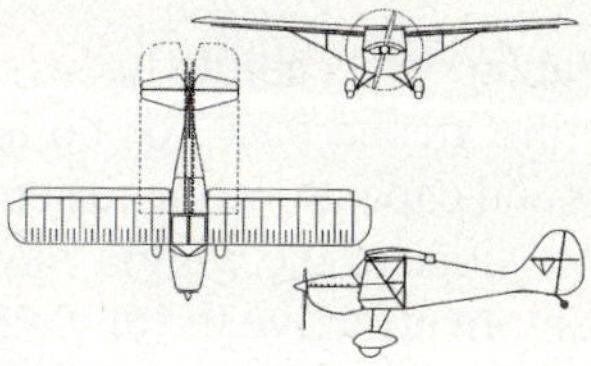

Most builders do it quicker. It took me four years to build my Avid Mark IV STOL. I'll admit that I am slow because I am so "fussy", but the result is the "Grand champion of Avids" according to the Avid Company owner.

I test flew my plane for the first time in September of 1996 and she proved to be a "Honey". The Rotax 582 drives a 74" culver prop through a 3-1 gearbox. This plane out climbs anything at our little airport—solo. It is responsive yet stable, has good ground visibility over the nose and excellent in-flight visibility.

My wife, Ginger strongly encouraged me throughout this project, in part because it got me out of the house. . . in fact she suggested I build another plane; lucky me! The only problem; I like my Avid and don't want another plane.

Fritz Bachem
1740 Riverside Dr.
Myrtle Creek, OR 97457

Brugar

Giancarlo Zanardo

A One-Man Aircraft Factory

After restoring a DeHavilland DH 82A Tiger Moth over twenty years ago, Giancarlo Zinardo progressed to building his own airplanes with the construction of a Fokker DR1 replica. He hasn't stopped since.

The prolific Italian builder's earlier accomplishments include a Bleriot X1-2 and a Wright Flyer. Moving from the "Romantic Era" machines into the WWII Warbird marques, Giancarlo recently rolled out a beautiful .7 scale P-51D Mustang based on Marcel Jurca's MJ 77 design. The Chevrolet 350 V-8 powered aircraft required 8,000 hours of labor over a period of four years.

Not one to rest on his accomplishments, Giancarlo is now planning to fill out his Flying Circus with two more replicas, a Supermarine Spitfire and a Messerschmitt BF 109 (both .9 scale).

Brugar

Lucio Turchet

Giancarlo Zanardo
via Cadore, 12
Conegliano (TV)
ITALIA

cal antagonists ability to "wiggle" its victim. In the case for both inductors and capacitors, the faster the rise time (slope of the wiggle waveform) the more effective the attenuation of any given filter will become. This is why small filter components can strip out the effects of VHF radio frequency interference (100 MHz = very fast wiggles) and have little or no effect on audio signals (1 KHz... a whole lot slower). Larger filter components can attenuate alternator whine (1 KHz) and have little effect on the 14 volt DC power which carries the noise into your audio system.

Ferrite Torroids and Beads are often suggested as potential cures for noise problems. As a general rule, they work only when the propagation path in question is CONDUCTED, only if the interfering signal is in the tens of megahertz or higher, and then only if the circuitry downstream of the ferrite is of sufficiently low impedance at the interfering frequency to let the ferrite do its job. The ferrite cores put around a wire are simply a single turn inductor with a higher inductance than if the cores were not present. Ferrite cores decrease rigidity in the conducting path, but we still need some downstream mass. Shunt capacitors are the first components to consider in any noise filtering problem with inductors to be added when still more attenuation is needed. I've found very few instances where adding ferrite cores to a wire or wire bundle made a problem go away.

Sizing filter components is a function of how much current flows in the path, strength of the interfering signal and its lowest frequency. Ninety percent of the amateur builder's noise problems will involve alternator noise, and most of those cases will be fixed by proper grounding or wire routing techniques. The remaining cases will benefit from filter assemblies and/or filter components that are readily available for automotive audio systems. For the more complicated cases, filter components will have to be sized to the task and fabricated from parts purchased from electronics suppliers. Let's talk either by e-mail or phone and we'll help you size both the task and the cure.

On a spectrum chart, the lowest frequency system is Loran, followed by ADF, marker beacons, VHF comm and omni nav, topped off by

CONTINUED ON PAGE 408

Engine Guide

Original photo courtesy Orenda Aerospace Corp.

Homebuilt "FAQ's"

Frequently Asked Questions posted to the Internet Newsgroup *rec. aviation.homebuilt.* Originally compiled by Steve Cornelius and maintained by Ron Wanttaja along with additional homebuilt aircraft information on his web page at http://www.halcyon.com/wanttaja/avlinks.html

Q: What types of engines can I use in a homebuilt?

A: Literally any. For years, homebuilders have relied on certified engines. But, as the cost of these engines rises, alternatives have been found. Several companies are building specialty engines specifically for experimental use, and others have adapted automotive engines.There are both two and four stroke engines availabe for experimental use and most are water cooled. All are fairly reasonable in cost compared to their certified counterparts. These products develop power in the 40 to 100hp range.

Some builders have taken the time to customize and tune auto engines to the application with good results. The all-aluminum small-block Chevrolet is becoming the poor-mans "Mini Merlin" and produces excellent power (not to mention sound!). Some are also experimenting with Wankel-type rotaries, as their power-to-weight ratio and physical compactness make them excellent candidates for aircraft use. Power in auto conversions can run from 80 to 400+ hp. Builders have used air cooled Volkswagen conversions for years, and several companies are now producing parts and completed engines. VW engines sound similar to their Lycoming or Continental counterparts but rev higher, in the 3100rpm range.The original crankshafts in these engines are usually replaced with stronger, custom crankshfts.

When choosing a certified engine, the engine must be equipped and maintained as it normally would be for a certified aircraft. This means that all AD and bulletins must be complied with. This can have some impact on your flight test period. If you make any mods to such an engine, it will be considered a non-certified engine by the FAA, which will increase the test period from 25 to 40 hours. It will also reduce the market value of an otherwise expensive engine.

Q: Why all the controversy regarding two stoke vs. four stroke engines?

A: The four stroke, air cooled engine has been the mainstay of light aviation for over 50 years and this shows no immediate signs of change.However, some two stroke designs have attracted a strong following in recent years.

Why has this happened? The two most obvious factors are cost and weight. A 65hp Rotax costs and weighs roughly 60% of its certified Continental A-65 counterpart. For the newer generation of "portable" lightplanes, this reduced weight is an obvious design advantage. In addition to the lower initial cost, these engines can be majored by the owner in his garage for a few hundred dollars.

There are pilots out there who think of the idea of a "chainsaw" engine as anathema to a proper aircraft. Improper mixture control in a two stroke can damage the engine quickly. The vibration through the airframe is of a higher pitch. The higher operating speed of the engine requires a geared speed reduction system. The sound of two-stroke aircraft is simply unpleasant to some eardrums.

Many of the reliability questions regarding two strokes may have resulted from improper installation, operation, or maintenance. Often the $10,000 certified engine understandably gets "fussed" over considerably more. Its care and feeding are well understood by most pilots since who are usually trained behind such engines.

The choice of two stroke or four stroke depends on your preferences and requirements. If you need more than 65hp, your decision is essentially made for now. If not, study your options and carefully evaluate your needs. Ride behind both and see which you're comfortable with.

AeroCrafter ENGINE GUIDE

The following data, supplied by the manufacturers, provides basic information on many of the powerplants suitable for homebuilt aircraft. Contact the manufacturers or distributors listed for more specific information regarding the requirements of your particular installation. Dealers for **Engine Accessories** ***are listed in the*** **SUPPLIER Section.**

2 Stroke International
Beaufort Industrial Park #8
Schein Loop
Beaufort SC 29902
Tel: 803-846-2167
Fax: 803-846-2169

• 230 F22
Single cylinder, two stroke, dual ignition, air cooled, carbureted
Max. HP: 21.9 hp @ 6500 rpm
Weight dry: 42 lbs.
Belt Reduction
Rotation: CCW
Fuel: 92 oct. min.
Fuel Burn @75%: 2.1 gal/hr

• 460 F40
Two cylinder, inline, two stroke, dual ignition, air cooled, carbureted
Max. HP: 41 hp @ 6500 rpm
Weight dry: 69.5 lbs
Rotation: CCW
Fuel: 87 oct. min.
Fuel Burn@ 75%: 3.7 gal/hr

• 460 F45
Two cylinder, inline, two stroke, dual ignition, air cooled, carbureted
Max. HP: 45.4 hp @ 6500 rpm
Weight dry: 72 lbs.
Rotation: CCW
Fuel: 92 oct. min.
Fuel Burn @ 75%: 4.0 gal/hr

• 540 L70
Two cylinder, inline, two stroke dual ignition, liquid cooled, carbureted
Max. HP: 70 hp @ 7000 rpm
Weight dry: 102 lbs.
Rotation: CW facing gearbox
Fuel: 92 oct. min.
Fuel Burn @ 75%: 5.6 gal/hr

• 808 L95
Three cylinder, two stroke, inline, dual ignition, liquid cooled, carbureted
Max. HP: 95 hp @ 7000 rpm
Weight dry: 130 lbs.
Rotation: CW facing gearbox
Fuel: 92 oct. min.
Fuel Burn @ 75%: 7.3 gal/hr

• 808 L100
Three cylinder, two stroke, inline, dual ignition, liquid cooled, fuel injected
Max. HP: 100 hp @ 7000 rpm
Weight dry: 130 lbs.
Rotation: CW facing gearbox
Fuel: 92 oct. min.
Fuel Burn @ 75%: 7.3 gal/hr

A.K.S. Inc.
P.O. Box 17161
Portland OR 97217-0161
Tel. 503-247-9207
Fax: 503-289-1491
e-mail: trikes@teleport.com
http://www.teleport.com/~trikes

• Verner SVS-1400
Two cylinder, opposed, four stroke, dual carbureted, dual CDI electronic ignition, air cooled
Max. HP: 80 hp @ 5000 rpm
Weight: 137 lbs. incl. reduction drive, starter, carbs, generator
Fuel: 92 oct. min. or 100LL
Fuel Burn: 1.9 @ 3500 rpm
TBO: est. 500+ hrs.

Alturair
660 Steel Street
El Cajon CA 92020
Tel: 619-449-1570
Fax: 619-442-0481

• A650-1
Single rotor rotary, solid state ignition, liquid cooled, carbureted
Max. HP: 100 hp @ 6000 rpm
Weight dry: 100 lbs.

• A650-2
Twin rotor rotary, solid state ignition, liquid cooled, carbureted
Max. HP: 200 hp @ 6000 rpm
Weight dry: 200 lbs.

Amtec Corporation
4808 Bradford Boulevard
Huntsville AL 35805-1949
Tel. 205-722-7200

• Buddy Twin (in development)
Two cylinder opposed, 1250cc, double-eccentric crankshaft
Max. HP: 40 hp @ 2900 rpm
Weight: 69 lbs.
TBO: 1000 hrs.

Arrow SNC
Via Badiaschi 25
Piacenza 29100 ITALY
Tel: 523-41932
Fax: 523-41340

• AE 270 AC
Single cylinder, two stroke, free air cooled
Max. HP: 35 hp @ 6800 rpm
Fuel: 96-98 oct. or 100 oct. Avgas
Weight: 13.2 lbs. including carb, fuel pump, starter, and alternator

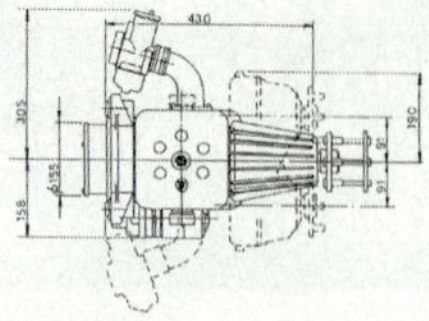

• Gp 500 AC
Two stroke, two cylinder opposed, electronic dual ignition, air cooled, carbureted
Max. HP: 60 hp @ 6800 rpm
Fuel: 96-98 oct. auto, 100 oct. Avgas
Rotation: CCW facing prop flange with reduction drive
Weight: 74.8 lbs. including carb, fuel pump, starter and alternator
TBO: 400 hrs.

• GP 1000
Four cylinder, horizontally opposed, two stroke, fuel-oil mix, free air cooled, reed valve induction
Max. HP: 100-120 hp @ 6800 rpm
Fuel: 96-98 oct. or 100 oct. Avgas
Weight: 124.5 lbs. including carb, fuel pumps, and starter
TBO: 300 hrs.

Atlas Motors, Inc.
685 North Loudoun St.
Winchester VA 22601
Tel: 800-336-9878
Fax: 703-665-2025

• Franklin/PZL Engines
120-250 hp certified aircraft engines

Belted Air Power
1408 Western Avenue
Las Vegas NV 89102
Tel: 702-384-8006
Fax: 702-384-8006

• Buick V-8
V-8, eight cylinder, point or electronic ignition, liquid cooled, carbureted or fuel injected
Max. HP: 150 hp @ 3500 rpm
Weight dry: 270 lbs.

• Vortec V-6
V-6, six cylinder, points or electronic ignition, liquid cooled, carbureted or fuel injected
Max. HP: 187 hp @ 3500 rpm
Weight dry: 290 lbs.

• Chevrolet 350
V-8, eight cylinder, liquid cooled, carbureted, points or electronic ignition
Max. HP: 246 hp @3500 rpm

Bradley Aerospace
236A W. East Ave. #260
Chico CA 95926
Tel: 916-899-7918
Fax: 916-899-2942
VW and Porsche Engine Conversions. Turbo Prop and Jet Engines.

• 1600cc
Max. HP: 53 hp
Weight: 128 lbs.
Direct Drive
TBO: 1000 hrs.
Price: $4,995

• 2000cc
Max. HP: 69 hp
Weight: 133 lbs.
Direct Drive
TBO: 1000 hrs.
Price: $5,800

• 2400cc
Max. HP: 92 hp
Weight: 178 lbs.
Direct Drive
TBO: 1000 hrs.
Price: $6,895

• 2800cc
Max. HP: 115 hp
Weight: 185 lbs. complete
Direct Drive
TBO: 1000 hrs.
Price: $7,800

The Cam-Fire Engine Group
9755 W Saanich Road
Sidney, British Columbia
V8L 5T5
CANADA
Tel: 604-656-4774
Fax: 604-656-8007

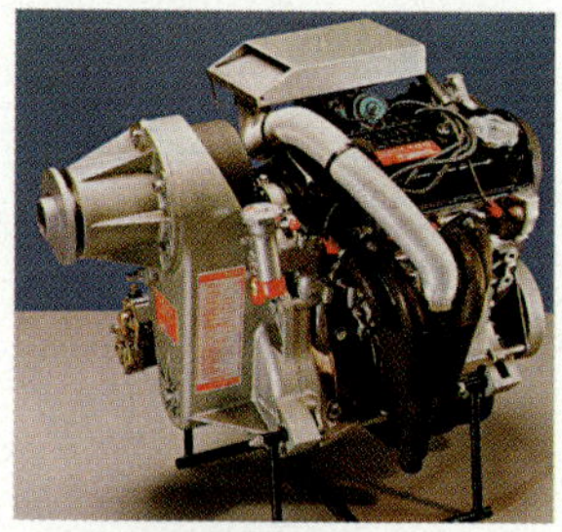

• Cam 100
Four cylinder, four stroke, liquid cooled, carbureted, or fuel injected
Max. HP: 100 hp
Weight dry: 224.3 lbs.
Fuel: 90 oct. min. or 100 LL
Fuel Burn: 3.5 to 4.5 gal/hr
Price: $7, 950

Christine Aero Engines
PO Box 216
Highway 178
Donalds SC 29638
Tel: 864-379-8677

• C1200
Two cylinder, four cycle, horizontal opposed, dual point or CDI, air cooled, carbureted
Max. HP: 48 hp @ 3200 rpm
Fuel: Minimum 100LL Aviation or 91 Auto
Weight dry: 82 lbs.
TBO: est. 1000 hrs.

• C2400-S
Two cylinder, four cycle, horizontal opposed, dual ignition, air cooled, single carburetor
Max. HP: 87 hp @ 3500 rpm
Fuel: Minimum 100LL Aviation or 91 Auto
Weight dry: 82 lbs.
TBO: est. 1000

Christine Aero Engines Cont,

• C2400-S
Two cylinder, four cycle, horizontal opposed, dual ignition, air cooled, dual carburetor
Max. HP: 93.8 hp @ 3500 rpm
Fuel: Minimum 100LL Aviation or 91 Auto
Weight dry: 124 lbs.
TBO: est. 1000 hrs.

Crossflow Aero Corp.

RR#1 Palgrave
Ontario L0N 1P0
CANADA
Tel: 519-942-9855
Fax: 519-942-8503

• CF4-20/20t/20tho
Four cylinder, four cycle, four cam, 2 liter, Multiport EFI, Direct Fire ignition. t=turbocharged, tho= turbocharged high output
Max. HP: 180/200/250 hp
Reduction Drive: Planetary gear
Fuel: Premium automotive/ 100LL
Fuel Burn: 7/7.9 gal/hr
Weight: <340 lbs. wet
Price: $17,000/$18,700/$24,000

• CF4-33/33t/33tho
Four cylinder, four cycle, four cam, 3.3 liter, Multiport EFI, Direct Fire ignition
Max. HP: 250/300/320 hp
Reduction Drive: Planetary gear
Fuel: Premium automotive/ 100LL
Fuel Burn: 6.9/8.5/9 gal/hr
Weight: <340 lbs. wet
Price: $19,500/$21,100/$22,700

DeltaHawk, Inc.

3003 Golf Ave.
Racine WI 53404
Tel: 414-634-9660
Fax: 414-634-9625
E-mail dhawk@execpc.com

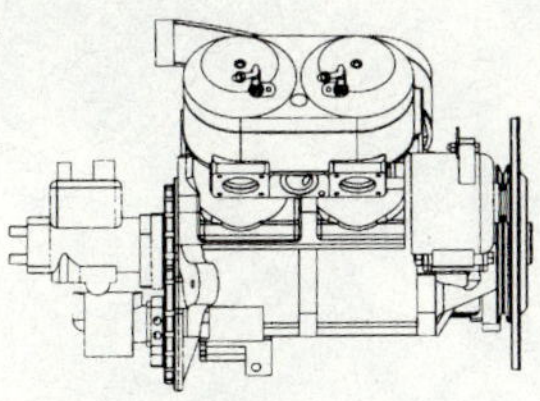

• V-4 Turbo Diesel
Upright 90 degree V-4, turbocharged, direct drive, two stroke, liquid cooled
Max. HP: 200 hp @ 2500 rpm
Fuel: Jet-A(JP5) or #2 diesel (>20 degrees F)
Fuel Burn: 7/7.9 gal/hr
Weight: < 300 lbs. total installed, wet

Dyna-Cam Aero Engine Corp.

PO Box 1159
Torrance CA 90505-0159
Tel: 310-791-4642
Fax: 310-791-4645

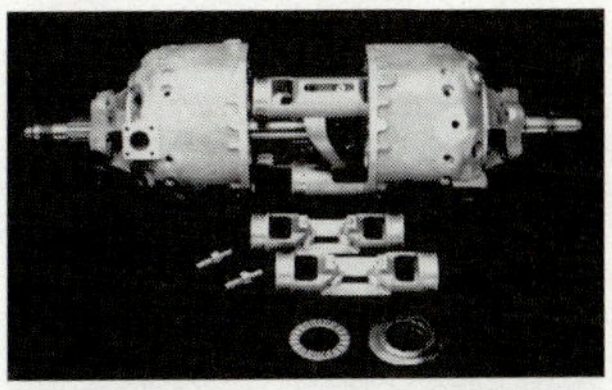

• DC-375
Turbine-shaped, twelve cylinder, six double acting pistons, fuel injected, magneto or electronic ignition, liquid cooled
Max. HP: 200 hp @ 1900 rpm
Weight dry: 265 lbs.
TBO: 2000 hrs.

Eggenfellner Advanced

Aircraft, Inc.
427-3 Amherst St., Ste. 230
Nashua NH 03063
Tel: 800-840-4620

• Subaru 2.2 L
Four cylinder, opposed, four cycle, fuel injected
Max. HP: 135 hp
Reduction Drive: Geared
Weight: 276 lbs.
Price: $7,900

• Subaru 2.5 L
Four cylinder, opposed, four cycle, fuel injected
Max. HP: 155 & 165 hp
Reduction Drive: Geared
Weight: 306 lbs.
Price: $8,900

Engineair Inc.

209 Cessna Blvd.
Daytona Beach FL 32124
Tel: 904-788-2852

• GTS IV
V-8, eight cylinder, electronic ignition, liquid cooled
Max. HP: 420 hp @ 2375 rpm
Weight dry: 580 lbs.

Flite-Lite, Inc.

200 Coventry Drive
Lexington SC 29072
Tel: 803-356-8587
Fax: 803-951-3398

• EMDAIR CF 122
Two cylinder, opposed, fuel injected, electronic ignition, air cooled
Max. HP: 80 hp @ 3000 rpm
Weight dry: 129 lbs.
Price: $9,400

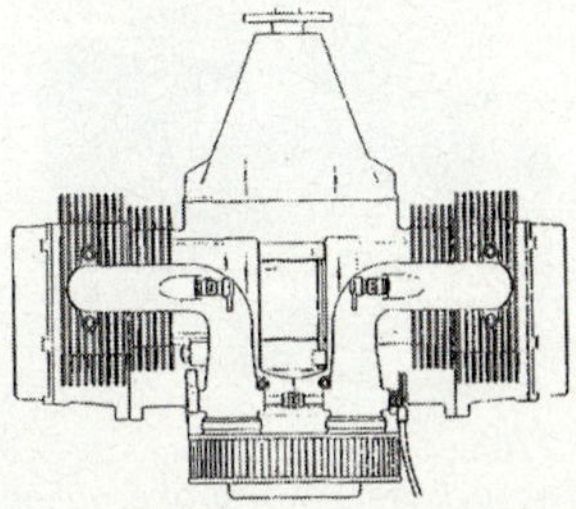

• EMDAIR MP60
Two cylinder, four stroke, opposed, fuel injected, electronic ignition, air cooled
Max. HP: 60 hp @ 5750 rpm
Fuel: 100LL or Mogas
Weight dry: 95 lbs.
Price: $5,400

Formula Power

1060 Baldy Park
PO Box 2168
Sandpoint ID 83864
Tel: 208-263-6637

• Subaru EA 81
Opposed, four cylinder, carbureted or fuel injected, liquid cooled
Max. HP: 80 hp @ 4800 rpm

• Subaru EA22
Opposed, four cylinder, fuel injected, electronic dual ignition, liquid cooled
Max. HP: 120 hp @ 4800 rpm

• SVX
Opposed, six cylinder, fuel injected, electronic ignition, liquid cooled
Max. HP: 260 hp @ 5400 rpm

Gateway Aviation

7124 Lanai Street
Long Beach CA 90808
Tel: 562-425-1016
Fax: 562-425-8407

• T-62
Turbine engines available with Ross Aeronautical Engineering reduction drives geared to 1892, 2105, 2765, or 3243 rpm
Max. HP: 225 Shaft Horsepower

Göbler-Hirthmotoren

Postfach 62
71726 Benningen/Neckar
GERMANY
Tel: 07144 6074
Fax: 07144 5415

U.S. Distributor
Recreational Power Engineering
5479 E. Country Road 38
Tiffin OH 44883
Ph:419-585-7002
Fax: 419-585-6004

• F-33
Single cylinder, carbureted, dual CDI, air cooled
Max. HP: 25 hp @ 5200 rpm
Fuel: 93 oct. unleaded auto
Rotation: CCW facing PTO
Weight: 35 lbs. with std. equip.

• F-23
Two cylinder, two stroke, piston ported, carbureted, electronic dual ignition, air cooled
Max. HP: 40 hp @ 5500 rpm
Fuel: 93 oct. unleaded auto
Rotation: CCW facing PTO.
Gearbox reverses rotation.
Weight: 70 lbs. with std. equip.
TBO: 1000 hrs.

• 2702
Two cylinder, two stroke, inline, carbureted, dual CDI, air cooled
Max. HP: 40 hp @ 5500 rpm
Fuel: 93 oct. unleaded auto
Rotation: CCW facing PTO.
Gearbox reverses rotation.
Weight: 71 lbs. with std. equip.

• 2703
Two cylinder, two stroke, inline, piston ported, electronic dual ignition, fan cooled
Max. HP: 55 hp @ 6200 rpm
Fuel: 93 oct. unleaded auto
Rotation: CCW facing PTO.
Gearbox reverses rotation.
Weight: 82 lbs. with std. equip.
TBO: 1000 hrs.

• 2704
Two cylinder, two stroke, inline, carbureted, dual CDI, air cooled
Max. HP: 50 hp @ 5000 rpm
Fuel: 93 oct. unleaded auto
Rotation: CCW facing PTO.
Gearbox reverses rotation.
Weight: 82 lbs. with std. equip.
TBO: 1000 hrs.

• 2706
Two cylinder, two stroke, inline, carbureted, dual CDI, air cooled
Max. HP: 65 hp @ 6300 rpm
Fuel: 93 oct. unleaded auto
Rotation: CCW facing PTO.
Gearbox reverses rotation.
Weight: 82 lbs. with std. equip.
TBO: 1000 hrs.

• F-30
Four cylinder, two stroke, opposed, carbureted, dual CDI, air cooled
Max. HP: 110 hp @ 6200 rpm
Fuel: 93 oct. unleaded auto
Weight: 109 lbs. with std. equip.
TBO: 1000 hrs.

Great Plains Aircraft

P.O. Box 545
Boys Town NE 68010
Tel: 402-493-6507
Fax: 402-333-7750

• Type 1 1600 VW
Four cylinder, horizontally opposed, air cooled
Max. HP: 55 hp @ 3600 rpm
Ignition: Slick magneto
Fuel: 100 oct. min.
Weight: 159.2 lbs.

• Type 1 1835 VW
Four cylinder, horizontally opposed, air cooled
Max. HP: 65 hp@3600 rpm
Ignition: Slick magneto
Fuel: 100 oct. min.
Weight: 160 lbs.

• Type 4 2400 VW
Max. HP: 83 hp @ 3600 rpm
Ignition: Slick magneto
Fuel: 100 oct. min.
Weight: 188 lbs.

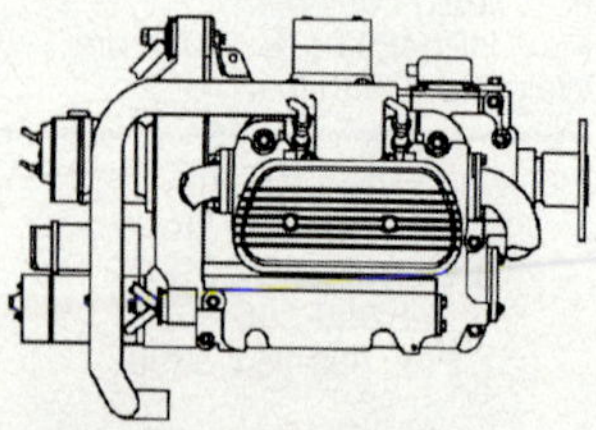

• Type 4 2600 VW
Max. HP: 96 hp @ 3400 rpm
Ignition: Slick magneto
Fuel: 100 oct. min.
Weight: 190 lbs.

HKS

Shizouka, Japan

U.S. Distributor
HPower Ltd.
P.O. Box 690
Ellington, CT 06029
Tel: 860-875-8185
Fax: 860-870-5499
http://www.hpower-ltd.com

• 700-E
Two cylinder, horizontally opposed, four stroke, dual carburetor, dual electronic ignition, air cooled
Max. HP: 60 hp@ 5900 rpm
Weight dry: 103 lbs. including gearbox and starter
Fuel: unleaded premium auto
Fuel Burn: 3.4 gal/hr @ 4750 rpm
TBO: 500 hrs
Price: $6,200

Hummel Aviation

509 E. Butler Street
Bryan OH 43506-0880
Tel: 419-636-3390

• 1/2 VW
Two cylinder, opposed, carbureted, magneto ignition, air cooled
Max. HP: 37 hp@ 3000 rpm
Weight dry: 80 lbs.

JPX
Z1 Nord
72320 Vibraye
FRANCE
Tel: 43 93 61 74
Fax: 43 93 62 71

U.S. Distributor
Leroy Enterprises
Rt. 3 Box 7150
Wilcox AZ 85643
Tel: 520-384-2611
Fax: 520-384-2612
http://www.leroyent.com

• 4TX75/A
Four cylinder, horizontally opposed, air cooled
Max. HP: 85 hp@ 3000 rpm
Rotation: CW from front
Direct drive
Fuel: UL 95 Oct. or 100 LL
Fuel Burn: 2.5 to 3 gal/hr @ cruise
Weight: 172 lbs. totally equipped less exhaust.
Price: $9,985

Jabiru Aircraft Pty Ltd
PO Box 5186
Bundaberg West
Queensland 4670
AUSTRALIA
Tel: 61-71-551-778
Fax: 61-71-552-669

North American Distributor
Tiger Aviation
134 Cessna Drive
Trenton SC 29847
Tel: 1-800-522-4781
Fax: 803-663-6655
Email tremfl@aol.com
http://www.tigeraviation.com

• 2200
Four cylinder, four stroke, horizontally opposed, dual transistorized magneto ignition, carbureted, wet sump lubrication, ram air cooled
Max. HP: 80 hp @ 3300 rpm
Fuel: 100/130 Avgas or Mogas
Fuel Burn: 3.75 gal/hr @ 75%
Prop Rotation: clockwise (pilot's view)
Weight: 123 lbs. complete
TBO: 1000 hrs.

Jones Custom Aircraft Services
4010 SW 4th Street
PO Box 423487
Kissimmee FL 34742
Tel: 407-846-1244
Fax: 407-846-1244
E-mail:
SJONESFLY@AOL.COM

• Chevrolet V-8 350
V-8, eight cylinder, carbureted or fuel injected, distributor or electronic ignition, liquid cooled
Max. HP: 340 hp @ 4500 rpm
Weight dry: 525 lbs.

• Chevrolet V-8 502
V-8, eight cylinder, carbureted or fuel injected, distributor or electronic ignition, liquid cooled
Max. HP: 440 hp @ 4500 rpm
Weight dry: 625 lbs.

König Motorenbau KG
Friedrich Albricht Damm 72
1000 Berlin 13
WEST GERMANY
Tel: 030-344 30 71
Fax: 030-344 15 95

• SC 430
Two stroke, rotary valve induction, three cylinder radial, carbureted, 3 CDI, air cooled
Max. HP: 24 hp @ 4200 rpm
Reduction: Belt
Fuel Burn: .5 liter/HP/hr
Weight: 35.2 lbs.

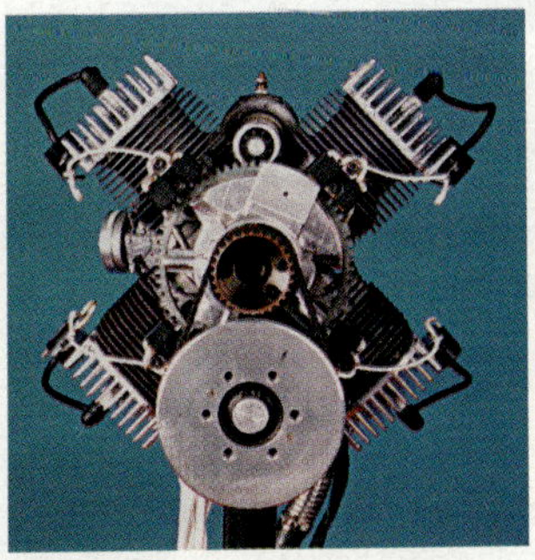

• SD 570
Two stroke, rotary valve induction, four cylinder radial, carbureted, 4 CDI, air cooled
Max. HP: 28 hp @ 4200 rpm
Reduction (optional): Belt
Fuel Burn: .5 liter/HP/hr
Weight: 35.2 lbs.

LOM Prague Aircraft Engines
100 38 Praha 10
Malesice, Cernokostelecka 270
CZECH REPUBLIC
Tel: 70 36 38
Fax: 70 65 23

North American Distributors

Krybus Aviation
822 Santa Maria Street
PO Box 14
Santa Paula CA 93060
Tel: 805-525-8764
Fax: 805-525-1147

Moravia, Inc.
P.O. Box 942
Thunder Bay, Ontario
P7C 4X8
CANADA
Tel: 807-475-5113
Fax: 807-475-9587

• LOM 132A
Four cylinder, inverted, inline, four stroke, fuel injected, air cooled
Max. HP: 120 hp @ 2700 rpm
Fuel Burn: 7.1 gal/hr @ 2400 rpm
Direct Drive
Rotation: CCW - crankshaft
Weight dry: 231 lbs.
TBO: 2000 hrs.

• LOM 332A
Four cylinder, four stroke, inverted, inline, fuel injected, air cooled
Max. HP: 140 hp @ 2700 rpm
Direct Drive
Fuel Burn: 7.6 gal/hr @ 2400 rpm
Rotation: CCW - crankshaft
Weight dry: 249 lbs.
TBO: 2000 hrs.

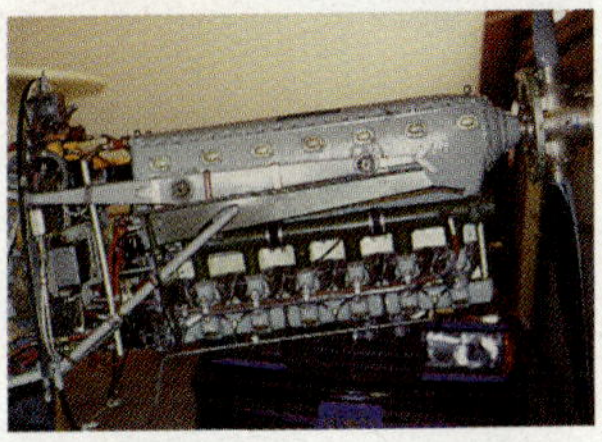

• LOM 137
Six cylinder, inverted, inline, four stroke, fuel injected, air cooled
Max. HP: 179 hp @ 2750 rpm
Fuel Burn: 7.6 gal/hr @ 2400 rpm
Rotation: CCW - crankshaft
Direct Drive
Weight dry: 325 lbs.
TBO: 2000 hrs.

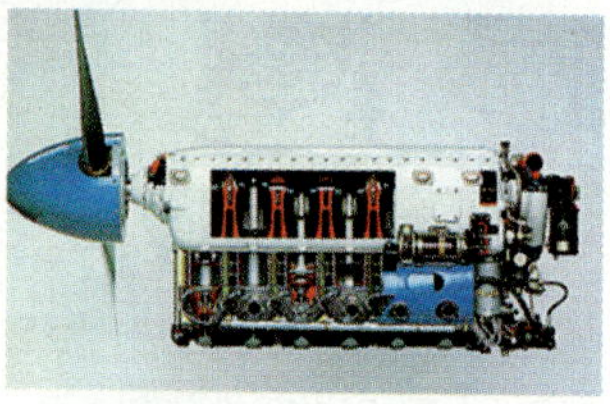

• LOM 337
Six cylinder, inverted, inline, four stroke, fuel injected, air cooled
Max. HP: 210 hp @ 2750 rpm
Fuel Burn: 7.6 gal/hr @ 2400 rpm
Rotation: CCW - crankshaft
Direct Drive
Weight dry: 350 lbs.
TBO: 2000 hrs.

Light Power Engine Corp.
PO Drawer 3350
Morgantown WV 26503
Tel: 304-291-3843
Fax: 304-292-1902

• III-200
Four cylinder, inline, inverted, fuel injected, dual electronic ignition, liquid cooled
Max. HP: 150 hp @ 2700 rpm
Weight dry: 229 lbs.
Direct drive
TBO: 2200 hrs.

• TIII-200
Four cylinder, inline, turbocharged, fuel injected, dual electronic ignition, liquid cooled
Max. HP: 175 hp @ 2700 rpm
Weight dry: 246 lbs.
Direct drive
TBO: 2200 hrs.

• IIV-400A
V-8, fuel injected, dual electronic ignition, liquid cooled
Max. HP: 230 hp @ 2700 rpm
Direct drive
Weight dry: 400 lbs.
TBO: 2200 hrs.

• IIV-400B
V-8, fuel injected, dual electronic ignition, liquid cooled
Max. HP: 250 hp @ 2700 rpm
Weight dry: 400 lbs.
Direct drive
TBO: 2200 hrs.

Light Power Engine Corp. Continued

• IIV-500
V-8, fuel injected, dual electronic ignition, liquid cooled
Max. HP: 320 hp @ 2700 rpm
Weight dry: 480 lbs.
Direct drive
TBO: 2200 hrs.

• TIIV-500
V-8, fuel injected, turbocharged, dual electronic ignition, liquid cooled
Max. HP: 320 hp @ 2700 rpm
Weight dry: 568 lbs.
Direct drive
TBO: 2200 hrs.

• IVG-500B
V-8, geared, fuel injected, liquid cooled
Max. HP: 500 hp @ 4300 rpm
Weight dry: 550 lbs.
TBO: 2200 hrs.

• IIV-560
V-8, fuel injected, dual electronic ignition, liquid cooled
Max. HP: 350 hp @ 2700 rpm
Weight dry: 480 lbs.
TBO: 2200 hrs.

• IVG-560
V-8, geared, fuel injected, liquid cooled
Max. HP: 513 hp @ 4300 rpm
TBO: 2200 hrs.

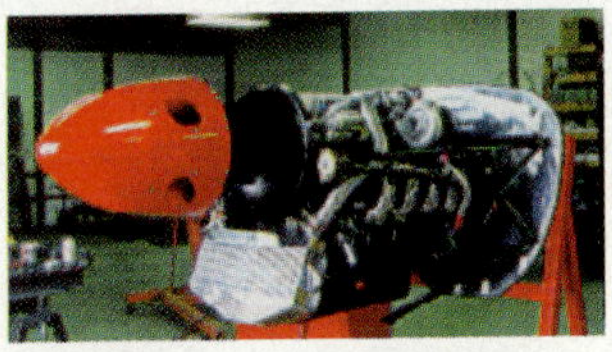

• IIV-600
V-8, fuel injected, dual electronic ignition, liquid cooled
Max. HP: 380 hp @ 2700 rpm
Weight dry: 517 lbs.
Direct drive
TBO: 2200 hrs.

• TIIV-600
V-8, fuel injected, turbocharged, dual electronic ignition, liquid cooled
Max. HP: 380 hp @ 2700 rpm
Weight dry: 578 lbs.
Direct drive
TBO: 2200 hrs.

• IVG-600
V-8, fuel injected, geared, turbocharged, dual electronic ignition, liquid cooled
Max. HP: 600 hp @ 4300 rpm
Weight dry: 562 lbs.
TBO: 2200 hrs.

• TIVG-600
V-8, fuel injected, geared, dual electronic ignition, liquid cooled
Max. HP: 600 hp @ 4300 rpm
Weight dry: 580 lbs.
TBO: 2200 hrs.

• IIV-650
V-8, fuel injected, dual electronic ignition, liquid cooled
Max. HP: 450 hp @ 2700 rpm
Weight dry: 531 lbs.
Direct drive
TBO: 2200 hrs.

• TIIV-650
V-8, fuel injected, turbocharged, dual electronic ignition, liquid cooled
Max. HP: 500 hp @ 2700 rpm
Weight dry: 596 lbs.
Direct drive
TBO: 2200 hrs.

Lightning Engines

4099 Brick Church Pike
White Creek TN 37189
Tel: 615-865-1802
Fax: 615-865-9164

• Chevrolet V-8
V-8, carbureted or fuel injected, liquid cooled
Max. HP: 600 hp @ 6000 rpm
Weight dry: 460 lbs.

Limbach Flugmotoren GmbH
Kotthausener Strasse 5
D-53639
Köenigswinter-Sassenberg
GERMANY
Tel: 224492010
Fax: 224492013

Limbach Aircraft Engines (USA)
3114 North Sunset Ave.
Sand Springs, OK 74063
Tel: 918-241-5468
Fax: 918-245-6910

• SL 1700 EA
Four cylinder, horizontally opposed, air cooled, single magneto ignition, wet sump pressure lubrication
Max. HP: 60 hp @ 3550 rpm
Fuel: 96 oct.
Direct Drive
Rotation: CCW
Weight: 150 lbs.
Fuel Burn: 2.6 gal/hr.
TBO: 1200 hrs

• L 1800 EA.X
Four cylinder, horizontally opposed, air cooled, single magneto ignition, wet sump pressure lubrication
Max. HP: 64 hp @ 3400 rpm
Fuel: 96 oct.
Direct Drive
Rotation: CCW
Weight: 150 lbs.
Fuel Burn: 2.6 gal/hr.
TBO: 1200 hrs.

• L 2000 DA
Four cylinder, horizontally opposed, air cooled, double magneto ignition, wet sump pressure lubrication
Max. HP: 75 hp @ 3200 rpm
Fuel: 96 oct.
Direct Drive
Rotation: CCW
Weight: 164 lbs.
Fuel Burn: 3.1 gal/hr.
TBO: 1000 hrs.

• L 2000 EA
Four cylinder, horizontally opposed, single magneto ignition, wet sump pressure lubrication
Max. HP: 80 hp @ 3400 rpm
Fuel: 96 oct.
Direct Drive
Rotation: CCW
Weight: 152 lbs.
Fuel Burn: 3.1 gal/hr.
TBO: 1000 hrs.

• L 2000 EO, EC
Four cylinder, horizontally opposed, air cooled, single magneto ignition, wet sump pressure lubrication
Max. HP: 80 hp @ 3400 rpm
Fuel: 96 oct.
Direct Drive
Rotation: CCW
Weight: 163 lbs.
Fuel Burn: 3.1 gal/hr.
TBO: 1000 hrs.

• L 2000 EB 1
Four cylinder, horizontally opposed, air cooled, single magneto ignition, wet sump pressure lubrication
Max. HP: 80 hp @ 3400 rpm
Cont. HP: 72 @ 3000 rpm
Fuel: 96 oct.
Direct Drive
Rotation: CCW
Weight: 159.5 lbs.
Fuel Burn: 3.1 gal/hr.
TBO: 1000 hrs.

• L 2400 DE3.X
Four cylinder, horizontally opposed, air cooled, dual magneto ignition, wet sump pressure lubrication
Max. HP: 90 hp @ 3200 rpm
Fuel: 96 oct.
Direct Drive
Rotation: CCW
Weight: 192.5 lbs.
Fuel Burn: 3.6 gal/hr.
TBO: 1400 hrs.

• L 2400 EO.X
Four cylinder, horizontally opposed air cooled engine, with single magneto ignition, wet sump pressure lubrication
Max. HP: 87 hp @ 3200 rpm
Fuel: 96 oct.
Direct Drive
Rotation: CCW
Weight: 178 lbs.
Fuel Burn: 3.6 gal/hr.
TBO: 1400 hrs.

• L 2400 EB
Four cylinder, horizontally opposed, air cooled, single magneto ignition, wet sump pressure lubrication
Max. HP: 87 hp @ 3200 rpm
Fuel: 96 oct.
Direct Drive
Rotation: CCW
Weight: 180 lbs.
Fuel Burn: 3.6 gal/hr.
TBO: 1400 hrs.

• L 2400 EFI
Four cylinder, horizontally opposed, air cooled, fuel injected, fully electronic system
Max. HP: 100 hp @ 3000 rpm
Fuel: 95 oct.
Direct Drive
Rotation: CCW
Weight: 187 lbs.
Fuel Burn: 3.9 gal/hr.
TBO: 1400 hrs.

MDB Flugtechnik AG
CH-3368 Bleienbach
Langenthal, SWITZERLAND
Tel: 41 62 923 45 45
Fax: 41 62 923 24 29

U.S. Distributor
MDB Aerospace Corp.
13420 Reese Blvd. West
Huntersville NC 28078
Tel: 704-948-1234
Fax: 704-875-0781

• FM2600
Four cylinder, horizontally opposed, fuel injected, liquid cooled
Max. HP: 160 hp @ 4500 rpm

Motavia Ltd.
Redlands Airfield
Lower Wanborough
Swindon, Wiltshire SN4 0AA
ENGLAND
Tel: 44 1793 791133
Fax: 44 1793 791155
Email sales@motavia.com

Motavia USA
P.O. Box 388
Port Angeles WA 98362
Tel/Fax: 360-452-0117
Email motavia@olypen.com

• Ultratec 60/80
Two cylinder, four stroke, fuel injected, liquid cooled, dry sump
Max. HP: 81 hp @ 7000 rpm
(Engine Management Unit upgrade to 60 hp version)
Fuel: Unleaded
Fuel Burn: 2 gal/hr
Weight dry: 125 lbs. complete

Nelson Engine of Florida, Inc.
420 Harbor Drive
Naples FL 33940
Tel: 813-263-1670
Fax: 813-261-8345

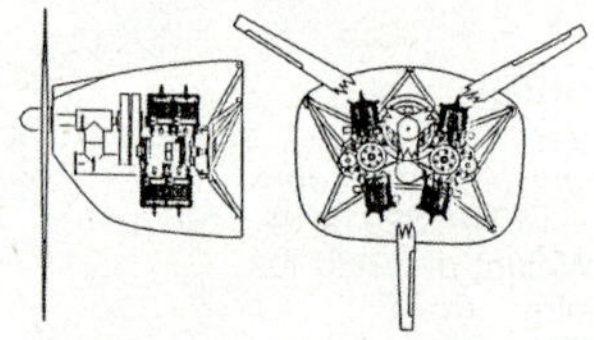

• H-63-CP-2
Twin engine, single prop, dual ignition, also available as a helicopter engine
Max. HP: 90 hp @ 2135 rpm
Fuel: Unleaded auto
Weight: 190 lbs. incl. prop and accessories
TBO: 800 hrs.

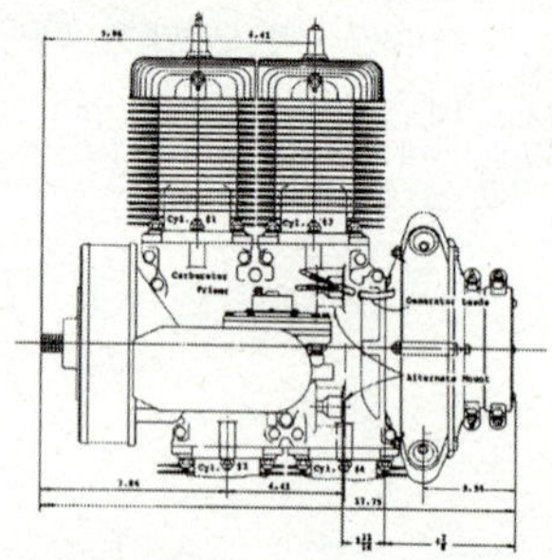

• H-63-CP
Two cylinder, two cycle, air cooled, FAA certified
Max. HP: 48 hp @ 4400 rpm
Fuel: Unleaded auto
Weight: 68 lbs. incl. starter, carb, and generator
TBO: 800 hrs.

Northwest Aero Products
2801 Academy Dr. S.E.
Auburn WA 98092
Tel: 206-735-5022
Fax: 206-939-1578
E-mail:
nwaero@ix.netcom.com
Web: HTTP://
www.halcyon.com/wpowers/

• Ford 3.8L V-6
V-6, carbureted, electronic ignition, liquid cooled
Max. HP: 200 hp @ 4800 rpm
Weight: 387 lbs.

• Chevrolet HO 350 V-8
V-8, carbureted, dual electronic ignition, liquid cooled
Max. HP: 375 hp

• Chevrolet 4.3L V-6 "Vortec"
V-6, carbureted, dual electronic ignition, liquid cooled
Max. HP: 230-240 hp @ 4800 rpm
Weight: 398 lbs.

NSI Propulsion Systems

19132 59th Dr. NE
Arlington WA 98223
Tel: 360-435-1055
Fax: 360-435-4554

• EA81-98-TBI
Four cylinder, horizontally opposed, fuel injected, NSI dual ignition, liquid cooled
Max. HP: 100 hp @ 5400 rpm
Weight dry: 188 lbs.
TBO: 1200 hrs.

• EA81-118-TBI
Four cylinder, horizontally opposed, fuel injected, NSI dual ignition, liquid cooled
Max. HP: 118 hp @ 5800 rpm
Weight dry: 190 lbs.
TBO: 1200 hrs.

• Turbo EA81-130-EFI
Four cylinder, horizontally opposed, fuel injected, NSI dual ignition, liquid cooled
Max. HP: 130 hp @ 4800 rpm
Weight dry: 238 lbs.
TBO: 1200 hrs.

• EJ22- 160-EFI
Four cylinder, horizontally opposed, fuel injected, NSI dual ignition, liquid cooled
Max. HP: 160 hp @ 5500 rpm
Weight dry: 272 lbs
TBO: 1200 hrs.

• Turbo EJ22-200-EFI
Four cylinder, horizontally opposed, fuel injected, NSI dual ignition, liquid cooled
Max. HP: 200 hp @ 5200 rpm
TBO: 1200 hrs.

Omega Aerospace Corp.

9724 - 32nd Place West,
Bldg C-84
Everett WA 98204
Tel: 206-513-6177
Fax: 206-513-8653

• OAC-228A
Single cylinder. carbureted, air cooled
Max. HP: 25 hp
Weight: 25 lbs.

O'Neill Airplane Company

791 Livington Street
Carlyle IL 62231
Tel: 618-594-2681
Fax: 618-594-2681
E-mail: teroneill@aol.com

• 351 W
V-8, carbureted, liquid cooled
Max. HP: 380 hp @ 5000 rpm

Orenda Aerospace Corp.

3160 Derry Road East
Mississauga, Ontario L4T 1A9
CANADA
Tel: 905-673-5319
Fax: 905-673-4049
http://www.orenda.com

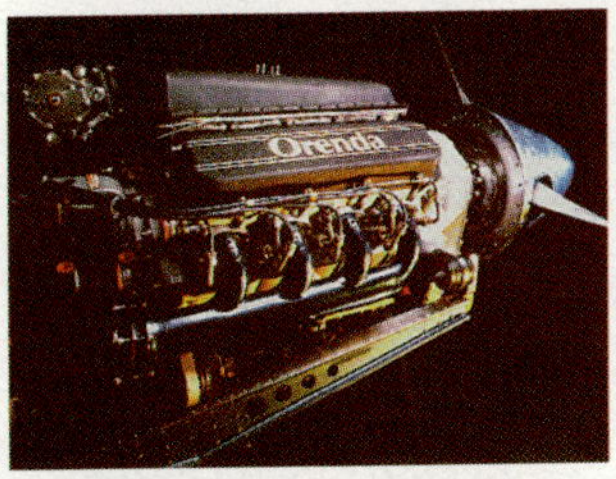

• OE 600
Eight cylinder, aluminum, 90 degree V, turbocharged, twin magneto ignition, liquid cooled
Max. HP: 600hp @ 4200 rpm
Cont. HP: 500 hp
Fuel: 100 LL
Reduction: geared
Weight: 720 lbs.
TBO: 1500 hrs.

PowerKits-Technical Materials, Inc.

263 Quillayute Place
PO Box 80
La Conner WA 98257
Tel: 360-466-1071
Fax: 360-466-1071

• Suzuki G10 Sprint
Three cylinder, inline, turbo, magneto ignition, liquid cooled
Max. HP: 104 hp @ 5500 rpm

Powersport, Inc.

12294 Hatch Lane
Aumsville OR 97325
Tel: 503-769-5469

• Mazda 13B
Twin rotor, rotary, fuel injected, dual electronic ignition, liquid cooled
Max. HP: 200 hp @ 5900 rpm
Weight dry: 260 lbs.

Precision Aerial Services

207-2 Cessna Blvd.
Daytona Beach FL 32124
Tel: 904-761-4122
Fax: 904-788-0150

• Chevy Corvair 2.7 L
Six cylinder, horizontally opposed, carbureted, air cooled
Max. HP: 90 hp @ 3000 rpm

Radne Motor AB

P.O. Box 3035
136 03 Haninge, SWEDEN
Tel: 46 8 777 15 31
Fax: 46 8 745 39 39
Email info@radne.se

• Raket 120
Single cylinder, two stroke, carbureted, air cooled
Max. HP: 14.9 hp @ 9000 rpm
Weight: 14.9 lbs.

Raven Redrives, Inc.

1025 Rosewood Ave. Suite 104
Boulder CO 80304
Tel: 303-440-6234
Fax: 303-440-4085

• Model 1000RS
Geo Metro three cylinder, four stroke, electronic ignition, throttle body fuel injected, liquid cooled
Max. HP: 65 hp @ 5700 rpm
Weight: 146 lbs. dry
Reduction Drive: Belt
Fuel: Unleaded auto
Fuel Burn: 2.2 gal/hr @ 75%
TBO: 1000 hrs.

• Model 1300GS
Geo Metro/Suzuki four cylinder, four stroke, electronic ignition, throttle body fuel injected, liquid cooled
Max. HP: 80 hp @ 5700 rpm
Weight: 170 lbs. dry
Reduction Drive: Belt
Fuel: Unleaded auto
Fuel Burn: 3.5 gal/hr @ 75%
TBO: 1000 hrs.

Rebel Experimental Engines, Inc.

P.O. Box 805
3920 South Arkansas Ave.
Russellville AR 72811-0805
Tel: 501-968-5284
Fax: 501-968-5444
Email: rebeleng@mail.cswnet.com

• Rebel Twin Model 2-215
Two cylinder opposed, two stroke, carbureted, air cooled
Max. HP: 25 hp @ 3000 rpm
Weight: 35 lbs. dry
Rotation: CW facing prop end
Direct drive
Fuel: 92 oct. min.

• Model 2-40
Two cylinder, horizontal opposed, four stroke, air cooled
Max. HP: 45 hp @ 2900 rpm
Weight: 85 lbs. dry
Rotation: CCW
Direct drive
Fuel: 90 oct. min.
TBO: 1500 hrs. estimated

• Model 4-40
Two cylinder, horizontal opposed, four stroke, air cooled
Max. HP: 90 hp @ 2900 rpm
Weight: 160 lbs. dry
Rotation: CCW
Direct drive
Fuel: 90 oct. min.
TBO: 1500 hrs. estimated

ROTAX
Bombardier-Rotax GmbH
Postach 5
A-4623 Gunskirchen
AUSTRIA
Tel. 072-462-710

For North American Distributors
See ENGINES in Supplier Section

• 447 UL-2V
Two cylinder, two stroke, fan cooled, electronic single ignition
Max. HP: 41.6 hp @ 6500 rpm
Weight: 71.9 lbs. incl. carb and exhaust
Fuel: 91 oct. RON
Fuel Burn: 5.5 gal/hr
TBO: 300 hrs.

• 503 UL-2V
Two cylinder, two stroke, fan cooled engine with electronic dual ignition
Max. HP: 50 hp @ 6800 rpm
Fuel: 91 oct. RON
Fuel Burn: 6.4 gal/hr
Weight: 84.4 lbs. incl. carbs and exhaust
TBO: 300 hrs.

• 582 UL-2V
Two cylinder, two stroke, liquid cooled, electronic dual ignition
Max. HP: 65 hp @ 6800 rpm
Fuel: 91 oct. RON
Fuel Burn: 7.2 gal/hr
Weight: 75.6 lbs. incl. carbs and exhaust
TBO: 300 hrs

• 618 UL-2V
Two cylinder, two stroke, liquid cooled, electronic dual ignition
Max. HP: 74 hp @ 6800 rpm
Fuel: 95 oct. RON
Fuel Burn: 8.1 gal/hr
Weight: 91.7 lbs. incl. carbs and exhaust
TBO: 300 hrs.

• 912
Four cylinder, four stroke, liquid/air cooled, dual carburetor, electronic dual ignition
Max. HP: 81 hp @ 5800 rpm
Fuel: 95 oct RON
Fuel Burn: 5.1 gal/hr @ 75%
Weight dry: 127.8 lbs. with carbs
TBO: 1000 hrs.

• 914
Four cylinder, turbo charged, four stroke, liquid/air cooled, dual carburetor, and electronic dual ignition
Max. HP: 115 hp @ 5800 rpm
Fuel: 95 oct RON
Fuel Burn: 5.7 gal/hr @ 75%
Weight dry: 137.1 lbs. with carbs
TBO: 600 hrs.

Sadler Radial Engines, Inc.
PO Box 953
Coupeville WA 98239-0953
Tel: 206-678-6994
Fax: 206-678-2005

• Sadler R1765U
Six cylinder, radial, carbureted, dual electronic ignition, air cooled
Max. HP: 65 hp @ 3000 rpm
Weight: 108 lbs.

Shadetree Aviation
822 Hartz Way, Suite 202
Danville CA 94523
Tel: 510-743-1786
Fax: 510-743-9807
or
4331 Gonl Road
Carson City NV 89706
Tel: 702-841-1068
Fax: 702-841-1069

• Russian M-14P
Nine cylinder, radial, air cooled
Max. HP: 365 hp

SOLO Kleinmotoren GmbH
Stuttgarterstrasse 41
D-71069 Sindelfingen
GERMANY
Tel: 07031 3010
Fax: 07031 301149

• Solo
Single cylinder, two stroke, carbureted, air cooled
Max. HP: 20 hp @ 6900 rpm
Weight: 44 lbs.
Reduction drive: belt

Stratus, Inc.
7750 12th Avenue NW
Seattle WA 98117
Tel: 206-783-3845
Fax: 206-784-8511
E-mail: Reinerh@aol.com

• EA81
Four cylinder, opposed, four stroke, carbureted, dual electronic ignition, liquid cooled
Max. HP: 100 hp @ 5400 rpm
Weight: 185 lbs.

• EJ22
Four cylinder, opposed, four stroke, carbureted, dual electronic ignition, liquid cooled
Max. HP: 160 hp @ 5600 rpm
Weight: 265 lbs.

Teledyne Continental Motors
Aircraft Products
P.O. Box 90, Mobile, AL 36601
Tel: 205-438-3411

• TSIO-550
Six cylinder, opposed, air cooled, fuel injected, turbocharged, direct drive
Cont. HP: 350 hp @ 2700 rpm
Fuel: 100/100 LL Avgas
Weight dry: 425 lbs.
TBO: 1600 hrs.

• IO-550
Six cylinder, opposed, air cooled, direct drive
Cont. HP: 280-305 hp
Fuel: 100/100 LL Avgas
Weight dry: 415 to 433 lbs.
TBO: 1700 hrs.

• GTSIO-520
Six cylinder, opposed, air cooled, fuel injected, turbocharged, geared drive
Cont. HP: 340-435 hp @ 3200 to 3400 rpm
Fuel: 100/100 LL Avgas
Weight dry: 481-600 lbs.
TBO: 1200 hrs.

• TSIO-520
Six cylinder, opposed, air cooled, fuel injected, turbocharged, direct drive
Cont. HP: 285-325 hp @ 2600 to 2700 rpm
Weight dry: 412 to 436
TBO: 1400 hrs.

• IO-520
Six cylinder, opposed, air cooled, fuel injected, direct drive
Cont. HP: 285-300 hp @ 2700 to 2850 rpm
Fuel: 100/100 LL Avgas
Weight dry: 413 to 431 lbs.
TBO: 1400 hrs.

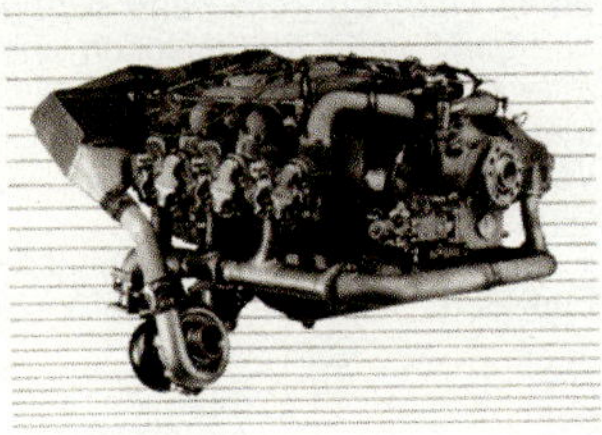

• TSIO-360
Six cylinder, opposed, air cooled, fuel injected, turbocharged, direct drive
Cont. HP: 200-225 hp @ 2575 to 2800 rpm
Fuel: 100/100 LL Avgas
Weight dry: 313 to 359 lbs.
TBO: 1400 hrs.

• IO-360
Six cylinder, opposed, air cooled, fuel injected, direct drive
Cont. HP: 195-210 hp @ 2600 to 2800 rpm
Fuel: 100/100 LL Avgas
Weight dry: 294 lbs.
TBO: 1500 hrs.

• O-200
Four cylinder, opposed, air cooled, direct drive
Cont. HP: 100 hp @ 2750 rpm
Fuel: 80/87 Avgas
Weight dry: 188 lbs.
TBO: 1800-2000 hrs.

TERF, Inc.
11411 Culver
Brighton MI 48114
Tel/Fax: 810-229-2082

• 13B & 20B Mazda
Two/three rotor, rotary engines, liquid cooled, electronic fuel and ignition, turbo and non turbo, brushless starter/alternator.
Max. HP: 400/600 hp @ 6500 rpm
150/200 continuous
Weight dry: 200 lbs. approx.

Textron Lycoming
Reciprocating Engine Div.
Subsidiary of Textrom Inc.
652 Oliver St.
Williamsport PA 17701
Tel: 717-327-7278
Fax: 717-327-7022
(TBO's based upon using genuine Textron Lycoming parts)

• 720 Series
Eight cylinder, horizontally opposed, fuel injected
Max. HP: 375-400 hp @ 2650 rpm
Direct drive
Weight: 558 lbs.
TBO: 1800 hrs.

• 540 Series
Six cylinder, opposed, carb or fuel injected (turbocharged & intercooling available)
Max. HP: 235-360 hp @ 2500 to 2700 rpm
Direct drive; right and left rotation
Weight: 356-595 lbs.
TBO: 2000 hrs.

• 360 Series
Four cylinder, opposed, carb or fuel injected(turbocharged models available)
Max. HP: 180-210 hp @ 2575 to 2700 rpm
Direct drive; right and left rotation
Weight: 255-348 lbs.
TBO: 2000 hrs.

• 320 Series
Four cylinder, horizontally opposed, carb or fuel injected
Max. HP: 140-160 hp @ 2450 to 2700 rpm
Direct drive; right and left rotation
Weight: 244-271 lbs.
TBO: 2000 hrs.

• 235 Series
Four cylinder, horizontally opposed, carbureted
Max. HP: 100-118 hp @ 2400 to 2800 rpm
Direct drive
Weight: 213-218 lbs.
TBO: 2400 hrs.

Total Engine Concepts
A Division of TEAM-38, Inc.
13205 SW Citrus Boulevard
Indiantown, FL 34956
Tel: 561-597-3801
Fax: 561-597-3804
www.team-38.com

• 502 Big-Block Chevy
Eight cylinder, V-configured, fuel injected, dual-ignition, liquid cooled, cast iron block, aluminum heads, bore: 4.470 inches, stroke: 4.000 inches, displacement: 8.2 liters (502 cu. in.), torque: 514 lb/ft @ 3200 rpm
HP: 500 @ 4800 rpm
Fuel Burn: 19 gal/hr at cruise
Weight: 610 lbs. dry
Lubrication: Dry sump

• 2165cc VW-Based Longblock
Four cylinder, horizontally opposed, magneto and electronic ignition, air cooled, Bore: 94 mm, Stroke: 78 mm, individual cylinder heads machined for dual ignition, solid valve lifters, forged 4130 crankshaft, massive front crank bearing
HP: 68 @ 3400 rpm
Weight: 180 lbs. with accessories
Direct drive

• 1915cc VW-Based Longblock
Four cylinder, horizontally opposed, magneto and electronic ignition, air cooled, bore: 94mm, stroke: 69 mm, forged 4130 crankshaft, massive front crank bearing, solid valve lifters
HP: 65 @ 3400 rpm
Weight: 175 lbs. with accessories
Direct drive

• 1083 Defiant
Two cylinder, horizontally opposed, magneto ignition, air cooled, solid valve lifters, custom-cast aluminum case, forged chrome-moly crankshaft, bore: 94 mm, stroke: 78 mm
HP: 35 @ 3400 rpm
Weight: 90 lbs.
Direct drive
TBO: 1000 hours
Fuel Burn: 1.5 gal/hr at cruise

Turbine Technologies
P.O. Box 105
430 Phillip Street
Chetek WI 54728
Tel: 715-924-4876
Fax: 715-924-2436
Email turbtech@discover-net.net
http://www.discover-net.net/turbtech

• SR-30
Turbojet, axial turbine, reverse flow annular combustor, centrifugal compressor, recirculating oil system
40 lbs. static thrust
Weight: < 10 lbs.

• TF-45
High bypass, geared turbofan

• TP-45
Turbo prop engine
HP: 55 Shaft Horse Power

• TRJ-200 (in development)
Two stage centrifugal compressor turbojet, axial turbine, reverse flow annular combustion chamber
Static Thrust: 321 lbs.
Weight: 72 lbs. dry
Fuel Burn: 34 gal/hr

Universal Engineering, Inc.
116 West Main
Winamac IN 46996
Tel: 219-946-6757
Fax: 219-946-7471
www.universalengineering.com/diesel.html

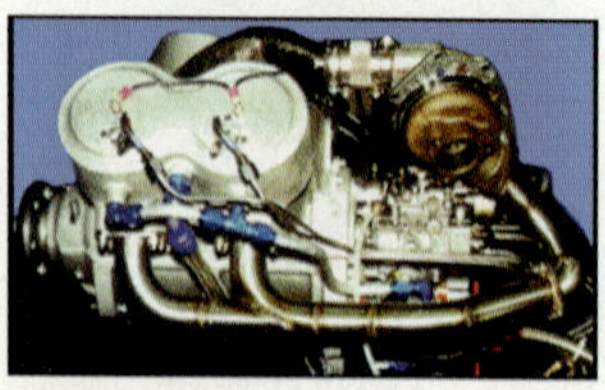

• V-4 Diesel
90 degree V-4, two stroke, liquid cooled, dry sump
HP: 200 @ 2500 rpm
Fuel Burn: 19 gal/hr at cruise
Weight: 197.5 lbs. dry
Direct drive
TBO: 2000 hrs.

Zenoah
142-1 Sakuragaoka-2
Higashiyamato, Tokyo
JAPAN
Tel: 0425-61-2141

U.S. Distributor
Tennessee Propellers Inc.
125 Deerfield Drive
Normandy TN 37360
Tel: 615-455-4516/455-1247
Fax: 615-455-1309

• Zenoah G50D
Two stroke, twin cylinder, horizontally opposed, CDI ignition, free air cooled
Max. HP: 45 hp @ 5800 rpm
Rotation: CCW viewed from PTO
Weight: 47.4 lbs.
TBO: 500 hrs.

• Zenoah G50C
Two stroke, twin cylinder, horizontally opposed, CDI ignition, fan cooled
Max. HP: 45 hp @ 5800 rpm
Rotation: CCW viewed from PTO
Weight: 51.8 lbs.
TBO: 500 hrs.

• Zenoah G25B-1
Two stroke, twin cylinder, horizontally opposed, magneto, CDI ignition, free air cooled
Max. HP: 22 hp @ 6600 rpm
Rotation: CCW viewed from PTO
TBO: 500 hrs.

Michael Zoche Antriebstechnik
Keferstr. 13
80802 Munich
GERMANY
Tel: 49 89 344591
Fax: 49 89 342451
http://193.26.97.194/

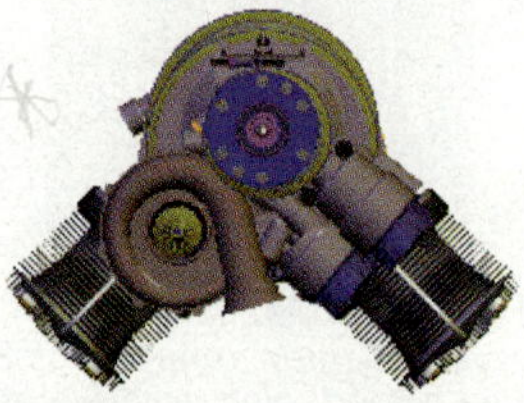

• ZO 03A
Two cylinder, two stroke, diesel, air cooled
Max. HP: 70 hp @ 2500 rpm
Weight: 121 lbs.
Direct drive
Fuel: Diesel #2, JP4, JP5, JP8, Jet A
Fuel Burn: 2.77 gal/hr @ 75%

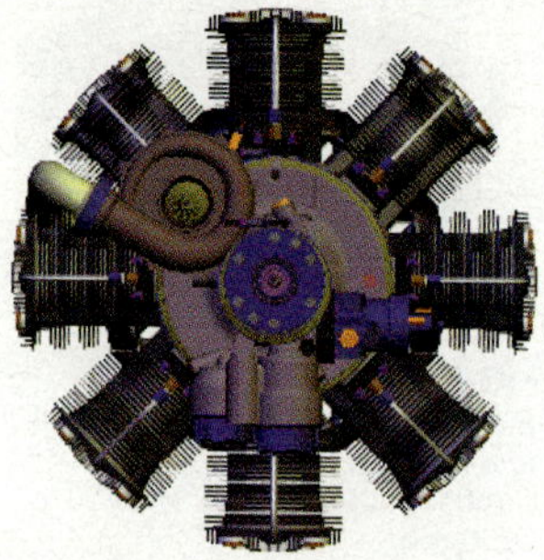

• ZO 02A
Eight cylinder radial (four cylinders per row), two stroke, diesel, two stage charging (turbo and supercharger), direct fuel injection and intercooling, air cooled
Max. HP: 300 hp @ 2500 rpm
Weight: 271 lbs.
Direct drive
Fuel: Diesel #2, JP4, JP5, JP8, Jet A
Fuel Burn: 11.43 gal/hr @ 75%

• ZO 01A
Four cylinder radial, two stroke, diesel, direct fuel injection, air cooled
Max. HP: 150 hp @ 2500 rpm
Weight: 185 lbs.
Direct drive
Fuel: Diesel #2, JP4, JP5, JP8, Jet A
Fuel Burn: 5.72 gal/hr @ 75%

AeroCrafter

Propeller Guide

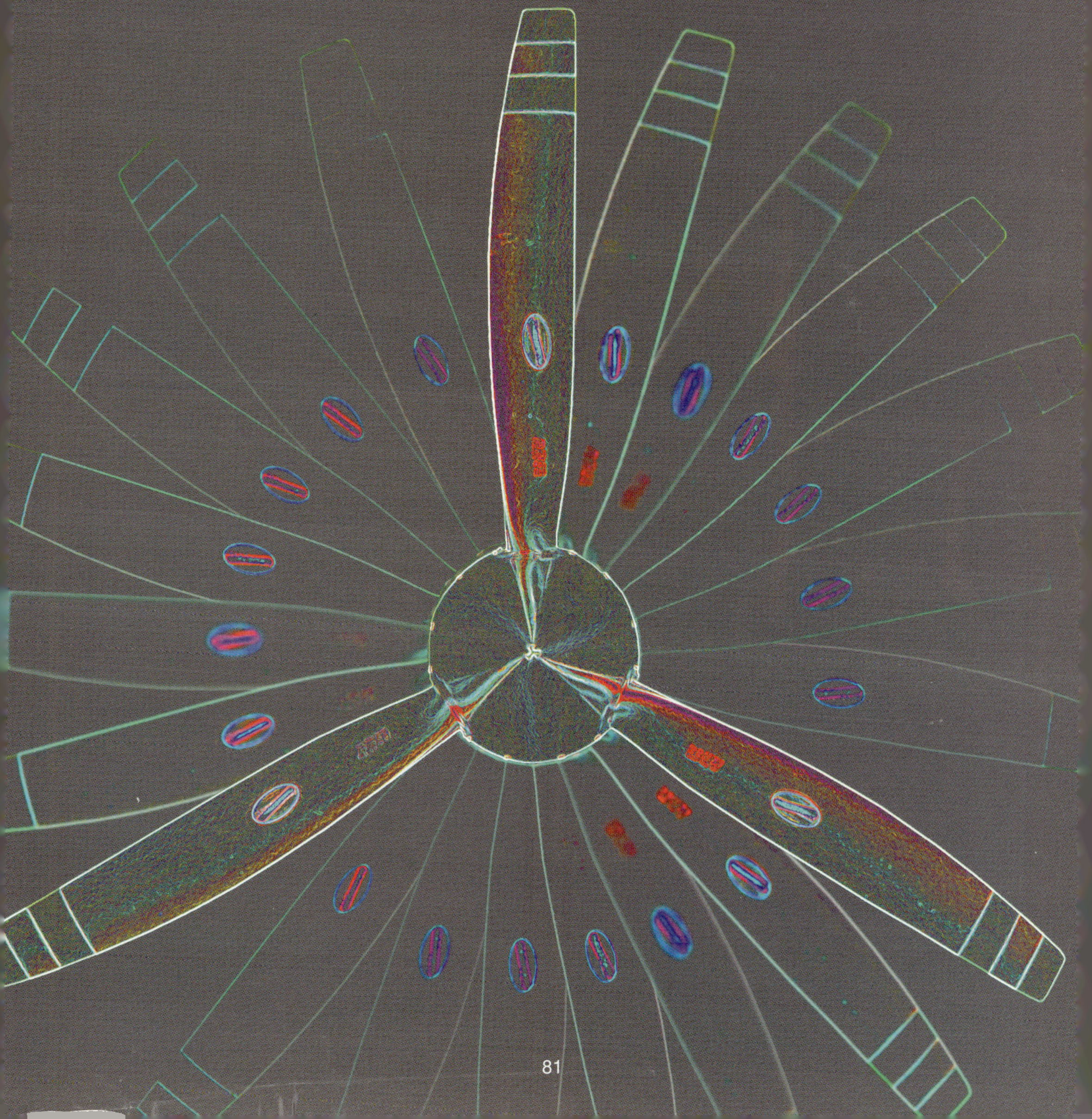

The following data, supplied by the manufacturers, provides basic information on many of the propellers suitable for homebuilt aircraft. Contact the manufacturers or distributors listed for more specific information regarding the requirements of your particular installation. See the listings under* Propellers *in the* SUPPLIER Section *for a more complete list of distributors.

A.K.S. Inc.

PO Box 17161
Portland OR 97217
Tel. 503-289-1491 • Fax 503-247-9207
E-Mail: trikes@teleport.com
Web: http://www.teleport.com/~trikes

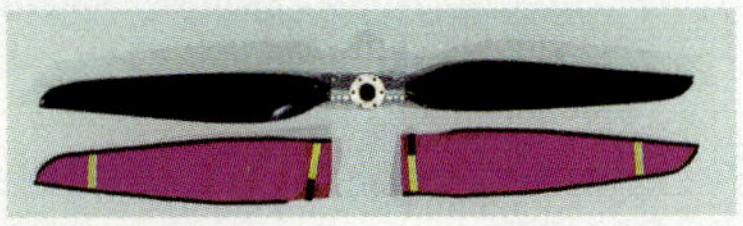

Techno Prop
Material: Composite
Diameter: 52" and 63"
Hub: 100 mm
Number of Blades: 2 or 3
Pitch: Ground Adjustable
Engines/Aircraft: Developed for Rotax 912/914, Subaru, and other 4-strokes.

Sport Prop
Material: Composite
Diameter: 62", 66", 68"
Hub Size: 75 mm
Number of Blades: 2, 3, 4
Pitch: Ground Adjustable
Engines/Aircraft: Rotax, SVS 1400, HKS, Hirth, 2si.

Avia-Hamilton Std. Aviation

1 Hamilton Rd., Windsor Locks CT
06096-1010, 860-654-3641

Material: Wood
Metal
Aluminum
Usable HP Range: 4-Cycle: 140 - 220
Number of Blades: 2 - 5
Pitch: Ground or In-Flight Adjustable
Method: Hydraulic
Application: All Aircraft

European Address: PO Box 22 • Hlavenecl
29 • Stara Boleslav 250 02
CZECH REPUBLIC
Tel. 42-202-811114 • Fax 42 202-811818

Aymar-DeMuth Propellers

8213 Elberta Drive • PO Box 853
Ellicott City MD 21041-0853
Tel. 410-461-4329 • Fax 410-461-4329

Material: Wood-Maple Laminate
Composite
Diameter: 86"
Hub Size: 5"
Usable HP Range: Up to 300
2 or 4-Cycle
Number of Blades: 2
Pitch: Fixed
Engines/Aircraft: RV, T-18, Glasair, Velocity & Other Experimental

Competition Aircraft, Inc.

11110 Gopher Mine Trail
Grass Valley CA 95949
Tel. 916-268-3048 • Fax 916-268-2321
E-Mail: ultraprop@ord.net

Material: Composite(Ultra-Prop™), injection molded; carbon/epoxy
Diameter: to 68"
Hub Size: 2.25"
Usable HP Range: 4-Cycle: to 100
2-Cycle: to 65
Number of Blades: 2 - 4
Pitch: Ground Adjustable
Engines/Aircraft: Rotax, Hirth, 2SI /Experimental, Ultralights, Gryos

Photo courtesy A.K.S. Inc.

Culver Props Inc.

RD1, Box 176
Galeton PA 16922
Tel. 814-435-1077 • Fax 814-435-6523
E-Mail: rmiller@conterra.com

Material: Wood- Birch, Maple, Walnut, Mahogany
Diameter: 20" - 96"
Hub Size: 1.5 - 5.25
Usable HP Range: 4-Cycle: to 200
2-Cycle: to 90
Number of Blades: 2
Pitch: Fixed
Engines/Aircraft: Powered Parachutes, Experimental, Ultralights, Gryos

Felix Propellers

W10508 Bell Rd.
Camp Douglas WI 54618
Tel. 800-776-7357 • Fax 608-427-6544

Material: Wood
Diameter: 50" - 90"
Hub Size: 1.5" - 4.5"
Usable HP Range: 4-Cycle: to 350
Number of Blades: 2 - 4
Pitch: Fixed
Engines/Aircraft: Experimental, Ultralights, Replicas, Antiques

GSC Systems Int'l., Ltd.

6255 Onanagan Landing Rd.
Vernon, B.C. V1T 6Y5 CANADA
Tel. 250-549-3772 • Fax 250-549-3769
E-Mail: asap@junction.net

Material: Wood- Maple
Composite
Usable HP Range: 4-Cycle: to 115
Pitch: Ground or In-Flight Adjustable
Method: Mechanical
Engines/Aircraft: Rotax bolt pattern standard. Other bolt patterns on request.

Hartzell Propeller Inc.

One Propeller Place
Factory Service Center
Piqua OH 45356-2634
Tel. 937-778-4201 • Fax 937-778-4202

Material: Wood
Composite- Kevlar
Metal- Aluminum
Diameter: 68"+
Usable HP Range: 4-Cycle: 150+
Number of Blades: 2 - 6
Pitch: In-Flight Adjustable (constant speed)
Method: Hydraulic
Application: Certified, Experimental

Hegy Propellers

407 W. Lincoln • Box 124
Marfa TX 79843
Tel. 915-729-4249

Material: Wood- Birch, Maple
Diameter: 12" - 120"
Hub Size: 2.25" - 7"
Number of Blades: 2, 4
Pitch: Fixed.
Engines/Aircraft: VW, Corvair, Continental, etc./Experimental, Ultralights Gryos

---

Hoffmann Propeller GmbH & Co. KG

Küpferlingstrasse 9
Rosenheim D-83022 GERMANY
Tel. 49-8031-1878-0 • Fax 49-8031-1878-78
U.S. Distributor: Steen Aero Lab
1210 Airport Rd., Marion NC 28752
Ph/Fax 704-652-7382

Material: Wood- Composite with FRP covering
Diameter: up to 118"
Usable HP Range: 2 or 4-Cycle: to 500 kW
Number of Blades: 2 - 5
Pitch: • Fixed or
• Adjustable: Ground, In-Flight
Method: Hydraulic, Electric, Mechanical
Engines/Aircraft: Katana, Rotax, and more/ All Types of Aircraft

---

Ivoprop Corp.

15903 Lakewood Blvd.
Bellflower CA 90706
Tel. 562-602-1451 • Fax 562-602-1374

Material: Composite
Size: Diameter: 36" - 76"
Usable HP Range: 2 or 4-Cycle: 15 - 700
Number of Blades: 2 - 3 - 6
Pitch: Ground or In-Flight Adjustable
Method: Electric
Engines/Aircraft: Experimental, Ultralights, Gryos

MT-Propeller Entwicklung GmbH

Flugplatz Straubing Wallmuehle
Atting D-94348 GERMANY
Tel. 49-9429-8433 • Fax 49-9429-8432
U.S. Distributor: Light Power Engine Corp.,
PO Drawer 3350, Morgantown WV 26505,
Tel. 305-291-3843, Fax 304-292-1902.

Material: Wood- Laminate
Composite- Wood/FRP
Diameter: 50" - 120"
Usable HP Range: 4-Cycle: 60 - 1000
2-Cycle: 50 - 150
Pitch: Ground or In-Flight Adjustable
Method: Hydraulic, Electric
Engines/Aircraft: Lycoming, Continental, Rotax, Franklin/ Certified or Experimental

---

McCauley Propeller Systems

3535 McCauley Drive
Vandalia OH 45377-5053
Tel. 800-621-7767 • Fax 937-890-6001

Material: Metal
Diameter: 68" to 114"
Usable HP Range: 4-Cycle: 65 to 1650 hp
Number of Blades: 2, 3, 4, 5
Pitch: • Fixed or
• Adjustable: In-Flight
Method: Hydraulic
Engines/Aircraft: Certified & Experimental

M.W. Propellers

3906 W. Ina Rd. #200-193
Tucson AZ 85741
Tel. 520-405-9309

Material: Wood- Maple Laminate
Usable HP Range: 65 to 260
Engines/Aircraft: Most Aircraft

P & K Enterprises

11470 W. 800N
Colfax IN 46035-9636
Tel. 765-324-2424 • Fax 765-324-2424
E-Mail: PKENTERPRISES@tctc.com
Web: http://pkenterprises/tctc/com/html

Material: Wood: Laminate
Composite
Diameter: Any
Hub Thickness: Any
Usable HP Range: 4-Cycle: to 90
2-Cycle: to 80
Number of Blades: 2 - 5
Pitch: • Fixed or
• Ground Adjustable
Engines/Aircraft: Rotax, Zenoah, Cayuna/ Experimental, Ultralights

---

Pacesetter Propeller Works, Ltd.

PO Box 1245
Hillsboro OR 97123
Tel.503-628-2797

Engines/Aircraft: RV-3, RV-4, RV-6, Glasair, Mustang II, Midget Mustang, T-18, and others.

Performance Propellers

P.O. Box 486, 63/27 Blue Haven • Nogales
International Airport
Patagonia AZ 85624
Tel. 520-394-2059

Material: Wood- Maple Laminate
Fiberglassed Tips
Diameter: to 76"
Hub: 3.5" to 4.5"
Usable HP Range: 4-Cycle: to 350
Number of Blades: 2 - 3
Pitch: Fixed
Engines/Aircraft: RV, Long-EZ, Cozy, Lancair and other Experimental
All designs are custom made.

Powerfin, Inc.

4700 188th St. NE #G
Arlington WA 98223
Tel. 800-581-8207 • Fax 360-403-0599
E-Mail: powerfin@supernal.net
Web: http://www.supernal.net/~powerfin

Material: Composite- Carbon Fiber/Kevlar
Diameter: 48" - 74"
Hub: 2-1/8"
Usable HP Range: 4-Cycle: 80
2-Cycle: 80
Number of Blades: 2 - 4
Pitch: Ground Adjustable
Method: Mechanical
Engines/Aircraft: All 2-stroke, VW, Subaru, Jabiru, HKS/ Experimental, Ultralights, Custom Manufacture.

Precision Propellers, Inc.

PO Box 1392
Vernal UT 84078
Tel. 801-789-3699 • Fax 801-789-6059

Material: Wood-Hard Rock Maple Laminate
Diameter: 20" to 74"
Hub: Any
Usable HP Range: 4-Cycle: to 1000
Number of Blades: 2 - 3
Pitch: Ground Adjustable
Engines/Aircraft: Experimental, Ultralights, Custom Manufacture

---

Prince Aircraft Co.

6774 Providence Street • PO Box 2669
Whitehouse OH 43571
Tel. 419-877-5557 • Fax 419-877-5564
E-Mail: 75557.31555@Compuserve.Com

Material: Wood- Hard Maple
Composite
Diameter: Any
Hub: Any
Usable HP Range: 4-Cycle: to 1000
Number of Blades: to 8
Pitch: • Fixed or
• In-Flight Adjustable
Engines/Aircraft: Custom manufacturer: Propellers of any shape, size or capability.

Props, Inc.

354 SE 2nd Street
Newport OR 97356
Tel. 541-265-3032 • Fax 541-265-3032

Material: Composite: Wood, Fiberglass
Diameter: to 96"
Hub: to 4.5"
Usable HP Range: 4-Cycle: 5 - 350
2-Cycle: 5 - 125
Number of Blades: 2 - 4
Pitch: • Fixed or
• Ground Adjustable
Engines/Aircraft: Experimental, U/L, Gryos

St. Croix Aircraft

1139 State Highway #148
Corning IA 50841
Tel. 515-322-4041

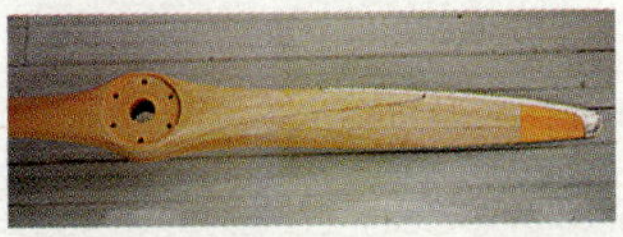

Material: Wood- Birch, Maple, Walnut
Diameter: to 120"
Hub: Any size
Usable HP Range: 4-Cycle: to 400
Number of Blades: 1 - 5
Pitch: Fixed
Engines/Aircraft: Experimental, Gryos, Certified; Custom Applications
Specially designed for antiques and homebuilts.

Sensenich Prop Manufacturing

14 Citation Lane
Lititz PA 17543
Tel. 717-569-0435 • Fax 717-560-3725
Web: http://WWW.SENSENICH.COM

Sensenich Wood Propeller Company

2008 Wood Ct.
Plant City FL 33567
Tel. 813-752-3711 • Fax 813-752-2818

Material: Wood- Birch Laminate
Composite- Wood Core w/ Fiber glass, Kevlar
Diameter: 60" to 98"
Hub: 2.75" to 6"
Usable HP Range: 4-Cycle: 40 to 275
Number of Blades: 2
Pitch: Fixed
Engines/Aircraft: Continental, Lycoming or others/ for most Experimental, Gryos, Certified

---

Sport Prop s.r.o.

PO Box 1 • 17100 Posta
Praha 71 CZECH REPUBLIC
E-Mail: zlinak@teleport.com

See A.K.S. Inc.

Ed Sterba Propellers

412 S. 5th St.
Delavan WI 53115
Tel. 414-728-1367

Material: Wood
Engines/Aircraft: Custom Carving for Homebuilders and Manufacturers

---

Tennessee Propellers Inc.

125 Deerfield Drive
Normandy TN 37360
Tel. 615-455-4516 • Fax 615-455-1309

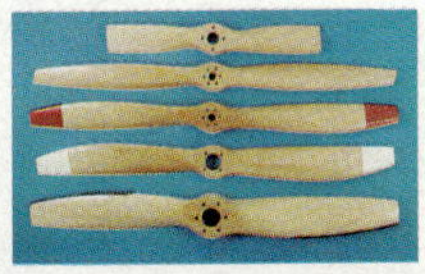

Material: Wood-Hard Rock Maple Laminate
Composite-2 Part Uralite
Diameter: 24" - 72"
Hub: 1" - 3"
Usable HP Range: 4-Cycle: to 100
2-Cycle: to 100
Number of Blades: 2
Pitch: Fixed
Engines/Aircraft: Zenoah/Experimental, Ultralights, Gryos

Vari-Prop International Inc.

P.O. Box 1856
Cottonwood CA 96022
Tel. 916-347-6122

Bernie Warnke

HC 2, Box 850484
Tucson AZ 85735
Tel./Fax 520-822-1166

Material: Wood- Maple Laminate
Usable HP Range: to 260
Engines/Aircraft: Experimental & Aerobatic

Warp Drive, Inc.

1207 Hwy 18 E. • PO Box 78
Ventura IA 50482
Tel. 800-833-9357 • Fax 515-357-7592

Material: Wood
Composite: Carbon Fiber
Number of Blades: 2 - 6
Pitch: Ground Adjustable

Whirlwind Propeller Corporation

Hangar H • 1860 Joe Crosson Drive
El Cajon CA 92020
Tel. 619-562-3725
E-Mail: wwpc@concentric.net

Material: Wood: Laminate
Composite: Carbon Fiber
Diameter: 68" - 102"
Usable HP Range: 4-Cycle: 80 - 450
Number of Blades: 2, 3
Pitch: In-Flight Adjustable
Method: Hydraulic
Engines/Aircraft: Rotax 912, Lyc. O-320, O-360, O-540, Cont. 550/ Experimental

Mike Hepperlen's

RV4

After 5-1/2 years of construction, I flew my RV-4 on February 19, 1996. As Van advertises, I am the proud owner of the RV grin. The aircraft is powered by a Lycoming O-360 engine which turns a 70/82 Sterba Propeller. It indicates 180 mph @ 7500 feet with full throttle and flies great. Some of the other accessories include Imron paint, a D.J. Lauritsen interior, an IFR panel, and radios. These accessories, plus the engine, helped to bring the gross weight up to 983 pounds. A very understanding and helpful wife, along with some good friends holding bucking bars and wrenches, made this project possible.

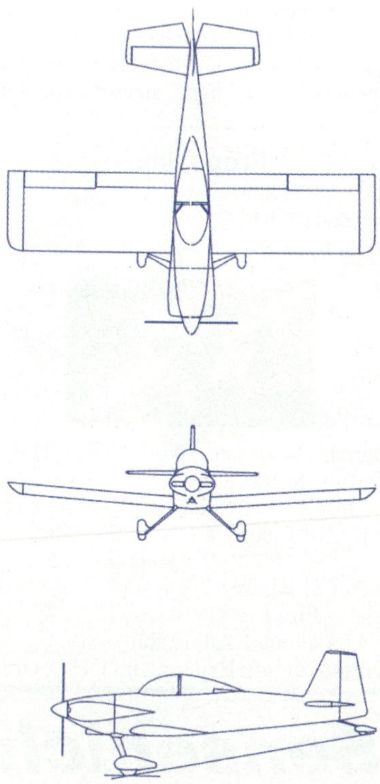

Mike Hepperlen
EAA 267043
6823 W. Hillside Rd.
Crystal Lake, IL 60012

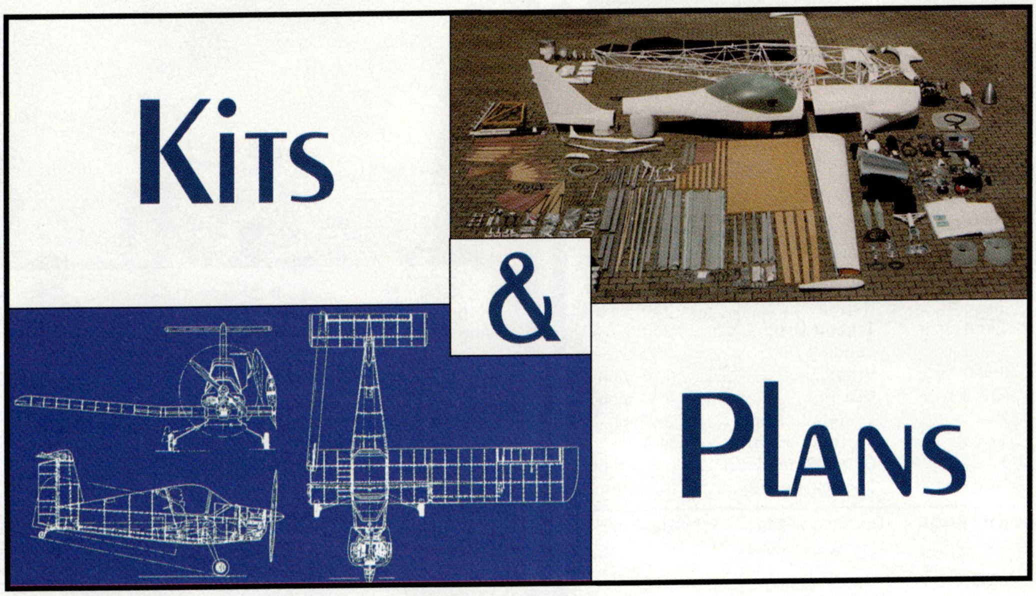

AeroCrafter presents a broad overview of the kits, plans, supplies and services offered to the homebuilt aircraft enthusiast and is intended solely for the reader's education and entertainment. All information has been provided by the manufacturers and other independent sources. The publisher cannot guarantee the accuracy or correctness of the information provided. A listing in AeroCrafter does not constitute an endorsement or recommendation of any kit, plan, supply or service.

It is important that you contact the manufacturer directly and make your own evaluation before deciding on a product or service.

Contact the manufacturer for any changes in prices or specifications.

- Kits and Plans are listed alphabetically by Company Name within the major categories:

AIRCRAFT NAME INDEX IS LOCATED ON PAGE 418

- Aircraft Performance Data is dependent upon manufacturer's testing and can vary with engine installation, weight, environmental factors, etc. Note: Some manufacturers include fuel weight in the "Payload" figure.
- N/A in the Kit and Plans listings indicates that the information was not available at the time of publication or is not applicable to the aircraft.
- Your comments on AeroCrafter and experiences with the companies listed are welcome. Send us email at AEROCFTR@BAICORP.COM or contact us at PO Box 1516; Benicia CA 94510-1516; 707-747-1509.

Baby Ace - Model D

Wing Span	26.50 ft.	**Range**	350 sm
Wing Area	112.3 sq. ft.	**Takeoff Dist.**	200 ft.
Length	17.60 ft.	**Landing Dist.**	250 ft.
Max Gross Wt.	950.0 lbs.	**Vmax**	110 mph
Empty Wt.	575 lbs.	**VCr**	100 mph
Payload (full fuel)	N/A	V_{S1} **(stall clean)**	34 mph
Fuel Capacity	17 gal.	V_{S0} **(lndg config)**	N/A
Seats	1	**Climb Rate @msl**	1200 fpm
Cabin Width	N/A	**Service Ceiling**	16,000 ft.

Std. Engine/HP	Usable HP Range	Construction Mtrl.	Est. Bldg. Time
Continental / 65 hp	N/A	Tube, Wood, Fabric	600 hrs.
Plans	**Kit**	**Info Packet**	**Video**
$70.00	$5,000.00	$5.00	N/A

A single seater that first flew in November of 1956. Has an open cockpit with a door on the starboard side. Fuselage is welded steel tube and fabric. Wings are a fabric covered two spar wood structure. Space for 10 lbs. of luggage. Engine and propeller not included in kit price.

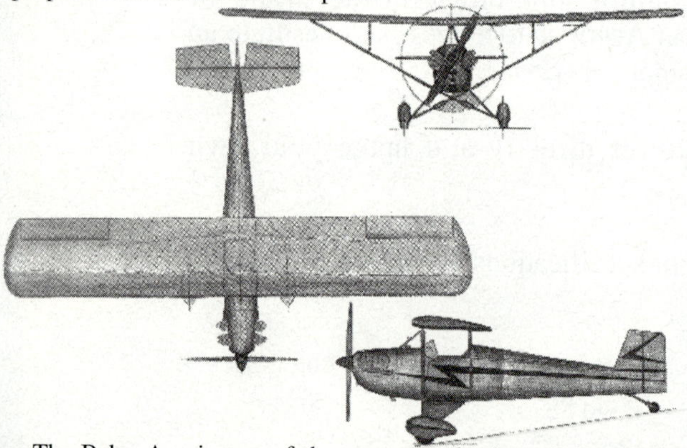

The Baby Ace is one of the most popular homebuilts of all time. It all started back in the Thirties when Mr. O. G. Corben designed the first of the Baby Ace series. There is a great deal of aeronautical talent and engineering skill behind this ship. Men such as O.G. Corben, Paul Poberezny, Stanley J. Dzik, Carl A. Walters, Cliff DuCharme and others, have all had a part in it. It is estimated that there are more than 350 Baby Aces flying throughout the world today, and hundreds more being built. The Baby Ace is a safe, sane, well engineered airplane; and one that you will not be afraid to fly when you have completed it. It is easy to build and easy to fly. It's among the most proven of all homebuilts and a real beauty when built according to the plans.

Henry Burdette, Barnesville OH

"My modified Baby Ace was built from a very complete and easy to read set of prints from Ace Aircraft. ...it's a straightforward, easy to build [aircraft]. Cruise is 85 mph."

Henry Burdette and his Baby Ace are pictured in the photo above.

ACE AIRCRAFT
05-134th St. • Chesapeake WV 25315
Tel. 304-949-3098

Junior Ace Model E

Wing Span	26.00 ft.	**Range**	350 sm
Wing Area	N/A	**Takeoff Dist.**	400 ft.
Length	18.00 ft.	**Landing Dist.**	600 ft.
Max Gross Wt.	1335 lbs.	**Vmax**	130 mph
Empty Wt.	809 lbs.	**VCr**	105 mph
Payload (full fuel)	N/A	V_{S1} **(stall clean)**	34 mph
Fuel Capacity	22 gal.	V_{S0} **(lndg config)**	N/A
Seats	2	**Climb Rate @msl**	N/A
Cabin Width	N/A	**Service Ceiling**	10,000 ft.

Std. Engine/HP	Usable HP Range	Construction Mtrl.	Est. Bldg. Time
Continental / 85 hp	N/A	Tube, Wood, Fabric	600 hrs.
Plans	**Kit**	**Info Packet**	**Video**
$85.00	$6,000.00	$5.00	N/A

The Junior Ace Model E differs from the Baby Ace Model D in being a side-by-side two seater. It is powered by the larger 85 hp C85 four-cylinder horizontally-opposed air cooled engine. It was first flown in August of 1966. Engine and propeller are not included in kit price.

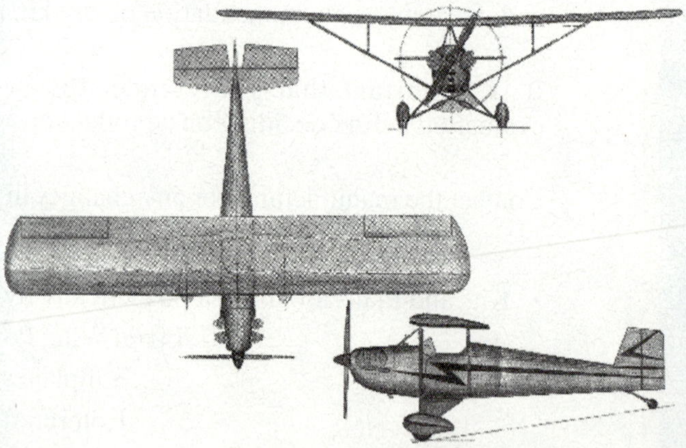

ACE AIRCRAFT
05-134th St. • Chesapeake WV 25315
Tel. 304-949-3098

George Reynolds' Corbin Baby Ace

Builders Story on Page 63

Cuby I

Wing Span	33.50 ft.	**Range**	N/A sm
Wing Area	161.00 sq. ft.	**Takeoff Dist.**	150 ft.
Length	18.25 ft.	**Landing Dist.**	300 ft.
Max Gross Wt.	850 lbs.	**Vmax**	95 mph
Empty Wt.	250 lbs.	**Vcr**	55 mph
Payload (full fuel)	N/A lbs.	**V_{S1} (stall clean)**	N/A mph
Fuel Capacity	5.0 gal.	**V_{S0} (lndg config)**	N/A mph
Seats	1	**Climb Rate @msl**	700 fpm
Cabin Width	N/A in.	**Serv. Ceiling**	N/A ft.

Std. Engine/HP	Usable HP Range	Construction Mtrl.	Est. Bldg. Time
Rotax 277 / 27 hp	27 - 60	Tube, Aluminum, Fabric	N/A
Plans	**Kit**	**Info Packet**	**Video**
No	$9,398	$5.00	No

Single seater version of the Cuby design. It is a high wing, welded steel fuselage monoplane that is Part 103 legal. It sports an enclosed cabin and can accept many different engines.

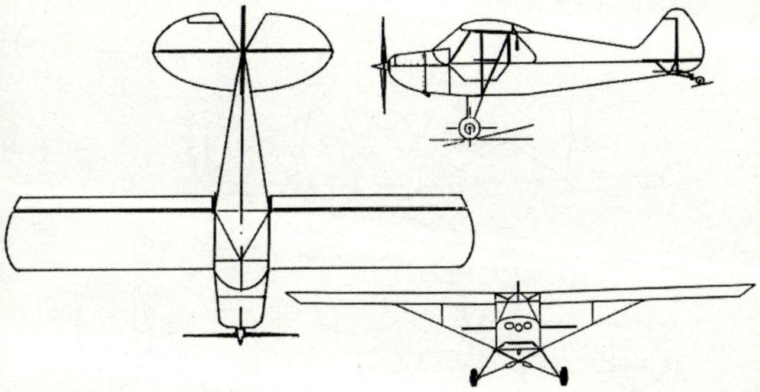

ACES HIGH LIGHT AIRCRAFT LTD.
RR 1 • London, Ontario N6A 4B5
CANADA
Tel. 519-652-3020 • Fax 519-652-3020

Cuby II

Wing Span	33.50 ft.	**Range**	240 sm
Wing Area	161.00 sq. ft.	**Takeoff Dist.**	125 ft.
Length	19.50 ft.	**Landing Dist.**	220 ft.
Max Gross Wt.	1,058 lbs.	**Vmax**	100 mph
Empty Wt.	530 lbs.	**Vcr**	80 mph
Payload (full fuel)	N/A lbs.	**V_{S1} (stall clean)**	30 mph
Fuel Capacity	12.0 gal.	**V_{S0} (lndg config)**	N/A mph
Seats	2	**Climb Rate @msl**	750 fpm
Cabin Width	N/A in.	**Serv. Ceiling**	N/A ft.

Std. Engine/HP	Usable HP Range	Construction Mtrl.	Est. Bldg. Time
Rotax 503 / 53 hp	47 - 65	Tube, Fabric, Alum. Wing	200 hrs.
Plans	**Kit**	**Info Packet**	**Video**
No	$17,484	$5.00	$10.50

Welded steel fuselage. Features include dual controls and it accepts a variety of engines. Options available include floats, folding wings and an agricultural spray package.

Completed & Flying: 200+

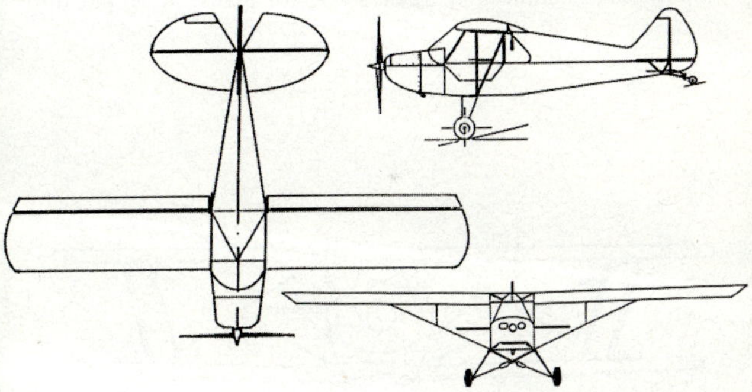

Joe Hampton, Thamesford Ontario

"The more I get into this aircraft the more convinced I am that this is probably the biggest sleeper of its type anywhere. I would highly recommend this aircraft to any pilot with sufficient training as it is far above the 'toy' category."

Chuck Gleason

"This plane is certainly my pride and joy. On my first flight, after attaining 1,000 feet, I trimmed for level flight and released the stick. The plane flew straight and level - no adjustments had to be made. Truly a wonder for a first-time builder."

ACES HIGH LIGHT AIRCRAFT LTD.
RR 1 • London, Ontario N6A 4B5
CANADA
Tel. 519-652-3020 • Fax 519-652-3020

Acro Sport I

Wing Span	19.60 ft.	**Range**	350 sm
Wing Area	117.0 sq. ft.	**Takeoff Dist.**	350 ft.
Length	17.50 ft.	**Landing Dist.**	800 ft.
Max Gross Wt.	1350 lbs.	**Vmax**	180 mph
Empty Wt.	739 lbs.	**VCr**	130 mph
Payload (full fuel)	N/A	**V_{S1} (stall clean)**	50 mph
Fuel Capacity	20 gal.	**V_{S0} (lndg config)**	N/A
Seats	1	**Climb Rate @msl**	3500 fpm
Cabin Width	N/A	**Service Ceiling**	25,000 ft.

Std. Engine/HP	Usable HP Range	Construction Mtrl.	Est. Bldg. Time
Lyc. IO-360 / 180 hp	85 - 200	Tube, Wood, Fabric	3500 hrs.
Plans	**Kit**	**Info Packet**	**Video**
$85.00	Yes	$8.00	N/A

Designed with the first time amateur builder in mind. Straight forward and uncomplicated. Chosen by hundreds of EAA Project Schoolflight programs at high schools around the country. Fully aerobatic. Welded steel tube fuselage, spruce wing construction, overall fabric covering. Baggage capacity of 35 lbs. Parts kits available from Wicks Aircraft Supply Co., 800-221-9425. Super Acro Wing Drawings for Acro Sport I $15.00. Shipping and handling charges: $4.50 for plans, $2.50 per info pack.

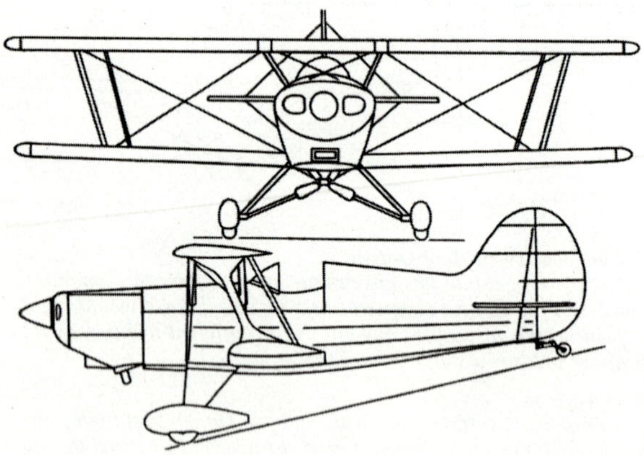

ACRO SPORT INC.
PO Box 462 • Hales Corners WI 53130
Tel. 414-529-2609

Acro Sport II

Wing Span	21.70 ft.	**Range**	300 sm
Wing Area	152.0 sq. ft.	**Takeoff Dist.**	350 ft.
Length	18.90 ft.	**Landing Dist.**	800 ft.
Max Gross Wt.	1520 lbs.	**Vmax**	152 mph
Empty Wt.	875 lbs.	**VCr**	123 mph
Payload (full fuel)	N/A	**V_{S1} (stall clean)**	53 mph
Fuel Capacity	26 gal.	**V_{S0} (lndg config)**	N/A
Seats	2	**Climb Rate @msl**	1200 fpm
Cabin Width	N/A	**Service Ceiling**	25,000 ft.

Std. Engine/HP	Usable HP Range	Construction Mtrl.	Est. Bldg. Time
Lycoming / 180 hp	108 - 200	Tube, Wood, Fabric	2000 hrs.
Plans	**Kit**	**Info Packet**	**Video**
$100.00	Yes	$8.00	N/A

A competitive, two place, open cockpit sport biplane. Capable of sportsman and intermediate aerobatics. Parts kits available from Wicks Aircraft Supply Co., 800-221-9425. Shipping and handling charges: $4.50 for plans, $2.50 per info pack.

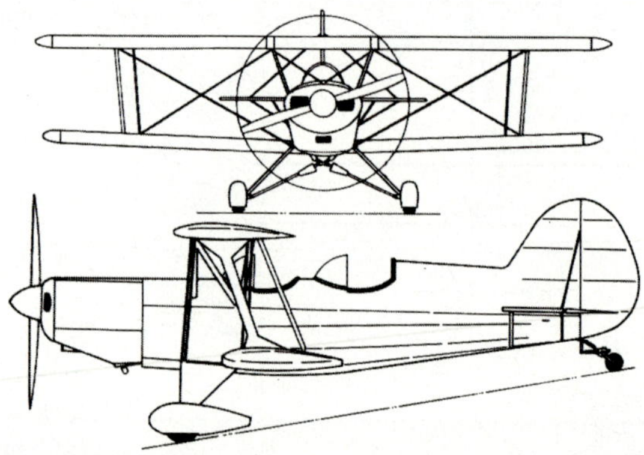

The Acro series was designed by EAA Founder Paul Poberezny with the first time amateur aircraft builder in mind. The Acros are easy to build aircraft for those who want to share the joy of open cockpit aerobatics. You won't find complicated or hard to fabricate fittings on these elegantly simple aerobatic biplanes. And, the larger, two place Acro II can be constructed for just a fraction more cost than its smaller, single placebrother.With its larger wheels, wider landing gear, light weight and responsive controls, the Acro II is a sport plane that is both docile on the ground and nimble in the air. Capable of sportsman and intermediate aerobatics, the Acro II is a competitive airplane. The Acro Sport series of aircraft have cockpits which are designed to be more comfortable for the larger pilot.

ACRO SPORT INC.
PO Box 462 • Hales Corners WI 53130
Tel. 414-529-2609

Nesmith Cougar

Wing Span	20.50 ft.	**Range**	600 sm
Wing Area	82.50 sq. ft.	**Takeoff Dist.**	800 ft.
Length	18.90 ft.	**Landing Dist.**	700 ft.
Max Gross Wt.	1250 lbs.	**Vmax**	195 mph
Empty Wt.	624 lbs.	**VCr**	135 mph
Payload (full fuel)	N/A	**V_{S1} (stall clean)**	53 mph
Fuel Capacity	25 gal.	**V_{S0} (lndg config)**	N/A
Seats	2	**Climb Rate @msl**	1000 fpm
Cabin Width	N/A	**Service Ceiling**	16,000 ft.

Std. Engine/HP	Usable HP Range	Construction Mtrl.	Est. Bldg. Time
Continental / 85 hp	65 - 125	Tube, Wood, Fabric	2000 hrs.
Plans	**Kit**	**Info Packet**	**Video**
$85.00	Yes	$8.00	N/A

A high, enclosed cabin wing design with a steel tube fuselage and wood and fabric wings. The cabin offers side-by-side seating. A folding wing option is available. Parts kits available from Wicks Aircraft Supply Co., 800-221-9425.

ACRO SPORT INC.
PO Box 462 • Hales Corners WI 53130
Tel. 414-529-2609

Pober Junior Ace

Wing Span	34.00 ft.	**Range**	250 sm
Wing Area	168.0 sq. ft.	**Takeoff Dist.**	350 ft.
Length	20.00 ft.	**Landing Dist.**	450 ft.
Max Gross Wt.	1320 lbs.	**Vmax**	130 mph
Empty Wt.	750 lbs.	**VCr**	80 mph
Payload (full fuel)	N/A	**V_{S1} (stall clean)**	38 mph
Fuel Capacity	24 gal.	**V_{S0} (lndg config)**	N/A
Seats	2	**Climb Rate @msl**	500 fpm
Cabin Width	N/A	**Service Ceiling**	25,000 ft.

Std. Engine/HP	Usable HP Range	Construction Mtrl.	Est. Bldg. Time
Continental / 85 hp	65 - 125	Tube, Wood, Fabric	1500 hrs.
Plans	**Kit**	**Info Packet**	**Video**
$125.00	Yes	$8.00	N/A

Side-by-side seating. Design has two doors and can be easily converted to an enclosed cabin version. This is an updated version of a 1930's design.Parts kits available from Wicks Aircraft Supply Co., 800-221-9425.

ACRO SPORT INC.
PO Box 462 • Hales Corners WI 53130
Tel. 414-529-2609

Pober Pixie

Wing Span	29.90 ft.	**Range**	290 sm
Wing Area	N/A	**Takeoff Dist.**	300 ft.
Length	17.30 ft.	**Landing Dist.**	300 ft.
Max Gross Wt.	900.0 lbs.	**Vmax**	130 mph
Empty Wt.	543 lbs.	**VCr**	83 mph
Payload (full fuel)	N/A	**V_{S1} (stall clean)**	30 mph
Fuel Capacity	12 gal.	**V_{S0} (lndg config)**	N/A
Seats	1	**Climb Rate @msl**	500 fpm
Cabin Width	N/A	**Service Ceiling**	13,000 ft.

Std. Engine/HP	Usable HP Range	Construction Mtrl.	Est. Bldg. Time
VW / 60 hp	N/A	Tube, Wood, Fabric	1500 hrs.
Plans	**Kit**	**Info Packet**	**Video**
$85.00	Yes	$8.00	N/A

For true light plane building and flying. The Pixie's style is reminiscent of the open cockpit flying of the helmet and goggle 1930's. Also utilizes Continental A-65 powerplant. J-3 style landing gear. Parts and kits are available from Wicks Aircraft Supply. 800-221-9425. Shipping and handling charges: $4.50 for plans, $2.50 per info pack.

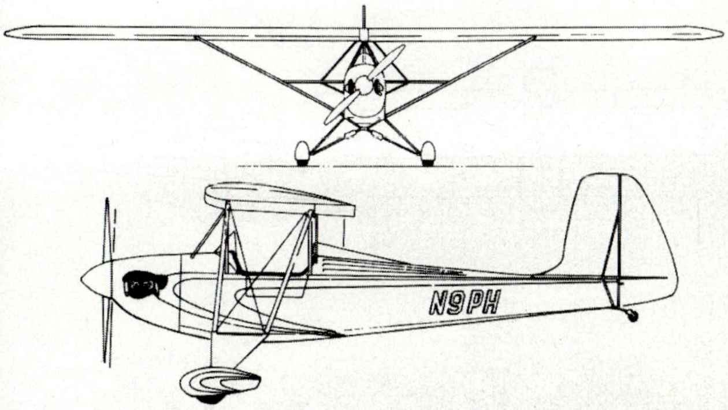

ACRO SPORT INC.
PO Box 462 • Hales Corners WI 53130
Tel. 414-529-2609

Pober Super Ace

Wing Span	27.20 ft.	**Range**	250 sm
Wing Area	N/A	**Takeoff Dist.**	350 ft.
Length	18.40 ft.	**Landing Dist.**	450 ft.
Max Gross Wt.	1030 lbs.	**Vmax**	160 mph
Empty Wt.	685 lbs.	**VCr**	110 mph
Payload (full fuel)	N/A	**V_{S1} (stall clean)**	N/A
Fuel Capacity	12 gal.	**V_{S0} (lndg config)**	N/A
Seats	1	**Climb Rate @msl**	500 fpm
Cabin Width	N/A	**Service Ceiling**	25,000 ft.

Std. Engine/HP	Usable HP Range	Construction Mtrl.	Est. Bldg. Time
Cont. C-85-8 / 85 hp	85 - 150	Tube, Wood, Fabric	1900 hrs.
Plans	**Kit**	**Info Packet**	**Video**
$100.00	Yes	$8.00	N/A

Standard 85 hp powerplant does not include starter or electrical system. Other powerplants up to 150 hp can be used. Constructed of 4130 steel tubing for the fuselage, tail group, landing gear, fittings, and wing struts. Wings are spruce spars and ribs. The complete aircraft is covered with Stits Poly-fiber and finished with Stits paints. Control system is cable operated. Shipping and handling charges: $4.50 for plans, $2.50 per info pack. Parts kits available from Wicks Aircraft Supply Co., 800-221-9425.

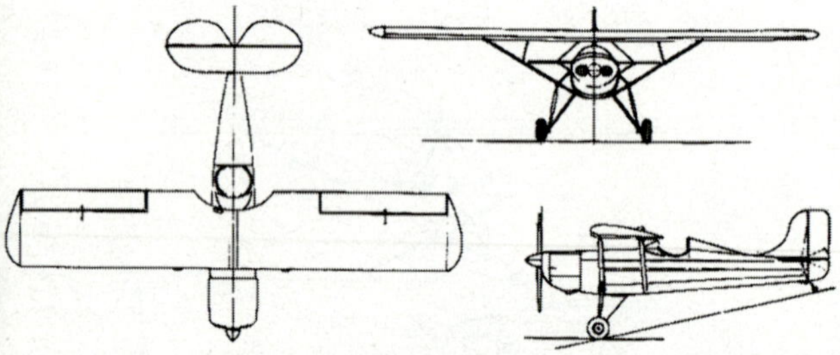

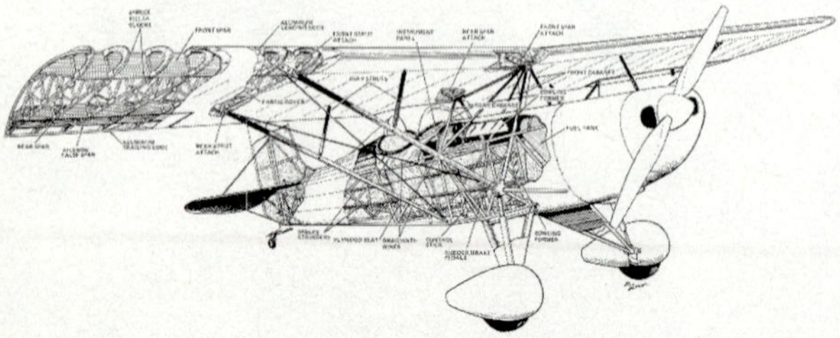

ACRO SPORT INC.
PO Box 462 • Hales Corners WI 53130
Tel. 414-529-2609

Mallard M2-A

Wing Span	29.67 ft.	**Range**	350 sm
Wing Area	156.0 sq. ft.	**Takeoff Dist.**	400 ft.
Length	22.33 ft.	**Landing Dist.**	650 ft.
Max Gross Wt.	1250 lbs.	**Vmax**	95 mph
Empty Wt.	750 lbs.	**VCr**	85 mph
Payload (full fuel)	N/A	**V_{S1} (stall clean)**	37 mph
Fuel Capacity	18 gal.	**V_{S0} (lndg config)**	34 mph
Seats	2	**Climb Rate @msl**	750 fpm
Cabin Width	46 in.	**Service Ceiling**	16,000 ft.

Std. Engine/HP	Usable HP Range	Construction Mtrl.	Est. Bldg. Time
Rotax 912 / 80 hp	66-110	Tube, Fabric, Composite	350 hrs.
Plans	**Kit**	**Info Packet**	**Video**
N/A	$19,000.00	$5.00	N/A

Available as a conversion kit for existing Buccaneers or as a completely new kit. Conversion can involve as many as 30 extensive modifications as well as numerous optional items. Large comfortable cabin. **Completed & Flying:** 5

ADVANCED AEROMARINE
152 East Eighth Street • Apopka Industrial Park
Apopka FL 32703
Tel. 407-877-7871 • Fax 407-877-7871

Carrera

Wing Span	29.67 ft.	**Range**	N/A
Wing Area	145.0 sq. ft.	**Takeoff Dist.**	150 ft.
Length	20.08 ft.	**Landing Dist.**	150 ft.
Max Gross Wt.	700.0 lbs.	**Vmax**	105 mph
Empty Wt.	350 lbs.	**VCr**	84 mph
Payload (full fuel)	350 lbs.	**V_{S1} (stall clean)**	30 mph
Fuel Capacity	12 gal.	**V_{S0} (lndg config)**	30 mph
Seats	1	**Climb Rate @msl**	1200 fpm
Cabin Width	N/A	**Service Ceiling**	15,000 ft.

Std. Engine/HP	Usable HP Range	Construction Mtrl.	Est. Bldg. Time
Rotax / 52 hp	45 - 115	Tube, Fabric, Composite	200 hrs.
Plans	**Kit**	**Info Packet**	**Video**
N/A	$7,995.00	$5.00	N/A

Available in several versions including single, two place and side-by-side seating. All-metal skins are optional. Easy to assemble basic structure features all-aluminum bolt/rivet construction. Optional floats and skis. **Completed:** 40

ADVANCED AEROMARINE
152 East Eighth Street • Apopka Industrial Park
Apopka FL 32703
Tel. 407-877-7871 • Fax 407-877-7871

Adventurer 2+2

Wing Span	35.85 ft.	**Range**	800 sm
Wing Area	179.0 sq. ft.	**Takeoff Dist.**	700 ft.
Length	24.50 ft.	**Landing Dist.**	600 ft.
Max Gross Wt.	3000 lbs.	**Vmax**	158 mph
Empty Wt.	1800 lbs.	**VCr**	115 mph
Payload (full fuel)	1000 lbs.	**V_{S1} (stall clean)**	54 mph
Fuel Capacity	60 gal.	**V_{S0} (lndg config)**	47 mph
Seats	4	**Climb Rate @msl**	900 fpm
Cabin Width	46 in.	**Service Ceiling**	12,500 ft.

Std. Engine/HP	Usable HP Range	Construction Mtrl.	Est. Bldg. Time
Lyc. IO-360 / 200 hp	200 - 350	Composite	1000 hrs.
Plans	**Kit**	**Info Packet**	**Video**
N/A	$35,200.00	$25.00	Yes

2+2 seating amphibian with a 7 ft. sleeping bunk. Price is for the basic kit and does not include engine or avionics. Also available in a Rapid Assembly kit that reduces build time from 1,000 hours to 400 - 600 hours. Options include larger wing tanks, dual controls, flaps, tricycle gear and preassembled wing. Various engine options. Takeoff/landing specs are for land. Water takeoff distance is 1,000 ft. Video included in Info Pack. See Ad on Page 7. **Kit Sold:** 120 **Completed & Flying:** 5

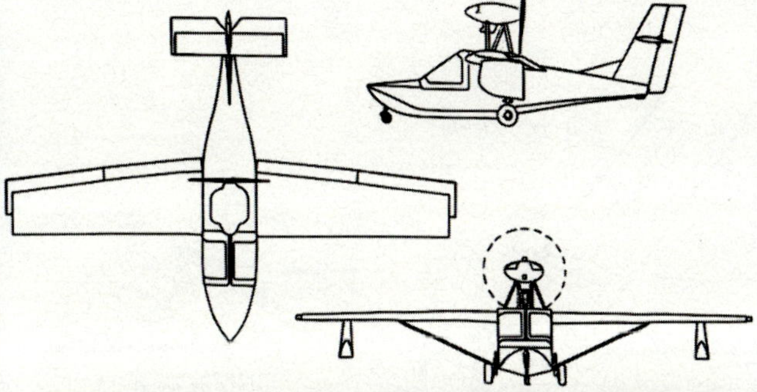

Emad R. Albana, MD, Builder
"... I am impressed with the simplicity and strength of the Adventurer. Your support and supply of parts is outstanding."

ADVENTUREAIR
PO Box 368 • Carroll County Airport
Berryville AR 72616-0368
Ph: 870-423-5350 • Fax: 870-423-5366
E-Mail: advair@cswnet.com
Web: http://www.pe.net/~advair

Adventurer 333

Wing Span	35.83 ft.	**Range**	850 sm
Wing Area	179.0 sq. ft.	**Takeoff Dist.**	600 ft.
Length	24.50 ft.	**Landing Dist.**	700 ft.
Max Gross Wt.	3333 lbs.	**Vmax**	158 mph
Empty Wt.	2000 lbs.	**VCr**	130 mph
Payload (full fuel)	1333 lbs.	**V_{S1} (stall clean)**	61 mph
Fuel Capacity	60 gal.	**V_{S0} (lndg config)**	54 mph
Seats	4	**Climb Rate @msl**	1200 fpm
Cabin Width	46 in.	**Service Ceiling**	15,000 ft.

Std. Engine/HP	Usable HP Range	Construction Mtrl.	Est. Bldg. Time
Chev HO350/ 333 hp	300 - 350	Composite	1000 hrs.
Plans	**Kit**	**Info Packet**	**Video**
N/A	$34,900.00	$25.00	Yes

333 hp Chevy engine powered. Price is for the basic kit and does not include engine or avionics. Also available in a Rapid Assembly kit that reduces build time from 1,000 hours to 400 - 600 hours. Options include 60,120, and 180 gallon wing tanks, dual controls, flaps, tricycle gear and preassembled wing. Various engine options. Takeoff/landing specs are for land. Video included in Info Pack. See Ad on Page 7. **Sold:** 12 **Flying:** 5

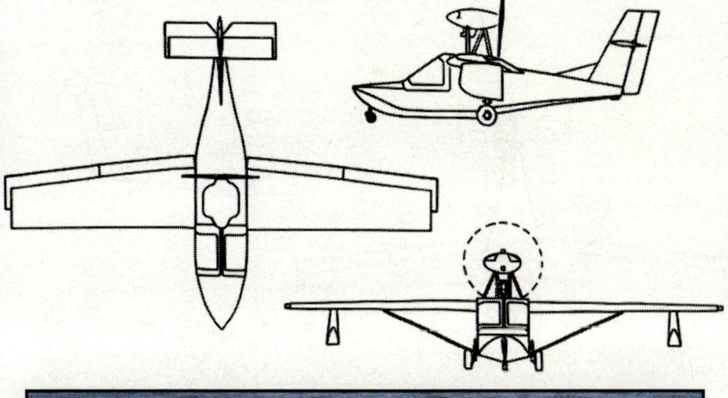

Paul Douglas, Builder
"...I was quite pleased with the quality of the fuselage, especially the bonded areas between all of the fuselage decks/bulkheads, and horizontal tail surfaces. ...The gas tanks are especially well made."

ADVENTUREAIR
PO Box 368 • Carroll County Airport
Berryville AR 72616-0368
Ph: 870-423-5350 • Fax: 870-423-5366
E-Mail: advair@cswnet.com
Web: http://www.pe.net/~advair

Adventurer 333 Heavy Hauler

Wing Span	35.83 ft.	**Range**	800 sm
Wing Area	179.0 sq. ft.	**Takeoff Dist.**	600 ft.
Length	24.50 ft.	**Landing Dist.**	700 ft.
Max Gross Wt.	4400 lbs.	**Vmax**	158 mph
Empty Wt.	2220 lbs.	**VCr**	130 mph
Payload (full fuel)	2180 lbs.	**V_{S1} (stall clean)**	61 mph
Fuel Capacity	60 gal.	**V_{S0} (lndg config)**	54 mph
Seats	4	**Climb Rate @msl**	1200 fpm
Cabin Width	46 in.	**Service Ceiling**	15,000 ft.

Std. Engine/HP	Usable HP Range	Construction Mtrl.	Est. Bldg. Time
Chev HO350/ 333 hp	200 - 350	Composite	1000 hrs.
Plans	**Kit**	**Info Packet**	**Video**
N/A	$35,200.00	$25.00	Yes

Long range, heavy hauler or military version of the 333. Price is for the basic kit and does not include engine or avionics. Options include larger fuel tanks (180, 216 or 260 gal.), dual controls, flaps, tricycle gear and preassembled wing. Seating is for 2, 3 or 2+2. Video included in Info Pack. See Ad on Page 7.

Kits/Plans Sold: 120 **Completed & Flying:** 5

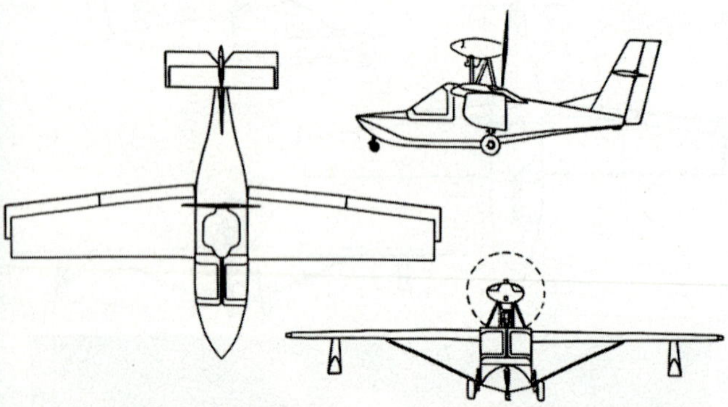

Ray Ackerman, Builder

"... had no real experience with composites but the instructions were simple enough so I didn't have any real problems. Besides, support was only a telephone call away."

ADVENTURE AIR
PO Box 368 • Carroll County Airport
Berryville AR 72616-0368
Ph: 870-423-5350 • Fax: 870-423-5366
E-Mail: advair@cswnet.com
Web: http://www.pe.net/~advair

Sirocco

Wing Span	33.00 ft.	**Range**	312 sm
Wing Area	148.0 sq. ft.	**Takeoff Dist.**	92 ft.
Length	19.00 ft.	**Landing Dist.**	N/A
Max Gross Wt.	546.0 lbs.	**Vmax**	75 mph
Empty Wt.	288 lbs.	**VCr**	62 mph
Payload (full fuel)	N/A	**V_{S1} (stall clean)**	28 mph
Fuel Capacity	10 gal.	**V_{S0} (lndg config)**	N/A
Seats	1	**Climb Rate @msl**	984 fpm
Cabin Width	N/A	**Service Ceiling**	N/A

Std. Engine/HP	Usable HP Range	Construction Mtrl.	Est. Bldg. Time
Rotax 447 / 35 hp	N/A	Metal & Composite	N/A
Plans	**Kit**	**Info Packet**	**Video**
N/A	$11,000.00	Yes	N/A

The first ultralight aircraft to have flown around the world. A three axis ultralight with specially designed spoilers that ensure extensive roll control. Body is a one piece construction, hand laminated of composite fiberglass/polyester. Wing is glass epoxy laminate and metal with a 3.8 oz. Dacron covering.

Completed & Flying: 160

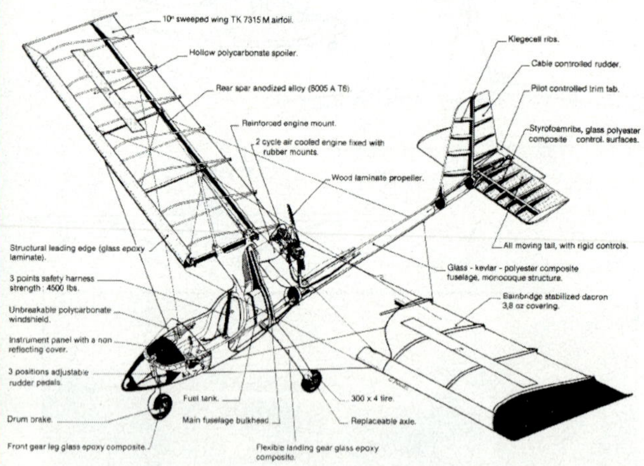

AERIANE
Rue de Poiriers, 7 • Gembloux B-5030
BELGIUM
Tel. 32 81 60 05 95 • Fax 32 81 60 05 96
E-Mail: Vincent.Piret@ping.be
Web: http://www.ping.be/~ping4026/aeriane.htm

Ryan ST

Wing Span	30.00 ft.	**Range**	350 sm
Wing Area	124.0 sq. ft.	**Takeoff Dist.**	525 ft.
Length	21.50 ft.	**Landing Dist.**	450 ft.
Max Gross Wt.	1575 lbs.	**Vmax**	150 mph
Empty Wt.	1023 lbs.	**VCr**	125 mph
Payload (full fuel)	N/A	**V_{S1} (stall clean)**	45 mph
Fuel Capacity	24 gal.	**V_{S0} (lndg config)**	N/A
Seats	2	**Climb Rate @msl**	1400 fpm
Cabin Width	N/A	**Service Ceiling**	17,500 ft.

Std. Engine/HP	Usable HP Range	Construction Mtrl.	Est. Bldg. Time
Menasco / 150 hp	125 - 150	Metal, Wood, Fabric	N/A
Plans	**Kit**	**Info Packet**	**Video**
$650.00	N/A	$12.00	N/A

A classic single engine design with an open cockpit. Three different engines can be used: the Menasco, the Walter LOM or the Casa-Tigre. Acceptable hp range is from 125 to 150. Photo: Ted Babbini, Sonoma, CA.

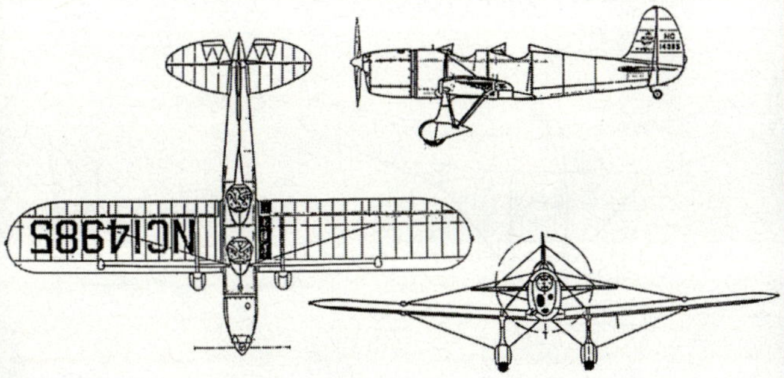

The original Ryan ST was designed in 1934 by T. Claude Ryan, and produced in San Diego, California by the Ryan Aeronautical Company from 1934 to 1940. This design was the outstanding, open cockpit monoplane of the "Golden Era", and considered to be a true classic then, as well as now.

The available plans have been redrawn, by hand, from original prints of the original manufacturer's drawings, with improved drafting techniques and clarity.

The drawings have been approved by the original manufacturer for either building up a "replica", or in restoring or repairing an original Ryan ST airplane. The plans are also available to model builders, who want to build that "ultimate" and super scale Ryan ST model, showing every rivet and other details.

AERO-DRAFT
430 Budding Ridge
Cheshire CT 06410
Tel. 203-272-2127

Cadet Model STF

Wing Span	27.00 ft.	**Range**	520 sm
Wing Area	120.0 sq. ft.	**Takeoff Dist.**	800 ft.
Length	18.00 ft.	**Landing Dist.**	835 ft.
Max Gross Wt.	1350 lbs.	**Vmax**	175 mph
Empty Wt.	789 lbs.	**VCr**	130 mph
Payload (full fuel)	N/A	**V_{S1} (stall clean)**	60 mph
Fuel Capacity	25 gal.	**V_{S0} (lndg config)**	55 mph
Seats	2	**Climb Rate @msl**	800 fpm
Cabin Width	39 in.	**Service Ceiling**	15,000 ft.

Std. Engine/HP	Usable HP Range	Construction Mtrl.	Est. Bldg. Time
Continental / 100 hp	85 - 120	Tube, Wood, Fabric	2500 hrs.
Plans	**Kit**	**Info Packet**	**Video**
$230.00	N/A	$10.00	N/A

Updated version of the 1941 Culver Cadet which was one of the better performing light planes ever built. The clean design, with retractable landing gear, and elliptical cantilever wing, give this airplane speed and docile handling characteristics. Lightly balanced controls and a comfortable cockpit.

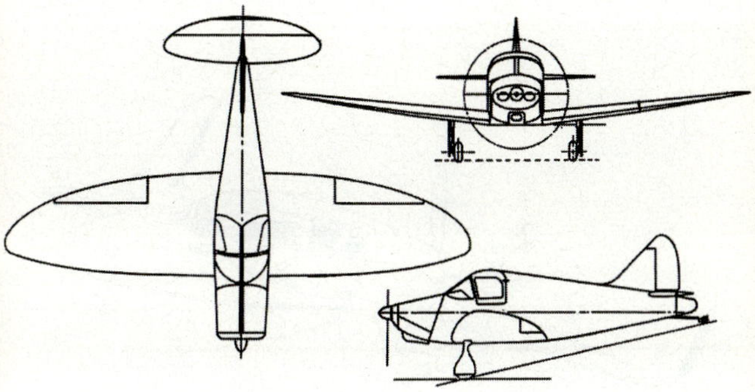

The predecessor M12 Cadet airplane was designed in the early 1940's by Al Mooney for the Culver Aircraft Company. Wartime shortages dictated that it be primarily constructed of wood. 368 were built in Wichita, Kansas. The low wing Cadet was powered by a 75 hp Continental aircraft engine. Even with this modest power, the Cadet cruised at 120 mph. The classic M12 was way ahead of its time. However, the years have taken their toll on these wood structured machines and very few examples are still airworthy.

Enter Neal La France, an avid aircraft homebuilder, racer, and designer who once owned a Culver Cadet. After extensive research, Neal found that there are quite a few of the old planes around. They usually have serviceable wings, tails, and equipment, but rotten plywood fuselages. So, Neal purchased one and, after much design and analysis, built a steel tube replacement fuselage. The new design has been documented and is now available as the plans only Cadet Model STF.

AERO SYSTEMS
9031 Suncrest St. • Wichita KS 67212
Tel. 316-722-2494

AeroCanard RG

Wing Span	28.10 ft.	**Range**	1000 sm
Wing Area	102.3 sq. ft.	**Takeoff Dist.**	800 ft.
Length	16.75 ft.	**Landing Dist.**	1500 ft.
Max Gross Wt.	2150 lbs.	**Vmax**	225 mph
Empty Wt.	1300 lbs.	**VCr**	210 mph
Payload (full fuel)	N/A	**V_{S1} (stall clean)**	N/A
Fuel Capacity	50 gal.	**V_{S0} (lndg config)**	N/A
Seats	4	**Climb Rate @msl**	1900 fpm
Cabin Width	N/A	**Service Ceiling**	25,000 ft.

Std. Engine/HP	Usable HP Range	Construction Mtrl.	Est. Bldg. Time
Lyc LIO-360 / 200 hp	160 - 200	Composite	1700 hrs.
Plans	**Kit**	**Info Packet**	**Video**
N/A	$27,341.00	$2.00	N/A

Canard pusher design with fully retractable tri-gear. Can be purchased in 16 sub-kits or complete. Kit can also be purchased without raw materials or hardware for $19,298. Also available in the "FG" version with fixed main gear and retractable nose wheel for $22,791.

Completed & Flying: 1

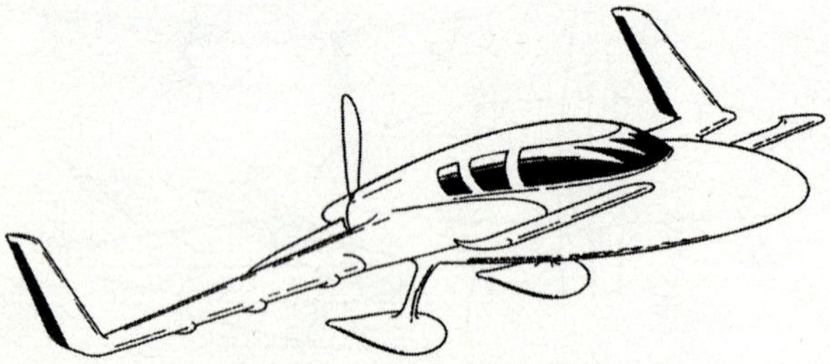

AEROCAD, INC.
PO Box 7307
Port St. Lucie FL 34985
Tel. 561-460-8020 • Fax 561-460-8020
E-Mail: jeff@aerocad.com
Web: http://www.aerocad.com

AeroCanard SB

Wing Span	28.10 ft.	**Range**	1000 sm
Wing Area	102.3 sq. ft.	**Takeoff Dist.**	1200 ft.
Length	16.75 ft.	**Landing Dist.**	1500 ft.
Max Gross Wt.	2050 lbs.	**Vmax**	230 mph
Empty Wt.	1150 lbs.	**VCr**	200 mph
Payload (full fuel)	N/A	**V_{S1} (stall clean)**	N/A
Fuel Capacity	52 gal.	**V_{S0} (lndg config)**	N/A
Seats	4	**Climb Rate @msl**	1500 fpm
Cabin Width	N/A	**Service Ceiling**	20,000 ft.

Std. Engine/HP	Usable HP Range	Construction Mtrl.	Est. Bldg. Time
Lyc 0-360 / 180 hp	160 - 200	Composite	1500 hrs.
Plans	**Kit**	**Info Packet**	**Video**
N/A	$20,786.00	$2.00	N/A

AeroCad's "Small Body" design. A version of the Cozy Mk IV licensed by Co-Z Development. A modern compact, efficient, high performance four place airplane of proven design. Seats two adults and two children. Available in 16 sub-kits. Kit can be ordered without the raw materials and hardware for $12,743.

Completed & Flying: 9

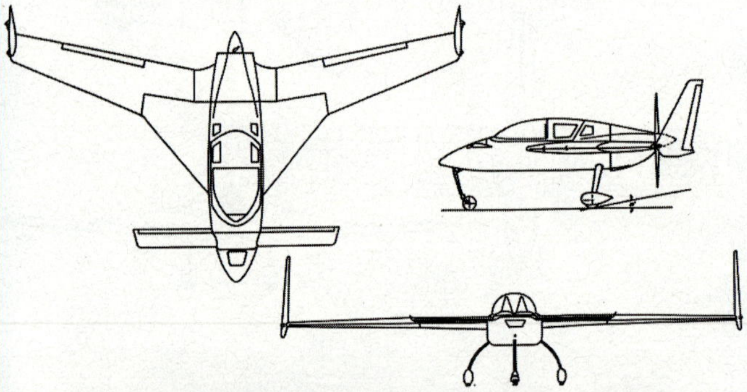

AEROCAD, INC.
PO Box 7307
Port St. Lucie FL 34985
Tel. 561-460-8020 • Fax 561-460-8020
E-Mail: jeff@aerocad.com
Web: http://www.aerocad.com

E-Z Flyer

Wing Span	26.00 ft.	**Range**	160 sm
Wing Area	165.0 sq. ft.	**Takeoff Dist.**	150 ft.
Length	21.00 ft.	**Landing Dist.**	150 ft.
Max Gross Wt.	1050 lbs.	**Vmax**	95 mph
Empty Wt.	469 lbs.	**VCr**	65 mph
Payload (full fuel)	527 lbs.	**V_{S1} (stall clean)**	N/A
Fuel Capacity	9 gal.	**V_{S0} (lndg config)**	30 mph
Seats	2	**Climb Rate @msl**	650 fpm
Cabin Width	N/A	**Service Ceiling**	N/A

Std. Engine/HP	Usable HP Range	Construction Mtrl.	Est. Bldg. Time
Rotax 503 / 52 hp	50 - 100	Steel, Fabric	200 hrs.
Plans	**Kit**	**Info Packet**	**Video**
N/A	$13,995.00	$10.00	N/A

Features rugged construction with 4130 Cr-Mo steel fuselage. Wings and tail are the same as those used on the Merlin GT. Exceptional visibility. Designed to be used by flying schools. Easy to build. Strong, simple, and easy to fly. Complete kit price includes Rotax 503 engine and a two bladed prop. Also available with Rotax 582, 618, or 912. Extended range fuel capacity is optional. See Ad on Page 40.

Completed & Flying: 15

Rich Wozniakowski, Munster, IN
"Thanks for designing a simple aircraft that a first time builder can handle."

AEROCOMP, INC.
2335 Newfound Harbor Drive
Merritt Island FL 32952
Tel. 407-453-6641 • Fax 407-453-6641
E-Mail: ral@nexusprime.org
Web: http://aerocompinc.com

Merlin GT

Wing Span	30.00 ft.	**Range**	350 sm
Wing Area	165.0 sq. ft.	**Takeoff Dist.**	110 ft.
Length	20.00 ft.	**Landing Dist.**	175 ft.
Max Gross Wt.	1300 lbs.	**Vmax**	120 mph
Empty Wt.	567 lbs.	**VCr**	93 mph
Payload (full fuel)	637 lbs.	**V_{S1} (stall clean)**	38 mph
Fuel Capacity	16 gal.	**V_{S0} (lndg config)**	N/A
Seats	2	**Climb Rate @msl**	1500 fpm
Cabin Width	41 in.	**Service Ceiling**	18,000 ft.

Std. Engine/HP	Usable HP Range	Construction Mtrl.	Est. Bldg. Time
Rotax 912 / 79 hp	50 - 125	Steel and Fabric	350 hrs.
Plans	**Kit**	**Info Packet**	**Video**
N/A	$19,995.00	$10.00	$15.00

Known for its rugged durability, superb ground handling stability, and simple construction. Excellent float plane. Available with wheels, floats, or as amphibious. A very rugged and roomy airplane with outstanding STOL and rough field capabilities. Wing main spar and leading edge are factory prebuilt. Price without engine is $12,995. In addition to the Rotax 912, other engine options include Rotax 582 and 618, Continental, and Subaru auto conversions. See Ad on Page 40.

Kits/Plans Sold: 350 **Completed & Flying:** 300

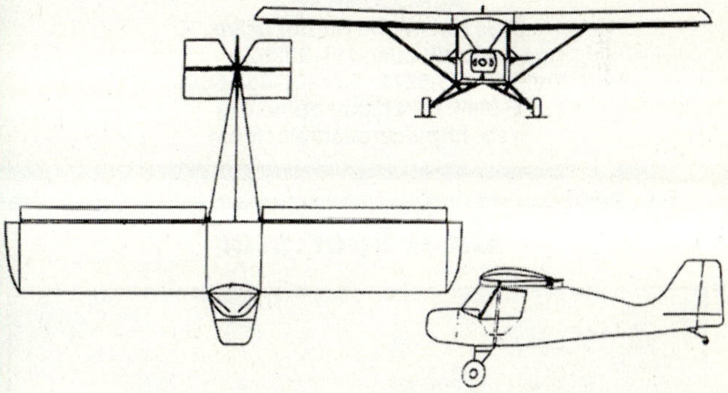

Steve Young, Merritt Island, FL
"This is a FUN airplane. It flies like a dream. With the doors open, it is like an open cockpit aircraft, only better."

Ron Wood, Jonesboro, AR
"The Merlin just wants to fly!"

Wayne Winters, Calgary, AB
"The hardest part of building my Merlin was getting the packing crate off the truck."

AEROCOMP, INC.
2335 Newfound Harbor Drive
Merritt Island FL 32952
Tel. 407-453-6641 • Fax 407-453-6641
E-Mail: ral@nexusprime.org
Web: http://aerocompinc.com

CA 6

Wing Span	34.50 ft.	**Range**	1000 sm
Wing Area	207.0 sq. ft.	**Takeoff Dist.**	450 ft.
Length	24.50 ft.	**Landing Dist.**	500 ft.
Max Gross Wt.	2850 lbs.	**Vmax**	200 mph
Empty Wt.	1500 lbs.	**VCr**	175 mph
Payload (full fuel)	N/A	**V_{S1} (stall clean)**	48 mph
Fuel Capacity	120 gal.	**V_{S0} (lndg config)**	N/A
Seats	6	**Climb Rate @msl**	1700 fpm
Cabin Width	43 in.	**Service Ceiling**	18,000 ft.

Std. Engine/HP	Usable HP Range	Construction Mtrl.	Est. Bldg. Time
Franklin / 220 hp	220-260	Composite	350 hrs.
Plans	**Kit**	**Info Packet**	**Video**
N/A	$26,900.00	N/A	$20.00

Six place, tapered or straight wing construction. Can accept a wide range of engines from 220-260 hp. Available in amphibious, floats, conventional gear, and tricycle gear configurations. Kit can be assembled in 350 hours with no special skills or tools needed. Kit price does not include engine, prop, or floats. Other kits are available that include engine and prop. Number of kits sold/flying includes all models. See Ad on Page 40.

Kits/Plans Sold: 53 **Completed & Flying:** 11

AEROCOMP, INC.
2335 Newfound Harbor Drive
Merritt Island FL 32952
Tel. 407-453-6641 • Fax 407-453-6641
E-Mail: ral@nexusprime.org
Web: http://aerocompinc.com

Comp Air Trainer

Wing Span	34.50 ft.	**Range**	600 sm
Wing Area	207.0 sq. ft.	**Takeoff Dist.**	450 ft.
Length	24.50 ft.	**Landing Dist.**	500 ft.
Max Gross Wt.	2400 lbs.	**Vmax**	150 mph
Empty Wt.	1375 lbs.	**VCr**	120 mph
Payload (full fuel)	N/A	**V_{S1} (stall clean)**	41 mph
Fuel Capacity	120 gal.	**V_{S0} (lndg config)**	N/A
Seats	4	**Climb Rate @msl**	800 fpm
Cabin Width	43 in.	**Service Ceiling**	15,500 ft.

Std. Engine/HP	Usable HP Range	Construction Mtrl.	Est. Bldg. Time
N/A	125-170	Composite	N/A
Plans	**Kit**	**Info Packet**	**Video**
N/A	$22,900.00	$10.00	$10.00

Four place composite. Available in float, conventional, and tricycle gear configurations. Kit can be assembled in 350 hours with no special skills or tools needed. Kit price does not include engine, prop, or floats. Other kits are available that include engine and prop. Number of kits sold/flying includes all models. See Ad on Page 40.

Kits/Plans Sold: 53 **Completed & Flying:** 11

AEROCOMP, INC.
2335 Newfound Harbor Drive
Merritt Island FL 32952
Tel. 407-453-6641 • Fax 407-453-6641
E-Mail: ral@nexusprime.org
Web: http://aerocompinc.com

CA 6 Sportster

AEROCOMP, INC.
2335 Newfound Harbor Drive
Merritt Island FL 32952
Tel. 407-453-6641 • Fax 407-453-6641
E-Mail: ral@nexusprime.org
Web: http://aerocompinc.com

Wing Span	34.50 ft.	**Range**	1000 sm
Wing Area	207.0 sq. ft.	**Takeoff Dist.**	450 ft.
Length	24.50 ft.	**Landing Dist.**	500 ft.
Max Gross Wt.	3600 lbs.	**Vmax**	195 mph
Empty Wt.	1850 lbs.	**VCr**	180 mph
Payload (full fuel)	N/A	**V_{S1} (stall clean)**	48 mph
Fuel Capacity	120 gal.	**V_{S0} (lndg config)**	N/A
Seats	6	**Climb Rate @msl**	1200 fpm
Cabin Width	43 in.	**Service Ceiling**	15,500 ft.

Std. Engine/HP	Usable HP Range	Construction Mtrl.	Est. Bldg. Time
N/A	250-300	Composite	500 hrs.
Plans	**Kit**	**Info Packet**	**Video**
N/A	$36,999.00	N/A	$20.00

Six place composite powered by 260-300 hp engine. Conventional gear only. Can be flown with two on board doing sport catagory aerobatics at 2600 lb. gross weight. It is rated at +5.2, -2.0 G's. Can also be flown cross country with six on board with a gross weight of 3200 lbs. Kit can be assembled in 500 hours. Kit price does not include engine or prop. Other kits are available that include engine and prop. Number of kits sold/flying includes all models. See Ad on Page 40.

Kits/Plans Sold: 53 **Completed & Flying:** 11

CA 6 WB

Wing Span	35.00 ft.	**Range**	1000 sm
Wing Area	212.0 sq. ft.	**Takeoff Dist.**	450 ft.
Length	8.00 ft.	**Landing Dist.**	500 ft.
Max Gross Wt.	3600 lbs.	**Vmax**	200 mph
Empty Wt.	1900 lbs.	**VCr**	185 mph
Payload (full fuel)	N/A	**V_{S1} (stall clean)**	50 mph
Fuel Capacity	60 gal.	**V_{S0} (lndg config)**	N/A
Seats	6	**Climb Rate @msl**	1700 fpm
Cabin Width	47 in.	**Service Ceiling**	18,000 ft.

Std. Engine/HP	Usable HP Range	Construction Mtrl.	Est. Bldg. Time
N/A	250 - 350	Composite	350 hrs.
Plans	**Kit**	**Info Packet**	**Video**
N/A	$36,995.00	$10.00	$20.00

Six place, 46.5" extra wide cabin, flush doors, tapered or straight wing, composite construction. Available in amphibious, float, conventional, or tri-gear configurations. Kit price does not include engine, prop, or floats. Other kits available that include engine and prop. Number of kits sold/flying includes all models.

Kits/Plans Sold: 53 **Completed & Flying:** 11

AEROCOMP, INC.
2335 Newfound Harbor Drive
Merritt Island FL 32952
Ph: 407-453-6641 • Fax: 407-453-6641
E-Mail: ral@nexusprime.org
Web: http://aerocompinc.com

CA 6 Camp

Wing Span	34.50 ft.	**Range**	1000 sm
Wing Area	207.0 sq. ft.	**Takeoff Dist.**	450 ft.
Length	24.50 ft.	**Landing Dist.**	500 ft.
Max Gross Wt.	3000 lbs.	**Vmax**	200 mph
Empty Wt.	1650 lbs.	**VCr**	175 mph
Payload (full fuel)	N/A	**V_{S1} (stall clean)**	48 mph
Fuel Capacity	120 gal.	**V_{S0} (lndg config)**	N/A
Seats	6	**Climb Rate @msl**	1700 fpm
Cabin Width	43 in.	**Service Ceiling**	18,000 ft.

Std. Engine/HP	Usable HP Range	Construction Mtrl.	Est. Bldg. Time
N/A	220-260	Composite	N/A
Plans	**Kit**	**Info Packet**	**Video**
N/A	$30,995.00	$10.00	$10.00

Six place, bulkhead free interior allowing folding seats to make a 42.5" x 78" flat sleeping area. Split seats can allow air ambulance configuration. Amphibious, float, tri or conventional gear configurations available. Number of kits sold/flying includes all models. See Ad on Page 40.

Kits/Plans Sold: 53 **Completed & Flying:** 11

AEROCOMP, INC.
2335 Newfound Harbor Drive
Merritt Island FL 32952
Tel. 407-453-6641 • Fax 407-453-6641
E-Mail: ral@nexusprime.org
Web: http://aerocompinc.com

CA 6 TW

AEROCOMP, INC.
2335 Newfound Harbor Drive
Merritt Island FL 32952
Tel. 407-453-6641 • Fax 407-453-6641
E-Mail: ral@nexusprime.org
Web: http://aerocompinc.com

Wing Span	34.50 ft.	**Range**	1000 sm
Wing Area	178.0 sq. ft.	**Takeoff Dist.**	450 ft.
Length	24.50 ft.	**Landing Dist.**	500 ft.
Max Gross Wt.	3000 lbs.	**Vmax**	200 mph
Empty Wt.	1650 lbs.	**VCr**	195 mph
Payload (full fuel)	N/A	**V_{S1} (stall clean)**	48 mph
Fuel Capacity	120 gal.	**V_{S0} (lndg config)**	N/A
Seats	6	**Climb Rate @msl**	1900 fpm
Cabin Width	43 in.	**Service Ceiling**	18,000 ft.

Std. Engine/HP	Usable HP Range	Construction Mtrl.	Est. Bldg. Time
Franklin / 220 hp	220-300	Composite	350 hrs.
Plans	**Kit**	**Info Packet**	**Video**
N/A	$34,995.00	N/A	$20.00

Six place, tapered wing, flush doors, composite construction aircraft that can accept a wide range of engines from 220-300 hp. Available in amphibious, float, conventional, or tri gear configurations. Kit can be assembled in 350 hours with no special skills or tools needed. Kit price does not include engine, prop, or floats. Other kits available that include engine and prop. Number of kits sold/flying includes all models. See Ad on Page 40.

Kits/Plans Sold: 53 **Completed & Flying:** 11

Comp Monster

Wing Span	34.50 ft.	**Range**	800 sm
Wing Area	207.0 sq. ft.	**Takeoff Dist.**	450 ft.
Length	24.80 ft.	**Landing Dist.**	500 ft.
Max Gross Wt.	2850 lbs.	**Vmax**	167 mph
Empty Wt.	1450 lbs.	**VCr**	132 mph
Payload (full fuel)	N/A	**V_{S1} (stall clean)**	41 mph
Fuel Capacity	120 gal.	**V_{S0} (lndg config)**	N/A
Seats	4	**Climb Rate @msl**	1000 fpm
Cabin Width	43 in.	**Service Ceiling**	15,500 ft.

Std. Engine/HP	Usable HP Range	Construction Mtrl.	Est. Bldg. Time
Lyc. 0-360 / 180 hp	125 - 250	Composite	350 hrs.
Plans	**Kit**	**Info Packet**	**Video**
N/A	$23,900.00	$10.00	N/A

Four place, composite construction aircraft that can accept a wide range of engines from 125 to 250 hp. Available in amphibious or standard gear configurations. Kit can be assembled in 350 hours with no special skills or tools needed. Kit price does not include engine, prop, or floats. Other kits available that include engine and prop. Number of kits sold/ flying includes all models. See Ad on Page 4

Kits/Plans Sold: 53 **Completed & Flying:** 11

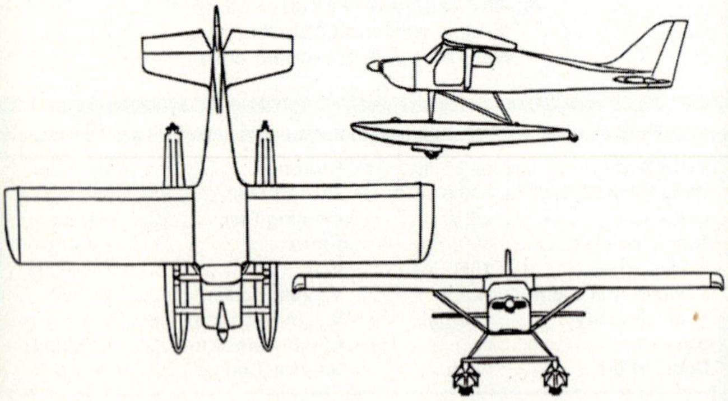

AEROCOMP, INC.
2335 Newfound Harbor Drive
Merritt Island FL 32952
Tel. 407-453-6641 • Fax 407-453-6641
E-Mail: ral@nexusprime.org
Web: http://aerocompinc.com

AeroMaster AG

Wing Span	28.70 ft.	**Range**	140 sm
Wing Area	146.00 sq. ft.	**Takeoff Dist.**	150 ft.
Length	17.50 ft.	**Landing Dist.**	150 ft.
Max Gross Wt.	975 lbs.	**Vmax**	95 mph
Empty Wt.	450 lbs.	**Vcr**	80 mph
Payload (full fuel)	510 lbs.	**V_{S1} (stall clean)**	23 mph
Fuel Capacity	10.0 gal.	**V_{S0} (lndg config)**	N/A mph
Seats	1	**Climb Rate @msl**	1,300 fpm
Cabin Width	22.0 in.	**Serv. Ceiling**	10,000 ft.

Std. Engine/HP	Usable HP Range	Construction Mtrl.	Est. Bldg. Time
Rotax 582 / 65 hp	50-80	Tube, Fabric, Composite	170 hrs.
Plans	**Kit**	**Info Packet**	**Video**
No	$21,075	$12.00	$20.00

Excellent visibility. Supplied with engine, propeller, instruments, and Spray Miser CDA agricultural spraying system with 30 gallon tank.

Kits/Plans Sold: 6 **Completed & Flying:** 4

AEROLITES, INC.
12104 David Rd. • Welsh LA 70591
Tel. 318-734-3865 • Fax 318-734-3865

Bearcat

Wing Span	30.75 ft.	**Range**	125 sm
Wing Area	162.00 sq. ft.	**Takeoff Dist.**	150 ft.
Length	17.50 ft.	**Landing Dist.**	150 ft.
Max Gross Wt.	750 lbs.	**Vmax**	75 mph
Empty Wt.	325 lbs.	**Vcr**	65 mph
Payload (full fuel)	360 lbs.	**V_{S1} (stall clean)**	26 mph
Fuel Capacity	5.0 gal.	**V_{S0} (lndg config)**	N/A mph
Seats	1	**Climb Rate @msl**	1,200 fpm
Cabin Width	24.0 in.	**Serv. Ceiling**	10,000 ft.

Std. Engine/HP	Usable HP Range	Construction Mtrl.	Est. Bldg. Time
Rotax 447 / 42 hp	42-65	Tube, Fabric, Composite	150 hrs.
Plans	**Kit**	**Info Packet**	**Video**
No	$11,495	$12.00	$20.00

Baby Ace replica. Supplied with engine and propeller. Ag Bearcat version includes Spray Master CDA spraying system.

Completed & Flying: 22 Bearcats; 9 Ag Bearcats

AEROLITES, INC.
12104 David Rd. • Welsh LA 70591
Tel. 318-734-3865 • Fax 318-734-3865

AeroSkiff

Wing Span	29.70 ft.	Range	180 sm
Wing Area	156.00 sq. ft.	Takeoff Dist.	550 ft.
Length	22.33 ft.	Landing Dist.	400 ft.
Max Gross Wt.	1125 lbs.	Vmax	90 mph
Empty Wt.	565 lbs.	Vcr	65 mph
Payload (full fuel)	N/A lbs.	V_{S1} (stall clean)	38 mph
Fuel Capacity	12.0 gal.	V_{S0} (lndg config)	N/A mph
Seats	2	Climb Rate @msl	600 fpm
Cabin Width	44.0 in.	Serv. Ceiling	10,000 ft.

Std. Engine/HP	Usable HP Range	Construction Mtrl.	Est. Bldg. Time
Rotax 582 / 65 hp	65-75	Fiberglass, Fabric	180 hrs.
Plans	**Kit**	**Info Packet**	**Video**
No	$19,990	Yes	No

Quick assembly; Dacron slip-on fabric. Fiberglass fuselage. Excellent visibility. Dual controls. Retractable landing gear. Removable wings take 30 minutes with 2 people. Rotax 582 65 hp water cooled engine supplied standard. Rotax 618, 75 hp optional. Takeoff distance is on water and landing distance is on land.

Kits Sold: 2 **Completed & Flying:** 2

AEROLITES, INC.
12104 David Rd. • Welsh LA 70591
Tel. 318-734-3865 • Fax 318-734-3865

Acroduster I

Wing Span	19.00 ft.	Range	300 sm
Wing Area	105.0 sq. ft.	Takeoff Dist.	N/A
Length	15.80 ft.	Landing Dist.	N/A
Max Gross Wt.	1190 lbs.	Vmax	180 mph
Empty Wt.	740 lbs.	VCr	165 mph
Payload (full fuel)	300 lbs.	V_{S1} (stall clean)	70 mph
Fuel Capacity	25 gal.	V_{S0} (lndg config)	70 mph
Seats	1	Climb Rate @msl	3000 fpm
Cabin Width	24 in.	Service Ceiling	12,000 ft.

Std. Engine/HP	Usable HP Range	Construction Mtrl.	Est. Bldg. Time
Lycoming / 200 hp	125 - 200	Aluminum, Wood, Fabric	1500 hrs.
Plans	**Kit**	**Info Packet**	**Video**
$120.00	N/A	$6.50	N/A

Easy to build, semi-monocoque, aluminum fuselage with aluminum tail surfaces and wood/fabric wings. Designed for competitive aerobatics with 180/200 hp engine. It is also an excellent sport aircraft with 125 and 150 hp engines. **Plans Sold:** 37

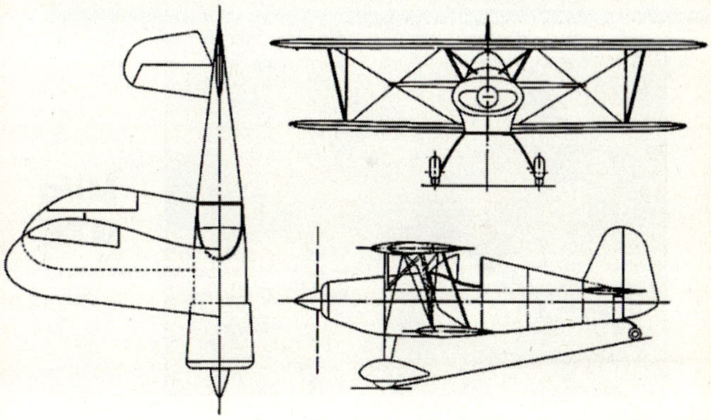

AEROVANT AIRCRAFT CORP.
2342 Jonquil Place • Rockford IL 61107
Tel. 815-877-4508

HOMEBUILT "FAQ'S" (See Page 68)

Q: How are registration numbers assigned, and can I choose a personalized one?

A: "N" numbers in the US are assigned by the FAA Aircraft Registry in Oklahoma City. You must get one assigned before you have your finished airplane inspected. It's sometimes suggested that you wait until about six months before estimated completion, since they will charge you an annual fee for reserving a number.

Requesting an assigned number will cost $5. If you want to request a special number, it costs $10. It is suggested that you submit a list of choices, five or so, in order of preference.

XP Clipper

Wing Span	33.00 ft.	**Range**	N/A
Wing Area	167.0 sq. ft.	**Takeoff Dist.**	135 ft.
Length	11.00 ft.	**Landing Dist.**	140 ft.
Max Gross Wt.	995 lbs.	**Vmax**	87 mph
Empty Wt.	424 lbs.	**VCr**	75 mph
Payload (full fuel)	571 lbs.	**V_{S1} (stall clean)**	33 mph
Fuel Capacity	15 gal.	**V_{S0} (lndg config)**	N/A
Seats	2	**Climb Rate @msl**	1044 fpm
Cabin Width	N/A	**Service Ceiling**	N/A

Std. Engine/HP	Usable HP Range	Construction Mtrl.	Est. Bldg. Time
Rotax 582 / 64 hp	N/A	Tube, Fabric	60 hrs.
Plans	**Kit**	**Info Packet**	**Video**
N/A	$20,500.00	Yes	N/A

The Clipper trike adds an aerodynamic pod to the open framed GTE trike, reducing drag and increasing performance. Bags, streamlining and instruments are standard features. Options include electric starter, towing system, and dual foot controls.

AIR CREATION
Aerodrome de Lanas•Aubenas F-07200
FRANCE
Tel. 75 93 66 66 • Fax 75 35 04 03
E-Mail: aircreation@bsi.f

Fun Twin

Wing Span	33.00 ft.	**Range**	N/A
Wing Area	187.0 sq. ft.	**Takeoff Dist.**	120 ft.
Length	11.00 ft.	**Landing Dist.**	130 ft.
Max Gross Wt.	849 lbs.	**Vmax**	65 mph
Empty Wt.	284 lbs.	**VCr**	46 mph
Payload (full fuel)	430 lbs.	**V_{S1} (stall clean)**	29 mph
Fuel Capacity	10 gal.	**V_{S0} (lndg config)**	N/A
Seats	2	**Climb Rate @msl**	807 fpm
Cabin Width	N/A	**Service Ceiling**	N/A

Std. Engine/HP	Usable HP Range	Construction Mtrl.	Est. Bldg. Time
Rotax 503 / 52 hp	N/A	Tube, Fabric	40 hrs.
Plans	**Kit**	**Info Packet**	**Video**
N/A	$12,000.00	Yes	$20.00

The two seat Twin trike combined with Air Creation's Fun wing creates a very forgiving aircraft with qualities similar to the Racer trike. Also offered in a 503 SL version.

AIR CREATION
Aerodrome de Lanas•Aubenas F-07200
FRANCE
Tel. 75 93 66 66 • Fax 75 35 04 03
E-Mail: aircreation@bsi.f

Mild GTE

Wing Span	33.00 ft.	**Range**	N/A
Wing Area	178.0 sq. ft.	**Takeoff Dist.**	120 ft.
Length	11.00 ft.	**Landing Dist.**	130 ft.
Max Gross Wt.	995 lbs.	**Vmax**	75 mph
Empty Wt.	392 lbs.	**VCr**	60 mph
Payload (full fuel)	601 lbs.	**V_{S1} (stall clean)**	33 mph
Fuel Capacity	15 gal.	**V_{S0} (lndg config)**	N/A
Seats	2	**Climb Rate @msl**	N/A
Cabin Width	N/A	**Service Ceiling**	N/A

Std. Engine/HP	Usable HP Range	Construction Mtrl.	Est. Bldg. Time
Rotax 582 / 64 hp	N/A	Tube, Fabric	50 hrs.
Plans	**Kit**	**Info Packet**	**Video**
N/A	$17,000.00	Yes	$20.00

The two seat GTE trike provides predictable flight behavior at all attitudes. It can withstand a frontal shock of 9G and an optional 3-point seat belt allows full freedom of movement. Designed/tested for British BCAR/S certification requirements.

XP Racer

Wing Span	30.00 ft.	**Range**	N/A
Wing Area	124.0 sq. ft.	**Takeoff Dist.**	100 ft.
Length	9.00 ft.	**Landing Dist.**	120 ft.
Max Gross Wt.	566 lbs.	**Vmax**	84 mph
Empty Wt.	255 lbs.	**VCr**	74 mph
Payload (full fuel)	274 lbs.	**V_{S1} (stall clean)**	28 mph
Fuel Capacity	5 gal.	**V_{S0} (lndg config)**	N/A
Seats	1	**Climb Rate @msl**	1378 fpm
Cabin Width	N/A	**Service Ceiling**	N/A

Std. Engine/HP	Usable HP Range	Construction Mtrl.	Est. Bldg. Time
Rotax 503 / 52 hp	N/A	Tube, Fabric	6 hrs.
Plans	**Kit**	**Info Packet**	**Video**
N/A	$12,500.00	Yes	$20.00

This combination of the Racer trike and the new 11 square meter XP wing offers higher performance. Air Creation wings are designed for motorized flight and are equipped with double lateral cables and tightening security systems. Also offered in 503 S and 503 SL versions.

North American Distributors:

AIR AMERICA
84 Boxborough Road
Stow MA 01775
Ph: 508-897-5220
Fax: 508-897-4231

CANADIAN POWERED PARACHUTES LTD.
Box 1769 • Vegreville, Alberta T9C 1S8 CANADA
Ph: 403-658-2044
Fax: 403-658-2387

KEMMERIES AVIATION
8710 W. Carefree Hwy.
Peoria AZ 85382
Tel.: 602-566-8068
Fax: 602-561-2287

AIR CREATION
Aerodrome de Lanas • Aubenas F-07200
FRANCE
Tel. 75 93 66 66 • Fax 75 35 04 03
E-Mail: aircreation@bsi.fr

Fun Racer

North American Distributors:

AIR AMERICA	CANADIAN POWERED PARACHUTES LTD.	KEMMERIES AVIATION
84 Boxborough Road	Box 1769 • Vegreville, Alberta T9C	8710 W. Carefree Hwy.
Stow MA 01775	1S8 CANADA	Peoria AZ 85382
Ph: 508-897-5220	Ph: 403-658-2044	Tel.: 602-566-8068
Fax: 508-897-4231	Fax: 403-658-2387	Fax: 602-561-2287

Wing Span	33.00 ft.	**Range**	N/A
Wing Area	145.0 sq. ft.	**Takeoff Dist.**	100 ft.
Length	9.00 ft.	**Landing Dist.**	110 ft.
Max Gross Wt.	555 lbs.	**Vmax**	55 mph
Empty Wt.	234 lbs.	**VCr**	47 mph
Payload (full fuel)	309 lbs.	**V_{S1} (stall clean)**	23 mph
Fuel Capacity	5 gal.	**V_{S0} (lndg config)**	N/A
Seats	1	**Climb Rate @msl**	1181 fpm
Cabin Width	N/A	**Service Ceiling**	N/A

Std. Engine/HP	Usable HP Range	Construction Mtrl.	Est. Bldg. Time
Rotax 447 / 42 hp	N/A	Tube and Fabric	N/A
Plans	**Kit**	**Info Packet**	**Video**
N/A	$9,900.00	Yes	$20.00

A Part 103 legal, weight-shift trike for the recreational pilot with entry level performance. A compact and easy to manage single seater. Options include rear wheel fairings, cargo bags, instrument console, ten gallon fuel capacity. In production since 1984. **Flying:** 110

AIR CREATION
Aerodrome de Lanas • Aubenas F-07200
FRANCE
Tel. 75 93 66 66 • Fax 75 35 04 03
E-Mail: aircreation@bsi.fr

Spitfire

Wing Span	30.00 ft.	**Range**	120 sm
Wing Area	152.0 sq. ft.	**Takeoff Dist.**	50 ft.
Length	17.75 ft.	**Landing Dist.**	70 ft.
Max Gross Wt.	550 lbs.	**Vmax**	65 mph
Empty Wt.	252 lbs.	**VCr**	55 mph
Payload (full fuel)	N/A	**V_{S1} (stall clean)**	28 mph
Fuel Capacity	5 gal.	**V_{S0} (lndg config)**	N/A
Seats	1	**Climb Rate @msl**	700 fpm
Cabin Width	32 in.	**Service Ceiling**	12,000 ft.

Std. Engine/HP	Usable HP Range	Construction Mtrl.	Est. Bldg. Time
Rotax 447 / 42 hp	N/A	Tube, Fabric	40 hrs.
Plans	**Kit**	**Info Packet**	**Video**
N/A	$7,995.00	Yes	N/A

A single seat ultralight utilizing tube and fabric wings with a fiberglass fuselage. Offered in 'Ultralight' and 'Super Sport' versions. The Super Sport comes with a larger engine, instruments, spring gear assembly, auxiliary fuel tank, and wheel pants.

Completed & Flying: 514

AIR MAGIC ULTRALIGHTS
4800 W. 34th Street • Building C-51
Houston TX 77092
Tel. 713-957-2300 • Fax 713-957-0066

Spitfire II

Wing Span	30.00 ft.	**Range**	100 sm
Wing Area	152.0 sq. ft.	**Takeoff Dist.**	125 ft.
Length	17.90 ft.	**Landing Dist.**	125 ft.
Max Gross Wt.	800 lbs.	**Vmax**	60 mph
Empty Wt.	350 lbs.	**VCr**	55 mph
Payload (full fuel)	N/A	**V_{S1} (stall clean)**	38 mph
Fuel Capacity	5 gal.	**V_{S0} (lndg config)**	N/A
Seats	2	**Climb Rate @msl**	800 fpm
Cabin Width	N/A	**Service Ceiling**	12,500 ft.

Std. Engine/HP	Usable HP Range	Construction Mtrl.	Est. Bldg. Time
Rotax 503 / 52 hp	N/A	Metal, Fabric	50 hrs.
Plans	**Kit**	**Info Packet**	**Video**
N/A	$9,995.00	$8.00	N/A

A high wing, two seat ultralight utilizing a fiberglass chassis along with tube and fabric wings. Available in an "Elite" version with bucket seats, a larger engine and 10 gallon fuel capacity. Offered in several packages both with, and without engine.

Completed & Flying: 68

AIR MAGIC ULTRALIGHTS
4800 W. 34th Street • Building C-51
Houston TX 77092
Tel. 713-957-2300 • Fax 713-957-0066

Wild Thing

Wing Span	30.00 ft.	**Range**	300 sm
Wing Area	150.0 sq. ft.	**Takeoff Dist.**	320 ft.
Length	21.40 ft.	**Landing Dist.**	145 ft.
Max Gross Wt.	992 lbs.	**Vmax**	122 mph
Empty Wt.	540 lbs.	**VCr**	100 mph
Payload (full fuel)	452 lbs.	**V_{S1} (stall clean)**	35 mph
Fuel Capacity	21 gal.	**V_{S0} (lndg config)**	28 mph
Seats	2	**Climb Rate @msl**	1100 fpm
Cabin Width	44 in.	**Service Ceiling**	N/A

Std. Engine/HP	Usable HP Range	Construction Mtrl.	Est. Bldg. Time
Gobler-Hirth / 95 hp	60 - 180	Aluminum	400 hrs.
Plans	**Kit**	**Info Packet**	**Video**
N/A	$17,990.00	$15.00	$30.00

Wings fold backward for towing and storage. Kit price is without engine. Prepared for the Gobler-Hirth F30 engine. Can also use Limbach 60 hp and Lycoming 180 hp powerplants. Available with Junker's Rescue System option. Complete ready to fly assembly available for $46,900 with German construction and registration. **Completed & Flying:** 75

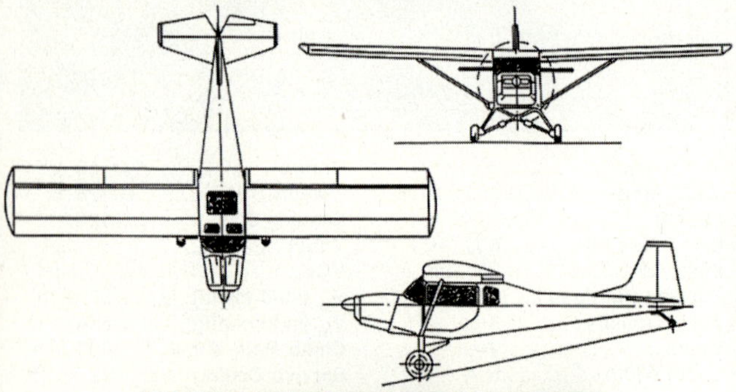

The aircraft was designed by R. Kurtz, an experienced pilot active in the German glider and sky diver scene. His design goal was to construct an all-aluminum ultralight aircraft for his own use and pleasure locally, but also suitable for long distance voyages over the rough terrain such as is found on the continent of Africa. Safety being a major consideration, the Junker's Rescue System is offered as an option.

AIR-MAX GMBH
Venatoriusstrasse 21
90455 Nuernberg
GERMANY
Tel. 09129-40 03 44 • Fax 09129-68 19

Vagabund

Wing Span	24.75 ft.	**Range**	250 sm
Wing Area	196.0 sq. ft.	**Takeoff Dist.**	330 ft.
Length	21.32 ft.	**Landing Dist.**	250 ft.
Max Gross Wt.	885 lbs.	**Vmax**	75 mph
Empty Wt.	500 lbs.	**VCr**	68 mph
Payload (full fuel)	385 lbs.	**V_{S1} (stall clean)**	40 mph
Fuel Capacity	13 gal.	**V_{S0} (lndg config)**	37 mph
Seats	2	**Climb Rate @msl**	500 fpm
Cabin Width	26 in.	**Service Ceiling**	N/A

Std. Engine/HP	Usable HP Range	Construction Mtrl.	Est. Bldg. Time
Limb L2000 / 80 hp	40 - 80	Polish pine, Fabric	300 hrs.
Plans	**Kit**	**Info Packet**	**Video**
N/A	$17,000.00	Yes	$25.00

An easy to fly biplane with no bad habits. The kit includes prefabricated parts (fuselage frame, spars, ribs, hardware, etc.) which allows a very short building time up to the firewall. The engine, engine mount, instruments, and covering are not included.

Kits/Plans Sold: 8 **Completed & Flying:** 4

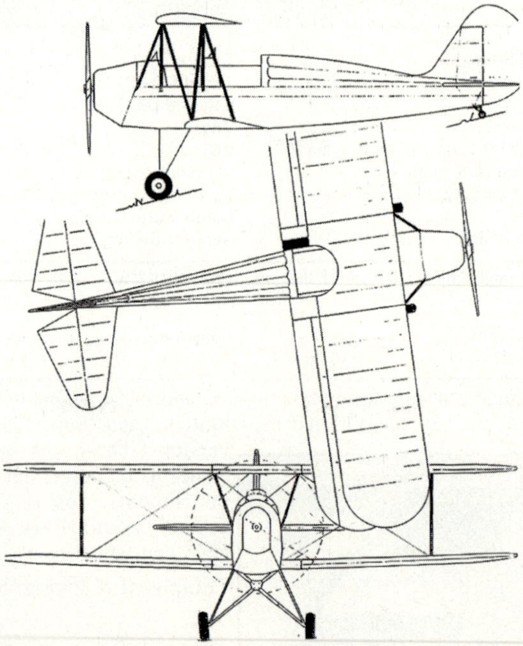

AIRCRAFT COVERINGS
Schloss Huennefeld 4
Bad Essen 49152
GERMANY
Tel. 00-49-5472-2413 • Fax 00-49-5472-2413

Super Stallion

Wing Span	35.00 ft.	**Range**	2600 sm
Wing Area	140.0 sq. ft.	**Takeoff Dist.**	1200 ft.
Length	24.50 ft.	**Landing Dist.**	700 ft.
Max Gross Wt.	3800 lbs.	**Vmax**	300 mph
Empty Wt.	2200 lbs.	**VCr**	235 mph
Payload (full fuel)	1600 lbs.	**V_{S1} (stall clean)**	80 mph
Fuel Capacity	180 gal.	**V_{S0} (lndg config)**	71 mph
Seats	6	**Climb Rate @msl**	1600 fpm
Cabin Width	48 in.	**Service Ceiling**	23,000 ft.

Std. Engine/HP	Usable HP Range	Construction Mtrl.	Est. Bldg. Time
Cont I / 300 hp	180 - 350	Composite, Steel	1500 hrs.
Plans	**Kit**	**Info Packet**	**Video**
$200.00	$59,050.00	$12.00	N/A

A clean, composite design with a drag coefficient equal to that of the Lancair IV. Large wings allow short distance, low speed takeoffs from dirt strips. Floats and skis optional. Kit price does not include engine, propeller or freight.

Kits/Plans Sold: 23 **Completed & Flying:** 1

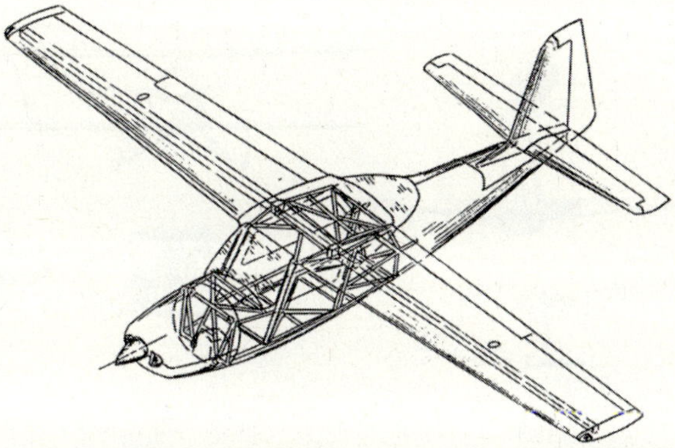

AIRCRAFT DESIGNS, INC.
5 Harris Court, Bldg. S
Monterey CA 93940
Tel. 408-649-6212 • Fax 408-649-5738

Chinook Plus

Wing Span	32.00 ft.	**Range**	350 sm
Wing Area	154.5 sq. ft.	**Takeoff Dist.**	250 ft.
Length	17.70 ft.	**Landing Dist.**	N/A
Max Gross Wt.	900 lbs.	**Vmax**	95 mph
Empty Wt.	380 lbs.	**VCr**	83 mph
Payload (full fuel)	N/A	**V_{S1} (stall clean)**	35 mph
Fuel Capacity	9 gal.	**V_{S0} (lndg config)**	32 mph
Seats	2	**Climb Rate @msl**	1200 fpm
Cabin Width	N/A	**Service Ceiling**	15,000 ft.

Std. Engine/HP	Usable HP Range	Construction Mtrl.	Est. Bldg. Time
Rotax 582 / 65 hp	50 - 65	Aluminum & Fabric	150 hrs.
Plans	**Kit**	**Info Packet**	**Video**
N/A	$7,250.00	$10.00	N/A

Modified and improved design that has sold more than 700 aircraft since its introduction. Pleasant flying characteristics. Suitable for both novice and experienced pilot. One seat or two seat tandem. Rotax 503 engine optional. **Kits Sold:** 700

AIRCRAFT SALES & PARTS
PO Box 995 • Vernon, B.C. V1T 6N2
CANADA
Tel. 250-549-1102 • Fax 250-549-3769
E-Mail: asap@junction.net • Web: http://www.ultralight.ca

Beaver RX550 Plus

Wing Span	32.00 ft.	**Range**	350 sm
Wing Area	154.5 sq. ft.	**Takeoff Dist.**	275 ft.
Length	20.70 ft.	**Landing Dist.**	N/A
Max Gross Wt.	1050 lbs.	**Vmax**	105 mph
Empty Wt.	450 lbs.	**VCr**	65 mph
Payload (full fuel)	N/A	**V_{S1} (stall clean)**	38 mph
Fuel Capacity	N/A	**V_{S0} (lndg config)**	N/A
Seats	2	**Climb Rate @msl**	900 fpm
Cabin Width	N/A	**Service Ceiling**	9,000 ft.

Std. Engine/HP	Usable HP Range	Construction Mtrl.	Est. Bldg. Time
Rotax 503 / 50 hp	50-65	Aluminum, Fabric	150 hrs.
Plans	**Kit**	**Info Packet**	**Video**
N/A	$9,000.00	$10.00	$10.00

Updated Beaver with new wing utilizing ladder-style aluminum tube frame with Ceconite covering and featuring full span ailerons. Wing retro kits available for older RX550's. Engine, instruments, and propeller not included in basic kit price.

AIRCRAFT SALES & PARTS
PO Box 995 • Vernon, B.C. V1T 6N2
CANADA
Tel. 250-549-1102 • Fax 250-549-3769
E-Mail: asap@junction.net • Web: http://www.ultralight.ca

Baby Lakes

Wing Span	16.70 ft.	Range	250 sm
Wing Area	88.00 sq. ft.	Takeoff Dist.	300 ft.
Length	13.80 ft.	Landing Dist.	300 ft.
Max Gross Wt.	850 lbs.	Vmax	135 mph
Empty Wt.	480 lbs.	VCr	118 mph
Payload (full fuel)	300 lbs.	V_{S1} (stall clean)	55 mph
Fuel Capacity	12 gal.	V_{S0} (lndg config)	N/A
Seats	1	Climb Rate @msl	2000 fpm
Cabin Width	24 in.	Service Ceiling	14,500 ft.

Std. Engine/HP	Usable HP Range	Construction Mtrl.	Est. Bldg. Time
Continental / 85 hp	65 - 90	Metal, Wood, Tube, Fabric	1200 hrs.
Plans	**Kit**	**Info Packet**	**Video**
$150.00	$12,000.00	$5.00	N/A

A classic wood, metal sheet, tube and fabric biplane. Typical of the 1930's era. Kit price does not include engine. See Ad on Page 24.

Kits/Plans Sold: 900 **Completed & Flying:** 50

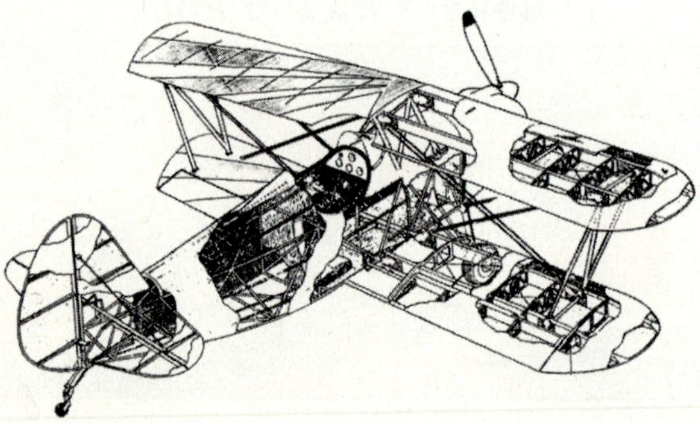

AIRCRAFT SPRUCE & SPECIALTY
PO Box 4000
Corona CA 91718-9961
Tel. 800-824-1930 • Fax 909-372-0555
E-Mail: info@aircraft-spruce.com
Web: http://www.aircraft-spruce.com

Super Baby Lakes

Wing Span	16.80 ft.	Range	250 sm
Wing Area	89.00 sq. ft.	Takeoff Dist.	300 ft.
Length	13.60 ft.	Landing Dist.	300 ft.
Max Gross Wt.	850 lbs.	Vmax	155 mph
Empty Wt.	510 lbs.	VCr	135 mph
Payload (full fuel)	250 lbs.	V_{S1} (stall clean)	55 mph
Fuel Capacity	12 gal.	V_{S0} (lndg config)	N/A
Seats	1	Climb Rate @msl	2000 fpm
Cabin Width	24 in.	Service Ceiling	15,000 ft.

Std. Engine/HP	Usable HP Range	Construction Mtrl.	Est. Bldg. Time
Lyc O-235 / 110 hp	100 - 125	Tubing, Wood, Fabric	1200 hrs.
Plans	**Kit**	**Info Packet**	**Video**
$175.00	$12,500.00	$5.00	N/A

An inexpensive metal, wood and fabric biplane. A first time builder's aerobatic aircraft. Design stressed to +9, -9 G's. Climb rate is initial performance only. Kit price does not include engine. See Ad on Page 24.

Kits/Plans Sold: 450 **Completed & Flying:** 50

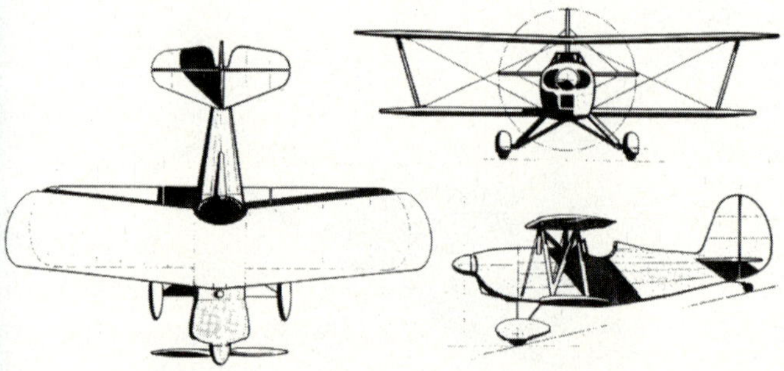

AIRCRAFT SPRUCE & SPECIALTY
PO Box 4000
Corona CA 91718-9961
Tel. 800-824-1930 • Fax 909-372-0555
E-Mail: info@aircraft-spruce.com
Web: http://www.aircraft-spruce.com

HOMEBUILT "FAQ'S" *(See Page 68)*

Q: Can I type-certify my airplane?

A: Not recommended. The costs are extremely prohibitive (which is often why the designers refuse to do it), and there is little benefit. Remember that the only major restriction on experimental operations is use of the aircraft for hire.

Buddy Baby Lakes

Wing Span	18.00 ft.	Range	250 sm
Wing Area	98.00 sq. ft.	Takeoff Dist.	350 ft.
Length	14.80 ft.	Landing Dist.	350 ft.
Max Gross Wt.	1000 lbs.	Vmax	135 mph
Empty Wt.	535 lbs.	VCr	118 mph
Payload (full fuel)	N/A	V_{S1} (stall clean)	55 mph
Fuel Capacity	15 gal.	V_{S0} (lndg config)	N/A
Seats	2	Climb Rate @msl	2000 fpm
Cabin Width	26 in.	Service Ceiling	12,500 ft.

Std. Engine/HP	Usable HP Range	Construction Mtrl.	Est. Bldg. Time
Continental / 100 hp	85 - 125	Tube, Wood, Aluminum	1500 hrs.
Plans	**Kit**	**Info Packet**	**Video**
$200.00	$14,500.00	$5.00	N/A

A unique design that provides great performance, low cost and holds two people. Design can also utilize Lycoming or Franklin engines. Climb rate is with pilot only. See Ad on Page 24.

Kits/Plans Sold: 125 **Completed & Flying:** 4

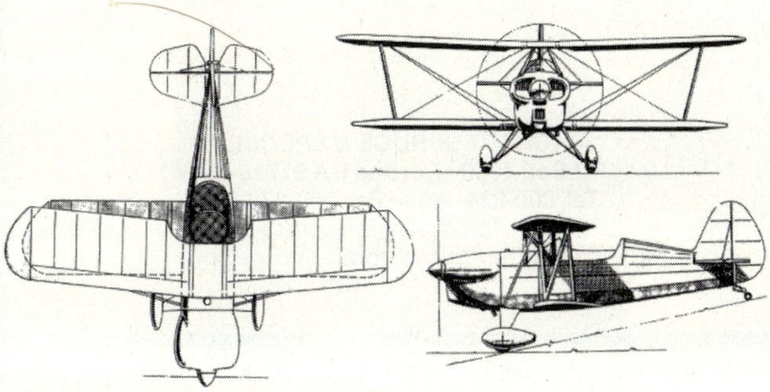

AIRCRAFT SPRUCE & SPECIALTY
PO Box 4000
Corona CA 91718-9961
Tel. 800-824-1930 • Fax 909-372-0555
E-Mail: info@aircraft-spruce.com
Web: http://www.aircraft-spruce.com

DR-107 One Design

Wing Span	19.50 ft.	Range	375 sm
Wing Area	75.55 sq. ft.	Takeoff Dist.	600 ft.
Length	17.00 ft.	Landing Dist.	1200 ft.
Max Gross Wt.	1150 lbs.	Vmax	184 mph
Empty Wt.	740 lbs.	VCr	160 mph
Payload (full fuel)	290 lbs.	V_{S1} (stall clean)	63 mph
Fuel Capacity	19 gal.	V_{S0} (lndg config)	63 mph
Seats	1	Climb Rate @msl	2000 fpm
Cabin Width	24 in.	Service Ceiling	22,000 ft.

Std. Engine/HP	Usable HP Range	Construction Mtrl.	Est. Bldg. Time
Lyc AEIO 320 / 160 hp	160 - 180	Wood, Steel and Fabric	2000 hrs.
Plans	**Kit**	**Info Packet**	**Video**
$300.00	Yes	$5.00	N/A

Fully aerobatic design, first flown in 1993. Created by Dan Rihn as an economical answer for pilots wishing to fly basic through advanced aerobatics. Designed for +10, -10 G's, IAC One Class aerobatic competition. 360 degree per second roll rate. Normally powered by an O-360 engine modified with an inverted oil system, high compression cylinders, and fuel injection. Specs are for a Lyc. AEIO-320 engine. A variety of materials kits as well as prefabricated kits, up to the full airframe, are available. Plans sold in 15 countries. See Ad on Page 24.

Kits/Plans Sold: 355 **Completed & Flying:** 5

AIRCRAFT SPRUCE & SPECIALTY
PO Box 4000
Corona CA 91718-9961
Tel. 800-824-1930 • Fax 909-372-0555
E-Mail: info@aircraft-spruce.com
Web: http://www.aircraft-spruce.com

Rihn DR-109

Wing Span	24.00 ft.	**Range**	534 sm
Wing Area	114.0 sq. ft.	**Takeoff Dist.**	600 ft.
Length	22.00 ft.	**Landing Dist.**	1500 ft.
Max Gross Wt.	2275 lbs.	**Vmax**	205 mph
Empty Wt.	1495 lbs.	**VCr**	178 mph
Payload (full fuel)	504 lbs.	**V_{S1} (stall clean)**	58 mph
Fuel Capacity	46 gal.	**V_{S0} (lndg config)**	N/A
Seats	2	**Climb Rate @msl**	2800 fpm
Cabin Width	48 in.	**Service Ceiling**	N/A

Std. Engine/HP	Usable HP Range	Construction Mtrl.	Est. Bldg. Time
AEIO-540-D4A5 / 260 hp	200-300	Tube, Metal, Fabric, Wood	1300 hrs.
Plans	**Kit**	**Info Packet**	**Video**
N/A	$50,000.00	$19.95	N/A

A two seat design suitable for aerobatic competition and training. Designed by Dan Rihn. Fuselage has a welded steel tube truss covered with sheet metal from the firewall aft to the cockpit, aft to the tail. Tail surfaces are fabric covered and wire braced with steel tube spars and bent up steel sheet metal ribs. The one piece wing is all-wood with Douglas fir box main spar, Douglas Fir spars, plywood ribs and plywood covered. Also offered with Lycoming 200 and 300 hp engines. Price does not include engine. See Ad on Page 24.

AIRCRAFT SPRUCE & SPECIALTY
PO Box 4000 • Corona CA 91718-9961
Tel. 800-824-1930 • Fax 909-372-0555
E-Mail: info@aircraft-spruce.com
Web: http://www.aircraft-spruce.com

W-10 Tailwind

Wing Span	24.00 ft.	**Range**	600 sm
Wing Area	92.00 sq. ft.	**Takeoff Dist.**	750 ft.
Length	19.60 ft.	**Landing Dist.**	650 ft.
Max Gross Wt.	1425 lbs.	**Vmax**	200 mph
Empty Wt.	878 lbs.	**VCr**	180 mph
Payload (full fuel)	N/A	**V_{S1} (stall clean)**	55 mph
Fuel Capacity	35 gal.	**V_{S0} (lndg config)**	N/A
Seats	2	**Climb Rate @msl**	1200 fpm
Cabin Width	N/A	**Service Ceiling**	20,000 ft.

Std. Engine/HP	Usable HP Range	Construction Mtrl.	Est. Bldg. Time
Cont. 0-300 / 145 hp	N/A	Wood, Fabric and Tubing	N/A
Plans	**Kit**	**Info Packet**	**Video**
$180.00	N/A	$5.00	N/A

A plans only, fixed gear taildragger. Introduced by S. J. Whittman in 1953. Offers outstanding performance, efficiency, and simplicity. Auto V-8 powerplant conversion plans available for $35. See Ad on Page 24.

AIRCRAFT SPRUCE & SPECIALTY
PO Box 4000 • Corona CA 91718-9961
Tel. 800-824-1930 • Fax 909-372-0555
E-Mail: info@aircraft-spruce.com
Web: http://www.aircraft-spruce.com

Acrolite

Wing Span	20.00 ft.	**Range**	250 sm
Wing Area	133.0 sq. ft.	**Takeoff Dist.**	500 ft.
Length	17.00 ft.	**Landing Dist.**	500 ft.
Max Gross Wt.	750.0 lbs.	**Vmax**	130 mph
Empty Wt.	495 lbs.	**VCr**	110 mph
Payload (full fuel)	N/A	**V_{S1} (stall clean)**	N/A
Fuel Capacity	8 gal.	**V_{S0} (lndg config)**	N/A
Seats	1	**Climb Rate @msl**	N/A
Cabin Width	N/A	**Service Ceiling**	12,000 ft.

Std. Engine/HP	Usable HP Range	Construction Mtrl.	Est. Bldg. Time
Rotax 912 / 80 hp	55-112	Metal, Wood, Tube	2000 hrs.
Plans	**Kit**	**Info Packet**	**Video**
Yes	N/A	$5.00	N/A

An entry level, aerobatic capable design. Can utilize wood or metal wings. See Ad on Page 24.

AIRCRAFT SPRUCE & SPECIALTY
PO Box 4000 • Corona CA 91718-9961
Tel. 800-824-1930 • Fax 909-372-0555
E-Mail: info@aircraft-spruce.com
Web: http://www.aircraft-spruce.com

Acro 1

Wing Span	20.00 ft.	**Range**	1000 sm
Wing Area	75.00 sq. ft.	**Takeoff Dist.**	N/A
Length	17.00 ft.	**Landing Dist.**	N/A
Max Gross Wt.	1250 lbs.	**Vmax**	245 mph
Empty Wt.	780 lbs.	**VCr**	220 mph
Payload (full fuel)	N/A	**V_{S1} (stall clean)**	58 mph
Fuel Capacity	47 gal.	**V_{S0} (lndg config)**	58 mph
Seats	1	**Climb Rate @msl**	3000 fpm
Cabin Width	N/A	**Service Ceiling**	25,000 ft.

Std. Engine/HP	Usable HP Range	Construction Mtrl.	Est. Bldg. Time
Lyc IO-360 / 200 hp	100 - 210	Graphite, Fiberglass	700 hrs.
Plans	**Kit**	**Info Packet**	**Video**
$225.00	$22,900.00	$24.00	Yes

A high speed, single seat design suitable for cross country flight. High-tech composite materials are used throughout the entire airframe. Fully aerobatic to +15, -15 G's. Accepts engines from 100 to 210 hp. Kit does not include engine or propeller. Info pack includes a video. International cost is $36.00.

Completed & Flying: 1

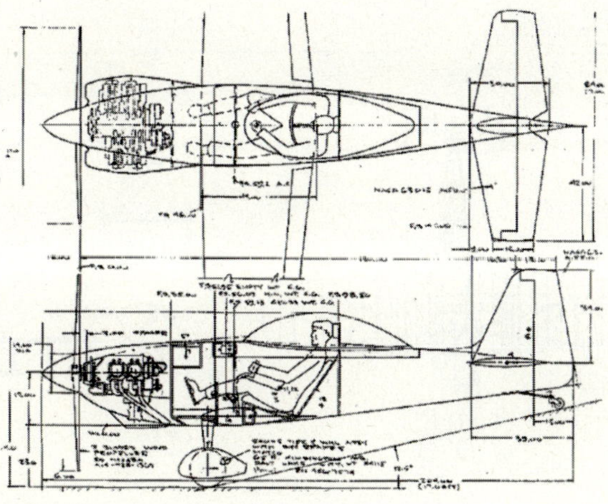

AIRCRAFT TECHNOLOGIES, INC.
4265 Lilburn Industrial Way
Lilburn GA 30047
Tel. 770-806-9098 • Fax 770-806-9098
E-Mail: fredmeyer@msn.com

Atlantis

Wing Span	24.50 ft.	**Range**	1000 sm
Wing Area	95.00 sq. ft.	**Takeoff Dist.**	N/A
Length	N/A	**Landing Dist.**	N/A
Max Gross Wt.	1900 lbs.	**Vmax**	200 mph
Empty Wt.	1100 lbs.	**VCr**	180 mph
Payload (full fuel)	800 lbs.	**V_{S1} (stall clean)**	65 mph
Fuel Capacity	60 gal.	**V_{S0} (lndg config)**	58 mph
Seats	2	**Climb Rate @msl**	1275 fpm
Cabin Width	42 in.	**Service Ceiling**	N/A

Std. Engine/HP	Usable HP Range	Construction Mtrl.	Est. Bldg. Time
Lyc IO-360 / 200 hp	150 - 210	Composite, Steel	700 hrs.
Plans	**Kit**	**Info Packet**	**Video**
$300.00	$10,365.00	$25.00	Yes

A strong, fast, cross country airplane which offers roomy comfort and an impressive 1,000 mile range. Fully aerobatic. Engineered to +15, -15 G's ultimate loading. Room for two adults and 80 pounds of baggage in 12 cubic feet of baggage space. Builders receive a prewelded fuselage with prewelded tail feathers ready to bolt on. Building time is further reduced by the use of large pre-molded components. Can be built in less than 700 hours. A Fast Build Kit is available for $36,800. Tricycle and landing gear option available.

Kits/Plans Sold: 6 **Completed & Flying:** 1

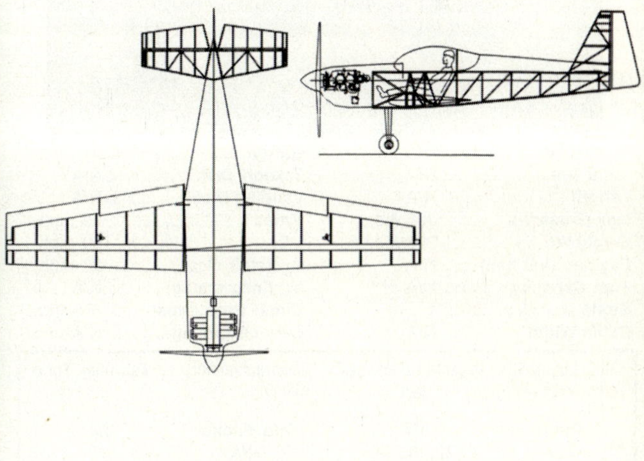

AIRCRAFT TECHNOLOGIES, INC.
4265 Lilburn Industrial Way
Lilburn GA 30047
Tel. 770-806-9098 • Fax 770-806-9098
E-Mail: fredmeyer@msn.com

E 3 Eindecker

Wing Span	24 ft.	**Range**	160 sm
Wing Area	89 sq. ft.	**Takeoff Dist.**	120 ft.
Length	15 ft.	**Landing Dist.**	280 ft.
Max Gross Wt.	548 lbs.	**Vmax**	80 mph
Empty Wt.	238 lbs.	**VCr**	N/A
Payload (full fuel)	310 lbs.	**V_{S1} (stall clean)**	26 mph
Fuel Capacity	9 gal.	**V_{S0} (lndg config)**	N/A
Seats	1	**Climb Rate @msl**	1100 fpm
Cabin Width	24 in.	**Service Ceiling**	N/A

Std. Engine/HP	Usable HP Range	Construction Mtrl.	Est. Bldg. Time
Rotax 503 / 52 hp	28 - 52	Aluminum tube	N/A
Plans	**Kit**	**Info Packet**	**Video**
N/A	$4,495.00	$9.95	$24.95

WWI fighter type. Aluminum tubing and rivet construction. CAD designed and detailed. Ridged box bracing. Quick build. Modern two stroke engine. Authentic appearance. **Flying:** 3

AIRDROME AEROPLANES
929 NW Road 1571
Holden MO 64040
Tel. 816-230-8585
Web: http://www.sunflower.org/~baron

Dream Classic

Wing Span	N/A	**Range**	N/A
Wing Area	N/A	**Takeoff Dist.**	N/A
Length	N/A	**Landing Dist.**	N/A
Max Gross Wt.	393 lbs.	**Vmax**	63 mph
Empty Wt.	223 lbs.	**VCr**	59 mph
Payload (full fuel)	N/A	**V_{S1} (stall clean)**	26 mph
Fuel Capacity	N/A	**V_{S0} (lndg config)**	N/A
Seats	1	**Climb Rate @msl**	875 fpm
Cabin Width	N/A	**Service Ceiling**	N/A

Std. Engine/HP	Usable HP Range	Construction Mtrl.	Est. Bldg. Time
Rotax 447 / 40 hp	N/A	Aluminum, Fabric	200 hrs.
Plans	**Kit**	**Info Packet**	**Video**
N/A	$2,995.00	N/A	N/A

Classic style ultralight available in complete airframe kit. Optional prebuilt aluminum tubing airframe ready for covering and engine installation.

AIRDROME AEROPLANES
929 NW Road 1571
Holden MO 64040
Tel. 816-230-8585
Web: http://www.sunflower.org/~baron

Fokker DR1

Wing Span	17.71 ft.	**Range**	250 sm
Wing Area	N/A	**Takeoff Dist.**	200 ft.
Length	N/A	**Landing Dist.**	300 ft.
Max Gross Wt.	634 lbs.	**Vmax**	85 mph
Empty Wt.	392 lbs.	**VCr**	70 mph
Payload (full fuel)	N/A	**V_{S1} (stall clean)**	36 mph
Fuel Capacity	5 gal.	**V_{S0} (lndg config)**	N/A
Seats	1	**Climb Rate @msl**	850 fpm
Cabin Width	N/A	**Service Ceiling**	10,000 ft.

Std. Engine/HP	Usable HP Range	Construction Mtrl.	Est. Bldg. Time
VW 1800 cc / 65 hp	N/A	Alum. Tubing & Riveted	400 hrs.
Plans	**Kit**	**Info Packet**	**Video**
N/A	$7,495.00	$9.95	$20.00

3/4 scale WWI Tri-Plane (Dreidecker) fighter replica. All-aluminum tubing and riveted construction. CAD designed and detailed. Ridged box bracing. Quick build. Inexpensive VW engine. Authentic appearance. Light responsive controls. Deluxe Airframe Kit includes basic kit plus engine mount and 5 gallon fuel tank. Optional 12 gallon fuel tank available.

Completed & Flying: 1

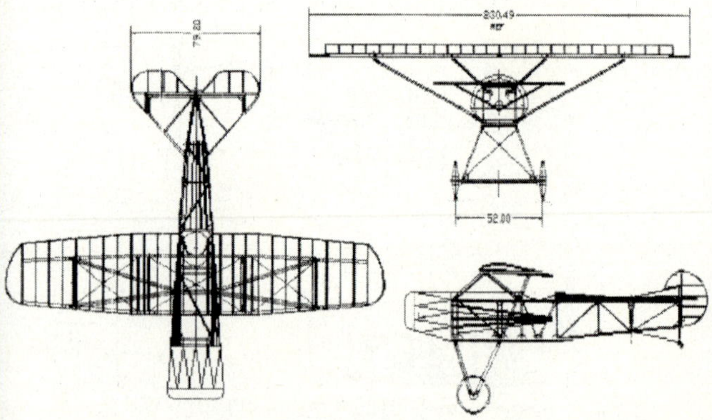

AIRDROME AEROPLANES
929 NW Road 1571
Holden MO 64040
Tel. 816-230-8585
Web: http://www.sunflower.org/~baron

Fokker D-VI

Wing Span	17.67 ft.	**Range**	90 sm
Wing Area	145.0 sq. ft.	**Takeoff Dist.**	95 ft.
Length	16.25 ft.	**Landing Dist.**	220 ft.
Max Gross Wt.	527 lbs.	**Vmax**	78 mph
Empty Wt.	297 lbs.	**VCr**	72 mph
Payload (full fuel)	N/A	**V_{S1} (stall clean)**	32 mph
Fuel Capacity	N/A	**V_{S0} (lndg config)**	N/A
Seats	1	**Climb Rate @msl**	750 fpm
Cabin Width	N/A	**Service Ceiling**	10,000 ft.

Std. Engine/HP	Usable HP Range	Construction Mtrl.	Est. Bldg. Time
Rotax / 46 hp	46 - 65	Alum. Tubing, Pop Rivet	350 hrs.
Plans	**Kit**	**Info Packet**	**Video**
N/A	$5,495.00	$9.95	$20.00

WWI fighter type. Aluminum tubing and rivet construction. CAD designed and detailed. Ridged box bracing. Quick build. Modern two stroke engine. Exciting authentic appearance. Light responsive controls. Horsepower range from 46 hp Rotax to 65 hp VW 1800 cc.

Completed & Flying: 3

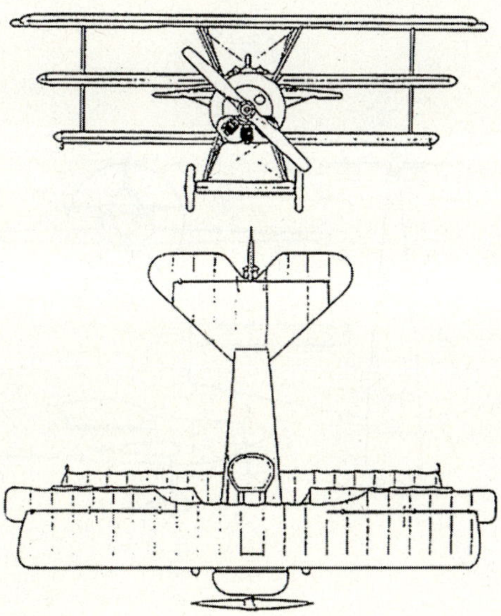

AIRDROME AEROPLANES
929 NW Road 1571
Holden MO 64040
Tel. 816-230-8585
Web: http://www.sunflower.org/~baron

Fokker D-VIII

Wing Span	24.00 ft.	**Range**	105 sm
Wing Area	74.00 sq. ft.	**Takeoff Dist.**	115 ft.
Length	15.20 ft.	**Landing Dist.**	300 ft.
Max Gross Wt.	505 lbs.	**Vmax**	92 mph
Empty Wt.	264 lbs.	**VCr**	80 mph
Payload (full fuel)	N/A	**V_{S1} (stall clean)**	40 mph
Fuel Capacity	5 gal.	**V_{S0} (lndg config)**	N/A
Seats	1	**Climb Rate @msl**	1100 fpm
Cabin Width	N/A	**Service Ceiling**	10,000 ft.

Std. Engine/HP	Usable HP Range	Construction Mtrl.	Est. Bldg. Time
Rotax / 52 hp	N/A	Alum. Tubing, Pop Rivet	300 hrs.
Plans	**Kit**	**Info Packet**	**Video**
N/A	$4,495.00	$9.95	$20.00

3/4 scale replica of famous WWI fighter. Aluminum tubing and pop rivet construction. Removable three-piece wings, easily transportable. Quick build. Economical operation. Modern two stroke engine. Light responsive controls. Deluxe kit includes engine mount and 5 gal. fuel tank. **Flying:** 2

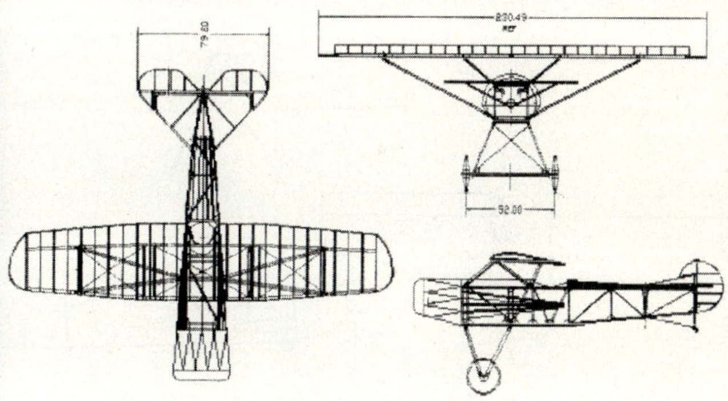

The Fokker D-VIII first flew in the fall of 1918 during the first World War. Although its engagement was limited to approximately one month, it was considered to be the most advanced fighter plane of the era, and was most feared by Allied pilots. It was nicknamed the "Flying Razor," a moniker earned because its aerodynamics were uncommon to the era.

This 3/4 scale replica is designed with the modern two stroke Rotax engine in mind. With its high power-to-weight ratio, coupled with the light rugged aluminum tubing and rivet construction, it is capable of delivering fighter-like performance. At an empty weight of 263 pounds, it can achieve speeds reaching 92 mph with as little as a 35 hp engine. And it's convenient to fly anywhere. In addition to its removable wings, the large wheels and taildragger configuration allow this aircraft to be operated from unimproved strips.

AIRDROME AEROPLANES
929 NW Road 1571
Holden MO 64040
Tel. 816-230-8585
Web: http://www.sunflower.org/~baron

Giles G-200

Wing Span	20.00 ft.	**Range**	750 sm
Wing Area	75.00 sq. ft.	**Takeoff Dist.**	N/A ft.
Length	18.00 ft.	**Landing Dist.**	N/A ft.
Max Gross Wt.	1,150 lbs.	**Vmax**	255 mph
Empty Wt.	750 lbs.	**Vcr**	215 mph
Payload (full fuel)	N/A lbs.	**V_{S1} (stall clean)**	66 mph
Fuel Capacity	36.0 gal.	**V_{S0} (lndg config)**	N/A mph
Seats	1	**Climb Rate @msl**	4,100 fpm
Cabin Width	23.0 in.	**Serv. Ceiling**	N/A

Std. Engine/HP	Usable HP Range	Construction Mtrl.	Est. Bldg. Time
Lycoming / 200 hp	150 - 220	Composite	1000 hrs.
Plans	**Kit**	**Info Packet**	**Video**
No	$39,500	Yes	$20.00

A single place, fully aerobatic design utilizing advanced composite construction. Cockpit accepts pilots from 5'1" and 100 lbs. to 6'5" and 245 lbs. Custom aircraft are available for other body sizes. Kit price is for the "E-Z Build" kit. Also available completely finished by AkroTech for $98,650.

Kits/Plans Sold: 26 **Completed & Flying:** 1

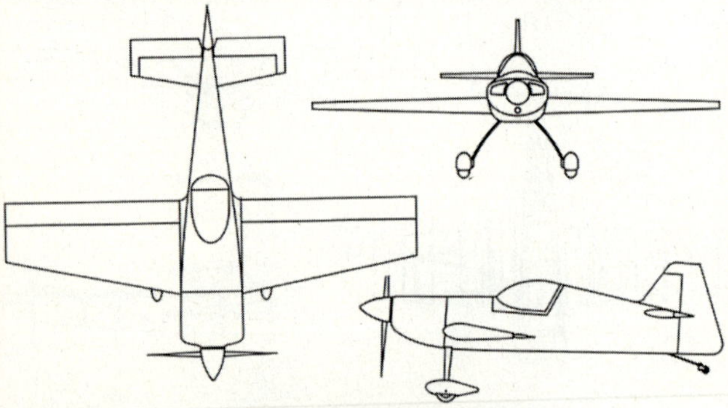

Giles G-202 in the foreground, G-200 in the background.

Mike Goulian

"Some airplanes have a great roll rate, however they have very little centering feel in the middle. But this airplane does. You can do a roll at maximum aileron input from either high or low speed and the aiplane still possesses a great centering feel. That's one of the greatest qualities you can have in an aerobatic plane."

AKROTECH AVIATION, INC
53774 Airport Rd. • Scappoose OR 97056
Tel. 503 543-7960 • Fax 503 543-7964

Giles G-202

Wing Span	22.00 ft.	**Range**	900 sm
Wing Area	90.00 sq. ft.	**Takeoff Dist.**	N/A
Length	19.75 ft.	**Landing Dist.**	N/A
Max Gross Wt.	1600 lbs.	**Vmax**	255 mph
Empty Wt.	950 lbs.	**VCr**	215 mph
Payload (full fuel)	N/A	**V_{S1} (stall clean)**	66 mph
Fuel Capacity	42 gal.	**V_{S0} (lndg config)**	N/A
Seats	2	**Climb Rate @msl**	3900 fpm
Cabin Width	28 in.	**Service Ceiling**	N/A

Std. Engine/HP	Usable HP Range	Construction Mtrl.	Est. Bldg. Time
Lyc 360 / 200 hp	150 - 220	Composite	1000 hrs.
Plans	**Kit**	**Info Packet**	**Video**
N/A	$54,500.00	Yes	$20.00

Two place, fully aerobatic airplane designed to compete in "unlimited" competitions. First flew December 1995. Wings are delivered with the spar and all ribs already attached to the top wing skin. Fuselage arrives in two halves with the firewall and bulkheads bonded to the bottom half. Also available completely finished by AkroTech for $120,000 using a Lycoming AEIO-360 engine and a two blade MT propeller. Other configurations available.

Kits/Plans Sold: 31 **Completed & Flying:** 1

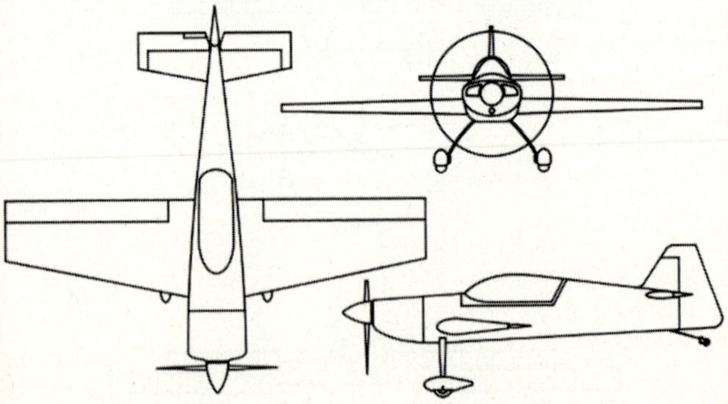

AKROTECH AVIATION, INC
53774 Airport Rd. • Scappoose OR 97056
Tel. 503 543-7960 • Fax 503 543-7964

BD-5G

Wing Span	17.00 ft.	**Range**	575 sm
Wing Area	N/A	**Takeoff Dist.**	590 ft.
Length	13.56 ft.	**Landing Dist.**	530 ft.
Max Gross Wt.	660 lbs.	**Vmax**	232 mph
Empty Wt.	355 lbs.	**VCr**	229 mph
Payload (full fuel)	N/A	**V_{S1} (stall clean)**	55 mph
Fuel Capacity	28 gal.	**V_{S0} (lndg config)**	N/A
Seats	1	**Climb Rate @msl**	1920 fpm
Cabin Width	N/A	**Service Ceiling**	26,000 ft.

Std. Engine/HP	Usable HP Range	Construction Mtrl.	Est. Bldg. Time
Zenoah	90 - 100	All-Metal	2400 hrs.
Plans	**Kit**	**Info Packet**	**Video**
$600.00	Yes	$20.00	$20.00

A high performance design that require above average skills by the builder and pilot. Other configurations include: the BD-5B, with a 21' wing span and a Honda 90 hp engine; the BD-5GR, with an Alturair 100 hp rotary engine; and the BD-5J using a Microturbo jet engine. +6, -6 G's stress limits.

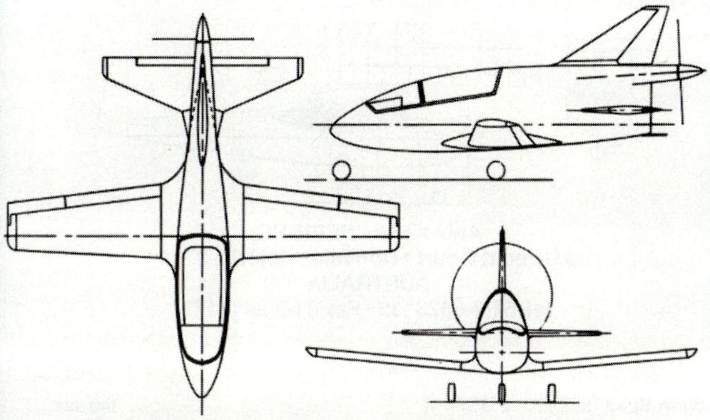

ALTURAIR
1405 N. Johnson
El Cajon CA 92020
Tel. 619-449-1570 • Fax 619-442-0481

Aviatika-890 U

Wing Span	26.57 ft.	**Range**	204 sm
Wing Area	153.9 sq. ft.	**Takeoff Dist.**	328 ft.
Length	18.05 ft.	**Landing Dist.**	N/A
Max Gross Wt.	990 lbs.	**Vmax**	75 mph
Empty Wt.	616 lbs.	**VCr**	68 mph
Payload (full fuel)	N/A	**V_{S1} (stall clean)**	40 mph
Fuel Capacity	10 gal.	**V_{S0} (lndg config)**	N/A
Seats	2	**Climb Rate @msl**	590 fpm
Cabin Width	N/A	**Service Ceiling**	13,000 ft.

Std. Engine/HP	Usable HP Range	Construction Mtrl.	Est. Bldg. Time
Rotax 912 / 80 hp	N/A	N/A	N/A
Plans	**Kit**	**Info Packet**	**Video**
N/A	Yes	N/A	N/A

A sturdy and reliable Ultralight (European) biplane. Seating for two in a side-by-side configuration. Meets JAR 23 guidelines.

ALIS.E.I., S.R.L.
Via dell'Artigianato, 25 • Camposanto (MO) 41031
ITALY
Tel. 0535 88149 • Fax 0535 88129

Eureka

Wing Span	26.00 ft.	**Range**	150 sm
Wing Area	104.0 sq. ft.	**Takeoff Dist.**	450 ft.
Length	13.00 ft.	**Landing Dist.**	250 ft.
Max Gross Wt.	460 lbs.	**Vmax**	75 mph
Empty Wt.	240 lbs.	**VCr**	60 mph
Payload (full fuel)	200 lbs.	**V_{S1} (stall clean)**	30 mph
Fuel Capacity	5 gal.	**V_{S0} (lndg config)**	30 mph
Seats	1	**Climb Rate @msl**	350 fpm
Cabin Width	24 in.	**Service Ceiling**	10,000 ft.

Std. Engine/HP	Usable HP Range	Construction Mtrl.	Est. Bldg. Time
2SI / 20 hp	20 - 30	Alum., Wood, Foam	60 hrs.
Plans	**Kit**	**Info Packet**	**Video**
N/A	$4,995.00	Yes	N/A

FAR 103 compliant. Aluminum skinned. Kit price includes engine. Prop and instruments extra. Kit components supplied precut, predrilled, prepainted. Assembly only required. Typically can be done in two weekends. **Sold:** 2 **Flying:** 1

AMERI-CANA ULTRALIGHTS
PO Box 360, Henricks Dr • Irricana, AB T0M 1B0
CANADA
Tel. 403-935-4859
E-Mail: wstark@compuserve.com

J-6 Karatoo

Wing Span	33.00 ft.	**Range**	340 sm
Wing Area	146.0 sq. ft.	**Takeoff Dist.**	300 ft.
Length	20.00 ft.	**Landing Dist.**	300 ft.
Max Gross Wt.	1300 lbs.	**Vmax**	115 mph
Empty Wt.	600 lbs.	**VCr**	85 mph
Payload (full fuel)	N/A	**V_{S1} (stall clean)**	35 mph
Fuel Capacity	16 gal.	**V_{S0} (Indg config)**	N/A
Seats	2	**Climb Rate @msl**	900 fpm
Cabin Width	N/A	**Service Ceiling**	14,000 ft.

Std. Engine/HP	Usable HP Range	Construction Mtrl.	Est. Bldg. Time
110 hp	85 - 110	Tube, Wood, Fabric	500 hrs.
Plans	**Kit**	**Info Packet**	**Video**
$250.00	$12,500.00	$5.00	Yes

Uses welded 4130 aircraft steel tube throughout the fuselage. Folding wings. Engine, propeller and instruments not included in kit.

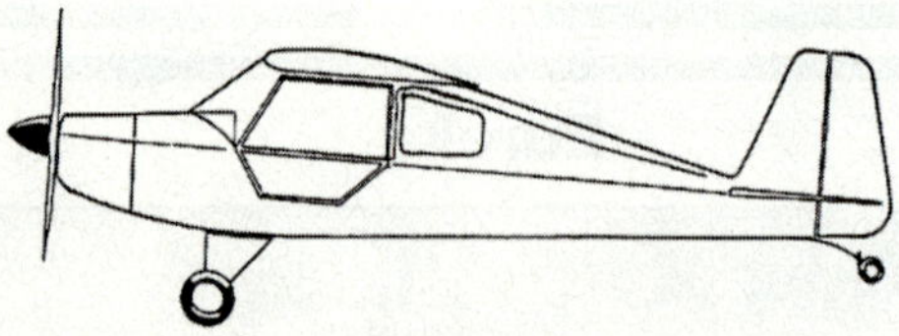

AMAX ENGINEERING
5 Utrecht Court • Donvale, Victoria, 3111
AUSTRALIA
Tel. 61-3-8423132 • Fax 61-3-8418177

Vixen 105

Wing Span	28.50 ft.	**Range**	350 sm
Wing Area	128.0 sq. ft.	**Takeoff Dist.**	300 ft.
Length	18.70 ft.	**Landing Dist.**	350 ft.
Max Gross Wt.	1200 lbs.	**Vmax**	130 mph
Empty Wt.	500 lbs.	**VCr**	100 mph
Payload (full fuel)	N/A	**V_{S1} (stall clean)**	40 mph
Fuel Capacity	16 gal.	**V_{S0} (Indg config)**	N/A
Seats	2	**Climb Rate @msl**	900 fpm
Cabin Width	N/A	**Service Ceiling**	14,000 ft.

Std. Engine/HP	Usable HP Range	Construction Mtrl.	Est. Bldg. Time
EA81 Subaru / 105 hp	N/A	Composite	300 hrs.
Plans	**Kit**	**Info Packet**	**Video**
N/A	$14,000.00	$5.00	N/A

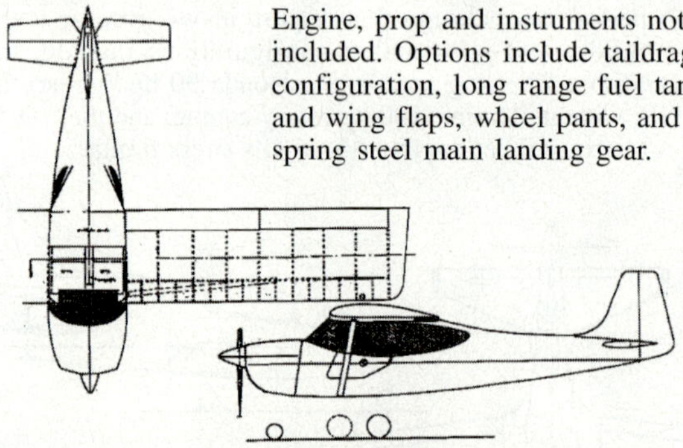

Engine, prop and instruments not included. Options include taildragger configuration, long range fuel tanks, and wing flaps, wheel pants, and spring steel main landing gear.

AMAX ENGINEERING
5 Utrecht Court • Donvale, Victoria, 3111
AUSTRALIA
Tel. 61-3-8423132 • Fax 61-3-8418177

Sport 1700

AMAX ENGINEERING
5 Utrecht Court • Donvale, Victoria, 3111
AUSTRALIA
Tel. 61-3-8423132 • Fax 61-3-8418177

Wing Span	33.00 ft.	**Range**	340 sm
Wing Area	146.0 sq. ft.	**Takeoff Dist.**	300 ft.
Length	20.00 ft.	**Landing Dist.**	300 ft.
Max Gross Wt.	1300 lbs.	**Vmax**	115 mph
Empty Wt.	600 lbs.	**VCr**	85 mph
Payload (full fuel)	N/A	**V_{S1} (stall clean)**	35 mph
Fuel Capacity	16 gal.	**V_{S0} (Indg config)**	N/A
Seats	2	**Climb Rate @msl**	900 fpm
Cabin Width	N/A	**Service Ceiling**	14,000 ft.

Std. Engine/HP	Usable HP Range	Construction Mtrl.	Est. Bldg. Time
110 hp	85 - 110	Tube, Wood, Fabric	500 hrs.
Plans	**Kit**	**Info Packet**	**Video**
$250.00	$12,500.00	$5.00	Yes

Uses welded 4130 aircraft steel tube throughout the fuselage. Folding wings. Engine, propeller, and instruments not included.

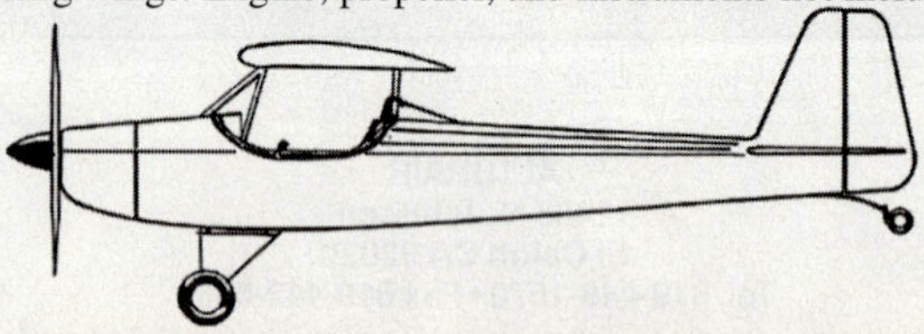

Vision

Wing Span	21.50 ft.	**Range**	N/A
Wing Area	85.00 sq. ft.	**Takeoff Dist.**	1500 ft.
Length	N/A	**Landing Dist.**	1200 ft.
Max Gross Wt.	1350 lbs.	**Vmax**	170 mph
Empty Wt.	850 lbs.	**VCr**	157 mph
Payload (full fuel)	368 lbs.	**V_{S1} (stall clean)**	60 mph
Fuel Capacity	22 gal.	**V_{S0} (lndg config)**	54 mph
Seats	2	**Climb Rate @msl**	1400 fpm
Cabin Width	44 in.	**Service Ceiling**	N/A

Std. Engine/HP	Usable HP Range	Construction Mtrl.	Est. Bldg. Time
Subaru / 100 hp	100 - 160	Composite	3000 hrs.
Plans	**Kit**	**Info Packet**	**Video**
$425.00	N/A	$17.00	N/A

The only plans built, non-canard, all-composite design available. Full sandwich open-bay construction. Incorporates the rugged, lightweight and proven construction methods of more expensive kit planes without the molds and complexity. Can be built with hand tools only. Some prefab parts available. First time builders can complete the airframe for less than $7,000.

Kits/Plans Sold: 21 **Completed & Flying:** 2

Stephen Upson
"I have just received my construction manual for the VISION and I am extremely delighted. The books are complete. Although I do hold an A&P certificate, I was not familiar with the "new" construction techniques used on this airplane. Any questions I had were answered before I could ask them. The whole process is well thought out and very well documented."

AMERICAN AFFORDABLE AIRCRAFT
210 Cessna Blvd., Unit 1-C
Daytona Beach FL 32124
Tel. 904-767-9691 • Fax 904-767-1217
E-Mail: visionflyr@aol.com
Web: http://members.aol.com/VisionFlyr

Ban-Bi MCR 01

Wing Span	22.00 ft.	**Range**	1435 sm
Wing Area	56.00 sq. ft.	**Takeoff Dist.**	900 ft.
Length	18.00 ft.	**Landing Dist.**	1000 ft.
Max Gross Wt.	990 lbs.	**Vmax**	200 mph
Empty Wt.	440 lbs.	**VCr**	180 mph
Payload (full fuel)	490 lbs.	**V_{S1} (stall clean)**	54 mph
Fuel Capacity	24 gal.	**V_{S0} (lndg config)**	N/A
Seats	2	**Climb Rate @msl**	1990 fpm
Cabin Width	43 in.	**Service Ceiling**	17,000 ft.

Std. Engine/HP	Usable HP Range	Construction Mtrl.	Est. Bldg. Time
Rotax 912 / 80 hp	66-80	Composite	400 hrs.
Plans	**Kit**	**Info Packet**	**Video**
N/A	$25,000.00	Yes	$20.00

Stress tested to +9, -4.5 G's. Capable of 180 mph cruise speed at 75% power with electrically adjustable inflight, variable pitch prop. Available Quick Build option saves hundreds of hours of building time. Available in Europe from Dyn' Aero, 19 rue de l'aviation, 21121 Darois, FRANCE **Sold:** 92 **Flying:** 22

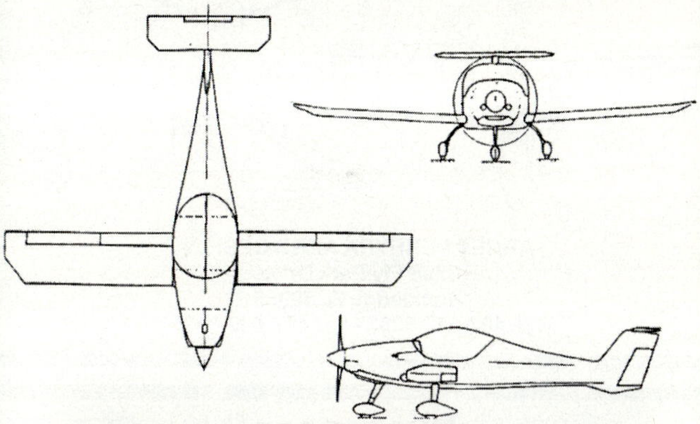

Mark Lindstrom former F4/F16/F117 pilot
"The control inputs are light and precise which leaves the pilot an impression that he or she is in the cockpit of a small fighter."

AMERICAN GHILES AIRCRAFT, INC.
522 East Washington St.
Orlando FL 32801
Tel. 407-839-0581 • Fax 407-872-3233

Aventura II

Wing Span	30.67 ft.	Range	150 sm
Wing Area	161.7 sq. ft.	Takeoff Dist.	200 ft.
Length	23.00 ft.	Landing Dist.	250 ft.
Max Gross Wt.	1200 lbs.	Vmax	90 mph
Empty Wt.	640 lbs.	VCr	70 mph
Payload (full fuel)	510 lbs.	V_{S1} (stall clean)	45 mph
Fuel Capacity	12 gal.	V_{S0} (lndg config)	32 mph
Seats	2	Climb Rate @msl	500 fpm
Cabin Width	44 in.	Service Ceiling	13,500 ft.

Std. Engine/HP	Usable HP Range	Construction Mtrl.	Est. Bldg. Time
Rotax 582 / 65 hp	65 - 80	Metal, Fiberglass, Fabric	200 hrs.
Plans	**Kit**	**Info Packet**	**Video**
N/A	$22,900.00	$5.00	N/A

Experimental two place amphibian. Past winner of Sun 'n Fun Best Lightplane Seabird award. Kit price includes BRS parachute, engine, Ivoprop, and instruments. See Ad on Page 9.
Flying: 25

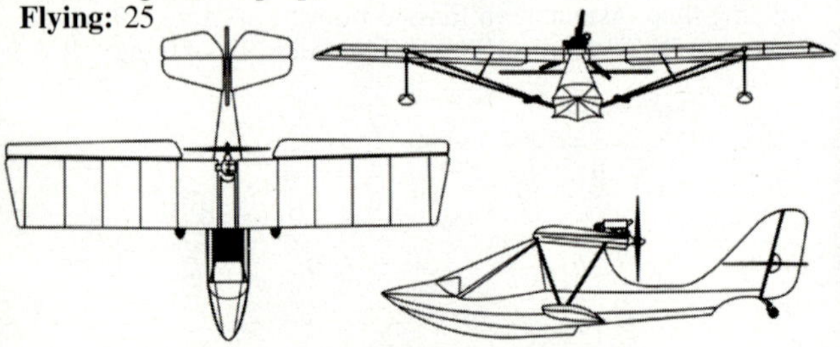

ARNET PEREYRA AERO DESIGN
3795 Fly Park Drive
Rockledge FL 32955
Tel. 407-635-8005 • Fax 407-635-8005

Aventura UL

Wing Span	21.67 ft.	Range	100 sm
Wing Area	156.0 sq. ft.	Takeoff Dist.	150 ft.
Length	20.50 ft.	Landing Dist.	125 ft.
Max Gross Wt.	650 lbs.	Vmax	63 mph
Empty Wt.	324 lbs.	VCr	55 mph
Payload (full fuel)	326 lbs.	V_{S1} (stall clean)	N/A
Fuel Capacity	5 gal.	V_{S0} (lndg config)	27 mph
Seats	1	Climb Rate @msl	500 fpm
Cabin Width	N/A	Service Ceiling	13,500 ft.

Std. Engine/HP	Usable HP Range	Construction Mtrl.	Est. Bldg. Time
Rotax 447 / 40 hp	28 - 40	Metal, Fiberglass, Fabric	90 hrs.
Plans	**Kit**	**Info Packet**	**Video**
N/A	$15,500.00	$5.00	N/A

Part 103 Ultralight amphibian flying boat. Past winner of the Sun 'n Fun Best UL Seabird award. Kit price includes BRS parachute, engine, Ivoprop, & instruments. See Ad on Page 9.
Flying: 9

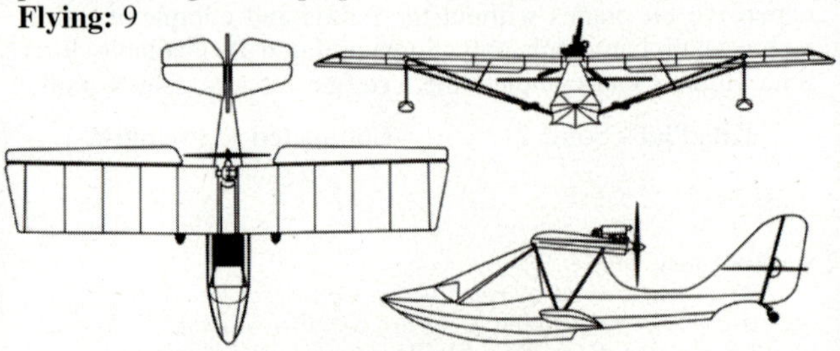

ARNET PEREYRA AERO DESIGN
3795 Fly Park Drive
Rockledge FL 32955
Tel. 407-635-8005 • Fax 407-635-8005

Buccaneer II

ARNET PEREYRA AERO DESIGN
3795 Fly Park Drive
Rockledge FL 32955
Tel. 407-635-8005 • Fax 407-635-8005

Wing Span	29.50 ft.	Range	150 sm
Wing Area	156.0 sq. ft.	Takeoff Dist.	300 ft.
Length	22.33 ft.	Landing Dist.	200 ft.
Max Gross Wt.	1200 lbs.	Vmax	90 mph
Empty Wt.	580 lbs.	VCr	70 mph
Payload (full fuel)	620 lbs.	V_{S1} (stall clean)	32 mph
Fuel Capacity	12 gal.	V_{S0} (lndg config)	N/A
Seats	2	Climb Rate @msl	500 fpm
Cabin Width	N/A	Service Ceiling	13,500 ft.

Std. Engine/HP	Usable HP Range	Construction Mtrl.	Est. Bldg. Time
Rotax 582 / 65 hp	65 - 80	Metal, Fiberglass, Fabric	200 hrs.
Plans	**Kit**	**Info Packet**	**Video**
N/A	$21,900.00	Yes	$10.00

Experimental two place amphibous flying boat. Kit price for this amphibian includes BRS parachute, engine, Ivoprop, and instruments.
See Ad on Page 9.
Completed & Flying: 300

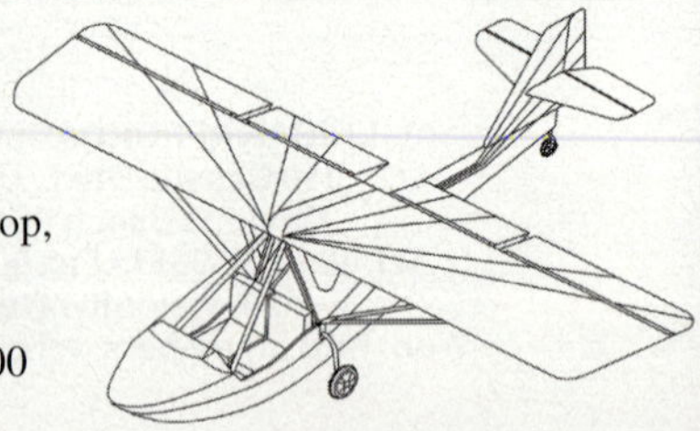

Buccaneer SX

Wing Span	29.67 ft.	**Range**	180 sm
Wing Area	143.0 sq. ft.	**Takeoff Dist.**	175 ft.
Length	20.10 ft.	**Landing Dist.**	150 ft.
Max Gross Wt.	810 lbs.	**Vmax**	90 mph
Empty Wt.	450 lbs.	**VCr**	65 mph
Payload (full fuel)	360 lbs.	**V_{S1} (stall clean)**	29 mph
Fuel Capacity	12 gal.	**V_{S0} (lndg config)**	26 mph
Seats	1	**Climb Rate @msl**	950 fpm
Cabin Width	N/A	**Service Ceiling**	15,000 ft.

Std. Engine/HP	Usable HP Range	Construction Mtrl.	Est. Bldg. Time
Rotax 503 / 52 hp	52 - 65	Aluminum, Fabric, Fiberglass	130 hrs.

Plans	Kit	Info Packet	Video
N/A	$16,800.00	$5.00	N/A

Experimental two place amphibian flying boat. Kit price includes BRS parachute, engine, Ivoprop, and instruments. See Ad on Page 9.
Flying: 200

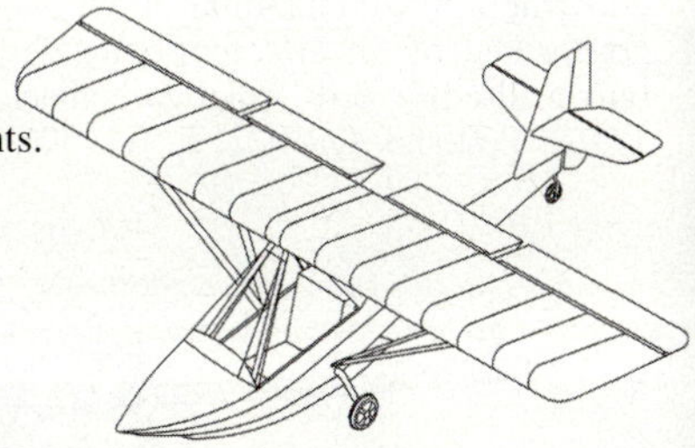

ARNET PEREYRA AERO DESIGN
3795 Fly Park Drive
Rockledge FL 32955
Tel. 407-635-8005 • Fax 407-635-8005

Sabre II

Wing Span	29.67 ft.	**Range**	150 sm
Wing Area	156.0 sq. ft.	**Takeoff Dist.**	150 ft.
Length	22.33 ft.	**Landing Dist.**	200 ft.
Max Gross Wt.	1100 lbs.	**Vmax**	90 mph
Empty Wt.	480 lbs.	**VCr**	75 mph
Payload (full fuel)	620 lbs.	**V_{S1} (stall clean)**	32 mph
Fuel Capacity	12 gal.	**V_{S0} (lndg config)**	29 mph
Seats	2	**Climb Rate @msl**	500 fpm
Cabin Width	N/A	**Service Ceiling**	14,000 ft.

Std. Engine/HP	Usable HP Range	Construction Mtrl.	Est. Bldg. Time
Rotax 582 / 65 hp	65 - 75	Metal, Fiberglass, Fabric	140 hrs.

Plans	Kit	Info Packet	Video
N/A	$18,600.00	Yes	N/A

Conventional tubing and fabric construction are used in this two place taildragger. Kit price includes BRS parachute, engine, Ivoprop, & instruments. See Ad on Page 9.
Sold: 10 **Flying:** 1

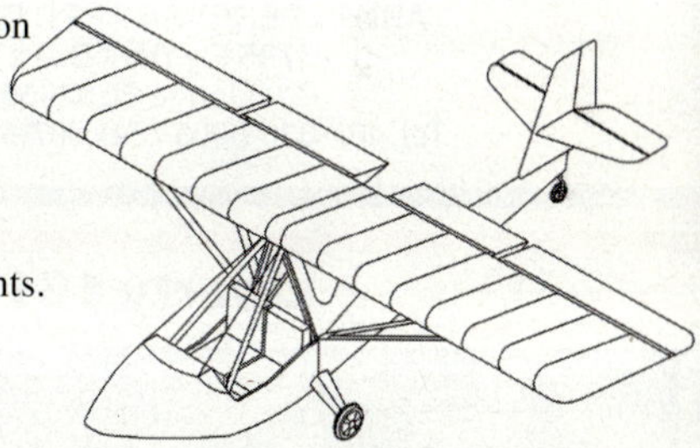

ARNET PEREYRA AERO DESIGN
3795 Fly Park Drive
Rockledge FL 32955
Tel. 407-635-8005 • Fax 407-635-8005

Zephyr II

ARNET PEREYRA AERO DESIGN
3795 Fly Park Drive
Rockledge FL 32955
Tel. 407-635-8005 • Fax 407-635-8005

Wing Span	29.67 ft.	**Range**	150 sm
Wing Area	156.0 sq. ft.	**Takeoff Dist.**	150 ft.
Length	22.50 ft.	**Landing Dist.**	200 ft.
Max Gross Wt.	1100 lbs.	**Vmax**	90 mph
Empty Wt.	480 lbs.	**VCr**	70 mph
Payload (full fuel)	620 lbs.	**V_{S1} (stall clean)**	33 mph
Fuel Capacity	12 gal.	**V_{S0} (lndg config)**	29 mph
Seats	2	**Climb Rate @msl**	800 fpm
Cabin Width	N/A	**Service Ceiling**	14,000 ft.

Std. Engine/HP	Usable HP Range	Construction Mtrl.	Est. Bldg. Time
Rotax 582 / 65 hp	65 - 75	Metal, Fiberglass, Fabric	140 hrs.

Plans	Kit	Info Packet	Video
N/A	$16,600.00	Yes	$10.00

Open cockpit, tandem two place taildragger. Kit price includes BRS parachute, & engine. See Ad on Page 9.

Sold: 40 **Flying:** 40

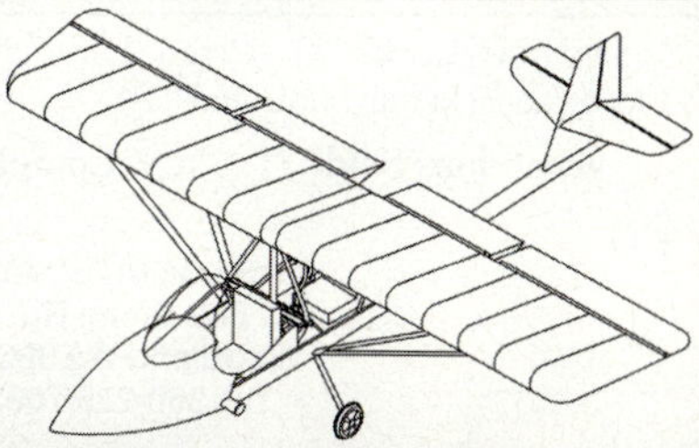

Aventura HP

Wing Span	29.67 ft.	**Range**	150 sm
Wing Area	156.0 sq. ft.	**Takeoff Dist.**	150 ft.
Length	20.50 ft.	**Landing Dist.**	150 ft.
Max Gross Wt.	950 lbs.	**Vmax**	90 mph
Empty Wt.	430 lbs.	**VCr**	75 mph
Payload (full fuel)	520 lbs.	**V_{S1} (stall clean)**	45 mph
Fuel Capacity	12 gal.	**V_{S0} (lndg config)**	32 mph
Seats	1	**Climb Rate @msl**	1100 fpm
Cabin Width	N/A	**Service Ceiling**	13,500 ft.

Std. Engine/HP	Usable HP Range	Construction Mtrl.	Est. Bldg. Time
Rotax 582 / 65 hp	52 - 65	Metal, Fiberglass, Fabric	110 hrs.
Plans	**Kit**	**Info Packet**	**Video**
N/A	$18,700.00	$5.00	N/A

Experimental single place amphibian flying boat. Kit price for this amphibian includes BRS parachute, engine, Ivoprop, and instruments. See Ad on Page 9. **Completed & Flying:** 8

ARNET PEREYRA AERO DESIGN
3795 Fly Park Drive
Rockledge FL 32955
Tel. 407-635-8005 • Fax 407-635-8005

Liberty 181

Wing Span	40.00 ft.	**Range**	1200 sm
Wing Area	200.0 sq. ft.	**Takeoff Dist.**	200 ft.
Length	24.00 ft.	**Landing Dist.**	250 ft.
Max Gross Wt.	3200 lbs.	**Vmax**	145 mph
Empty Wt.	1890 lbs.	**VCr**	130 mph
Payload (full fuel)	810 lbs.	**V_{S1} (stall clean)**	38 mph
Fuel Capacity	100 gal.	**V_{S0} (lndg config)**	N/A
Seats	4	**Climb Rate @msl**	2000 fpm
Cabin Width	46 in.	**Service Ceiling**	22,050 ft.

Std. Engine/HP	Usable HP Range	Construction Mtrl.	Est. Bldg. Time
Cont O-470 / 230 hp	230-300	Metal, Composite	2000 hrs.
Plans	**Kit**	**Info Packet**	**Video**
N/A	$24,495.00	Yes	N/A

Heavy duty workhorse design. Large doors for cargo handling. Floats and skis are optional.

Kits/Plans Sold: 7 **Completed & Flying:** 1

ASSOCIATED AIR
582 Niemi Rd.
Woodland WA 98674
Tel. 360-225-7093

Zephyr

Wing Span	34.70 ft.	**Range**	497 sm
Wing Area	N/A	**Takeoff Dist.**	N/A
Length	20.30 ft.	**Landing Dist.**	N/A
Max Gross Wt.	990 lbs.	**Vmax**	137 mph
Empty Wt.	616 lbs.	**VCr**	112 mph
Payload (full fuel)	N/A	**V_{S1} (stall clean)**	37 mph
Fuel Capacity	16 gal.	**V_{S0} (lndg config)**	N/A
Seats	2	**Climb Rate @msl**	N/A
Cabin Width	N/A	**Service Ceiling**	N/A

Std. Engine/HP	Usable HP Range	Construction Mtrl.	Est. Bldg. Time
Rotax 912 / 80 hp	50-80	Wood, Composite	N/A
Plans	**Kit**	**Info Packet**	**Video**
N/A	Yes	Yes	N/A

Designed by Oldrich Olsansky. Wing is cantilever, all wood construction with laminar flow UA-2 airfoil. Fuselage is comprised of an all composite shell braced by wooden bulkheads. In North America contact: A.K.S. Inc., PO Box 17161, Portland, OR 97217, Tel. 503-247-9207

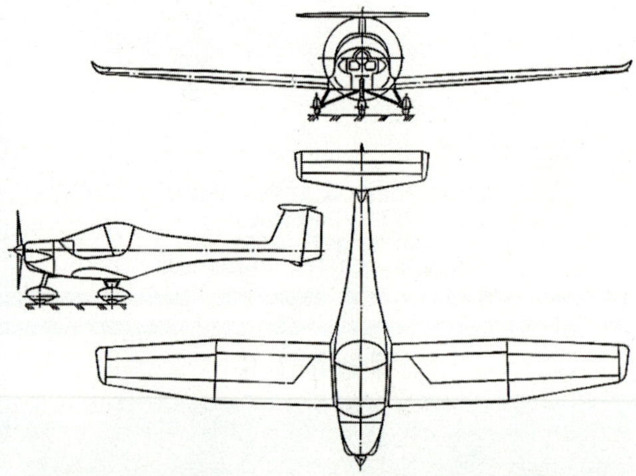

ATEC V.O.S.
Opolanska 171
289 07 Libice nad Cidlinou
CZECH REPUBLIC
Tel. 0042-324-77371 • Fax 0042-324-3002

Pitts S-1-11B

Wing Span	18.00 ft.	**Range**	N/A
Wing Area	110.3 sq. ft.	**Takeoff Dist.**	N/A
Length	18.10 ft.	**Landing Dist.**	N/A
Max Gross Wt.	1500 lbs.	**Vmax**	208 mph
Empty Wt.	1090 lbs.	**VCr**	N/A
Payload (full fuel)	N/A	**V_{S1} (stall clean)**	56 mph
Fuel Capacity	35 gal.	**V_{S0} (lndg config)**	N/A
Seats	1	**Climb Rate @msl**	4000 fpm
Cabin Width	N/A	**Service Ceiling**	N/A

Std. Engine/HP	Usable HP Range	Construction Mtrl.	Est. Bldg. Time
Lyc. IO-540 / 300 hp	N/A	Tube, Fabric	N/A
Plans	**Kit**	**Info Packet**	**Video**
$300.00	Yes	N/A	Yes

Designed specifically to compete in today's unlimited aerobatic environment. Utilizes revolutionary airfoil and aileron concepts. The simplicity of the design greatly increases reliability and ease of maintenance. Component kits available.

See Ad Inside Back Cover.

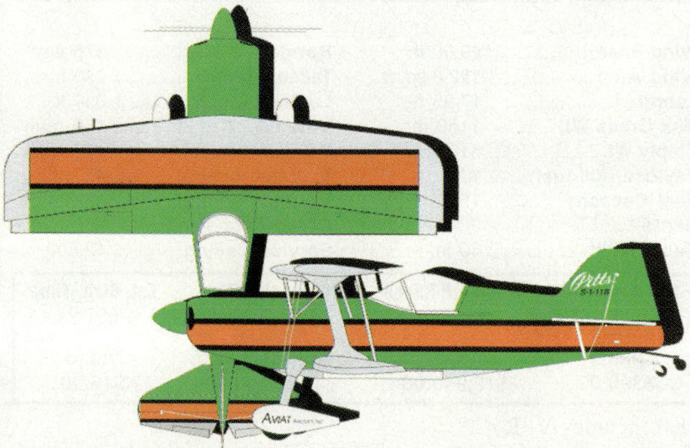

AVIAT AIRCRAFT, INC.
Box 1240 • 672 S. Washington
Afton WY 83110
Tel. 307-886-3151 • Fax 307-886-9674

Eagle II

Wing Span	19.92 ft.	**Range**	350 sm
Wing Area	126.0 sq. ft.	**Takeoff Dist.**	900 ft.
Length	18.50 ft.	**Landing Dist.**	1375 ft.
Max Gross Wt.	1600 lbs.	**Vmax**	184 mph
Empty Wt.	1050 lbs.	**VCr**	165 mph
Payload (full fuel)	N/A	**V_{S1} (stall clean)**	63 mph
Fuel Capacity	25 gal.	**V_{S0} (lndg config)**	N/A
Seats	2	**Climb Rate @msl**	2120 fpm
Cabin Width	N/A	**Service Ceiling**	22,000 ft.

Std. Engine/HP	Usable HP Range	Construction Mtrl.	Est. Bldg. Time
Lyc. AEIO-360 / 200 hp	180-200	Tube, Wood, Fabric	1800 hrs.
Plans	**Kit**	**Info Packet**	**Video**
N/A	$55,000.00	Yes	Yes

Proven two place design suitable for advanced aerobatic competition. +6, -3 G's limits. Engine and propeller not included. See Ad Inside Back Cover.
Completed & Flying: 350

AVIAT AIRCRAFT, INC.
Box 1240 • 672 S. Washington
Afton WY 83110
Tel. 307-886-3151 • Fax 307-886-9674

Pitts S-1S

Wing Span	17.33 ft.	**Range**	300 sm
Wing Area	98.50 sq. ft.	**Takeoff Dist.**	700 ft.
Length	15.00 ft.	**Landing Dist.**	1300 ft.
Max Gross Wt.	1150 lbs.	**Vmax**	154 mph
Empty Wt.	760 lbs.	**VCr**	143 mph
Payload (full fuel)	N/A	**V_{S1} (stall clean)**	64 mph
Fuel Capacity	20 gal.	**V_{S0} (lndg config)**	N/A
Seats	1	**Climb Rate @msl**	2400 fpm
Cabin Width	N/A	**Service Ceiling**	15,000 ft.

Std. Engine/HP	Usable HP Range	Construction Mtrl.	Est. Bldg. Time
Lyc. AEIO-360 / 180 hp	180-200	Tube, Wood, Fabric	N/A
Plans	**Kit**	**Info Packet**	**Video**
$250.00	N/A	Yes	N/A

Well proven, single place design. Highly successful in aerobatic competition for many years. Plans only. See Ad Inside Back Cover. **Completed & Flying:** 275

AVIAT AIRCRAFT, INC.
Box 1240 • 672 S. Washington
Afton WY 83110
Tel. 307-886-3151 • Fax 307-886-9674

Sky Rocket III

AVIABELLANCA AIRCRAFT CORP.
11800 Sunrise Valley Dr. • Suite 322
Reston VA 20191
Tel. 703-860-3823 • Fax 703-860-9144
E-Mail: AVBELLANCA@AOL. COM
Web: http://WWW.GAPTECH.COM/BELLANCA

Wing Span	35.00 ft.	**Range**	2300 sm
Wing Area	182.6 sq. ft.	**Takeoff Dist.**	680 ft.
Length	37.00 ft.	**Landing Dist.**	922 ft.
Max Gross Wt.	4200 lbs.	**Vmax**	340 mph
Empty Wt.	2490 lbs.	**VCr**	327 mph
Payload (full fuel)	900 lbs.	**V_{S1} (stall clean)**	76 mph
Fuel Capacity	176 gal.	**V_{S0} (lndg config)**	68 mph
Seats	6	**Climb Rate @msl**	2080 fpm
Cabin Width	45 in.	**Service Ceiling**	30,000 ft.

Std. Engine/HP	Usable HP Range	Construction Mtrl.	Est. Bldg. Time
Cont. GTSIO 520-435	325-500	Composite	3500 hrs.
Plans	**Kit**	**Info Packet**	**Video**
N/A	$85,000.00	Yes	N/A

August Bellanca's newest design.Advanced, all composite derivative of the world record holding Sky Rocket II with greater speed, payload, and range. Fast build, modular construction.
Sold: 11

Bandit

Wing Span	29.88 ft.	**Range**	340 sm
Wing Area	122.5 sq. ft.	**Takeoff Dist.**	125 ft.
Length	17.92 ft.	**Landing Dist.**	125 ft.
Max Gross Wt.	950 lbs.	**Vmax**	95 mph
Empty Wt.	425 lbs.	**VCr**	80 mph
Payload (full fuel)	450 lbs.	**V_{S1} (stall clean)**	35 mph
Fuel Capacity	15 gal.	**V_{S0} (lndg config)**	32 mph
Seats	2	**Climb Rate @msl**	800 fpm
Cabin Width	40 in.	**Service Ceiling**	12,500 ft.

Std. Engine/HP	Usable HP Range	Construction Mtrl.	Est. Bldg. Time
Rotax 503 / 50 hp	50 - 110	Tube, Fabric	500 hrs.
Plans	**Kit**	**Info Packet**	**Video**
$395.00	$11,515.00	$13.00	$16.00

Same airframe as the Avid Mark IV. 65 hp Rotax 582 optional. Wings fold in five minutes. Upholstered cloth seat. Disc brakes. VFR instruments. Trigear or tailwheel. See Ad on Page 27.
Sold: 130 **Flying:** 40

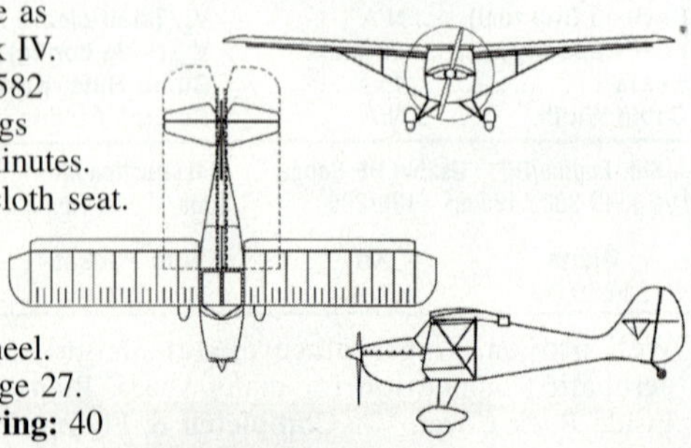

AVID AIRCRAFT, INC.
P.O. Box 728 • 4823 Aviation Way
Caldwell ID 83606
Tel. 208-454-2600 • Fax 208-454-8608
E-Mail: SALES@avidair.com
Web: http://www.avid.com

Mark IV Speedwing

Wing Span	29.90 ft.	**Range**	375 sm
Wing Area	122.8 sq. ft.	**Takeoff Dist.**	250 ft.
Length	17.85 ft.	**Landing Dist.**	350 ft.
Max Gross Wt.	1150 lbs.	**Vmax**	150 mph
Empty Wt.	510 lbs.	**VCr**	110 mph
Payload (full fuel)	N/A	**V_{S1} (stall clean)**	45 mph
Fuel Capacity	15 gal.	**V_{S0} (lndg config)**	42 mph
Seats	2	**Climb Rate @msl**	950 fpm
Cabin Width	40 in.	**Service Ceiling**	12,500 ft.

Std. Engine/HP	Usable HP Range	Construction Mtrl.	Est. Bldg. Time
Rotax 582 / 65 hp	N/A	Tube, Fabric	650 hrs.
Plans	**Kit**	**Info Packet**	**Video**
$395.00	$12,940.00	$13.00	$16.00

Kit includes VFR instruments. Specifications are for "Speedwing" version. Interchangeable "STOL Wing" version also available. The two types of wings are readily interchangeable. See Ad on Page 27.

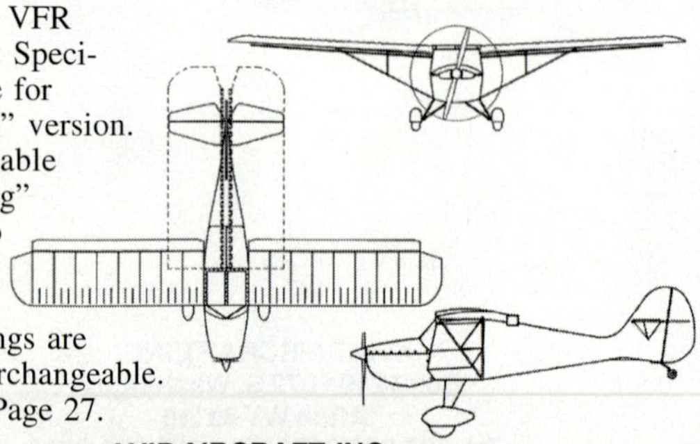

AVID AIRCRAFT, INC.
P.O. Box 728 • 4823 Aviation Way
Caldwell ID 83606
Tel. 208-454-2600 • Fax 208-454-8608
E-Mail: SALES@avidair.com
Web: http://www.avid.com

Mark IV STOL

Wing Span	23.96 ft.	**Range**	350 sm
Wing Area	97.31 sq. ft.	**Takeoff Dist.**	125 ft.
Length	17.85 ft.	**Landing Dist.**	125 ft.
Max Gross Wt.	1150 lbs.	**Vmax**	135 mph
Empty Wt.	510 lbs.	**VCr**	95 mph
Payload (full fuel)	N/A	V_{S1} **(stall clean)**	39 mph
Fuel Capacity	15 gal.	V_{S0} **(Indg config)**	36 mph
Seats	2	**Climb Rate @msl**	1000 fpm
Cabin Width	40 in.	**Service Ceiling**	12,500 ft.

Std. Engine/HP	Usable HP Range	Construction Mtrl.	Est. Bldg. Time
Rotax 582 / 65 hp	N/A	Tube, Fabric	650 hrs.
Plans	**Kit**	**Info Packet**	**Video**
$395.00	$12,940.00	$13.00	$16.00

Kit includes VFR instruments. Performance based on solo flight. Specifications are for "STOL Wing" version. "Speedwing" version also available. The two types of wings are readily interchangeable. Some owners use both. Wings fold in less than five minutes without disturbing control system. See Ad on Page 27. **Sold:** 425 **Completed & Flying:** 200

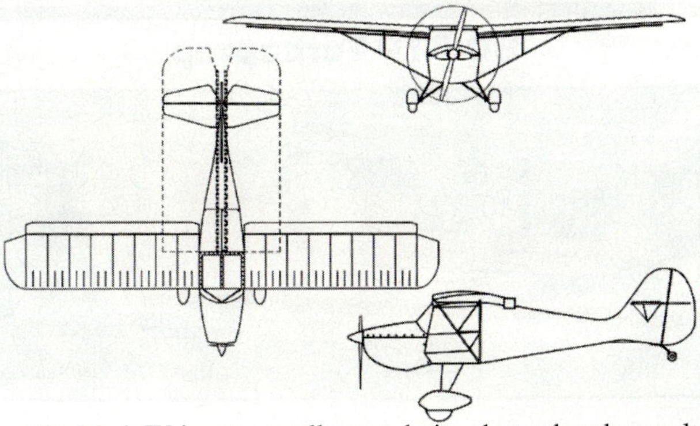

The Mark IV is not a totally new design, but rather the result of what Avid customers were doing in the field and suggestions made by them to the Avid management. It is a fourth-generation Avid Flyer with 12 additional features plus nine items added as standard features.

The most noticeable difference from earlier Flyer models is the engine cowl which has been completely redesigned with radiators moved to the very front of the cowling, and a larger spinner added. The landing gear is four inches longer to accommodate a larger propeller and the baggage compartment is now accessible from the outside.

David Schaffer, Branford, CT

"Building the Mark IV has been a truly marvelous experience from start to finish. Factory support was great and extra parts, if needed, were shipped promptly. The aircraft performs as advertised and I have enjoyed flying on warm summer days with the doors open."

AVID AIRCRAFT, INC.
P.O. Box 728 • 4823 Aviation Way
Caldwell ID 83606
Tel. 208-454-2600 • Fax 208-454-8608
E-Mail: SALES@avidair.com
Web: http://www.avid.com

Magnum

Wing Span	32.00 ft.	**Range**	500 sm
Wing Area	136.0 sq. ft.	**Takeoff Dist.**	150 ft.
Length	21.00 ft.	**Landing Dist.**	200 ft.
Max Gross Wt.	1750 lbs.	**Vmax**	150 mph
Empty Wt.	950 lbs.	**VCr**	130 mph
Payload (full fuel)	632 lbs.	V_{S1} **(stall clean)**	45 mph
Fuel Capacity	28 gal.	V_{S0} **(Indg config)**	36 mph
Seats	2	**Climb Rate @msl**	1800 fpm
Cabin Width	45 in.	**Service Ceiling**	17,500 ft.

Std. Engine/HP	Usable HP Range	Construction Mtrl.	Est. Bldg. Time
Lyc 0-320 / 160 hp	120 - 180	Tube, Fabric	800 hrs.
Plans	**Kit**	**Info Packet**	**Video**
N/A	$19,995.00	$13.00	$16.00

Price does not include engine. Performance based on a Lycoming 0-320 engine and solo flight. Options include 10 gallon add-on fuel tank, jump seat, and side baggage door. Tri-gear is now also available. Wings fold in less than five minutes without disturbing control system. See Ad on Page 27.

Kits/Plans Sold: 98 **Completed & Flying:** 17

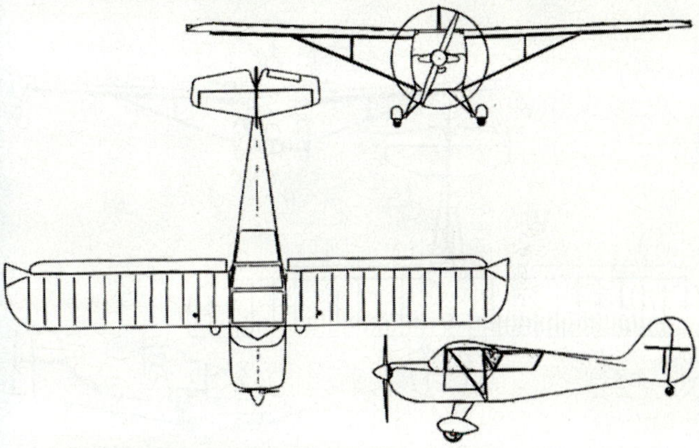

AVID AIRCRAFT, INC.
P.O. Box 728 • 4823 Aviation Way
Caldwell ID 83606
Tel. 208-454-2600 • Fax 208-454-8608
E-Mail: SALES@avidair.com
Web: http://www.avid.com

Catalina

Wing Span	36.00 ft.	**Range**	364 sm
Wing Area	150.0 sq. ft.	**Takeoff Dist.**	180 ft.
Length	19.40 ft.	**Landing Dist.**	300 ft.
Max Gross Wt.	1200 lbs.	**Vmax**	90 mph
Empty Wt.	600 lbs.	**VCr**	75 mph
Payload (full fuel)	N/A	**V_{S1} (stall clean)**	32 mph
Fuel Capacity	17 gal.	**V_{S0} (lndg config)**	N/A
Seats	3	**Climb Rate @msl**	1000 fpm
Cabin Width	N/A	**Service Ceiling**	12,500 ft.

Std. Engine/HP	Usable HP Range	Construction Mtrl.	Est. Bldg. Time
Rotax 582 / 65 hp	N/A	Tube, Fabric, Fiberglass	700 hrs.
Plans	**Kit**	**Info Packet**	**Video**
N/A	$16,030.00	$13.00	$16.00

A three place amphibian. Kit includes VFR instruments. Performance based on solo, water operations. 17.5 gallon fuel tanks are standard. 31.5 and 45.5 gallon tanks are optional. Wings fold in less than five minutes without disturbing control system. See Ad on Page 27.

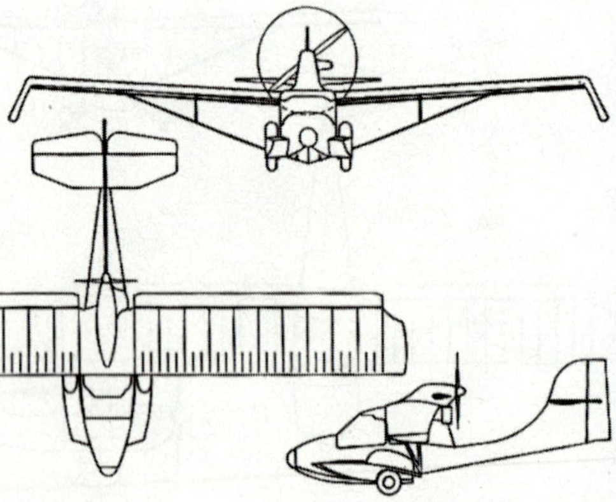

Rod Snider, Boise, ID

"I completed the land version of the Avid Amphibian in January of 1987. Since then I have flown this remarkable little plane for 1960 hours. It is really a superb plane for sightseeing and flying slow and camping in back country airstrips. Three people of medium build can be carried comfortably, as can two people and all necessary camping gear with full fuel. ..."

AVID AIRCRAFT, INC.
P.O. Box 728 • 4823 Aviation Way
Caldwell ID 83606
Tel. 208-454-2600 • Fax 208-454-8608
E-Mail: SALES@avidair.com
Web: http://www.avid.com

Jodel D 18 D 19

Wing Span	24.20 ft.	**Range**	525 sm
Wing Area	105.8 sq. ft.	**Takeoff Dist.**	833 ft.
Length	N/A	**Landing Dist.**	1477 ft.
Max Gross Wt.	1014 lbs.	**Vmax**	155 mph
Empty Wt.	550 lbs.	**VCr**	105 mph
Payload (full fuel)	N/A	**V_{S1} (stall clean)**	45 mph
Fuel Capacity	17 gal.	**V_{S0} (lndg config)**	N/A
Seats	N/A	**Climb Rate @msl**	590 fpm
Cabin Width	N/A	**Service Ceiling**	N/A

Std. Engine/HP	Usable HP Range	Construction Mtrl.	Est. Bldg. Time
VW 1600 / 58 hp	N/A	Wood, Fabric	N/A
Plans	**Kit**	**Info Packet**	**Video**
Yes	Yes	Yes	N/A

Simple construction, easy to fly, and good performance. All-wood with fabric covered wing.

AVIONS JODEL
37, route de Serres
F-21200 Beanue FRANCE
Tel. 80 22 01 51

BD-5TP Turboprop

Wing Span	19.00 ft.	**Range**	928 sm
Wing Area	41.80 sq. ft.	**Takeoff Dist.**	550 ft.
Length	14.68 ft.	**Landing Dist.**	750 ft.
Max Gross Wt.	910.0 lbs.	**Vmax**	240 mph
Empty Wt.	452 lbs.	**VCr**	225 mph
Payload (full fuel)	N/A	**V_{S1} (stall clean)**	63 mph
Fuel Capacity	38 gal.	**V_{S0} (lndg config)**	N/A
Seats	1	**Climb Rate @msl**	2200 fpm
Cabin Width	N/A	**Service Ceiling**	25,000 ft.

Std. Engine/HP	Usable HP Range	Construction Mtrl.	Est. Bldg. Time
Turboprop / 95 hp	N/A	All-Metal	800 hrs.
Plans	**Kit**	**Info Packet**	**Video**
N/A	$39,500.00	$20.00	$20.00

The turboprop version of the BD 5 using an economical turboprop engine which was designed for this airplane. The airframe and wings have been redesigned to take advantage of the higher cruising speed. Engine and prop are included in kit price.

BD MICRO TECHNOLOGIES, INC.
1260 Wade Road
Siletz OR 97380
Tel. 541-444-1343

BD-5B

Wing Span	21.50 ft.	**Range**	720 sm
Wing Area	47.40 sq. ft.	**Takeoff Dist.**	650 ft.
Length	13.99 ft.	**Landing Dist.**	700 ft.
Max Gross Wt.	830 lbs.	**Vmax**	210 mph
Empty Wt.	450 lbs.	**VCr**	180 mph
Payload (full fuel)	N/A	**V_{S1} (stall clean)**	59 mph
Fuel Capacity	26 gal.	**V_{S0} (lndg config)**	N/A
Seats	1	**Climb Rate @msl**	1900 fpm
Cabin Width	N/A	**Service Ceiling**	N/A

Std. Engine/HP	Usable HP Range	Construction Mtrl.	Est. Bldg. Time
Hirth 2706 / 68 hp	68 - 100	All-Metal	800 hrs.
Plans	**Kit**	**Info Packet**	**Video**
$450.00	$16,800.00	$20.00	$25.00

A high performance, single place, low wing, all-metal pusher sport design. Welded, machined, difficult-to-make parts are supplied. Easy-to-follow instructions. Excellent stability and high maneuverability. Kit price does not include engine.

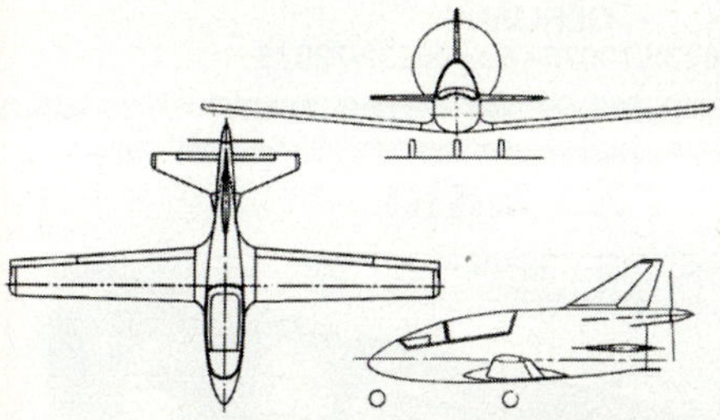

Completed: 160

BD MICRO TECHNOLOGIES, INC.
1260 Wade Road
Siletz OR 97380
Tel. 541-444-1343

BD-5J Microjet

Wing Span	17.00 ft.	**Range**	380 sm
Wing Area	37.80 sq. ft.	**Takeoff Dist.**	1800 ft.
Length	13.26 ft.	**Landing Dist.**	1000 ft.
Max Gross Wt.	860 lbs.	**Vmax**	290 mph
Empty Wt.	432 lbs.	**VCr**	240 mph
Payload (full fuel)	N/A	**V_{S1} (stall clean)**	67 mph
Fuel Capacity	36 gal.	**V_{S0} (lndg config)**	N/A
Seats	1	**Climb Rate @msl**	2400 fpm
Cabin Width	N/A	**Service Ceiling**	26,000 ft.

Std. Engine/HP	Usable HP Range	Construction Mtrl.	Est. Bldg. Time
Turbojet	N/A	All-Metal	600 hrs.
Plans	**Kit**	**Info Packet**	**Video**
N/A	$68,500.00	$20.00	$20.00

Similar in appearance to the BD-5B but powered by a turbojet engine generating 258 lbs. of thrust. Completely redesigned for the higher speeds and stresses generated by the turbojet engine. A popular airshow performer seen by thousands of spectators.

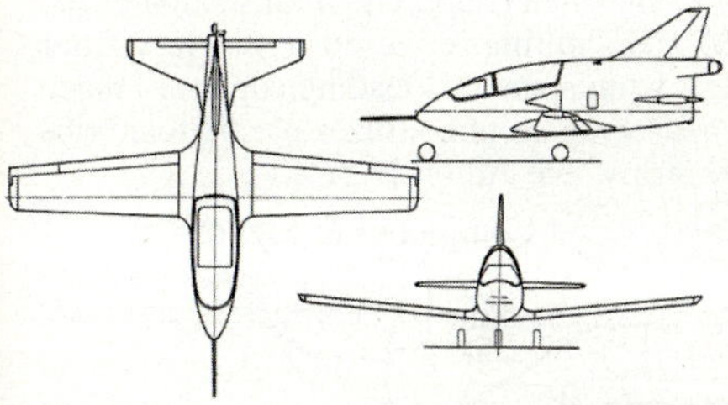

Completed: 12

BD MICRO TECHNOLOGIES, INC.
1260 Wade Road
Siletz OR 97380
Tel. 541-444-1343

BD-4 / BD-6

Wing Span	25.00 ft.	**Range**	750 sm
Wing Area	102.0 sq. ft.	**Takeoff Dist.**	450 ft.
Length	21.50 ft.	**Landing Dist.**	550 ft.
Max Gross Wt.	2500 lbs.	**Vmax**	205 mph
Empty Wt.	1150 lbs.	**VCr**	190 mph
Payload (full fuel)	N/A	**V_{S1} (stall clean)**	56 mph
Fuel Capacity	50 gal.	**V_{S0} (lndg config)**	N/A
Seats	4	**Climb Rate @msl**	1800 fpm
Cabin Width	N/A	**Service Ceiling**	22,000 ft.

Std. Engine/HP	Usable HP Range	Construction Mtrl.	Est. Bldg. Time
Lyc IO-360 / 180 hp	N/A	Metal	1600 hrs.
Plans	**Kit**	**Info Packet**	**Video**
$225.00	$18,500.00	$10.00	N/A

The BD-4 is a time proven, four place, all-metal design. Still going strong after more than 25 years. 50+ BD-4's flying. The single place BD-6 of similar design is also available.

Completed & Flying: 50

BD-4 AIRCRAFT
568 Larkspur Ct.
Medina OH 44256
Tel. 330-725-1382 • Fax 330-239-2739

FK 9

Wing Span	33.30 ft.	**Range**	340 sm
Wing Area	127.50 sq. ft.	**Takeoff Dist.**	300 ft.
Length	19.50 ft.	**Landing Dist.**	450 ft.
Max Gross Wt.	940 lbs.	**Vmax**	113 mph
Empty Wt.	357 lbs.	**Vcr**	104 mph
Payload (full fuel)	N/A lbs.	**V_{S1} (stall clean)**	30 mph
Fuel Capacity	10.6 gal.	**V_{S0} (lndg config)**	N/A mph
Seats	2	**Climb Rate @msl**	980 fpm
Cabin Width	42.5 in.	**Serv. Ceiling**	N/A ft.

Std. Engine/HP	Usable HP Range	Construction Mtrl.	Est. Bldg. Time
Rotax 503 / 46 hp	40 - 80	Steel, Alum., Carbon Fiber	200 hrs.
Plans	**Kit**	**Info Packet**	**Video**
No	$20,000	$5.00	No

Ultralight (European) taildragger test-flown and certified in Germany. Price does not include engine, instruments or seats. Design includes two 22 liter fuel tanks. The front section of the fuselage is made of steel tubing, covered with glass-fiber reinforced plastic. The wing spars and leading edges are made of carbon fiber. The rear wing section utilizes plastic foam ribs with pinewood cover strips. See Ad on Page 8.

Kits/Plans Sold: 40 **Completed & Flying:** 35

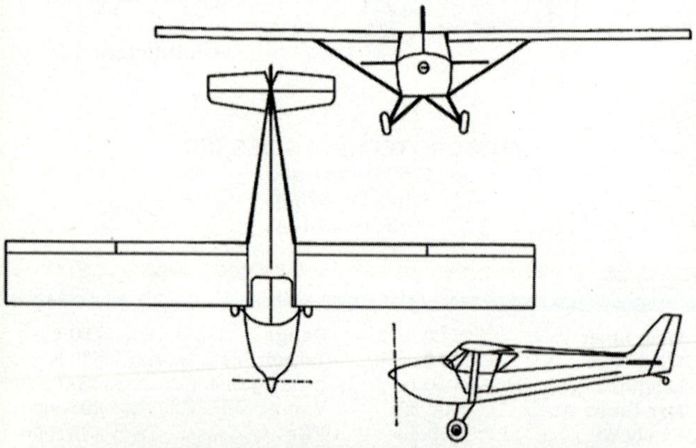

B&F TECHNIK V. GMBH.
Am Neuen Rheinhafen 10 • Speyer 67346
GERMANY
Tel. 06232/72076 • Fax 06232/72078

FK 9 - TG

Wing Span	33.30 ft.	**Range**	340 sm
Wing Area	127.5 sq. ft.	**Takeoff Dist.**	300 ft.
Length	19.50 ft.	**Landing Dist.**	450 ft.
Max Gross Wt.	940 lbs.	**Vmax**	113 mph
Empty Wt.	375 lbs.	**VCr**	104 mph
Payload (full fuel)	N/A	**V_{S1} (stall clean)**	43 mph
Fuel Capacity	11 gal.	**V_{S0} (lndg config)**	40 mph
Seats	2	**Climb Rate @msl**	N/A
Cabin Width	43 in.	**Service Ceiling**	N/A

Std. Engine/HP	Usable HP Range	Construction Mtrl.	Est. Bldg. Time
Rotax 503 / 46 hp	40 - 80	Alum., Steel, Carbon Fiber	200 hrs.
Plans	**Kit**	**Info Packet**	**Video**
N/A	$21,034.00	$5.00	N/A

Tricycle gear version of the FK 9. Ultralight (European) test-flown, manufactured and certified in Germany. See FK 9 listing for additional info. See Ad on Page 8.

Kits/Plans Sold: 35 **Completed & Flying:** 27

B&F TECHNIK V. GMBH.
Am Neuen Rheinhafen 10 • Speyer 67346
GERMANY
Tel. 06232/72076 • Fax 06232/72078

FK9 MKIII

Wing Span	32.10 ft.	**Range**	390 sm
Wing Area	38.00 sq. ft.	**Takeoff Dist.**	330 ft.
Length	19.00 ft.	**Landing Dist.**	450 ft.
Max Gross Wt.	990 lbs.	**Vmax**	150 mph
Empty Wt.	583 lbs.	**VCr**	116 mph
Payload (full fuel)	352 lbs.	**V_{S1} (stall clean)**	44 mph
Fuel Capacity	12 gal.	**V_{S0} (lndg config)**	40 mph
Seats	2	**Climb Rate @msl**	N/A
Cabin Width	43 in.	**Service Ceiling**	N/A

Std. Engine/HP	Usable HP Range	Construction Mtrl.	Est. Bldg. Time
Rotax 912 / 80 hp	65-120	Carbon Composite	200 hrs.
Plans	**Kit**	**Info Packet**	**Video**
N/A	Yes	$5.00	N/A

High performance version of the FK 9. Full composite airframe with welded steel truss in cockpit section. Fowler type flaps enable higher performance and a lower stall speed than FK 9. Folding wings standard. Certified as an ultralight in Germany. See Ad on Page 8.

B&F TECHNIK V. GMBH.
Am Neuen Rheinhafen 10 • Speyer 67346
GERMANY
Tel. 06232/72076 • Fax 06232/72078

FK 12 Comet

Wing Span	22.30 ft.	**Range**	295 sm
Wing Area	148.6 sq. ft.	**Takeoff Dist.**	350 ft.
Length	17.90 ft.	**Landing Dist.**	450 ft.
Max Gross Wt.	990 lbs.	**Vmax**	138 mph
Empty Wt.	550 lbs.	**VCr**	94 mph
Payload (full fuel)	374 lbs.	**V_{S1} (stall clean)**	43 mph
Fuel Capacity	12 gal.	**V_{S0} (Indg config)**	40 mph
Seats	2	**Climb Rate @msl**	1000 fpm
Cabin Width	28 in.	**Service Ceiling**	N/A

Std. Engine/HP	Usable HP Range	Construction Mtrl.	Est. Bldg. Time
Rotax 912 / 80 hp	70-120	Alum, Steel, Comp, Fabric	350 hrs.

Plans	Kit	Info Packet	Video
N/A	$22,000.00	$3.00	N/A

Fuselage and tail are tube structure covered with composite parts and fabric. Wings are composite with fabric covered ailerons. The wings can be folded by one person in three minutes. See Ad on Page 8.

Kits/Plans Sold: 5 **Completed & Flying:** 1

B&F TECHNIK V. GMBH.
Am Neuen Rheinhafen 10
Speyer 67346
GERMANY
Tel. 06232/72076 • Fax 06232/72078

Cherry BX-2

Wing Span	23.00 ft.	**Range**	650 sm
Wing Area	92.00 sq. ft.	**Takeoff Dist.**	940 ft.
Length	17.50 ft.	**Landing Dist.**	600 ft.
Max Gross Wt.	1211 lbs.	**Vmax**	163 mph
Empty Wt.	705 lbs.	**VCr**	128 mph
Payload (full fuel)	374 lbs.	**V_{S1} (stall clean)**	50 mph
Fuel Capacity	22 gal.	**V_{S0} (Indg config)**	45 mph
Seats	2	**Climb Rate @msl**	800 fpm
Cabin Width	40 in.	**Service Ceiling**	12,000 ft.

Std. Engine/HP	Usable HP Range	Construction Mtrl.	Est. Bldg. Time
Rotax / 80 hp	65 - 100	Composite, Wood	3000 hrs.

Plans	Kit	Info Packet	Video
$660.00	N/A	$10.00	N/A

Low wing, two seater with good flight performance. Features economy and low noise. Easy handling with folding and loading to trailer in 10 minutes. Manually operated retractable gear. Recommended engine is the Rotax 912, 80 hp.

Kits/Plans Sold: 190 **Completed & Flying:** 30

G. Dubener, Switzerland
"The Cherry BX-2 is one of the well designed, amateur-built airplanes. The good outside shape gives it an economic flight speed with a low powered engine as well as good performance."

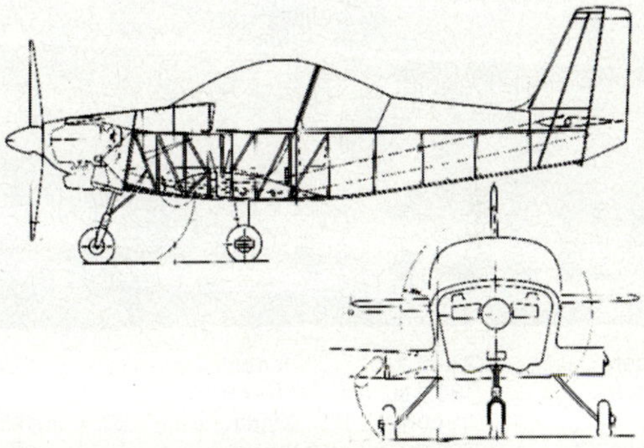

BX-AVIATION
Hoeheweg 2 • Safnern CH 2553
SWITZERLAND
Tel. 032/551823 • Fax 032/551823

Bagalini

Wing Span	35.00 ft.	Range	100 sm
Wing Area	153.0 sq. ft.	Takeoff Dist.	250 ft.
Length	21.00 ft.	Landing Dist.	250 ft.
Max Gross Wt.	450 lbs.	Vmax	60 mph
Empty Wt.	254 lbs.	VCr	40 mph
Payload (full fuel)	N/A	V_{S1} (stall clean)	27 mph
Fuel Capacity	3 gal.	V_{S0} (lndg config)	N/A
Seats	1	Climb Rate @msl	1200 fpm
Cabin Width	N/A	Service Ceiling	N/A

Std. Engine/HP	Usable HP Range	Construction Mtrl.	Est. Bldg. Time
Rotax 447 / 40 hp	N/A	Wood, Metal	700 hrs.
Plans	**Kit**	**Info Packet**	**Video**
$170.00	N/A	$5.00	N/A

Available with optional 35 hp Rotax 377 engine. Wood and fabric wings with wood and metal frame. Load factor of +7, -3.5 G's.

BAGALINI
via delle Rose 6 • Milan 20147
ITALY
Tel. 39 2 4121784

Baganfibio

Wing Span	35.00 ft.	Range	130 sm
Wing Area	167.0 sq. ft.	Takeoff Dist.	262 ft.
Length	22.00 ft.	Landing Dist.	328 ft.
Max Gross Wt.	759 lbs.	Vmax	62 mph
Empty Wt.	385 lbs.	VCr	43 mph
Payload (full fuel)	N/A	V_{S1} (stall clean)	28 mph
Fuel Capacity	7 gal.	V_{S0} (lndg config)	N/A
Seats	2	Climb Rate @msl	600 fpm
Cabin Width	N/A	Service Ceiling	N/A

Std. Engine/HP	Usable HP Range	Construction Mtrl.	Est. Bldg. Time
Rotax 503 / 48 hp	N/A	All Wood	700 hrs.
Plans	**Kit**	**Info Packet**	**Video**
$260.00	N/A	$5.00	N/A

All-wood, flying boat. Tandem seating. Optional Rotax 447 40 hp engine.

BAGALINI
via delle Rose 6 • Milan 20147
ITALY
Tel. 39 2 4121784

Colombo

Wing Span	34.40 ft.	Range	120 sm
Wing Area	180.0 sq. ft.	Takeoff Dist.	165 ft.
Length	21.00 ft.	Landing Dist.	165 ft.
Max Gross Wt.	700 lbs.	Vmax	62 mph
Empty Wt.	330 lbs.	VCr	41 mph
Payload (full fuel)	N/A	V_{S1} (stall clean)	26 mph
Fuel Capacity	6 gal.	V_{S0} (lndg config)	N/A
Seats	2	Climb Rate @msl	800 fpm
Cabin Width	N/A	Service Ceiling	N/A

Std. Engine/HP	Usable HP Range	Construction Mtrl.	Est. Bldg. Time
Rotax 447 / 40 hp	N/A	Wood, Metal	700 hrs.
Plans	**Kit**	**Info Packet**	**Video**
$220.00	N/A	$5.00	N/A

Fixed landing gear with tricycle or taildragger options.

BAGALINI
via delle Rose 6 • Milan 20147
ITALY
Tel. 39 2 4121784

Souricette Mono / Two Place

Wing Span	26.50 ft.	Range	250 sm
Wing Area	98.50 sq. ft.	Takeoff Dist.	490 ft.
Length	16.40 ft.	Landing Dist.	N/A
Max Gross Wt.	440 lbs.	Vmax	76 mph
Empty Wt.	220 lbs.	VCr	65 mph
Payload (full fuel)	190 lbs.	V_{S1} (stall clean)	31 mph
Fuel Capacity	10 gal.	V_{S0} (lndg config)	N/A
Seats	1	Climb Rate @msl	350 fpm
Cabin Width	22 in.	Service Ceiling	13,000 ft.

Std. Engine/HP	Usable HP Range	Construction Mtrl.	Est. Bldg. Time
JPX / 15 hp	12 - 25	Wood	450 hrs.
Plans	**Kit**	**Info Packet**	**Video**
$200.00	N/A	N/A	N/A

Single place version of the Souricette. A docile, charming and economical all-wood, mid wing design. Powered by a JPX 15 hp engine. Also available in a two-place version.

Kits/Plans Sold: 110 **Completed & Flying:** 48

MICHEL BARRY
171, av.du General Leclerc • F-91190 Gif/Yvette
FRANCE
Tel. 46 45 77 72

E-Z Flyer

Wing Span	26.00 ft.	**Range**	120 sm
Wing Area	157.0 sq. ft.	**Takeoff Dist.**	150 ft.
Length	21.00 ft.	**Landing Dist.**	300 ft.
Max Gross Wt.	1200 lbs.	**Vmax**	100 mph
Empty Wt.	469 lbs.	**VCr**	N/A
Payload (full fuel)	731 lbs.	**V_{S1} (stall clean)**	33 mph
Fuel Capacity	9 gal.	**V_{S0} (Indg config)**	N/A
Seats	2	**Climb Rate @msl**	500 fpm
Cabin Width	24 in.	**Service Ceiling**	10,000 ft.

Std. Engine/HP	Usable HP Range	Construction Mtrl.	Est. Bldg. Time
Rotax 582 / 65 hp	52 - 100	Steel, Alum.	250 hrs.
Plans	**Kit**	**Info Packet**	**Video**
N/A	$13,995.00	$5.00	$10.00

Two seat tandem, open cockpit design. Steel and aluminum construction. Uses the same wing and tail as the Merlin GT. Kit price includes Rotax 503 52 hp engine, Hegar wheels and brakes, tundra tires, and instruments. **Kits Sold:** 16 **Flying:** 8

BLUE YONDER AVIATION, INC.
Box 12, Site 9, RR 5 • Calgary, AL T2P 2G6
CANADA
Ph: 403-936-5701 • E-Mail: EZFlyer@sportflyer.com

Bounsall Super Prospector

Wing Span	29.67 ft.	**Range**	300 sm
Wing Area	120.8 sq. ft.	**Takeoff Dist.**	300 ft.
Length	19.30 ft.	**Landing Dist.**	250 ft.
Max Gross Wt.	800 lbs.	**Vmax**	100 mph
Empty Wt.	440 lbs.	**VCr**	90 mph
Payload (full fuel)	N/A	**V_{S1} (stall clean)**	36 mph
Fuel Capacity	12 gal.	**V_{S0} (Indg config)**	N/A
Seats	1	**Climb Rate @msl**	600 fpm
Cabin Width	N/A	**Service Ceiling**	12,000 ft.

Std. Engine/HP	Usable HP Range	Construction Mtrl.	Est. Bldg. Time
HAPI 1835cc / 60 hp	50-85	Tube, Wood, Fabric	500 hrs.
Plans	**Kit**	**Info Packet**	**Video**
$54.00	$7,800.00	$5.00	N/A

Rugged STOL bush plane. Excellent all-around visibility. Prefabricated fuselage availabe at extra cost. Supplied with standard instruments. **Kits/Plans Sold:** 3

BOUNSALL AIRCRAFT
PO Box 506
Mesquite NV 89024
Tel. 702-248-1175

MJ-7 P51 Mustang

Wing Span	23.10 ft.	**Range**	975 sm
Wing Area	112.0 sq. ft.	**Takeoff Dist.**	800 ft.
Length	22.00 ft.	**Landing Dist.**	1000 ft.
Max Gross Wt.	2400 lbs.	**Vmax**	300 mph
Empty Wt.	1800 lbs.	**VCr**	255 mph
Payload (full fuel)	520 lbs.	**V_{S1} (stall clean)**	69 mph
Fuel Capacity	50 gal.	**V_{S0} (Indg config)**	N/A
Seats	2	**Climb Rate @msl**	2000 fpm
Cabin Width	26 in.	**Service Ceiling**	16,500 ft.

Std. Engine/HP	Usable HP Range	Construction Mtrl.	Est. Bldg. Time
Chevy / 350 hp	N/A	Wood, Fiberglass	650 hrs.
Plans	**Kit**	**Info Packet**	**Video**
$350.00	$35,995.00	$5.00	N/A

P-51 Mustang replica designed by Marcel Jurca. Kit price does not include engine and propeller. Also uses Ford V-8, 351 powerplant. Retractable gear, detachable wing, aerobatic. Easy replacement of parts. Surface easy to sand and paint a favorite color scheme. **Completed & Flying:** 11

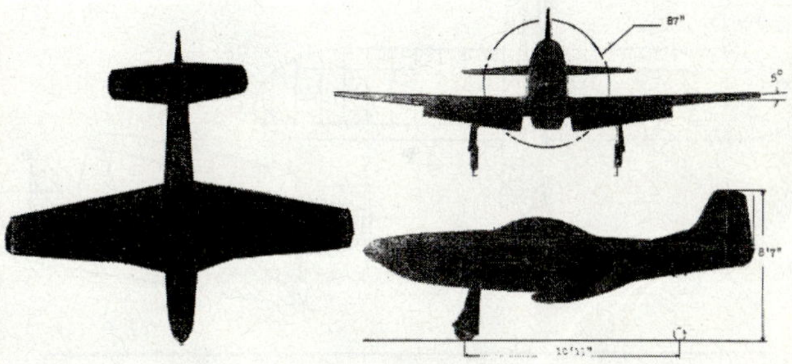

Designed by Marcel Jurca. Plans sold by Ken Heit of Flint, Michigan. Prototype first built in Canada and flown to Chicago, Illinois in 1970, powered by a Ranger Six 200 hp. In 1972 Boeve Fiberglass built a new belly scoop, radiator tunnels, wing fairings, cowling and spinner. A Ford 351 was also installed. Boeve Fiberglass made the first molded fiberglass aircraft.

BOEVE FIBERGLASS COMPONENTS
37 W. 31 Street
Holland MI 49423
Tel. 616-392-1329

Fly Baby 1A

Wing Span	28.00 ft.	**Range**	N/A
Wing Area	120.0 sq. ft.	**Takeoff Dist.**	350 ft.
Length	18.50 ft.	**Landing Dist.**	450 ft.
Max Gross Wt.	925 lbs.	**Vmax**	120 mph
Empty Wt.	605 lbs.	**VCr**	95 mph
Payload (full fuel)	N/A	**V_{S1} (stall clean)**	45 mph
Fuel Capacity	16 gal.	**V_{S0} (lndg config)**	N/A
Seats	1	**Climb Rate @msl**	1100 fpm
Cabin Width	N/A	**Service Ceiling**	15,000 ft.

Std. Engine/HP	Usable HP Range	Construction Mtrl.	Est. Bldg. Time
Continental / 100 hp	65 - 100	Wood, Fabric	1200 hrs.
Plans	**Kit**	**Info Packet**	**Video**
$65.00	N/A	$4.00	N/A

Classic, low wing, wire-braced monoplane that uses Piper J-3 Cub items: engine mounts, cowling, air scoops, gas tank, wheels and brakes.

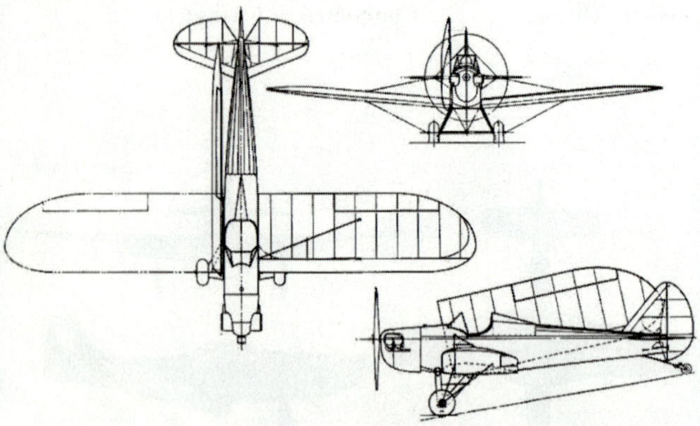

Jim Young, Carrolton TX
"I found [the Fly Baby] to be about the most honest thing I've handled on the ground."

William Burg, Sturgeon Bay WI
"Building the Fly Baby gave me so much pleasure it is hard to express it. While flying it was fun, it was the building that I really reveled in."

PETER M. BOWERS CO.
10458 16th Avenue South
Seattle WA 98168
Tel. 206-242-2582

Fly Baby 1B

Wing Span	28.00 ft.	**Range**	N/A
Wing Area	150.0 sq. ft.	**Takeoff Dist.**	350 ft.
Length	18.10 ft.	**Landing Dist.**	450 ft.
Max Gross Wt.	925 lbs.	**Vmax**	115 mph
Empty Wt.	605 lbs.	**VCr**	90 mph
Payload (full fuel)	N/A	**V_{S1} (stall clean)**	45 mph
Fuel Capacity	16 gal.	**V_{S0} (lndg config)**	N/A
Seats	1	**Climb Rate @msl**	1100 fpm
Cabin Width	N/A	**Service Ceiling**	15,000 ft.

Std. Engine/HP	Usable HP Range	Construction Mtrl.	Est. Bldg. Time
Continental / 100 hp	N/A	Wood, Fabric	1200 hrs.
Plans	**Kit**	**Info Packet**	**Video**
$80.00	N/A	$4.00	N/A

The biplane version of the Fly Baby. The biplane wing drawings are a supplement to the 1A drawings. The supplement alone is $15.00. The design uses Piper J-3 cub items: engine mount, cowling, air scoops, gas tank, wheels and brakes.

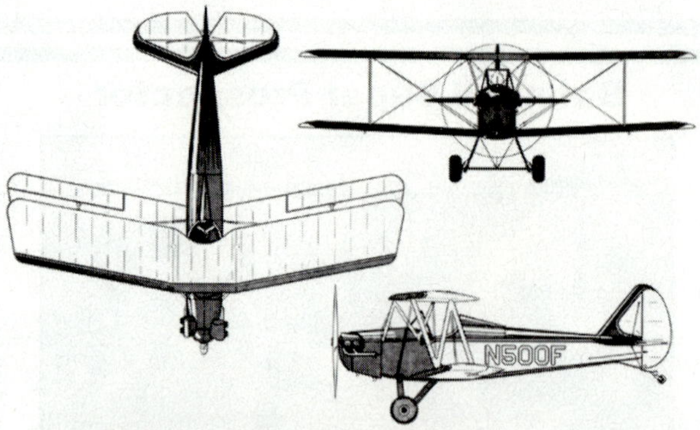

W. Gordon Standing, Nepean Canada
"...it is a joy to fly, very forgiving and is everything that you advertised."

David O. Henderson
"It is a very stable and airworthy airplane. The design of the Fly Baby is excellent and will provide any builder with many hours of enjoyment and safe flying."

PETER M. BOWERS CO.
10458 16th Avenue South
Seattle WA 98168
Tel. 206-242-2582

Bradley Aerobat BA100

Wing Span	18.40 ft.	**Range**	400 sm
Wing Area	63.00 sq. ft.	**Takeoff Dist.**	300 ft.
Length	14.50 ft.	**Landing Dist.**	400 ft.
Max Gross Wt.	620 lbs.	**Vmax**	200 mph
Empty Wt.	348 lbs.	**VCr**	150 mph
Payload (full fuel)	272 lbs.	**V_{S1} (stall clean)**	50 mph
Fuel Capacity	10 gal.	**V_{S0} (lndg config)**	N/A
Seats	1	**Climb Rate @msl**	2100 fpm
Cabin Width	22 in.	**Service Ceiling**	20,000 ft.

Std. Engine/HP	Usable HP Range	Construction Mtrl.	Est. Bldg. Time
VW Conv. 2000 / 70 hp	70 - 100	Metal	150 hrs.
Plans	**Kit**	**Info Packet**	**Video**
N/A	$14,500.00	$15.00	$25.00

Plane is fully aerobatic and stressed to +9, -9 G's. Executes aerobatic maneuvers with ease. Optional taildragger conversion available. Kits include engine, instruments, and prop. STOL option is available. Overseas information packs are $30. See Ad on Page 40.

Kits/Plans Sold: 33 **Completed & Flying:** 28

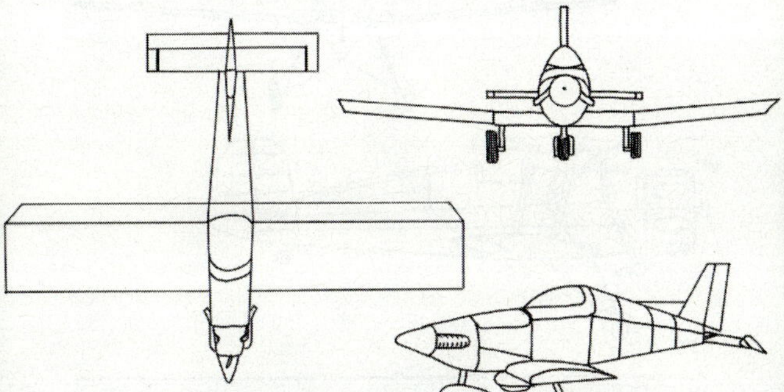

BRADLEY AEROSPACE
236A West East Avenue #260
Chico CA 95926
Tel. 916-899-7918 • Fax 916-899-2942
Web: http://www.kgmsoftworks.com/bradley

Freedom 40

Wing Span	28.00 ft.	**Range**	N/A
Wing Area	112.0 sq. ft.	**Takeoff Dist.**	N/A
Length	18.00 ft.	**Landing Dist.**	N/A
Max Gross Wt.	600 lbs.	**Vmax**	87 mph
Empty Wt.	330 lbs.	**VCr**	N/A
Payload (full fuel)	270 lbs.	**V_{S1} (stall clean)**	30 mph
Fuel Capacity	10 gal.	**V_{S0} (lndg config)**	N/A
Seats	1	**Climb Rate @msl**	1040 fpm
Cabin Width	28 in.	**Service Ceiling**	18,500 ft.

Std. Engine/HP	Usable HP Range	Construction Mtrl.	Est. Bldg. Time
Hirth 2702 / 40 hp	28 - 42	Aluminum	0 hrs.
Plans	**Kit**	**Info Packet**	**Video**
$250.00	Yes	Yes	N/A

Aluminum sheet metal monocoque ultralight. Formed metal parts can be made with simple hand tools - no compound metal parts required. Partial kit available. Materials cost for plans-builders is estimated at $3000 not including engine and instruments. See Ad on Page 64.

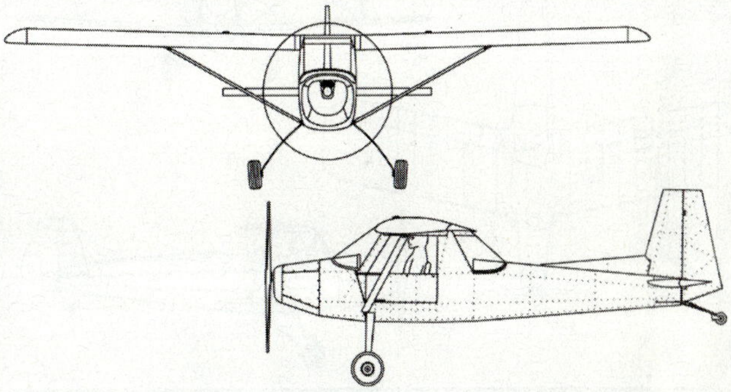

BRUTSCHE AIRCRAFT CORP.
475 East 900 S. m/s 67
Salt Lake City UT 84111
Ph: 801-355-8060 • Fax: 801-328-2060

Jungmann

Wing Span	24.27 ft.	**Range**	222 sm
Wing Area	142.5 sq. ft.	**Takeoff Dist.**	N/A
Length	22.06 ft.	**Landing Dist.**	N/A
Max Gross Wt.	1587 lbs.	**Vmax**	120 mph
Empty Wt.	993 lbs.	**VCr**	111 mph
Payload (full fuel)	N/A	**V_{S1} (stall clean)**	N/A
Fuel Capacity	23 gal.	**V_{S0} (lndg config)**	51 mph
Seats	2	**Climb Rate @msl**	N/A
Cabin Width	N/A	**Service Ceiling**	14,500 ft.

Std. Engine/HP	Usable HP Range	Construction Mtrl.	Est. Bldg. Time
Various	N/A	Tube, Wood, Fabric	N/A
Plans	**Kit**	**Info Packet**	**Video**
N/A	$34,500.00	Yes	N/A

Full scale reproduction of the classic, WWII-era, German trainer, first flown in 1934. Two place, tandem configuration. Engine, mounts, and cowlings not included in the basic kit but can be supplied.

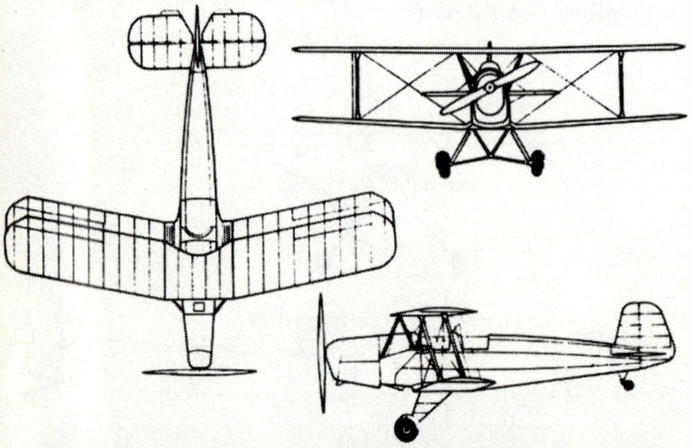

BUCKER PRADO S.L.
P.O. Box 1239
Albacete 02080
SPAIN
Ph: 34-67-508325 • Fax: 34-67-210581

Jungmeister

Wing Span	21.66 ft.	**Range**	300 sm
Wing Area	140.0 sq. ft.	**Takeoff Dist.**	N/A
Length	19.75 ft.	**Landing Dist.**	N/A
Max Gross Wt.	1342 lbs.	**Vmax**	135 mph
Empty Wt.	924 lbs.	**VCr**	123 mph
Payload (full fuel)	410 lbs.	**V_{S1} (stall clean)**	N/A
Fuel Capacity	25 gal.	**V_{S0} (lndg config)**	56 mph
Seats	1	**Climb Rate @msl**	2100 fpm
Cabin Width	N/A	**Service Ceiling**	18,000 ft.

Std. Engine/HP	Usable HP Range	Construction Mtrl.	Est. Bldg. Time
Lyc. 0-360 / 180 hp	N/A	Tube, wood, fabric	N/A
Plans	**Kit**	**Info Packet**	**Video**
N/A	$35,500.00	Yes	Yes

Famous, single seat, aerobatic biplane that first flew in 1935. Lycoming O-360 engine installation kit offered. Custom engine mount and round cowling ensure proper cooling while retaining the appearance of the original aircraft.

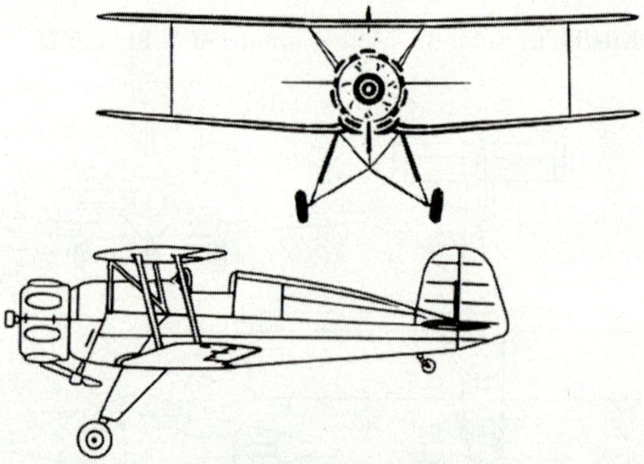

BUCKER PRADO S.L.
P.O. Box 1239
Albacete 02080
SPAIN
Ph: 34-67-508325 • Fax: 34-67-210581

U.S. Distributor:
KRYBUS AVIATION
822 Santa Maria Street • PO Box 14
Santa Paula CA 93061
Tel. 805-525-8764
Fax 805-525-1147

Barracuda

Wing Span	24.90 ft.	**Range**	800 sm
Wing Area	120.0 sq. ft.	**Takeoff Dist.**	1000 ft.
Length	21.60 ft.	**Landing Dist.**	1200 ft.
Max Gross Wt.	2300 lbs.	**Vmax**	210 mph
Empty Wt.	1500 lbs.	**VCr**	200 mph
Payload (full fuel)	600 lbs.	**V_{S1} (stall clean)**	62 mph
Fuel Capacity	52 gal.	**V_{S0} (lndg config)**	60 mph
Seats	2	**Climb Rate @msl**	2000 fpm
Cabin Width	42 in.	**Service Ceiling**	16,000 ft.

Std. Engine/HP	Usable HP Range	Construction Mtrl.	Est. Bldg. Time
Lyc IO-540 / 250 hp	200 - 300	Wood	1500 hrs.
Plans	**Kit**	**Info Packet**	**Video**
$200.00	N/A	$15.00	N/A

Responsive handling, docile stall, very comfortable cockpit facilities. Limited aerobatic capabilities. Good cross country aircraft. Although it is not offered as a kit, some components, such as the landing gear and engine mount, can be ordered. Plans include 32 sheets of drawings on 24 x 36 paper and a builders manual. See Ad on Page 17.

Kits/Plans Sold: 600 **Completed & Flying:** 30

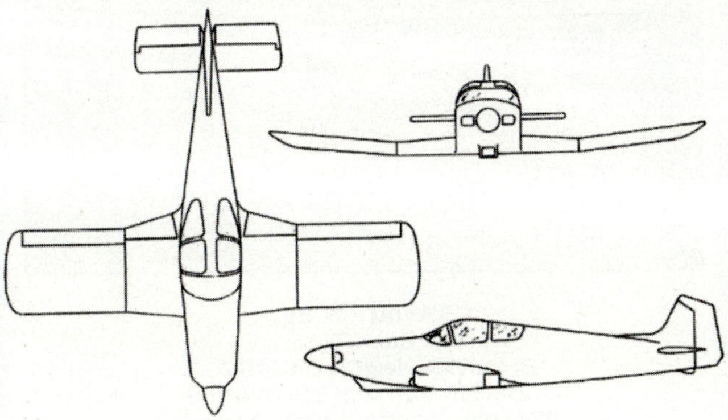

BUETHE ENTERPRISES, INC.
P.O. Box 486
Cathedral City CA 92235
Tel. 760-324-9455 • Fax 760-324-9454
E-Mail: wbbuethe@earthlink.net
Web: http://www.flybarracuda.com

Kimbrel Banty

Wing Span	32.00 ft.	**Range**	90 sm
Wing Area	128.0 sq. ft.	**Takeoff Dist.**	220 ft.
Length	18.80 ft.	**Landing Dist.**	200 ft.
Max Gross Wt.	500 lbs.	**Vmax**	60 mph
Empty Wt.	237 lbs.	**VCr**	50 mph
Payload (full fuel)	N/A	**V_{S1} (stall clean)**	25 mph
Fuel Capacity	5 gal.	**V_{S0} (lndg config)**	N/A
Seats	1	**Climb Rate @msl**	400 fpm
Cabin Width	22 in.	**Service Ceiling**	9,000 ft.

Std. Engine/HP	Usable HP Range	Construction Mtrl.	Est. Bldg. Time
Rotax 277 / 28 hp	28 - 36	Wood, Fabric	500 hrs.
Plans	**Kit**	**Info Packet**	**Video**
$25.00	N/A	$1.00	N/A

A plans only, parasol ultralight. Plans include detail drawings of all parts. A materials list and instructions for the inexperienced builder are also included.

Kits/Plans Sold: 1820 **Completed & Flying:** 30

BUTTERFLY AERO
1333 Garrard Creek Road
Oakville WA 98568
Tel. 360-273-9202 • Fax 360-273-5082

Free Spirit Mk II

Wing Span	30.30 ft.	**Range**	1200 sm
Wing Area	86.00 sq. ft.	**Takeoff Dist.**	750 ft.
Length	20.50 ft.	**Landing Dist.**	800 ft.
Max Gross Wt.	2150 lbs.	**Vmax**	300 mph
Empty Wt.	1250 lbs.	**Vcr**	240 mph
Payload (full fuel)	600 lbs.	**V_{S1} (stall clean)**	58 mph
Fuel Capacity	52.0 gal.	**V_{S0} (lndg config)**	52 mph
Seats	3	**Climb Rate @msl**	1,600 fpm
Cabin Width	44.5 in.	**Serv. Ceiling**	24,000 ft.

Std. Engine/HP	Usable HP Range	Construction Mtrl.	Est. Bldg. Time
Cont. IO-360 / 210 hp	150 - 220	Composite	N/A hrs.
Plans	**Kit**	**Info Packet**	**Video**
No	$24,950	No	No

High speed. High rate of climb. Long range. Lightweight premolded composite construction. Two seats plus a jump seat. First 10 basic kits will be sold at $24,950; thereafter kit price will be $29,680. **Completed & Flying:** 2

CEI
345 Woodside Way
Auburn CA 95603
Tel. 916-878-6867 • Fax 916-878-6967

Hawk Arrow

Wing Span	28.90 ft.	**Range**	150 sm
Wing Area	135.0 sq. ft.	**Takeoff Dist.**	150 ft.
Length	21.25 ft.	**Landing Dist.**	150 ft.
Max Gross Wt.	650 lbs.	**Vmax**	100 mph
Empty Wt.	282 lbs.	**VCr**	75 mph
Payload (full fuel)	N/A	**V_{S1} (stall clean)**	34 mph
Fuel Capacity	5 gal.	**V_{S0} (lndg config)**	N/A
Seats	1	**Climb Rate @msl**	1100 fpm
Cabin Width	N/A	**Service Ceiling**	12,000 ft.

Std. Engine/HP	Usable HP Range	Construction Mtrl.	Est. Bldg. Time
Rotax 447 / 40 hp	N/A	Aluminum, Dacron	175 hrs.
Plans	**Kit**	**Info Packet**	**Video**
N/A	$11,850.00	$10.00	$15.00

General utility aircraft. Cross country cruiser based on Hawk Classic design. Tricycle or taildragger. Standard equipment: streamlined wing struts, curved windshield, curved overhead. Price includes engine. Options include Rotax 503, or 582 and Hirth 2703 or 2706. Also, instruments, ballistic chute, brakes, floats, and electric starter. See Ad on Page 34.

Completed & Flying: 50

CGS AVIATION, INC.
P.O. Box 470635
Broadview Heights OH 44147
Tel. 216-632-1424 • Fax 216-632-1207
Web: http://www.cgsaviation.com

Hawk Classic

Wing Span	28.90 ft.	**Range**	100 sm
Wing Area	135.0 sq. ft.	**Takeoff Dist.**	125 ft.
Length	20.70 ft.	**Landing Dist.**	150 ft.
Max Gross Wt.	575 lbs.	**Vmax**	95 mph
Empty Wt.	250 lbs.	**VCr**	70 mph
Payload (full fuel)	N/A	**V_{S1} (stall clean)**	30 mph
Fuel Capacity	5 gal.	**V_{S0} (lndg config)**	N/A
Seats	1	**Climb Rate @msl**	900 fpm
Cabin Width	N/A	**Service Ceiling**	12,000 ft.

Std. Engine/HP	Usable HP Range	Construction Mtrl.	Est. Bldg. Time
Rotax 447 / 40 hp	N/A	Aluminum, Dacron	175 hrs.
Plans	**Kit**	**Info Packet**	**Video**
N/A	$8,559.00	$10.00	$15.00

General utility, single seat aircraft. Twelve years in production. Cross country cruiser, all-season design. Fully enclosed with doors. Tricycle or taildragger. Price includes engine. Engine options include Rotax 503 or 582 and Hirth 2703 and 2706. Other options are instruments, ballistic chute, brakes, floats, electric starter, and streamlined wing struts. See Ad on Page 34.

Completed & Flying: 900

CGS AVIATION, INC.
P.O. Box 470635
Broadview Heights OH 44147
Tel. 216-632-1424 • Fax 216-632-1207
Web: http://www.cgsaviation.com

HOMEBUILT "FAQ'S" *(See Page 68)*

Q: What operations are illegal in my homebuilt that might not be in a Type-Certified airplane?

A: Operations for hire are "expressly" forbidden. No paid cargo or passengers are permitted. Also, flight over densely populated areas is against the regs, but this does not seem to be enforced strongly unless the "reckless operation" clause in FAR Part 91 is called into play.
This is not as restrictive as you might think. Often clubs are formed around homebuilt aircraft, and it is legal for you to pay an instructor to give you a checkout or BFR in your airplane.
So in general, you can do most anything with your homebuilt that you can do with normal private pilot privileges in a certified aircraft.

Hawk II Classic

Wing Span	34.00 ft.	Range	180 sm
Wing Area	159.0 sq. ft.	Takeoff Dist.	200 ft.
Length	21.40 ft.	Landing Dist.	300 ft.
Max Gross Wt.	800 lbs.	Vmax	95 mph
Empty Wt.	340 lbs.	VCr	70 mph
Payload (full fuel)	N/A	V_{S1} (stall clean)	35 mph
Fuel Capacity	5 gal.	V_{S0} (lndg config)	N/A
Seats	2	Climb Rate @msl	1100 fpm
Cabin Width	N/A	Service Ceiling	12,000 ft.

Std. Engine/HP	Usable HP Range	Construction Mtrl.	Est. Bldg. Time
Rotax 503 / 52 hp	N/A	Aluminum, Dacron	200 hrs.
Plans	**Kit**	**Info Packet**	**Video**
N/A	$13,688.00	$10.00	$15.00

Proven, two place, general utility aircraft for flight training and cross country cruising, all-season design. Full enclosure with doors. Price includes engine. Options include Rotax 582, Hirth 2703, Hirth 2706, instruments, ballistic chute, brakes, floats, electric starter, streamlined wing struts and long wing. See Ad on Page 34. **Completed & Flying:** 280

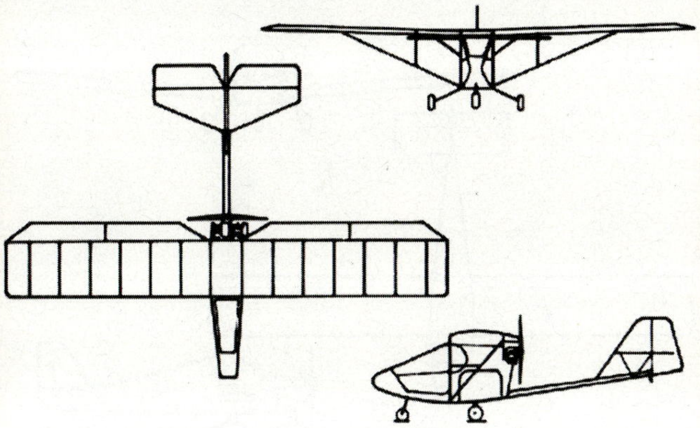

CGS AVIATION, INC.
P.O. Box 470635
Broadview Heights OH 44147
Tel. 216-632-1424 • Fax 216-632-1207
Web: http://www.cgsaviation.com

AG-Hawk

Wing Span	28.90 ft.	Range	150 sm
Wing Area	135.0 sq. ft.	Takeoff Dist.	100 ft.
Length	20.70 ft.	Landing Dist.	150 ft.
Max Gross Wt.	800 lbs.	Vmax	95 mph
Empty Wt.	275 lbs.	VCr	70 mph
Payload (full fuel)	N/A	V_{S1} (stall clean)	30 mph
Fuel Capacity	5 gal.	V_{S0} (lndg config)	N/A
Seats	1	Climb Rate @msl	950 fpm
Cabin Width	N/A	Service Ceiling	12,000 ft.

Std. Engine/HP	Usable HP Range	Construction Mtrl.	Est. Bldg. Time
Rotax 447 / 40 hp	N/A	Aluminum, Dacron	150 hrs.
Plans	**Kit**	**Info Packet**	**Video**
N/A	$12,495.00	$10.00	$15.00

General utility aircraft with reinforced wing and boom designed for agricultural uses. All season design. Options include Rotax 582 L/C, crop spray unit, instruments, parachute, brakes, floats, electric starter. See Ad on Page 34. **Completed & Flying:** 30

CGS AVIATION, INC.
P.O. Box 470635
Broadview Heights OH 44147
Tel. 216-632-1424 • Fax 216-632-1207
Web: http://www.cgsaviation.com

Hawk II Arrow

Wing Span	34.00 ft.	Range	150 sm
Wing Area	159.0 sq. ft.	Takeoff Dist.	200 ft.
Length	22.07 ft.	Landing Dist.	300 ft.
Max Gross Wt.	950 lbs.	Vmax	95 mph
Empty Wt.	395 lbs.	VCr	70 mph
Payload (full fuel)	N/A	V_{S1} (stall clean)	35 mph
Fuel Capacity	5 gal.	V_{S0} (lndg config)	N/A
Seats	2	Climb Rate @msl	750 fpm
Cabin Width	N/A	Service Ceiling	12,000 ft.

Std. Engine/HP	Usable HP Range	Construction Mtrl.	Est. Bldg. Time
Rotax 503 / 52 hp	N/A	Aluminum, Dacron	150 hrs.
Plans	**Kit**	**Info Packet**	**Video**
N/A	$14,936.00	$10.00	$15.00

General utility aircraft for flight training and cross country cruising. Tricycle or taildragger. Full enclosure with doors. Price includes engine. Options include instruments, ballistic chute, floats, long wing, and electric starter. See Ad on Page 34. **Flying:** 60+

CGS AVIATION, INC.
P.O. Box 470635
Broadview Heights OH 44147
Tel. 216-632-1424 • Fax 216-632-1207
Web: http://www.cgsaviation.com

Corby Starlet CJ-1

Wing Span	18.50 ft.	**Range**	425 sm
Wing Area	68.50 sq. ft.	**Takeoff Dist.**	420 ft.
Length	14.80 ft.	**Landing Dist.**	750 ft.
Max Gross Wt.	750 lbs.	**Vmax**	160 mph
Empty Wt.	450 lbs.	**VCr**	130 mph
Payload (full fuel)	230 lbs.	**V_{S1} (stall clean)**	46 mph
Fuel Capacity	11 gal.	**V_{S0} (lndg config)**	35 mph
Seats	N/A	**Climb Rate @msl**	1000 fpm
Cabin Width	22 in.	**Service Ceiling**	14,500 ft.

Std. Engine/HP	Usable HP Range	Construction Mtrl.	Est. Bldg. Time
VW / 75 hp	45-75	Wood	1400 hrs.
Plans	**Kit**	**Info Packet**	**Video**
$175.00	N/A	$8.00	$15.00

All-wood, ply and fabric covered, single seat, low wing, cantilever monoplane. Can use most existing VW powerplants from 45 to 75 hp. Well proven design with Type Certificate. Highly maneuverable. Can operate out of cropdusting strips and grass runways. Low cost to build and affordable. Starlet aircraft are being built and flown in 26 countries and 36 of the 50 states. Newsletter $24 US, $27 Canada, $30 Others. **Flying:** 67

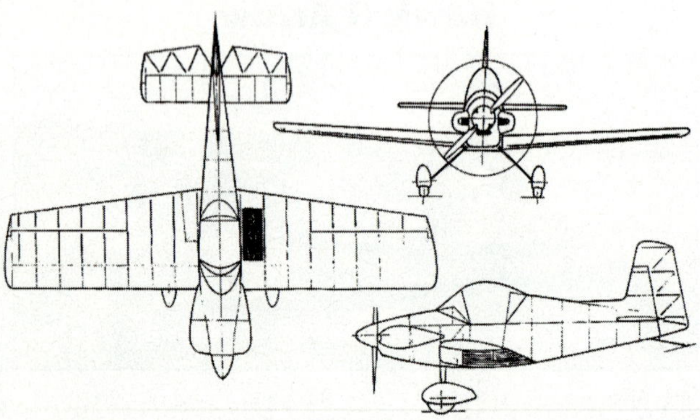

The Starlet was designed in the late sixties by John Corby, an aeronautical engineer, with assistance from Barrie Bishton who handled the engine and propeller components. While John intended it to be his own, low cost homebuilt aircraft, he had always felt that Australia should have an aircraft designed by Australians for Australians. So, he was easily persuaded to provide plans for his aircraft to other builders. Some Starlets have flown over 1000 hours while others have performed aerobatics at air shows for more than 15 years. It is considered to be a true "pilot toy."

Barry Morris, Spokane, WA

"I have researched plans/kitplanes for the last 4 years. I own a set of RV-4 plans also, but the Starlet and its wood construction and cost to build were too much to pass up."

C.S.N.
510 N.W. 46th Terrace
Plantation FL 33317-2044
Tel. 954-581-8835

Campana

Wing Span	36.40 ft.	**Range**	1118 sm
Wing Area	155.0 sq. ft.	**Takeoff Dist.**	165 ft.
Length	23.30 ft.	**Landing Dist.**	200 ft.
Max Gross Wt.	1324 lbs.	**Vmax**	132 mph
Empty Wt.	547 lbs.	**VCr**	121 mph
Payload (full fuel)	534 lbs.	**V_{S1} (stall clean)**	45 mph
Fuel Capacity	42 gal.	**V_{S0} (lndg config)**	40 mph
Seats	2	**Climb Rate @msl**	1200 fpm
Cabin Width	44 in.	**Service Ceiling**	20,000 ft.

Std. Engine/HP	Usable HP Range	Construction Mtrl.	Est. Bldg. Time
Rotax 912 / 80 hp	50 - 115	Composite	250 hrs.
Plans	**Kit**	**Info Packet**	**Video**
N/A	$26,023.00	$10.00	$15.00

A composite, two seat design available in ultralight and experimental versions. Specs are for the experimental version. The Campana 912 version uses an 80 hp Rotax 912 engine and has a gross weight of 993 lbs. Also available in the "Sport" version with a 98.7 ft. wing span and stressed to +6, -4 G's.

Kits/Plans Sold: 12 **Completed & Flying:** 7

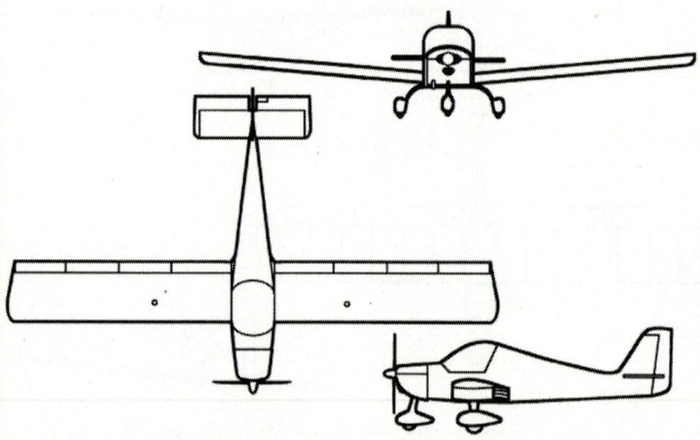

The designer, Bela Nogrady, is a French aeronautical engineer with degrees from the University of Toulouse. He began his first project, a wooden flying wing, when he was eighteen and has been flying since 1973. After fourteen years in academia and five years performing R&D for several small aeronautical companies, Bela formed his own company specializing in ultralights.

Bela designed the Campana to be easy to fly while maintaining high speeds and short field capabilities. Development began in October 1992 and the steel-tube prototype flew on May 25, 1993. The second, composite prototype flew in November 1994 and full production began in July 1995.

CAMPANA AVIATION
Aeroport de Tarbes • Pyrenees Airpole
F65000 Tarbes
FRANCE
Tel. 33-62-32-73-67 • Fax 33-62-32-73-69

ARV Griffin Mk III

Wing Span	35.50 ft.	**Range**	700 sm
Wing Area	136.0 sq. ft.	**Takeoff Dist.**	300 ft.
Length	21.00 ft.	**Landing Dist.**	450 ft.
Max Gross Wt.	1500 lbs.	**Vmax**	135 mph
Empty Wt.	840 lbs.	**VCr**	121 mph
Payload (full fuel)	438 lbs.	**V_{S1} (stall clean)**	50 mph
Fuel Capacity	37 gal.	**V_{S0} (lndg config)**	45 mph
Seats	2	**Climb Rate @msl**	1030 fpm
Cabin Width	47 in.	**Service Ceiling**	13,000 ft.

Std. Engine/HP	Usable HP Range	Construction Mtrl.	Est. Bldg. Time
Cont. 0-200 / 100 hp	65 - 150	All-Metal	800 hrs.
Plans	**Kit**	**Info Packet**	**Video**
N/A	$15,874.00	$10.00	N/A

A two place, side-by-side, high wing aircraft. Tricycle, taildragger or convertible landing gear. The all-metal structure features superior aerodynamic efficiency, 43" to 47" wide cabin, excellent visibility and adjustable, removable seats. Large baggage compartment. Wing detaches quickly for easy storage. Winglets are offered as an option which improve low-speed handling, lateral control and increases STOL characteristics. Other engine options include Subaru, CAM 100 and Rotax 912. Price does not include engine, propeller or avionics. See Ad on Page 40.

Kits/Plans Sold: 18 **Completed & Flying:** 3

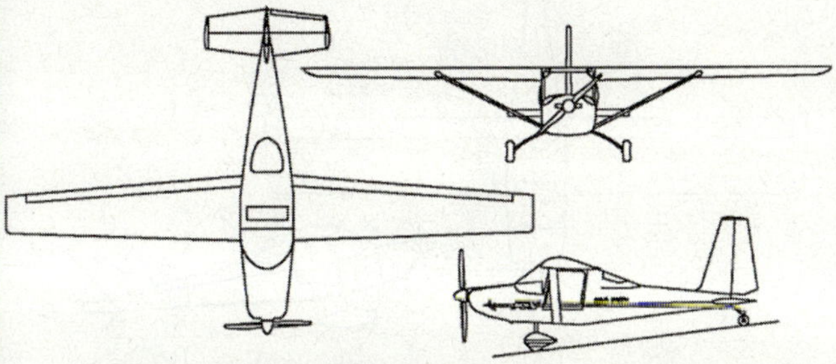

CANADA AIR R.V., INC.
PO Box 34013 • Kingsway Garden Mall PO
Edmonton, Alberta T5G 3G4
CANADA
Tel. 403-944-9210 • Fax 403-479-1002
E-Mail: sales@arvcorp.com • Web: http://www.arvcorp.com

Capella Fastback

Wing Span	28.50 ft.	**Range**	550 sm
Wing Area	114.0 sq. ft.	**Takeoff Dist.**	155 ft.
Length	18.50 ft.	**Landing Dist.**	150 ft.
Max Gross Wt.	1200 lbs.	**Vmax**	135 mph
Empty Wt.	560 lbs.	**VCr**	120 mph
Payload (full fuel)	640 lbs.	**V_{S1} (stall clean)**	41 mph
Fuel Capacity	22 gal.	**V_{S0} (lndg config)**	38 mph
Seats	2	**Climb Rate @msl**	1450 fpm
Cabin Width	43 in.	**Service Ceiling**	12,000 ft.

Std. Engine/HP	Usable HP Range	Construction Mtrl.	Est. Bldg. Time
Rotax 912 / 80 hp	N/A	Metal, Fabric	350 hrs.
Plans	**Kit**	**Info Packet**	**Video**
N/A	$16,670.00	$5.00	N/A

The newest addition to the Capella line. Equipped with a high rear turtledeck with side and upper aft window panels instead the of the standard wrap-around aft windscreen. The result is a styling reminiscent of the early Champ, Cub, or Taylorcraft designs. See Ad on Page 51.

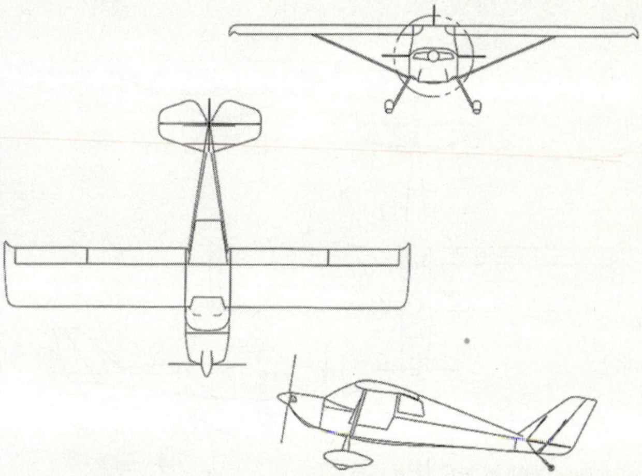

CAPELLA AIRCRAFT CORP.
4211-C Todd Lane
Austin TX 78744
Tel. 512-441-8844 • Fax 512-441-1997
E-Mail: webmaster@capellakitplanes.com
Web: http://www.capellakitplanes.com

Capella SS

Wing Span	27.50 ft.	**Range**	158 sm
Wing Area	110.0 sq. ft.	**Takeoff Dist.**	200 ft.
Length	17.33 ft.	**Landing Dist.**	200 ft.
Max Gross Wt.	625 lbs.	**Vmax**	112 mph
Empty Wt.	375 lbs.	**VCr**	102 mph
Payload (full fuel)	N/A	**V_{S1} (stall clean)**	38 mph
Fuel Capacity	10 gal.	**V_{S0} (lndg config)**	34 mph
Seats	1	**Climb Rate @msl**	980 fpm
Cabin Width	24 in.	**Service Ceiling**	10,000 ft.

Std. Engine/HP	Usable HP Range	Construction Mtrl.	Est. Bldg. Time
Rotax 503 / 50 hp	N/A	Aluminum, Steel	300 hrs.
Plans	**Kit**	**Info Packet**	**Video**
N/A	$13,285.00	$5.00	N/A

A top of the line single seater. Also available with the Rotax 447 engine. Folding wing option available. Specs are for FW1C50 airframe with optional 3:1 propeller redrive and speed fairings. Price is for taildragger airframe kit and does not include engine or instruments. See Ad on Page 51.

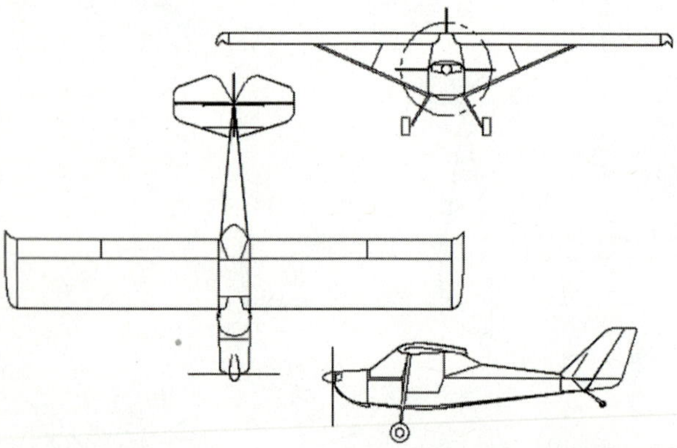

CAPELLA AIRCRAFT CORP.
4211-C Todd Lane
Austin TX 78744
Tel. 512-441-8844 • Fax 512-441-1997
E-Mail: webmaster@capellakitplanes.com
Web: http://www.capellakitplanes.com

Capella XS

Wing Span	28.50 ft.	**Range**	275 sm
Wing Area	114.0 sq. ft.	**Takeoff Dist.**	175 ft.
Length	18.41 ft.	**Landing Dist.**	250 ft.
Max Gross Wt.	1200 lbs.	**Vmax**	135 mph
Empty Wt.	480 lbs.	**VCr**	111 mph
Payload (full fuel)	720 lbs.	**V_{S1} (stall clean)**	41 mph
Fuel Capacity	11 gal.	**V_{S0} (lndg config)**	36 mph
Seats	2	**Climb Rate @msl**	1200 fpm
Cabin Width	43 in.	**Service Ceiling**	10,000 ft.

Std. Engine/HP	Usable HP Range	Construction Mtrl.	Est. Bldg. Time
Rotax 582 / 65 hp	N/A	Aluminum, Steel	350 hrs.
Plans	**Kit**	**Info Packet**	**Video**
N/A	$13,195.00	$5.00	N/A

The Rotax 582 powered Capella XS can be equipped to qualify as a two seat ultralight trainer. Folding wing and tri-gear options available. Performance specs are for solo flight. Price is for taildragger airframe kit and does not include engine or instruments. See Ad on Page 51.

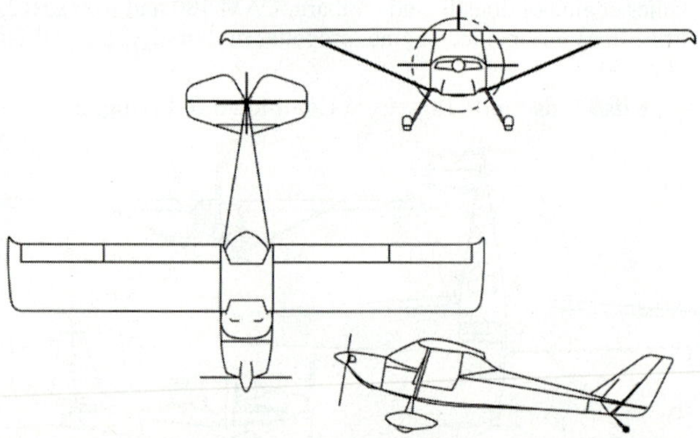

CAPELLA AIRCRAFT CORP.
4211-C Todd Lane
Austin TX 78744
Tel. 512-441-8844 • Fax 512-441-1997
E-Mail: webmaster@capellakitplanes.com
Web: http://www.capellakitplanes.com

Capella XLS

Wing Span	28.50 ft.	**Range**	550 sm
Wing Area	114.0 sq. ft.	**Takeoff Dist.**	200 ft.
Length	18.41 ft.	**Landing Dist.**	150 ft.
Max Gross Wt.	1200 lbs.	**Vmax**	135 mph
Empty Wt.	580 lbs.	**VCr**	120 mph
Payload (full fuel)	620 lbs.	**V_{S1} (stall clean)**	41 mph
Fuel Capacity	22 gal.	**V_{S0} (lndg config)**	38 mph
Seats	2	**Climb Rate @msl**	1450 fpm
Cabin Width	43 in.	**Service Ceiling**	12,000 ft.

Std. Engine/HP	Usable HP Range	Construction Mtrl.	Est. Bldg. Time
Rotax 912 / 80 hp	N/A	Aluminum, Fabric	350 hrs.
Plans	**Kit**	**Info Packet**	**Video**
N/A	$16,295.00	$5.00	N/A

Complete kit with high level of prefab. All-aluminum wing and FAR Part 23 Utility category strength and stability. Folding wing. Price is for taildragger airframe kit and does not include engine or instruments. See Ad on Page 51.

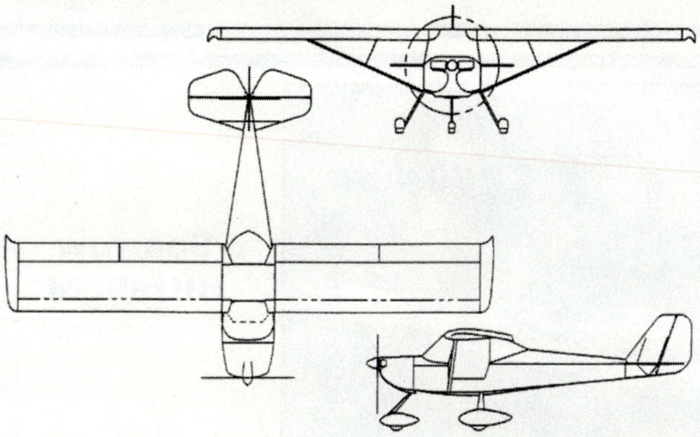

Duncan Carlton, August 1997
"After a long, hard look at the market for two seat moderate performace high wing designs, I decided on the Capella. The plane is stable, yet has a nice "tossable" feel to it that makes it a delight to fly compared to some certified aircraft such as the Cessna 150.

CAPELLA AIRCRAFT CORP.
4211-C Todd Lane
Austin TX 78744
Tel. 512-441-8844 • Fax 512-441-1997
E-Mail: webmaster@capellakitplanes.com
Web: http://www.capellakitplanes.com

Javelin I

Wing Span	28.00 ft.	**Range**	120 sm
Wing Area	112.0 sq. ft.	**Takeoff Dist.**	75 ft.
Length	18.00 ft.	**Landing Dist.**	100 ft.
Max Gross Wt.	575 lbs.	**Vmax**	85 mph
Empty Wt.	240 lbs.	**VCr**	60 mph
Payload (full fuel)	N/A	**V_{S1} (stall clean)**	26 mph
Fuel Capacity	5 gal.	**V_{S0} (lndg config)**	26 mph
Seats	1	**Climb Rate @msl**	700 fpm
Cabin Width	41 in.	**Service Ceiling**	10,000 ft.

Std. Engine/HP	Usable HP Range	Construction Mtrl.	Est. Bldg. Time
Rotax 277 / 28 hp	25 - 35	Steel, Alum., Fabric	80 hrs.
Plans	**Kit**	**Info Packet**	**Video**
N/A	$5,495.00	$5.00	N/A

A high wing aircraft of conventional configuration that's an exercise in simplicity. Low cost, lightweight, easy to assemble. Low operating cost. Price is for airframe kit and does not include engine or instruments. See Ad on Page 51.

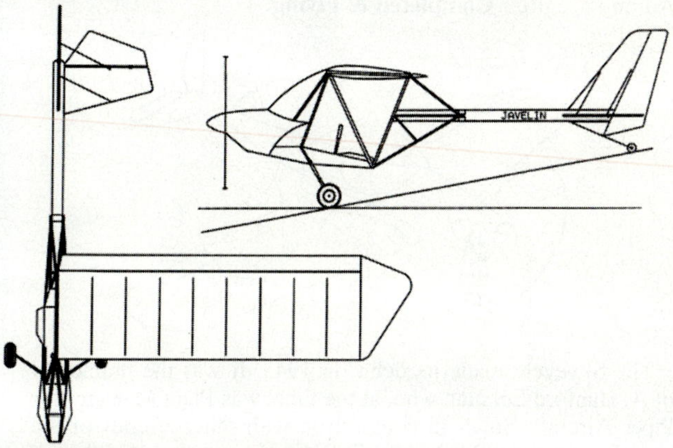

CAPELLA AIRCRAFT CORP.
4211-C Todd Lane
Austin TX 78744
Tel. 512-441-8844 • Fax 512-441-1997
E-Mail: webmaster@capellakitplanes.com
Web: http://www.capellakitplanes.com

Skycycle

Wing Span	20.00 ft.	**Range**	200 sm
Wing Area	95.00 sq. ft.	**Takeoff Dist.**	375 ft.
Length	15.90 ft.	**Landing Dist.**	N/A
Max Gross Wt.	N/A	**Vmax**	109 mph
Empty Wt.	295 lbs.	**VCr**	100 mph
Payload (full fuel)	N/A	**V_{S1} (stall clean)**	40 mph
Fuel Capacity	8 gal.	**V_{S0} (lndg config)**	37 mph
Seats	1	**Climb Rate @msl**	875 fpm
Cabin Width	29 in.	**Service Ceiling**	N/A

Std. Engine/HP	Usable HP Range	Construction Mtrl.	Est. Bldg. Time
Lycoming / 55 hp	N/A	Steel, Fabric	N/A
Plans	**Kit**	**Info Packet**	**Video**
N/A	$13,900.00	N/A	N/A

A modern redesign of the small, affordable, post-WW II aircraft designed by A. Hanford Eckman which first flew in 1945. Has a fiberglass exterior and plexiglas canopy. Tail cone uses 2024-T3 aluminum and includes formed bulkheads for strength. Kit price does not include the "factory assist" package and requires special tooling. Parts are available prefabricated. See Ad on Page 46. **Completed & Flying:** 1

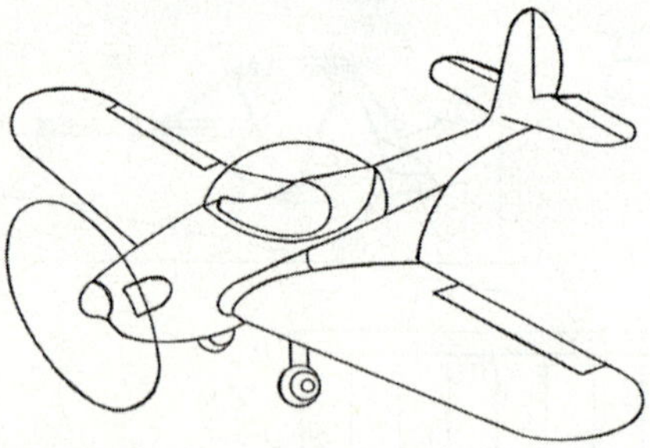

The Skycycle made its debut in 1945. It was the brainchild of A. Hanford Eckman who, at the time, was Plant Manager for Piper Aircraft. Small in comparison with other aircraft of the era, the diminutive "class act" was truly an advanced design for its day. Its performance was considered to be phenomenal.

The current Skycycle closely resembles the original Piper prototype. Design parameters for size and structure have been duplicated as nearly as possible from available information. Sketches, a few original photographs, and a three view drawing appear to be the only surviving accurate (and authentic) paper confirmation of the original design.

Design performance has not been altered. Only the materials used have been changed to include modern composites and engines designed for today's light aircraft.

After a thorough period of flight testing, the first "reproduction" prototype was donated to the Piper Museum in honor of the original design and the people who created it.

CARLSON AIRCRAFT, INC.
50643 S.R. 14, P.O. Box 88
East Palestine OH 44413-0088
Tel. 303-426-3934 • Fax 303-426-1144

Sparrow Sport Special

Wing Span	26.00 ft.	**Range**	300 sm
Wing Area	117.0 sq. ft.	**Takeoff Dist.**	85 ft.
Length	16.20 ft.	**Landing Dist.**	175 ft.
Max Gross Wt.	710 lbs.	**Vmax**	120 mph
Empty Wt.	360 lbs.	**VCr**	85 mph
Payload (full fuel)	N/A	**V_{S1} (stall clean)**	31 mph
Fuel Capacity	10 gal.	**V_{S0} (lndg config)**	N/A
Seats	1	**Climb Rate @msl**	1750 fpm
Cabin Width	N/A	**Service Ceiling**	N/A

Std. Engine/HP	Usable HP Range	Construction Mtrl.	Est. Bldg. Time
Rotax 503 / 52 hp	N/A	Aluminum, Fabric	N/A
Plans	**Kit**	**Info Packet**	**Video**
N/A	$8,995.00	Yes	N/A

A single seat, high wing taildragger with a fully enclosed cockpit that is designed for short field performance. Welded fuselage and extruded aluminum rib and spar wings. Kit price does not include engine. See Ad on Page 46. **Completed & Flying:** 9

CARLSON AIRCRAFT, INC.
50643 S.R. 14, P.O. Box 88
East Palestine OH 44413-0088
Tel. 303-426-3934 • Fax 303-426-1144

Sparrow Ultralight

Wing Span	30.20 ft.	**Range**	200 sm
Wing Area	120.0 sq. ft.	**Takeoff Dist.**	140 ft.
Length	16.75 ft.	**Landing Dist.**	195 ft.
Max Gross Wt.	504.0 lbs.	**Vmax**	63 mph
Empty Wt.	254 lbs.	**VCr**	58 mph
Payload (full fuel)	N/A	**V_{S1} (stall clean)**	26 mph
Fuel Capacity	5 gal.	**V_{S0} (lndg config)**	N/A
Seats	1	**Climb Rate @msl**	750 fpm
Cabin Width	22	**Service Ceiling**	N/A

Std. Engine/HP	Usable HP Range	Construction Mtrl.	Est. Bldg. Time
Rotax 277 / 28 hp	N/A	Steel, Fabric	N/A
Plans	**Kit**	**Info Packet**	**Video**
N/A	$7,875.00	Yes	N/A

Three-axis controlled, tricycle design. Part 103 legal. Requires minimum runway. Features include factory completed ribs, round aluminum struts, bench seat with cushion and riveted aluminum tail surfaces. Engine not included. See Ad on Page 46. **Flying:** 23

CARLSON AIRCRAFT, INC.
50643 S.R. 14, P.O. Box 88
East Palestine OH 44413-0088
Tel. 303-426-3934 • Fax 303-426-1144

Sparrow II

Wing Span	31.19 ft.	**Range**	300 sm
Wing Area	140.0 sq. ft.	**Takeoff Dist.**	100 ft.
Length	18.00 ft.	**Landing Dist.**	150 ft.
Max Gross Wt.	1050 lbs.	**Vmax**	130 mph
Empty Wt.	510 lbs.	**VCr**	95 mph
Payload (full fuel)	N/A	**V_{S1} (stall clean)**	36 mph
Fuel Capacity	10 gal.	**V_{S0} (lndg config)**	N/A
Seats	1	**Climb Rate @msl**	1750 fpm
Cabin Width	N/A	**Service Ceiling**	N/A

Std. Engine/HP	Usable HP Range	Construction Mtrl.	Est. Bldg. Time
Rotax 582 / 66 hp	N/A	Steel, Fabric	N/A
Plans	**Kit**	**Info Packet**	**Video**
N/A	$13,095.00	Yes	N/A

Two seat version of the Sparrow UL. Tricycle design with side-by-side seating. 4130 steel fuselage with aluminum rib and spar wings.Dual controls, streamline struts, independent ailerons. Engine not included. See Ad on Page 46.

Kits Sold: 32

CARLSON AIRCRAFT, INC.
50643 S.R. 14, P.O. Box 88
East Palestine OH 44413-0088
Tel. 303-426-3934 • Fax 303-426-1144

Sparrow II XTC

Wing Span	31.19 ft.	**Range**	400 sm
Wing Area	140.0 sq. ft.	**Takeoff Dist.**	150 ft.
Length	18.00 ft.	**Landing Dist.**	250 ft.
Max Gross Wt.	1250 lbs.	**Vmax**	130 mph
Empty Wt.	600 lbs.	**VCr**	110 mph
Payload (full fuel)	N/A	**V_{S1} (stall clean)**	39 mph
Fuel Capacity	22 gal.	**V_{S0} (lndg config)**	N/A
Seats	2	**Climb Rate @msl**	1100 fpm
Cabin Width	N/A	**Service Ceiling**	N/A

Std. Engine/HP	Usable HP Range	Construction Mtrl.	Est. Bldg. Time
Continental / 100 hp	N/A	Metal, Fabric	500 hrs.
Plans	**Kit**	**Info Packet**	**Video**
N/A	$14,875.00	Yes	N/A

Higher power/performance version of the Sparrow UL side-by-side seating and dual controls. 22 gallon wing tanks come standard. Other features include factory completed ribs, and streamline struts. Engine not included. See Ad on Page 46.

CARLSON AIRCRAFT, INC.
50643 S.R. 14, P.O. Box 88
East Palestine OH 44413-0088
Tel. 303-426-3934 • Fax 303-426-1144

H-2 Honey Bee

Wing Span	19.00 ft.	**Range**	120 sm
Wing Area	140.0 sq. ft.	**Takeoff Dist.**	200 ft.
Length	15.80 ft.	**Landing Dist.**	275 ft.
Max Gross Wt.	700 lbs.	**Vmax**	80 mph
Empty Wt.	495 lbs.	**VCr**	68 mph
Payload (full fuel)	205 lbs.	**V_{S1} (stall clean)**	35 mph
Fuel Capacity	12 gal.	**V_{S0} (lndg config)**	N/A
Seats	1	**Climb Rate @msl**	850 fpm
Cabin Width	21 in.	**Service Ceiling**	9,000 ft.

Std. Engine/HP	Usable HP Range	Construction Mtrl.	Est. Bldg. Time
Hirth 2706 / 65 hp	40 - 95	Aluminum, Fabric	800 hrs.
Plans	**Kit**	**Info Packet**	**Video**
$250.00	N/A	$5.00	N/A

Lightweight aerobatic biplane designed by Bert Howland who first used the unique construction technique of fabricating the fuselage frame from 3/4" square aluminum 6061-T6 tubing with TIG welding instead of gussets and rivets. The H-2 is the standard model with a single set of ailerons. The H-2A aerobatic (up to IAC Intermediate level) model utilizes the Hirth F30 95hp engine, has a double set of ailerons, inverted fuel system, and extra length on the aileron control arm. Plans consist of 40 engineering drawings in 17"x11" format and the designer's construction notes. Materials kits available from Aircraft Spruce & Specialty, 800-831-2949 and Wicks Aircraft Supply, 618-654-7447.

Kits/Plans Sold: 10 **Completed & Flying:** 5

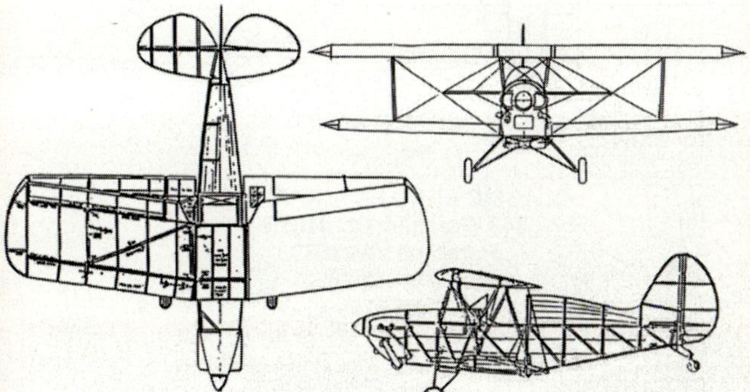

CLASSIC AERO ENTERPRISES
343 Wrexham Ct., #101B
Hampton VA 23669
Tel. 757-851-2856

H-3 Pegasus

Wing Span	25.00 ft.	**Range**	120 sm
Wing Area	110.0 sq. ft.	**Takeoff Dist.**	200 ft.
Length	15.00 ft.	**Landing Dist.**	250 ft.
Max Gross Wt.	595 lbs.	**Vmax**	95 mph
Empty Wt.	345 lbs.	**VCr**	80 mph
Payload (full fuel)	218 lbs.	**V_{S1} (stall clean)**	35 mph
Fuel Capacity	8 gal.	**V_{S0} (lndg config)**	N/A
Seats	1	**Climb Rate @msl**	700 fpm
Cabin Width	21 in.	**Service Ceiling**	9,000 ft.

Std. Engine/HP	Usable HP Range	Construction Mtrl.	Est. Bldg. Time
Hirth 2704 / 55 hp	40 - 55	Aluminum, Fabric	700 hrs.
Plans	**Kit**	**Info Packet**	**Video**
$250.00	N/A	$5.00	N/A

Low-wing, open cockpit monoplane by noted lightweight sportplane designer Bert Howland. Utilizes similar construction techniques to the H-2 Honeybee. Cantilevered wing design eliminates struts and cables. Weight of the welded aluminum fuselage is 18 lbs. Design load is +5, -3 G's. Plans only available Classic Aero Enterprises. Materials kits are available from Aircraft Spruce & Specialty, 800-831-2949 and Wicks Aircraft Supply, 618-654-7447.

Completed & Flying: 20

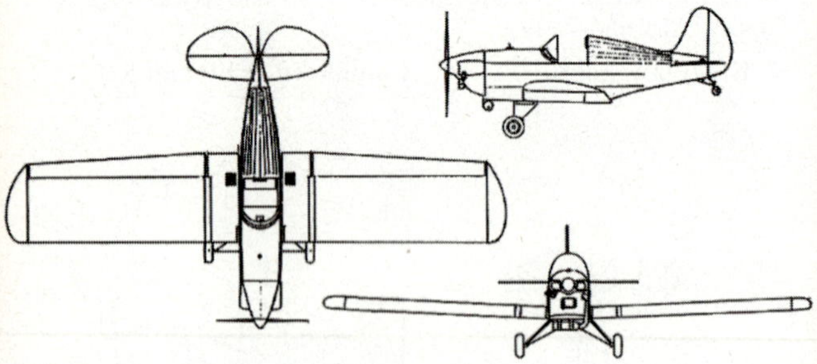

CLASSIC AERO ENTERPRISES
343 Wrexham Ct., #101B
Hampton VA 23669
Tel. 757-851-2856

Dipper

Wing Span	33.40 ft.	**Range**	400 sm
Wing Area	160.0 sq. ft.	**Takeoff Dist.**	600 ft.
Length	25.40 ft.	**Landing Dist.**	700 ft.
Max Gross Wt.	1760 lbs.	**Vmax**	125 mph
Empty Wt.	1100 lbs.	**VCr**	115 mph
Payload (full fuel)	N/A	**V_{S1} (stall clean)**	50 mph
Fuel Capacity	39 gal.	**V_{S0} (lndg config)**	N/A
Seats	2	**Climb Rate @msl**	1400 fpm
Cabin Width	N/A	**Service Ceiling**	18,000 ft.

Std. Engine/HP	Usable HP Range	Construction Mtrl.	Est. Bldg. Time
Lyc 0-360 / 180 hp	N/A	Fiberglass	1000 hrs.
Plans	**Kit**	**Info Packet**	**Video**
$200.00	N/A	$10.00	N/A

A two seat pusher amphibious conversion of a Cessna 150. Fiberglass hull. Instrument panel and windshield are moved forward three feet with a fiberglass boat hull and nose cowling added. Nose wheel retracts forward and the main gear goes into the rear windows. Design was started in 1964; first flight in 1982.

Kits/Plans Sold: 13 **Completed & Flying:** 1

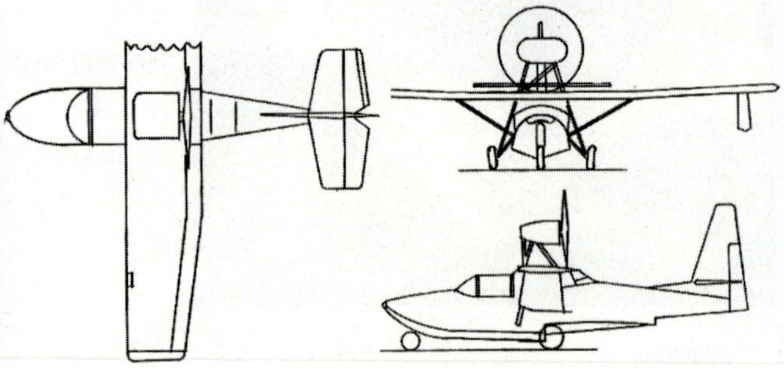

COLLINS AERO
238 Fairville Road
Chadds Ford PA 19317
Tel. 610-388-2393

S-18

Wing Span	20.84 ft.	Range	600 sm
Wing Area	86.00 sq. ft.	Takeoff Dist.	900 ft.
Length	19.33 ft.	Landing Dist.	1200 ft.
Max Gross Wt.	1600 lbs.	Vmax	213 mph
Empty Wt.	923 lbs.	VCr	170 mph
Payload (full fuel)	677 lbs.	V_{S1} (stall clean)	63 mph
Fuel Capacity	29 gal.	V_{S0} (lndg config)	60 mph
Seats	2	Climb Rate @msl	1200 fpm
Cabin Width	40 in.	Service Ceiling	18,000 ft.

Std. Engine/HP	Usable HP Range	Construction Mtrl.	Est. Bldg. Time
Lyc. 0-320 / 150 hp	135 - 180	Metal	2200 hrs.
Plans	**Kit**	**Info Packet**	**Video**
$195.00	$13,500.00	$5.00	N/A

All-metal, low wing design featuring a folding wing that meets highway width requirements. T-18 was first homebuilt to fly (1) around the world, (2) over the North Pole, and (3) to Australia/New Zealand and back. The S-18 is an improved aircraft with longer fuselage and improved airfoil. Structure meets requirements of CAM 3 "Aerobatic" category with +6, -3 G's.

Kits/Plans Sold: 187 **Completed & Flying:** 40

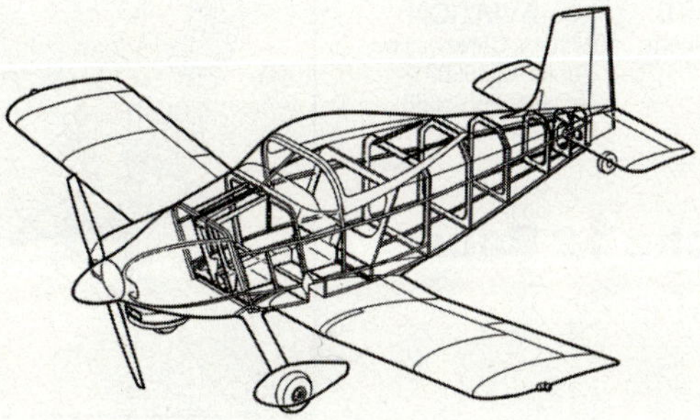

CLASSIC SPORT AIRCRAFT
19426 Campbell Creek Drive
Springville CA 93265
Tel. 209-539-2755 • Fax 209-539-2755
E-Mail: t18thorp@aol.com

Cozy Mark IV

Wing Span	28.10 ft.	Range	1000 sm
Wing Area	103.0 sq. ft.	Takeoff Dist.	1200 ft.
Length	16.90 ft.	Landing Dist.	1200 ft.
Max Gross Wt.	2050 lbs.	Vmax	228 mph
Empty Wt.	1050 lbs.	VCr	210 mph
Payload (full fuel)	700 lbs.	V_{S1} (stall clean)	N/A
Fuel Capacity	52 gal.	V_{S0} (lndg config)	N/A
Seats	4	Climb Rate @msl	2000 fpm
Cabin Width	42 in.	Service Ceiling	20,000 ft.

Std. Engine/HP	Usable HP Range	Construction Mtrl.	Est. Bldg. Time
Lyc. 0-360 / 180 hp	160 - 220	Composite	2500 hrs.
Plans	**Kit**	**Info Packet**	**Video**
$500.00	$14,000.00	$10.00	N/A

Canard pusher design with retractable nose gear. Stall resistant. Full-bubble canopy combined with rear engine and wings offer excellent visibility. Easy entrance and exit. Full dual controls. Specs are for solo flight. See Ad on Page 22.

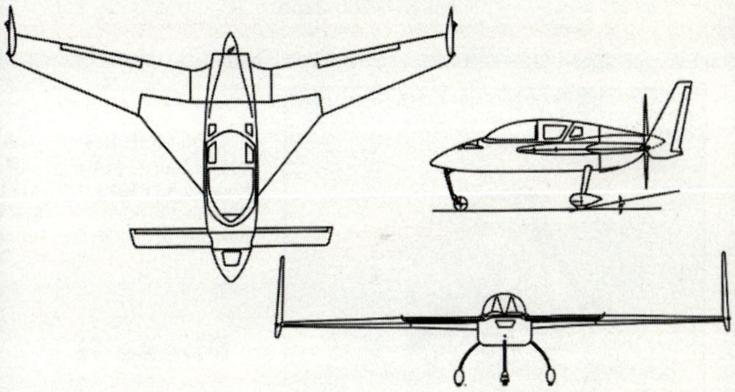

The Cozy concept originated in 1980 when the designer, Nat Puffer modified a Long EZ by adding side-by-side seating in the front. So many others wanted to copy his idea that he obtained permission to publish plans for the side-by-side Long EZ, which he named the Cozy. The design was originally a three place, but in 1988 a larger, four place version was prototyped, which was named the Cozy Mark IV, to differentiate it from the three place. Plans were published in 1991. The first proof-of-plans model was completed in 1992, and it was extensively flight tested in 1993 and the results published.

Dewey Davis, Warrenton, VA

"The Cozy is a delight to fly. It met all the performance objectives as advertised. I have flown this aircraft all over the United States and the Bahamas ... The Rutan aircraft are easy to build and they are getting easier every day because there is a large builder's network with helpful ideas and new techniques. ... I don't think there are very many alternatives that even come close in terms of performance per dollar invested. I can heartily recommend the Cozy to anyone that enjoys building and wants a fast, efficient cross country airplane."

CO-Z DEVELOPMENT CORPORATION
2046 N. 63rd Place
Mesa AZ 85215
Tel. 602-981-6401 • Fax 602-981-6401

FRED

Wing Span	22.60 ft.	Range	N/A
Wing Area	110.0 sq. ft.	Takeoff Dist.	400 ft.
Length	16.00 ft.	Landing Dist.	150 ft.
Max Gross Wt.	820 lbs.	Vmax	85 mph
Empty Wt.	550 lbs.	VCr	75 mph
Payload (full fuel)	210 lbs.	V_{S1} (stall clean)	N/A
Fuel Capacity	10 gal.	V_{S0} (lndg config)	N/A
Seats	1	Climb Rate @msl	700 fpm
Cabin Width	21 in.	Service Ceiling	N/A

Std. Engine/HP	Usable HP Range	Construction Mtrl.	Est. Bldg. Time
A-65 / 65 hp	N/A	Wood	N/A
Plans	**Kit**	**Info Packet**	**Video**
$40.00	N/A	$4.00	N/A

A plans-only parasol monoplane. Acceptable powerplants include VW, Subaru, and Continental engines up to 220 lbs. Designed to be road-towed with no trailer up to 70 mph. Wings fold easily by one person in five minutes. Folded width is 4.5 feet. No stall design. Simple wood and fabric structure. Prototype first flew in 1963 and is still flying. **Sold:** 370 **Flying:** 25

ERIC CLUTTON CO.
913 Cedar Lane • Tullahoma TN 37388
Tel. 615-455-2256

Samba Topless

Wing Span	32.00 ft.	Range	80 sm
Wing Area	150.6 sq. ft.	Takeoff Dist.	50 ft.
Length	6.00 ft.	Landing Dist.	50 ft.
Max Gross Wt.	450 lbs.	Vmax	55 mph
Empty Wt.	205 lbs.	VCr	40 mph
Payload (full fuel)	225 lbs.	V_{S1} (stall clean)	25 mph
Fuel Capacity	2 gal.	V_{S0} (lndg config)	N/A
Seats	1	Climb Rate @msl	500 fpm
Cabin Width	N/A	Service Ceiling	17,999 ft.

Std. Engine/HP	Usable HP Range	Construction Mtrl.	Est. Bldg. Time
Zenoah G22 / 22 hp	22	Tube, Fabric	10 hrs.
Plans	**Kit**	**Info Packet**	**Video**
N/A	$10,300.00	$5.00	N/A

A soaring trike that is part 103 legal. Design incorporates the 'Topless' hang gliding wing which removes the king-post and upper rigging and replaces them with a composite cross-spar and an extra internal washout strut.

COSMOS ULM
1, rue du stade • Fontaine-les-Dijon F-21121
FRANCE
Tel. 80 57 47 47 • Fax 80 57 47 50 • E-Mail: info@cosmos-ulm.com
Web: http://www.cosmos-ulm.com

Distributed By:

COSMOS US
501 W. San Juan Ave.
Phoenix AZ 85013
Ph: 602-285-4094
Fax: 602-530-2938

CANADIAN POWERED PARACHUTES LTD.
Box 1769 • Vegreville, Alberta
T9C 1S8 CANADA
Ph: 403-658-2044
Fax: 403-658-2387

KEMMERIES AVIATION
8710 W. Carefree Hwy.
Peoria AZ 85382
Tel.: 602-566-8068
Fax: 602-561-2287

Phase II

Wing Span	33.00 ft.	Range	325 sm
Wing Area	164.0 sq. ft.	Takeoff Dist.	200 ft.
Length	7.00 ft.	Landing Dist.	200 ft.
Max Gross Wt.	800 lbs.	Vmax	80 mph
Empty Wt.	300 lbs.	VCr	65 mph
Payload (full fuel)	500 lbs.	V_{S1} (stall clean)	28 mph
Fuel Capacity	14 gal.	V_{S0} (lndg config)	N/A
Seats	2	Climb Rate @msl	1000 fpm
Cabin Width	N/A	Service Ceiling	17,999 ft.

Std. Engine/HP	Usable HP Range	Construction Mtrl.	Est. Bldg. Time
Rotax 503 ES / 50 hp	40 - 75	Tube, Fabric	10 hrs.
Plans	**Kit**	**Info Packet**	**Video**
N/A	N/A	$5.00	N/A

A weight shift trike offered with a variety of wings. Multi-purpose trainer with one or two occupants. Standard features include a tool bag, removable seat-cushions, adjustable gas shock absorbers and hanging point safety wire.

COSMOS ULM
1, rue du stade • Fontaine-les-Dijon F-21121
FRANCE
Tel. 80 57 47 47 • Fax 80 57 47 50 • E-Mail: info@cosmos-ulm.com
Web: http://www.cosmos-ulm.com

Echo 12 Racer

Wing Span	32.00 ft.	Range	325 sm
Wing Area	141.0 sq. ft.	Takeoff Dist.	50 ft.
Length	6.00 ft.	Landing Dist.	50 ft.
Max Gross Wt.	550 lbs.	Vmax	100 mph
Empty Wt.	254 lbs.	VCr	65 mph
Payload (full fuel)	300 lbs.	V_{S1} (stall clean)	28 mph
Fuel Capacity	10 gal.	Service Ceiling	17,999 ft.
Seats	1	Climb Rate @msl	800 fpm

Std. Engine/HP	Usable HP Range	Construction Mtrl.	Est. Bldg. Time
Rotax 503 / 50 hp	40 - 65	Tube, Fabric	10 hrs.
Plans	**Kit**	**Info Packet**	**Video**
N/A	$12,995.00	$5.00	N/A

A single seat, weight-shift trike offered in "Fun" and "Racer" versions. The Echo Fun is equipped with a foam cushion seat. The Echo Racer has wheel pants, a front fairing and a composite fuel tank/seat. Uses a variety of Cosmos wings.

COSMOS ULM
1, rue du stade • Fontaine-les-Dijon F-21121
FRANCE
Tel. 80 57 47 47 • Fax 80 57 47 50 • E-Mail: info@cosmos-ulm.com
Web: http://www.cosmos-ulm.com

Cosy Classic - Kit

Wing Span	26.10 ft.	**Range**	1000 sm
Wing Area	96.00 sq. ft.	**Takeoff Dist.**	1200 ft.
Length	17.53 ft.	**Landing Dist.**	1000 ft.
Max Gross Wt.	1750 lbs.	**Vmax**	230 mph
Empty Wt.	1000 lbs.	**VCr**	200 mph
Payload (full fuel)	N/A	**V_{S1} (stall clean)**	67 mph
Fuel Capacity	52 gal.	**V_{S0} (Indg config)**	N/A
Seats	3	**Climb Rate @msl**	1500 fpm
Cabin Width	N/A	**Service Ceiling**	20,000 ft.

Std. Engine/HP	Usable HP Range	Construction Mtrl.	Est. Bldg. Time
Lyc. 0-320 / 160 hp	N/A	Composite	1500 hrs.
Plans	**Kit**	**Info Packet**	**Video**
N/A	$25,000.00	$15.00	N/A

Kit-built version of the Cosy Classic with front hinged canopy, 44 inch wide fuselage, R1145 Roncz canard and Lycoming engine. Price includes plans and drawings, serial number, one year subscription to newsletter, and shipping. Mix and match with plans-built version. Exact prices dependant on current exchange rate. Also contact MAP, La Verrerie Basse, F-81140, Vaour, France. Telephone 33-6356-3824 Fax 33-6356-3046. **Completed & Flying:** 3

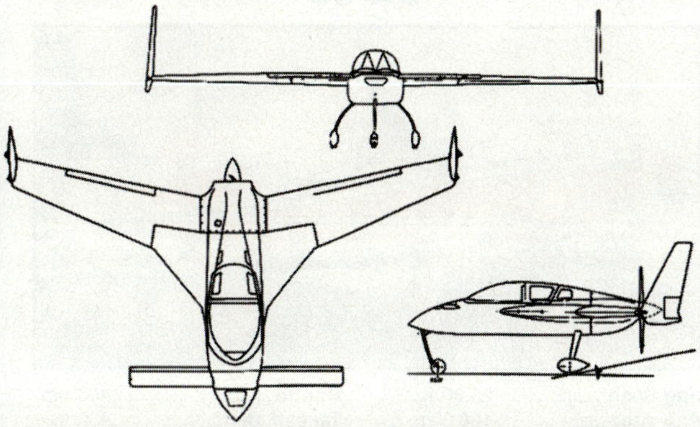

While stationed at Sheppard AFB in Wichita Falls, Texas, in the ENJJPT program, as an instructor pilot, Uli Wolter heard that one of the American officers was building a Long-EZ. He couldn't wait to look over the project. It didn't take long and he was hooked. A few months later at Oshkosh, he saw Nat Puffer's new "Cozy", side-by-side seating version of the Long-EZ. It was love at first sight.

Uli built and flew the first plans built Cozy in 1985. He became very enthusiastic about its great flying qualities. In 1986, it was time to return home. To avoid the high cost of shipping, Uli flew his Cozy via the North Atlantic to Germany. Shortly after his arrival, he set up Cosy-Europe. He bought the exclusive design and marketing rights for the Cozy in 1989, and introduced the new improved "Cosy Classic" which is available as Kit or Plans. See next listing.

COSY-EUROPE
Ahornstrasse 10 • D-86510 Ried
GERMANY
Tel. 49-8233-60594 • Fax 49-8233-20150

Cosy Classic - Plans

Wing Span	25.90 ft.	**Range**	1000 sm
Wing Area	N/A	**Takeoff Dist.**	1200 ft.
Length	17.50 ft.	**Landing Dist.**	1000 ft.
Max Gross Wt.	1750 lbs.	**Vmax**	210 mph
Empty Wt.	1000 lbs.	**VCr**	190 mph
Payload (full fuel)	N/A	**V_{S1} (stall clean)**	67 mph
Fuel Capacity	52 gal.	**V_{S0} (Indg config)**	N/A
Seats	3	**Climb Rate @msl**	1500 fpm
Cabin Width	N/A	**Service Ceiling**	20,000 ft.

Std. Engine/HP	Usable HP Range	Construction Mtrl.	Est. Bldg. Time
Lyc 0-320 / 160 hp	N/A	Composite	2500 hrs.
Plans	**Kit**	**Info Packet**	**Video**
$500.00	N/A	$15.00	N/A

Plans-built version of the Cosy Classic. Mix and match with kit-built version detailed in previous listing. Also contact MAP, La Verrerie Basse, F-81140, Vaour, France. Telephone 33-6356-3824 Fax 33-6356-3046 **Completed & Flying:** 3

COSY-EUROPE
Ahornstrasse 10 • D-86510 Ried
GERMANY
Tel. 49-8233-60594 • Fax 49-8233-20150

Criquet L

Wing Span	26.20 ft.	**Range**	200 sm
Wing Area	189.6 sq. ft.	**Takeoff Dist.**	130 ft.
Length	N/A	**Landing Dist.**	150 ft.
Max Gross Wt.	880 lbs.	**Vmax**	81 mph
Empty Wt.	385 lbs.	**VCr**	68 mph
Payload (full fuel)	N/A	**V_{S1} (stall clean)**	31 mph
Fuel Capacity	11 gal.	**V_{S0} (Indg config)**	N/A
Seats	2	**Climb Rate @msl**	500 fpm
Cabin Width	N/A	**Service Ceiling**	N/A

Std. Engine/HP	Usable HP Range	Construction Mtrl.	Est. Bldg. Time
Rotax 503 / 50 hp	N/A	Wood	N/A
Plans	**Kit**	**Info Packet**	**Video**
$360.00	N/A	N/A	N/A

A unique plans-only, staggered wing design. The Criquet L sports an enclosed cabin and all-wood construction. Can be fitted with two or three axis controls. Approximate plans price is in U.S. funds.

EMILIEN CROSES
63 Route de Davaye, aerodrome
Carnay-les-Macon 71000
FRANCE
Tel. 85 34 16 11

North Star

Wing Span	36.33 ft.	Range	650 sm
Wing Area	190.0 sq. ft.	Takeoff Dist.	300 ft.
Length	22.50 ft.	Landing Dist.	200 ft.
Max Gross Wt.	2200 lbs.	Vmax	135 mph
Empty Wt.	1170 lbs.	VCr	110 mph
Payload (full fuel)	N/A	V_{S1} (stall clean)	35 mph
Fuel Capacity	52 gal.	V_{S0} (lndg config)	N/A
Seats	2	Climb Rate @msl	1000 fpm
Cabin Width	N/A	Service Ceiling	15,000 ft.

Std. Engine/HP	Usable HP Range	Construction Mtrl.	Est. Bldg. Time
Lyc. 0-320 / 150 hp	100 - 180	Tube, Fabric	1000 hrs.
Plans	**Kit**	**Info Packet**	**Video**
N/A	$30,000.00	Yes	N/A

A derivative of the Super Cub. Good performance with floats. Features 2-1/2" wider cockpit, doors on both sides with handles inside and out, 90 lbs. baggage capacity with baggage door, and adjustable front seat with tilting back. Kit price does not include engine, covering, wheels, brakes, upholstery, hardware or instruments. Simplified version (Bright Star) has no flaps, one door. Available for $20,000.

When Morgan Williams wanted a Super Cub floatplane, he couldn't find one that was suitable. So, he built one. Williams is a lifelong airplane junkie who has spent much of his adult life maintaining, rebuilding and restoring airplanes. His firm, Custom Flight Components, Ltd., has assisted homebuilders in various projects and specializes in custom-built aerobatic wings for Pitts biplanes. Through it all, Williams took note of where certain airplanes fell short in the durability and accessibility departments.

He knew Piper's PA-18 Super Cub was an excellent performing airplane, but he also knew it could be improved, even after the aircraft's 40 years. Rather than work over an existing hulk - something he has done for clients far too often - Williams opted to start afresh and design his own Cub.

CUSTOM FLIGHT COMPONENTS LTD.
R.R. #1, • Perkinsfield
Ontario L0L 2J0
CANADA
Tel. 705 526-9626 • Fax 705 526-2529

Culp Special

Wing Span	24.00 ft.	Range	600 sm
Wing Area	N/A	Takeoff Dist.	400 ft.
Length	21.00 ft.	Landing Dist.	1000 ft.
Max Gross Wt.	2100 lbs.	Vmax	220 mph
Empty Wt.	1480 lbs.	VCr	150 mph
Payload (full fuel)	N/A	V_{S1} (stall clean)	70 mph
Fuel Capacity	70 gal.	V_{S0} (lndg config)	N/A
Seats	2	Climb Rate @msl	3500 fpm
Cabin Width	N/A	Service Ceiling	N/A

Std. Engine/HP	Usable HP Range	Construction Mtrl.	Est. Bldg. Time
M-14P / 360 hp	N/A	Tube, Wood, Fabric	2500 hrs.
Plans	**Kit**	**Info Packet**	**Video**
$480.00	$47,500.00	N/A	N/A

Everything you need in an airplane...two wings and a round engine! Reminiscent of early era biplanes, the Culp Special is fully aerobatic and utilizes the Russian M-14P radial engine.

Completed & Flying: 1

CULPS SPECIALTIES
PO Box 7542
Shreveport LA 71107
Tel. 318-222-0850

CA-65

Wing Span	25.00 ft.	Range	400 sm
Wing Area	108.0 sq. ft.	Takeoff Dist.	450 ft.
Length	19.00 ft.	Landing Dist.	600 ft.
Max Gross Wt.	1500 lbs.	Vmax	180 mph
Empty Wt.	900 lbs.	VCr	155 mph
Payload (full fuel)	N/A	V_{S1} (stall clean)	55 mph
Fuel Capacity	28 gal.	Service Ceiling	15,000 ft.
Seats	2	Climb Rate @msl	1000 fpm

Std. Engine/HP	Usable HP Range	Construction Mtrl.	Est. Bldg. Time
Lycoming / 125 hp	108 - 160	Metal, Wood	N/A hrs.
Plans	**Kit**	**Info Packet**	**Video**
$180.00	N/A	$5.00	$45.00

Low wing design with retractable landing gear. Gear is lighter than some fixed gear types. Optional folding wings. Dual controls. Streamlined canopy, completely enclosed engine, and tapered wing offer economic performance at all speeds. A 160 hp Lyc. engine gives good climb and speed over 190 mph.

ANTON CVJETKOVIC
5324 W. 121st Street • Hawthorne CA 90250
Ph: 310-643-6931 • Fax: 310-643-6931

Ascender III-B

Wing Span	33.00 ft.	**Range**	120 sm
Wing Area	173.0 sq. ft.	**Takeoff Dist.**	100 ft.
Length	16.80 ft.	**Landing Dist.**	50 ft.
Max Gross Wt.	465 lbs.	**Vmax**	55 mph
Empty Wt.	215 lbs.	**VCr**	40 mph
Payload (full fuel)	250 lbs.	**V_{S1} (stall clean)**	N/A
Fuel Capacity	5 gal.	**V_{S0} (lndg config)**	N/A
Seats	1	**Climb Rate @msl**	1000 fpm
Cabin Width	N/A	**Service Ceiling**	18,000 ft.

Std. Engine/HP	Usable HP Range	Construction Mtrl.	Est. Bldg. Time
Cuyuna / 35 hp	22 - 45	Alum. Tube, Dacron	100 hrs.
Plans	**Kit**	**Info Packet**	**Video**
N/A	$7,400.00	Yes	N/A

Advanced copy of the Ascender II developed in 1980 with an open cockpit, low speed, and excellent rough field capabilities. Most components are supplied prefabricated. Wing covering completely pre-sewn. **Sold:** 1000 **Flying:** 1000

DFE ULTRALIGHTS, INC.
RD 1, Box 185
Vanderbilt PA 15486
Tel. 412-529-0450 • Fax 412-529-0596

Ascender III-C

Wing Span	33.00 ft.	**Range**	120 sm
Wing Area	173.0 sq. ft.	**Takeoff Dist.**	100 ft.
Length	16.80 ft.	**Landing Dist.**	50 ft.
Max Gross Wt.	535 lbs.	**Vmax**	55 mph
Empty Wt.	235 lbs.	**VCr**	40 mph
Payload (full fuel)	300 lbs.	**V_{S1} (stall clean)**	N/A
Fuel Capacity	5 gal.	**V_{S0} (lndg config)**	N/A
Seats	1	**Climb Rate @msl**	900 fpm
Cabin Width	N/A	**Service Ceiling**	18,000 ft.

Std. Engine/HP	Usable HP Range	Construction Mtrl.	Est. Bldg. Time
Cuyuna / 35 hp	22 - 45	Alum. Tube, Dacron	100 hrs.
Plans	**Kit**	**Info Packet**	**Video**
N/A	$7,600.00	Yes	N/A

Heavy duty version of the Ascender III-B. For pilots over 225 lbs. Can also be fitted with a sidecar for two place operation. Ascender III-T (TUG) for aerial towing also available. The TUG is built and tested before delivery and not available as a kit.

DFE ULTRALIGHTS, INC.
RD 1, Box 185
Vanderbilt PA 15486
Tel. 412-529-0450 • Fax 412-529-0596

D-201 Sportwing

Wing Span	27.00 ft.	**Range**	350 sm
Wing Area	184.5 sq. ft.	**Takeoff Dist.**	420 ft.
Length	21.65 ft.	**Landing Dist.**	550 ft.
Max Gross Wt.	1845 lbs.	**Vmax**	132 mph
Empty Wt.	1185 lbs.	**VCr**	122 mph
Payload (full fuel)	N/A	**V_{S1} (stall clean)**	47 mph
Fuel Capacity	28 gal.	**V_{S0} (lndg config)**	N/A
Seats	2	**Climb Rate @msl**	1000 fpm
Cabin Width	N/A	**Service Ceiling**	19,000 ft.

Std. Engine/HP	Usable HP Range	Construction Mtrl.	Est. Bldg. Time
Lyc.10-320 / 160 hp	125 - 200	Tube, Fabric	2500 hrs.
Plans	**Kit**	**Info Packet**	**Video**
$400.00	N/A	$5.00	N/A

Incorporates more than 20 years of design improvements. 25 built, 20 in the works. Drawings for metal center section with extra fuel tanks available. Wide flange, hydrapress formed ribs available. Wide rib flanges permit pop riveting of fabric. This method used on prototype and PJ-260/D-260 predecessors, still with original cecconite after 28 years. **Flying:** 20

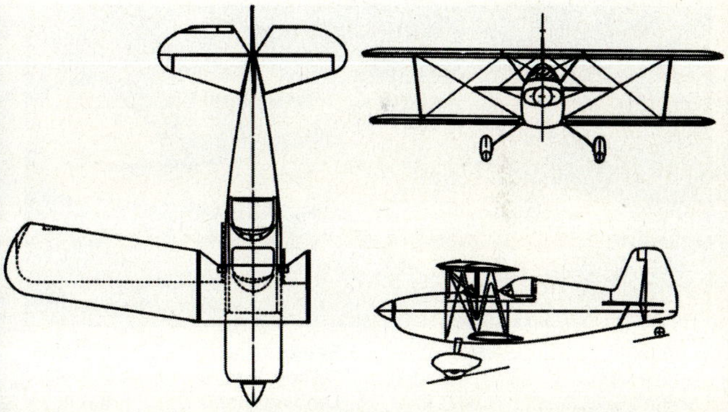

The Sportwing is the latest model in the PJ-260/D-260 series of biplanes; an airplane not as demanding of flying skills as the Pitts or Christen Eagle. It incorporates a significant number of design improvements and meritorious suggestions recorded over the more than 20 years.

The design has been analyzed to make it compatible with engines of lower horsepower, 125 hp on the low end to a maximum of 200 hp on the high end. Both the empty and gross weights are substantially reduced over those of the earlier models, primarily due to the lighter weights of the smaller engines suitable for this model.

NICK D'APUZZO AIRPLANE DESIGN
1029 Blue Rock Lane
Blue Bell PA 19422-2003
Tel. 215-646-4792 • Fax 215-646-8122

E-Racer

Wing Span	26.20 ft.	**Range**	1000 sm
Wing Area	94.00 sq. ft.	**Takeoff Dist.**	1200 ft.
Length	16.80 ft.	**Landing Dist.**	1500 ft.
Max Gross Wt.	1800 lbs.	**Vmax**	250 mph
Empty Wt.	1000 lbs.	**VCr**	225 mph
Payload (full fuel)	N/A	**V_{S1} (stall clean)**	N/A
Fuel Capacity	46 gal.	**V_{S0} (lndg config)**	N/A
Seats	2	**Climb Rate @msl**	2500 fpm
Cabin Width	N/A	**Service Ceiling**	25,000 ft.

Std. Engine/HP	Usable HP Range	Construction Mtrl.	Est. Bldg. Time
Buick V-8 / 240 hp	N/A	Composite	2000 hrs.
Plans	**Kit**	**Info Packet**	**Video**
$250.00	N/A	$9.00	$20.00

A high speed, canard aircraft designed around the proven aerodynamics of the Rutan Long EZ. Available in a Mk. I and a Mk. II configuration. Fully retractable landing gear. Pre-built wings and canards are available from Glassic Composites. **Completed & Flying:** 4

SHIRL DICKEY ENTERPRISES
8631 W. College Dr. • Phoenix AZ 85037
Tel. 602-691-0515

Durand Mark V

Wing Span	24.50 ft.	**Range**	400 sm
Wing Area	144.0 sq. ft.	**Takeoff Dist.**	550 ft.
Length	20.25 ft.	**Landing Dist.**	450 ft.
Max Gross Wt.	1840 lbs.	**Vmax**	135 mph
Empty Wt.	1210 lbs.	**VCr**	125 mph
Payload (full fuel)	483 lbs.	**V_{S1} (stall clean)**	N/A
Fuel Capacity	25 gal.	**V_{S0} (lndg config)**	N/A
Seats	2	**Climb Rate @msl**	1200 fpm
Cabin Width	43 in.	**Service Ceiling**	15,000 ft.

Std. Engine/HP	Usable HP Range	Construction Mtrl.	Est. Bldg. Time
Lyc 0-320 / 150 hp	125 - 180	Tube, Aluminum	2500 hrs.
Plans	**Kit**	**Info Packet**	**Video**
$275.00	N/A	$5.00	N/A

Two place, negative stagger, biplane with a fully enclosed cabin and side-by-side seating. Dual controls standard. Optional leading edge tanks available for additional range. Not intended for aerobatic flight. **Kits/Plans Sold:** 91 **Completed & Flying:** 9

DURAND ASSOCIATES, INC.
11317 N. 84th St. • McKinley Road
Omaha NE 68122
Tel. 402-571-7058

Ultrababy

Wing Span	27.50 ft.	**Range**	150 sm
Wing Area	121.5 sq. ft.	**Takeoff Dist.**	250 ft.
Length	16.00 ft.	**Landing Dist.**	400 ft.
Max Gross Wt.	620 lbs.	**Vmax**	75 mph
Empty Wt.	300 lbs.	**VCr**	60 mph
Payload (full fuel)	280 lbs.	**V_{S1} (stall clean)**	35 mph
Fuel Capacity	5 gal.	**V_{S0} (lndg config)**	N/A
Seats	1	**Climb Rate @msl**	600 fpm
Cabin Width	N/A	**Service Ceiling**	10,000 ft.

Std. Engine/HP	Usable HP Range	Construction Mtrl.	Est. Bldg. Time
1/2 VW / 48 hp	35-52	Wood, Fabric	700 hrs.
Plans	**Kit**	**Info Packet**	**Video**
$75.00	N/A	$3.00	$15.00

Plans-only 75% scale Fly Baby replica. Can qualify as an Ultralight or Experimental depending on engine. Plans have been available for more than a year.
Kits/Plans Sold: 40 **Completed /Flying:** 1

DUANE'S HANGAR
226 Melton Rd.
Liberty SC 29657
Tel. 864-843-3732

Dyke Delta

Wing Span	22.18 ft.	**Range**	650 sm
Wing Area	177.5 sq. ft.	**Takeoff Dist.**	800 ft.
Length	19.00 ft.	**Landing Dist.**	1600 ft.
Max Gross Wt.	1950 lbs.	**Vmax**	210 mph
Empty Wt.	1080 lbs.	**VCr**	175 mph
Payload (full fuel)	N/A	**V_{S1} (stall clean)**	70 mph
Fuel Capacity	47 gal.	**V_{S0} (lndg config)**	N/A
Seats	4	**Climb Rate @msl**	1800 fpm
Cabin Width	47 in.	**Service Ceiling**	14,000 ft.

Std. Engine/HP	Usable HP Range	Construction Mtrl.	Est. Bldg. Time
Lyc 0-360 / 180 hp	N/A	Tube, Fiberglass	N/A
Plans	**Kit**	**Info Packet**	**Video**
$200.00	N/A	$5.00	N/A

Proven delta design. Has folding wings, retractable gear and is towable. Folded width is 7' 4". Maximum towing speed is 60 mph. **Completed & Flying:** 45

DYKE AIRCRAFT
2840 Old Yellow Spring Road
Fairborn OH 45324
Tel. 937-878-9832

CR.100

Wing Span	27.80 ft.	**Range**	N/A
Wing Area	N/A	**Takeoff Dist.**	N/A
Length	N/A	**Landing Dist.**	N/A
Max Gross Wt.	1870 lbs.	**Vmax**	198 mph
Empty Wt.	1210 lbs.	**VCr**	155 mph
Payload (full fuel)	N/A	**V_{S1} (stall clean)**	N/A
Fuel Capacity	21 gal.	**V_{S0} (lndg config)**	N/A
Seats	2	**Climb Rate @msl**	N/A
Cabin Width	N/A	**Service Ceiling**	N/A

Std. Engine/HP	Usable HP Range	Construction Mtrl.	Est. Bldg. Time
Lyc. IO-360 / 180 hp	N/A	Metal	9500 hrs.
Plans	**Kit**	**Info Packet**	**Video**
N/A	$14,600.00	N/A	N/A

All-metal forerunner of the MCR 01. Price is for basic kit excluding engine, prop, avionics, instruments, upholstery, paint, and wiring.

DYN' AERO SA
19 rue de l'Aviation
F21121 Darois 21121
FRANCE
Tel. (33) 3 80 35 60 6 • Fax (33) 3 80 35 60 6

Early Bird 80% SPAD 13

Wing Span	20.17 ft.	**Range**	240 sm
Wing Area	142.0 sq. ft.	**Takeoff Dist.**	250 ft.
Length	16.25 ft.	**Landing Dist.**	350 ft.
Max Gross Wt.	800 lbs.	**Vmax**	90 mph
Empty Wt.	550 lbs.	**VCr**	80 mph
Payload (full fuel)	200 lbs.	**V_{S1} (stall clean)**	45 mph
Fuel Capacity	9 gal.	**V_{S0} (lndg config)**	N/A
Seats	1	**Climb Rate @msl**	800 fpm
Cabin Width	24 in.	**Service Ceiling**	14,000 ft.

Std. Engine/HP	Usable HP Range	Construction Mtrl.	Est. Bldg. Time
GEO / 85 hp	80 - 100	Steel, Aluminum	600 hrs.
Plans	**Kit**	**Info Packet**	**Video**
$125.00	$14,900.00	$2.00	N/A

Quick and responsive — flys like the little fighter she is supposed to be. The multipoint fuel injected Geo Tracker engine burns three gallons per hour. Kit includes engine.

Kits/Plans Sold: 35 **Completed & Flying:** 1

EARLY BIRD AIRCRAFT CO.
125 Stearman Court
Erie CO 80516
Tel. 303-665-5169

Early Bird Jenny

Wing Span	27.50 ft.	**Range**	130 sm
Wing Area	175.0 sq. ft.	**Takeoff Dist.**	100 ft.
Length	18.33 ft.	**Landing Dist.**	250 ft.
Max Gross Wt.	800 lbs.	**Vmax**	70 mph
Empty Wt.	419 lbs.	**VCr**	65 mph
Payload (full fuel)	327 lbs.	**V_{S1} (stall clean)**	35 mph
Fuel Capacity	9 gal.	**V_{S0} (lndg config)**	N/A
Seats	2	**Climb Rate @msl**	800 fpm
Cabin Width	24 in.	**Service Ceiling**	14,000 ft.

Std. Engine/HP	Usable HP Range	Construction Mtrl.	Est. Bldg. Time
Rotax 503 / 46 hp	46 - 65	Steel, Aluminum	500 hrs.
Plans	**Kit**	**Info Packet**	**Video**
$98.00	$10,245.00	$2.00	$28.00

Qualifies as an ultralight trainer. Kit includes Rotax 503 engine and detachable wings. Plans include material list for those who want to do their own shopping. Now flying with a 60 hp Geo Metro engine.

Kits/Plans Sold: 530 **Completed & Flying:** 24

EARLY BIRD AIRCRAFT CO.
125 Stearman Court
Erie CO 80516
Tel. 303-665-5169

Thunder Gull J

Wing Span	20.00 ft.	**Range**	165 sm
Wing Area	95.00 sq. ft.	**Takeoff Dist.**	125 ft.
Length	17.60 ft.	**Landing Dist.**	75 ft.
Max Gross Wt.	550 lbs.	**Vmax**	120 mph
Empty Wt.	252 lbs.	**VCr**	63 mph
Payload (full fuel)	268 lbs.	**V_{S1} (stall clean)**	27 mph
Fuel Capacity	5 gal.	**V_{S0} (lndg config)**	25 mph
Seats	1	**Climb Rate @msl**	700 fpm
Cabin Width	27 in.	**Service Ceiling**	14,000 ft.

Std. Engine/HP	Usable HP Range	Construction Mtrl.	Est. Bldg. Time
Rotax 277 / 27 hp	27 - 53	Tube, Composite, Alum.	150 hrs.
Plans	**Kit**	**Info Packet**	**Video**
N/A	$8,980.00	$5.00	$25.00

Single seat, high wing pusher. The first of the Gull series flown in 1976. Many have been flown from coast to coast and have more than 300 hours flying time on them. Very refined handling. Meets FAR Part 103 regulations. Available as kit or ready to fly.

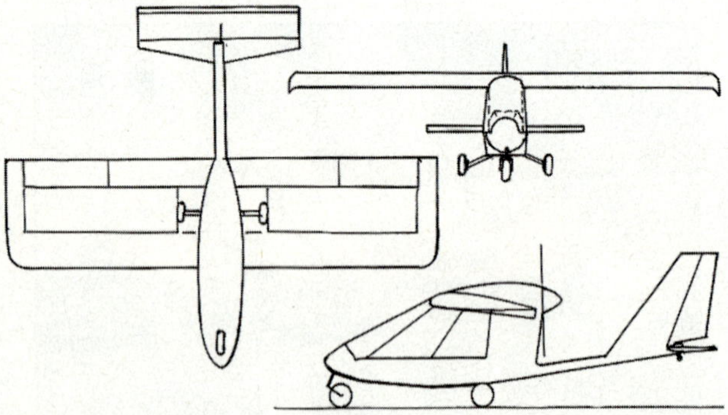

Mark Beierle has been refining this design since 1976 in order to give the Thunder Gull control harmony and agility not found in any other aircraft in its class. The prototype is powered with a 26 hp Rotax 277 engine. It has been flown across the continent seven times with 100 lbs. of baggage and a 180 lb. pilot through strong winds and heavy rain.

Tim Stormer, Falls Church, VA
"... The flight performance of the Thunder Gull is astonishing and the ground tracking punishment given the flying machine certainly says it all about the structural soundness of the vehicle. Wow!!!"

Bill Hanline, Ventura, CA, J Kit Builder
"I shopped for three years before choosing which ultralight was for me. I decided on a Thunder Gull. It's the best plane out there in its class."

EARTHSTAR AIRCRAFT, INC.
Star Route 313
Santa Margarita CA 93453
Tel. 805-438-5235 • Fax 805-438-5235

Thunder Gull JT2

Wing Span	24.00 ft.	**Range**	275 sm
Wing Area	114.0 sq. ft.	**Takeoff Dist.**	200 ft.
Length	17.60 ft.	**Landing Dist.**	175 ft.
Max Gross Wt.	850 lbs.	**Vmax**	120 mph
Empty Wt.	398 lbs.	**VCr**	87 mph
Payload (full fuel)	398 lbs.	**V_{S1} (stall clean)**	39 mph
Fuel Capacity	10 gal.	**V_{S0} (lndg config)**	37 mph
Seats	2	**Climb Rate @msl**	800 fpm
Cabin Width	27 in.	**Service Ceiling**	14,500 ft.

Std. Engine/HP	Usable HP Range	Construction Mtrl.	Est. Bldg. Time
Rotax 503 / 53 hp	40 - 80	Tube, Composite, Alum.	300 hrs.
Plans	**Kit**	**Info Packet**	**Video**
N/A	$12,980.00	$5.00	$25.00

High wing pusher with tandem seating and dual controls. No forming or welding required. Leading edge section of wing from main spar forward comes preassembled, as do control surfaces and stabilizers. Available as kit or ready to fly.

EARTHSTAR AIRCRAFT, INC.
Star Route 313
Santa Margarita CA 93453
Tel. 805-438-5235 • Fax 805-438-5235

Thunder Gull Odyssey

Wing Span	26.00 ft.	**Range**	300 sm
Wing Area	124.0 sq. ft.	**Takeoff Dist.**	200 ft.
Length	17.80 ft.	**Landing Dist.**	175 ft.
Max Gross Wt.	1000 lbs.	**Vmax**	120 mph
Empty Wt.	420 lbs.	**VCr**	87 mph
Payload (full fuel)	520 lbs.	**V_{S1} (stall clean)**	37 mph
Fuel Capacity	10 gal.	**V_{S0} (lndg config)**	34 mph
Seats	2	**Climb Rate @msl**	900 fpm
Cabin Width	41 in.	**Service Ceiling**	15,500 ft.

Std. Engine/HP	Usable HP Range	Construction Mtrl.	Est. Bldg. Time
Rotax 503 / 53 hp	40 - 80	Tube, Composite, Alum.	350 hrs.
Plans	**Kit**	**Info Packet**	**Video**
N/A	$14,980.00	$5.00	$25.00

Two seat, side-by-side, staggered seating with dual controls. Has a single center stick. 26 foot wing is standard. Introduced at Sun 'n Fun in 1995. Register as an Ultralight Trainer or Experimental. Available as kit or ready to fly.

EARTHSTAR AIRCRAFT, INC.
Star Route 313
Santa Margarita CA 93453
Tel. 805-438-5235 • Fax 805-438-5235

Coach II S

Wing Span	34.40 ft.	**Range**	260 sm
Wing Area	161.4 sq. ft.	**Takeoff Dist.**	330 ft.
Length	17.40 ft.	**Landing Dist.**	280 ft.
Max Gross Wt.	993 lbs.	**Vmax**	77 mph
Empty Wt.	552 lbs.	**VCr**	62 mph
Payload (full fuel)	441 lbs.	**V_{S1} (stall clean)**	31 mph
Fuel Capacity	11 gal.	**V_{S0} (lndg config)**	N/A
Seats	2	**Climb Rate @msl**	700 fpm
Cabin Width	39 in.	**Service Ceiling**	12,000 ft.

Std. Engine/HP	Usable HP Range	Construction Mtrl.	Est. Bldg. Time
Sauer 2200 cc / 70 hp	40 - 80	Steel, Wood, Fabric	450 hrs.
Plans	**Kit**	**Info Packet**	**Video**
N/A	$18,400.00	$10.00	N/A

Side-by-side seating with dual controls. +5, -3 G's limit. Detachable wings for storage.Kit does not include engine or propeller. A "ready to fly" version is available for $43,750. Designed by Egon Scheibe. Test flown and certified as an ultralight in Germany.

Completed & Flying: 15

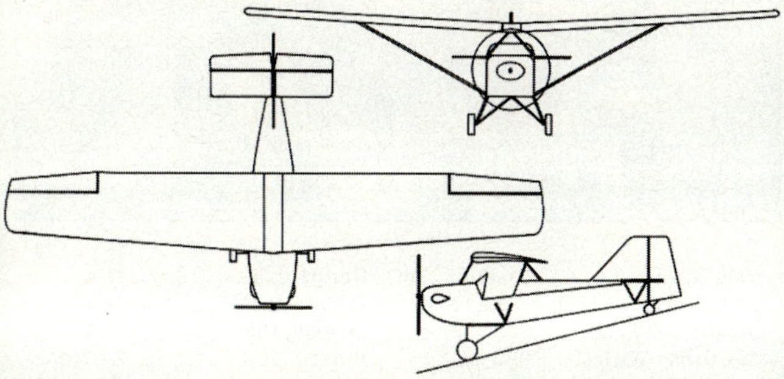

EHROFLUG GMBH
Winkelacker
Altnau CH-8595
SWITZERLAND
Tel. 01141 71 695 2808 • Fax 01141 71 695 2808

Thorp T-18

Wing Span	20.80 ft.	**Range**	530 sm
Wing Area	86.00 sq. ft.	**Takeoff Dist.**	1000 ft.
Length	18.90 ft.	**Landing Dist.**	1800 ft.
Max Gross Wt.	1500 lbs.	**Vmax**	210 mph
Empty Wt.	923 lbs.	**VCr**	201 mph
Payload (full fuel)	400 lbs.	**V_{S1} (stall clean)**	68 mph
Fuel Capacity	29 gal.	**V_{S0} (lndg config)**	59 mph
Seats	2	**Climb Rate @msl**	1200 fpm
Cabin Width	36 in.	**Service Ceiling**	18,000 ft.

Std. Engine/HP	Usable HP Range	Construction Mtrl.	Est. Bldg. Time
Lycoming / 180 hp	125-180	Metal	2000 hrs.
Plans	**Kit**	**Info Packet**	**Video**
$250.00	N/A	$15.00	N/A

Simple to build, high performance sport aircraft. Durable, all-metal design with low maintenance. The first easy to build all-metal homebuilt. Laser cut aileron kits available. Other sub-assembly kits under development. Now 30+ years old but still out-performs many contemporary designs. Performance specs are with 180 hp engine. **Plans Sold:** 1500 **Flying:** 400

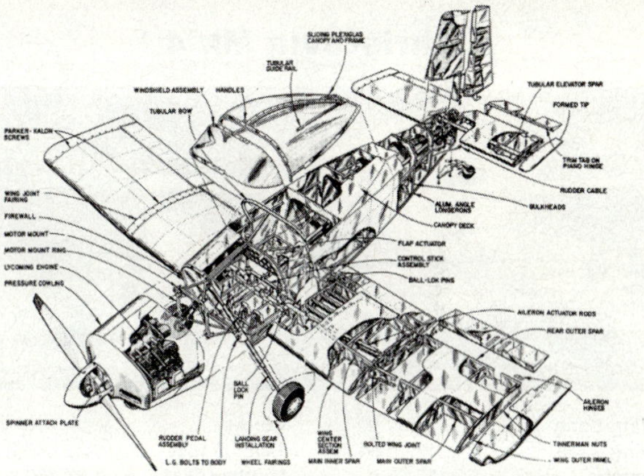

EKLUND ENGINEERING, INC.
19960 Elliott Road • PO Box 1510
Lockeford CA 95237-1510
Tel. 209-727-0318 • Fax 209-727-0873
E-Mail: ThorpT18@compuserve.com

Christavia Mk 1

Wing Span	32.50 ft.	**Range**	300 sm
Wing Area	146.3 sq. ft.	**Takeoff Dist.**	300 ft.
Length	21.25 ft.	**Landing Dist.**	550 ft.
Max Gross Wt.	1500 lbs.	**Vmax**	120 mph
Empty Wt.	750 lbs.	**VCr**	105 mph
Payload (full fuel)	N/A	**V_{S1} (stall clean)**	45 mph
Fuel Capacity	22 gal.	**V_{S0} (Indg config)**	N/A
Seats	2	**Climb Rate @msl**	850 fpm
Cabin Width	N/A	**Service Ceiling**	14,000 ft.

Std. Engine/HP	Usable HP Range	Construction Mtrl.	Est. Bldg. Time
Continental / 65 hp	N/A	Wood, Fabric, Tube	2200 hrs.
Plans	**Kit**	**Info Packet**	**Video**
$200.00	N/A	$10.00	N/A

Two place taildragger from Canada. Conventional tube, wood, and fabric construction.

Completed & Flying: 200

ELMWOOD AVIATION
RR4, Elmwood Drive
Belleville, Ont. K8N 4Z4
CANADA
Tel. 613-967-1853

Basic Fox

Wing Span	31.48 ft.	**Range**	N/A
Wing Area	173.2 sq. ft.	**Takeoff Dist.**	N/A
Length	20.33 ft.	**Landing Dist.**	N/A
Max Gross Wt.	N/A	**Vmax**	75 mph
Empty Wt.	418 lbs.	**VCr**	50 mph
Payload (full fuel)	N/A	**V_{S1} (stall clean)**	25 mph
Fuel Capacity	N/A	**V_{S0} (Indg config)**	N/A
Seats	N/A	**Climb Rate @msl**	980 fpm
Cabin Width	N/A	**Service Ceiling**	N/A

Std. Engine/HP	Usable HP Range	Construction Mtrl.	Est. Bldg. Time
Rotax 503 / 50 hp	N/A	Metal, Tube, Fabric	N/A
Plans	**Kit**	**Info Packet**	**Video**
N/A	N/A	Yes	N/A

A simple, sturdy pusher design with side-by-side seating. Perfect for training. Various configurations available.

EURO-FLY
Via Ca Onorai
Galliera Veneta (Padova) 50-35015
ITALY
Tel. 049 596 5464 • Fax 049 596 5464

Christavia Mk 4

Wing Span	35.50 ft.	**Range**	400 sm
Wing Area	177.3 sq. ft.	**Takeoff Dist.**	450 ft.
Length	22.25 ft.	**Landing Dist.**	750 ft.
Max Gross Wt.	2200 lbs.	**Vmax**	130 mph
Empty Wt.	1100 lbs.	**VCr**	120 mph
Payload (full fuel)	N/A	**V_{S1} (stall clean)**	48 mph
Fuel Capacity	41 gal.	**V_{S0} (Indg config)**	N/A
Seats	4	**Climb Rate @msl**	800 fpm
Cabin Width	N/A	**Service Ceiling**	19,000 ft.

Std. Engine/HP	Usable HP Range	Construction Mtrl.	Est. Bldg. Time
Lycoming / 150 hp	N/A	Wood, Fabric, Tube	2600 hrs.
Plans	**Kit**	**Info Packet**	**Video**
$200.00	N/A	$10.00	N/A

Four place taildragger from Canada. Longer fuselage and wider wingspan than the Mk 1 to accommodate additional passengers. Currently being redesigned with metal wing spars. Plans are temporarily not available. Call. **Completed & Flying:** 250

ELMWOOD AVIATION
RR4, Elmwood Drive
Belleville, Ont. K8N 4Z4
CANADA
Tel. 613-967-1853

Fire Fox

Wing Span	31.48 ft.	**Range**	N/A
Wing Area	173.2 sq. ft.	**Takeoff Dist.**	N/A
Length	20.33 ft.	**Landing Dist.**	N/A
Max Gross Wt.	990 lbs.	**Vmax**	81 mph
Empty Wt.	484 lbs.	**VCr**	56 mph
Payload (full fuel)	N/A	**V_{S1} (stall clean)**	28 mph
Fuel Capacity	13 gal.	**V_{S0} (Indg config)**	N/A
Seats	2	**Climb Rate @msl**	984 fpm
Cabin Width	N/A	**Service Ceiling**	N/A

Std. Engine/HP	Usable HP Range	Construction Mtrl.	Est. Bldg. Time
Rotax 503 / 50 hp	N/A	Metal, Tube, Fabric	N/A
Plans	**Kit**	**Info Packet**	**Video**
N/A	N/A	N/A	N/A

A sleek, tandem two seat, three-axis ultralight. ALS 500 Steel structure with Alloy 6082 T6 wings. Waterproof, fireproof, anti-U.V. covering is standard.

EURO-FLY
Via Ca Onorai
Galliera Veneta (Padova) 50-35015
ITALY
Tel. 049 596 5464 • Fax 049 596 5464

Fire Cat

Wing Span	31.50 ft.	**Range**	N/A
Wing Area	172.1 sq. ft.	**Takeoff Dist.**	164 ft.
Length	20.33 ft.	**Landing Dist.**	229 ft.
Max Gross Wt.	990 lbs.	**Vmax**	81 mph
Empty Wt.	480 lbs.	**VCr**	56 mph
Payload (full fuel)	N/A	**V_{S1} (stall clean)**	28 mph
Fuel Capacity	13 gal.	**V_{S0} (Indg config)**	N/A
Seats	2	**Climb Rate @msl**	984 fpm
Cabin Width	N/A	**Service Ceiling**	N/A

Std. Engine/HP	Usable HP Range	Construction Mtrl.	Est. Bldg. Time
Rotax 503 / 50 hp	N/A	Metal, Tube, Fabric	N/A
Plans	**Kit**	**Info Packet**	**Video**
N/A	N/A	Yes	N/A

A two seat, side-by-side, pusher design. Electric starter standard.

EURO-FLY
Via Ca Onorai
Galliera Veneta (Padova) 50-35015
ITALY
Tel. 049 596 5464 • Fax 049 596 5464

Flash

Wing Span	28.53 ft.	**Range**	N/A
Wing Area	140.4 sq. ft.	**Takeoff Dist.**	262 ft.
Length	21.32 ft.	**Landing Dist.**	295 ft.
Max Gross Wt.	990 lbs.	**Vmax**	118 mph
Empty Wt.	583 lbs.	**VCr**	99 mph
Payload (full fuel)	N/A	**V_{S1} (stall clean)**	36 mph
Fuel Capacity	13 gal.	**V_{S0} (Indg config)**	N/A
Seats	2	**Climb Rate @msl**	1377 fpm
Cabin Width	N/A	**Service Ceiling**	N/A

Std. Engine/HP	Usable HP Range	Construction Mtrl.	Est. Bldg. Time
Rotax 912 / 80 hp	N/A	Metal, Fabric	N/A
Plans	**Kit**	**Info Packet**	**Video**
N/A	N/A	Yes	N/A

An easy to use advanced ultralight design. Steel fuselage. Alloy wing ribs. Other engine options include the Rotax 582 and Rotax 618. Specs are for the Rotax 912. Available in kit form or fully assembled.

EURO-FLY
Via Ca Onorai
Galliera Veneta (Padova) 50-35015
ITALY
Tel. 049 596 5464 • Fax 049 596 5464

Europa

Wing Span	26.00 ft.	**Range**	550 sm
Wing Area	95.00 sq. ft.	**Takeoff Dist.**	350 ft.
Length	19.20 ft.	**Landing Dist.**	800 ft.
Max Gross Wt.	1375 lbs.	**Vmax**	188 mph
Empty Wt.	700 lbs.	**VCr**	150 mph
Payload (full fuel)	445 lbs.	**V_{S1} (stall clean)**	48 mph
Fuel Capacity	19 gal.	**V_{S0} (Indg config)**	42 mph
Seats	2	**Climb Rate @msl**	800 fpm
Cabin Width	44 in.	**Service Ceiling**	15,000 ft.

Std. Engine/HP	Usable HP Range	Construction Mtrl.	Est. Bldg. Time
Rotax 912 / 80 hp	80-120	Composite	850 hrs.
Plans	**Kit**	**Info Packet**	**Video**
N/A	$19,985.00	$15.00	$15.00

Able to operate from unprepared grass strips with retractable, single landing wheel and outriggers. Also available with tricycle landing gear. Rigs/de-rigs in under 5 minutes. Ergonomically designed cockpit and instrument module. Prices depend on current exchange rate. The Europa Club has a web page at http://www.avnet.co.uk/europa. **Sold:** 300+ **Flying:** 10

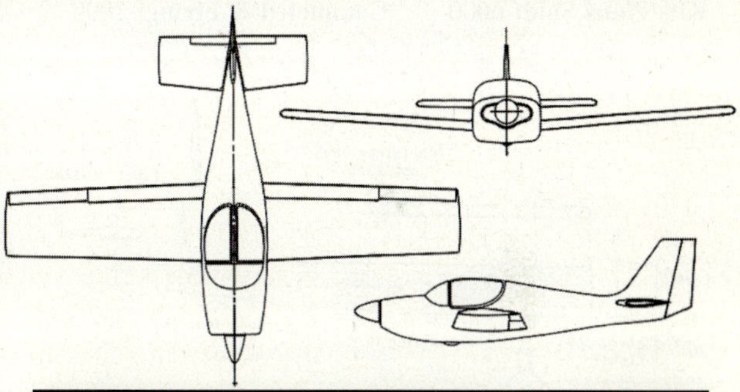

Optional glider wings available. Wingspan of 43 feet gives a glide ratio of approximately 25:1 and lowers stall speed to 39 mph.

U.S. DISTRIBUTOR:
Europa Aviation Inc. • 3400 Airfield Dr. West • Suite 99
Lakeland FL 33811 • Tel.: 941-647-5355 • Fax: 941-646-2877
E-Mail: europa@gate.net

EUROPA AVIATION
Unit 2A Dove Way • Kirby Mills Industrial Estate
Kirbymoorside, N. Yorks YO6 6NR
ENGLAND
Tel. 44 01751 431773 • Fax 44 01751 431706

Volksplane VP-1

Wing Span	24.00 ft.	**Range**	200 sm
Wing Area	100.0 sq. ft.	**Takeoff Dist.**	500 ft.
Length	18.00 ft.	**Landing Dist.**	400 ft.
Max Gross Wt.	750 lbs.	**Vmax**	95 mph
Empty Wt.	475 lbs.	**VCr**	75 mph
Payload (full fuel)	225 lbs.	**V_{S1} (stall clean)**	45 mph
Fuel Capacity	8 gal.	**V_{S0} (lndg config)**	N/A
Seats	1	**Climb Rate @msl**	600 fpm
Cabin Width	22 in.	**Service Ceiling**	10,000 ft.

Std. Engine/HP	Usable HP Range	Construction Mtrl.	Est. Bldg. Time
VW 1600 / 50 hp	50 - 65	Wood, Fabric	700 hrs.
Plans	**Kit**	**Info Packet**	**Video**
$44.00	N/A	$3.00	N/A

Low wing, single place taildragger. Designed as the simplest possible, basic aircraft for the novice homebuilder with a Cub's easy handling and flight safety characteristics. Functional lines have a beauty of their own. Good STOL performance. Simple, easy to build construction. Very low cost. Introduced in 1969, hundreds are flying in many countries.

Kits/Plans Sold: 6000 **Completed & Flying:** 1000

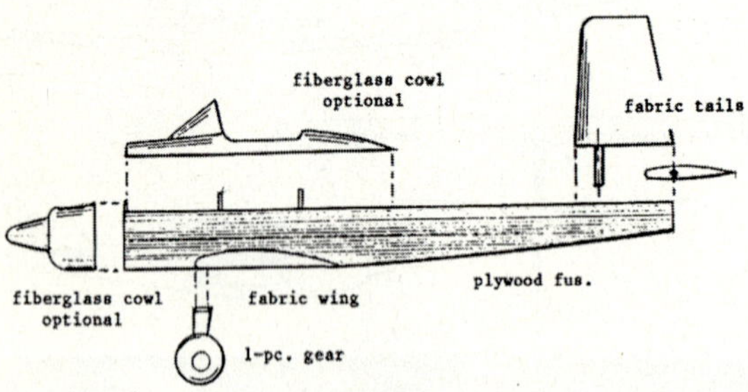

EVANS AIRCRAFT
Box 744-A
La Jolla CA 92038
Web: http://www.evanair.com

Express Series 90

Wing Span	31.00 ft.	**Range**	1360 sm
Wing Area	130.0 sq. ft.	**Takeoff Dist.**	N/A
Length	25.00 ft.	**Landing Dist.**	N/A
Max Gross Wt.	3200 lbs.	**Vmax**	N/A
Empty Wt.	1840 lbs.	**VCr**	210 mph
Payload (full fuel)	800 lbs.	**V_{S1} (stall clean)**	N/A
Fuel Capacity	92 gal.	**V_{S0} (lndg config)**	57 mph
Seats	4	**Climb Rate @msl**	N/A
Cabin Width	46 in.	**Service Ceiling**	N/A

Std. Engine/HP	Usable HP Range	Construction Mtrl.	Est. Bldg. Time
Lyc. IO-540 / 300 hp	N/A	Composite	2000 hrs.
Plans	**Kit**	**Info Packet**	**Video**
N/A	$31,980.00	Yes	Yes

Update of the Express design providing enhancements in performance, ease of construction and price. 20% larger conventional tail gives a wider CG range and enhanced pitch and directional stability. Variety of build packages available.

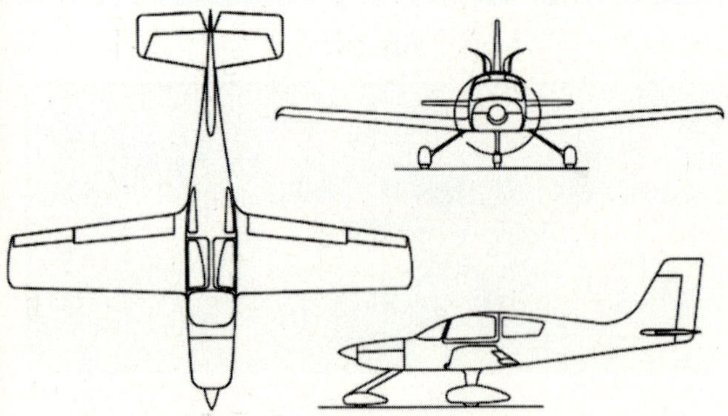

EXPRESS AIRCRAFT COMPANY L.L.C.
7825 Old Highway 99 SE, Suite • Olympia Airport
Olympia WA 98501
Tel. 360-352-3554 • Fax 360-352-3231
Web: http://www.express-aircraft.com

HOMEBUILT "FAQ'S" (See Page 68)

Q: Can I insure a homebuilt airplane?

A: Yes you can. The insurance company may have their own requirements in addition to the FARs, but they normally will insure a homebuilt. You may find lower limits on passenger liability coverage though.

Where fast, high performance kit aircraft are concerned, one major aviation insurance company is insisting on periodic construction inspections by company reps, and thorough checkouts in type (+10hrs) before they will underwrite a policy for that aircraft.

Express

Wing Span	31.50 ft.	**Range**	1800 sm
Wing Area	130.0 sq. ft.	**Takeoff Dist.**	700 ft.
Length	26.00 ft.	**Landing Dist.**	1200 ft.
Max Gross Wt.	3200 lbs.	**Vmax**	225 mph
Empty Wt.	1825 lbs.	**VCr**	220 mph
Payload (full fuel)	823 lbs.	**V_{S1} (stall clean)**	N/A
Fuel Capacity	92 gal.	**V_{S0} (lndg config)**	54 mph
Seats	4	**Climb Rate @msl**	1800 fpm
Cabin Width	46 in.	**Service Ceiling**	24,000 ft.

Std. Engine/HP	Usable HP Range	Construction Mtrl.	Est. Bldg. Time
Lyc. IO540 / 250 hp	N/A	Composite	N/A
Plans	**Kit**	**Info Packet**	**Video**
N/A	$31,950.00	Yes	Yes

All composite performance cruiser. Available in three kits: Empennage, Fuselage, and Wing.

EXPRESS AIRCRAFT COMPANY L.L.C.
7825 Old Highway 99 SE, Suite • Olympia Airport
Olympia WA 98501
Tel. 360-352-3554 • Fax 360-352-3231
Web: http://www.express-aircraft.com

Peregrine Falcon

Wing Span	36.00 ft.	**Range**	240 sm
Wing Area	187.0 sq. ft.	**Takeoff Dist.**	150 ft.
Length	14.25 ft.	**Landing Dist.**	90 ft.
Max Gross Wt.	500 lbs.	**Vmax**	130 mph
Empty Wt.	254 lbs.	**VCr**	55 mph
Payload (full fuel)	225 lbs.	**V_{S1} (stall clean)**	24 mph
Fuel Capacity	5 gal.	**V_{S0} (lndg config)**	35 mph
Seats	1	**Climb Rate @msl**	800 fpm
Cabin Width	24 in.	**Service Ceiling**	15,000 ft.

Std. Engine/HP	Usable HP Range	Construction Mtrl.	Est. Bldg. Time
Rotax 447 / 42 hp	N/A	Alum. Tube, Composite, Fabric	50 hrs.
Plans	**Kit**	**Info Packet**	**Video**
N/A	$12,500.00	Yes	Yes

Part 103 UL legal with Rotax 447 engine. Enclosed cockpit canard design. L/D of 14:1. Very short building time.
Kits/Plans Sold: 2 **Completed & Flying:** 2

FALCON EAST
155 Oscawana Lake Road • Putnam Valley NY 10579
Tel. 888-528-8940 • Fax 914-528-8940
E-Mail: Falconeast@aol.com
Web: http://ul-flyer.com/aircraft/falcon.html

Ellipse

Wing Span	36.62 ft.	**Range**	650 sm
Wing Area	128.0 sq. ft.	**Takeoff Dist.**	1100 ft.
Length	20.00 ft.	**Landing Dist.**	1100 ft.
Max Gross Wt.	2200 lbs.	**Vmax**	N/A
Empty Wt.	1177 lbs.	**VCr**	150 mph
Payload (full fuel)	1023 lbs.	**V_{S1} (stall clean)**	55 mph
Fuel Capacity	41 gal.	**V_{S0} (lndg config)**	N/A
Seats	4	**Climb Rate @msl**	950 fpm
Cabin Width	43 in.	**Service Ceiling**	15,200 ft.

Std. Engine/HP	Usable HP Range	Construction Mtrl.	Est. Bldg. Time
Lyc. 0-320 / 150 hp	125-200	Tube, Fabric	1000 hrs.
Plans	**Kit**	**Info Packet**	**Video**
N/A	$35,000.00	$10.00	N/A

Fast, high-wing design with elliptical wings. High payload . Folding rear seats with access to baggage area. Wings foldable in five minutes by one person. Pontoon and ski fittings are standard. Engine & propeller not included. **Sold:** 3 **Flying:** 1

EXPLORER AVIATION
Idaho County Airport • 120 E. South 6th St.
Grangeville ID 83530
Tel. 208-983-2423 • Fax 208-983-3767

Mignet HM290/HM293

Wing Span	20.00 ft.	**Range**	150 sm
Wing Area	120.0 sq. ft.	**Takeoff Dist.**	175 ft.
Length	11.50 ft.	**Landing Dist.**	175 ft.
Max Gross Wt.	580 lbs.	**Vmax**	72 mph
Empty Wt.	360 lbs.	**VCr**	60 mph
Payload (full fuel)	190 lbs.	**V_{S1} (stall clean)**	90 mph
Fuel Capacity	10 gal.	**V_{S0} (lndg config)**	85 mph
Seats	1	**Climb Rate @msl**	500 fpm
Cabin Width	22 in.	**Service Ceiling**	N/A

Std. Engine/HP	Usable HP Range	Construction Mtrl.	Est. Bldg. Time
VW / 60 hp	30 - 70	Wood, fabric	500 hrs.
Plans	**Kit**	**Info Packet**	**Video**
$75.00	$4,450.00	$10.00	$20.00

Specs are for HM290. HM293 is larger. E models of both designs offer lighter weight and more wing area to meet ultralight vehicle rules. Engine options are available for the four models.
Kits/Plans Sold: 650 **Completed & Flying:** 35

FALCONAR AVIA, INC.
7739 - 81 Ave. • Edmonton, Alberta T6C 0V4
CANADA
Tel. 403-465-2024 • Fax 403-465-2029
E-Mail: falconar@tic.ab.ca

SAL P-51D Mustang

Wing Span	24.83 ft.	**Range**	500 sm
Wing Area	110.0 sq. ft.	**Takeoff Dist.**	700 ft.
Length	22.50 ft.	**Landing Dist.**	700 ft.
Max Gross Wt.	1985 lbs.	**Vmax**	258 mph
Empty Wt.	1420 lbs.	**VCr**	176 mph
Payload (full fuel)	536 lbs.	**V_{S1} (stall clean)**	67 mph
Fuel Capacity	31 gal.	**V_{S0} (lndg config)**	60 mph
Seats	1	**Climb Rate @msl**	1850 fpm
Cabin Width	24 in.	**Service Ceiling**	17,400 ft.

Std. Engine/HP	Usable HP Range	Construction Mtrl.	Est. Bldg. Time
Ranger / 175 hp	150-250	Wood, Fiberglass	2500 hrs.
Plans	**Kit**	**Info Packet**	**Video**
$500.00	$25,500.00	$10.00	N/A

2/3 scale. Introduced in 1969. Two seat version available. Engine choices include Ranger, Javelin Ford 230 V-6B 200 hp, other small V-6 or V-8, Avia M337, Lycoming IO-360, Continental 0-470. Wood and plywood with some fiberglass and metal parts. Flight proven by several aircraft over several hundred hours of flying with Ranger engine. Features include jettisonable canopy, removable belly scoop assembly, electric or manual flaps, electric or manual tailwheel retract, and a close to scale cockpit layout. **Kits/Plans Sold:** 825 **Flying:** 16

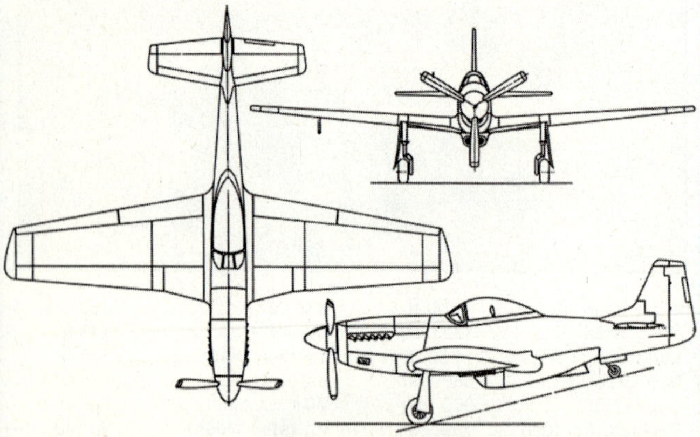

Joe Broeders, Pilot
"Real good on the ground - no problem keeping it straight for landing or takeoff ... turns excellent ...pitch excellent ... feels very good at speed 110 mph and above."

Bill Green, Pilot
"You have a very nice flying airplane ... the nearest thing to the real Mustang I have flown."

FALCONAR AVIA, INC.
7739 - 81 Ave. • Edmonton, Alberta T6C 0V4
CANADA
Tel. 403-465-2024 • Fax 403-465-2029
E-Mail: falconar@tic.ab.ca

Cubmajor

Wing Span	35.00 ft.	**Range**	235 sm
Wing Area	163.0 sq. ft.	**Takeoff Dist.**	300 ft.
Length	23.75 ft.	**Landing Dist.**	250 ft.
Max Gross Wt.	1400 lbs.	**Vmax**	115 mph
Empty Wt.	900 lbs.	**VCr**	110 mph
Payload (full fuel)	500 lbs.	**V_{S1} (stall clean)**	N/A
Fuel Capacity	13 gal.	**V_{S0} (lndg config)**	40 mph
Seats	2	**Climb Rate @msl**	1000 fpm
Cabin Width	24 in.	**Service Ceiling**	N/A

Std. Engine/HP	Usable HP Range	Construction Mtrl.	Est. Bldg. Time
Cont C-85 / 85 hp	75 - 125	Wood, Fabric	1400 hrs.
Plans	**Kit**	**Info Packet**	**Video**
$150.00	$8,200.00	N/A	$20.00

Tandem seating. Generous wing area for short takeoff and landing without need of flaps. Constructed primarily of wood with steel tube cabin section and steel fittings. Recent improvements include self lubricating Delrin bearings for aileron and tail hinges, sealed ailerons for improved cruise effectiveness, lower stall, and ease of construction. Adaptable to skis or floats. Folding wings for ease of storage and transportation.

Completed & Flying: 6

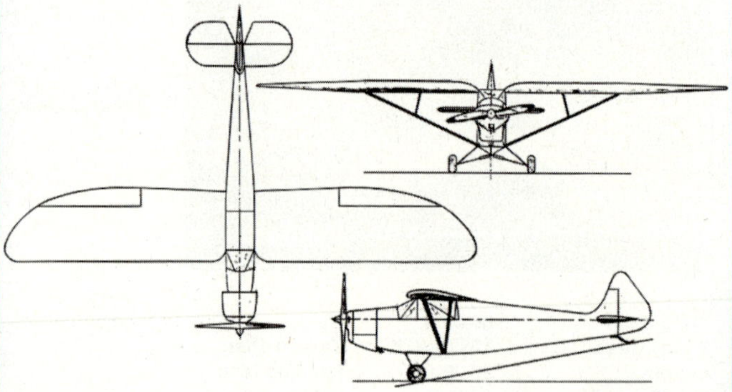

The "Major" is a time proven design. It was originally designed in England by Mr. C. H. Latimer-Needham to Air Ministry requirements, as laid down in A.P. 1208 but was re-stressed in 1961 to current A.R.B. requirements. The opportunity was taken to increase the permitted engine power and the gross weight to 1,100 pounds. Although the 62 hp Walter Mikron was used for the pre-war prototype in England, recommended power is now 85 to 125. Continental and Lycoming engines in this power range are suitable.

FALCONAR AVIA, INC.
7739 - 81 Ave. • Edmonton, Alberta T6C 0V4
CANADA
Tel. 403-465-2024 • Fax 403-465-2029
E-Mail: falconar@tic.ab.ca

ARV-1K Golden Hawk

Wing Span	34.00 ft.	Range	200 sm
Wing Area	168.0 sq. ft.	Takeoff Dist.	250 ft.
Length	14.00 ft.	Landing Dist.	250 ft.
Max Gross Wt.	990 lbs.	Vmax	105 mph
Empty Wt.	500 lbs.	VCr	90 mph
Payload (full fuel)	N/A	V_{S1} (stall clean)	50 mph
Fuel Capacity	9 gal.	V_{S0} (lndg config)	40 mph
Seats	2	Climb Rate @msl	1100 fpm
Cabin Width	30 in.	Service Ceiling	N/A

Std. Engine/HP	Usable HP Range	Construction Mtrl.	Est. Bldg. Time
Hirth / 55 hp	50 - 70	Composite, Metal	500 hrs.
Plans	**Kit**	**Info Packet**	**Video**
N/A	$19,700.00	$10.00	N/A

Canard design, high wing, strut braced. Spinproof. Good visibility. Highly maneuverable. Composite fuselage. Metal wings, canard, and fins. Fabric covered aluminum wings and canard. Lightweight. Tandem seating. Assemble/disassemble in 10 minutes. Kit is complete with engine, prop, fuselage pod, fuel tank, hydraulic brakes, instruments, HIPEC covering materials. Available in three sub-kits. **Completed & Flying:** 4

FALCONAR AVIA, INC.
7739 - 81 Ave. • Edmonton, Alberta T6C 0V4
CANADA
Tel. 403-465-2024 • Fax 403-465-2029
E-Mail: falconar@tic.ab.ca

Falconar F-11A

Wing Span	27.25 ft.	Range	550 sm
Wing Area	138.0 sq. ft.	Takeoff Dist.	200 ft.
Length	22.00 ft.	Landing Dist.	300 ft.
Max Gross Wt.	1300 lbs.	Vmax	185 mph
Empty Wt.	780 lbs.	VCr	120 mph
Payload (full fuel)	515 lbs.	V_{S1} (stall clean)	35 mph
Fuel Capacity	17 gal.	V_{S0} (lndg config)	N/A
Seats	2	Climb Rate @msl	1000 fpm
Cabin Width	40 in.	Service Ceiling	14,000 ft.

Std. Engine/HP	Usable HP Range	Construction Mtrl.	Est. Bldg. Time
Continental / 85 hp	65 - 130	Wood, Fabric	1000 hrs.
Plans	**Kit**	**Info Packet**	**Video**
$200.00	$7,000.00	$10.00	N/A

Offers large area, low aspect ratio control surfaces for good "feel" and positive control. Good longitudinal, directional, and lateral stability for comfortable and easy flying. Variety of engine options include Continental 65-100 hp, Lyc 100-125 hp, Franklin 125-130 hp, and others. **Sold:** 800 **Flying:** 50

[illegible] Allinott, Builder/Pilot
"Easy to build ... really stable and easy to fly."

FALCONAR AVIA, INC.
7739 - 81 Ave. • Edmonton, Alberta T6C 0V4
CANADA
Tel. 403-465-2024 • Fax 403-465-2029
E-Mail: falconar@tic.ab.ca

Falconar F-12A

Wing Span	28.00 ft.	Range	400 sm
Wing Area	140.0 sq. ft.	Takeoff Dist.	450 ft.
Length	24.00 ft.	Landing Dist.	500 ft.
Max Gross Wt.	1800 lbs.	Vmax	185 mph
Empty Wt.	1150 lbs.	VCr	160 mph
Payload (full fuel)	650 lbs.	V_{S1} (stall clean)	57 mph
Fuel Capacity	26 gal.	V_{S0} (lndg config)	57 mph
Seats	2	Climb Rate @msl	1100 fpm
Cabin Width	44 in.	Service Ceiling	17,000 ft.

Std. Engine/HP	Usable HP Range	Construction Mtrl.	Est. Bldg. Time
Lyc. 0-320 / 150 hp	125 - 200	Wood, Fabric	1200 hrs.
Plans	**Kit**	**Info Packet**	**Video**
$220.00	$7,700.00	$10.00	N/A

A robust aircraft designed for years of rough field operation. The 44" wide cockpit is comfortable for two large adults. A third seat is optional. It can cruise at 150 mph with a 150 hp engine.

Completed & Flying: 25

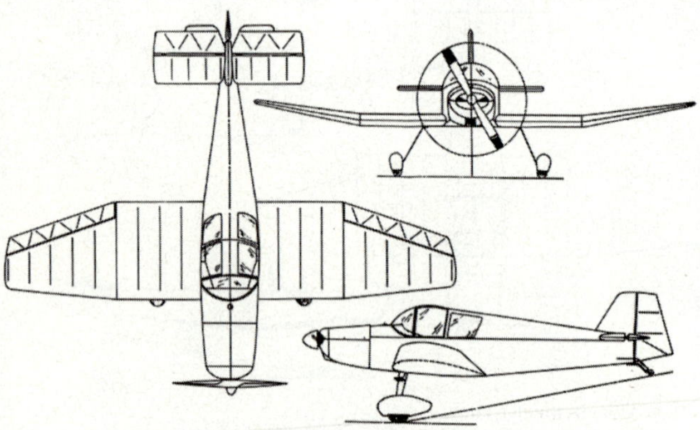

FALCONAR AVIA, INC.
7739 - 81 Ave. • Edmonton, Alberta T6C 0V4
CANADA
Tel. 403-465-2024 • Fax 403-465-2029
E-Mail: falconar@tic.ab.ca

AMF-S14 Maranda

Wing Span	32.00 ft.	Range	400 sm
Wing Area	140.0 sq. ft.	Takeoff Dist.	200 ft.
Length	22.00 ft.	Landing Dist.	180 ft.
Max Gross Wt.	1850 lbs.	Vmax	150 mph
Empty Wt.	1100 lbs.	VCr	125 mph
Payload (full fuel)	750 lbs.	V_{S1} (stall clean)	39 mph
Fuel Capacity	52 gal.	V_{S0} (lndg config)	N/A
Seats	2	Climb Rate @msl	1300 fpm
Cabin Width	46 in.	Service Ceiling	17,000 ft.

Std. Engine/HP	Usable HP Range	Construction Mtrl.	Est. Bldg. Time
Lyc 0-320 / 150 hp	N/A	Wood, Fabric	1200 hrs.
Plans	**Kit**	**Info Packet**	**Video**
$170.00	$9,400.00	$10.00	N/A

A versatile aircraft for the sportsman. Excellent visibility, high wing, and long fuselage. Simple, robust, all-wood construction, optional flaps, folding wings, large cabin and larger baggage compartment. Top loading door. Adaptable to floats or skis. Introduced in 1961. Recent improvements include a 46" extra wide cabin and heavy duty shock struts for higher gross weights. More than 90 completions reported.

Kits/Plans Sold: 850 **Completed & Flying:** 90

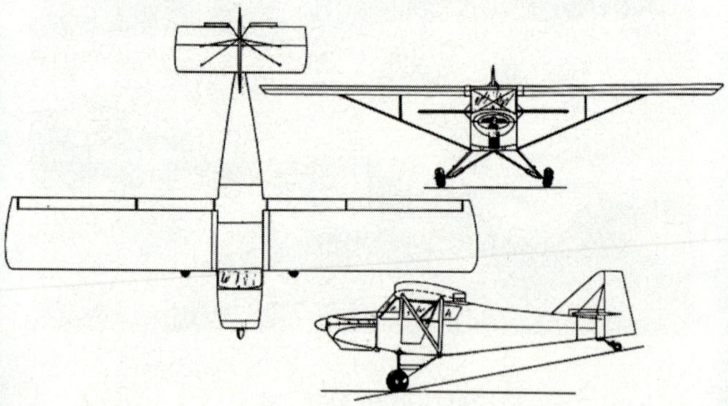

Design influences of the AMF-S14 Maranda include the Adam RA-14 from France, Fleet Canuck from Canada, the Jodel D11 from France, and the Taylorcraft BC12D from the U.S. The first prototype flew in 1961 from the Del-Van airpark near Vancouver, B.C. Since then, the gross weight has been increased and numerous design refinements have been incorporated. The latest to fly incorporates a wide cabin and Nissan diesel power.

Bob Pearce, EAA Chapter, Regina, Sask.
"One of the nicest flying aircraft I have ever flown."

Gavin Grimmer, Hastings, New Zealand
"Super smooth in flight - a dream to fly."

FALCONAR AVIA, INC.
7739 - 81 Ave. • Edmonton, Alberta T6C 0V4
CANADA
Tel. 403-465-2024 • Fax 403-465-2029
E-Mail: falconar@tic.ab.ca

Falconar F9A

Wing Span	23.00 ft.	**Range**	300 sm
Wing Area	99.00 sq. ft.	**Takeoff Dist.**	260 ft.
Length	17.90 ft.	**Landing Dist.**	300 ft.
Max Gross Wt.	660 lbs.	**Vmax**	140 mph
Empty Wt.	405 lbs.	**VCr**	100 mph
Payload (full fuel)	195 lbs.	V_{S1} **(stall clean)**	43 mph
Fuel Capacity	10 gal.	V_{S0} **(lndg config)**	38 mph
Seats	1	**Climb Rate @msl**	1020 fpm
Cabin Width	21 in.	**Service Ceiling**	N/A

Std. Engine/HP	Usable HP Range	Construction Mtrl.	Est. Bldg. Time
VW / 60 hp	40 - 90	Wood, Fabric	700 hrs.
Plans	**Kit**	**Info Packet**	**Video**
$190.00	$5,300.00	N/A	$20.00

Small, simple, single seater. Larger, more comfortable cockpit than the older F9. Engine options include Hirth and VW. Introduced in 1965. **Completed & Flying:** 30

FALCONAR AVIA, INC.
7739 - 81 Ave. • Edmonton, Alberta T6C 0V4
CANADA
Tel. 403-465-2024 • Fax 403-465-2029
E-Mail: falconar@tic.ab.ca

Mignet HM380 Ladybug

Wing Span	27.00 ft.	**Range**	470 sm
Wing Area	180.5 sq. ft.	**Takeoff Dist.**	500 ft.
Length	16.00 ft.	**Landing Dist.**	500 ft.
Max Gross Wt.	1100 lbs.	**Vmax**	124 mph
Empty Wt.	580 lbs.	**VCr**	113 mph
Payload (full fuel)	N/A	V_{S1} **(stall clean)**	28 mph
Fuel Capacity	23 gal.	V_{S0} **(lndg config)**	N/A
Seats	2	**Climb Rate @msl**	900 fpm
Cabin Width	38 in.	**Service Ceiling**	13,000 ft.

Std. Engine/HP	Usable HP Range	Construction Mtrl.	Est. Bldg. Time
Cont / 100 hp	60 - 100	Wood, Tube, Fabric	1200 hrs.
Plans	**Kit**	**Info Packet**	**Video**
$175.00	$6,286.00	$10.00	N/A

A Mignet type tandem wing aircraft. Info pack covers three models: the single place HM360, the two place, wood and fabric HM380, and the two place, tube and fabric HM380L. Kit price does not include engine. **Completed & Flying:** 1

FALCONAR AVIA, INC.
7739 - 81 Ave. • Edmonton, Alberta T6C 0V4
CANADA
Tel. 403-465-2024 • Fax 403-465-2029
E-Mail: falconar@tic.ab.ca

Falconar F-11E

Wing Span	27.25 ft.	**Range**	560 sm
Wing Area	138.0 sq. ft.	**Takeoff Dist.**	300 ft.
Length	21.00 ft.	**Landing Dist.**	250 ft.
Max Gross Wt.	1100 lbs.	**Vmax**	140 mph
Empty Wt.	650 lbs.	**VCr**	110 mph
Payload (full fuel)	460 lbs.	V_{S1} **(stall clean)**	35 mph
Fuel Capacity	18 gal.	V_{S0} **(lndg config)**	N/A
Seats	2	**Climb Rate @msl**	1000 fpm
Cabin Width	40 in.	**Service Ceiling**	15,000 ft.

Std. Engine/HP	Usable HP Range	Construction Mtrl.	Est. Bldg. Time
VW / 85 hp	50 - 95	Wood, Fabric	1000 hrs.
Plans	**Kit**	**Info Packet**	**Video**
$210.00	$5,900.00	N/A	$20.00

Similar to F11A but 65 pounds lighter empty weight and significantly less gross weight for better performance. Introduced in 1987. **Completed & Flying:** 1

FALCONAR AVIA, INC.
7739 - 81 Ave. • Edmonton, Alberta T6C 0V4
CANADA
Tel. 403-465-2024 • Fax 403-465-2029
E-Mail: falconar@tic.ab.ca

Minihawk

Wing Span	25.50 ft.	**Range**	570 sm
Wing Area	106.0 sq. ft.	**Takeoff Dist.**	270 ft.
Length	18.00 ft.	**Landing Dist.**	300 ft.
Max Gross Wt.	1070 lbs.	**Vmax**	128 mph
Empty Wt.	645 lbs.	**VCr**	115 mph
Payload (full fuel)	N/A	V_{S1} **(stall clean)**	56 mph
Fuel Capacity	15 gal.	V_{S0} **(lndg config)**	50 mph
Seats	2	**Climb Rate @msl**	1350 fpm
Cabin Width	N/A	**Service Ceiling**	N/A

Std. Engine/HP	Usable HP Range	Construction Mtrl.	Est. Bldg. Time
Cont O-200 / 100 hp	65 - 100	Wood, Fabric	1500 hrs.
Plans	**Kit**	**Info Packet**	**Video**
$235.00	$11,600.00	$10.00	N/A

Wood and fabric construction. Can be built in either tricycle or taildragger configurations.

FALCONAR AVIA, INC.
7739 - 81 Ave. • Edmonton, Alberta T6C 0V4
CANADA
Tel. 403-465-2024 • Fax 403-465-2029
E-Mail: falconar@tic.ab.ca

F-II

Wing Span	29.50 ft.	**Range**	200 sm
Wing Area	140.0 sq. ft.	**Takeoff Dist.**	200 ft.
Length	22.00 ft.	**Landing Dist.**	200 ft.
Max Gross Wt.	900 lbs.	**Vmax**	80 mph
Empty Wt.	400 lbs.	**VCr**	70 mph
Payload (full fuel)	475 lbs.	V_{S1} **(stall clean)**	32 mph
Fuel Capacity	13 gal.	V_{S0} **(lndg config)**	28 mph
Seats	2	**Climb Rate @msl**	1200 fpm
Cabin Width	41 in.	**Service Ceiling**	10,000 ft.

Std. Engine/HP	Usable HP Range	Construction Mtrl.	Est. Bldg. Time
Rotax 503 / 52 hp	46-70	Aluminum, Fabric, Steel	250 hrs.
Plans	**Kit**	**Info Packet**	**Video**
N/A	$9,850.00	$3.00	N/A

1991 Sun 'n Fun Grand Champion. Easy entry and exit through big gull wing doors. Fully enclosed and vented cabin. Easy fuel fill, baggage bin, foldable wings and tail. Rugged but light construction. Light and precise handling, flaperons, two axis trim, dual lift struts and tail wires. Can be flown from either seat. New Quick Build Kit includes some prefab components.

Kits Sold: 21 **Completed & Flying:** 14

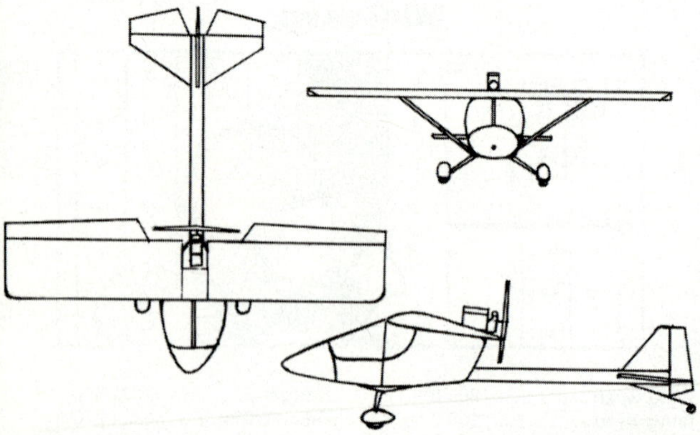

The F-II has a fully enclosed and vented two seat, side-by-side cabin made of fiberglass and fabric. Dual rudder controls and a centrally located control stick allow the plane to be flown from either seat. The wing has dual support struts for more rigidity and this, coupled with the boost tabs, makes for smooth, light and responsive handling. Visibility out of the Lexan windows is excellent.

B.V. Lawson, Nashville TN

"I really like the performance and handling. The F-II turned out to be my dream plane and it now sits in my hanger."

Bob Leatherwood, Dallas, GA

"In September 1993, I received my Ferguson F-II kit and the following March 1994 I made my first flight without a hitch. ... After 80 hours I am very pleased ... The Ferguson was my first kit construction project and it was very easy to build. ... Flying the Ferguson is a real pleasure."

FERGUSON AIRCRAFT, INC.
2431 Ferguson Place
Dallas GA 30132
Tel. 770-443-2747 • Fax 770-443-2747

Tucano X2

Wing Span	33.35 ft.	**Range**	N/A
Wing Area	182.6 sq. ft.	**Takeoff Dist.**	N/A
Length	20.66 ft.	**Landing Dist.**	N/A
Max Gross Wt.	N/A	**Vmax**	209 mph
Empty Wt.	625 lbs.	**VCr**	143 mph
Payload (full fuel)	N/A	V_{S1} **(stall clean)**	121 mph
Fuel Capacity	N/A	V_{S0} **(lndg config)**	N/A
Seats	2	**Climb Rate @msl**	N/A
Cabin Width	N/A	**Service Ceiling**	N/A

Std. Engine/HP	Usable HP Range	Construction Mtrl.	Est. Bldg. Time
2-Rotax 582	N/A	Tube, Fabric	N/A
Plans	**Kit**	**Info Packet**	**Video**
N/A	N/A	N/A	N/A

A unique design that uses two engines in a push-pull configuration. Fuel consumption at cruise speed is 4 gallons per hour.

FERRARI, U.L.M.
Via Garibaldi 104/a
Castelbaldo (PD) 35040
ITALY
Tel. 0425 57316

Youngster

Wing Span	18.00 ft.	**Range**	N/A
Wing Area	126.0 sq. ft.	**Takeoff Dist.**	200 ft.
Length	15.50 ft.	**Landing Dist.**	250 ft.
Max Gross Wt.	650 lbs.	**Vmax**	110 mph
Empty Wt.	400 lbs.	**VCr**	85 mph
Payload (full fuel)	N/A	V_{S1} **(stall clean)**	33 mph
Fuel Capacity	8 gal.	V_{S0} **(lndg config)**	N/A
Seats	1	**Climb Rate @msl**	700 fpm
Cabin Width	N/A	**Service Ceiling**	N/A

Std. Engine/HP	Usable HP Range	Construction Mtrl.	Est. Bldg. Time
Great Plains / 50 hp	50-85	Wood, Fabric	400 hrs.
Plans	**Kit**	**Info Packet**	**Video**
$200.00	$3,400.00	$5.00	N/A

An affordable, nostalgic beauty. With a +4, -2.5 G's design and ultimate of +6, -3.75 G's, the properly trained pilot can perform sportsman category aerobatics safely. Can accommodate a four cycle VW or Suburu engine up to 85 hp and 160 pounds.

FISHER AERO CORP.
7118 S.R. 335
Portsmouth OH 45662
Tel. 614-820-2219 • Fax 614-820-2219

Celebrity

Wing Span	22.00 ft.	**Range**	200 sm
Wing Area	176.0 sq. ft.	**Takeoff Dist.**	350 ft.
Length	17.60 ft.	**Landing Dist.**	300 ft.
Max Gross Wt.	1050 lbs.	**Vmax**	110 mph
Empty Wt.	525 lbs.	**VCr**	95 mph
Payload (full fuel)	N/A	**V_{S1} (stall clean)**	40 mph
Fuel Capacity	13 gal.	**V_{S0} (lndg config)**	N/A
Seats	2	**Climb Rate @msl**	700 fpm
Cabin Width	26 in.	**Service Ceiling**	N/A

Std. Engine/HP	Usable HP Range	Construction Mtrl.	Est. Bldg. Time
Cont. / 115 hp	70-125	Metal, Wood, Fabric	600 hrs.
Plans	**Kit**	**Info Packet**	**Video**
N/A	$7,600.00	$8.00	N/A

Two place biplane. Easy access to both cockpits. Wings are detachable for storage. Continental engines available.

FISHER AERO CORP.
7118 S.R. 335
Portsmouth OH 45662
Tel. 614-820-2219 • Fax 614-820-2219

Horizon II

Wing Span	26.00 ft.	**Range**	250 sm
Wing Area	114.0 sq. ft.	**Takeoff Dist.**	250 ft.
Length	19.80 ft.	**Landing Dist.**	250 ft.
Max Gross Wt.	1050 lbs.	**Vmax**	105 mph
Empty Wt.	570 lbs.	**VCr**	95 mph
Payload (full fuel)	N/A	**V_{S1} (stall clean)**	37 mph
Fuel Capacity	13 gal.	**V_{S0} (lndg config)**	N/A
Seats	2	**Climb Rate @msl**	900 fpm
Cabin Width	27 in.	**Service Ceiling**	N/A

Std. Engine/HP	Usable HP Range	Construction Mtrl.	Est. Bldg. Time
Limbach / 115 hp	N/A	Metal/Fabric or Wood/Fabric	600 hrs.
Plans	**Kit**	**Info Packet**	**Video**
$350.00	$7,500.00	$8.00	N/A

Two place tandem seating. Look and feel of the L-19. Folding wings. Continental engines available. All-wood or welded fuselage.

FISHER AERO CORP.
7118 S.R. 335
Portsmouth OH 45662
Tel. 614-820-2219 • Fax 614-820-2219

Horizon I

Wing Span	25.30 ft.	**Range**	250 sm
Wing Area	114.0 sq. ft.	**Takeoff Dist.**	[illegible] ft.
Length	18.80 ft.	**Landing Dist.**	250 ft.
Max Gross Wt.	1050 lbs.	**Vmax**	97 mph
Empty Wt.	520 lbs.	**VCr**	90 mph
Payload (full fuel)	N/A	**V_{S1} (stall clean)**	39 mph
Fuel Capacity	13 gal.	**V_{S0} (lndg config)**	N/A
Seats	2	**Climb Rate @msl**	900 fpm
Cabin Width	27 in.	**Service Ceiling**	N/A

Std. Engine/HP	Usable HP Range	Construction Mtrl.	Est. Bldg. Time
115 hp	N/A	Wood, Fabric	500 hrs.
Plans	**Kit**	**Info Packet**	**Video**
$275.00	$5,800.00	$8.00	N/A

Easy to fly. Tandem seating. Look and feel of the Champ. Folding wings. Variety of engines available.

FISHER AERO CORP.
7118 S.R. 335
Portsmouth OH 45662
Tel. 614-820-2219 • Fax 614-820-2219

Mariah

Wing Span	25.00 ft.	**Range**	400 sm
Wing Area	125.0 sq. ft.	**Takeoff Dist.**	400 ft.
Length	20.00 ft.	**Landing Dist.**	500 ft.
Max Gross Wt.	1300 lbs.	**Vmax**	150 mph
Empty Wt.	800 lbs.	**VCr**	130 mph
Payload (full fuel)	N/A	**V_{S1} (stall clean)**	N/A
Fuel Capacity	22 gal.	**V_{S0} (lndg config)**	N/A
Seats	2	**Climb Rate @msl**	1200 fpm
Cabin Width	N/A	**Service Ceiling**	N/A

Std. Engine/HP	Usable HP Range	Construction Mtrl.	Est. Bldg. Time
Continental / 85 hp	85 - 125	Wood, Fabric	1000 hrs.
Plans	**Kit**	**Info Packet**	**Video**
$350.00	$8,900.00	N/A	N/A

All-wood with tandem seating. Easy to assemble Quick Build Kit available for $12,500.

FISHER AERO CORP.
7118 S.R. 335
Portsmouth OH 45662
Tel. 614-820-2219 • Fax 614-820-2219

Avenger

Wing Span	27.00 ft.	**Range**	N/A
Wing Area	121.0 sq. ft.	**Takeoff Dist.**	200 ft.
Length	16.25 ft.	**Landing Dist.**	200 ft.
Max Gross Wt.	600 lbs.	**Vmax**	63 mph
Empty Wt.	250 lbs.	**VCr**	60 mph
Payload (full fuel)	N/A	**V_{S1} (stall clean)**	26 mph
Fuel Capacity	5 gal.	**V_{S0} (lndg config)**	N/A
Seats	1	**Climb Rate @msl**	900 fpm
Cabin Width	24 in.	**Service Ceiling**	N/A

Std. Engine/HP	Usable HP Range	Construction Mtrl.	Est. Bldg. Time
Rotax / 40 hp	40 - 65	Wood, Fabric	300 hrs.
Plans	**Kit**	**Info Packet**	**Video**
$150.00	$2,495.00	$5.00	N/A

Designed for ultralight and light plane builders and flyers. However, can accommodate a 6'4", 270 lb. pilot. Quite stable for hands-off flying. Kit includes everything except engine and mount. Can also use a 65 hp engine. **Sold:** 150 **Flying:** 30

FISHER AERO CORP.
7118 S.R. 335
Portsmouth OH 45662
Tel. 614-820-2219 • Fax 614-820-2219

Super Koala

Wing Span	31.00 ft.	**Range**	200 sm
Wing Area	140.0 sq. ft.	**Takeoff Dist.**	150 ft.
Length	18.08 ft.	**Landing Dist.**	200 ft.
Max Gross Wt.	830 lbs.	**Vmax**	95 mph
Empty Wt.	400 lbs.	**VCr**	75 mph
Payload (full fuel)	N/A	**V_{S1} (stall clean)**	30 mph
Fuel Capacity	10 gal.	**V_{S0} (lndg config)**	N/A
Seats	2	**Climb Rate @msl**	800 fpm
Cabin Width	34 in.	**Service Ceiling**	12,000 ft.

Std. Engine/HP	Usable HP Range	Construction Mtrl.	Est. Bldg. Time
Rotax 503 / 50 hp	50 - 65	Wood, Fabric	300 hrs.
Plans	**Kit**	**Info Packet**	**Video**
N/A	$6,575.00	$10.00	$15.00

Two place version of one of the most popular aircraft ever built. Detachable wings. Field assembly time is 20 minutes. Quick Build airframe, wings, and fuselage kits available. See low cost, single seat FP-202 on next page. **Sold:** 150 **Flying:** 75

FISHER FLYING PRODUCTS
P.O. Box 468 • Industrial Park • Edgeley ND 58433
Tel. 701-493-2286 • Fax 701-493-2539
E-Mail: ffpjac@daktel.com
Web: http://www.fisherflying.com

R-80 Tiger Moth

Wing Span	23.00 ft.	**Range**	250 sm
Wing Area	170.0 sq. ft.	**Takeoff Dist.**	300 ft.
Length	19.00 ft.	**Landing Dist.**	300 ft.
Max Gross Wt.	1150 lbs.	**Vmax**	110 mph
Empty Wt.	560 lbs.	**VCr**	90 mph
Payload (full fuel)	N/A	**V_{S1} (stall clean)**	30 mph
Fuel Capacity	12 gal.	**V_{S0} (lndg config)**	N/A
Seats	2	**Climb Rate @msl**	800 fpm
Cabin Width	26 in.	**Service Ceiling**	N/A

Std. Engine/HP	Usable HP Range	Construction Mtrl.	Est. Bldg. Time
Norton Rot. / 100 hp	75 - 100	Wood, Fabric	500 hrs.
Plans	**Kit**	**Info Packet**	**Video**
N/A	$9,300.00	$10.00	$15.00

80% scale replica of the original DeHaviland DH82A Tiger Moth built in England from 1931-1944. Quick Build components and 7.5 gallon center section auxiliary tank available.
Kits/Plans Sold: 40 **Completed & Flying:** 1

FISHER FLYING PRODUCTS
P.O. Box 468 • Industrial Park • Edgeley ND 58433
Tel. 701-493-2286 • Fax 701-493-2539
E-Mail: ffpjac@daktel.com
Web: http://www.fisherflying.com

FP-202 Koala

Wing Span	29.83 ft.	**Range**	135 sm
Wing Area	120.0 sq. ft.	**Takeoff Dist.**	125 ft.
Length	17.75 ft.	**Landing Dist.**	150 ft.
Max Gross Wt.	500 lbs.	**Vmax**	75 mph
Empty Wt.	250 lbs.	**VCr**	60 mph
Payload (full fuel)	N/A	**V_{S1} (stall clean)**	26 mph
Fuel Capacity	5 gal.	**V_{S0} (lndg config)**	N/A
Seats	1	**Climb Rate @msl**	800 fpm
Cabin Width	22 in.	**Service Ceiling**	12,000 ft.

Std. Engine/HP	Usable HP Range	Construction Mtrl.	Est. Bldg. Time
Rotax 277 / 28 hp	28 - 50	Wood, Fabric	300 hrs.
Plans	**Kit**	**Info Packet**	**Video**
N/A	$5,150.00	$10.00	$15.00

A faithful reproduction of one of the most popular aircraft ever built. With large, totally enclosed cockpit. Detachable wings. The two place Super Koala is detailed above.Quick Build component kits are available. **Sold:** 450 **Flying:** 300

FISHER FLYING PRODUCTS
P.O. Box 468 • Industrial Park • Edgeley ND 58433
Tel. 701-493-2286 • Fax 701-493-2539
E-Mail: ffpjac@daktel.com
Web: http://www.fisherflying.com

Classic

Wing Span	22.00 ft.	Range	225 sm
Wing Area	154.0 sq. ft.	Takeoff Dist.	150 ft.
Length	16.75 ft.	Landing Dist.	250 ft.
Max Gross Wt.	850 lbs.	Vmax	100 mph
Empty Wt.	400 lbs.	VCr	85 mph
Payload (full fuel)	400 lbs.	V_{S1} (stall clean)	35 mph
Fuel Capacity	8 gal.	V_{S0} (Indg config)	N/A
Seats	2	Climb Rate @msl	1300 fpm
Cabin Width	N/A	Service Ceiling	10,000 ft.

Std. Engine/HP	Usable HP Range	Construction Mtrl.	Est. Bldg. Time
Rotax 582 / 65 hp	65-80	Wood, Fabric	400 hrs.
Plans	**Kit**	**Info Packet**	**Video**
N/A	$6,775.00	$10.00	$15.00

A two place biplane with similar flight characteristics to the FP-404. Dual controls allow front or back seat operation. 20 minute assembly. Specs are for solo flight. Quick Build options available for airframe, wing, and fuselage. **Sold:** 200

FISHER FLYING PRODUCTS
P.O. Box 468 • Industrial Park • Edgeley ND 58433
Tel. 701-493-2286 • Fax 701-493-2539
E-Mail: ffpjac@daktel.com
Web: http://www.fisherflying.com

Dakota Hawk

Wing Span	28.50 ft.	Range	250 sm
Wing Area	128.0 sq. ft.	Takeoff Dist.	300 ft.
Length	19.00 ft.	Landing Dist.	400 ft.
Max Gross Wt.	1150 lbs.	Vmax	100 mph
Empty Wt.	600 lbs.	VCr	90 mph
Payload (full fuel)	N/A	V_{S1} (stall clean)	35 mph
Fuel Capacity	12 gal.	V_{S0} (Indg config)	N/A
Seats	2	Climb Rate @msl	800 fpm
Cabin Width	38 in.	Service Ceiling	N/A

Std. Engine/HP	Usable HP Range	Construction Mtrl.	Est. Bldg. Time
Rotax 912 / 80 hp	65 - 100	Wood, Fabric	500 hrs.
Plans	**Kit**	**Info Packet**	**Video**
N/A	$9,300.00	$10.00	$15.00

Two place, side-by-side seating. For high performance flight or cross country trips. Dual controls, steerable tail wheel, folding wings. Quick Build kit reduces build time to 300-400 hours. **Kits Sold:** 40 **Completed and Flying:** 4

FISHER FLYING PRODUCTS
P.O. Box 468 • Industrial Park • Edgeley ND 58433
Tel. 701-493-2286 • Fax 701-493-2539
E-Mail: ffpjac@daktel.com
Web: http://www.fisherflying.com

FP-303

Wing Span	27.67 ft.	Range	85 sm
Wing Area	111.0 sq. ft.	Takeoff Dist.	125 ft.
Length	16.50 ft.	Landing Dist.	130 ft.
Max Gross Wt.	450 lbs.	Vmax	70 mph
Empty Wt.	235 lbs.	VCr	60 mph
Payload (full fuel)	N/A	V_{S1} (stall clean)	N/A
Fuel Capacity	5 gal.	V_{S0} (Indg config)	25 mph
Seats	1	Climb Rate @msl	700 fpm
Cabin Width	20 in.	Service Ceiling	10,000 ft.

Std. Engine/HP	Usable HP Range	Construction Mtrl.	Est. Bldg. Time
Rotax 277 / 28 hp	28 - 40	Wood, Fabric	250 hrs.
Plans	**Kit**	**Info Packet**	**Video**
$175.00	$2,850.00	$10.00	$15.00

An economical ultralight designed to be highly portable. Field assembly time is less than 10 minutes. Cockpit positioned for good visibility. Quick Build Kits available for airframe, wing, and fuselage. **Kits/Plans Sold:** 400 **Flying:** 300

FISHER FLYING PRODUCTS
P.O. Box 468 • Industrial Park • Edgeley ND 58433
Tel. 701-493-2286 • Fax 701-493-2539
E-Mail: ffpjac@daktel.com
Web: http://www.fisherflying.com

FP-404

Wing Span	18.00 ft.	Range	140 sm
Wing Area	120.0 sq. ft.	Takeoff Dist.	125 ft.
Length	14.50 ft.	Landing Dist.	[illegible] ft
Max Gross Wt.	540 lbs.	Vmax	80 mph
Empty Wt.	270 lbs.	VCr	72 mph
Payload (full fuel)	270 lbs.	V_{S1} (stall clean)	30 mph
Fuel Capacity	5 gal.	V_{S0} (Indg config)	N/A
Seats	1	Climb Rate @msl	700 fpm
Cabin Width	22 in.	Service Ceiling	10,000 ft.

Std. Engine/HP	Usable HP Range	Construction Mtrl.	Est. Bldg. Time
Rotax 447 / 40 hp	40-50	Wood, Fabric	400 hrs.
Plans	**Kit**	**Info Packet**	**Video**
N/A	$5,100.00	$10.00	$15.00

Single seat wood and fabric biplane. Shares characteristics of the Classic. Experimental version with the Rotax 447 or 503 engine has a max speed of 80 mph, cruise of 72 mph. Detachable wings. Quick Build options. **Sold:** 350 **Flying:** 300

FISHER FLYING PRODUCTS
P.O. Box 468 • Industrial Park • Edgeley ND 58433
Tel. 701-493-2286 • Fax 701-493-2539
E-Mail: ffpjac@daktel.com
Web: http://www.fisherflying.com

FP-505 Skeeter

Wing Span	28.00 ft.	**Range**	180 sm
Wing Area	112.0 sq. ft.	**Takeoff Dist.**	150 ft.
Length	16.50 ft.	**Landing Dist.**	150 ft.
Max Gross Wt.	500 lbs.	**Vmax**	62 mph
Empty Wt.	245 lbs.	**VCr**	55 mph
Payload (full fuel)	N/A	**V_{S1} (stall clean)**	26 mph
Fuel Capacity	5 gal.	**V_{S0} (lndg config)**	N/A
Seats	1	**Climb Rate @msl**	750 fpm
Cabin Width	22 in.	**Service Ceiling**	13,000 ft.

Std. Engine/HP	Usable HP Range	Construction Mtrl.	Est. Bldg. Time
Rotax 277 / 28 hp	28 - 50	Wood, Fabric	250 hrs.
Plans	**Kit**	**Info Packet**	**Video**
N/A	$3,650.00	$10.00	$15.00

A single seat, open cockpit, parasol ultralight. The pylon-mounted high wing leaves the cockpit essentially unenclosed. Sturdy design utilizing shock-absorbing gear and geodetic construction. Quick Build Kits available. **Sold:** 50 **Flying:** 25

FISHER FLYING PRODUCTS
P.O. Box 468 • Industrial Park • Edgeley ND 58433
Tel. 701-493-2286 • Fax 701-493-2539
E-Mail: ffpjac@daktel.com
Web: http://www.fisherflying.com

Hercules Cruiser

Wing Span	34.00 ft.	**Range**	180 sm
Wing Area	190.0 sq. ft.	**Takeoff Dist.**	200 ft.
Length	10.00 ft.	**Landing Dist.**	200 ft.
Max Gross Wt.	790 lbs.	**Vmax**	55 mph
Empty Wt.	254 lbs.	**VCr**	40 mph
Payload (full fuel)	446 lbs.	**V_{S1} (stall clean)**	22 mph
Fuel Capacity	10 gal.	**V_{S0} (lndg config)**	N/A
Seats	2	**Climb Rate @msl**	700 fpm
Cabin Width	22 in.	**Service Ceiling**	10,000 ft.

Std. Engine/HP	Usable HP Range	Construction Mtrl.	Est. Bldg. Time
Cuyuna ULII-02/ 40 hp	N/A	Dacron, Alum. Tube	10 hrs.
Plans	**Kit**	**Info Packet**	**Video**
N/A	$8,995.00	$3.00	N/A

Designed by A. W. Harrison. Features a wing mount/dismount system which is easily worked by one person. Also offered in a higher performance Hercules Cross Country version powered by a Rotax 503, 50 hp engine at $13,314. **Sold:** 16 **Flying:** 16

FLETCHER'S ULTRALIGHTS
2165 Xavier Ave.
Turlock CA 95382
Tel. 209-667-4397

FP-606 Sky Baby

Wing Span	28.83 ft.	**Range**	130 sm
Wing Area	116.0 sq. ft.	**Takeoff Dist.**	125 ft.
Length	17.33 ft.	**Landing Dist.**	175 ft.
Max Gross Wt.	500 lbs.	**Vmax**	65 mph
Empty Wt.	250 lbs.	**VCr**	55 mph
Payload (full fuel)	N/A	**V_{S1} (stall clean)**	26 mph
Fuel Capacity	5 gal.	**V_{S0} (lndg config)**	N/A
Seats	1	**Climb Rate @msl**	800 fpm
Cabin Width	22 in.	**Service Ceiling**	N/A

Std. Engine/HP	Usable HP Range	Construction Mtrl.	Est. Bldg. Time
Rotax 277 / 28 hp	28 - 50	Wood, Fabric	250 hrs.
Plans	**Kit**	**Info Packet**	**Video**
$250.00	$5,200.00	$10.00	$15.00

An economical ultralight. Replica of the Cessna 150 and 152. Fully enclosed cockpit. Steerable nosewheel. Rotax 447 or 503 engines optional. Choice of tri-gear or taildragger. Detachable wings. Quick Build sub-kits available. **Sold:** 25 **Complete:** 20

FISHER FLYING PRODUCTS
P.O. Box 468 • Industrial Park • Edgeley ND 58433
Tel. 701-493-2286 • Fax 701-493-2539
E-Mail: ffpjac@daktel.com
Web: http://www.fisherflying.com

Flightstar Formula

Wing Span	30.00 ft.	**Range**	180 sm
Wing Area	144.0 sq. ft.	**Takeoff Dist.**	120 ft.
Length	16.50 ft.	**Landing Dist.**	150 ft.
Max Gross Wt.	750 lbs.	**Vmax**	96 mph
Empty Wt.	320 lbs.	**VCr**	75 mph
Payload (full fuel)	N/A	**V_{S1} (stall clean)**	N/A
Fuel Capacity	10 gal.	**V_{S0} (lndg config)**	28 mph
Seats	1	**Climb Rate @msl**	1100 fpm
Cabin Width	29 in.	**Service Ceiling**	14,000 ft.

Std. Engine/HP	Usable HP Range	Construction Mtrl.	Est. Bldg. Time
Rotax 503 / 48 hp	42-48	Composite, Aluminum	80 hrs.
Plans	**Kit**	**Info Packet**	**Video**
N/A	$12,190.00	Yes	N/A

Single seat, high performance kit. Rugged airframe and high lift wing. Options includes flap system, brakes, instruments, floats, BRS chute, electric starter. See Ads on Pages 47 and 68.
Kits/Plans Sold:26 **Completed & Flying:** 20

FLIGHTSTAR INC.
Ellington Airport • P.O. Box 760 • Ellington CT 06029
Tel. 203-875-8185 • Fax 203-870-5499
Web: http://www.webtoad.com/fs/

Flightstar IISL

Wing Span	32.50 ft.	**Range**	165 sm
Wing Area	N/A	**Takeoff Dist.**	200 ft.
Length	19.00 ft.	**Landing Dist.**	250 ft.
Max Gross Wt.	900 lbs.	**Vmax**	N/A
Empty Wt.	365 lbs.	**VCr**	55 mph
Payload (full fuel)	N/A	**V_{S1} (stall clean)**	34 mph
Fuel Capacity	10 gal.	**V_{S0} (lndg config)**	32 mph
Seats	2	**Climb Rate @msl**	600 fpm
Cabin Width	42 in.	**Service Ceiling**	N/A

Std. Engine/HP	Usable HP Range	Construction Mtrl.	Est. Bldg. Time
Rotax 503 / 48 hp	48-54	Aluminum, Fabric, Composite	130 hrs.

Plans	Kit	Info Packet	Video
N/A	$12,950.00	N/A	N/A

Sport-ultralight version of the Flightstar II. Designed for sport flying and training. Fitted with full dual controls. Can be upgraded to the full cabin configuration with the addition of a modular kit. See Ads on Pages 47 and 68.

Kits/Plans Sold: 120 **Completed & Flying:** 100

FLIGHTSTAR INC.
Ellington Airport • P.O. Box 760
Ellington CT 06029
Tel. 203-875-8185 • Fax 203-870-5499
Web: http://www.webtoad.com/fs/

C-42 Cyclone

Wing Span	31.00 ft.	**Range**	300 sm
Wing Area	135.0 sq. ft.	**Takeoff Dist.**	258 ft.
Length	20.33 ft.	**Landing Dist.**	N/A
Max Gross Wt.	992 lbs.	**Vmax**	118 mph
Empty Wt.	540 lbs.	**VCr**	105 mph
Payload (full fuel)	N/A	**V_{S1} (stall clean)**	N/A
Fuel Capacity	10 gal.	**V_{S0} (lndg config)**	39 mph
Seats	2	**Climb Rate @msl**	960 fpm
Cabin Width	48 in.	**Service Ceiling**	N/A

Std. Engine/HP	Usable HP Range	Construction Mtrl.	Est. Bldg. Time
Rotax 912 / 80 hp	N/A	Aluminum Tube, Composite, Fabr	N/A

Plans	Kit	Info Packet	Video
N/A	$29,995.00	Yes	N/A

Kit version of the German Ikarus C-42. Wing features one man folding system which can be completed without tools in minutes. Central mounted control stick and dual throttle handles. Price is for complete kit which includes engine, installation package, Culver wood prop, all fuel system equipment, fastener package, all electrical components,wheels, tires, brake system, presewn dacron covering set, composite door kit, and trim system. See Ads on Pages 47 and 68.

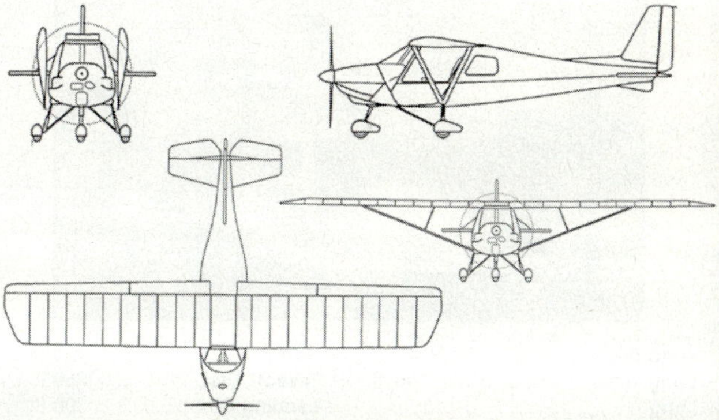

FLIGHTSTAR INC.
Ellington Airport • P.O. Box 760
Ellington CT 06029
Tel. 203-875-8185 • Fax 203-870-5499
Web: http://www.webtoad.com/fs/

Flightstar Spyder

Wing Span	30.00 ft.	**Range**	200 sm
Wing Area	144.0 sq. ft.	**Takeoff Dist.**	100 ft.
Length	16.50 ft.	**Landing Dist.**	150 ft.
Max Gross Wt.	650 lbs.	**Vmax**	96 mph
Empty Wt.	280 lbs.	**VCr**	60 mph
Payload (full fuel)	370 lbs.	**V_{S1} (stall clean)**	26 mph
Fuel Capacity	10 gal.	**V_{S0} (lndg config)**	26 mph
Seats	1	**Climb Rate @msl**	1000 fpm
Cabin Width	29 in.	**Service Ceiling**	14,000 ft.

Std. Engine/HP	Usable HP Range	Construction Mtrl.	Est. Bldg. Time
Rotax 477 / 42 hp	N/A	Composite, Aluminum, Fabric	75 hrs.
Plans	**Kit**	**Info Packet**	**Video**
N/A	$10,695.00	Yes	N/A

Spyder UL can be equipped with with one of several lightweight engines and operated as an unlicensed ultralight. Optional equipment includes brakes, instruments, floats, BRS chute, and electric starter. See Ads on Pages 47 and 68.

Sold: 46
Flying: 39

FLIGHTSTAR INC.
Ellington Airport • P.O. Box 760
Ellington CT 06029
Tel. 203-875-8185 • Fax 203-870-5499
Web: http://www.webtoad.com/fs/

Flightstar II

Wing Span	32.50 ft.	**Range**	180 sm
Wing Area	162.0 sq. ft.	**Takeoff Dist.**	250 ft.
Length	19.00 ft.	**Landing Dist.**	300 ft.
Max Gross Wt.	970 lbs.	**Vmax**	100 mph
Empty Wt.	420 lbs.	**VCr**	65 mph
Payload (full fuel)	550 lbs.	**V_{S1} (stall clean)**	36 mph
Fuel Capacity	10 gal.	**V_{S0} (lndg config)**	34 mph
Seats	2	**Climb Rate @msl**	700 fpm
Cabin Width	42 in.	**Service Ceiling**	14,000 ft.

Std. Engine/HP	Usable HP Range	Construction Mtrl.	Est. Bldg. Time
Rotax 503 / 54 hp	N/A	Composite, Aluminum, Fabric	130 hrs.
Plans	**Kit**	**Info Packet**	**Video**
N/A	$14,950.00	Yes	N/A

Two seat lightplane. No fabrication required. Wide stance gear enhance crosswind landing performance. Price includes engine and fully enclosed cockpit. Options include brakes, instruments, floats, BRS chute, & electric starter. See Ads on Pages 47 & 68.

Sold: 120
Flying: 100

FLIGHTSTAR INC.
Ellington Airport • P.O. Box 760
Ellington CT 06029
Tel. 203-875-8185 • Fax 203-870-5499
Web: http://www.webtoad.com/fs/

Storch

Wing Span	33.27 ft.	**Range**	200 sm
Wing Area	142.0 sq. ft.	**Takeoff Dist.**	290 ft.
Length	20.83 ft.	**Landing Dist.**	310 ft.
Max Gross Wt.	992 lbs.	**Vmax**	99 mph
Empty Wt.	529 lbs.	**VCr**	80 mph
Payload (full fuel)	402 lbs.	**V_{S1} (stall clean)**	36 mph
Fuel Capacity	10 gal.	**V_{S0} (lndg config)**	N/A
Seats	2	**Climb Rate @msl**	850 fpm
Cabin Width	43 in.	**Service Ceiling**	15,000 ft.

Std. Engine/HP	Usable HP Range	Construction Mtrl.	Est. Bldg. Time
Rotax 582 / 60 hp	N/A	Glass, Epoxy and Aluminum	150 hrs.
Plans	**Kit**	**Info Packet**	**Video**
N/A	$32,712.00	$2.00	$10.00

A sturdy, two seat ultralight (European) with over 100 currently flying. Folding wings take only four minutes to deploy by a single person. Standard features include dual controls, dual ignition, ballistic parachute and strong landing gear. Specs are for the standard version. The A.A. De Lux and S.A. De Lux versions have slightly greater empty weights. Price includes Rotax engine.

Kits/Plans Sold: 20 **Completed & Flying:** 150

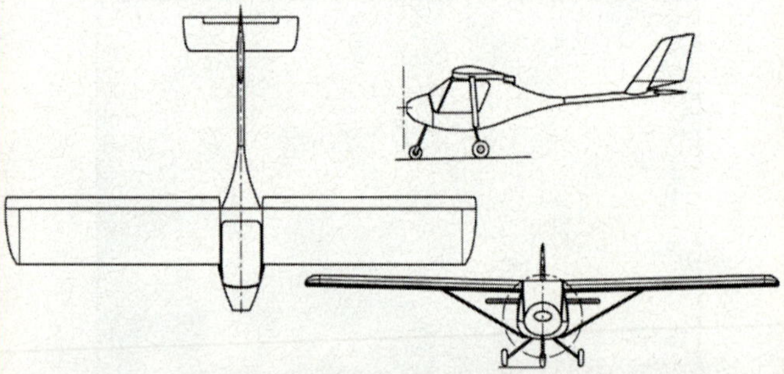

Lucio Turchet Photo

FLY SYNTHESIS
Via Gorizia No.63 • 33050 Gonars/Udine
ITALY
Tel. 0432 992482 • Fax 0432 993557
E-Mail: guabais@mbox.vol.it
Web: http://www.compouse.com/FlySynthesis/

Texan

Wing Span	26.90 ft.	**Range**	495 sm
Wing Area	123.0 sq. ft.	**Takeoff Dist.**	327 ft.
Length	22.30 ft.	**Landing Dist.**	N/A
Max Gross Wt.	992 lbs.	**Vmax**	118 mph
Empty Wt.	551 lbs.	**VCr**	106 mph
Payload (full fuel)	440 lbs.	**V_{S1} (stall clean)**	107 mph
Fuel Capacity	N/A	**V_{S0} (lndg config)**	35 mph
Seats	2	**Climb Rate @msl**	1200 fpm
Cabin Width	N/A	**Service Ceiling**	N/A

Std. Engine/HP	Usable HP Range	Construction Mtrl.	Est. Bldg. Time
Rotax 912 / 80 hp	N/A	Composite	N/A
Plans	**Kit**	**Info Packet**	**Video**
N/A	N/A	N/A	N/A

New all composite sport ultralight (European). The design emphasis is on performance and safety.

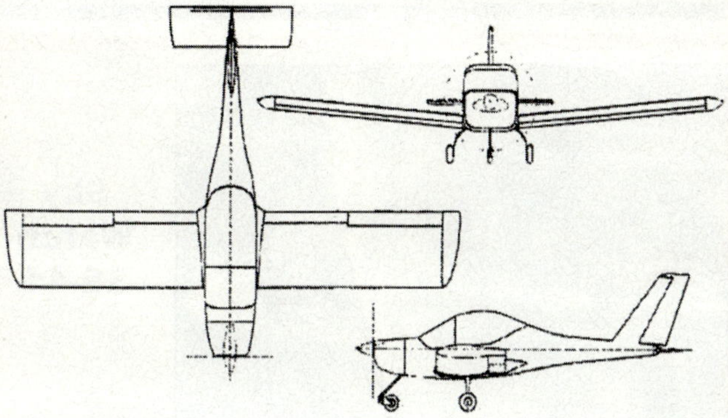

Lucio Turchet Photo

FLY SYNTHESIS
Via Gorizia No.63 • 33050 Gonars/Udine
ITALY
Tel. 0432 992482 • Fax 0432 993557
E-Mail: guabais@mbox.vol.it
Web: http://www.compouse.com/FlySynthesis/

Flying Flea HM-14

Wing Span	18.00 f.	**Range**	180 sm
Wing Area	100.0 sq. ft.	**Takeoff Dist.**	150 ft.
Length	13.50 ft.	**Landing Dist.**	150 ft.
Max Gross Wt.	500 lbs.	**Vmax**	70 mph
Empty Wt.	260 lbs.	**VCr**	55 mph
Payload (full fuel)	N/A	**V_{S1} (stall clean)**	25 mph
Fuel Capacity	6 gal.	**V_{S0} (lndg config)**	N/A
Seats	1	**Climb Rate @msl**	500 fpm
Cabin Width	20 in.	**Service Ceiling**	N/A

Std. Engine/HP	Usable HP Range	Construction Mtrl.	Est. Bldg. Time
Various / 35 hp	25-45	Wood, Fabric	300 hrs.
Plans	**Kit**	**Info Packet**	**Video**
$45.00	N/A	Yes	$49.00

The Flying Flea was designed in 1934 by Henri Mignet. It features an 18 ft. pivoting front wing and 13 ft. fixed, tandem rear wing, and two-axis controls. Originally powered by low horsepower motorcycle and auto engines. Modern Fleas can utilize a variety of engines including Rotax, Citroen, DAF, etc.

FLYING FLEA ARCHIVE USA
P.O. Box 892
Wooster OH 44691-0892
Tel. 330-377-5013 • Fax 330-377-5013

Flying Flea HM-160

Wing Span	15.00 ft.	**Range**	N/A
Wing Area	77.00 sq. ft.	**Takeoff Dist.**	100 ft.
Length	10.50 ft.	**Landing Dist.**	150 ft.
Max Gross Wt.	472 lbs.	**Vmax**	80 mph
Empty Wt.	242 lbs.	**VCr**	65 mph
Payload (full fuel)	N/A	**V_{S1} (stall clean)**	20 mph
Fuel Capacity	6 gal.	**V_{S0} (lndg config)**	20 mph
Seats	1	**Climb Rate @msl**	500 fpm
Cabin Width	20 in.	**Service Ceiling**	N/A

Std. Engine/HP	Usable HP Range	Construction Mtrl.	Est. Bldg. Time
Various / 25 hp	20-35	Wood, Fabric	400 hrs.
Plans	**Kit**	**Info Packet**	**Video**
$125.00	N/A	$5.00	N/A

Based on Henri Mignet's favorite HM-16 "Baby Pou", this is at US ultralight weight. The HM-16 was designed and built after the lessons he learned with original HM-14 "Flying Flea." Plans include tri-gear version and trailering configuration.

FLYING FLEA ARCHIVE USA
P.O. Box 892
Wooster OH 44691-0892
Tel. 330-377-5013 • Fax 330-377-5013

Sky Raider

Wing Span	26.20 ft.	**Range**	N/A
Wing Area	N/A	**Takeoff Dist.**	100 ft.
Length	17.00 ft.	**Landing Dist.**	100 ft.
Max Gross Wt.	N/A	**Vmax**	100 mph
Empty Wt.	240 lbs.	**VCr**	63 mph
Payload (full fuel)	N/A	**V_{S1} (stall clean)**	27 mph
Fuel Capacity	N/A	**V_{S0} (lndg config)**	23 mph
Seats	N/A	**Climb Rate @msl**	600 fpm
Cabin Width	24 in.	**Service Ceiling**	N/A

Std. Engine/HP	Usable HP Range	Construction Mtrl.	Est. Bldg. Time
277 Rotax / 27 hp	N/A	Tube, Fabric	200 hrs.
Plans	**Kit**	**Info Packet**	**Video**
N/A	N/A	Yes	N/A

Enclosed chrome-moly steel fuselage. Folding wings with flaps and ailerons that require no disconecting of controls. Cockpit accomodates large pilots. Can meet Part 103 requirements.

FLYING K ENTERPRISES
3403 Arthur St.
Caldwell ID 83605
Tel. 208-455-7529 • Fax 208-455-7678

Freebird

Wing Span	29.00 ft.	**Range**	200 sm
Wing Area	132.0 sq. ft.	**Takeoff Dist.**	125 ft.
Length	17.50 ft.	**Landing Dist.**	150 ft.
Max Gross Wt.	870 lbs.	**Vmax**	110 mph
Empty Wt.	385 lbs.	**VCr**	80 mph
Payload (full fuel)	425 lbs.	**V_{S1} (stall clean)**	38 mph
Fuel Capacity	10 gal.	**V_{S0} (lndg config)**	32 mph
Seats	2	**Climb Rate @msl**	800 fpm
Cabin Width	42 in.	**Service Ceiling**	12,000 ft.

Std. Engine/HP	Usable HP Range	Construction Mtrl.	Est. Bldg. Time
Rotax 503 / 52 hp	N/A	Tube, Aluminum	100 hrs.
Plans	**Kit**	**Info Packet**	**Video**
N/A	$9,995.00	$5.00	$10.00

Performance specs are with optional enclosed two place cabin. Engine not included. Single seat and twin engine versions available. Wings are removable by one person in five minutes. A trailer kit is offered at $795. **Sold:** 42 **Flying:** 33

FREEBIRD AIRPLANE COMPANY
1916 Pleasant Grove Church Rd.
Marshville NC 28103
Tel. 800-850-3708

Sky Watch SS-11

Wing Span	33.00 ft.	**Range**	405 sm
Wing Area	155.0 sq. ft.	**Takeoff Dist.**	200 ft.
Length	21.50 ft.	**Landing Dist.**	N/A
Max Gross Wt.	950 lbs.	**Vmax**	90 mph
Empty Wt.	420 lbs.	**VCr**	80 mph
Payload (full fuel)	N/A	**V_{S1} (stall clean)**	N/A
Fuel Capacity	13 gal.	**V_{S0} (lndg config)**	29 mph
Seats	2	**Climb Rate @msl**	1100 fpm
Cabin Width	N/A	**Service Ceiling**	13,000 ft.

Std. Engine/HP	Usable HP Range	Construction Mtrl.	Est. Bldg. Time
Rotax 503 / 50 hp	50-75	Tube, fabric	N/A
Plans	**Kit**	**Info Packet**	**Video**
N/A	N/A	N/A	N/A

Tandem seat, full dual control aircraft with ergonomically designed cockpit, a compact, upright engine package mounted on a chrome-moly dynafocal mount, and components manufactured with computer numeric controlled matching centers.

FREEDOM LITE, INC.
Box 51 • Walton, Ontario N0K 1Z0
CANADA
Tel. 519-887-9966 • Fax 519-887-6465

Frontier MD-II

Wing Span	29.50 ft.	Range	690 sm
Wing Area	177.0 sq. ft.	Takeoff Dist.	750 ft.
Length	27.50 ft.	Landing Dist.	400 ft.
Max Gross Wt.	2000 lbs.	Vmax	184 mph
Empty Wt.	1150 lbs.	VCr	140 mph
Payload (full fuel)	850 lbs.	V_{S1} (stall clean)	35 mph
Fuel Capacity	43 gal.	V_{S0} (lndg config)	30 mph
Seats	2	Climb Rate @msl	1200 fpm
Cabin Width	41 in.	Service Ceiling	16,000 ft.

Std. Engine/HP	Usable HP Range	Construction Mtrl.	Est. Bldg. Time
LOM / 140 hp	65 - 200	2024T-3 Aluminum	1500 hrs.
Plans	**Kit**	**Info Packet**	**Video**
N/A	$19,500.00	$5.00	N/A

Combines characteristics of several well proven designs. US35B airfoil selected for extremely gentle stall and very short takeoff roll. Partially pre-assembled. Assembly hardware provided. Four sub-kits can be ordered separately.

Kits Sold: 2 **Completed & Flying:** 2

FRONTIER AIRCRAFT, INC.
225 Wall Street
Vail CO 81657
Tel. 970 476-9300 • Fax 970 476-6232

SQ2000 XP

Wing Span	26.58 ft.	Range	851 sm
Wing Area	104.0 sq. ft.	Takeoff Dist.	1600 ft.
Length	17.50 ft.	Landing Dist.	1500 ft.
Max Gross Wt.	2250 lbs.	Vmax	260 mph
Empty Wt.	1200 lbs.	VCr	230 mph
Payload (full fuel)	688 lbs.	V_{S1} (stall clean)	N/A
Fuel Capacity	42 gal.	V_{S0} (lndg config)	N/A
Seats	4	Climb Rate @msl	2200 fpm
Cabin Width	48 in.	Service Ceiling	25,000 ft.

Std. Engine/HP	Usable HP Range	Construction Mtrl.	Est. Bldg. Time
Lyc. IO-360 / 200 hp	160-260	Composite	700 hrs.
Plans	**Kit**	**Info Packet**	**Video**
N/A	$35,450.00	$10.00	$18.00

Features gull wing doors, electro-hydraulic retractable gear. Cutting edge technology rib and spar construction for lighter weight & higher strength. See Ad on Page 9. **Sold:** 2 **Flying:** 1

GLASSIC COMPOSITES, LLC
15425 Dayton Pike, Box 716 • Sale Creek TN 37373-0716
Tel. 423-332-8300 • Fax 423-332-4080
E-Mail: GlassiComp@aol.com
Web: http://www.glassic.com

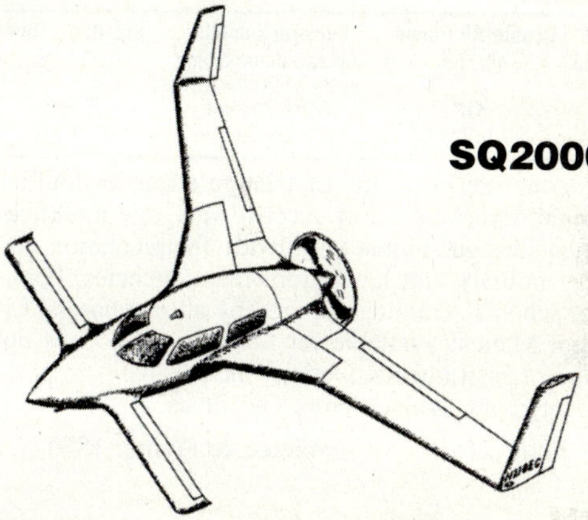

SQ2000 ES

Wing Span	26.58 ft.	Range	850 sm
Wing Area	100.0 sq. ft.	Takeoff Dist.	1600 ft.
Length	17.52 ft.	Landing Dist.	1500 ft.
Max Gross Wt.	1900 lbs.	Vmax	250 mph
Empty Wt.	1200 lbs.	VCr	225 mph
Payload (full fuel)	688 lbs.	V_{S1} (stall clean)	N/A
Fuel Capacity	39 gal.	V_{S0} (lndg config)	N/A
Seats	4	Climb Rate @msl	2200 fpm
Cabin Width	48 in.	Service Ceiling	25,000 ft.

Std. Engine/HP	Usable HP Range	Construction Mtrl.	Est. Bldg. Time
Lyc. IO-360 / 200 hp	N/A	Composite	750 hrs.
Plans	**Kit**	**Info Packet**	**Video**
N/A	$25,450.00	$10.00	Yes

Fixed gear model still features gull wing doors. The gear are oleo struts, mounted in the same stable outboard position as the retract version. The wings and canard come as pre-shaped foam cores, which the builder completes utilizing the simple moldless composite method. See Ad on Page 9.

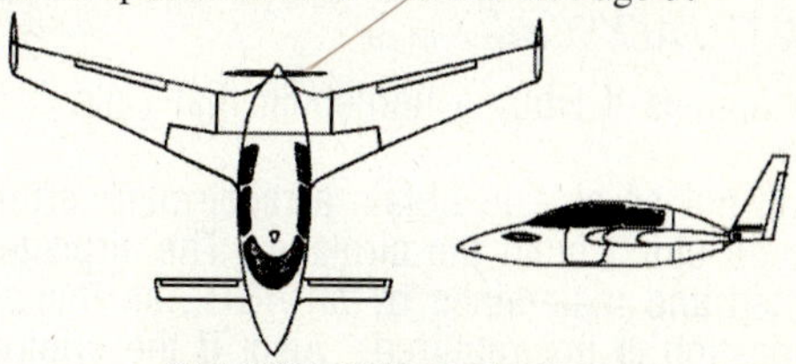

GLASSIC COMPOSITES, LLC
15425 Dayton Pike, Box 716 • Sale Creek TN 37373-0716
Ph: 423-332-8300 • Fax: 423-332-4080
E-Mail: GlassiComp@aol.com
Web: http://www.glassic.com

Fun Flight 550

Wing Span	35.00 ft.	**Range**	150 sm
Wing Area	163.0 sq. ft.	**Takeoff Dist.**	100 ft.
Length	20.67 ft.	**Landing Dist.**	150 ft.
Max Gross Wt.	950 lbs.	**Vmax**	85 mph
Empty Wt.	395 lbs.	**VCr**	50 mph
Payload (full fuel)	505 lbs.	**V_{S1} (stall clean)**	32 mph
Fuel Capacity	10 gal.	**V_{S0} (lndg config)**	32 mph
Seats	2	**Climb Rate @msl**	950 fpm
Cabin Width	48 in.	**Service Ceiling**	13,000 ft.

Std. Engine/HP	Usable HP Range	Construction Mtrl.	Est. Bldg. Time
Rotax 503 / 48 hp	48 - 65	Aircraft grade tubing	100 hrs.
Plans	**Kit**	**Info Packet**	**Video**
N/A	$9,850.00	$5.00	$12.00

Very simple construction. Lots of triangulation in double-sleeved airframe. Excellent safety record. Very few modifications to original design. Flown worldwide for recreation and surveilance by military and law enforcement agencies. Popular with flight schools. Training required by all purchasers. License required. Manual available for $85. Kit price does not include engine or instruments. Options include ballistic parachute, radio, intercom, 3-blade prop, and floats.

Kits/Plans Sold: 2154 **Completed & Flying:** 1850

Vance Lorenzana, Oswego IL 60543
"Just wanted to let you know how pleased I am with my Beaver RX550. ... I have put over 75 hours on it without any problems. The plane is everything you said it would be and more. It is a well built, solid aircraft with stable, forgiving flight characteristics, but most of all it is fun to fly!"

Glenn Rounsevell, 550 Owner
"... Ultralight flying was a conservative way to go. I found it much less expensive than other types of flying and the purchase of my plane through Fun Flight was made easy and pleasant throughout. Flying low and slow has been most enjoyable."

FUN FLIGHT INC.
Box 4410
Alexandria VA 22303
Tel. 703-768-6496

T-Bird Tandem TBT06

Wing Span	31.00 ft.	**Range**	225 sm
Wing Area	154.0 sq. ft.	**Takeoff Dist.**	300 ft.
Length	21.25 ft.	**Landing Dist.**	200 ft.
Max Gross Wt.	1000 lbs.	**Vmax**	95 mph
Empty Wt.	400 lbs.	**VCr**	75 mph
Payload (full fuel)	500 lbs.	**V_{S1} (stall clean)**	38 mph
Fuel Capacity	12 gal.	**V_{S0} (lndg config)**	N/A
Seats	2	**Climb Rate @msl**	800 fpm
Cabin Width	27 in.	**Service Ceiling**	10,000 ft.

Std. Engine/HP	Usable HP Range	Construction Mtrl.	Est. Bldg. Time
Rotax 582 / 65 hp	N/A	Tube, Fabric	100 hrs.
Plans	**Kit**	**Info Packet**	**Video**
N/A	$18,110.00	$5.00	$10.00

A high wing pusher with fully enclosable, tandem seating. Strong triangular construction. Amphibian floats with electrical retract available. Dual controls. Agricultral spray unit available. Other options include strut fairings, wheel pants, 800x6 tundra tires and more. Instruments not included in kit. See Ad on Back Cover.

Kits/Plans Sold: 38 **Completed & Flying:** 21

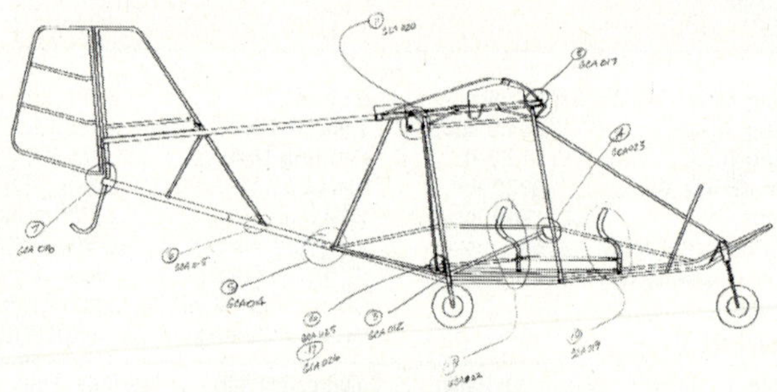

GOLDEN CIRCLE AIR INC.
PO Box 676 • #10 Ellefson Drive
De Soto IA 50069-0676
Ph: 515-834-2225 • Fax: 515-834-2152
E-Mail: bobdorly@netins.net
Web: http://www.netins.net/showcase/t-birds/

HOMEBUILT "FAQ'S" (See Page 68)

Q: What happens if I buy a homebuilt that I didn't build?

A: You will not be able to obtain a repairman certificate, since you didn't build the airplane. The airplane can be inspected and maintained by an A&P... an Inspection Authorization is not required. Also, if the original owner retains his or her Repairman Certificate they can continue working on your plane..

The FAA allows the owner of a homebuilt to perform maintenance on the aircraft, as long as an A&P signs off the work within a year. In other words, the A&P's annual inspection not only covers the inspection of the air plane, but it counts as the yearly signoff for work done by the owner in the past year. An EAA article on this agreement can be found at http://www.halcyon.com/wanttaja/maint.html

Parts for kit aircraft may or may not be available de pending on the source. For either kit or plans-built, one thing to insist upon is having the plans in your posses sion or available. This way, should a part need to be fabricated later, you have the specs for it.

T-Bird I

Wing Span	31.00 ft.	**Range**	120 sm
Wing Area	154.0 sq. ft.	**Takeoff Dist.**	90 ft.
Length	18.00 ft.	**Landing Dist.**	120 ft.
Max Gross Wt.	570 lbs.	**Vmax**	88 mph
Empty Wt.	250 lbs.	**VCr**	65 mph
Payload (full fuel)	300 lbs.	**V_{S1} (stall clean)**	28 mph
Fuel Capacity	5 gal.	**V_{S0} (lndg config)**	N/A
Seats	1	**Climb Rate @msl**	900 fpm
Cabin Width	26 in.	**Service Ceiling**	6,000 ft.

Std. Engine/HP	Usable HP Range	Construction Mtrl.	Est. Bldg. Time
Rotax 503 / 48 hp	28 - 66	Tube, Fabric	60 hrs.
Plans	**Kit**	**Info Packet**	**Video**
N/A	$8,992.00	$5.00	$10.00

High wing taildragger with tricycle option. Full enclosure and floats available. Fuselage is a cage constructed from metal tubing. Completely strut-braced, very solid. The wings utilize a Mylar covering. Engine options include Rotax 477, 503 and 582. Other options include enlarged cabin, aluminum wheels and 5 and 10 gallon seat tanks. Instruments not included in kit. See Ad on Back Cover.

Kits/Plans Sold: 3200 **Completed & Flying:** 2000

GOLDEN CIRCLE AIR INC.
PO Box 676 • #10 Ellefson Drive
De Soto IA 50069-0676
Ph: 515-834-2225 • Fax: 515-834-2152
E-Mail: bobdorly@netins.net
Web: http://www.netins.net/showcase/t-birds/

T-Bird II

Wing Span	36.50 ft.	**Range**	180 sm
Wing Area	190.0 sq. ft.	**Takeoff Dist.**	300 ft.
Length	18.00 ft.	**Landing Dist.**	250 ft.
Max Gross Wt.	1000 lbs.	**Vmax**	88 mph
Empty Wt.	400 lbs.	**VCr**	68 mph
Payload (full fuel)	500 lbs.	**V_{S1} (stall clean)**	31 mph
Fuel Capacity	12 gal.	**V_{S0} (lndg config)**	N/A
Seats	2	**Climb Rate @msl**	800 fpm
Cabin Width	42 in.	**Service Ceiling**	10,000 ft.

Std. Engine/HP	Usable HP Range	Construction Mtrl.	Est. Bldg. Time
Rotax 582 / 66 hp	N/A	Tube, Fabric	60 hrs.
Plans	**Kit**	**Info Packet**	**Video**
N/A	$13,487.00	$5.00	$10.00

Two seat, side-by-side version of the T-Bird. Widest cabin in the industry. High wing pusher with dual controls and strong triangular construction. Other options include strut fairings. See Ad on Back Cover. **Sold:** 2000 **Completed & Flying:** 2000

GOLDEN CIRCLE AIR INC.
PO Box 676 • #10 Ellefson Drive
De Soto IA 50069-0676
Ph: 515-834-2225 • Fax: 515-834-2152
E-Mail: bobdorly@netins.net
Web: http://www.netins.net/showcase/t-birds/

T-Bird III

GOLDEN CIRCLE AIR INC.
PO Box 676 • #10 Ellefson Drive
De Soto IA 50069-0676
Ph: 515-834-2225 • Fax: 515-834-2152
E-Mail: bobdorly@netins.net
Web: http://www.netins.net/showcase/t-birds/

Wing Span	36.50 ft.	**Range**	150 sm
Wing Area	190.0 sq. ft.	**Takeoff Dist.**	350 ft.
Length	20.83 ft.	**Landing Dist.**	300 ft.
Max Gross Wt.	1230 lbs.	**Vmax**	85 mph
Empty Wt.	500 lbs.	**VCr**	68 mph
Payload (full fuel)	650 lbs.	**V_{S1} (stall clean)**	39 mph
Fuel Capacity	12 gal.	**V_{S0} (lndg config)**	N/A
Seats	3	**Climb Rate @msl**	650 fpm
Cabin Width	42 in.	**Service Ceiling**	8,500 ft.

Std. Engine/HP	Usable HP Range	Construction Mtrl.	Est. Bldg. Time
Rotax 582 / 66 hp	N/A	Tube, Fabric	100 hrs.
Plans	**Kit**	**Info Packet**	**Video**
N/A	$18,475.00	$5.00	$10.00

Three seat T-Bird. A high wing, strut-braced, pusher design. Options include ag sprayer, strut fairings, wheel pants, 800x6 tundra tires and more. Instruments not included in kit. See Ad on Back Cover.

Sold: 19 **Flying:** 19

Great Lakes Sport Trainer

Wing Span	26.67ft.	**Range**	340sm
Wing Area	187.5 sq. ft.	**Takeoff Dist.**	400 ft.
Length	20.33 ft.	**Landing Dist.**	500 ft.
Max Gross Wt.	1618 lbs.	**Vmax**	138 mph
Empty Wt.	1025 lbs.	**VCr**	125 mph
Payload (full fuel)	500 lbs.	**V_{S1} (stall clean)**	55 mph
Fuel Capacity	25 gal.	**V_{S0} (lndg config)**	N/A
Seats	2	**Climb Rate @msl**	1200 fpm
Cabin Width	24 in.	**Service Ceiling**	14,000 ft.

Std. Engine/HP	Usable HP Range	Construction Mtrl.	Est. Bldg. Time
Lyc O-360 / 180 hp	125 - 200	Wood, Tube	2000 hrs.
Plans	**Kit**	**Info Packet**	**Video**
$300.00	$23,500.00	$10.00	N/A

Two place trainer with the look and feel of the 1930's. Representative of the famed Great Lakes aircraft of that era. **Sold:** 475 **Flying:** 40

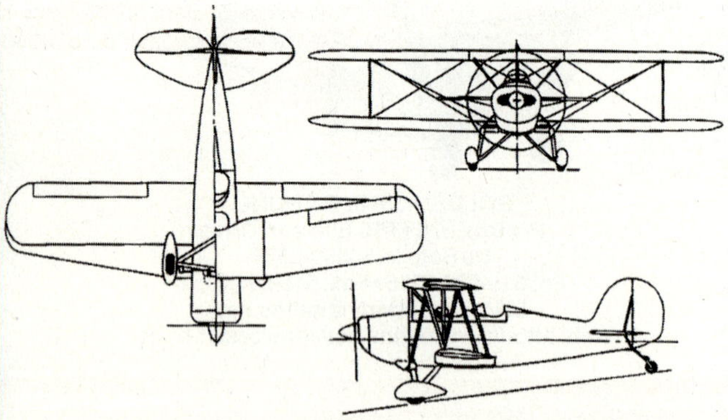

GREAT AIRPLANES INC.
P.O. Box 615
Needham Heights MA 02194-0005
Tel. 617-444-5480 • Fax 617-444-5480
E-Mail: hswack@aol.com
Web: http://aso.solid.com/greatair

Great Lakes Biplane 2T-1A

Wing Span	26.50 ft.	**Range**	300 sm
Wing Area	187.6 sq. ft.	**Takeoff Dist.**	475 ft.
Length	20.50 ft.	**Landing Dist.**	400 ft.
Max Gross Wt.	1800 lbs.	**Vmax**	131 mph
Empty Wt.	1200 lbs.	**VCr**	118 mph
Payload (full fuel)	600 lbs.	**V_{S1} (stall clean)**	50 mph
Fuel Capacity	26 gal.	**V_{S0} (lndg config)**	50 mph
Seats	2	**Climb Rate @msl**	1200 fpm
Cabin Width	26 in.	**Service Ceiling**	17,000 ft.

Std. Engine/HP	Usable HP Range	Construction Mtrl.	Est. Bldg. Time
Lycoming / 180 hp	100 - 200	Tube, Fabric	1500 hrs.
Plans	**Kit**	**Info Packet**	**Video**
$250.00	N/A	N/A	N/A

A classic aerobatic biplane. One of the most popular designs of all time. Great handling, no bad habits, and a joy to fly. High resale value, low operating costs. **Sold:** 500 **Flying:** 50

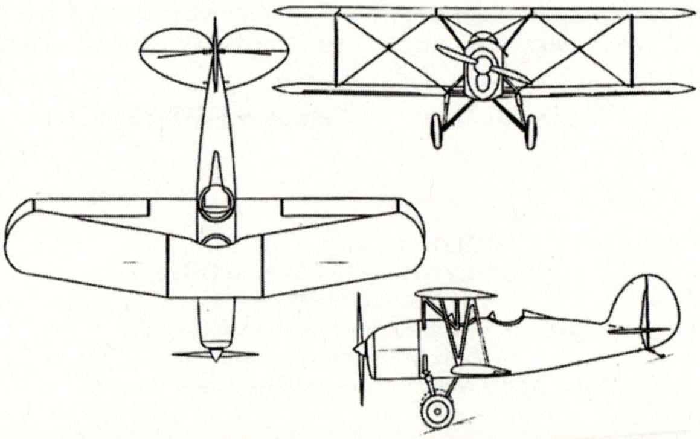

Great Lakes originally designed this plane in 1929 - 1930 as a trainer. However, because the Luftwaffe trained on the almost identical Bucker Jungman, political considerations dictated that the U.S. Air Corps select the Boeing Stearman for that purpose. Some Great Lakes reportedly were exported to South America where they were fitted with machine guns and took part in various brushfire wars. Between World Wars, the Great Lakes was the premier aerobatic aircraft in the U.S.

Current plans have been redrawn for legibility and ease of use from hard to get originals. Construction is straight forward, with no surprises, difficult process, or exotic materials. Several hundred Great Lakes have been built and most are still flying.

John A. Hern Jr., Coeur d'Alene, Idaho
"The Great Lakes is a joy to fly, very forgiving and yet very responsive. ... It's a great all around fun aircraft combining nostalgia, simplicity, and performance."

GREAT LAKES PLANS CO.
1900 Millview Lane
Coeur D'Alene ID 83814-7622
Tel. 800-228-7794 • Fax 208-664-8540

Sonerai I

Wing Span	16.67 ft.	**Range**	300 sm
Wing Area	75.00 sq. ft.	**Takeoff Dist.**	600 ft.
Length	16.67 ft.	**Landing Dist.**	800 ft.
Max Gross Wt.	700 lbs.	**Vmax**	180 mph
Empty Wt.	440 lbs.	**VCr**	150 mph
Payload (full fuel)	260 lbs.	**V_{S1} (stall clean)**	45 mph
Fuel Capacity	16 gal.	**V_{S0} (Indg config)**	N/A
Seats	1	**Climb Rate @msl**	1000 fpm
Cabin Width	N/A	**Service Ceiling**	10,000 ft.

Std. Engine/HP	Usable HP Range	Construction Mtrl.	Est. Bldg. Time
VW / 70 hp	N/A	Tube, Aluminum, Fabric	800 hrs.
Plans	**Kit**	**Info Packet**	**Video**
$99.95	$5,500.00	$3.00	$10.95

Mid wing racer designed to meet all Professional Race Pilots Assoc. (PRPA) and Formula Vee Racing requirements for 1600cc Volkswagen powered aircraft. Can use engine displacements from 1600cc through 2180cc. Conventional or tricycle gear options. All-aluminum folding wings. See Ad on Pg. 80.

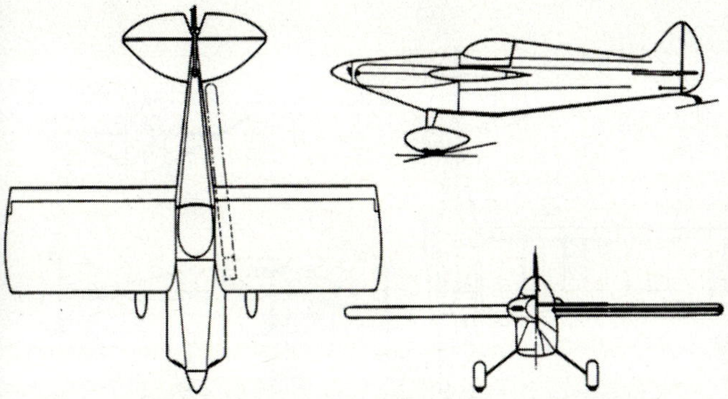

Sonerai homebuilt aircraft were designed by John Monnett. They have been on the market since the early 70's. The Sonerai I was first, followed by the very popular Sonerai II which, including its various versions, is among the most popular homebuilt aircraft ever.

Fuselage construction of the Sonerai Series is 4130 steel tubing with fabric covering. The wings are made of 2024-T3 aluminum. You will need simple metal working tools and access to a gas welder to complete the aircraft. One of the attractive features of the Sonerai is that the wings can easily be folded back. You can keep it at home and tow it to the airport on its own wheels.

The Sonerai I enjoys an excellent reputation for strength, speed, and agility, especially in the Formula VEE racing circuit. The Sonerai I can handle + or - 6 G's with the "S" wing. It can accommodate a 6 foot, 200 lb. pilot. The Sonerai I will do 170 mph flat out, full throttle on a Great Plains 2180cc engine.

GREAT PLAINS AIRCRAFT
P.O. Box 545 • Boys Town NE 68010
Tel. 402-493-6507 • Fax 402-333-7750
E-Mail: gpasc@earthlink.net
Web: http://www.greatplainsas.com

Sonerai II

Wing Span	18.67 ft.	**Range**	245 sm
Wing Area	84.00 sq. ft.	**Takeoff Dist.**	900 ft.
Length	18.83 ft.	**Landing Dist.**	800 ft.
Max Gross Wt.	950 lbs.	**Vmax**	180 mph
Empty Wt.	520 lbs.	**VCr**	140 mph
Payload (full fuel)	430 lbs.	**V_{S1} (stall clean)**	45 mph
Fuel Capacity	10 gal.	**V_{S0} (Indg config)**	N/A
Seats	2	**Climb Rate @msl**	500 fpm
Cabin Width	N/A	**Service Ceiling**	10,000 ft.

Std. Engine/HP	Usable HP Range	Construction Mtrl.	Est. Bldg. Time
VW / 70 hp	N/A	Tube, Aluminum, Fabric	800 hrs.
Plans	**Kit**	**Info Packet**	**Video**
$99.95	$5,500.00	$3.00	$10.95

Three versions: II, II-L, and II-LT. The II (Original) is a mid wing, tandem aircraft that is flown from the rear seat solo. The II-L is the same plane with a low wing. The II-LT is the same as II-L but has a steerable nosewheel added. See Ad on Page 80.

GREAT PLAINS AIRCRAFT
P.O. Box 545 • Boys Town NE 68010
Tel. 402-493-6507 • Fax 402-333-7750
E-Mail: gpasc@earthlink.net
Web: http://www.greatplainsas.com

Sonerai II Stretch

Wing Span	18.67 ft.	**Range**	245 sm
Wing Area	84.00 sq. ft.	**Takeoff Dist.**	900 ft.
Length	20.33 ft.	**Landing Dist.**	800 ft.
Max Gross Wt.	1150 lbs.	**Vmax**	180 mph
Empty Wt.	540 lbs.	**VCr**	140 mph
Payload (full fuel)	610 lbs.	**V_{S1} (stall clean)**	50 mph
Fuel Capacity	10 gal.	**V_{S0} (Indg config)**	N/A
Seats	2	**Climb Rate @msl**	700 fpm
Cabin Width	N/A	**Service Ceiling**	10,000 ft.

Std. Engine/HP	Usable HP Range	Construction Mtrl.	Est. Bldg. Time
VW 2180cc / 70 hp	70+ hp	Tube, Aluminum, Fabric	800 hrs.
Plans	**Kit**	**Info Packet**	**Video**
$99.95	$5,500.00	$3.00	$10.95

The Sonerai low wing stretch II-LS and II-LTS tri-gear make up the rest of the Sonerai family. Both have all-aluminum folding wings. Note that the Sonerai II are among the most popular homebuilt aircraft ever. See Ad on Page 80.

GREAT PLAINS AIRCRAFT
P.O. Box 545 • Boys Town NE 68010
Tel. 402-493-6507 • Fax 402-333-7750
E-Mail: gpasc@earthlink.net
Web: http://www.greatplainsas.com

Micro Mong

Wing Span	19.50 ft.	**Range**	200 sm
Wing Area	110.0 sq. ft.	**Takeoff Dist.**	200 ft.
Length	14.00 ft.	**Landing Dist.**	400 ft.
Max Gross Wt.	650 lbs.	**Vmax**	110 mph
Empty Wt.	340 lbs.	**VCr**	85 mph
Payload (full fuel)	250 lbs.	**V_{S1} (stall clean)**	35 mph
Fuel Capacity	9 gal.	**V_{S0} (lndg config)**	N/A
Seats	1	**Climb Rate @msl**	1000 fpm
Cabin Width	21 in.	**Service Ceiling**	15,000 ft.

Std. Engine/HP	Usable HP Range	Construction Mtrl.	Est. Bldg. Time
Rotax 503 / 50 hp	N/A	Metal, Tube, Fabric	400 hrs.
Plans	**Kit**	**Info Packet**	**Video**
$130.00	$14,534.00	$7.00	N/A

Classic biplane lines packaged into a very compact design. Agile, docile handling. Kit price includes engine.

Kits/Plans Sold: 200 **Completed & Flying:** 5

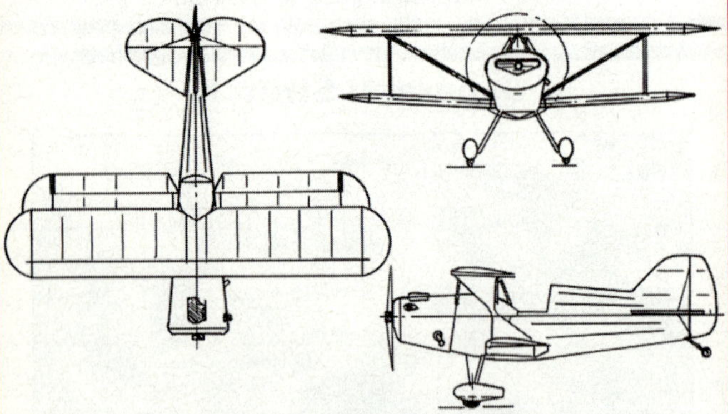

GREEN SKY ADVENTURES, INC.
2395 Cream Ridge Rd.
Orwell OH 44076
Tel. 216-293-6624 • Fax 216-293-6321
Web: http://www.greensky-adventures.com

Zippy Sport

Wing Span	26.30 ft.	**Range**	300 sm
Wing Area	100.0 sq. ft.	**Takeoff Dist.**	350 ft.
Length	17.90 ft.	**Landing Dist.**	550 ft.
Max Gross Wt.	680 lbs.	**Vmax**	120 mph
Empty Wt.	421 lbs.	**VCr**	100 mph
Payload (full fuel)	220 lbs.	**V_{S1} (stall clean)**	40 mph
Fuel Capacity	6 gal.	**V_{S0} (lndg config)**	40 mph
Seats	5	**Climb Rate @msl**	700 fpm
Cabin Width	N/A	**Service Ceiling**	15,000 ft.

Std. Engine/HP	Usable HP Range	Construction Mtrl.	Est. Bldg. Time
Rotax 503 / 50 hp	N/A	Metal, Wood, Fabric	650 hrs.
Plans	**Kit**	**Info Packet**	**Video**
$125.00	N/A	$5.00	N/A

Building cost is about $4,000 not including engine. Light controls but not overly sensitive. Roomy cockpit for the larger pilot. Comprehensive builder's manual available for $10.00.

Kits/Plans Sold: 84 **Completed & Flying:** 4

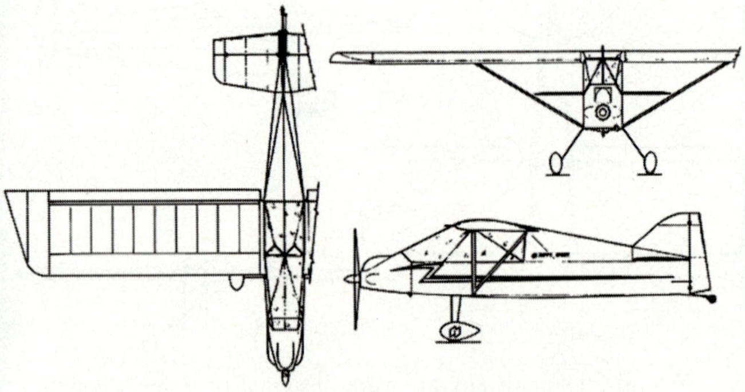

GREEN SKY ADVENTURES, INC.
2395 Cream Ridge Rd.
Orwell OH 44076
Tel. 216-293-6624 • Fax 216-293-6321
Web: http://www.greensky-adventures.com

GN-1 Aircamper

Wing Span	29.00 ft.	**Range**	358 sm
Wing Area	145.0 sq. ft.	**Takeoff Dist.**	300 ft.
Length	18.10 ft.	**Landing Dist.**	300 ft.
Max Gross Wt.	1100 lbs.	**Vmax**	115 mph
Empty Wt.	650 lbs.	**VCr**	87 mph
Payload (full fuel)	N/A	**V_{S1} (stall clean)**	25 mph
Fuel Capacity	18 gal.	**V_{S0} (lndg config)**	N/A
Seats	2	**Climb Rate @msl**	500 fpm
Cabin Width	N/A	**Service Ceiling**	12,000 ft.

Std. Engine/HP	Usable HP Range	Construction Mtrl.	Est. Bldg. Time
Continental / 65 hp	65 - 90	Tube, Wood	3000 hrs.
Plans	**Kit**	**Info Packet**	**Video**
$25.00	Yes	$1.00	N/A

Classic Pietenpol. Plans ordered from overseas are $25.00 plus $6.00 in U.S. funds.

Completed & Flying: 500

Thomas Vogt, Sr., Builder/Pilot
"... The plane is airborne before I know it. It's unbelievable. I keep my climb-outs shallow, my banks flat. ... Time and time again, I show an indicated 90 airspeed at 2,000 RPM ... I have a 75 hp Continental with a Sensenich metal prop."

Tom Mathes, Builder/Pilot
"...SN 932 is alive and well. First flew her in Sept. '95 and have 61 hours to date. With a Hegy 72x40 prop she cruises at about 72 mph. Thank you for a good set of plans and help when needed."

JOHN W. GREGA
355 Grand Blvd.
Bedford OH 44146
Tel. 216-232-5790

Lionheart

Wing Span	31.00 ft.	**Range**	1500 sm
Wing Area	238.0 sq. ft.	**Takeoff Dist.**	800 ft.
Length	26.60 ft.	**Landing Dist.**	1400 ft.
Max Gross Wt.	5200 lbs.	**Vmax**	230 mph
Empty Wt.	2900 lbs.	**VCr**	210 mph
Payload (full fuel)	940 lbs.	**V_{S1} (stall clean)**	61 mph
Fuel Capacity	160 gal.	**V_{S0} (lndg config)**	56 mph
Seats	7	**Climb Rate @msl**	2500 fpm
Cabin Width	51 in.	**Service Ceiling**	20,000 ft.

Std. Engine/HP	Usable HP Range	Construction Mtrl.	Est. Bldg. Time
P&W R-985 / 450 hp	350 - 600	Composite	2000 hrs.
Plans	**Kit**	**Info Packet**	**Video**
N/A	$89,900.00	$10.00	Yes

A spirited, 200+ mph, six or seven place, long distance cruiser. Lionheart combines the classic look and sound of the 30's with materials and aerodynamics of the 90's. The aircraft's primary structure consists of high temperature glass and carbon composites molded in production tooling. The prototype made its first flight the end of July '97 and was flown daily at Oshkosh '97. Current testing results indicate the aircraft meets or exceeds all structural, handling, and performance objectives. Four customer aircraft are under construction.

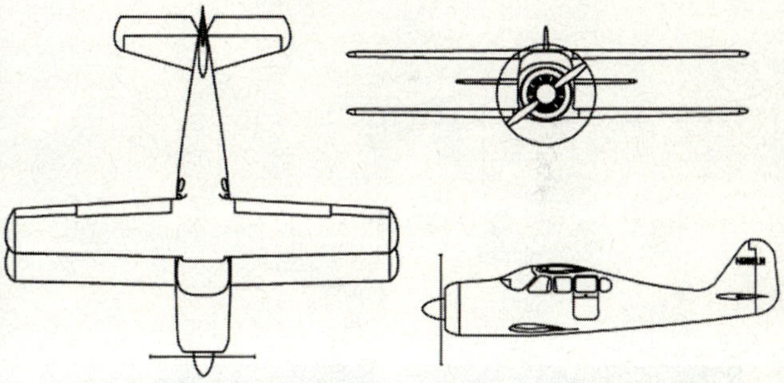

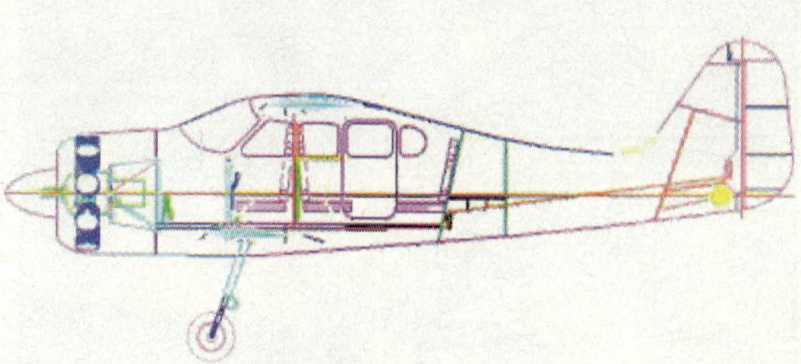

GRIFFON AEROSPACE, INC.
920 Yarbrough Rd.
Harvest AL 35749
Tel. 205-721-9055 • Fax 205-837-1694
E-Mail: griffon@traveller.com
Web: http://www.griffon-aerospace.com

ALFA

Wing Span	29.50 ft.	**Range**	560 sm
Wing Area	31.00 sq. ft.	**Takeoff Dist.**	N/A
Length	19.50 ft.	**Landing Dist.**	N/A
Max Gross Wt.	1540 lbs.	**Vmax**	185 mph
Empty Wt.	N/A	**VCr**	140 mph
Payload (full fuel)	572 lbs.	**V_{S1} (stall clean)**	61 mph
Fuel Capacity	N/A	**V_{S0} (lndg config)**	54 mph
Seats	2	**Climb Rate @msl**	N/A
Cabin Width	44 in.	**Service Ceiling**	N/A

Std. Engine/HP	Usable HP Range	Construction Mtrl.	Est. Bldg. Time
Porsche / 110 hp	N/A	Aluminum and fiberglass	N/A
Plans	**Kit**	**Info Packet**	**Video**
N/A	N/A	N/A	N/A

Porsche engine drives a quiet, geared, five-bladed prop at 1500 RPM in cruise. The sliding canopy offers easy entrance and provides excellent visability. The ALFA is easy to disassemble and can be rigged on a trailer in just a few minutes.

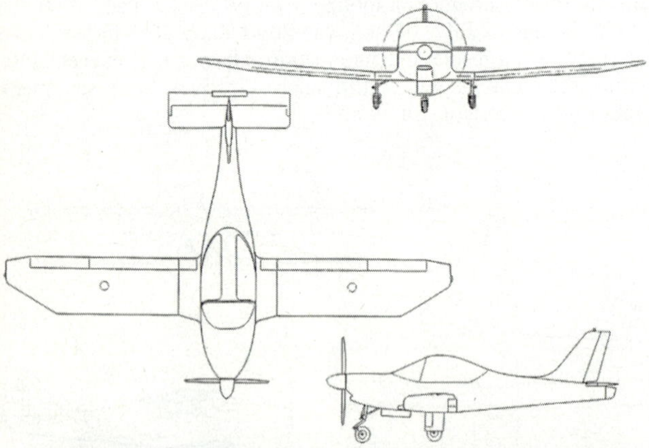

HB FLUGTECHNIK GMBH
Dr.-Adolf-Scharf-Str. 42, Pf.7
A-4053 Haid AUSTRIA
Tel. 07229/79104 • Fax 07229/79104-15

Highlander

Wing Span	29.60 ft.	**Range**	530 sm
Wing Area	92.50 sq. ft.	**Takeoff Dist.**	600 ft.
Length	19.50 ft.	**Landing Dist.**	700 ft.
Max Gross Wt.	1200 lbs.	**Vmax**	145 mph
Empty Wt.	700 lbs.	**VCr**	110 mph
Payload (full fuel)	500 lbs.	**V_{S1} (stall clean)**	52 mph
Fuel Capacity	20 gal.	**V_{S0} (lndg config)**	50 mph
Seats	2	**Climb Rate @msl**	800 fpm
Cabin Width	40 in.	**Service Ceiling**	10,000 ft.

Std. Engine/HP	Usable HP Range	Construction Mtrl.	Est. Bldg. Time
Rotax 912 / 80 hp	N/A	Metal, Aluminum	800 hrs.
Plans	**Kit**	**Info Packet**	**Video**
$350.00	$14,500.00	$5.00	N/A

A two seat, light tourer of British design (completed in 1987). Derivative of the ARV Super 2 using a Rotax engine. Single spar all-aluminum construction with dual stick and rudder. Canopy hinges up and back. Excellent all-around vision. Completely stress and load tested and flight proven. Manufactured in the U.S. since 1995. A lower cost economy kit with raw materials plus hardware is offered for experienced builders. **Completed & Flying:** 32

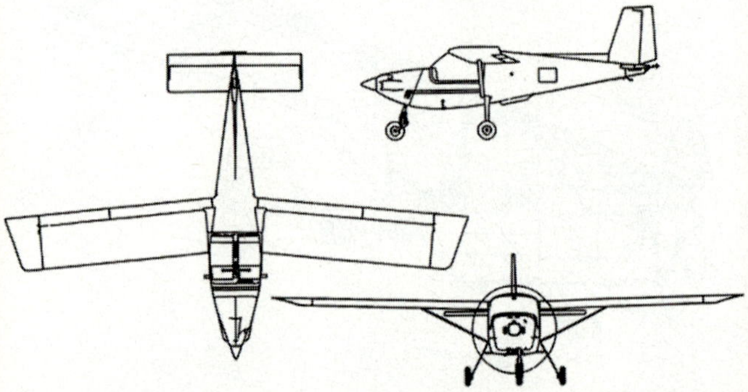

The Highlander is a derivative of ARV Super 2 originally produced by ARV Aviation Ltd. The ARV (Aviation Recreational Vehicle) was the brain child of Richard Noble who attained the world land speed record at Black Lake, NV in 1983. On returning to the United Kingdom, he learned that nobody was manufacturing a small modern light aircraft. Richard engaged a first class design team to do the job. But, after producing 30 aircraft, further research and development costs forced the company to cease it's manufacturing operations.

In the spring of 1990 a new company, Aviation (Scotland) Ltd, was formed to take up the challenge. In September 1993 A.S.L. established a second facility in Sweden in conjunction with a major Swedish industrial conglomerate.This increased production capability and provided additional capitol to fund numerous improvements to the ARV. The kit version is produced entirely in the U.S. and is named "Highlander."

HIGHLANDER AIRCRAFT CORP.
2255 Orkla Drive
Golden Valley MN 55427
Tel. 612-593-0341 • Fax 612-525-1566

Henderson Little Bear

Wing Span	35.00 ft.	Range	200 sm
Wing Area	178.0 sq. ft.	Takeoff Dist.	175 ft.
Length	22.50 ft.	Landing Dist.	275 ft.
Max Gross Wt.	1220 lbs.	Vmax	122 mph
Empty Wt.	700 lbs.	VCr	N/A
Payload (full fuel)	420 lbs.	V_{S1} (stall clean)	38 mph
Fuel Capacity	12 gal.	V_{S0} (lndg config)	N/A
Seats	2	Climb Rate @msl	500 fpm
Cabin Width	28 in.	Service Ceiling	10,500 ft.

Std. Engine/HP	Usable HP Range	Construction Mtrl.	Est. Bldg. Time
Continental / 65 hp	65 - 100	Tube, Fabric	400 hrs.
Plans	**Kit**	**Info Packet**	**Video**
Yes	$24,500.00	N/A	N/A

Super Cub replica. Prefab parts available.

Kits/Plans Sold: 5 **Completed & Flying:** 2

HENDERSON AERO SPECIALTIES
7431 Canterbury Rd.
Felton DE 19943
Tel. 302-284-8107 • Fax 302-284-8184

J-4 Sportster

Wing Span	28.00 ft.	Range	150 sm
Wing Area	120.0 sq. ft.	Takeoff Dist.	50 ft.
Length	16.25 ft.	Landing Dist.	N/A
Max Gross Wt.	500 lbs.	Vmax	63 mph
Empty Wt.	242 lbs.	VCr	60 mph
Payload (full fuel)	N/A	V_{S1} (stall clean)	24 mph
Fuel Capacity	5 gal.	V_{S0} (lndg config)	N/A
Seats	1	Climb Rate @msl	750 fpm
Cabin Width	N/A	Service Ceiling	N/A

Std. Engine/HP	Usable HP Range	Construction Mtrl.	Est. Bldg. Time
Rotax 277 / 28 hp	N/A	Wood, Tube, Fabric	350 hrs.
Plans	**Kit**	**Info Packet**	**Video**
$150.00	$6,884.00	$5.00	N/A

Ultralight. Close cousin to the Kitten. Received Grand Champion, Charles Lindbergh award. Plans build time is about 800 hrs. Available in the higher performance Super Sportster version. Kit price does not include engine. **Flying:** 50

HIPP'S SUPERBIRDS, INC.
P.O.Box 266
Saluda NC 28773
Tel. 704-749-9134 • Fax 704-748-9803

BLT-ARA

Wing Span	33.00 ft.	Range	1000 sm
Wing Area	129.0 sq. ft.	Takeoff Dist.	1300 ft.
Length	22.00 ft.	Landing Dist.	1300 ft.
Max Gross Wt.	1600 lbs.	Vmax	168 mph
Empty Wt.	1000 lbs.	VCr	158 mph
Payload (full fuel)	430 lbs.	V_{S1} (stall clean)	53 mph
Fuel Capacity	26 gal.	V_{S0} (lndg config)	N/A
Seats	2	Climb Rate @msl	800 fpm
Cabin Width	40 in.	Service Ceiling	18,000 ft.

Std. Engine/HP	Usable HP Range	Construction Mtrl.	Est. Bldg. Time
Lim. L2400 / 95 hp	90 - 110	Fiberglass	3000 hrs.
Plans	**Kit**	**Info Packet**	**Video**
$1,000.00	N/A	N/A	N/A

Sleek, two place design. Slightly modified version of the earlier BL1-KEA (shown in photo) which was first flown in 1989. Detachable wings and elevator. Stores in a trailer like a glider. Can be built without molds. Plans supplied with a 300 page construction manual. **Kits/Plans Sold:** 13

L AND B HINZ
Bei der Kirche 17 • Filderstadt D-70794 • GERMANY
Tel. 0049-7158-65441 • Fax 0049-7158-65441

J-3 Kitten

Wing Span	30.00 ft.	Range	150 sm
Wing Area	120.0 sq. ft.	Takeoff Dist.	50 ft.
Length	16.25 ft.	Landing Dist.	N/A
Max Gross Wt.	500 lbs.	Vmax	63 mph
Empty Wt.	250 lbs.	VCr	60 mph
Payload (full fuel)	N/A	V_{S1} (stall clean)	24 mph
Fuel Capacity	5 gal.	V_{S0} (lndg config)	N/A
Seats	1	Climb Rate @msl	750 fpm
Cabin Width	N/A	Service Ceiling	N/A

Std. Engine/HP	Usable HP Range	Construction Mtrl.	Est. Bldg. Time
Rotax 277 / 28 hp	N/A	Tube, Fabric, Wood	350 hrs.
Plans	**Kit**	**Info Packet**	**Video**
$100.00	$6,684.00	$5.00	N/A

Experimental well proven design. Estimated build time is for kit construction. Plans construction is estimated at 800 hours. Price is for firewall back without engine.

Completed & Flying: 50

HIPP'S SUPERBIRDS, INC.
P.O.Box 266
Saluda NC 28773
Tel. 704-749-9134 • Fax 704-748-9803

Reliant

Wing Span	30.00 ft.	**Range**	150 sm
Wing Area	N/A	**Takeoff Dist.**	75 ft.
Length	16.33 ft.	**Landing Dist.**	200 ft.
Max Gross Wt.	500 lbs.	**Vmax**	63 mph
Empty Wt.	254 lbs.	**VCr**	60 mph
Payload (full fuel)	N/A	**V_{S1} (stall clean)**	24 mph
Fuel Capacity	5 gal.	**V_{S0} (lndg config)**	N/A
Seats	1	**Climb Rate @msl**	800 fpm
Cabin Width	N/A	**Service Ceiling**	12,000 ft.

Std. Engine/HP	Usable HP Range	Construction Mtrl.	Est. Bldg. Time
Rotax 277 / 28 hp	N/A	Wood, Metal	350 hrs.
Plans	**Kit**	**Info Packet**	**Video**
N/A	$6,884.00	$5.00	N/A

First introduced at Oshkosh 1987. Received Grand Champion, Charles Lindbergh award. Price does not include engine. Kit does include instruments and most everything else to make it ready-to-fly except final paint colors and thinner. **Flying:** 50

HIPP'S SUPERBIRDS, INC.
P.O.Box 266
Saluda NC 28773
Tel. 704-749-9134 • Fax 704-748-9803

J-5 Super Kitten

Wing Span	30.00 ft.	**Range**	350 sm
Wing Area	120.0 sq. ft.	**Takeoff Dist.**	50 ft.
Length	16.25 ft.	**Landing Dist.**	N/A
Max Gross Wt.	600 lbs.	**Vmax**	90 mph
Empty Wt.	290 lbs.	**VCr**	77 mph
Payload (full fuel)	N/A	**V_{S1} (stall clean)**	31 mph
Fuel Capacity	9 gal.	**V_{S0} (lndg config)**	N/A
Seats	1	**Climb Rate @msl**	800 fpm
Cabin Width	N/A	**Service Ceiling**	N/A

Std. Engine/HP	Usable HP Range	Construction Mtrl.	Est. Bldg. Time
Rotax 447 / 40 hp	N/A	Wood, Tube, Fabric	350 hrs.
Plans	**Kit**	**Info Packet**	**Video**
$100.00	$9,287.00	$5.00	N/A

The Super Kitten has all the options: extra fuel capacity, shocks, wheel pants, brakes and a 40 hp Rotax engine.

Completed & Flying: 50

HIPP'S SUPERBIRDS, INC.
P.O.Box 266
Saluda NC 28773
Tel. 704-749-914 • Fax 704-748-9803

Reliant SX

Wing Span	30.00 ft.	**Range**	190 sm
Wing Area	120.0 sq. ft.	**Takeoff Dist.**	150 ft.
Length	16.33 ft.	**Landing Dist.**	200 ft.
Max Gross Wt.	575 lbs.	**Vmax**	95 mph
Empty Wt.	285 lbs.	**VCr**	85 mph
Payload (full fuel)	N/A	**V_{S1} (stall clean)**	30 mph
Fuel Capacity	5 gal.	**V_{S0} (lndg config)**	N/A
Seats	1	**Climb Rate @msl**	1150 fpm
Cabin Width	N/A	**Service Ceiling**	N/A

Std. Engine/HP	Usable HP Range	Construction Mtrl.	Est. Bldg. Time
Rotax 447 / 40 hp	N/A	Wood, Tube, Fabric	350 hrs.
Plans	**Kit**	**Info Packet**	**Video**
N/A	$9,487.00	$5.00	N/A

Hipp's "Classy" plane, with sleek lines. Experimental. A clean, fast design. Available with an additional 5 gal. tank. Kit includes engine, shocks and wheel pants.

Completed & Flying: 50

HIPP'S SUPERBIRDS, INC.
P.O.Box 266
Saluda NC 28773
Tel. 704-749-9134 • Fax 704-748-9803

F4U Corsair

Wing Span	24.80 ft.	**Range**	650 sm
Wing Area	127.5 sq. ft.	**Takeoff Dist.**	1200 ft.
Length	21.20 ft.	**Landing Dist.**	1500 ft.
Max Gross Wt.	2004 lbs.	**Vmax**	256 mph
Empty Wt.	1405 lbs.	**VCr**	225 mph
Payload (full fuel)	330 lbs.	**V_{S1} (stall clean)**	65 mph
Fuel Capacity	45 gal.	**V_{S0} (lndg config)**	55 mph
Seats	1	**Climb Rate @msl**	2235 fpm
Cabin Width	21 in.	**Service Ceiling**	26,250 ft.

Std. Engine/HP	Usable HP Range	Construction Mtrl.	Est. Bldg. Time
Ford V-6 / 230 hp	N/A	Tube, Composite	2000 hrs.
Plans	**Kit**	**Info Packet**	**Video**
N/A	Yes	$3.50	N/A

60% scale replica with exact external detail. Complete kit equipped for VFR cross country flight. Utilizes 230 HP converted auto engine.

HISTORICAL AIRCRAFT CORP.
PO Box 888
Nucla CO 81424
Tel. 970-864-7328 • Fax 970-864-7624
E-Mail: hac@montrose.net

P-51D Mustang

Wing Span	24.00 ft.	**Range**	650 sm
Wing Area	100.0 sq. ft.	**Takeoff Dist.**	1200 ft.
Length	20.60 ft.	**Landing Dist.**	1500 ft.
Max Gross Wt.	1954 lbs.	**Vmax**	255 mph
Empty Wt.	1354 lbs.	**VCr**	228 mph
Payload (full fuel)	330 lbs.	**V_{S1} (stall clean)**	72 mph
Fuel Capacity	41 gal.	**V_{S0} (lndg config)**	57 mph
Seats	1	**Climb Rate @msl**	2290 fpm
Cabin Width	21 in.	**Service Ceiling**	27,750 ft.

Std. Engine/HP	Usable HP Range	Construction Mtrl.	Est. Bldg. Time
Ford V-6 / 230 hp	N/A	Tube, Composite	2000 hrs.
Plans	**Kit**	**Info Packet**	**Video**
N/A	Yes	$3.50	N/A

62.5% scale Mustang. Prefabricated steel internal structure to handle aerobatic loads. External shell is shaped out with urethane foam and fiberglass to give the original's metal appearance when painted. Complete with prefabricated assemblies, engine, scale fixed blade prop, instruments, avionics, etc.

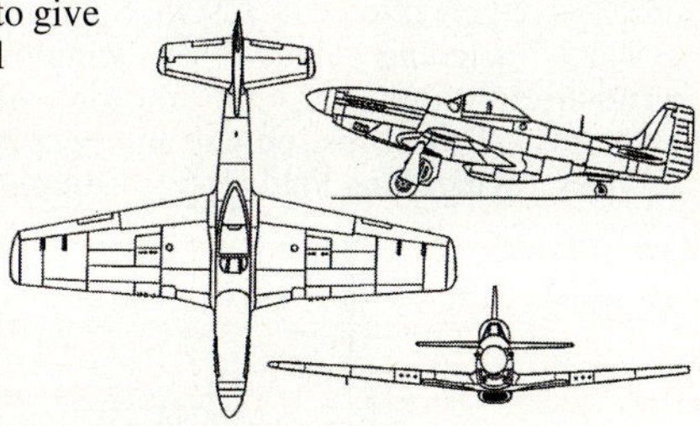

HISTORICAL AIRCRAFT CORP.
PO Box 888
Nucla CO 81424
Ph: 970-864-7328 • Fax: 970-864-7624
E-Mail: hac@montrose.net

P-40C Tomahawk

Wing Span	24.50 ft.	**Range**	650 sm
Wing Area	100.0 sq. ft.	**Takeoff Dist.**	1200 ft.
Length	21.27 ft.	**Landing Dist.**	1500 ft.
Max Gross Wt.	1938 lbs.	**Vmax**	250 mph
Empty Wt.	1347 lbs.	**VCr**	224 mph
Payload (full fuel)	330 lbs.	**V_{S1} (stall clean)**	71 mph
Fuel Capacity	42 gal.	**V_{S0} (lndg config)**	61 mph
Seats	1	**Climb Rate @msl**	2365 fpm
Cabin Width	21 in.	**Service Ceiling**	27,250 ft.

Std. Engine/HP	Usable HP Range	Construction Mtrl.	Est. Bldg. Time
Ford V-6 / 230 hp	N/A	Tube, Composite	2000 hrs.
Plans	**Kit**	**Info Packet**	**Video**
N/A	Yes	$3.50	N/A

62.5% scale P-40C Tomahawk. Pre-fabricated steel internal structure to handle aerobatic loads. External shell is shaped out with urethane foam and fiberglass to give the original's metal appearance when painted. Kit comes complete with prefabricated assemblies, engine, scale fixed blade prop, instruments, avionics, foam, fiberglass, paint, etc.

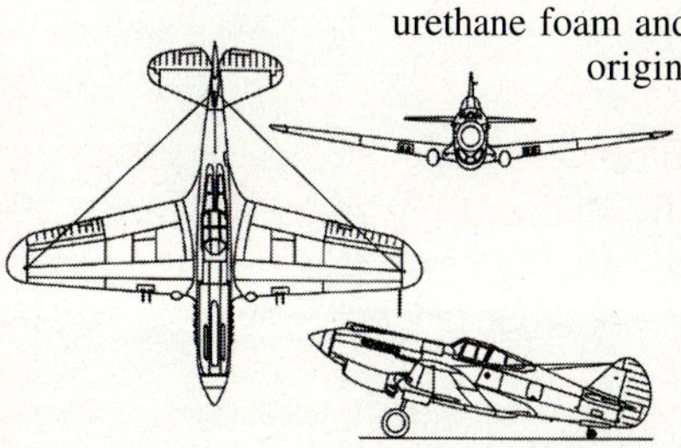

HISTORICAL AIRCRAFT CORP.
PO Box 888
Nucla CO 81424
Ph: 970-864-7328 • Fax: 970-864-7624
E-Mail: hac@montrose.net

PZL P.11c

Wing Span	30.00 ft.	**Range**	300 sm
Wing Area	84.00 sq. ft.	**Takeoff Dist.**	800 ft.
Length	18.30 ft.	**Landing Dist.**	1000 ft.
Max Gross Wt.	1100 lbs.	**Vmax**	105 mph
Empty Wt.	800 lbs.	**VCr**	90 mph
Payload (full fuel)	180 lbs.	**V_{S1} (stall clean)**	55 mph
Fuel Capacity	11 gal.	**V_{S0} (lndg config)**	N/A
Seats	1	**Climb Rate @msl**	800 fpm
Cabin Width	22 in.	**Service Ceiling**	12,500 ft.

Std. Engine/HP	Usable HP Range	Construction Mtrl.	Est. Bldg. Time
CAM / 100 hp	N/A	Tube, Wood, Fabric	1400 hrs.
Plans	**Kit**	**Info Packet**	**Video**
N/A	Yes	$3.50	N/A

66% scale replica of the Polish fighter used in the early days of WWII. Low cost beginner kit equipped for local VFR flying. Complete kit includes prefabricated assemblies, engine, prop, basic VFR instrumentation, fabric, paint, etc. Also includes scale control stick with MG triggers, dummy 7.9mm MGs, and ring-and-bead gunsight.

HISTORICAL AIRCRAFT CORP.
PO Box 888
Nucla CO 81424
Tel. 970-864-7328 • Fax 970-864-7624
E-Mail: hac@montrose.net

Ryan STA PT-16/PT-20

Wing Span	26.00 ft.	**Range**	490 sm
Wing Area	112.6 sq. ft.	**Takeoff Dist.**	800 ft.
Length	20.10 ft.	**Landing Dist.**	1200 ft.
Max Gross Wt.	1275 lbs.	**Vmax**	125 mph
Empty Wt.	725 lbs.	**VCr**	110 mph
Payload (full fuel)	412 lbs.	**V_{S1} (stall clean)**	53 mph
Fuel Capacity	23 gal.	**V_{S0} (lndg config)**	53 mph
Seats	2	**Climb Rate @msl**	930 fpm
Cabin Width	22 in.	**Service Ceiling**	16,300 ft.

Std. Engine/HP	Usable HP Range	Construction Mtrl.	Est. Bldg. Time
CAM / 100 hp	N/A	Tube, Wood, Fabric, F/G	1500 hrs.
Plans	**Kit**	**Info Packet**	**Video**
N/A	$35,500.00	$3.50	N/A

85% scale Ryan Trainer is designed as a simple and complete aircraft kit for beginners. With prefabricated fuselage frame, tail section, gear legs, and engine mount. Wings have wood spars and built up wood ribs.

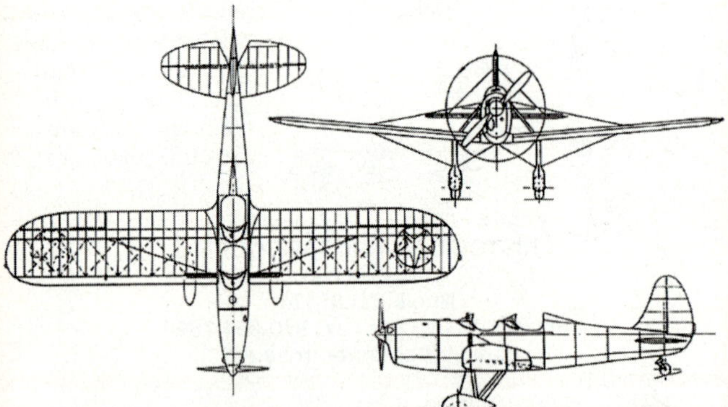

HISTORICAL AIRCRAFT CORP.
PO Box 888
Nucla CO 81424
Tel. 970-864-7328 • Fax 970-864-7624
E-Mail: hac@montrose.net

Hummel Bird

Wing Span	21.00 ft.	**Range**	250 sm
Wing Area	63.00 sq. ft.	**Takeoff Dist.**	300 ft.
Length	13.30 ft.	**Landing Dist.**	600 ft.
Max Gross Wt.	520 lbs.	**Vmax**	135 mph
Empty Wt.	268 lbs.	**VCr**	115 mph
Payload (full fuel)	250 lbs.	**V_{S1} (stall clean)**	38 mph
Fuel Capacity	7 gal.	**V_{S0} (lndg config)**	32 mph
Seats	1	**Climb Rate @msl**	1000 fpm
Cabin Width	22 in.	**Service Ceiling**	13,000 ft.

Std. Engine/HP	Usable HP Range	Construction Mtrl.	Est. Bldg. Time
1/2 VW / 35 hp	N/A	Metal	600 hrs.
Plans	**Kit**	**Info Packet**	**Video**
$180.00	N/A	$6.00	$29.95

Attractive choice for first time builders. Low cost, fast build, good performance. Tricycle and taildragger options. 60-minute VHS descriptive tape offered at $29.95. Some prefab components are available including bulkhead sets, wing root sets, and fairings, windshield and canopy, motor mount, and wheel assembly. Also props, engine kits, engine plans, parts, and services. **Kits/Plans Sold:** 1800 **Completed & Flying:** 150

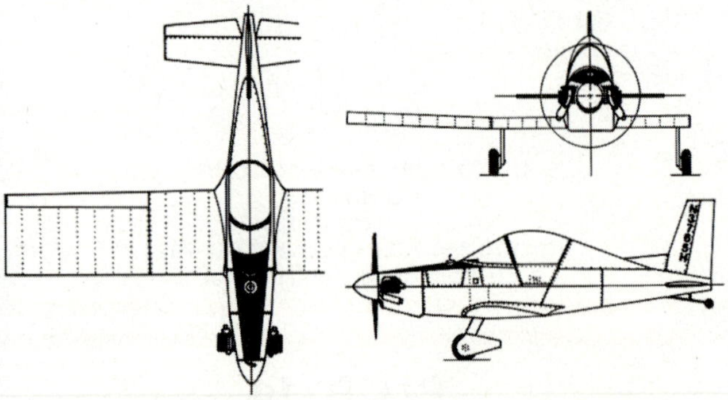

The Hummel Bird was developed using the Watson Wingwagon aircraft as a basic design. It was simple and quick to build including the engine. However, the tri-gear, open cockpit and no shoulder harness were seen as negatives by Morry Hummel. So, he raised the two front bulkheads 2" providing more foot space and the seat bulkheads by 5" to allow for a shoulder harness. Also, the weight was reduced and it was stressed for 6 G's instead of 3.5 G's. Along with other improvements, this has provided a much sturdier aircraft suitable for larger pilots.

Terry Hamilton, Adams, Mn

"I enjoyed building the little bird as much as I do flying it! ...I like everything about the plane and would recommend it to others who are looking for a low cost fun plane to own and fly."

Brian D. Klenoski, Vicksburg, MI

"... The engine and airplane have performed above my expectations."

HUMMEL AVIATION
509 E. Butler Street
Bryan OH 43506-0880
Tel. 419-636-3390

CA-2

Wing Span	26.00 ft.	**Range**	125 sm
Wing Area	117.0 sq. ft.	**Takeoff Dist.**	100 ft.
Length	16.50 ft.	**Landing Dist.**	200 ft.
Max Gross Wt.	500 lbs.	**Vmax**	63 mph
Empty Wt.	250 lbs.	**VCr**	50 mph
Payload (full fuel)	250 lbs.	**V_{S1} (stall clean)**	26 mph
Fuel Capacity	5 gal.	**V_{S0} (lndg config)**	N/A
Seats	1	**Climb Rate @msl**	600 fpm
Cabin Width	24 in.	**Service Ceiling**	N/A

Std. Engine/HP	Usable HP Range	Construction Mtrl.	Est. Bldg. Time
Rotax / 28 hp	22 - 40	Aluminum	500 hrs.
Plans	**Kit**	**Info Packet**	**Video**
$125.00	N/A	$5.00	N/A

New design. Only all-metal legal ultralight available (when powered with Rotax 277 engine). Total cost between $5,000 and $6,000. Other engine options include the 1/2 VW, 25 hp Hirth, and the Rotax 337 and 477. **Completed & Flying:** 3

HUMMEL AVIATION
509 E. Butler Street
Bryan OH 43506-0880
Tel. 419-636-3390

Hurricane Clip Wing HP

Wing Span	26.00 ft.	**Range**	100 sm
Wing Area	132.0 sq. ft.	**Takeoff Dist.**	100 ft.
Length	17.00 ft.	**Landing Dist.**	150 ft.
Max Gross Wt.	625 lbs.	**Vmax**	105 mph
Empty Wt.	275 lbs.	**VCr**	70 mph
Payload (full fuel)	275 lbs.	**V_{S1} (stall clean)**	30 mph
Fuel Capacity	10 gal.	**V_{S0} (lndg config)**	N/A
Seats	1	**Climb Rate @msl**	900 fpm
Cabin Width	20 in.	**Service Ceiling**	12,500 ft.

Std. Engine/HP	Usable HP Range	Construction Mtrl.	Est. Bldg. Time
Rotax 503 / 52 hp	40 - 75	Tube, Fabric	100 hrs.
Plans	**Kit**	**Info Packet**	**Video**
N/A	$10,229.00	$5.00	$15.00

High performance ultralight for sport flying. The HP series is available in standard and "Clipwing" versions. +8, -6 G's loading. Experimental type FAA registration. See Ad on Page 45.
Kits/Plans Sold: 125 **Completed & Flying:** 125

HY-TEK HURRICANE
23055 Airport Road NE #1
Aurora OR 97002
Tel. 503-678-5740 • Fax 503-678-2771
E-Mail: mkhytek@aol.com

Hurricane Hauler

Wing Span	28.00 ft.	**Range**	100 sm
Wing Area	142.0 sq. ft.	**Takeoff Dist.**	150 ft.
Length	17.00 ft.	**Landing Dist.**	150 ft.
Max Gross Wt.	750 lbs.	**Vmax**	100 mph
Empty Wt.	285 lbs.	**VCr**	65 mph
Payload (full fuel)	390 lbs.	**V_{S1} (stall clean)**	30 mph
Fuel Capacity	10 gal.	**V_{S0} (lndg config)**	N/A
Seats	1	**Climb Rate @msl**	800 fpm
Cabin Width	26 in.	**Service Ceiling**	12,500 ft.

Std. Engine/HP	Usable HP Range	Construction Mtrl.	Est. Bldg. Time
Rotax 503 / 52 hp	52 - 75	Tube, Fabric	100 hrs.
Plans	**Kit**	**Info Packet**	**Video**
N/A	$10,399.00	$5.00	$15.00

For sport and cross country flying. Extra wide cockpit for large pilots. Heavy duty landing gear. Experimental type FAA registration. See Ad on Page 45. **Kits Sold:** 125 **Complete & Flying:** 15

HY-TEK HURRICANE
23055 Airport Road NE #1
Aurora OR 97002
Tel. 503-678-5740 • Fax 503-678-2771
E-Mail: mkhytek@aol.com

Hurricane 103 Ultralight

Wing Span	28.00 ft.	**Range**	80 sm
Wing Area	142.0 sq. ft.	**Takeoff Dist.**	100 ft.
Length	17.00 ft.	**Landing Dist.**	150 ft.
Max Gross Wt.	525 lbs.	**Vmax**	100 mph
Empty Wt.	252 lbs.	**VCr**	65 mph
Payload (full fuel)	243 lbs.	**V_{S1} (stall clean)**	27 mph
Fuel Capacity	5 gal.	**V_{S0} (lndg config)**	N/A
Seats	1	**Climb Rate @msl**	800 fpm
Cabin Width	20 in.	**Service Ceiling**	12,500 ft.

Std. Engine/HP	Usable HP Range	Construction Mtrl.	Est. Bldg. Time
Rotax 447 / 40 hp	40	Tube, Fabric	100 hrs.
Plans	**Kit**	**Info Packet**	**Video**
N/A	$9,199.00	$5.00	$15.00

High performance ultralight for sport flying. FAR 103 legal. Price is for the "Standard" version. Also sold in the "Ultra 103" package for $8699 which has slightly reduced performance. See Ad on Page 45. **Kits Sold:** 125 **Completed & Flying:** 15

HY-TEK HURRICANE
23055 Airport Road NE #1
Aurora OR 97002
Tel. 503-678-5740 • Fax 503-678-2771
E-Mail: mkhytek@aol.com

Savannah

Wing Span	29.50 ft.	**Range**	400 sm
Wing Area	139.2 sq. ft.	**Takeoff Dist.**	260 ft.
Length	20.02 ft.	**Landing Dist.**	190 ft.
Max Gross Wt.	972 lbs.	**Vmax**	108 mph
Empty Wt.	605 lbs.	**VCr**	95 mph
Payload (full fuel)	N/A	**V_{S1} (stall clean)**	N/A
Fuel Capacity	19 gal.	**V_{S0} (lndg config)**	28 mph
Seats	2	**Climb Rate @msl**	1200 fpm
Cabin Width	N/A	**Service Ceiling**	13,000 ft.

Std. Engine/HP	Usable HP Range	Construction Mtrl.	Est. Bldg. Time
Rotax 912 / 79 hp	79-118	Metal	N/A
Plans	**Kit**	**Info Packet**	**Video**
N/A	Yes	N/A	N/A

Advanced kit includes finished and painted aircraft, upholstered seats, pre-cut instrument panel, engine mount for Rotax 912, assembled and painted engine cowling, assembled radiator, wheels brakes, and tires. Engine, instruments and prop not included.

I.C.P. SRL
via Torinon 12 • 14020 Piova Massaia (AT)
ITALY
Tel. 39 (0141) 996261 • Fax 39 (0141) 996353
E-Mail: info@icp.it

Amigo

Wing Span	26.40 ft.	**Range**	460 sm
Wing Area	138.0 sq. ft.	**Takeoff Dist.**	300 ft.
Length	19.80 ft.	**Landing Dist.**	330 ft.
Max Gross Wt.	972 lbs.	**Vmax**	127 mph
Empty Wt.	605 lbs.	**VCr**	110 mph
Payload (full fuel)	N/A	**V_{S1} (stall clean)**	N/A
Fuel Capacity	32 gal.	**V_{S0} (lndg config)**	40 mph
Seats	2	**Climb Rate @msl**	1000 fpm
Cabin Width	N/A	**Service Ceiling**	13,000 ft.

Std. Engine/HP	Usable HP Range	Construction Mtrl.	Est. Bldg. Time
Rotax / 79 hp	79-118	Metal	N/A
Plans	**Kit**	**Info Packet**	**Video**
N/A	N/A	N/A	N/A

Advanced kit includes finished and painted aircraft, upholstered seats, pre-cut instrument panel, engine mount for Rotax 912, assembled and painted engine cowling, and heating system, wheels brakes, and tires. Engine, instruments and prop not included.

I.C.P. SRL
via Torinon 12 • 14020 Piova Massaia (AT)
ITALY
Tel. 39 (0141) 996261 • Fax 39 (0141) 996353
E-Mail: info@icp.it

Genesis

Wing Span	26.66 ft.	**Range**	225 sm
Wing Area	155.0 sq. ft.	**Takeoff Dist.**	150 ft.
Length	19.33 ft.	**Landing Dist.**	100 ft.
Max Gross Wt.	1000 lbs.	**Vmax**	120 mph
Empty Wt.	490 lbs.	**VCr**	90 mph
Payload (full fuel)	515 lbs.	**V_{S1} (stall clean)**	32 mph
Fuel Capacity	14 gal.	**V_{S0} (lndg config)**	N/A
Seats	2	**Climb Rate @msl**	800 fpm
Cabin Width	43 in.	**Service Ceiling**	15,000 ft.

Std. Engine/HP	Usable HP Range	Construction Mtrl.	Est. Bldg. Time
Rotax 503 / 53 hp	53 - 100	Metal, Fiberglass, Fabric	200 hrs.
Plans	**Kit**	**Info Packet**	**Video**
N/A	$11,495.00	$5.00	$10.00

Two place, side-by-side. Engine, accessories and electric start are standard. Options include 20 gallon add-on fuel tank, floats and snow skis, electric flaperons/trim, adjustable seats, wheel pants and ballistic parachute. 6.08 ft wide XL wing option also offered.

Kits/Plans Sold: 82 **Completed & Flying:** 32

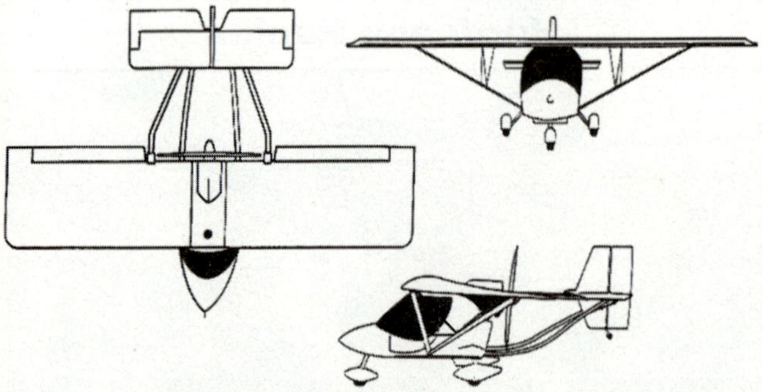

INNOVATION ENGINEERING, INC.
Davenport Municipal Airport • 8970 Harrison Street
Davenport IA 52804
Tel. 800-464-3664 • Fax 319-386-4569

Sky Quest

Wing Span	31.50 ft.	**Range**	N/A
Wing Area	155.0 sq. ft.	**Takeoff Dist.**	350 ft.
Length	19.40 ft.	**Landing Dist.**	150 ft.
Max Gross Wt.	1400 lbs.	**Vmax**	120 mph
Empty Wt.	550 lbs.	**VCr**	90 mph
Payload (full fuel)	575 lbs.	**V_{S1} (stall clean)**	40 mph
Fuel Capacity	20 gal.	**V_{S0} (lndg config)**	N/A
Seats	2	**Climb Rate @msl**	1200 fpm
Cabin Width	43 in.	**Service Ceiling**	N/A

Std. Engine/HP	Usable HP Range	Construction Mtrl.	Est. Bldg. Time
2xRotax 503/106 hp	N/A	Metal, Fiberglass, Fabric	300 hrs.
Plans	**Kit**	**Info Packet**	**Video**
N/A	$23,500.00	$5.00	N/A

Twin engine, high wing, pusher configuration. 20 gallon auxiliary fuel tank is optional.

Kits/Plans Sold: 1

INNOVATION ENGINEERING, INC.
Davenport Municipal Airport • 8970 Harrison Street
Davenport IA 52804
Ph: 800-464-3664 • Fax: 319-386-4569

Griffon

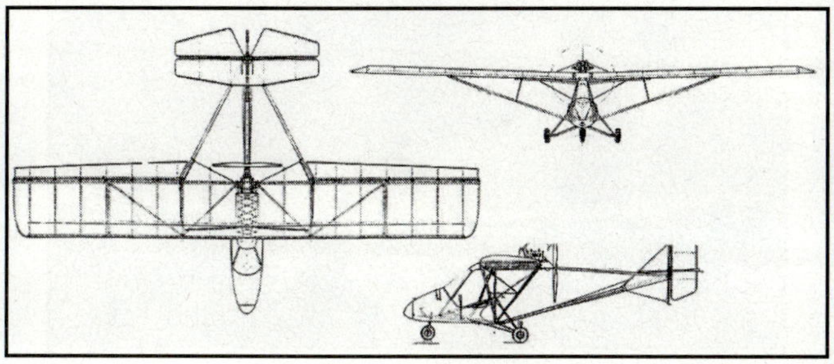

Wing Span	32.90 ft.	**Range**	124 sm
Wing Area	150.3 sq. ft.	**Takeoff Dist.**	165 ft.
Length	18.90 ft.	**Landing Dist.**	115 ft.
Max Gross Wt.	606 lbs.	**Vmax**	87 mph
Empty Wt.	342 lbs.	**VCr**	59 mph
Payload (full fuel)	N/A	**V_{S1} (stall clean)**	30 mph
Fuel Capacity	N/A	**V_{S0} (lndg config)**	N/A
Seats	1	**Climb Rate @msl**	1181 fpm
Cabin Width	N/A	**Service Ceiling**	N/A

Std. Engine/HP	Usable HP Range	Construction Mtrl.	Est. Bldg. Time
Rotax 503 / 50 hp	40-50	Aluminum tube, fabric, composi	N/A
Plans	**Kit**	**Info Packet**	**Video**
N/A	$7,400.00	N/A	N/A

Single seat, braced, high wing design. Two laminated plastic half-shells form the crew position. Wing struts and tail surfaces can be folded for transport and storage.

Kits/Plans Sold: 15 **Completed & Flying:** 15

INTERPLANE SRO
84 05 Letiste Zbraslavice
CZECHOSLOVAKIA
Tel. 0327/92145 • Fax 0327/92145

Skyboy

Wing Span	34.40 ft.	**Range**	186 sm
Wing Area	163.6 sq. ft.	**Takeoff Dist.**	330 ft.
Length	20.90 ft.	**Landing Dist.**	265 ft.
Max Gross Wt.	904 lbs.	**Vmax**	85 mph
Empty Wt.	489 lbs.	**VCr**	53 mph
Payload (full fuel)	N/A	**V_{S1} (stall clean)**	33 mph
Fuel Capacity	N/A	**V_{S0} (lndg config)**	28 mph
Seats	2	**Climb Rate @msl**	492 fpm
Cabin Width	N/A	**Service Ceiling**	N/A

Std. Engine/HP	Usable HP Range	Construction Mtrl.	Est. Bldg. Time
Rotax 503 / 50 hp	50-64	Aluminum tube, laminated plast	500 hrs.
Plans	**Kit**	**Info Packet**	**Video**
N/A	$14,400.00	N/A	N/A

Side-by-side, adjustable seats in fully enclosed cabin with upward opening window/door on each side. Wing struts and tail surfaces can be folded for transportation and storage. Skyboy S version is reinforced for higher Max Gross Takeoff Weight and speeds; utilizes more powerful Rotax 582 engine.

Kits/Plans Sold: 22 **Completed & Flying:** 22

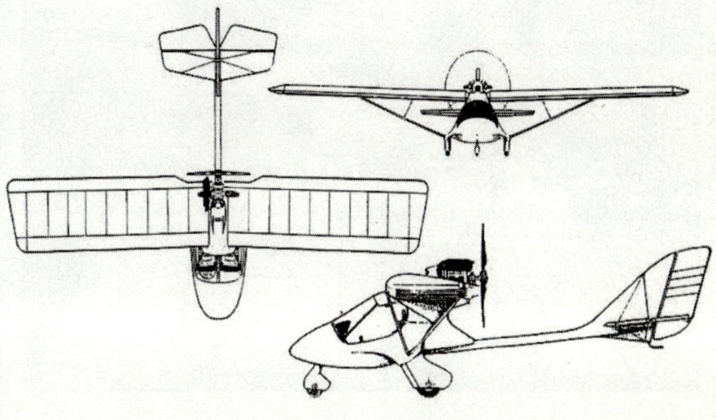

U.S. Distributor:
INTERPLANE USA, INC.
International Center • 30 Lee Gate
Grosse Pointe Farms MI 48236
Ph: 313-882-3400
Fax: 313-882-3401
E-Mail: InPlaneUSA@aol.com

INTERPLANE SRO
84 05 Letiste Zbraslavice
CZECHOSLOVAKIA
Ph: 0327/92145 • Fax: 0327/92145

Jabiru ST

Wing Span	26.33 ft.	**Range**	600 sm
Wing Area	85.00 sq. ft.	**Takeoff Dist.**	480 ft.
Length	16.40 ft.	**Landing Dist.**	300 ft.
Max Gross Wt.	1060 lbs.	**Vmax**	130 mph
Empty Wt.	520 lbs.	**VCr**	115 mph
Payload (full fuel)	540 lbs.	**V_{S1} (stall clean)**	45 mph
Fuel Capacity	17 gal.	**V_{S0} (lndg config)**	39 mph
Seats	2	**Climb Rate @msl**	1200 fpm
Cabin Width	43 in.	**Service Ceiling**	15,000 ft.

Std. Engine/HP	Usable HP Range	Construction Mtrl.	Est. Bldg. Time
Jabiru / 80 hp	N/A	Composite	600 hrs.
Plans	**Kit**	**Info Packet**	**Video**
N/A	$19,950.00	N/A	N/A

All composite construction with parts bonding only required. Quickbuild Kit price does not include engine.

Completed & Flying: 130.

U.S. Distributor:
TIGER AVIATION, INC.
134 Cessna Drive
Trenton SC 29847
Ph: 803-663-6200 • Fax: 803-663-6655
E-Mail: TREMFL@AOL.COM
Web: http://www.tigeraviation.com

JABIRU AIRCRAFT PTY LTD
PO Box 5186 • Bundaberg West
Queensland 4670
AUSTRALIA
Tel. 61-71-551-778 • Fax 61-71-552-669

V6 STOL

Wing Span	32.00 ft.	**Range**	700 sm
Wing Area	168.0 sq. ft.	**Takeoff Dist.**	150 ft.
Length	22.25 ft.	**Landing Dist.**	300 ft.
Max Gross Wt.	2200 lbs.	**Vmax**	140 mph
Empty Wt.	1200 lbs.	**VCr**	130 mph
Payload (full fuel)	N/A	**V_{S1} (stall clean)**	52 mph
Fuel Capacity	36 gal.	**V_{S0} (lndg config)**	N/A
Seats	4	**Climb Rate @msl**	1800 fpm
Cabin Width	N/A	**Service Ceiling**	18,000 ft.

Std. Engine/HP	Usable HP Range	Construction Mtrl.	Est. Bldg. Time
Ford V6 / 230 hp	N/A	Metal, Tube, Fabric	400 hrs.
Plans	**Kit**	**Info Packet**	**Video**
$89.00	N/A	$6.00	N/A

Combines a Piper Pacer airframe with a Ford V-6 engine for STOL performance and economical operation. Uses converted auto engines up to 300 hp.

JAVELIN AIRCRAFT COMPANY
662 S. Governour
Wichita KS 67207
Ph: 316-686-8500

Ultima

Wing Span	26.90 ft.	**Range**	700 sm
Wing Area	113.0 sq. ft.	**Takeoff Dist.**	560 ft.
Length	19.69 ft.	**Landing Dist.**	690 ft.
Max Gross Wt.	992 lbs.	**Vmax**	162 mph
Empty Wt.	573 lbs.	**VCr**	130 mph
Payload (full fuel)	330 lbs.	**V_{S1} (stall clean)**	47 mph
Fuel Capacity	32 gal.	**V_{S0} (lndg config)**	40 mph
Seats	2	**Climb Rate @msl**	N/A
Cabin Width	40 in.	**Service Ceiling**	16,400 ft.

Std. Engine/HP	Usable HP Range	Construction Mtrl.	Est. Bldg. Time
Rotax 912 / 80 hp	50-80	N/A	400 hrs.
Plans	**Kit**	**Info Packet**	**Video**
N/A	Yes	N/A	N/A

Aerobatic. Stressed to +6, -4 G's. Designed by Andre Konig. Introduced in 1993.

JUNKERS PROFLY
Georg-Nutzel-Str.10
95961 Kodnitz
GERMANY
Tel. 499221/84464 • Fax 499221/1842

IBIS - RJ03

Wing Span	20.66 ft.	**Range**	500 sm
Wing Area	52.30 sq. ft.	**Takeoff Dist.**	1000 ft.
Length	16.10 ft.	**Landing Dist.**	1050 ft.
Max Gross Wt.	1034 lbs.	**Vmax**	155 mph
Empty Wt.	572 lbs.	**VCr**	126 mph
Payload (full fuel)	374 lbs.	**V_{S1} (stall clean)**	59 mph
Fuel Capacity	16 gal.	**V_{S0} (lndg config)**	N/A
Seats	2	**Climb Rate @msl**	700 fpm
Cabin Width	26 in.	**Service Ceiling**	N/A

Std. Engine/HP	Usable HP Range	Construction Mtrl.	Est. Bldg. Time
VW 1830cc / 60 hp	60 - 80	Wood, Styrofoam	2000 hrs.
Plans	**Kit**	**Info Packet**	**Video**
$575.00	N/A	Yes	N/A

Two seat, energy-saving, canard design combines modern lines and traditional materials. Manufacturer claims operating costs are 20 to 25% lower than for other "classic" aircraft designs with the same engine under similar conditions. Totally enclosed in wood. Steerable nose wheel. The design features low weight, easy assembly. Other engine options include several VW models from 60 to 80 hp, JPX-4T75, Limbach-2000, and the Norton P64 rotary engine. Plans available in English or French. **Kits/Plans Sold:** 55 **Completed & Flying:** 4

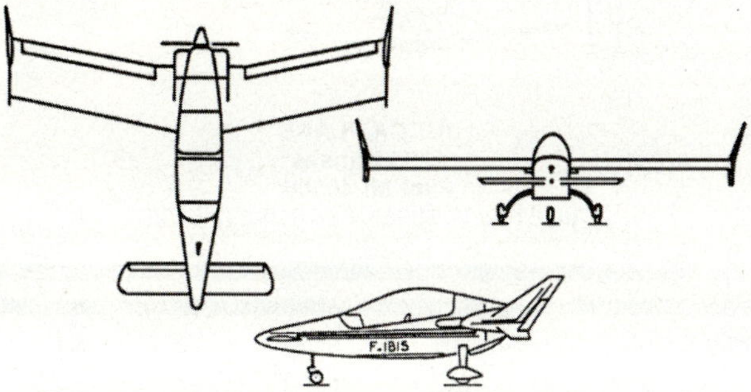

The IBIS is the product of technical advances made during research-work carried out on its forerunner, the Volucelle-RJ.02. Construction cost of the RJ03 is that of a one seat airplane. By utilizing a combination of "Styrodur" (polystyrene) with identically-matching plywood ribs and depression-bonding of surface-coatings, the manufacturer has been able to scrupulously respect the design's lines and give it a surface finish matching that of top synthetic-fiber aircraft.

Fitted with "flaperons" on the wings, the RJ03's approach speed is identical to that of classic aircraft bearing the same load per square meter. With little nosing-up during approach, rearward and forward visibility is excellent. Flight is not affected by rainy weather because of a non-laminar surface on the canard. Together, these features give the IBIS exceptional performance.

JUNQUA-DIFFUSION INTERNATIONAL
c/o S. C. A. M. • 69 rue Garibaldi
94100 Saint-Maur
FRANCE
Tel. 33 1 42 834579 • Fax 33 1 42 830065

MJ-2 Tempete

Wing Span	19.52 ft.	**Range**	350 sm
Wing Area	86.00 sq. ft.	**Takeoff Dist.**	825 ft.
Length	18.50 ft.	**Landing Dist.**	800 ft.
Max Gross Wt.	950 lbs.	**Vmax**	233 mph
Empty Wt.	639 lbs.	**VCr**	155 mph
Payload (full fuel)	N/A	**V_{S1} (stall clean)**	62 mph
Fuel Capacity	16 gal.	**V_{S0} (lndg config)**	N/A
Seats	1	**Climb Rate @msl**	600 fpm
Cabin Width	N/A	**Service Ceiling**	11,500 ft.

Std. Engine/HP	Usable HP Range	Construction Mtrl.	Est. Bldg. Time
Lycoming / 160 hp	65 - 180	Wood, Fabric	1250 hrs.
Plans	**Kit**	**Info Packet**	**Video**
$500.00	N/A	$29.00	N/A

Unit shown was built by Mr. Hendrycks near Paris, France. Highly aerobatic. +6, -4 G's. 0-200 with Ellison throttle body injection. Suitable for inverted flight. One of only two Jurca designs which are not WWII replicas. **Flying:** 54

JURCA PLANS
1733 Kansas
Flint MI 48506
Tel. 810-232-5395 • Fax 810-232-5395

MJ-5 Sirocco

Wing Span	21.00 ft.	**Range**	500 sm
Wing Area	112.0 sq. ft.	**Takeoff Dist.**	800 ft.
Length	21.00 ft.	**Landing Dist.**	1500 ft.
Max Gross Wt.	1850 lbs.	**Vmax**	225 mph
Empty Wt.	1250 lbs.	**VCr**	160 mph
Payload (full fuel)	600 lbs.	**V_{S1} (stall clean)**	55 mph
Fuel Capacity	35 gal.	**V_{S0} (lndg config)**	N/A
Seats	2	**Climb Rate @msl**	1800 fpm
Cabin Width	32 in.	**Service Ceiling**	15,000 ft.

Std. Engine/HP	Usable HP Range	Construction Mtrl.	Est. Bldg. Time
Lycoming / 160 hp	N/A	Wood, Metal, Composite	3500 hrs.
Plans	**Kit**	**Info Packet**	**Video**
$350.00	N/A	$5.00	N/A

This aircraft was built and static flight tested by the French Air Ministry. Makes an excellent aerobatic trainer. +6, -4 G's aerobatic. Nosegear is optional. Like all Jurca designs, the MJ-5 has been professionally stress analyzed. **Sold:** 500 **Flying:** 50

JURCA PLANS
1733 Kansas
Flint MI 48506
Tel. 810-232-5395 • Fax 810-232-5395

MJ-8 FW 190

Wing Span	24.00 ft.	**Range**	500 sm
Wing Area	112.0 sq. ft.	**Takeoff Dist.**	1200 ft.
Length	29.33 ft.	**Landing Dist.**	1800 ft.
Max Gross Wt.	2500 lbs.	**Vmax**	330 mph
Empty Wt.	1700 lbs.	**VCr**	225 mph
Payload (full fuel)	N/A	**V_{S1} (stall clean)**	65 mph
Fuel Capacity	50 gal.	**V_{S0} (lndg config)**	N/A
Seats	1	**Climb Rate @msl**	2500 fpm
Cabin Width	N/A	**Service Ceiling**	15,000 ft.

Std. Engine/HP	Usable HP Range	Construction Mtrl.	Est. Bldg. Time
Lycoming / 290 hp	N/A	Wood, Metal, Composite	5000 hrs.
Plans	**Kit**	**Info Packet**	**Video**
$500.00	N/A	$29.00	$29.00

75% scale size. Uses any radial engine from 200 to 400 hp up to one meter in diameter or six cylinder opposed. Some prefab components available. Also available 100% scale as the MJ-80 utilizing various 1200 hp powerplants.

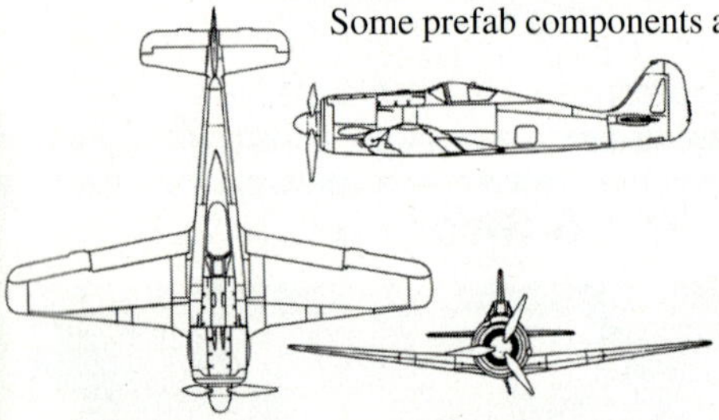

Sold: 300
Flying: 3

JURCA PLANS
1733 Kansas
Flint MI 48506
Tel. 810-232-5395 • Fax 810-232-5395

MJ-12 P-40

Wing Span	27.00 ft.	**Range**	500 sm
Wing Area	135.0 sq. ft.	**Takeoff Dist.**	900 ft.
Length	25.00 ft.	**Landing Dist.**	1500 ft.
Max Gross Wt.	2860 lbs.	**Vmax**	330 mph
Empty Wt.	2300 lbs.	**VCr**	230 mph
Payload (full fuel)	N/A	**V_{S1} (stall clean)**	65 mph
Fuel Capacity	35 gal.	**V_{S0} (lndg config)**	N/A
Seats	1	**Climb Rate @msl**	2500 fpm
Cabin Width	N/A	**Service Ceiling**	15,000 ft.

Std. Engine/HP	Usable HP Range	Construction Mtrl.	Est. Bldg. Time
Various / 400 hp	200 - 400	Wood, Metal, Composite	5000 hrs.
Plans	**Kit**	**Info Packet**	**Video**
$500.00	N/A	$29.00	$29.00

75% scale size. Like all Jurca designs, the MJ-12 has been professionally stress analyzed. It is possible to use the V-12 Ranger engine but designer must approve plan modifications to support additional weight and horsepower. Some prefab components available.

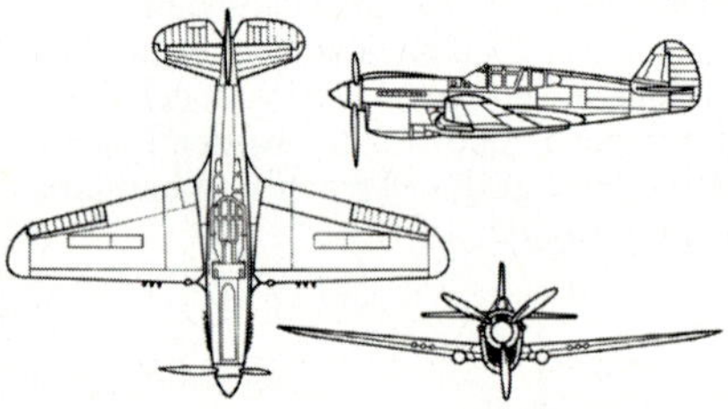

Sold: 300
Flying: 1

JURCA PLANS
1733 Kansas
Flint MI 48506
Tel. 810-232-5395 • Fax 810-232-5395

MJ-77 Mustang

Wing Span	27.00 ft.	**Range**	500 sm
Wing Area	135.0 sq. ft.	**Takeoff Dist.**	900 ft.
Length	21.00 ft.	**Landing Dist.**	1500 ft.
Max Gross Wt.	3300 lbs.	**Vmax**	330 mph
Empty Wt.	2300 lbs.	**VCr**	230 mph
Payload (full fuel)	700 lbs.	**V_{S1} (stall clean)**	60 mph
Fuel Capacity	50 gal.	**V_{S0} (lndg config)**	57 mph
Seats	2	**Climb Rate @msl**	2600 fpm
Cabin Width	N/A	**Service Ceiling**	25,000 ft.

Std. Engine/HP	Usable HP Range	Construction Mtrl.	Est. Bldg. Time
Geschwender/ 400 hp	N/A	Wood, Metal, Composite	5000 hrs.
Plans	**Kit**	**Info Packet**	**Video**
$500.00	N/A	$29.00	$29.00

75% scale model. Two versions can be built from same plans; standard or reinforced, 2860 lbs. or 3300 lbs. gross. 200 to 400 hp or 400 to 650 hp. Horsepower will dictate performance. Like all Jurca designs, the MJ-77 Mustang has been professionally stress analyzed. Some prefab components available.

Kits/Plans Sold: 500 **Completed & Flying:** 10

JURCA PLANS
1733 Kansas
Flint MI 48506
Tel. 810-232-5395 • Fax 810-232-5395

MJ-90 ME 109

Wing Span	32.00 ft.	**Range**	500 sm
Wing Area	176.0 sq. ft.	**Takeoff Dist.**	1000 ft.
Length	29.67 ft.	**Landing Dist.**	1800 ft.
Max Gross Wt.	5060 lbs.	**Vmax**	375 mph
Empty Wt.	3900 lbs.	**VCr**	250 mph
Payload (full fuel)	N/A	**V_{S1} (stall clean)**	70 mph
Fuel Capacity	150 gal.	**V_{S0} (lndg config)**	N/A
Seats	1	**Climb Rate @msl**	3000 fpm
Cabin Width	N/A	**Service Ceiling**	30,000 ft.

Std. Engine/HP	Usable HP Range	Construction Mtrl.	Est. Bldg. Time
V12 / 1200 hp	N/A	Wood, Metal, Composite	7000 hrs.
Plans	**Kit**	**Info Packet**	**Video**
$3,750.00	N/A	$29.00	$29.00

100% scale. Note that these full scale models can be expensive to build. Estimate total cost of this model from $100,000 to $150,000. Like all Jurca designs, the MJ-90 ME109 has been professionally stress analyzed. Inverted V-12 engines from 600 to 1500 hp currently not available for this model. V-12 Ranger can be used but requires added weight to correct for center of gravity. Some prefab components available.

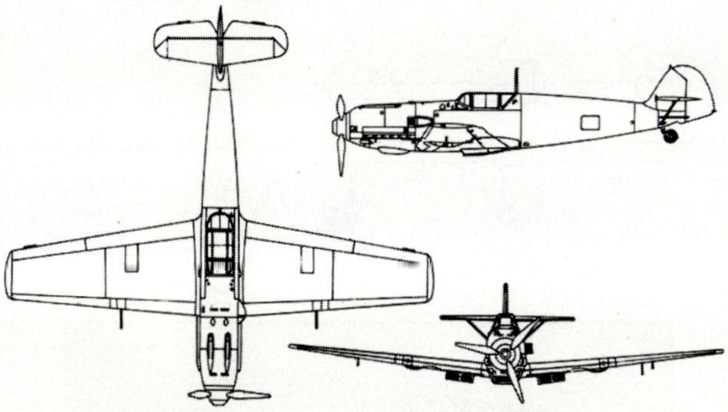

JURCA PLANS
1733 Kansas
Flint MI 48506
Tel. 810-232-5395 • Fax 810-232-5395

MJ-10 Spitfire

Wing Span	27.00 ft.	**Range**	500 sm
Wing Area	135.0 sq. ft.	**Takeoff Dist.**	1000 ft.
Length	23.50 ft.	**Landing Dist.**	1500 ft.
Max Gross Wt.	2860 lbs.	**Vmax**	330 mph
Empty Wt.	1700 lbs.	**VCr**	225 mph
Payload (full fuel)	N/A	**V_{S1} (stall clean)**	60 mph
Fuel Capacity	35 gal.	**V_{S0} (lndg config)**	60 mph
Seats	1	**Climb Rate @msl**	2500 fpm
Cabin Width	N/A	**Service Ceiling**	15,000 ft.

Std. Engine/HP	Usable HP Range	Construction Mtrl.	Est. Bldg. Time
Various / 400 hp	200 - 400	Wood, Fabric	5000 hrs.
Plans	**Kit**	**Info Packet**	**Video**
$500.00	N/A	$29.00	$29.00

75% scale size. Use any reliable engine from 200 to 400 hp. Those which have actually flown are the six cylinder Ranger, Potez, and V-12 Jaguar. Performance will depend on horsepower. Some prefab components available. Like all Jurca designs, the MJ-10 has been professionally stress analyzed. Also available 100% scale as the MJ-100 utilizing any V-12 650 to 1800 hp Allison or Merlin engine.

Kits/Plans Sold: 400 **Flying:** 6

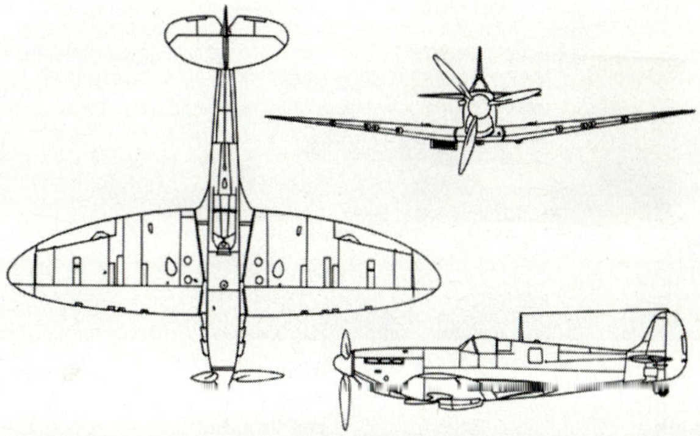

100% Scale MJ-100

JURCA PLANS
1733 Kansas
Flint MI 48506
Tel. 810-232-5395 • Fax 810-232-5395

KP 2U - Sova (Owl)

Wing Span	32.50 ft.	Range	600 sm
Wing Area	12.67 sq. ft.	Takeoff Dist.	330 ft.
Length	23.60 ft.	Landing Dist.	390 ft.
Max Gross Wt.	990 lbs.	Vmax	150 mph
Empty Wt.	594 lbs.	VCr	125 mph
Payload (full fuel)	303 lbs.	V_{S1} (stall clean)	N/A
Fuel Capacity	25 gal.	V_{S0} (lndg config)	32 mph
Seats	2	Climb Rate @msl	1320 fpm
Cabin Width	42 in.	Service Ceiling	N/A

Std. Engine/HP	Usable HP Range	Construction Mtrl.	Est. Bldg. Time
Rotax 618 / 74 hp	65-80	Aluminum	N/A
Plans	**Kit**	**Info Packet**	**Video**
N/A	Yes	N/A	N/A

Side-by-side, two seat ultralight (European). Electrically retractable tricycle landing gear with steerable nose wheel. In North America contact: A.K.S. Inc., PO Box 17161, Portland, OR 97217, Phone 503-247-9207.

KAPPA, SPOL. S. R.O.
Brtnicka 21 • CZ - 586 01 Jihlava
CZECH REPUBLIC
Ph: 00420-66-28121 • Fax: 00420-66-28122
E-Mail: KAPPA77@mbox.vol.CZ

Kelly-D

Wing Span	26.30 ft.	Range	275 sm
Wing Area	230.0 sq. ft.	Takeoff Dist.	400 ft.
Length	19.25 ft.	Landing Dist.	600 ft.
Max Gross Wt.	1500 lbs.	Vmax	100 mph
Empty Wt.	925 lbs.	VCr	90 mph
Payload (full fuel)	N/A	V_{S1} (stall clean)	40 mph
Fuel Capacity	24 gal.	V_{S0} (lndg config)	N/A
Seats	2	Climb Rate @msl	825 fpm
Cabin Width	N/A	Service Ceiling	N/A

Std. Engine/HP	Usable HP Range	Construction Mtrl.	Est. Bldg. Time
Lyc 0-235 / 115 hp	N/A	Wood, Fabric, Tube	4000 hrs.
Plans	**Kit**	**Info Packet**	**Video**
$150.00	N/A	$8.00	N/A

Classic two place biplane design. Larger than the CB-1, more space and comfort for the larger pilot. Detachable wings.
Completed & Flying: 20

DUDLEY R. KELLY CO.
Route 4, Box 194
Versailles KY 40383
Tel. 606-873-5253

Hatz CB-1

Wing Span	25.25 ft.	Range	200 sm
Wing Area	N/A	Takeoff Dist.	400 ft.
Length	19.00 ft.	Landing Dist.	600 ft.
Max Gross Wt.	1400 lbs.	Vmax	100 mph
Empty Wt.	850 lbs.	VCr	85 mph
Payload (full fuel)	N/A	V_{S1} (stall clean)	40 mph
Fuel Capacity	18 gal.	V_{S0} (lndg config)	N/A
Seats	2	Climb Rate @msl	750 fpm
Cabin Width	N/A	Service Ceiling	N/A

Std. Engine/HP	Usable HP Range	Construction Mtrl.	Est. Bldg. Time
Cont 0-200 / 100 hp	N/A	Tube, Wood, Fabric	3500 hrs.
Plans	**Kit**	**Info Packet**	**Video**
$150.00	N/A	$8.00	N/A

Classic two place biplane. A straightforward design with simple details and a minimum of time consuming jig work. Drawings include full size rib and spar fitting patterns. Hatz newsletter includes a great deal of information about various machines that have been built and flown.

Materials kit available from Makelan Corp. PO Box 312142, New Braunfels, TX 78130 (210) 905-7832 **Completed & Flying:** 95

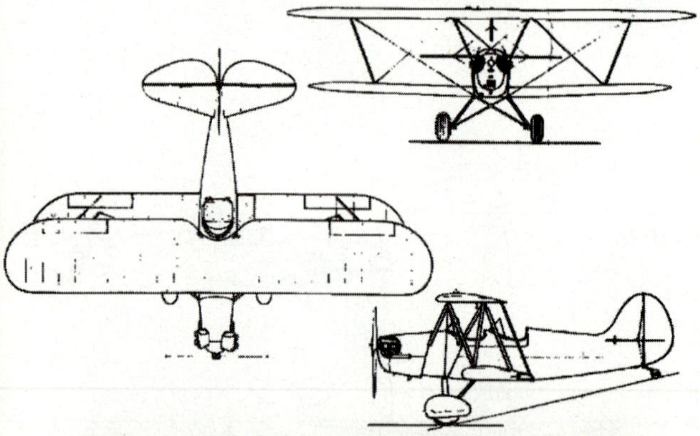

DUDLEY R. KELLY CO.
Route 4, Box 194
Versailles KY 40383
Tel. 606-873-5253

Tukan Gulf

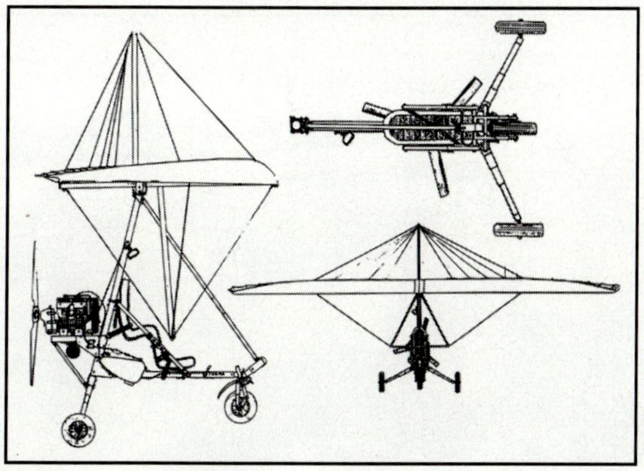

Wing Span	35.00 ft.	**Range**	N/A
Wing Area	209.0 sq. ft.	**Takeoff Dist.**	50 ft.
Length	N/A	**Landing Dist.**	N/A
Max Gross Wt.	769 lbs.	**Vmax**	70 mph
Empty Wt.	254 lbs.	**VCr**	61 mph
Payload (full fuel)	N/A	**V_{S1} (stall clean)**	23 mph
Fuel Capacity	5 gal.	**V_{S0} (lndg config)**	N/A
Seats	2	**Climb Rate @msl**	N/A
Cabin Width	N/A	**Service Ceiling**	N/A

Std. Engine/HP	Usable HP Range	Construction Mtrl.	Est. Bldg. Time
Rotax 447 / 42 hp	N/A	Tube & Fabric	N/A
Plans	**Kit**	**Info Packet**	**Video**
$125.00	$10,960.00	Yes	N/A

A weight shift trike. Part 103 legal. Accepts the "Trikezilla" upgrade. Anodized aluminum chariot, folding cantilever, tricycle. 19 meter wing. Options: 10 gallon fuel, tow and release package, ag spray unit, floats and skis.

KEMMERIES AVIATION
8710 W. Carefree Hwy.
Peoria AZ 85382
Tel. 602-566-8068 • Fax 602-561-2287
Web: http://www.sportplanes.com/kemries.htm

Firestar I

Wing Span	27.67 ft.	**Range**	N/A
Wing Area	143.0 sq. ft.	**Takeoff Dist.**	125 ft.
Length	20.25 ft.	**Landing Dist.**	125 ft.
Max Gross Wt.	550 lbs.	**Vmax**	63 mph
Empty Wt.	254 lbs.	**VCr**	53 mph
Payload (full fuel)	N/A	**V_{S1} (stall clean)**	N/A
Fuel Capacity	5 gal.	**V_{S0} (lndg config)**	N/A
Seats	1	**Climb Rate @msl**	650 fpm
Cabin Width	N/A	**Service Ceiling**	N/A

Std. Engine/HP	Usable HP Range	Construction Mtrl.	Est. Bldg. Time
Rotax 447 / 41 hp	N/A	Tube, Fabric	450 hrs.
Plans	**Kit**	**Info Packet**	**Video**
N/A	$8,193.00	$5.00	N/A

Proven single seat ultralight. Clean, lightweight design with good takeoff and climb performance. Higher performance available with optional Rotax 503 engine. Lower cost and more economical operation available with optional Rotax 277 engine. Removable doors and windows for summer/winter flying. 360 degree visibility. Folding wings and tail. Eight minutes from trailer to air, one person and no tools required. See Ad on Page 52.

Woody Spurlock, Percival, IA
"Building the little jewel has been a fun project. ... we are very pleased with the flying characteristics, it handles beautifully."

KOLB AIRCRAFT, INC.
RD 3, Box 38
Phoenixville PA 19460
Tel. 610-948-4136 • Fax 610-948-6727

Firestar II

Wing Span	27.67 ft.	**Range**	N/A
Wing Area	143.0 sq. ft.	**Takeoff Dist.**	200 ft.
Length	21.25 ft.	**Landing Dist.**	100 ft.
Max Gross Wt.	725 lbs.	**Vmax**	80 mph
Empty Wt.	300 lbs.	**VCr**	70 mph
Payload (full fuel)	N/A	**V_{S1} (stall clean)**	N/A
Fuel Capacity	5 gal.	**V_{S0} (lndg config)**	N/A
Seats	2	**Climb Rate @msl**	1000 fpm
Cabin Width	N/A	**Service Ceiling**	N/A

Std. Engine/HP	Usable HP Range	Construction Mtrl.	Est. Bldg. Time
Rotax 503 / 53 hp	N/A	Tube, Fabric	450 hrs.
Plans	**Kit**	**Info Packet**	**Video**
N/A	$8,845.00	$5.00	N/A

Two seat, ultralight experimental version of the FireStar. Clean, lightweight design with good takeoff and climb performance. Lower cost and more economical performance available with optional Rotax 447 engine. Removable doors and windows for summer and winter flying. 360 degree visibility. Folding wings and tail. Eight minutes from trailer to air, one person and no tools required. See Ad on Page 52.

KOLB AIRCRAFT, INC.
RD 3, Box 38
Phoenixville PA 19460
Tel. 610-948-4136 • Fax 610-948-6727

Mark III

Wing Span	30.00 ft.	**Range**	N/A
Wing Area	160.0 sq. ft.	**Takeoff Dist.**	200 ft.
Length	22.50 ft.	**Landing Dist.**	250 ft.
Max Gross Wt.	850 lbs.	**Vmax**	85 mph
Empty Wt.	395 lbs.	**VCr**	75 mph
Payload (full fuel)	N/A	**V_{S1} (stall clean)**	N/A
Fuel Capacity	10 gal.	**V_{S0} (lndg config)**	N/A
Seats	2	**Climb Rate @msl**	1150 fpm
Cabin Width	N/A	**Service Ceiling**	N/A

Std. Engine/HP	Usable HP Range	Construction Mtrl.	Est. Bldg. Time
Rotax 582 / 65 hp	N/A	Tube, Fabric	500 hrs.
Plans	**Kit**	**Info Packet**	**Video**
N/A	$10,954.00	$5.00	N/A

Features lower landing speed, higher top speed, and greater climb and payload. Large effective flaps offer STOL class performance. Two higher hp and one lower hp engine options available. Numerous other options offered. Performance data is with pilot only. 45" wide cabin. Folding wings. Eight minute set-up by one person. See Ad on Page 52.

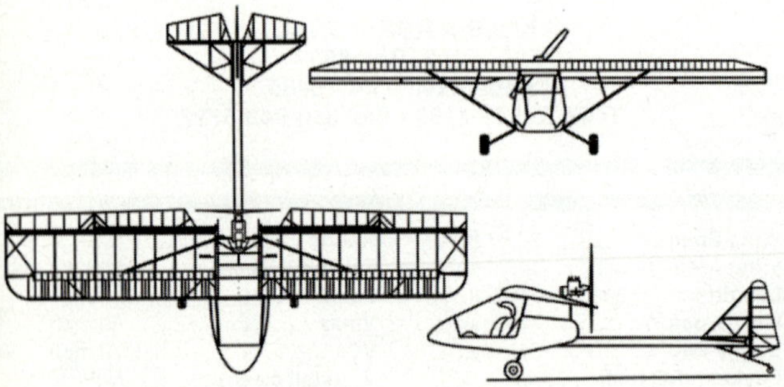

KOLB AIRCRAFT, INC.
RD 3, Box 38
Phoenixville PA 19460
Tel. 610-948-4136 • Fax 610-948-6727

Slingshot

Wing Span	22.00 ft.	**Range**	N/A
Wing Area	110.0 sq. ft.	**Takeoff Dist.**	100 ft.
Length	19.00 ft.	**Landing Dist.**	N/A
Max Gross Wt.	850 lbs.	**Vmax**	100 mph
Empty Wt.	345 lbs.	**VCr**	87 mph
Payload (full fuel)	N/A	**V_{S1} (stall clean)**	39 mph
Fuel Capacity	10 gal.	**V_{S0} (lndg config)**	N/A
Seats	2	**Climb Rate @msl**	1300 fpm
Cabin Width	N/A	**Service Ceiling**	N/A

Std. Engine/HP	Usable HP Range	Construction Mtrl.	Est. Bldg. Time
Rotax 582 / 65 hp	N/A	Tube, Fabric	400 hrs.
Plans	**Kit**	**Info Packet**	**Video**
N/A	$11,454.00	$5.00	N/A

High horsepower to weight ratio for high performance. Delivers the cruise speeds of GA aircraft but with the fun and economy of an ultralight. Complete kit price is with the 65 hp Rotax 582 engine. Kit with Rotax 503 engine is $10,341. Total cost for three airframe only sub-kits without engine is $7,265. See Ad on Page 52.

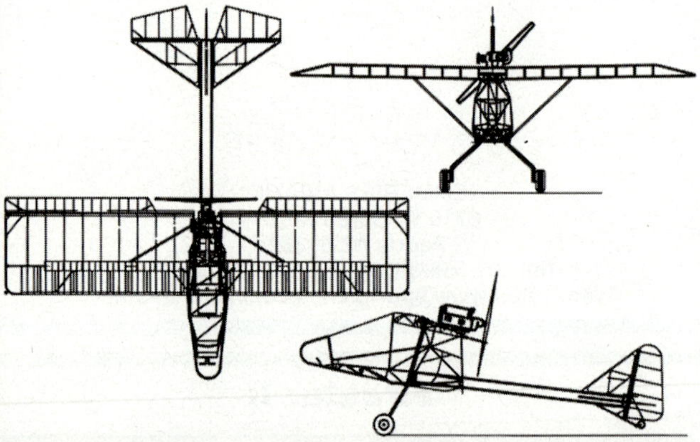

KOLB AIRCRAFT, INC.
RD 3, Box 38
Phoenixville PA 19460
Tel. 610-948-4136 • Fax 610-948-6727

Lancair 360 Mk II

Wing Span	23.50 ft.	**Range**	940 sm
Wing Area	76.00 sq. ft.	**Takeoff Dist.**	690 ft.
Length	21.00 ft.	**Landing Dist.**	1200 ft.
Max Gross Wt.	1685 lbs.	**Vmax**	260 mph
Empty Wt.	1090 lbs.	**VCr**	240 mph
Payload (full fuel)	343 lbs.	**V_{S1} (stall clean)**	N/A
Fuel Capacity	42 gal.	**V_{S0} (lndg config)**	63 mph
Seats	2	**Climb Rate @msl**	2700 fpm
Cabin Width	43 in.	**Service Ceiling**	18,000 ft.

Std. Engine/HP	Usable HP Range	Construction Mtrl.	Est. Bldg. Time
Lyc. IO-360 / 180 hp	160-180	Composite	2500 hrs.
Plans	**Kit**	**Info Packet**	**Video**
$425.00	$27,950.00	$15.00	$20.00

Two place, high performance design. Excellent for fun and long cross country flights. Range listed is with standard tanks. Also available with O-320,160 hp engine. Can be assembled in a standard 20' x 20' two-car garage. Estimated build time is 1,500 hours with Fast Build Kit.

Kits/Plans Sold: 400 **Completed & Flying:** 200

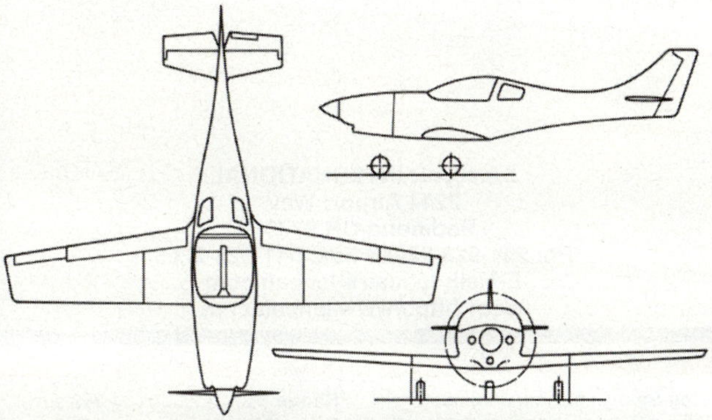

LANCAIR INTERNATIONAL
2244 Airport Way
Redmond OR 97756
Ph: 541-923-2244 • Fax: 541-923-2255
E-Mail: lancair@lancair.com
Web: http://www.lancair.com

Lancair Super ES

Wing Span	35.50 ft.	**Range**	1200 sm
Wing Area	140.0 sq. ft.	**Takeoff Dist.**	600 ft.
Length	25.00 ft.	**Landing Dist.**	800 ft.
Max Gross Wt.	3000 lbs.	**Vmax**	240 mph
Empty Wt.	1800 lbs.	**VCr**	192 mph
Payload (full fuel)	750 lbs.	**V_{S1} (stall clean)**	57 mph
Fuel Capacity	75 gal.	**V_{S0} (lndg config)**	54 mph
Seats	4	**Climb Rate @msl**	2000 fpm
Cabin Width	46 in.	**Service Ceiling**	18,000 ft.

Std. Engine/HP	Usable HP Range	Construction Mtrl.	Est. Bldg. Time
Cont. IO-360 / 210 hp	200 - 300	Composite	2400 hrs.
Plans	**Kit**	**Info Packet**	**Video**
$425.00	$34,950.00	$15.00	$20.00

The Lancair Super ES is an economical four place airplane. It has the speed of a Bonanza, economy of a 172, and the load carrying capacity of a 182. Expanded fuel capacity option available. Advanced composite construction. Estimated build time is 1,400 hours with the Fast Build Kit. With advanced composites, epoxy prepregs, Nomex honeycomb cores, and carbon fiber components.

Kits/Plans Sold: 100 **Completed & Flying:** 12

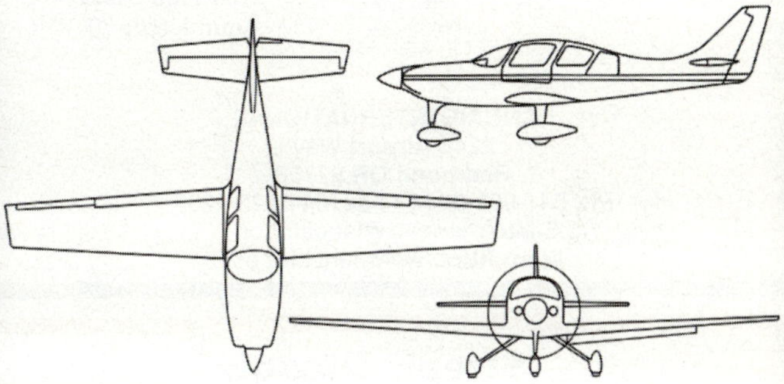

The Lancair was first made available in 1985.The design philosophy was to stress simplicity in what is a relatively complex high performance aircraft. Due to aerodynamic and pure styling considerations, the only suitable materials were composite. Until this date, the only composite airframes on the emerging kit plane market were standard room temperature, wet lay up systems - a derivative from common "boat technology."

From the beginning, these materials were not suitable for the Lancair's goals of high material standards, repeatable quality assurance and airframe longevity. Thus Lancair became the first kit plane to use 100% high temperature, epoxy based, advanced composites throughout its primary structural airframe.

LANCAIR INTERNATIONAL
2244 Airport Way
Redmond OR 97756
Ph: 541-923-2244 • Fax: 541-923-2255
E-Mail: lancair@lancair.com
Web: http://www.lancair.com

Lancair IV & IV-P

Wing Span	30.20 ft.	**Range**	1450 sm
Wing Area	98.00 sq. ft.	**Takeoff Dist.**	1200 ft.
Length	25.00 ft.	**Landing Dist.**	1600 ft.
Max Gross Wt.	3200 lbs.	**Vmax**	315 mph
Empty Wt.	1900 lbs.	**VCr**	300 mph
Payload (full fuel)	808 lbs.	**V_{S1} (stall clean)**	N/A
Fuel Capacity	82 gal.	**V_{S0} (lndg config)**	75 mph
Seats	4	**Climb Rate @msl**	2600 fpm
Cabin Width	46 in.	**Service Ceiling**	29,000 ft.

Std. Engine/HP	Usable HP Range	Construction Mtrl.	Est. Bldg. Time
Cont. 550 / 350 hp	N/A	Composite	3000 hrs.
Plans	**Kit**	**Info Packet**	**Video**
$625.00	$49,900.00	$15.00	$20.00

Fast, four place design. The prototype set an official N.A.A. world airspeed record of 362 mph on February 20, 1991 for the San Francisco to Denver run. Easily removable 135 lb. wing sections. Useful load 1300 lbs. The IV-P model includes cabin pressurization. Fast Build Kit available.

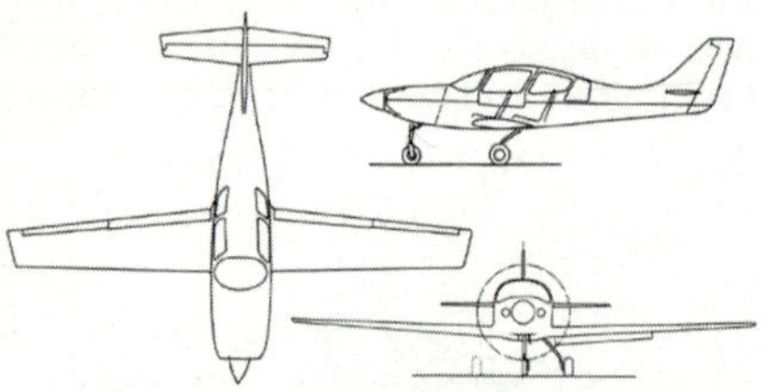

Kits/Plans Sold: 400
Completed: 70

LANCAIR INTERNATIONAL
2244 Airport Way
Redmond OR 97756
Ph: 541-923-2244 • Fax: 541-923-2255
E-Mail: lancair@lancair.com
Web: http://www.lancair.com

Tigress

Wing Span	30.20 ft.	**Range**	1450 sm
Wing Area	98.00 sq. ft.	**Takeoff Dist.**	N/A
Length	25.00 ft.	**Landing Dist.**	N/A
Max Gross Wt.	3400 lbs.	**Vmax**	N/A
Empty Wt.	2400 lbs.	**VCr**	405 mph
Payload (full fuel)	1000 lbs.	**V_{S1} (stall clean)**	N/A
Fuel Capacity	115 gal.	**V_{S0} (lndg config)**	N/A
Seats	N/A	**Climb Rate @msl**	N/A
Cabin Width	N/A	**Service Ceiling**	N/A

Std. Engine/HP	Usable HP Range	Construction Mtrl.	Est. Bldg. Time
Orenda / 600 hp	N/A	Composite	N/A
Plans	**Kit**	**Info Packet**	**Video**
N/A	$109,500.00	Yes	N/A

Based on the Lancair IV-P, this new design offers greater performance levels. Although many structural changes have been made to accomodate higher horsepower and speeds, the Tigress shares proven wing and empennage planforms with the IV-P. It also utilizes the custom pressurization system of the IV-P. Introductory Fast-Build Kit price listed.

LANCAIR INTERNATIONAL
2244 Airport Way
Redmond OR 97756
Ph: 541-923-2244 • Fax: 541-923-2255
E-Mail: lancair@lancair.com
Web: http://www.lancair.com

Streak Shadow

LARON AVIATION TECH., INC.
PO Box 5026
Borger TX 79008-5026
Tel. 806-273-8513 • Fax 806-273-8375
E-Mail: laron@arn.net
Web: http://www.thebook.com/laronaviation

Wing Span	28.00 ft.	**Range**	363 sm
Wing Area	140.0 sq. ft.	**Takeoff Dist.**	100 ft.
Length	21.00 ft.	**Landing Dist.**	200 ft.
Max Gross Wt.	900 lbs.	**Vmax**	121 mph
Empty Wt.	388 lbs.	**VCr**	100 mph
Payload (full fuel)	N/A	**V_{S1} (stall clean)**	31 mph
Fuel Capacity	14 gal.	**V_{S0} (lndg config)**	N/A
Seats	2	**Climb Rate @msl**	1800 fpm
Cabin Width	N/A	**Service Ceiling**	29,000 ft.

Std. Engine/HP	Usable HP Range	Construction Mtrl.	Est. Bldg. Time
Rotax 582 / 65 hp	N/A	Metal, Wood, Fabric	500 hrs.
Plans	**Kit**	**Info Packet**	**Video**
N/A	$13,500.00	$10.00	$15.00

Clipped wing Shadow uses a higher horsepower engine. Span is reduced by five feet and thickness reduced by one inch. These changes produce an aircraft with impressive handling while retaining safety features of the Shadow. Engine is not included in price but instruments are. Detachable wings. Quick Build, starter, and staged kits available. See Ad on Page 49.

Kits/Plans Sold: 200 **Completed & Flying:** 180

Shadow

Wing Span	31.11 ft.	**Range**	193 sm
Wing Area	162.0 sq. ft.	**Takeoff Dist.**	100 ft.
Length	21.00 ft.	**Landing Dist.**	100 ft.
Max Gross Wt.	824 lbs.	**Vmax**	102 mph
Empty Wt.	349 lbs.	**VCr**	90 mph
Payload (full fuel)	436 lbs.	**V_{S1} (stall clean)**	28 mph
Fuel Capacity	7 gal.	**V_{S0} (lndg config)**	25 mph
Seats	2	**Climb Rate @msl**	1100 fpm
Cabin Width	26 in.	**Service Ceiling**	25,000 ft.

Std. Engine/HP	Usable HP Range	Construction Mtrl.	Est. Bldg. Time
Rotax 503 / 50 hp	40 - 65	Metal, Wood, Fabric	500 hrs.
Plans	**Kit**	**Info Packet**	**Video**
N/A	$13,500.00	$10.00	$15.00

Safe and fun design. Unique wing design has no conventional stall characteristics and is spin resistant. Kit includes instruments, propeller and brakes. Quick Build, starter, and staged kits available. Wings are detachable. Can be fitted with one or two seats. Engine is not included in kit. Choices include Hirth 2704 and Rotax 582. A G Series Glider version of the Shadow is available for $11,500. See Ad on Page 49. **Kits Sold:** 100 **Flying:** 90

John Hutchison, British Airways Concorde pilot
"I was really most impressed by the docile handling qualities and lively performance of your little aircraft; she is great fun to fly and I felt entirely safe and 'at home' in her."

LARON AVIATION TECH., INC.
PO Box 5026
Borger TX 79008-5026
Tel. 806-273-8513 • Fax 806-273-8375
E-Mail: laron@arn.net
Web: http://www.thebook.com/laronaviation

Star Streak

Wing Span	30.00 ft.	**Range**	310 sm
Wing Area	N/A	**Takeoff Dist.**	100 ft.
Length	21.00 ft.	**Landing Dist.**	200 ft.
Max Gross Wt.	900 lbs.	**Vmax**	141 mph
Empty Wt.	400 lbs.	**VCr**	122 mph
Payload (full fuel)	N/A	**V_{S1} (stall clean)**	35 mph
Fuel Capacity	14 gal.	**V_{S0} (lndg config)**	N/A
Seats	2	**Climb Rate @msl**	2800 fpm
Cabin Width	N/A	**Service Ceiling**	30,000 ft.

Std. Engine/HP	Usable HP Range	Construction Mtrl.	Est. Bldg. Time
Hirth F30 / 80 hp	N/A	Wood, Metal, Fiberglass	500 hrs.
Plans	**Kit**	**Info Packet**	**Video**
N/A	$13,500.00	$10.00	$15.00

The final word in the Shadow Series of powered aircraft. A higher aspect ratio and smaller thickness ratio improve performance. Design uses a higher horsepower Hirth F-30 engine. Kit price does not include engine. Quick Build, starter, and staged kits available. All machining, welding, and forming of metal parts is done at the factory. Molded fiberglass parts come completed. See Ad on Page 49.

Kits/Plans Sold: 4 **Completed & Flying:** 3

LARON AVIATION TECH., INC.
PO Box 5026
Borger TX 79008-5026
Tel. 806-273-8513 • Fax 806-273-8375
E-Mail: laron@arn.net
Web: http://www.thebook.com/laronaviation

1/2 TUN

LARON AVIATION TECH., INC.
PO Box 5026
Borger TX 79008-5026
Tel. 806-273-8513 • Fax 806-273-8375
E-Mail: laron@arn.net
Web: http://www.thebook.com/laronaviation

Wing Span	26.00 ft.	**Range**	100 sm
Wing Area	117.0 sq. ft.	**Takeoff Dist.**	150 ft.
Length	21.00 ft.	**Landing Dist.**	100 ft.
Max Gross Wt.	550 lbs.	**Vmax**	65 mph
Empty Wt.	250 lbs.	**VCr**	55 mph
Payload (full fuel)	300 lbs.	**V_{S1} (stall clean)**	25 mph
Fuel Capacity	5 gal.	**V_{S0} (lndg config)**	N/A
Seats	1	**Climb Rate @msl**	1000 fpm
Cabin Width	N/A	**Service Ceiling**	10,000 ft.

Std. Engine/HP	Usable HP Range	Construction Mtrl.	Est. Bldg. Time
Rotax 447 / 30 hp	N/A	Metal, Tube, Fabric	90 hrs.
Plans	**Kit**	**Info Packet**	**Video**
N/A	$4,900.00	N/A	N/A

Low cost, Part 103 Legal based on the Tundra design. Fast build. See Ad on Page 49.

Completed & Flying: 3

Wizard

Wing Span	30.50 ft.	**Range**	270 sm
Wing Area	140.0 sq. ft.	**Takeoff Dist.**	170 ft.
Length	19.60 ft.	**Landing Dist.**	N/A
Max Gross Wt.	925 lbs.	**Vmax**	120 mph
Empty Wt.	462 lbs.	**VCr**	90 mph
Payload (full fuel)	N/A	**V_{S1} (stall clean)**	N/A
Fuel Capacity	10 gal.	**V_{S0} (lndg config)**	N/A
Seats	2	**Climb Rate @msl**	1350 fpm
Cabin Width	N/A	**Service Ceiling**	15,000 ft.

Std. Engine/HP	Usable HP Range	Construction Mtrl.	Est. Bldg. Time
Rotax 582 / 65 hp	N/A	Comp., Tube, Fabric	450 hrs.
Plans	**Kit**	**Info Packet**	**Video**
N/A	$14,500.00	$5.00	$15.00

A high wing, two place model with a unique twin-tail design. Good STOL performance. Engine not included in kit price. Choices include Rotax 618, Rotax 912, and Hirth 2706. Quick Build, starter, and staged kits available. See Ad on Page 49.

Kits/Plans Sold: 240 **Completed & Flying:** 220

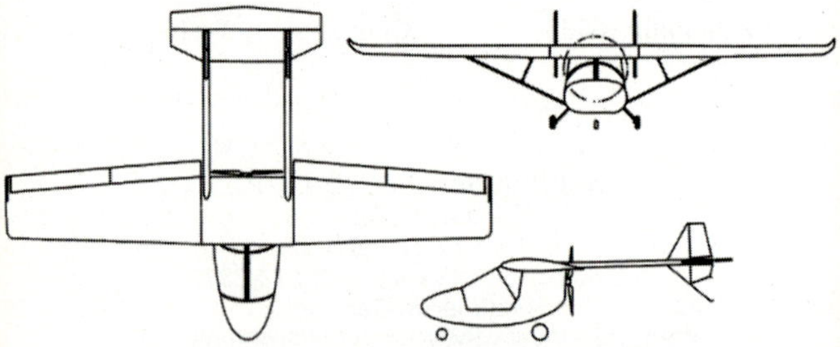

LARON AVIATION TECH., INC.
PO Box 5026
Borger TX 79008-5026
Tel. 806-273-8513 • Fax 806-273-8375
E-Mail: laron@arn.net
Web: http://www.thebook.com/laronaviation

Tundra

Wing Span	32.00 ft.	**Range**	N/A
Wing Area	170.0 sq. ft.	**Takeoff Dist.**	100 ft.
Length	21.00 ft.	**Landing Dist.**	100 ft.
Max Gross Wt.	900 lbs.	**Vmax**	120 mph
Empty Wt.	400 lbs.	**VCr**	85 mph
Payload (full fuel)	500 lbs.	**V_{S1} (stall clean)**	N/A
Fuel Capacity	11 gal.	**V_{S0} (lndg config)**	35 mph
Seats	2	**Climb Rate @msl**	1200 fpm
Cabin Width	N/A	**Service Ceiling**	15,000 ft.

Std. Engine/HP	Usable HP Range	Construction Mtrl.	Est. Bldg. Time
Rotax 503 / 50 hp	50-80	Tube, Fabric	250 hrs.
Plans	**Kit**	**Info Packet**	**Video**
N/A	$7,900.00	$5.00	N/A

Rugged all-metal, three-axis airplane capable of withstanding the abuse expected in a training environment. Meets Transport Canada TP 10141. Very good load carrying capability and speed make it well suited for floats, cross country flying or just short trips around the patch. Quick Build, starter, and staged kits available. Engine not included in kit price. Choices include Rotax 618, Rotax 582, and Hirth 2706. See Ad on Page 49.

Kits/Plans Sold: 20 **Completed & Flying:** 15

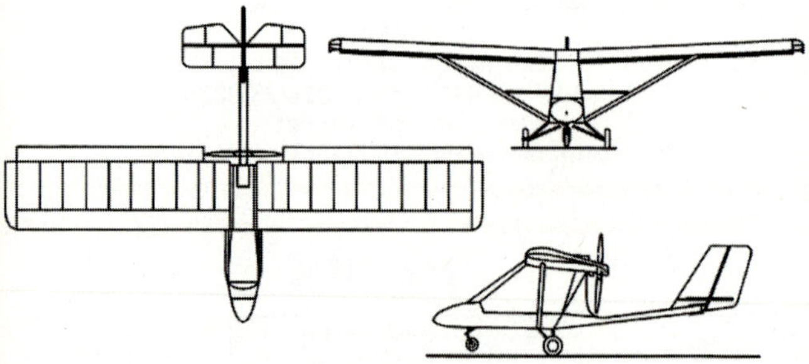

Ian Davis, Former Bush Pilot

"The Tundra is agile, tough, and roomy. Even loaded down with a couple of guys my size, the Tundra's performance is amazing. I can't believe you can get all this for an incredibly low price and minimal assembly time. And to think, this is an ultralight!"

LARON AVIATION TECH., INC.
PO Box 5026
Borger TX 79008-5026
Tel. 806-273-8513 • Fax 806-273-8375
E-Mail: laron@arn.net
Web: http://www.thebook.com/laronaviation

MoHawk

Wing Span	30.00 ft.	**Range**	320 sm
Wing Area	124.5 sq. ft.	**Takeoff Dist.**	100 ft.
Length	18.00 ft.	**Landing Dist.**	150 ft.
Max Gross Wt.	1000 lbs.	**Vmax**	130 mph
Empty Wt.	450 lbs.	**VCr**	85 mph
Payload (full fuel)	550 lbs.	**V_{S1} (stall clean)**	35 mph
Fuel Capacity	12 gal.	**V_{S0} (lndg config)**	30 mph
Seats	2	**Climb Rate @msl**	1200 fpm
Cabin Width	N/A	**Service Ceiling**	12,000 ft.

Std. Engine/HP	Usable HP Range	Construction Mtrl.	Est. Bldg. Time
Rotax 503 / 55 hp	N/A	Tube & Fabric	900 hrs.
Plans	**Kit**	**Info Packet**	**Video**
$229.00	N/A	N/A	N/A

A plans-only high wing aircraft that can be built for about $6,500 in parts. Standard features include hydraulic brakes, folding wings, stearable tail wheel.

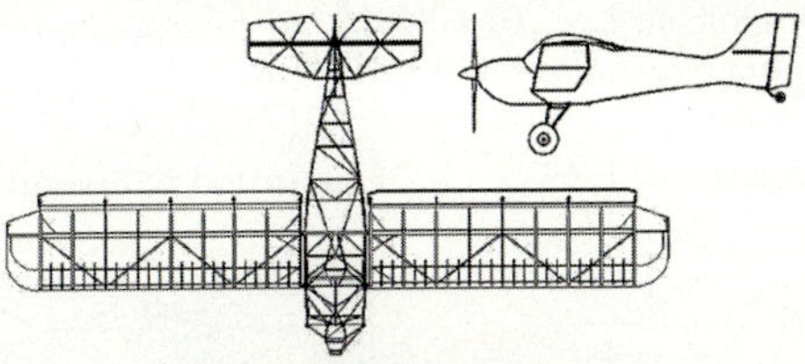

LAS BRISAS SALES
188 Tower View Lane
Ozark MO 65721
Tel. 417-443-2100 • Fax 417-443-2100
E-Mail: mohawk@katman2.com
Web: http://katman2.com/mohawk.htm

Early Bird Jenny

Wing Span	27.50 ft.	**Range**	180 sm
Wing Area	175.0 sq. ft.	**Takeoff Dist.**	100 ft.
Length	18.33 ft.	**Landing Dist.**	250 ft.
Max Gross Wt.	800 lbs.	**Vmax**	80 mph
Empty Wt.	419 lbs.	**VCr**	60 mph
Payload (full fuel)	381 lbs.	**V_{S1} (stall clean)**	35 mph
Fuel Capacity	10 gal.	**V_{S0} (lndg config)**	N/A
Seats	2	**Climb Rate @msl**	800 fpm
Cabin Width	24 in.	**Service Ceiling**	14,000 ft.

Std. Engine/HP	Usable HP Range	Construction Mtrl.	Est. Bldg. Time
Rotax 582 / 64 hp	46 - 74	Metal, Wood Spars	400 hrs.
Plans	**Kit**	**Info Packet**	**Video**
$98.00	$12,500.00	$10.00	$28.00

2/3 scale classic biplane designed by Dennis Wiley. The Jenny was probably the first barnstorming airplane and the first airplane that many people ever saw. Engine is included with the kit. Choices include a Rotax 503, 532, 582, or 618. Price shown is with Rotax 582. May also be purchased without engine and in sections. See Ad on Page 34.

Kits/Plans Sold: 485 **Completed & Flying:** 18

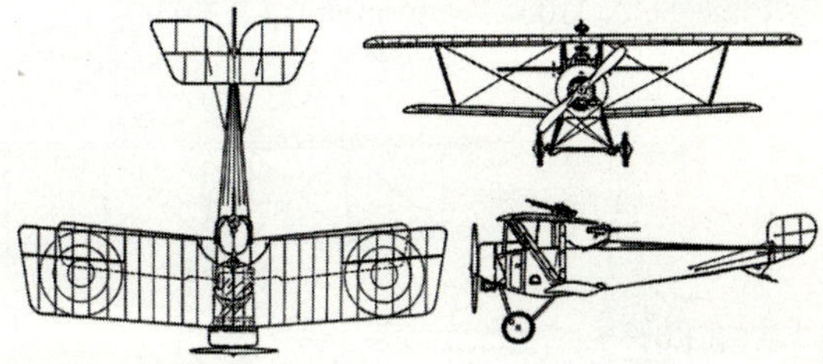

One of the best of the classic biplanes was the Curtis Jenny. It was a conglomeration of sticks and wires that, if pushed to its limits, could barely outrun a horse and couldn't beat a train for very long. Nevertheless, the Jenny was the first barnstorming airplane and the first replica aircraft to be successfully marketed in kit form.

Dennis Wiley's 2/3 scale replica is a light weight, easy to build, and inexpensive opportunity to experience those early years of aviation. From the inside-out, the airplane looks and feels like a Jenny. It includes a wooden instrument panel and wooden floor boards in a cockpit from which the pilot looks out on lots of struts and wires, not as many as a real Jenny but a sufficient number to create the ambiance.

Massimo Stucchi, via San Francesco, Italy
"... flying behavior and performances are satisfying and up to my expectations."

Dennis Wiley, Erie, CO
"I've flown more than 1,000 miles of cross country in my small open-cockpit biplane and have a great story to tell my grandkids someday..."

LEADING EDGE AIR FOILS
Meadowlake Airport • 8242 Cessna Drive
Peyton CO 80831
Tel. 719-683-5323 • Fax 719-683-5333
E-Mail: leafkp@kktv.com
Web: http://www.leadingedgeairfoils.com

Nieuport 11

Wing Span	21.50 ft.	**Range**	165 sm
Wing Area	114.0 sq. ft.	**Takeoff Dist.**	160 ft.
Length	16.50 ft.	**Landing Dist.**	275 ft.
Max Gross Wt.	550 lbs.	**Vmax**	83 mph
Empty Wt.	250 lbs.	**VCr**	65 mph
Payload (full fuel)	200 lbs.	**V_{S1} (stall clean)**	27 mph
Fuel Capacity	12 gal.	**V_{S0} (lndg config)**	N/A
Seats	1	**Climb Rate @msl**	750 fpm
Cabin Width	24 in.	**Service Ceiling**	15,000 ft.

Std. Engine/HP	Usable HP Range	Construction Mtrl.	Est. Bldg. Time
Rotax 503 / 48 hp	30 -50	Aluminum Tubing	700 hrs.
Plans	**Kit**	**Info Packet**	**Video**
$139.00	$8,500.00	$10.00	$39.95

7/8 scale size. Engine included in kit. Specs and price are for Rotax 503. Available with Rotax 447 engines. May also be purchased without engine and in sections. Steel fuselage is optional. This plane is also available in the Nieuport 17 version with 2 ft. wider wing span and round cowl. Call for information. Plans also available from the designer Graham Lee, Box 889, Lamont, Alberta, T0B 2R0, Canada, 403-895-2975. See Ad on Page 34.

Kits/Plans Sold: 1100 **Completed & Flying:** 100

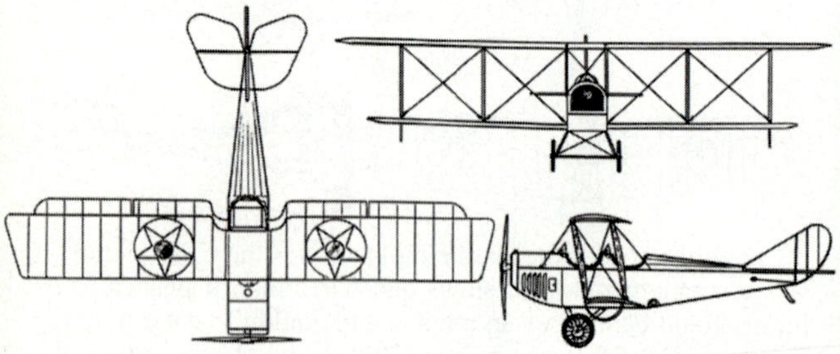

The Nieuport 11 has been described by WWI pilots as one of the most delightful warplanes they have flown. For some of us, to own and fly such an aeroplane has been a long standing dream. Until now, the choices were a either a five or six figure purchase, or a five or six year building project. Now, Leading Edge Air Foils offers a practical way to satisfy these desires. The 7/8 scale Nieuport can be built from plans or a kit and flown with all the sight, sound and emotions that an enthusiast could desire. For the pilot looking for the pleasure of "flying" a plane, rather than driving it; the challenge of taming a taildragger, and the pride in having built a thing of beauty, this 7/8 scale Nieuport reproduction can be one of life's more fulfilling experiences.

LEADING EDGE AIR FOILS
Meadowlake Airport • 8242 Cessna Drive
Peyton CO 80831
Tel. 719-683-5323 • Fax 719-683-5333
E-Mail: leafkp@kktv.com
Web: http://www.leadingedgeairfoils.com

Nieuport 12

Wing Span	26.50 ft.	**Range**	300 sm
Wing Area	189.0 sq. ft.	**Takeoff Dist.**	350 ft.
Length	21.50 ft.	**Landing Dist.**	250 ft.
Max Gross Wt.	1000 lbs.	**Vmax**	94 mph
Empty Wt.	536 lbs.	**VCr**	75 mph
Payload (full fuel)	450 lbs.	**V_{S1} (stall clean)**	33 mph
Fuel Capacity	12 gal.	**V_{S0} (lndg config)**	N/A
Seats	2	**Climb Rate @msl**	350 fpm
Cabin Width	26 in.	**Service Ceiling**	N/A

Std. Engine/HP	Usable HP Range	Construction Mtrl.	Est. Bldg. Time
Rotax 582 / 64 hp	60 - 90	Alum. Tubing, Fabric	1400 hrs.
Plans	**Kit**	**Info Packet**	**Video**
$250.00	$13,695.00	$10.00	Yes

A very forgiving and attractive design. The test pilot described it as a "creampuff" after the first flight. Kit price includes tubing kit, cable/rigging, hardware, wheels, brakes, covering and Rotax 582 engine. Plans may be purchased separately from the designer, Graham Lee, Box 889, Lamont, Alberta, T0B-2R0, Canada, 403-895-2975. Plans include a 4.5 hour NTSC video. See Ad on Page 34.

Kits/Plans Sold: 67 **Completed & Flying:** 5

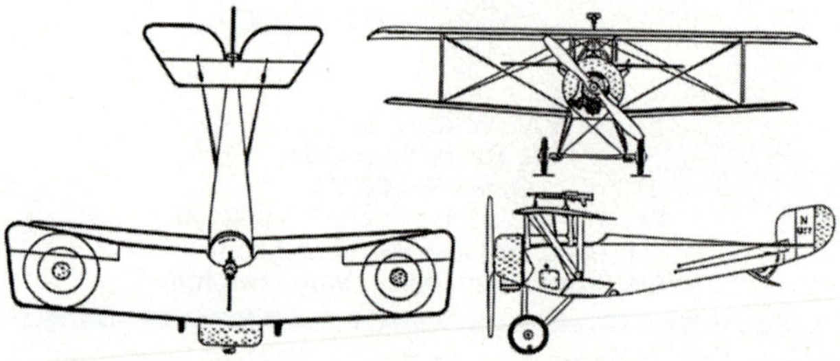

This biplane has been a favorite flying machine for its inherent strength and attractive appearance. Originally designed as a stable observation and gun platform, it combines light weight and strength with the stable yet responsive flight handing. It's an ideal two place, fun flying machine.

A wide range of power plants may be adapted; from the water cooled Rotax 582 to small Continental and Lycoming engines. With the addition of a reduction system, 1700 to 2100 VW engines can also be used.

Rick Giles, Freeport, IL

"My Nieuport is powered with a Yamaha 1100 cc motorcycle engine. ...Takeoffs and landings are made in 200 to 250 feet. Cruise is 60 to 70 mph, and top speed is 85 mph. A handmade 80 x 70 inch black walnut propeller turning at 900 rpm results in 350 pounds of static thrust."

Nieuport 11 & 12 designs also available from:
GRAHAM LEE
Box 889 • General Delivery
Lamont, Alberta T0B 2R0 CANADA
Ph: 403-895-2975
E-Mail: cara@freenet.edmonton.ab.ca

LEADING EDGE AIR FOILS
Meadowlake Airport • 8242 Cessna Drive
Peyton CO 80831
Tel. 719-683-5323 • Fax 719-683-5333
E-Mail: leafkp@kktv.com
Web: http://www.leadingedgeairfoils.com

Piper UL

Wing Span	30.84 ft.	Range	N/A
Wing Area	136.7 sq. ft.	Takeoff Dist.	200 ft.
Length	19.95 ft.	Landing Dist.	130 ft.
Max Gross Wt.	990 lbs.	Vmax	100 mph
Empty Wt.	550 lbs.	VCr	75 mph
Payload (full fuel)	N/A	V_{S1} (stall clean)	39 mph
Fuel Capacity	N/A	V_{S0} (lndg config)	34 mph
Seats	2	Climb Rate @msl	500 fpm
Cabin Width	28 in.	Service Ceiling	N/A

Std. Engine/HP	Usable HP Range	Construction Mtrl.	Est. Bldg. Time
Rotax 503 / 52 hp	N/A	Tube, Fabric	800 hrs.
Plans	**Kit**	**Info Packet**	**Video**
N/A	$18,000.00	Yes	N/A

Piper replica. Ultralight by European standards. Stress limits of +6, -6 G's. Price is approximate in U.S. funds for complete finished aircraft. Available in the U.S. as a kit. **Flying:** 25

LET-MONT, S.R.O.
sidlo Vikyrovice, cp. 226
CZECH REPUBLIC
Tel. 0649/6043,4081 • Fax 38 77

Tulak

Wing Span	31.50 ft.	Range	N/A
Wing Area	140.0 sq. ft.	Takeoff Dist.	200 ft.
Length	N/A	Landing Dist.	130 ft.
Max Gross Wt.	990 lbs.	Vmax	93 mph
Empty Wt.	507 lbs.	VCr	68 mph
Payload (full fuel)	N/A	V_{S1} (stall clean)	31 mph
Fuel Capacity	N/A	V_{S0} (lndg config)	27 mph
Seats	2	Climb Rate @msl	550 fpm
Cabin Width	N/A	Service Ceiling	N/A

Std. Engine/HP	Usable HP Range	Construction Mtrl.	Est. Bldg. Time
Rotax 503 / 52 hp	N/A	Tube, Fabric	700 hrs.
Plans	**Kit**	**Info Packet**	**Video**
N/A	Yes	N/A	N/A

Two place lightweight. Tube and fabric construction. Stress limits of +6, -6 G's. Although available in Europe as a finished aircraft, it is available in the U.S. in kit form.

Completed & Flying: 10

LET-MONT, S.R.O.
sidlo Vikyrovice, cp. 226
CZECH REPUBLIC
Tel. 0649/6043,4081 • Fax 38 77

LK-2M Sluka

Wing Span	30.20 ft.	Range	218 sm
Wing Area	143.5 sq. ft.	Takeoff Dist.	196 ft.
Length	16.70 ft.	Landing Dist.	492 ft.
Max Gross Wt.	605 lbs.	Vmax	80 mph
Empty Wt.	341 lbs.	VCr	60 mph
Payload (full fuel)	N/A	V_{S1} (stall clean)	32 mph
Fuel Capacity	9 gal.	V_{S0} (lndg config)	N/A
Seats	1	Climb Rate @msl	885 fpm
Cabin Width	N/A	Service Ceiling	10,000 ft.

Std. Engine/HP	Usable HP Range	Construction Mtrl.	Est. Bldg. Time
Rotax 447 / 47 hp	40 - 47	Alum. Tube, Dacron, Comp.	60 hrs.
Plans	**Kit**	**Info Packet**	**Video**
N/A	$11,765.00	$10.00	N/A

Conventionally designed, reliable and safe, Ultralight (European) aircraft with tractor propeller.

Kits/Plans Sold: 84 **Completed & Flying:** 62

LETOV AIR DIVISION
Beranovych 65
Praha 9-Letnany 19902
CZECH REPUBLIC
Tel. (+42-2) 8581653 • Fax (+42-2) 8587175

ST-4 Aztek

Wing Span	34.10 ft.	Range	212 sm
Wing Area	163.7 sq. ft.	Takeoff Dist.	196 ft.
Length	19.40 ft.	Landing Dist.	480 ft.
Max Gross Wt.	990 lbs.	Vmax	82 mph
Empty Wt.	484 lbs.	VCr	60 mph
Payload (full fuel)	420 lbs.	V_{S1} (stall clean)	31 mph
Fuel Capacity	15 gal.	V_{S0} (lndg config)	33 mph
Seats	2	Climb Rate @msl	750 fpm
Cabin Width	N/A	Service Ceiling	9,000 ft.

Std. Engine/HP	Usable HP Range	Construction Mtrl.	Est. Bldg. Time
Rotax 503 / 50 hp	45 - 50	Alum. Tube, Dacron, Comp.	80 hrs.
Plans	**Kit**	**Info Packet**	**Video**
N/A	$12,970.00	$10.00	N/A

Two place Ultralight (European). Standard kit includes instruments and engine.

Kits/Plans Sold: 26 **Completed & Flying:** 15

LETOV AIR DIVISION
Beranovych 65
Praha 9-Letnany 19902
CZECH REPUBLIC
Tel. (+42-2) 8581653 • Fax (+42-2) 8587175

North American Distributor:
BOHEMICAN INTERNATIONAL
36 Buchanan Road • St. Catharines, Ontario
L2M 4R6 CANADA • Ph: 905-935-4938
Fax: 905-935-4938 • E-Mail: pdokter@niagara.com

Air Cam

Wing Span	36.00 ft.	**Range**	250 sm
Wing Area	203.0 sq. ft.	**Takeoff Dist.**	170 ft.
Length	27.00 ft.	**Landing Dist.**	200 ft.
Max Gross Wt.	1500 lbs.	**Vmax**	100 mph
Empty Wt.	880 lbs.	**VCr**	80 mph
Payload (full fuel)	450 lbs.	**V_{S1} (stall clean)**	35 mph
Fuel Capacity	28 gal.	**V_{S0} (lndg config)**	35 mph
Seats	2	**Climb Rate @msl**	1200 fpm
Cabin Width	25 in.	**Service Ceiling**	19,500 ft.

Std. Engine/HP	Usable HP Range	Construction Mtrl.	Est. Bldg. Time
2-Rotax 582/64hp/128	128 - 230	Aluminum, Steel, Fabric	600 hrs.
Plans	**Kit**	**Info Packet**	**Video**
N/A	$38,500.00	Yes	$20.00

A twin engine, open cockpit design intended for aerial photography. A utility aircraft for many applications. Very good STOL performance. The Air Cam will fly and climb with only one engine operating. Each engine has its own separate fuel and electrical system. Also available with dual Rotax 912 and 914 engines. Floats and cockpit enclosure are optional.

Kits/Plans Sold: 56 **Completed & Flying:** 4

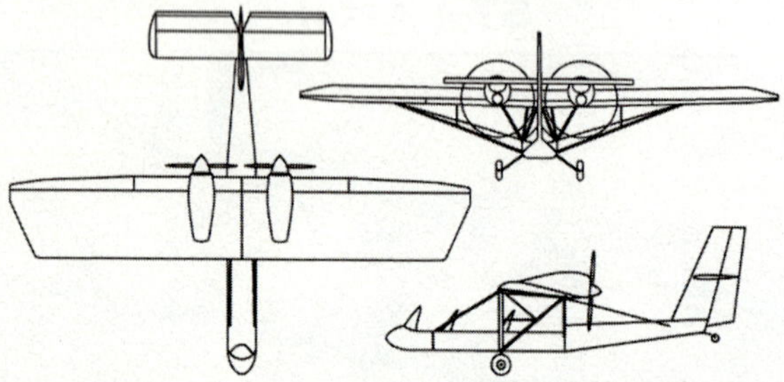

The Air Cam was originally designed and constructed for use as an aerial camera platform for The National Geographic Society. It was highly successful in meeting their requirements for a rugged and reliable aircraft capable of operating from small clearings.

A fully loaded Air Cam has the ability to takeoff with one engine. The positioning of the 64 hp liquid cooled powerplants makes it one of the easiest twins to fly on one engine. With both engines (128 hp), the Air Cam has an exceptional power-to-weight ratio for quick acceleration, short takeoffs, and breathtaking climb-outs. It is now available with pairs of larger Rotax 81 hp (162 total hp) and 115 hp (230 total hp) engines for higher performance.

Experimenter, July '94

"... the aircraft is very stable and docile to handle which of course adds to it's utility as a photo platform ... capable of getting in and out of those places where all the great shots tend to hide but which frequently have less than hospitable terrain for flying."

LEZA-LOCKWOOD CORP.
1 Leza Drive
Sebring FL 33870
Tel. 941-655-4242 • Fax 941-655-0310
E-Mail: aircam@ct.net
Web: http://www.sebring-airport.com/aircam

LM-1-2P Puddle Jumper

Wing Span	30.00 ft.	**Range**	225 sm
Wing Area	130.0 sq. ft.	**Takeoff Dist.**	200 ft.
Length	19.30 ft.	**Landing Dist.**	300 ft.
Max Gross Wt.	875 lbs.	**Vmax**	85 mph
Empty Wt.	430 lbs.	**VCr**	75 mph
Payload (full fuel)	N/A	**V_{S1} (stall clean)**	32 mph
Fuel Capacity	9 gal.	**V_{S0} (lndg config)**	N/A
Seats	2	**Climb Rate @msl**	750 fpm
Cabin Width	N/A	**Service Ceiling**	N/A

Std. Engine/HP	Usable HP Range	Construction Mtrl.	Est. Bldg. Time
65 hp	N/A	Wood, Dacron	400 hrs.
Plans	**Kit**	**Info Packet**	**Video**
$175.00	$4,450.00	$7.50	N/A

An 85% scale J-3 replica, two place version. Also available in the LM-1A 85% scale wood/Dacron, one place version and LM-1X 75% scale alum/Dacron, one place version. Engine, instruments, and paint not included in kit price. **Flying:** 2

LIGHT MINIATURE AIRCRAFT
19695 NW 80th Dr.
Okeechobee FL 34972
Tel. 941-467-0933 • Fax 941-467-0933

LM-3X-W Aeronca Champ

Wing Span	30.00 ft.	**Range**	N/A
Wing Area	120.0 sq. ft.	**Takeoff Dist.**	175 ft.
Length	18.60 ft.	**Landing Dist.**	200 ft.
Max Gross Wt.	600 lbs.	**Vmax**	85 mph
Empty Wt.	335 lbs.	**VCr**	75 mph
Payload (full fuel)	N/A	**V_{S1} (stall clean)**	28 mph
Fuel Capacity	5 gal.	**V_{S0} (lndg config)**	N/A
Seats	1	**Climb Rate @msl**	650 fpm
Cabin Width	N/A	**Service Ceiling**	N/A

Std. Engine/HP	Usable HP Range	Construction Mtrl.	Est. Bldg. Time
Rotax 503 / 52 hp	40 - 65	Wood	450 hrs.
Plans	**Kit**	**Info Packet**	**Video**
$175.00	$6,950.00	$7.50	N/A

A 75% scale size, all-wood construction, Aeronca replica.

Completed & Flying: 12

LIGHT MINIATURE AIRCRAFT
19695 NW 80th Dr.
Okeechobee FL 34972
Ph: 941-467-0933 • Fax: 941-467-0933

LM-2X-2P-W Taylorcraft

Wing Span	32.90 ft.	**Range**	160 sm
Wing Area	135.0 sq. ft.	**Takeoff Dist.**	350 ft.
Length	17.80 ft.	**Landing Dist.**	400 ft.
Max Gross Wt.	925 lbs.	**Vmax**	95 mph
Empty Wt.	475 lbs.	**VCr**	85 mph
Payload (full fuel)	N/A	**V_{S1} (stall clean)**	38 mph
Fuel Capacity	10 gal.	**V_{S0} (Indg config)**	N/A
Seats	2	**Climb Rate @msl**	650 fpm
Cabin Width	N/A	**Service Ceiling**	N/A

Std. Engine/HP	Usable HP Range	Construction Mtrl.	Est. Bldg. Time
Hirth / 65 hp	N/A	Wood, Dacron	500 hrs.
Plans	**Kit**	**Info Packet**	**Video**
$175.00	$4,950.00	$7.50	N/A

A 7/8 scale Taylorcraft utilizing all-wood and Dacron construction. Also available in the LM-2X-2P aluminum and Dacron design as well as the LM-2X single place aluminum and Dacron kit. Engine, engine accessories, instruments, and final coats of paint are not included. **Completed & Flying:** 1

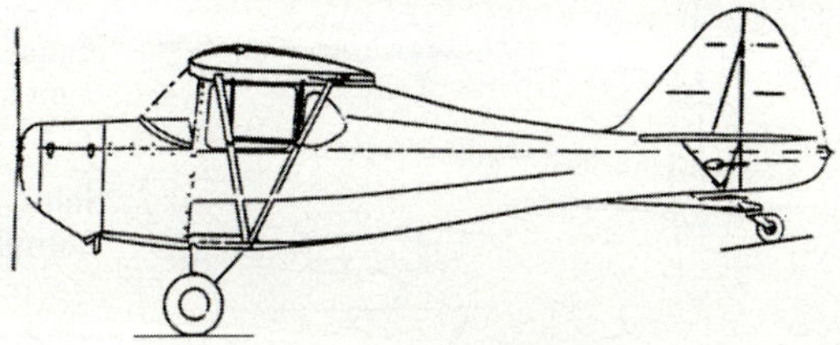

LIGHT MINIATURE AIRCRAFT
19695 NW 80th Dr.
Okeechobee FL 34972
Tel. 941-467-0933 • Fax 941-467-0933

LM-5X-W Super Cub

Wing Span	35.20 ft.	**Range**	160 sm
Wing Area	180.0 sq. ft.	**Takeoff Dist.**	400 ft.
Length	22.70 ft.	**Landing Dist.**	500 ft.
Max Gross Wt.	1200 lbs.	**Vmax**	95 mph
Empty Wt.	700 lbs.	**VCr**	80 mph
Payload (full fuel)	N/A	**V_{S1} (stall clean)**	36 mph
Fuel Capacity	10 gal.	**V_{S0} (Indg config)**	N/A
Seats	2	**Climb Rate @msl**	600 fpm
Cabin Width	N/A	**Service Ceiling**	N/A

Std. Engine/HP	Usable HP Range	Construction Mtrl.	Est. Bldg. Time
Rotax 582 / 66 hp	N/A	Wood, Fabric	500 hrs.
Plans	**Kit**	**Info Packet**	**Video**
$275.00	$11,500.00	$10.00	N/A

A full scale, two place, wood and Dacron Super Cub replica. See the LM-5X for the aluminum and fabric version. Accepts a wide range of engines including the less expensive VWs. Engine, engine accessories, instruments, and final coats of paint are not included. **Completed & Flying:** 1

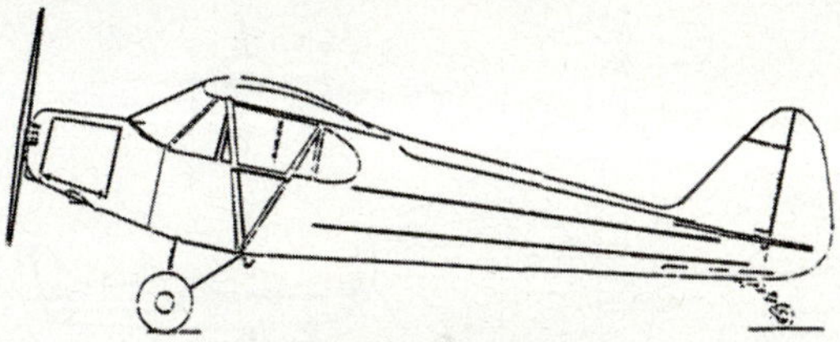

LIGHT MINIATURE AIRCRAFT
19695 NW 80th Dr.
Okeechobee FL 34972
Tel. 941-467-0933 • Fax 941-467-0933

LM-5X Super Cub

Wing Span	35.20 ft.	**Range**	160 sm
Wing Area	180.0 sq. ft.	**Takeoff Dist.**	350 ft.
Length	22.70 ft.	**Landing Dist.**	400 ft.
Max Gross Wt.	1024 lbs.	**Vmax**	95 mph
Empty Wt.	620 lbs.	**VCr**	80 mph
Payload (full fuel)	N/A	**V_{S1} (stall clean)**	35 mph
Fuel Capacity	10 gal.	**V_{S0} (Indg config)**	N/A
Seats	2	**Climb Rate @msl**	750 fpm
Cabin Width	N/A	**Service Ceiling**	N/A

Std. Engine/HP	Usable HP Range	Construction Mtrl.	Est. Bldg. Time
Rotax 582 / 66 hp	N/A	Aluminum, Fabric	500 hrs.
Plans	**Kit**	**Info Packet**	**Video**
$275.00	$11,500.00	$10.00	N/A

A full scale, two place, bonded aluminum and Dacron, Super Cub replica. See the LM-5X-W for the wood and fabric version. Engine, engine accessories, instruments, and final coats of paint are not included.

LIGHT MINIATURE AIRCRAFT
19695 NW 80th Dr.
Okeechobee FL 34972
Tel. 941-467-0933 • Fax 941-467-0933

C.P. 60 Diamant

Wing Span	30.50 ft.	**Range**	1200 sm
Wing Area	143.1 sq. ft.	**Takeoff Dist.**	600 ft.
Length	21.80 ft.	**Landing Dist.**	800 ft.
Max Gross Wt.	1880 lbs.	**Vmax**	155 mph
Empty Wt.	1085 lbs.	**VCr**	142 mph
Payload (full fuel)	475 lbs.	**V_{S1} (stall clean)**	56 mph
Fuel Capacity	42 gal.	**V_{S0} (lndg config)**	51 mph
Seats	3	**Climb Rate @msl**	1000 fpm
Cabin Width	41 in.	**Service Ceiling**	15,000 ft.

Std. Engine/HP	Usable HP Range	Construction Mtrl.	Est. Bldg. Time
Lycoming / 150 hp	100 - 160	All Wood	3500 hrs.
Plans	**Kit**	**Info Packet**	**Video**
$300.00	Yes	$6.00	N/A

Cantilevered elliptical wing. An extrapolation of the basic Emeraude. Reinforced for aerobatics. Achieved 193 mph speed during flight test. Plans include both rectractable tri-gear and conventional taildragger designs. Complete info pack on all eight aircraft is $15.00. **Sold:** 146 **Completed & Flying** 18

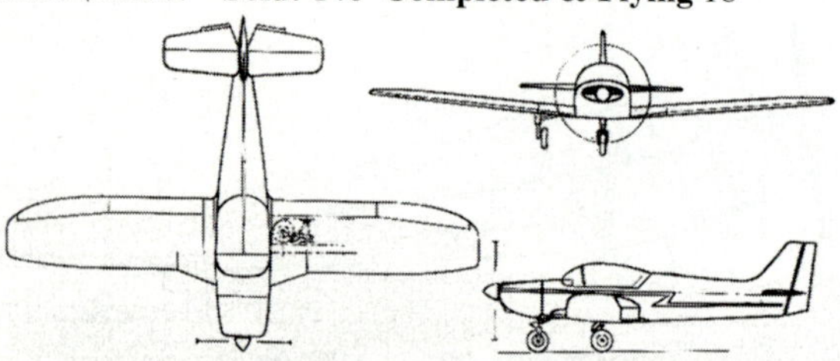

E. (GENE) LITTNER
118 Notre Dame • Le Gardeur • Quebec J5Z 3C3
CANADA
Tel. 514-585-4059 • Fax 514-585-7565

C.P. 1320 Saphir

Wing Span	26.00 ft.	**Range**	950 sm
Wing Area	119.0 sq. ft.	**Takeoff Dist.**	650 ft.
Length	22.00 ft.	**Landing Dist.**	980 ft.
Max Gross Wt.	2073 lbs.	**Vmax**	200 mph
Empty Wt.	1190 lbs.	**VCr**	162 mph
Payload (full fuel)	830 lbs.	**V_{S1} (stall clean)**	62 mph
Fuel Capacity	42 gal.	**V_{S0} (lndg config)**	53 mph
Seats	2	**Climb Rate @msl**	1900 fpm
Cabin Width	42 in.	**Service Ceiling**	13,000 ft.

Std. Engine/HP	Usable HP Range	Construction Mtrl.	Est. Bldg. Time
Lycoming / 150 hp	N/A	All Wood	3000 hrs.
Plans	**Kit**	**Info Packet**	**Video**
$275.00	N/A	$6.00	N/A

Cantilevered elliptical wing. Designed with ribs and frames cut out of plywood instead of trusses and gussets to reduce construction time. Normally a two seater, a third passenger of up to 130 lbs. can fit on the back bench seat. Sliding canopy. Literature on all eight aircraft for $15.00.

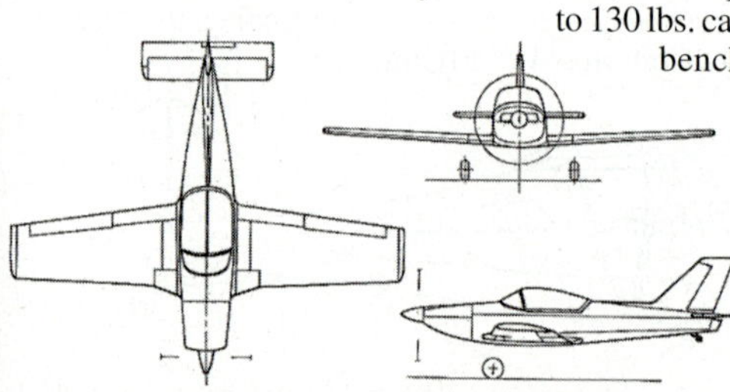

Sold: 40
Completed: 13

E. (GENE) LITTNER
118 Notre Dame • Le Gardeur • Quebec J5Z 3C3
CANADA
Tel. 514-585-4059 • Fax 514-585-7565

C.P. 750 Beryl

Wing Span	26.50 ft.	**Range**	700 sm
Wing Area	117.0 sq. ft.	**Takeoff Dist.**	620 ft.
Length	22.70 ft.	**Landing Dist.**	900 ft.
Max Gross Wt.	1760 lbs.	**Vmax**	185 mph
Empty Wt.	1000 lbs.	**VCr**	160 mph
Payload (full fuel)	565 lbs.	**V_{S1} (stall clean)**	65 mph
Fuel Capacity	36 gal.	**V_{S0} (lndg config)**	56 mph
Seats	2	**Climb Rate @msl**	1200 fpm
Cabin Width	27 in.	**Service Ceiling**	15,000 ft.

Std. Engine/HP	Usable HP Range	Construction Mtrl.	Est. Bldg. Time
Lycoming / 150 hp	115 - 160	Wood, Steel Tube	2500 hrs.
Plans	**Kit**	**Info Packet**	**Video**
$250.00	Yes	$6.00	Yes

Cantilevered elliptical wing, identical, including flight characteristics, to the Super Emeraude C.P. 328 except for a stronger spar. Wooden fuselage option included in plans. Tandem seating. Aircraft is fully aerobatic as a single seater. Can be flown solo from either front or rear seat. Sliding canopy. First presented to the public at the 1955 French Fly-in. Complete info pack on all eight aircraft is $15.00.

Sold: 128 **Completed & Flying** 10

E. (GENE) LITTNER
118 Notre Dame • Le Gardeur • Quebec J5Z 3C3
CANADA
Tel. 514-585-4059 • Fax 514-585-7565

C.P. 80 Zephyr

Wing Span	19.70 ft.	**Range**	300 sm
Wing Area	66.70 sq. ft.	**Takeoff Dist.**	650 ft.
Length	17.30 ft.	**Landing Dist.**	650 ft.
Max Gross Wt.	835 lbs.	**Vmax**	190 mph
Empty Wt.	575 lbs.	**VCr**	155 mph
Payload (full fuel)	205 lbs.	**V_{S1} (stall clean)**	60 mph
Fuel Capacity	10 gal.	**V_{S0} (Indg config)**	55 mph
Seats	1	**Climb Rate @msl**	2300 fpm
Cabin Width	24 in.	**Service Ceiling**	10,000 ft.

Std. Engine/HP	Usable HP Range	Construction Mtrl.	Est. Bldg. Time
Continental / 100 hp	65 - 100	All-Wood	2000 hrs.
Plans	**Kit**	**Info Packet**	**Video**
$200.00	N/A	$6.00	N/A

Designed primarily for Formula I racing. Optional flaps are included in plans. Cessna type landing gear. Complete info pack for all eight airplanes $15.00. **Sold:** 42 **Completed:** 15

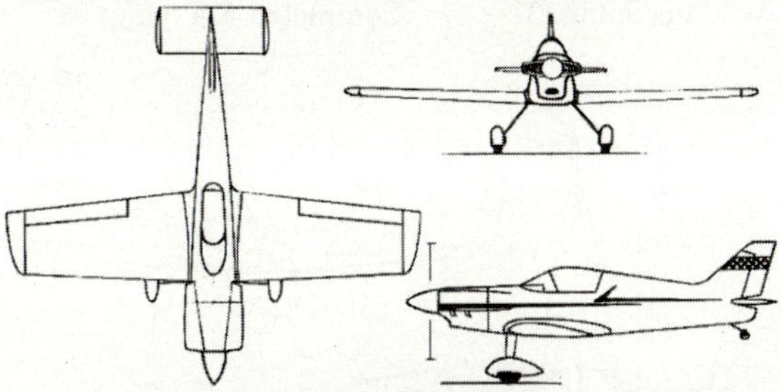

E. (GENE) LITTNER
118 Notre Dame • Le Gardeur • Quebec J5Z 3C3
CANADA
Tel. 514-585-4059 • Fax 514-585-7565

C.P. 90 Pinocchio

Wing Span	23.50 ft.	**Range**	350 sm
Wing Area	104.0 sq. ft.	**Takeoff Dist.**	580 ft.
Length	19.70 ft.	**Landing Dist.**	525 ft.
Max Gross Wt.	1015 lbs.	**Vmax**	150 mph
Empty Wt.	700 lbs.	**VCr**	143 mph
Payload (full fuel)	220 lbs.	**V_{S1} (stall clean)**	45 mph
Fuel Capacity	15 gal.	**V_{S0} (Indg config)**	N/A
Seats	1	**Climb Rate @msl**	1100 fpm
Cabin Width	30 in.	**Service Ceiling**	12,000 ft.

Std. Engine/HP	Usable HP Range	Construction Mtrl.	Est. Bldg. Time
Lycoming / 115 hp	65 - 115	All Wood	2000 hrs.
Plans	**Kit**	**Info Packet**	**Video**
$200.00	N/A	$6.00	N/A

Cantilevered elliptical wing. A reduced design of the C.P. 328 (Super-Emeraude). Complete info pack on all eight aircraft available for $15.00. **Sold:** 15 **Completed:** 5

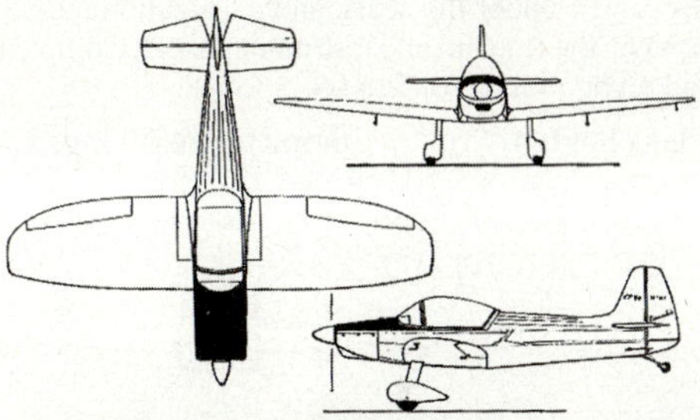

E. (GENE) LITTNER
118 Notre Dame • Le Gardeur • Quebec J5Z 3C3
CANADA
Tel. 514-585-4059 • Fax 514-585-7565

Champion V

E. (GENE) LITTNER
118 Notre Dame • Le Gardeur • Quebec J5Z 3C3
CANADA
Tel. 514-585-4059 • Fax 514-585-7565

Wing Span	27.30 ft.	**Range**	450 sm
Wing Area	120.0 sq. ft.	**Takeoff Dist.**	425 ft.
Length	20.00 ft.	**Landing Dist.**	425 ft.
Max Gross Wt.	1058 lbs.	**Vmax**	140 mph
Empty Wt.	600 lbs.	**VCr**	125 mph
Payload (full fuel)	78 lbs.	**V_{S1} (stall clean)**	45 mph
Fuel Capacity	13 gal.	**V_{S0} (Indg config)**	40 mph
Seats	2	**Climb Rate @msl**	800 fpm
Cabin Width	42 in.	**Service Ceiling**	N/A

Std. Engine/HP	Usable HP Range	Construction Mtrl.	Est. Bldg. Time
VW / 75 hp	75 - 100	Wood, Dacron Cover	1500 hrs.
Plans	**Kit**	**Info Packet**	**Video**
$400.00	N/A	$6.00	N/A

Low wing airplane looks like a downsized "Falco." Designed in Italy by Giuseppe Vidor. First flown June 10, 1995. More than 50 now under construction in Europe. Meets F.A.R. 23 normal category with two people on board and the utility category in solo configurations. Can be taken apart in about 15 minutes. Features fixed or retractable tri-gear. Complete info packet on all eight aircraft for $15.00.

Kits/Plans Sold: 15 **Completed & Flying:** 3

Vega ST 87

Wing Span	25.00 ft.	**Range**	750 sm
Wing Area	108.0 sq. ft.	**Takeoff Dist.**	800 ft.
Length	23.00 ft.	**Landing Dist.**	800 ft.
Max Gross Wt.	1600 lbs.	**Vmax**	150 mph
Empty Wt.	1035 lbs.	**VCr**	125 mph
Payload (full fuel)	180 lbs.	**V_{S1} (stall clean)**	60 mph
Fuel Capacity	30 gal.	**V_{S0} (lndg config)**	50 mph
Seats	2	**Climb Rate @msl**	700 fpm
Cabin Width	43 in.	**Service Ceiling**	13,000 ft.

Std. Engine/HP	Usable HP Range	Construction Mtrl.	Est. Bldg. Time
Lycoming / 108 hp	85 - 125	Wood, Dacron Cover	2000 hrs.
Plans	**Kit**	**Info Packet**	**Video**
$400.00	N/A	$6.00	N/A

Cross country design from France by Rene Stern. First flown in July, 1992. Wing built into fuselage for great rigidity. A 25 gallon fuel tank can be installed behind the seats on the rear spar; two 15 gallon tanks can fit under the seats, and a 15 gallon tank can be installed between the firewall and instrument panel. Complete info packet on all eight aircraft for $15.00.

Kits/Plans Sold: 4 **Completed & Flying:** 2

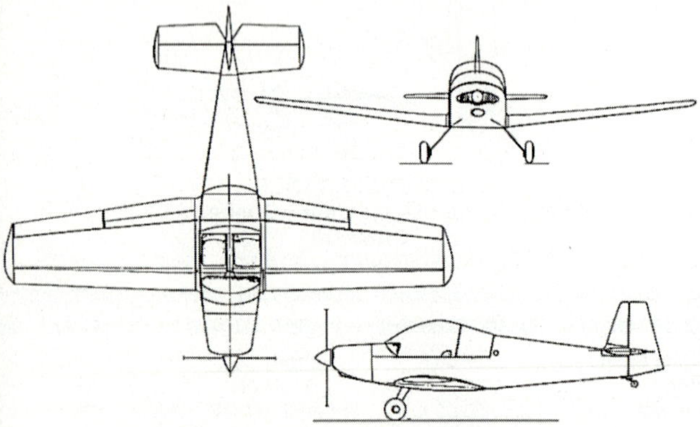

E. (GENE) LITTNER
118 Notre Dame • Le Gardeur • Quebec J5Z 3C3
CANADA
Tel. 514-585-4059 • Fax 514-585-7565

C.P. 328 Super Emeraude

Wing Span	26.50 ft.	**Range**	620 sm
Wing Area	117.0 sq. ft.	**Takeoff Dist.**	450 ft.
Length	21.00 ft.	**Landing Dist.**	600 ft.
Max Gross Wt.	1540 lbs.	**Vmax**	145 mph
Empty Wt.	850 lbs.	**VCr**	125 mph
Payload (full fuel)	500 lbs.	**V_{S1} (stall clean)**	60 mph
Fuel Capacity	30 gal.	**V_{S0} (lndg config)**	55 mph
Seats	2	**Climb Rate @msl**	900 fpm
Cabin Width	41 in.	**Service Ceiling**	14,000 ft.

Std. Engine/HP	Usable HP Range	Construction Mtrl.	Est. Bldg. Time
Lycoming / 115 hp	N/A	All Wood	3000 hrs.
Plans	**Kit**	**Info Packet**	**Video**
$275.00	N/A	$6.00	N/A

Cantilevered elliptical wing. Limited aerobatics possible as a single seater. Sliding canopy. Complete info packet on all eight aircraft for $15.00.

Kits/Plans Sold: 232 **Completed & Flying:** 45

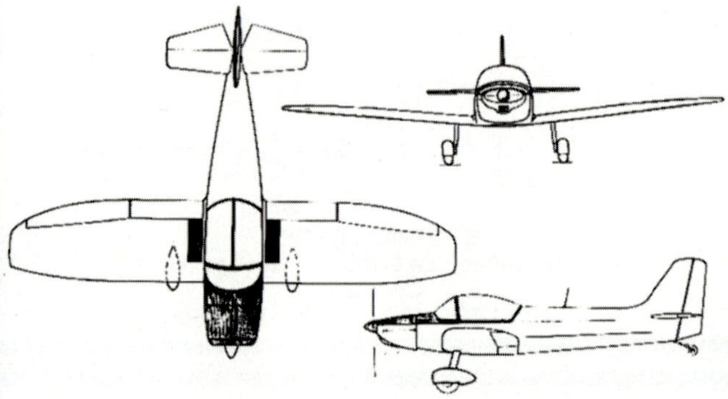

The C.P. 30 Emeraude, equipped with a Continental 90 hp engine flew for the first time in 1955. Flight and static tested by a French government agency, it was then certified for production as the C.P. 301 Emeraude. More than 300 were built in the late 50's and 60's.

In 1963, the designer C. Piel beefed up the structure to take more power. Gross weight was increased from 1450 to 1550 lbs. A sliding canopy was added. The new model was called the C.P. 320 Super-Emeraude. It was certified in September, 1988. More than 200 were produced.

Later, C. Piel released the C.P. 320 plans for use by amateur builders. The design was modified by replacing the oleo with a spring gear and adding a swept back vertical tail as an option. This plans-only version is designated the C.P. 328.

E. (GENE) LITTNER
118 Notre Dame • Le Gardeur • Quebec J5Z 3C3
CANADA
Tel. 514-585-4059 • Fax 514-585-7565

Drifter MU582

Wing Span	30.00 ft.	**Range**	180 sm
Wing Area	150.0 sq. ft.	**Takeoff Dist.**	250 ft.
Length	21.00 ft.	**Landing Dist.**	350 ft.
Max Gross Wt.	900 lbs.	**Vmax**	85 mph
Empty Wt.	420 lbs.	**VCr**	70 mph
Payload (full fuel)	N/A	**V_{S1} (stall clean)**	38 mph
Fuel Capacity	10 gal.	**V_{S0} (lndg config)**	38 mph
Seats	2	**Climb Rate @msl**	850 fpm
Cabin Width	N/A	**Service Ceiling**	18,000 ft.

Std. Engine/HP	Usable HP Range	Construction Mtrl.	Est. Bldg. Time
Rotax 582 / 65 hp	65 - 80	Alum., Steel, Fabric	150 hrs.
Plans	**Kit**	**Info Packet**	**Video**
N/A	$15,695.00	Yes	N/A

Two seat, pusher configuration. Designed to handle the higher gross weights associated with special applications such as float flying and agricultural spraying. Optional STOL flap kit available. Standard features include dual controls, tandem seating, 3 axis controls, cable braced for superior strength, and a center-mounted joy stick. Length includes nose fairing. Basic price includes engine, 3-blade Ivoprop, and instruments.

Kits/Plans Sold: 1300 **Completed & Flying:** 1000

LOCKWOOD AVIATION SUPPLY, INC.
280 Hendricks Way • Suite A
Sebring FL 33870
Tel. 941-655-5100 • Fax 941-655-6225
E-Mail: lockwood@digital.net
Web: http://www.lockwood-aviation.com

Drifter XP503

Wing Span	30.00 ft.	**Range**	180 sm
Wing Area	150.0 sq. ft.	**Takeoff Dist.**	250 ft.
Length	21.00 ft.	**Landing Dist.**	350 ft.
Max Gross Wt.	785 lbs.	**Vmax**	75 mph
Empty Wt.	385 lbs.	**VCr**	65 mph
Payload (full fuel)	N/A	**V_{S1} (stall clean)**	36 mph
Fuel Capacity	10 gal.	**V_{S0} (lndg config)**	36 mph
Seats	2	**Climb Rate @msl**	750 fpm
Cabin Width	N/A	**Service Ceiling**	18,000 ft.

Std. Engine/HP	Usable HP Range	Construction Mtrl.	Est. Bldg. Time
Rotax 503UL / 50 hp	N/A	Alum. Steel, Fabric	150 hrs.
Plans	**Kit**	**Info Packet**	**Video**
N/A	$13,595.00	Yes	N/A

Two seat, pusher configuration with open cockpit. Low operating costs and stable flying characteristics. A number of XP503's now have more than 3,000 hours. Optional STOL flap kit available. Standard features include dual controls, tandem seating, 3 axis controls, cable braced for superior strength, and a center-mounted joy stick. Length includes nose fairing. Basic price includes engine, 2-blade Ivoprop, and instruments.

Kits/Plans Sold: 1300 **Completed & Flying:** 1000

LOCKWOOD AVIATION SUPPLY, INC.
280 Hendricks Way • Suite A
Sebring FL 33870
Tel. 941-655-5100 • Fax 941-655-6225
E-Mail: lockwood@digital.net
Web: http://www.lockwood-aviation.com

Drifter DR447

LOCKWOOD AVIATION SUPPLY, INC.
280 Hendricks Way • Suite A
Sebring FL 33870
Tel. 941-655-5100 • Fax 941-655-6225
E-Mail: lockwood@digital.net
Web: http://www.lockwood-aviation.com

Wing Span	30.00 ft.	**Range**	180 sm
Wing Area	150.0 sq. ft.	**Takeoff Dist.**	250 ft.
Length	21.00 ft.	**Landing Dist.**	350 ft.
Max Gross Wt.	785 lbs.	**Vmax**	75 mph
Empty Wt.	320 lbs.	**VCr**	65 mph
Payload (full fuel)	N/A	**V_{S1} (stall clean)**	36 mph
Fuel Capacity	10 gal.	**V_{S0} (lndg config)**	36 mph
Seats	1	**Climb Rate @msl**	700 fpm
Cabin Width	N/A	**Service Ceiling**	18,000 ft.

Std. Engine/HP	Usable HP Range	Construction Mtrl.	Est. Bldg. Time
Rotax 447UL / 42 hp	N/A	Aluminum, Metal, Fabric	150 hrs.
Plans	**Kit**	**Info Packet**	**Video**
N/A	$12,595.00	Yes	N/A

Sing seat, pusher configuration with open cockpit. Optional STOL flap kit available. Standard features include 3 axis controls, cable braced for superior strength, and a center-mounted joy stick. Length includes nose fairing. Basic price includes engine, 2-blade Ivoprop, and instruments.

Kits/Plans Sold: 1300 **Completed & Flying:** 1000

5151 Mustang

Wing Span	27.50 ft.	**Range**	300 sm
Wing Area	130.0 sq. ft.	**Takeoff Dist.**	150 ft.
Length	22.10 ft.	**Landing Dist.**	250 ft.
Max Gross Wt.	895 lbs.	**Vmax**	105 mph
Empty Wt.	525 lbs.	**VCr**	85 mph
Payload (full fuel)	N/A	**V_{S1} (stall clean)**	30 mph
Fuel Capacity	5 gal.	**V_{S0} (lndg config)**	N/A
Seats	1	**Climb Rate @msl**	900 fpm
Cabin Width	N/A	**Service Ceiling**	12,500 ft.

Std. Engine/HP	Usable HP Range	Construction Mtrl.	Est. Bldg. Time
Rotax 582 / 65 hp	N/A	Wood, Fabric	500 hrs.
Plans	**Kit**	**Info Packet**	**Video**
N/A	$7,995.00	$10.00	$25.00

Scale warbird replica with retractable gear. Kit price does not include engine. Wings detach easily. Fixed gear, Rotax 503 powered version available. See Ad on Page 36.

Completed & Flying: 18

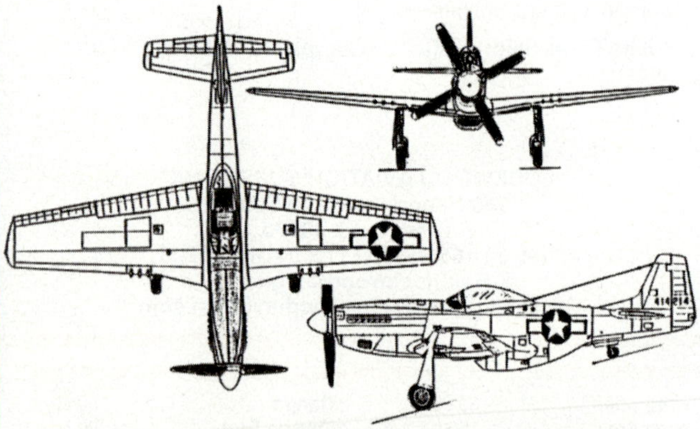

LOEHLE AIRCRAFT CORP.
380 Shipmans Creek Road, No. K
Wartrace TN 37183
Tel. 615-857-3419 • Fax 615-857-3908

P-40

Wing Span	28.80 ft.	**Range**	300 sm
Wing Area	136.0 sq. ft.	**Takeoff Dist.**	150 ft.
Length	22.10 ft.	**Landing Dist.**	250 ft.
Max Gross Wt.	885 lbs.	**Vmax**	100 mph
Empty Wt.	550 lbs.	**VCr**	85 mph
Payload (full fuel)	N/A	**V_{S1} (stall clean)**	30 mph
Fuel Capacity	5 gal.	**V_{S0} (lndg config)**	N/A
Seats	1	**Climb Rate @msl**	1200 fpm
Cabin Width	N/A	**Service Ceiling**	12,500 ft.

Std. Engine/HP	Usable HP Range	Construction Mtrl.	Est. Bldg. Time
Rotax 582 / 66 hp	N/A	Wood, Fabric	500 hrs.
Plans	**Kit**	**Info Packet**	**Video**
N/A	$7,995.00	$10.00	$25.00

Warbird replica. A realistic 75% scale design using simple wooden construction. Price does not include engine. Includes fixed gear. Simple wing and tail removal for easy transport and storage. Engine package and retractable gear available. Other options include electric start, custom wing tanks, ballistic parachute, and custom upholstery. See Ad on Page 34.

Completed & Flying: 1

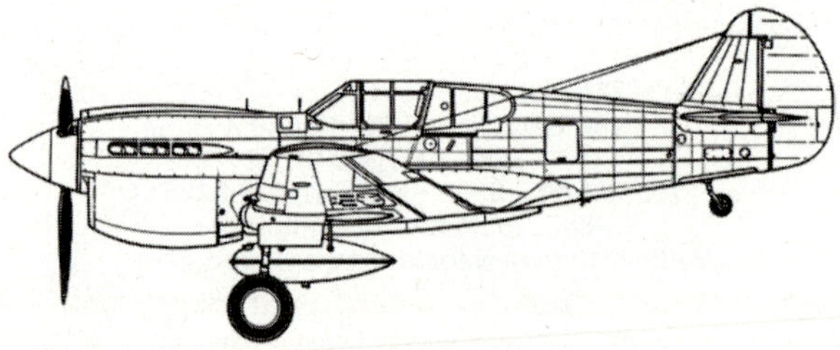

LOEHLE AIRCRAFT CORP.
380 Shipmans Creek Road, No. K
Wartrace TN 37183
Tel. 615-857-3419 • Fax 615-857-3908

Spad XIII

Wing Span	24.00 ft.	**Range**	80 sm
Wing Area	168.0 sq. ft.	**Takeoff Dist.**	90 ft.
Length	16.50 ft.	**Landing Dist.**	150 ft.
Max Gross Wt.	525 lbs.	**Vmax**	65 mph
Empty Wt.	260 lbs.	**VCr**	52 mph
Payload (full fuel)	N/A	**V_{S1} (stall clean)**	21 mph
Fuel Capacity	5 gal.	**V_{S0} (Indg config)**	N/A
Seats	1	**Climb Rate @msl**	900 fpm
Cabin Width	N/A	**Service Ceiling**	12,000 ft.

Std. Engine/HP	Usable HP Range	Construction Mtrl.	Est. Bldg. Time
Rotax 277 / 28 hp	N/A	Tube, Fabric	400 hrs.
Plans	**Kit**	**Info Packet**	**Video**
N/A	$9,995.00	$10.00	$25.00

WW I replica. Tube and fabric construction. Covering and cable rigging required. Other WWI replicas available include the Fokker D-VII and the SE5a. Contact manufacturer for details. See Ad on Page 34. **Completed & Flying:** 1

LOEHLE AIRCRAFT CORP.
380 Shipmans Creek Road, No. K
Wartrace TN 37183
Tel. 615-857-3419 • Fax 615-857-3908

Sport Parasol

Wing Span	25.60 ft.	**Range**	200 sm
Wing Area	114.0 sq. ft.	**Takeoff Dist.**	100 ft.
Length	18.50 ft.	**Landing Dist.**	150 ft.
Max Gross Wt.	600 lbs.	**Vmax**	85 mph
Empty Wt.	252 lbs.	**VCr**	70 mph
Payload (full fuel)	N/A	**V_{S1} (stall clean)**	22 mph
Fuel Capacity	5 gal.	**V_{S0} (Indg config)**	N/A
Seats	1	**Climb Rate @msl**	750 fpm
Cabin Width	N/A	**Service Ceiling**	12,500 ft.

Std. Engine/HP	Usable HP Range	Construction Mtrl.	Est. Bldg. Time
Rotax 503 / 50 hp	N/A	Wood, Fabric	350 hrs.
Plans	**Kit**	**Info Packet**	**Video**
N/A	$3,495.00	$10.00	$25.00

Simple wood and fabric construction. Price does not include engine. Wings are detachable. Meets part 103 UL requirements. See Ad on Page 34. **Completed & Flying:** 38

LOEHLE AIRCRAFT CORP.
380 Shipmans Creek Road, No. K
Wartrace TN 37183
Tel. 615-857-3419 • Fax 615-857-3908

Lucas L-5

Wing Span	30.00 ft.	**Range**	680 sm
Wing Area	130.0 sq. ft.	**Takeoff Dist.**	1300 ft.
Length	20.00 ft.	**Landing Dist.**	900 ft.
Max Gross Wt.	2150 lbs.	**Vmax**	170 mph
Empty Wt.	1230 lbs.	**VCr**	140 mph
Payload (full fuel)	830 lbs.	**V_{S1} (stall clean)**	47 mph
Fuel Capacity	50 gal.	**V_{S0} (Indg config)**	N/A
Seats	4	**Climb Rate @msl**	1200 fpm
Cabin Width	42 in.	**Service Ceiling**	18,000 ft.

Std. Engine/HP	Usable HP Range	Construction Mtrl.	Est. Bldg. Time
Various / 108 hp	108-200	Metal	4000 hrs.
Plans	**Kit**	**Info Packet**	**Video**
$400.00	N/A	N/A	N/A

All-metal successor to the early 1960's wood and fabric L-4. Seating for three or four. Uses a variety of engines. The L5 is the only all-metal three or four seat aircraft with retractable landing gear offered in Europe. Selected as a typical all-metal aircraft for teaching sheet metal workers in the aircraft industry.

Kits/Plans Sold: 60 **Completed & Flying:** 12

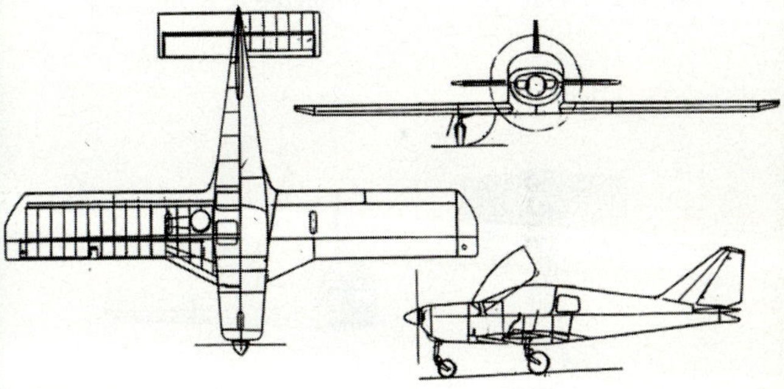

EMILE LUCAS
7 allee des Acacias
Lagny le Sec 60330
FRANCE
Tel. 44 60 87 45 • Fax 16 44 60 04 81

Lucas L-6B

Wing Span	29.00 ft.	**Range**	1100 sm
Wing Area	140.0 sq. ft.	**Takeoff Dist.**	900 ft.
Length	26.00 ft.	**Landing Dist.**	800 ft.
Max Gross Wt.	1590 lbs.	**Vmax**	120 mph
Empty Wt.	1060 lbs.	**VCr**	108 mph
Payload (full fuel)	530 lbs.	**V_{S1} (stall clean)**	N/A
Fuel Capacity	40 gal.	**V_{S0} (lndg config)**	N/A
Seats	2	**Climb Rate @msl**	N/A
Cabin Width	28 in.	**Service Ceiling**	20,000 ft.

Std. Engine/HP	Usable HP Range	Construction Mtrl.	Est. Bldg. Time
Lycoming / 108 hp	108-180	Metal	4000 hrs.
Plans	**Kit**	**Info Packet**	**Video**
$400.00	N/A	N/A	N/A

All-metal, folding wing, tandem sport plane version of L-6. See listing for the Lucas L-6A under SAILPLANES. Taildragger equipped with retractable landing gear and variable pitch propeller. Wing center section contains retractable wheel, flaps, and leading edge fuel tanks. The outer wing panels include ailerons that fold upward. Excellent lift/drag ratio. Choice of Lycoming engines from O-235 to O-360. Note 20,000 ft. service ceiling and 1,100 mile range.

Kits/Plans Sold: 5 **Completed & Flying:** 2

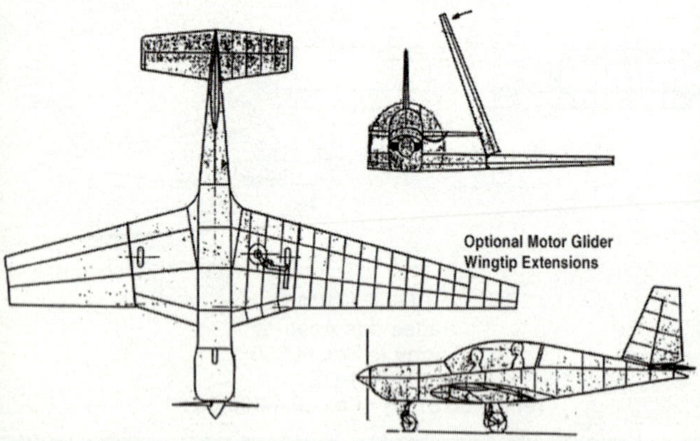

EMILE LUCAS
7 allee des Acacias
Lagny le Sec 60330
FRANCE
Tel. 44 60 87 45 • Fax 16 44 60 04 81

Lucas L-7

Wing Span	33.50 ft.	**Range**	450 sm
Wing Area	135.0 sq. ft.	**Takeoff Dist.**	1300 ft.
Length	21.00 ft.	**Landing Dist.**	1000 ft.
Max Gross Wt.	1720 lbs.	**Vmax**	115 mph
Empty Wt.	1150 lbs.	**VCr**	105 mph
Payload (full fuel)	570 lbs.	**V_{S1} (stall clean)**	49 mph
Fuel Capacity	30 gal.	**V_{S0} (lndg config)**	40 mph
Seats	2	**Climb Rate @msl**	550 fpm
Cabin Width	43 in.	**Service Ceiling**	14,000 ft.

Std. Engine/HP	Usable HP Range	Construction Mtrl.	Est. Bldg. Time
Lyc. 0235 / 108 hp	108 - 180	All Metal	3200 hrs.
Plans	**Kit**	**Info Packet**	**Video**
$15.00	N/A	$15.00	N/A

An all-metal design featuring a large 5.5' x 2.5' door underneath the fuselage allowing easy storage of bicycles, skis, etc. Can also be used for parachuting. Retractable nose-wheel.

Kits/Plans Sold: 15 **Completed & Flying:** 2

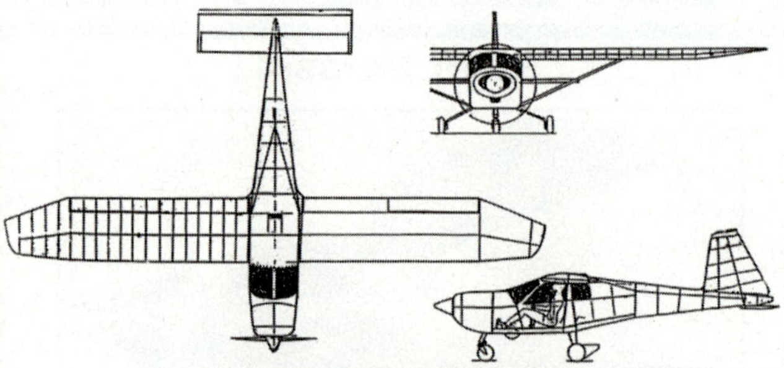

EMILE LUCAS
7 allee des Acacias
Lagny le Sec 60330
FRANCE
Tel. 44 60 87 45 • Fax 16 44 60 04 81

Mainair Blade

Wing Span	34.60 ft.	**Range**	200 sm
Wing Area	168.0 sq. ft.	**Takeoff Dist.**	N/A
Length	N/A	**Landing Dist.**	N/A
Max Gross Wt.	858 lbs.	**Vmax**	101 mph
Empty Wt.	380 lbs.	**VCr**	76 mph
Payload (full fuel)	478 lbs.	**V_{S1} (stall clean)**	30 mph
Fuel Capacity	12 gal.	**V_{S0} (Indg config)**	N/A
Seats	2	**Climb Rate @msl**	850 fpm
Cabin Width	N/A	**Service Ceiling**	10,000 ft.

Std. Engine/HP	Usable HP Range	Construction Mtrl.	Est. Bldg. Time
Rotax 582 / 65 hp	50 - 80	Aluminum Tube, Fabric	75 hrs.
Plans	**Kit**	**Info Packet**	**Video**
N/A	$17,150.00	Yes	N/A

Introduced in 1994, the Blade is proving to be exceptionally stable, even in rough air. Great for long flights or local cruising. Kit is supplied complete with all components except instruments but including in-flight trimmer. Minimal drilling required. Rotax 912 engine also available. A wing building service is available. Available ready built. Meets British BCAR/S specifications with CAA approval. Kit price is approximate in U.S. funds and includes Rotax 582 engine, electric starter, C-type gearbox and 4-blade carbon propeller.

Kits/Plans Sold: 120 **Completed & Flying:** 120

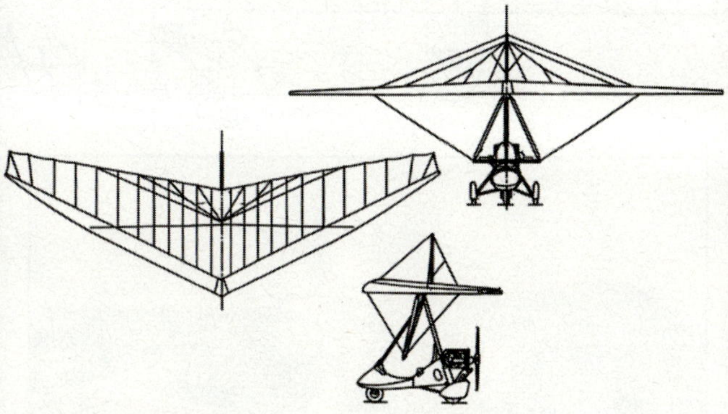

MAINAIR SPORTS LTD.
Alma Industrial Estate • Regent Street
Rochedale, Lancashire 0L12 0HQ
UNITED KINGDOM
Tel. +44(0)1706-55134 • Fax +44(0)1706-31561
E-Mail: rpattrick@compuserve.com
Web: http://ourworld.compuserve.com/homepages/RPattrick

Mainair Rapier

Paul Tomlin Photo

Wing Span	34.60 ft.	**Range**	200 sm
Wing Area	168.0 sq. ft.	**Takeoff Dist.**	N/A
Length	N/A	**Landing Dist.**	N/A
Max Gross Wt.	814 lbs.	**Vmax**	101 mph
Empty Wt.	334 lbs.	**VCr**	73 mph
Payload (full fuel)	480 lbs.	**V_{S1} (stall clean)**	30 mph
Fuel Capacity	12 gal.	**V_{S0} (Indg config)**	N/A
Seats	2	**Climb Rate @msl**	650 fpm
Cabin Width	N/A	**Service Ceiling**	10,000 ft.

Std. Engine/HP	Usable HP Range	Construction Mtrl.	Est. Bldg. Time
Rotax 503 / 50 hp	50 - 65	Aluminum Tube, Fabric	70 hrs.
Plans	**Kit**	**Info Packet**	**Video**
N/A	$12,400.00	Yes	N/A

A flexwing trike utilizing the Mainair Blade wing. Kit is supplied pre-drilled and a wing building service is available. Kit price is approximate in U.S. funds and includes Rotax 503 engine, electric starter, C-type gearbox and 4-blade carbon propeller.

MAINAIR SPORTS LTD.
Alma Industrial Estate • Regent Street
Rochedale, Lancashire 0L12 0HQ UNITED KINGDOM
Tel. +44(0)1706-55134 • Fax +44(0)1706-31561
E-Mail: rpattrick@compuserve.com

Hatz Classic

Wing Span	25.00 ft.	**Range**	280 sm
Wing Area	N/A	**Takeoff Dist.**	400 ft.
Length	19.00 ft.	**Landing Dist.**	600 ft.
Max Gross Wt.	1600 lbs.	**Vmax**	150 mph
Empty Wt.	1000 lbs.	**VCr**	105 mph
Payload (full fuel)	450 lbs.	**V_{S1} (stall clean)**	42 mph
Fuel Capacity	26 gal.	**V_{S0} (Indg config)**	42 mph
Seats	2	**Climb Rate @msl**	1000 fpm
Cabin Width	26 in.	**Service Ceiling**	10,000 ft.

Std. Engine/HP	Usable HP Range	Construction Mtrl.	Est. Bldg. Time
Lycoming / 150 hp	150 - 160	Wood, Tube, Fabric	1000 hrs.
Plans	**Kit**	**Info Packet**	**Video**
$250.00	$12,500.00	$7.00	$12.00

A refined model of the Hatz CB-1, an award winning design similar to the style and easy flying qualities of the Waco of the 1930's. Aerobatic. Engine, spars, and covering are not included.

Kits/Plans Sold: 5 **Completed & Flying:** 3

MAKELAN CORPORATION
P.O. Box 312142
New Braunfels TX 78130
Tel. 210-905-7832 • Fax 210-899-5407

MA-5 Charger

Wing Span	24.00 ft.	**Range**	400 sm
Wing Area	170.0 sq. ft.	**Takeoff Dist.**	600 ft.
Length	20.00 ft.	**Landing Dist.**	750 ft.
Max Gross Wt.	1500 lbs.	**Vmax**	120 mph
Empty Wt.	1000 lbs.	**VCr**	110 mph
Payload (full fuel)	N/A	**V_{S1} (stall clean)**	48 mph
Fuel Capacity	27 gal.	**V_{S0} (lndg config)**	N/A
Seats	2	**Climb Rate @msl**	1000 fpm
Cabin Width	N/A	**Service Ceiling**	12,500 ft.

Std. Engine/HP	Usable HP Range	Construction Mtrl.	Est. Bldg. Time
Lycoming / 125 hp	125 -180	Tube, Fabric	2500 hrs.
Plans	**Kit**	**Info Packet**	**Video**
$135.00	N/A	$5.00	N/A

Plans-only biplane. Similar to the classic Great Lakes Trainer series. Various Lycoming engine options available from 125 to 180 hp.

Kits/Plans Sold: 400 **Completed & Flying:** 80

ED MARQUART
PO Box 3032
Riverside CA 92509
Tel. 909-683-9582

Sport 1000

Wing Span	32.75 ft.	**Range**	90 sm
Wing Area	180.0 sq. ft.	**Takeoff Dist.**	150 ft.
Length	21.00 ft.	**Landing Dist.**	300 ft.
Max Gross Wt.	1000 lbs.	**Vmax**	100 mph
Empty Wt.	475 lbs.	**VCr**	65 mph
Payload (full fuel)	525 lbs.	**V_{S1} (stall clean)**	26 mph
Fuel Capacity	16 gal.	**V_{S0} (lndg config)**	N/A
Seats	2	**Climb Rate @msl**	600 fpm
Cabin Width	N/A	**Service Ceiling**	12,000 ft.

Std. Engine/HP	Usable HP Range	Construction Mtrl.	Est. Bldg. Time
Rotax 502 / 65 hp	65 - 100	Tube, Fabric	125 hrs.
Plans	**Kit**	**Info Packet**	**Video**
N/A	$15,995.00	Yes	N/A

High wing pusher. Tube and fabric construction.

Completed & Flying: 16

M-SQUARED, INC.
1300 Marshall Smith Rd.
Lucedale MS 39452
Tel. 601-947-9414 • Fax 601-947-4959

Corsario MK-III

Wing Span	33.80 ft.	**Range**	222 sm
Wing Area	177.4 sq. ft.	**Takeoff Dist.**	260 ft.
Length	22.44 ft.	**Landing Dist.**	220 ft.
Max Gross Wt.	1100 lbs.	**Vmax**	100 mph
Empty Wt.	540 lbs.	**VCr**	75 mph
Payload (full fuel)	N/A	**V_{S1} (stall clean)**	30 mph
Fuel Capacity	17 gal.	**V_{S0} (lndg config)**	N/A
Seats	2	**Climb Rate @msl**	800 fpm
Cabin Width	N/A	**Service Ceiling**	10,000 ft.

Std. Engine/HP	Usable HP Range	Construction Mtrl.	Est. Bldg. Time
Rotax 912 / 80 hp	66 - 80	Alum., Composite, Fabric	300 hrs.
Plans	**Kit**	**Info Packet**	**Video**
N/A	$15,285.00	$10.00	N/A

A high wing amphibian with side-by-side seating. Retractable tri-gear or tailwheel landing gear. Detachable wings. Engine options include Rotax 582, 618 and 912. Kit price does not include engine, propeller or instruments. The original Mk. I was designed in 1988.

Completed & Flying: 350

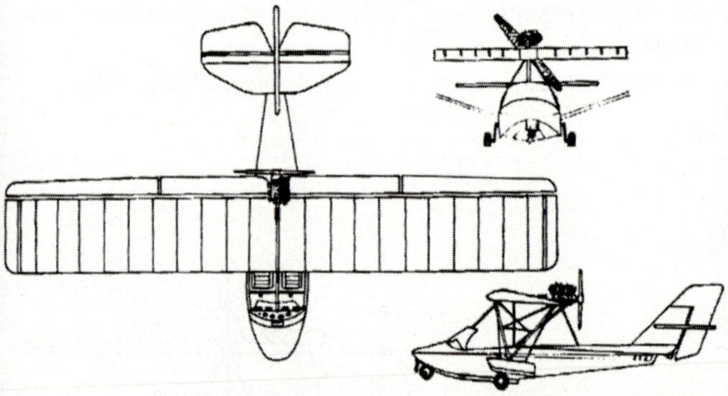

MICROLEVE
Avenida Ayrton Senna 2541 • Aeroporto de Jacarepagua, Hang
Rio de Janeiro, RJ 22 775 00 BRAZIL
Ph: 55 21-325-1020 • Fax: 55 21-325-1835
E-Mail: clark@cyberhome.com.br
Web: http://www.asasonline.com.br/vitrine/microlev

ML450

Wing Span	33.80 ft.	**Range**	222 sm
Wing Area	177.4 sq. ft.	**Takeoff Dist.**	200 ft.
Length	22.44 ft.	**Landing Dist.**	120 ft.
Max Gross Wt.	980 lbs.	**Vmax**	100 mph
Empty Wt.	484 lbs.	**VCr**	70 mph
Payload (full fuel)	N/A	**V_{S1} (stall clean)**	20 mph
Fuel Capacity	17 gal.	**V_{S0} (lndg config)**	N/A
Seats	2	**Climb Rate @msl**	1000 fpm
Cabin Width	N/A	**Service Ceiling**	10,000 ft.

Std. Engine/HP	Usable HP Range	Construction Mtrl.	Est. Bldg. Time
Rotax 618 / 74 hp	55 - 74	Metal, Composite, Fabric	250 hrs.
Plans	**Kit**	**Info Packet**	**Video**
N/A	$12,275.00	$10.00	N/A

A side-by-side design available in either tri-gear or taildragger configurations. Available with enclosed or open cockpit. Engine options include Rotax 503, 582 and 618 engines. Kit price does not include engine, propeller or instruments.

Completed & Flying: 250

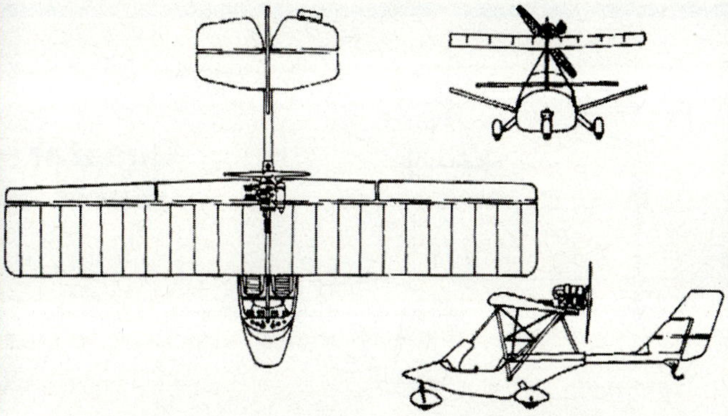

MICROLEVE
Avenida Ayrton Senna 2541 • Aeroporto de Jacarepagua, Hang
Rio de Janeiro, RJ 22 775 00 BRAZIL
Ph: 55 21-325-1020 • Fax: 55 21-325-1835
E-Mail: clark@cyberhome.com.br
Web: http://www.asasonline.com.br/vitrine/microlev

Pitts Model 12 "Pitts Monster"

Wing Span	23.00 ft.	**Range**	525 sm
Wing Area	154.0 sq. ft.	**Takeoff Dist.**	N/A
Length	21.70 ft.	**Landing Dist.**	N/A
Max Gross Wt.	2250 lbs.	**Vmax**	200 mph
Empty Wt.	1500 lbs.	**VCr**	175 mph
Payload (full fuel)	420 lbs.	**V_{S1} (stall clean)**	63 mph
Fuel Capacity	55 gal.	**V_{S0} (lndg config)**	N/A
Seats	2	**Climb Rate @msl**	2500 fpm
Cabin Width	N/A	**Service Ceiling**	N/A

Std. Engine/HP	Usable HP Range	Construction Mtrl.	Est. Bldg. Time
M-14P / 360 hp	250 - 360	Wood, Tube, Fabric	3000 hrs.
Plans	**Kit**	**Info Packet**	**Video**
$300.00	Yes	Yes	N/A

Also called the "Macho Stinker," the Pitts Monster is a high-powered, aerobatic aircraft designed by Curtis Pitts. A variety of component packages are available including welded fuselage, empennage, engine and other smaller sub-components.

Kits/Plans Sold: 70 **Completed & Flying:** 1

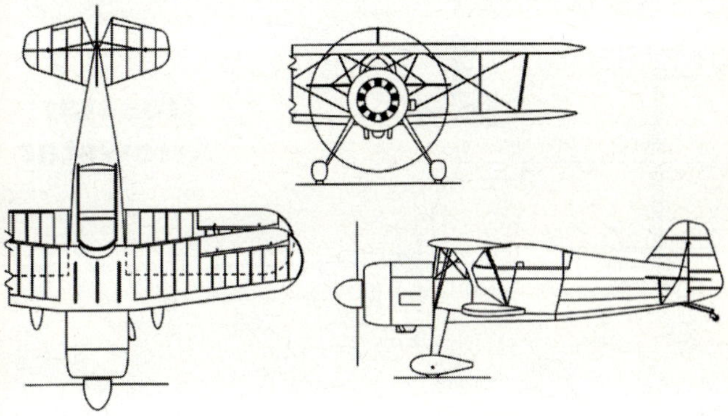

MID AMERICA AIRCRAFT, INC.
528 S. Belmont
Wichita KS 67218
Ph: 316-683-6537 • Fax: 316-682-5028

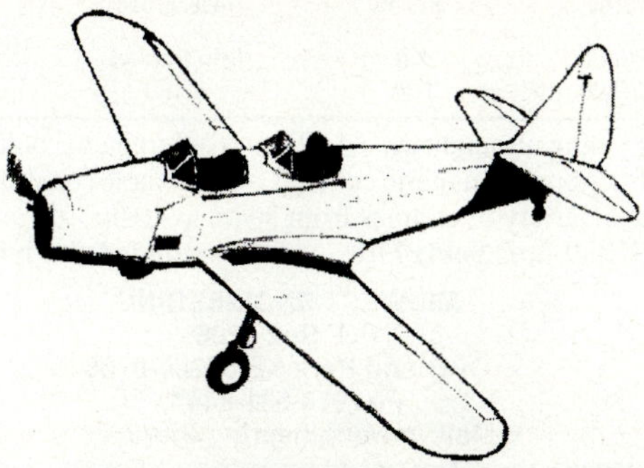

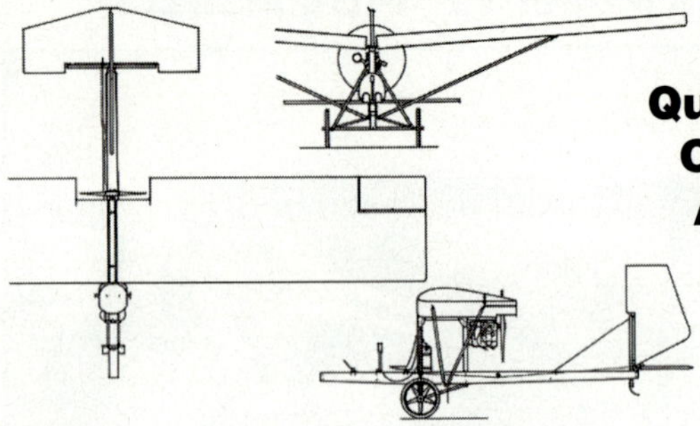

Questar Open Aire

Wing Span	28.66 ft.	**Range**	N/A
Wing Area	124.0 sq. ft.	**Takeoff Dist.**	100 ft.
Length	16.66 ft.	**Landing Dist.**	100 ft.
Max Gross Wt.	450 lbs.	**Vmax**	60 mph
Empty Wt.	220 lbs.	**VCr**	52 mph
Payload (full fuel)	230 lbs.	**V_{S1} (stall clean)**	25 mph
Fuel Capacity	3 gal.	**V_{S0} (lndg config)**	N/A
Seats	N/A	**Climb Rate @msl**	600 fpm
Cabin Width	N/A	**Service Ceiling**	10,000 ft.

Std. Engine/HP	Usable HP Range	Construction Mtrl.	Est. Bldg. Time
30 hp	30 - 40	Wood, Tube, Fabric	N/A
Plans	**Kit**	**Info Packet**	**Video**
$30.00	N/A	$5.00	N/A

A unique ultralight taildragger. Utilizes "bolt-together" construction with a fabric covered wing. Full panoramic view while in flight. Excellent for aerial photography.

Kits/Plans Sold: 50 **Completed & Flying:** 10

MIDWEST ENGINEERING
PO Box 4706
Overland Park KS 66204-0706
Tel. 913-631-8447
E-Mail: mwengineering@juno.com

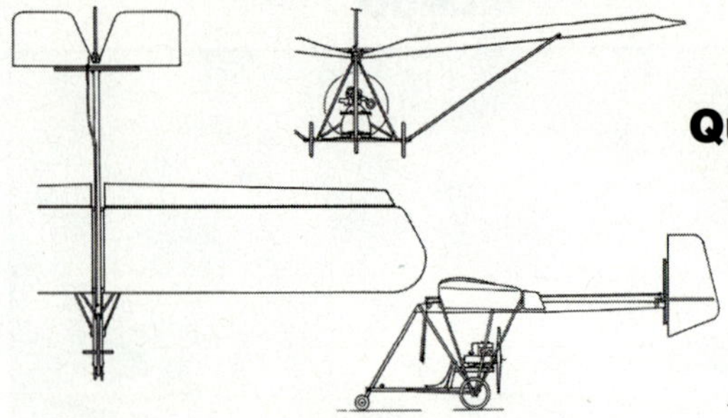

Questar XLS

Wing Span	31.30 ft.	**Range**	N/A
Wing Area	146.0 sq. ft.	**Takeoff Dist.**	100 ft.
Length	17.00 ft.	**Landing Dist.**	80 ft.
Max Gross Wt.	500 lbs.	**Vmax**	54 mph
Empty Wt.	240 lbs.	**VCr**	45 mph
Payload (full fuel)	260 lbs.	**V_{S1} (stall clean)**	24 mph
Fuel Capacity	5 gal.	**V_{S0} (lndg config)**	N/A
Seats	1	**Climb Rate @msl**	580 fpm
Cabin Width	N/A	**Service Ceiling**	N/A

Std. Engine/HP	Usable HP Range	Construction Mtrl.	Est. Bldg. Time
30 hp	30 - 40	Tube, Fabric	200 hrs.
Plans	**Kit**	**Info Packet**	**Video**
$30.00	N/A	$5.00	N/A

A rear-engine ultralight trike. Rear-engine design increases visiblity. Utilizes T6 aluminum "bolt-together" airframe and a Fabric covered wing.

Kits/Plans Sold: 70 **Completed & Flying:** 10

MIDWEST ENGINEERING
PO Box 4706
Overland Park KS 66204-0706
Tel. 913-631-8447
E-Mail: mwengineering@juno.com

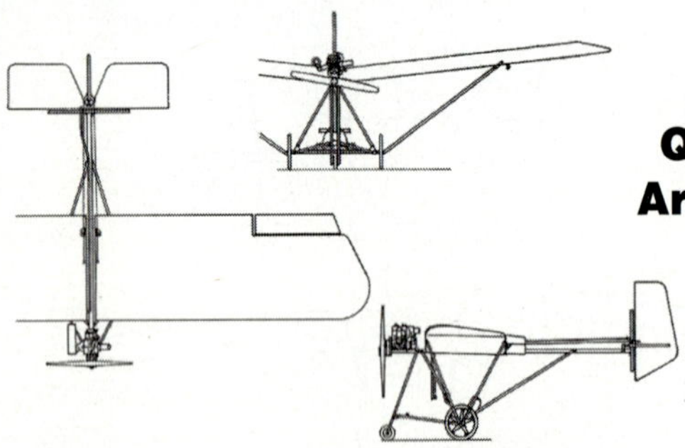

Questar Arrowstar

Wing Span	27.25 ft.	**Range**	N/A
Wing Area	130.0 sq. ft.	**Takeoff Dist.**	100 ft.
Length	15.00 ft.	**Landing Dist.**	80 ft.
Max Gross Wt.	450 lbs.	**Vmax**	58 mph
Empty Wt.	230 lbs.	**VCr**	48 mph
Payload (full fuel)	220 lbs.	**V_{S1} (stall clean)**	24 mph
Fuel Capacity	3 gal.	**V_{S0} (lndg config)**	N/A
Seats	1	**Climb Rate @msl**	600 fpm
Cabin Width	N/A	**Service Ceiling**	10,000 ft.

Std. Engine/HP	Usable HP Range	Construction Mtrl.	Est. Bldg. Time
30 hp	30 - 40	Tube, Fabric	80 hrs.
Plans	**Kit**	**Info Packet**	**Video**
$30.00	N/A	$5.00	N/A

A high-wing ultralight trike. Utilizes T6 aluminum "bolt-together" construction with a fabric covered wing. Simple construction. Can utilize a variety of engines from 30 hp to 40 hp.

Kits/Plans Sold: 170 **Completed & Flying:** 60

MIDWEST ENGINEERING
PO Box 4706
Overland Park KS 66204-0706
Ph: 913-631-8447
E-Mail: mwengineering@juno.com

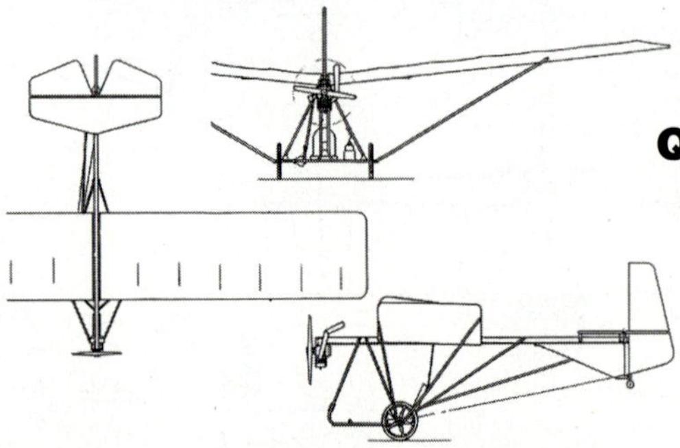

Questar Sport

Wing Span	33.25 ft.	**Range**	60 sm
Wing Area	165.0 sq. ft.	**Takeoff Dist.**	120 ft.
Length	17.50 ft.	**Landing Dist.**	80 ft.
Max Gross Wt.	400 lbs.	**Vmax**	48 mph
Empty Wt.	160 lbs.	**VCr**	38 mph
Payload (full fuel)	225 lbs.	**V_{S1} (stall clean)**	20 mph
Fuel Capacity	3 gal.	**V_{S0} (lndg config)**	N/A
Seats	1	**Climb Rate @msl**	350 fpm
Cabin Width	N/A	**Service Ceiling**	10,000 ft.

Std. Engine/HP	Usable HP Range	Construction Mtrl.	Est. Bldg. Time
30 hp	30 - 40	Tube, Fabric	80 hrs.
Plans	**Kit**	**Info Packet**	**Video**
$30.00	N/A	$5.00	N/A

An easily transportable ultralight featuring folding wings. 10 minute setup time. T6 aluminum "bolt-together" construction, fabric covered wing. Can be stored in a normal garage.

Kits/Plans Sold: 120 **Completed & Flying:** 30

MIDWEST ENGINEERING
PO Box 4706
Overland Park KS 66204-0706
Tel. 913-631-8447
E-Mail: mwengineering@juno.com

Celerity

Wing Span	25.00 ft.	Range	900 sm
Wing Area	100.0 sq. ft.	Takeoff Dist.	800 ft.
Length	22.80 ft.	Landing Dist.	600 ft.
Max Gross Wt.	1825 lbs.	Vmax	220 mph
Empty Wt.	1169 lbs.	VCr	205 mph
Payload (full fuel)	356 lbs.	V_{S1} (stall clean)	60 mph
Fuel Capacity	50 gal.	V_{S0} (lndg config)	55 mph
Seats	2	Climb Rate @msl	1800 fpm
Cabin Width	40 in.	Service Ceiling	20,000 ft.

Std. Engine/HP	Usable HP Range	Construction Mtrl.	Est. Bldg. Time
Lyc. O-300 / 160 hp	150 - 180	Wood, Fiberglass	3000 hrs.
Plans	**Kit**	**Info Packet**	**Video**
$245.00	$11,500.00	$7.50	$25.00

"Fast with class" at low cost. High performance, side-by-side seating, clean all-retractable tailwheel configuration. Easy to build with wood, foam and fiberglass. Options include factory molded leading edge fuel tanks, molded sliding canopy, and landing gear ready to install. Accepts Lycoming, Subaru, Mazda, Chevy V-6 and Ford V-6 engines. Comfortable upright seating, spacious baggage compartment and quiet cockpit due to sound attenuation of wood, foam, and fiberglass. Full 3-axis trim is autopilot compatible. Cruising range up to 1,100 miles. Newsletter $7.50 per year.

Kits/Plans Sold: 55 **Completed & Flying:** 2

MIRAGE AIRCRAFT CORP.
1091 W. Sahara Palms Dr.
Tucson AZ 85704
Tel. 520-797-2161
E-Mail: celerity@primemet.com

Marathon

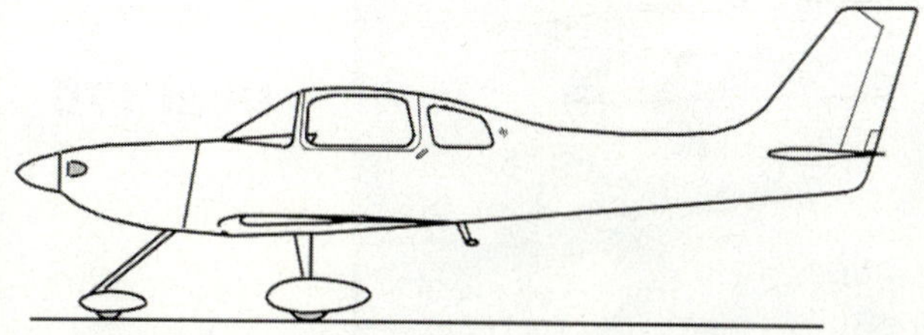

Wing Span	25.00 ft.	Range	800 sm
Wing Area	100.0 sq. ft.	Takeoff Dist.	800 ft.
Length	22.80 ft.	Landing Dist.	1000 ft.
Max Gross Wt.	1825 lbs.	Vmax	200 mph
Empty Wt.	1119 lbs.	VCr	185 mph
Payload (full fuel)	406 lbs.	V_{S1} (stall clean)	60 mph
Fuel Capacity	50 gal.	V_{S0} (lndg config)	55 mph
Seats	2	Climb Rate @msl	1500 fpm
Cabin Width	40 in.	Service Ceiling	20,000 ft.

Std. Engine/HP	Usable HP Range	Construction Mtrl.	Est. Bldg. Time
Chevy V6 / 160 hp	150 - 180	Wood, Foam, Fiberglass	N/A
Plans	**Kit**	**Info Packet**	**Video**
$245.00	$8,500.00	$7.50	$25.00

Economical cross country luxury traveler. The Mirage Marathon is a fixed gear adaptation of the Celerity. With side-by-side comfort, ergonomic seat design. Fixed gear (tricycle or taildragger), wood, foam and fiberglass construction. Options include factory molded leading edge fuel tanks, molded canopy, and complete landing gear ready to install. Accepts Lycoming, Subaru, Mazda, Chevy V-6 and Ford V-6 engines. Low noise levels. Roomy baggage compartment, 3-axis trim, "Easy-step" entry and exit. Marathon is designed for anyone with high performance needs, average builders' skills, and limited funds. Newsletter $7.50 per year. **Kits/Plans Sold:** 10

MIRAGE AIRCRAFT CORP.
1091 W. Sahara Palms Dr.
Tucson AZ 85704
Tel. 520-797-2161
E-Mail: celerity@primemet.com

Morrisey 2000

MORRISEY AIRCRAFT CO.
P.O. Box 27889
Las Vegas NV 89126
Tel. 702-251-5551

Wing Span	30.00 ft.	Range	375 sm
Wing Area	142.0 sq. ft.	Takeoff Dist.	500 ft.
Length	20.00 ft.	Landing Dist.	450 ft.
Max Gross Wt.	1600 lbs.	Vmax	120 mph
Empty Wt.	1050 lbs.	VCr	80 mph
Payload (full fuel)	430 lbs.	V_{S1} (stall clean)	50 mph
Fuel Capacity	20 gal.	V_{S0} (lndg config)	50 mph
Seats	2	Climb Rate @msl	1000 fpm
Cabin Width	30 in.	Service Ceiling	16,000 ft.

Std. Engine/HP	Usable HP Range	Construction Mtrl.	Est. Bldg. Time
Lyc. O-235 / 115 hp	90 - 180	Aluminum, Steel	N/A
Plans	**Kit**	**Info Packet**	**Video**
N/A	$28,000.00	N/A	N/A

All-metal, low wing, tricycle gear design. The Morrisey 2000 is the predecessor of the Varga Kachina which was first known as the Morrisey 2150. The Morrisey 2000 was first certified as a factory built aircraft in 1955 under CAR 03 rules. However, liability considerations dictated that it be offered in kit form for the homebuilt market.

Completed & Flying: 200

Dual 175

Wing Span	33.33 ft.	**Range**	N/A
Wing Area	175.0 sq. ft.	**Takeoff Dist.**	250 ft.
Length	11.00 ft.	**Landing Dist.**	150 ft.
Max Gross Wt.	750 lbs.	**Vmax**	67 mph
Empty Wt.	500 lbs.	**VCr**	50 mph
Payload (full fuel)	N/A	**V_{S1} (stall clean)**	27 mph
Fuel Capacity	10 gal.	**V_{S0} (lndg config)**	N/A
Seats	2	**Climb Rate @msl**	900 fpm
Cabin Width	N/A	**Service Ceiling**	10,000 ft.

Std. Engine/HP	Usable HP Range	Construction Mtrl.	Est. Bldg. Time
Rotax 503DC/52 hp	N/A	Tube, Fabric	8 hrs.
Plans	**Kit**	**Info Packet**	**Video**
N/A	$12,850.00	$8.00	N/A

Trike Ultralight. Standard size, two place. TIG welded Cr-Mo steel engine mount. Cantilevered fiberglass main axles. Molded, tear drop fuel tank. Olite bronze main hinges. TIG welded 6061-T6 aluminum front fork. Free instructions with kit purchase.

Completed & Flying: 25

MOUNTAINEER TRIKES
PO Box 557, Finkle Rd. • Millerton NY 12546-0557
Ph: 518-789-6550 • Fax: 518-789-6775

Solo 175

Wing Span	33.33 ft.	**Range**	N/A
Wing Area	175.0 sq. ft.	**Takeoff Dist.**	150 ft.
Length	11.00 ft.	**Landing Dist.**	50 ft.
Max Gross Wt.	750 lbs.	**Vmax**	63 mph
Empty Wt.	505 lbs.	**VCr**	45 mph
Payload (full fuel)	N/A	**V_{S1} (stall clean)**	25 mph
Fuel Capacity	5 gal.	**V_{S0} (lndg config)**	N/A
Seats	1	**Climb Rate @msl**	900 fpm
Cabin Width	N/A	**Service Ceiling**	10,000 ft.

Std. Engine/HP	Usable HP Range	Construction Mtrl.	Est. Bldg. Time
Rotax 447SC / 40 hp	N/A	Tube, Fabric	8 hrs.
Plans	**Kit**	**Info Packet**	**Video**
N/A	$12,250.00	$8.00	N/A

Trike Ultralight. Standard size. Single seat. Steel engine mount. Cantilevered fiberglass main axles. Molded tear drop fuel tank. Olite bronze main hinges. TIG welded 6061-T6 aluminum front fork. Free instructions with kit purchase. **Flying:** 10

MOUNTAINEER TRIKES
PO Box 557, Finkle Rd. • Millerton NY 12546-0557
Tel. 518-789-6550 • Fax 518-789-6775

Rebel

Wing Span	30.00 ft.	**Range**	733 sm
Wing Area	150.0 sq. ft.	**Takeoff Dist.**	300 ft.
Length	21.40 ft.	**Landing Dist.**	400 ft.
Max Gross Wt.	1650 lbs.	**Vmax**	140 mph
Empty Wt.	950 lbs.	**VCr**	120 mph
Payload (full fuel)	700 lbs.	**V_{S1} (stall clean)**	44 mph
Fuel Capacity	44 gal.	**V_{S0} (lndg config)**	40 mph
Seats	3	**Climb Rate @msl**	1200 fpm
Cabin Width	N/A	**Service Ceiling**	15,000 ft.

Std. Engine/HP	Usable HP Range	Construction Mtrl.	Est. Bldg. Time
Lyc.0-320 / 160 hp	80-160	Metal, Fabric	800 hrs.
Plans	**Kit**	**Info Packet**	**Video**
N/A	$13,950.00	$15.00	$20.00

STOL performance. Ample room for two full size adults and 150 lbs. of baggage. Engine not included in kit. Engine options include the O-235 116 hp and the model 912, 80 hp. Load limits are +5.7, -3.8 G's. Detachable wings. Floats and skis are optional. See Ad on Page 55.

Completed & Flying: 90

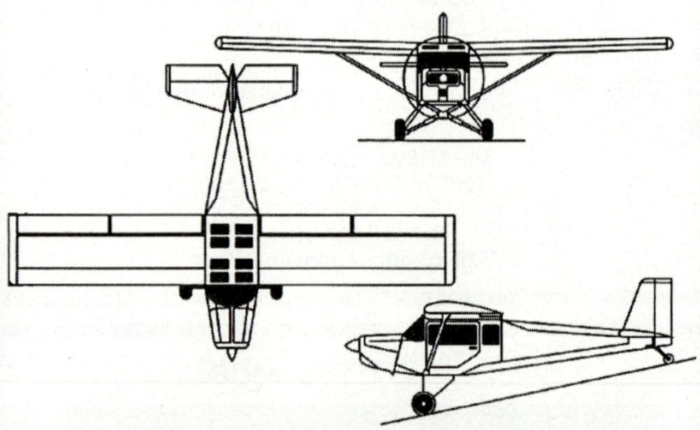

MURPHY AIRCRAFT MFG. LTD.
Unit 1 - 8155 Aitken Rd. • Chilliwack BC V2R 4H5
CANADA
Tel. 604-792-5855 • Fax 604-792-7006
E-Mail: mursales@murphyair.com
Web: http://www.murphyair.com

SR 2500 Super Rebel

Wing Span	36.00 ft.	**Range**	620 sm
Wing Area	185.0 sq. ft.	**Takeoff Dist.**	550 ft.
Length	23.00 ft.	**Landing Dist.**	460 ft.
Max Gross Wt.	3000 lbs.	**Vmax**	160 mph
Empty Wt.	1450 lbs.	**VCr**	145 mph
Payload (full fuel)	1200 lbs.	**V_{S1} (stall clean)**	43 mph
Fuel Capacity	60 gal.	**V_{S0} (lndg config)**	40 mph
Seats	4	**Climb Rate @msl**	1150 fpm
Cabin Width	44 in.	**Service Ceiling**	12,000 ft.

Std. Engine/HP	Usable HP Range	Construction Mtrl.	Est. Bldg. Time
Lyc 0-540 / 250 hp	150 - 210	Aluminum	1500 hrs.
Plans	**Kit**	**Info Packet**	**Video**
N/A	$19,995.00	$20.00	N/A

Newest and largest member of the Murphy family. Kit price does not include engine. The all-metal, four seater offers great utility with impressive performance. With dual controls, removable rear seats, large cargo area, and a double reinforced floor. Float attach points. Adjustable front seats. Electric trim. See Ad on Page 55. **Completed & Flying:** 1

The management at Murphy Aircraft was well aware that there were some fine four place homebuilts already available in the market. They knew that any new aircraft of this type would need to be something special. They feel that the Super Rebel 2500 is that and more. In addition to being the only all-aluminum four place, they claim that it goes to the head of the class for low cost and large payload.

The Super Rebel 2500 was first shown at Sun 'n Fun Fly-In '95. The prototype is powered by a 210 hp engine, though the aircraft will have the option of 150 to 210 hp powerplants. The aircraft will be offered as either a tricycle or taildragger. The landing gear configuration can be easily converted from one format to the other. Super Rebels adapt quickly to straight or amphibious floats, to skis, or to tundra tires for bush-flying. One of the options offered allows the aircraft to be used as a camper.

MURPHY AIRCRAFT MFG. LTD.
Unit 1 - 8155 Aitken Rd. • Chilliwack BC V2R 4H5
CANADA
Tel. 604-792-5855 • Fax 604-792-7006
E-Mail: mursales@murphyair.com
Web: http://www.murphyair.com

Maverick

Wing Span	29.50 ft.	**Range**	264 sm
Wing Area	147.0 sq. ft.	**Takeoff Dist.**	100 ft.
Length	20.80 ft.	**Landing Dist.**	200 ft.
Max Gross Wt.	950 lbs.	**Vmax**	110 mph
Empty Wt.	450 lbs.	**VCr**	85 mph
Payload (full fuel)	N/A	**V_{S1} (stall clean)**	24 mph
Fuel Capacity	11 gal.	**V_{S0} (lndg config)**	N/A
Seats	2	**Climb Rate @msl**	900 fpm
Cabin Width	N/A	**Service Ceiling**	10,000 ft.

Std. Engine/HP	Usable HP Range	Construction Mtrl.	Est. Bldg. Time
Rotax 582 / 65 hp	53-65	Metal, Fabric	850 hrs.
Plans	**Kit**	**Info Packet**	**Video**
N/A	$13,275.00	$15.00	N/A

Lightest member of the Murphy family. Side-by-side seating. Rugged all-metal fuselage. Fabric covered wing and tail section. Semi-monocoque construction. Single strut. Dual load paths. Complete kit price is with Rotax 503/DC 53 hp engine. Can be purchased with or without engine. Extended wings option available. See Ad on Page 55.

Completed & Flying: 18

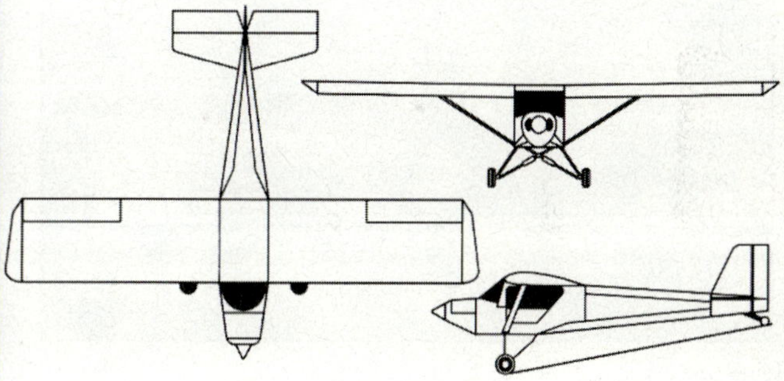

MURPHY AIRCRAFT MFG. LTD.
Unit 1 - 8155 Aitken Rd. • Chilliwack BC V2R 4H5
CANADA
Tel. 604-792-5855 • Fax 604-792-7006
E-Mail: mursales@murphyair.com
Web: http://www.murphyair.com

Renegade Spirit

Wing Span	21.30 ft.	**Range**	300 sm
Wing Area	153.8 sq. ft.	**Takeoff Dist.**	100 ft.
Length	18.50 ft.	**Landing Dist.**	250 ft.
Max Gross Wt.	950 lbs.	**Vmax**	105 mph
Empty Wt.	500 lbs.	**VCr**	85 mph
Payload (full fuel)	N/A	V_{S1} **(stall clean)**	38 mph
Fuel Capacity	14 gal.	V_{S0} **(lndg config)**	N/A
Seats	2	**Climb Rate @msl**	900 fpm
Cabin Width	N/A	**Service Ceiling**	10,000 ft.

Std. Engine/HP	Usable HP Range	Construction Mtrl.	Est. Bldg. Time
Rotax 912 / 80 hp	53-80	Wood, Fabric	550 hrs.
Plans	**Kit**	**Info Packet**	**Video**
$275.00	$13,995.00	$15.00	$20.00

Capable of limited aerobatics. Stressed to +10, -6 G's. Very quick roll rate. Semi-symmetrical NACA 23012 airfoil for extra performance. Engine not included in kit. Several engine options. Lower performance Renegade II version also available at lower cost. See Ad on Page 55. **Completed & Flying:** 250

MURPHY AIRCRAFT MFG. LTD.
Unit 1-8155 Aitken Rd. • Chilliwack BC V2R 4H5 CANADA
Tel. 604-792-5855 • Fax 604-792-7006 • E-Mail: mursales@murphyair.com
Web: http://www.murphyair.com

Elite

Wing Span	30.33 ft.	**Range**	685 sm
Wing Area	152.0 sq. ft.	**Takeoff Dist.**	470 ft.
Length	22.25 ft.	**Landing Dist.**	400 ft.
Max Gross Wt.	1800 lbs.	**Vmax**	145 mph
Empty Wt.	980 lbs.	**VCr**	132 mph
Payload (full fuel)	820 lbs.	V_{S1} **(stall clean)**	46 mph
Fuel Capacity	44 gal.	V_{S0} **(lndg config)**	42 mph
Seats	3	**Climb Rate @msl**	1500 fpm
Cabin Width	44 in.	**Service Ceiling**	15,000 ft.

Std. Engine/HP	Usable HP Range	Construction Mtrl.	Est. Bldg. Time
Lyc. 0-360 / 180 hp	116-180	Metal, Fabric	1200 hrs.
Plans	**Kit**	**Info Packet**	**Video**
N/A	$14,500.00	$15.00	$20.00

The new 180 hp, tricycle gear, Elite is an evolution of the very popular Murphy Rebel. First shown at Sun 'n Fun '96. Load limits are +5.7, -3.8 G's. Use with floats, skis, and oversize tires. Estimated price is in US funds. See Ad on Page 55.

MURPHY AIRCRAFT MFG. LTD.
Unit 1-8155 Aitken Rd. • Chilliwack BC V2R 4H5 CANADA
Tel. 604-792-5855 • Fax 604-792-7006 • E-Mail: mursales@murphyair.com
Web: http://www.murphyair.com

Midget Mustang

Wing Span	18.60 ft.	**Range**	450 sm
Wing Area	68.00 sq. ft.	**Takeoff Dist.**	550 ft.
Length	16.50 ft.	**Landing Dist.**	680 ft.
Max Gross Wt.	900 lbs.	**Vmax**	205 mph
Empty Wt.	610 lbs.	**VCr**	180 mph
Payload (full fuel)	N/A	V_{S1} **(stall clean)**	62 mph
Fuel Capacity	15 gal.	V_{S0} **(lndg config)**	57 mph
Seats	1	**Climb Rate @msl**	1600 fpm
Cabin Width	20 in.	**Service Ceiling**	22,000 ft.

Std. Engine/HP	Usable HP Range	Construction Mtrl.	Est. Bldg. Time
Cont. 0-200 / 100 hp	85-150	Aluminum	950 hrs.
Plans	**Kit**	**Info Packet**	**Video**
$120.00	$5,950.00	$10.00	$16.00

Also known as "Long Midget." The Midget Mustang is well known for its air racing legacy during the late '40s and early '50s. Design limit is +6, -6 G's, 300 degree per second roll rate. Variety of engine options include Continental C-85 to the Lycoming 0-320. See Ad on Page 9. **Sold: 2,000 Flying:** 300+

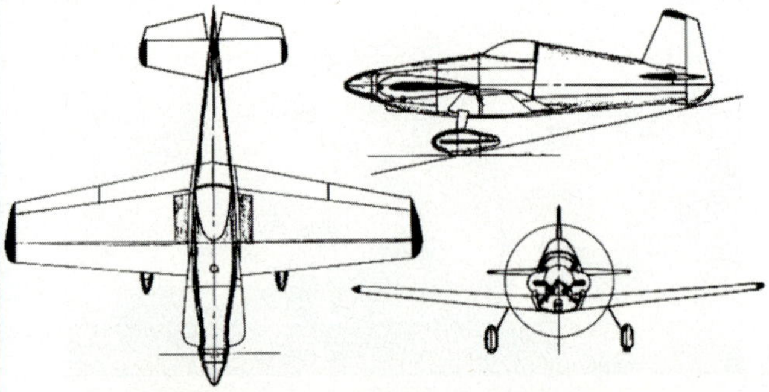

Stephen Jones' Midget Mustang

MUSTANG AERONAUTICS
1470 Temple City
Troy MI 48084
Tel. 248-589-9277 • Fax 248-588-6788
E-Mail: mustair@wwnet.com
Web: http://www.sportflyer.com\mustang

Mustang II

Wing Span	24.40 ft.	**Range**	800 sm
Wing Area	97.00 sq. ft.	**Takeoff Dist.**	450 ft.
Length	19.50 ft.	**Landing Dist.**	650 ft.
Max Gross Wt.	1600 lbs.	**Vmax**	220 mph
Empty Wt.	1000 lbs.	**VCr**	210 mph
Payload (full fuel)	N/A	**V_{S1} (stall clean)**	60 mph
Fuel Capacity	60 gal.	**V_{S0} (lndg config)**	57 mph
Seats	2	**Climb Rate @msl**	1900 fpm
Cabin Width	N/A	**Service Ceiling**	22,000 ft.

Std. Engine/HP	Usable HP Range	Construction Mtrl.	Est. Bldg. Time
Lyc 0-360 / 180 hp	135-200	Aluminum	1300 hrs.
Plans	**Kit**	**Info Packet**	**Video**
$195.00	$8,600.00	$10.00	$16.00

All-metal, high performance aircraft. Designed for both sport and cross country flying with side-by-side seating. 200+ mph cruising speed at altitude. Folding wing and tri-gear options. Engine and avionics are not included in the price. Use any 320 or 360 Series Lycoming. 1,600 plan sets sold around the world. See Ad on Page 9. **Completed & Flying:** 350+.

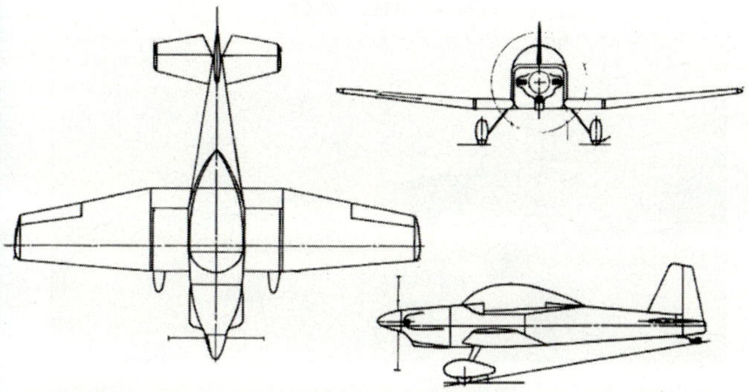

George Linkis' Mustang II

MUSTANG AERONAUTICS
1470 Temple City
Troy MI 48084
Tel. 248-589-9277 • Fax 248-588-6788
E-Mail: mustair@wwnet.com
Web: http://www.sportflyer.com\mustang

Spruce Coupe

Wing Span	22.00 ft.	**Range**	80 sm
Wing Area	88.00 sq. ft.	**Takeoff Dist.**	75 ft.
Length	16.33 ft.	**Landing Dist.**	100 ft.
Max Gross Wt.	600 lbs.	**Vmax**	75 mph
Empty Wt.	300 lbs.	**VCr**	60 mph
Payload (full fuel)	N/A	**V_{S1} (stall clean)**	35 mph
Fuel Capacity	6 gal.	**V_{S0} (lndg config)**	N/A
Seats	1	**Climb Rate @msl**	1200 fpm
Cabin Width	23 in.	**Service Ceiling**	N/A

Std. Engine/HP	Usable HP Range	Construction Mtrl.	Est. Bldg. Time
Zenoah / 45 hp	40 - 50	Wood, Fabric	500 hrs.
Plans	**Kit**	**Info Packet**	**Video**
$75.00	$6,995.00	$5.00	$15.00

Original design minimal aircraft that can be built and flow inexpensively. Features a spacious cockpit. Excellent STOL performance. Good visibility. Use spruce or pinewood and fir, birch, and mahogany plywood and Dacron fabric. Kit price includes engine and prop. **Flying:** 1

NWT CO.
RR1 Box 4730 • Charleston ME 04422
Tel. 207-285-7829 • E-Mail: flyco1@aol.com
Web: http://members.aol.com/flyco1/homepage.htm

Cassutt IIIM

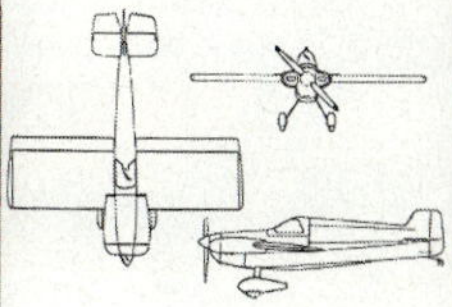

Wing Span	17.00 ft.	**Range**	490 sm
Wing Area	76.50 sq. ft.	**Takeoff Dist.**	1200 ft.
Length	16.00 ft.	**Landing Dist.**	800 ft.
Max Gross Wt.	800 lbs.	**Vmax**	200 mph
Empty Wt.	525 lbs.	**VCr**	180 mph
Payload (full fuel)	N/A	**V_{S1} (stall clean)**	68 mph
Fuel Capacity	14 gal.	**V_{S0} (lndg config)**	N/A
Seats	1	**Climb Rate @msl**	1500 fpm
Cabin Width	N/A	**Service Ceiling**	20,000 ft.

Std. Engine/HP	Usable HP Range	Construction Mtrl.	Est. Bldg. Time
Continental / 100 hp	N/A	Tube, Wood	1500 hrs.
Plans	**Kit**	**Info Packet**	**Video**
$125.00	$11,850.00	$5.00	N/A

Proven design, one place, full cantilever, high performance aircraft. Fully aerobatic and meets specs for competitive racing, "G" loading, safety characteristics, and visibility. Price is for the Racing Model with a 15' wing span. Sport Model, with 17' span, is available for $11,600. Prices do not include engine, instruments, skin, paint, or propeller. **Flying:** 225

NATIONAL AERONAUTICS CO.
5611 Kendall Court, Unit 4 • Arvada CO 80002
Tel. 303-940-8442 • Fax 303-940-8442

Tipsy Nipper Mk. III

Wing Span	19.67 ft.	**Range**	210 sm
Wing Area	80.70 sq. ft.	**Takeoff Dist.**	1110 ft.
Length	15.00 ft.	**Landing Dist.**	1500 ft.
Max Gross Wt.	750 lbs.	**Vmax**	107 mph
Empty Wt.	465 lbs.	**VCr**	93 mph
Payload (full fuel)	285 lbs.	**V_{S1} (stall clean)**	38 mph
Fuel Capacity	8 gal.	**V_{S0} (lndg config)**	38 mph
Seats	1	**Climb Rate @msl**	600 fpm
Cabin Width	18 in.	**Service Ceiling**	12,000 ft.

Std. Engine/HP	Usable HP Range	Construction Mtrl.	Est. Bldg. Time
VW / 60 hp	N/A	Steel Tube, Wood	N/A
Plans	**Kit**	**Info Packet**	**Video**
$105.00	$5,800.00	$14.00	N/A

A compact, single seat design with simple, effective systems. Primarily of wood construction. Aerobatic. Stressed to +6, -3 G's. Optional wing tip tanks increase fuel capacity to 14 imperial gallons. Note that prices are approximate in US funds.

Kits/Plans Sold: 100 **Completed & Flying:** 100

NIPPER KITS AND COMPONENTS
Foxley, Blackness Lane • Keston, Kent BR2 6HL
UNITED KINGDOM
Tel. 01689-858351 • Fax 01689-858351

Nordic VI

Wing Span	33.00 ft.	**Range**	400 sm
Wing Area	154.0 sq. ft.	**Takeoff Dist.**	200 ft.
Length	19.00 ft.	**Landing Dist.**	350 ft.
Max Gross Wt.	1058 lbs.	**Vmax**	100 mph
Empty Wt.	620 lbs.	**VCr**	95 mph
Payload (full fuel)	N/A	**V_{S1} (stall clean)**	38 mph
Fuel Capacity	12 gal.	**V_{S0} (lndg config)**	N/A
Seats	2	**Climb Rate @msl**	600 fpm
Cabin Width	N/A	**Service Ceiling**	10,000 ft.

Std. Engine/HP	Usable HP Range	Construction Mtrl.	Est. Bldg. Time
Subaru / 90 hp	60 - 100	Tube, Wood, Fabric	300 hrs.
Plans	**Kit**	**Info Packet**	**Video**
N/A	$13,900.00	$5.00	N/A

Taildragger, UL design. Options include Rotax 65/80 hp and Subaru 71/100 hp. Ready-to-fly a price is $27,700. Also available utilizing a Rotax 912 powerplant as the Nordic VI-912. The 912 increases Vmax to 130 and Vcr to 103.

NORMAN AVIATION INTERNATIONAL INC.
8130-1 Boul De La Rive Sud • CP #61032 • Levis, Que G6V 8X3
CANADA
Fax 418-833-7057 • E-Mail: norman.aviation@sympatico.ca
Tel. 418-833-4337 • Web: http://www3.sympatico.ca/norman.aviation

Nordic II

Wing Span	33.00 ft.	**Range**	350 sm
Wing Area	154.0 sq. ft.	**Takeoff Dist.**	150 ft.
Length	19.00 ft.	**Landing Dist.**	300 ft.
Max Gross Wt.	1058 lbs.	**Vmax**	100 mph
Empty Wt.	620 lbs.	**VCr**	85 mph
Payload (full fuel)	428 lbs.	**V_{S1} (stall clean)**	35 mph
Fuel Capacity	12 gal.	**V_{S0} (lndg config)**	N/A
Seats	2	**Climb Rate @msl**	600 fpm
Cabin Width	39 in.	**Service Ceiling**	10,000 ft.

Std. Engine/HP	Usable HP Range	Construction Mtrl.	Est. Bldg. Time
Subaru / 90 hp	65 - 100	Tube, Wood, Fabric	300 hrs.
Plans	**Kit**	**Info Packet**	**Video**
$500.00	$18,000.00	$5.00	N/A

Engine options include Rotax 65 to 80 hp and Subaru 71 to 100 hp. Complete ready-to-fly airplane price is $27,700 Prices are approximate in US funds.

NORMAN AVIATION INTERNATIONAL INC.
8130-1 Boul De La Rive Sud • CP #61032 • Levis, Que G6V 8X3
CANADA
Fax 418-833-7057 • E-Mail: norman.aviation@sympatico.ca
Tel. 418-833-4337 • Web: http://www3.sympatico.ca/norman.aviation

Nordic VII

Wing Span	33.80 ft.	**Range**	350 sm
Wing Area	135.0 sq. ft.	**Takeoff Dist.**	250 ft.
Length	20.17 ft.	**Landing Dist.**	350 ft.
Max Gross Wt.	1058 lbs.	**Vmax**	115 mph
Empty Wt.	625 lbs.	**VCr**	105 mph
Payload (full fuel)	433 lbs.	**V_{S1} (stall clean)**	40 mph
Fuel Capacity	12 gal.	**V_{S0} (lndg config)**	N/A
Seats	2	**Climb Rate @msl**	500 fpm
Cabin Width	41 in.	**Service Ceiling**	10,000 ft.

Std. Engine/HP	Usable HP Range	Construction Mtrl.	Est. Bldg. Time
Subaru / 100 hp	90 - 100	Tube, Wood, Fabric	300 hrs.
Plans	**Kit**	**Info Packet**	**Video**
N/A	$18,000.00	$5.00	N/A

Complete, ready-to-fly airplane price is $32,375. Prices are approximate in US funds.

NORMAN AVIATION INTERNATIONAL INC.
8130-1 Boul De La Rive Sud • CP #61032 • Levis, Que G6V 8X3
CANADA
Fax 418-833-7057 • E-Mail: norman.aviation@sympatico.ca
Tel. 418-833-4337 • Web: http://www3.sympatico.ca/norman.aviation

Karatoo J-6

Wing Span	33.00 ft.	Range	350 sm
Wing Area	169.0 sq. ft.	Takeoff Dist.	200 ft.
Length	19.00 ft.	Landing Dist.	350 ft.
Max Gross Wt.	1058 lbs.	Vmax	100 mph
Empty Wt.	630 lbs.	VCr	90 mph
Payload (full fuel)	N/A	V_{S1} (stall clean)	38 mph
Fuel Capacity	12 gal.	V_{S0} (Indg config)	N/A
Seats	2	Climb Rate @msl	650 fpm
Cabin Width	N/A	Service Ceiling	10,000 ft.

Std. Engine/HP	Usable HP Range	Construction Mtrl.	Est. Bldg. Time
Subaru / 90 hp	N/A	Tube, Wood, Fabric	300 hrs.
Plans	**Kit**	**Info Packet**	**Video**
N/A	$13,900.00	$5.00	N/A

Engine options include Rotax 65 to 80 hp and Subaru 71 to 100 hp. Ready-to-fly complete airplane price is $22,650. Prices are approximate in US funds.

NORMAN AVIATION INTERNATIONAL INC.
8130-1 Boul De La Rive Sud • CP #61032 • Levis, Que G6V 8X3
CANADA
Fax 418-833-7057 • E-Mail: norman.aviation@sympatico.ca
Tel. 418-833-4337 • Web: http://www3.sympatico.ca/norman.aviation

Magnum V8

Wing Span	36.00 ft.	Range	800 sm
Wing Area	189.0 sq. ft.	Takeoff Dist.	900 ft.
Length	25.00 ft.	Landing Dist.	900 ft.
Max Gross Wt.	3800 lbs.	Vmax	185 mph
Empty Wt.	1900 lbs.	VCr	140 mph
Payload (full fuel)	1900 lbs.	V_{S1} (stall clean)	76 mph
Fuel Capacity	110 gal.	V_{S0} (Indg config)	61 mph
Seats	2	Climb Rate @msl	1300 fpm
Cabin Width	50 in.	Service Ceiling	20,000 ft.

Std. Engine/HP	Usable HP Range	Construction Mtrl.	Est. Bldg. Time
Ford 351 HO / 385 hp	300 - 500	Aluminum, Steel	2000 hrs.
Plans	**Kit**	**Info Packet**	**Video**
$250.00	$8,000.00	$15.00	N/A

A bush type design with swing tail for loading cargo. Designed to 7G's at 3800 lbs. gross. Kit price is without engine.

Kits/Plans Sold: 45 **Completed & Flying:** 1

O'NEILL AIRPLANE COMPANY
791 Livingston
Carlyle IL 62231
Tel. 618-594-2681 • Fax 618-594-2681
E-Mail: troneill@accessus.net

Desert Eagle

Wing Span	33.00 ft.	Range	255 sm
Wing Area	165.0 sq. ft.	Takeoff Dist.	200 ft.
Length	21.50 ft.	Landing Dist.	150 ft.
Max Gross Wt.	1200 lbs.	Vmax	110 mph
Empty Wt.	495 lbs.	VCr	85 mph
Payload (full fuel)	600 lbs.	V_{S1} (stall clean)	38 mph
Fuel Capacity	10 gal.	V_{S0} (Indg config)	35 mph
Seats	2	Climb Rate @msl	600 fpm
Cabin Width	50 in.	Service Ceiling	16,000 ft.

Std. Engine/HP	Usable HP Range	Construction Mtrl.	Est. Bldg. Time
Motavia / 80 hp	65 - 90	Tube, Fabric	240 hrs.
Plans	**Kit**	**Info Packet**	**Video**
N/A	$14,900.00	$5.00	$20.00

Tough, two seat side-by-side, high wing pusher. Optional 20 and 55 gal.fuel tank configurations. Kit includes instruments, doors and vortex generators. Engine and prop not included. Desert Eagle/Ag-Wagon variation currently in development. See Ad on Page 54.

OLYMPIC ULTRALIGHTS
PO Box 388
Port Angeles WA 98362
Tel. 360-452-0117 • Fax 360-452-0117

Deland Travel Air 2000

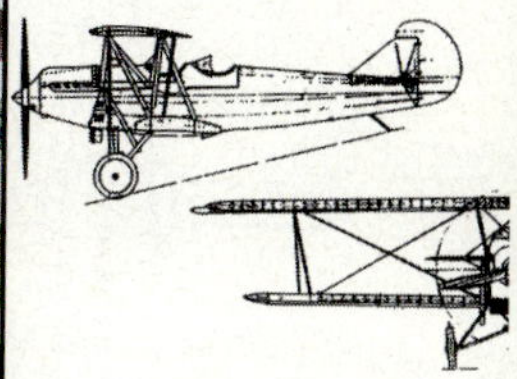

Wing Span	34.70 ft.	Range	500 sm
Wing Area	296.0 sq. ft.	Takeoff Dist.	200 ft.
Length	24.20 ft.	Landing Dist.	200 ft.
Max Gross Wt.	2590 lbs.	Vmax	120 mph
Empty Wt.	1450 lbs.	VCr	90 mph
Payload (full fuel)	1140 lbs.	V_{S1} (stall clean)	N/A
Fuel Capacity	30 gal.	V_{S0} (Indg config)	40 mph
Seats	3	Climb Rate @msl	1200 fpm
Cabin Width	N/A	Service Ceiling	18,000 ft.

Std. Engine/HP	Usable HP Range	Construction Mtrl.	Est. Bldg. Time
Thrust MK VI / 300 hp	N/A	Metal, Wood, Fabric	N/A
Plans	**Kit**	**Info Packet**	**Video**
Yes	$55,900.00	Yes	Yes

Vintage design. Called "Old Elephant Ears" because of the balanced ailerons on the upper wings. Available again with a modern, reliable powerplant and numerous improvements. **Completed & Flying:** 1

ORLANDO HELICOPTER AIRWAYS, INC.
1380 Flight Line Blvd. • Deland FL 32724
Tel. 904-943-8999 • Fax 904-943-9496

GP-4

Wing Span	24.80 ft.	**Range**	1200 sm
Wing Area	104.0 sq. ft.	**Takeoff Dist.**	600 ft.
Length	21.60 ft.	**Landing Dist.**	1200 ft.
Max Gross Wt.	2100 lbs.	**Vmax**	255 mph
Empty Wt.	1260 lbs.	**VCr**	240 mph
Payload (full fuel)	500 lbs.	**V_{S1} (stall clean)**	70 mph
Fuel Capacity	54 gal.	**V_{S0} (lndg config)**	62 mph
Seats	2	**Climb Rate @msl**	2200 fpm
Cabin Width	41 in.	**Service Ceiling**	N/A

Std. Engine/HP	Usable HP Range	Construction Mtrl.	Est. Bldg. Time
Lycoming / 200 hp	N/A	Wood	4000 hrs.
Plans	**Kit**	**Info Packet**	**Video**
$385.00	N/A	$15.00	N/A

A high performance cross country design with retractable landing gear. Design stress at full gross is +8, -6 G's.
Plans Sold: 360 **Flying:** 7

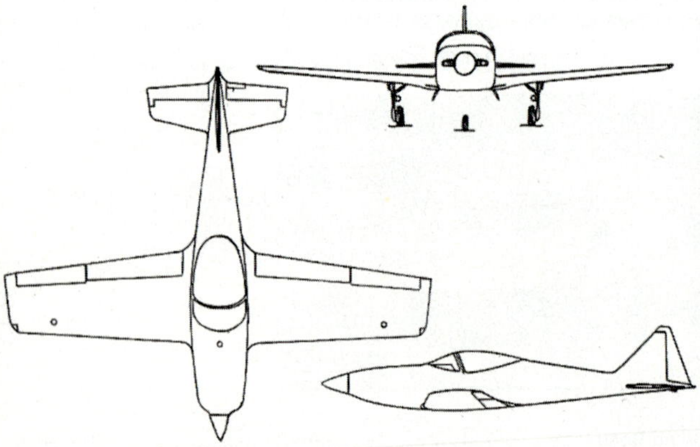

OSPREY AIRCRAFT
3741 El Ricon Way • Sacramento CA 95864
Tel. 916-483-3004 • E-Mail: gp-4@juno.com

Osprey 2

Wing Span	26.00 ft.	**Range**	N/A
Wing Area	130.0 sq. ft.	**Takeoff Dist.**	400 ft.
Length	21.00 ft.	**Landing Dist.**	700 ft.
Max Gross Wt.	1560 lbs.	**Vmax**	150 mph
Empty Wt.	960 lbs.	**VCr**	130 mph
Payload (full fuel)	420 lbs.	**V_{S1} (stall clean)**	65 mph
Fuel Capacity	32 gal.	**V_{S0} (lndg config)**	60 mph
Seats	2	**Climb Rate @msl**	1200 fpm
Cabin Width	43 in.	**Service Ceiling**	N/A

Std. Engine/HP	Usable HP Range	Construction Mtrl.	Est. Bldg. Time
Lycoming / 150 hp	N/A	Wood	2000 hrs.
Plans	**Kit**	**Info Packet**	**Video**
$210.00	N/A	$12.00	N/A

Proven amphibian design. Good water performance at gross weight. Takeoff distance in water is 550 ft. and landing distance in water is 800 ft. **Plans Sold:** 1480 **Flying:** 525

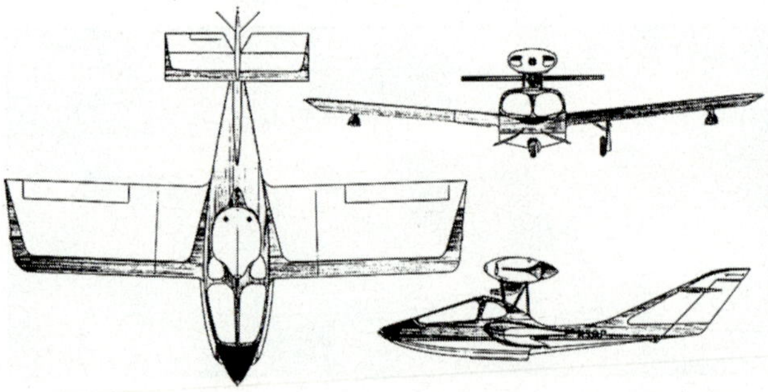

OSPREY AIRCRAFT
3741 El Ricon Way • Sacramento CA 95864
Tel. 916-483-3004 • E-Mail: gp-4@juno.com

Sky Arrow 1450L

Wing Span	31.49 ft.	**Range**	500 sm
Wing Area	144.5 sq. ft.	**Takeoff Dist.**	500 ft.
Length	24.93 ft.	**Landing Dist.**	250 ft.
Max Gross Wt.	1450 lbs.	**Vmax**	151 mph
Empty Wt.	770 lbs.	**VCr**	115 mph
Payload (full fuel)	566 lbs.	V_{S1} **(stall clean)**	46 mph
Fuel Capacity	18 gal.	V_{S0} **(lndg config)**	40 mph
Seats	2	**Climb Rate @msl**	850 fpm
Cabin Width	32 in.	**Service Ceiling**	18,000 ft.

Std. Engine/HP	Usable HP Range	Construction Mtrl.	Est. Bldg. Time
Rotax 912 / 80 hp	80 - 115	Composite	500 hrs.
Plans	**Kit**	**Info Packet**	**Video**
N/A	$37,000.00	$15.00	$20.00

Land version of the Sky Arrow. Has detachable wings and is designed to be car-towable without a trailer. It offers limited aerobatic capability. The airframe has been stress tested to +7.6, -3.8 G's. Engine and propeller included in price. The "Turbo" version, using Rotax 914 engines, is also available. Available as certified production aircraft. See Ad on Page 25.

Kits Sold: 180 **Completed & Flying:** 150

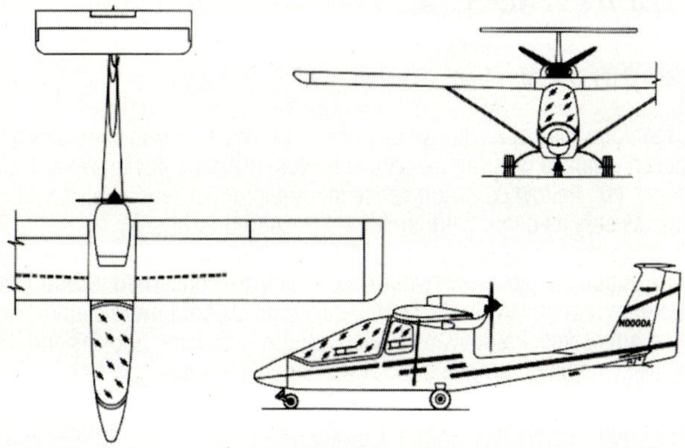

From 1977 until 1990, Pacific Aerosystems, Inc. (PAI), was involved in research, development and production of military products. In 1991, PAI changed its activities from military to commercial products, mainly devoted to the general aviation market.

The Sky Arrow is a 1,450 lb gross weight, high wing, two place (tandem configuration) aircraft with fixed landing gear capable of being converted to an amphibious configuration for land/sea operations or for water operations only, by installing a kit also available from the manfacturer. The high wing tandem configuration allows the pilot and passenger to have excellent spherical visibility of over 300 degrees. The airframe is manufactured from all composite materials consisting mainly of Kevlar and carbon sandwich construction.

PACIFIC AEROSYSTEM, INC.
5760 Chesapeake Court
San Diego CA 92123
Tel. 619-571-1441 • Fax 619-571-0803
E-Mail: pacaero@compuserve.com

Sky Arrow 1450A

Wing Span	31.49 ft.	**Range**	360 sm
Wing Area	144.5 sq. ft.	**Takeoff Dist.**	600 ft.
Length	24.93 ft.	**Landing Dist.**	250 ft.
Max Gross Wt.	1450 lbs.	**Vmax**	151 mph
Empty Wt.	920 lbs.	**VCr**	106 mph
Payload (full fuel)	430 lbs.	V_{S1} **(stall clean)**	47 mph
Fuel Capacity	19 gal.	V_{S0} **(lndg config)**	43 mph
Seats	2	**Climb Rate @msl**	600 fpm
Cabin Width	32 in.	**Service Ceiling**	13,000 ft.

Std. Engine/HP	Usable HP Range	Construction Mtrl.	Est. Bldg. Time
Rotax 912 / 80 hp	80 - 115	Composite	500 hrs.
Plans	**Kit**	**Info Packet**	**Video**
N/A	$52,000.00	$15.00	$20.00

Amphibious version of this composite design. Has detachable wings and is designed to be car-towable without a trailer. Limited aerobatic capability, airframe stress tested to +7.6, -3.8 G's. Engine and propeller included in price. "Turbo" version using Rotax 914 engine is also available. See Ad on Page 25.

Kits/Plans Sold: 21 **Completed & Flying:** 14

In Europe:
INIZIATIVE INDUSTRIALI ITALIANE
150 Corso Trieste
Rome 00198 ITALY
Ph: 6-8546341 • Fax: 6-8557162
Literature/Catalog: $35.00

PACIFIC AEROSYSTEM, INC.
5760 Chesapeake Court
San Diego CA 92123
Tel. 619-571-1441 • Fax 619-571-0803
E-Mail: pacaero@compuserve.com

Thunder Mustang

Wing Span	23.80 ft.	**Range**	1300 sm
Wing Area	104.0 sq. ft.	**Takeoff Dist.**	N/A
Length	24.20 ft.	**Landing Dist.**	N/A
Max Gross Wt.	3200 lbs.	**Vmax**	375 mph
Empty Wt.	2200 lbs.	**VCr**	345 mph
Payload (full fuel)	400 lbs.	**V_{S1} (stall clean)**	77 mph
Fuel Capacity	102 gal.	**V_{S0} (lndg config)**	68 mph
Seats	2	**Climb Rate @msl**	5200 fpm
Cabin Width	N/A	**Service Ceiling**	25,000 ft.

Std. Engine/HP	Usable HP Range	Construction Mtrl.	Est. Bldg. Time
Falconer V-12 / 640 hp	N/A	Composite	N/A
Plans	**Kit**	**Info Packet**	**Video**
N/A	$195,000.00	$10.00	$15.00

A unique blend of speed, maneuverability, and mystique, in a 3/4 scale P-51 developed by Dan Denney. With 640 hp, performance comparable to an original Mustang. Kits are complete including everything from the spinner, prop, and engine to the tailwheel.

The Thunder Mustang was developed using state of the art design and analysis methods currently in use in the aerospace industry. Computer technology played a major roll in the development of the airplane.

The thunder Mustang was scaled from the original P-51 drawings obtained on microfilm and then re-lofted using Peter Garrison's Loftsman program to recreate the various aerodynamic shapes. With accurate definition of the external shape of the aircraft through CAD (computer aided drafting), precise placement of all internal features such as bulkheads, seats, controls, engine placement, etc. was possible.

PAPA 51 CORPORATION
102 North King's Road
Nampa ID 83687
Tel. 208 466-5204 • Fax 208 466-1385

Viper

Wing Span	27.00 ft.	**Range**	585 sm
Wing Area	102.0 sq. ft.	**Takeoff Dist.**	300 ft.
Length	20.50 ft.	**Landing Dist.**	400 ft.
Max Gross Wt.	1050 lbs.	**Vmax**	125 mph
Empty Wt.	585 lbs.	**VCr**	110 mph
Payload (full fuel)	N/A	**V_{S1} (stall clean)**	38 mph
Fuel Capacity	17 gal.	**V_{S0} (lndg config)**	N/A
Seats	2	**Climb Rate @msl**	1500 fpm
Cabin Width	N/A	**Service Ceiling**	N/A

Std. Engine/HP	Usable HP Range	Construction Mtrl.	Est. Bldg. Time
Suzuki / 100 hp	65 - 110	Wood, Fabric	500 hrs.
Plans	**Kit**	**Info Packet**	**Video**
N/A	$15,000.00	$12.00	N/A

A wood and fabric taildragger design. Kit price includes engine.

Completed & Flying: 1

PAXMAN'S NORTHERN AIRCRAFT
PO Box 1155
Glenwood, AB T0K 2R0
CANADA
Tel. 403-626-3490 • Fax 403-626-3490

HOMEBUILT "FAQ'S" *(See Page 68)*

Q: What health hazards might be involved while building?

A: The most obvious hazards can be prevented with common shop safety practices, such as wearing protective lenses, handling power tools properly, etc. Follow common sense in shop practice, and you should be just as safe as if you were building household furniture.

A more subtle danger exists where chemicals are concerned. Composite structures require handling of chemical resins that are more exotic than simple adhesives. Paint systems also require extra care. Epoxies and Polyurethane finishes pose the worst problem.

Epoxies emit fumes that, while annoying, seem generally harmless at first. But after exposure, your body builds an allergic reaction to the substance. Once that threshold has been crossed, you will be "sensitized" for the rest of your life. The isocyanate content of polyurethane paint can trigger severe respiratory spasms once you become sensitized to them. Again, the reaction potential never goes away. Some paint and primer products have carcinogenic potential as well.

The solution is skin and respiratory protection, and good ventilation of the shop. *All* paints require at least filter respiration Isocyanate based paints require a fresh-air system as well. Wear protective gloves and eyewear. Above all, put an exhaust fan in your shop and use it, so as not to affect the whole household.

Before you open an unfamiliar substance, read "all" the supplier's warnings about protection. If you don't understand them, the supplier should be happy to explain the requirements. Whatever you do, don't ignore those precautions.

You may invest years and thousands of dollars in your airplane. Make sure you're still healthy enough to fly it when the time comes.

PL-2

Wing Span	28.50 ft.	**Range**	380 sm
Wing Area	116.0 sq. ft.	**Takeoff Dist.**	700 ft.
Length	19.30 ft.	**Landing Dist.**	600 ft.
Max Gross Wt.	1447 lbs.	**Vmax**	153 mph
Empty Wt.	902 lbs.	**VCr**	136 mph
Payload (full fuel)	380 lbs.	**V_{S1} (stall clean)**	60 mph
Fuel Capacity	25 gal.	**V_{S0} (lndg config)**	54 mph
Seats	2	**Climb Rate @msl**	1500 fpm
Cabin Width	40 in.	**Service Ceiling**	15,000 ft.

Std. Engine/HP	Usable HP Range	Construction Mtrl.	Est. Bldg. Time
Lyc. 0-320 / 150 hp	90 -150	Metal	3000 hrs.
Plans	**Kit**	**Info Packet**	**Video**
$400.00	N/A	$5.00	N/A

Side-by-side seating. Plans only. Stressed to +6 G's. Bubble canopy opens backward on ball bearing slides. Baggage compartment behind seats holds 40 lbs. This design has been used as a military trainer by several Latin American, African, and Asiatic nations including China. **Plans Sold:** 407 **Flying:** 200

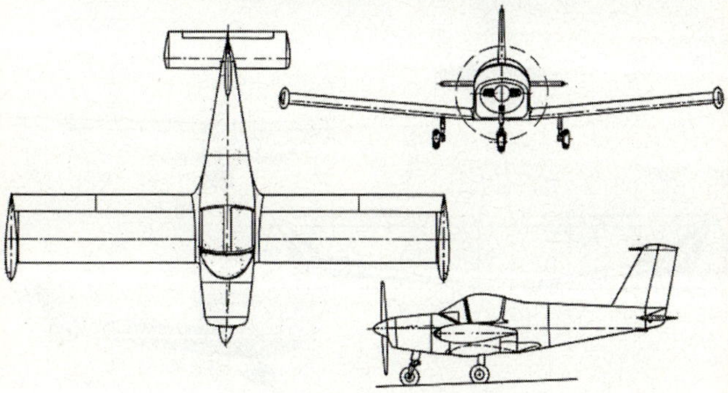

The PL-2 is an advanced version of the PL-1. The prototype was built in California by Mr. Harold Pio and was flown for the first time in April 1969. The external shape and flight characteristics are almost identical to the PL-1 equipped with the same engine. However, the construction has been simplified. Other improvements incorporated in the PL-2 include a wider cockpit, better canopy and fuselage lines, simplified fuselage construction, larger rudder, increased wing dihedral, and the capability of utilizing a larger variety of engines from 90 to 150 hp.

Trevor Parker, Aroha, New Zealand
"To be honest I was a bit apprehensive about [aluminum construction] but now I wouldn't build another aircraft any other way. The distortion caused by the heat treatment is minimal and is easily rectified afterwards by refitting the parts to the mold and reworking the flange to relieve any stress."

PAZMANY AIRCRAFT CORP.
P.O. Box 80051
San Diego CA 92138
Tel. 619-224-7330 • Fax 619-224-7358

PL-4A

Wing Span	26.67 ft.	**Range**	400 sm
Wing Area	89.00 sq. ft.	**Takeoff Dist.**	486 ft.
Length	16.42 ft.	**Landing Dist.**	436 ft.
Max Gross Wt.	850 lbs.	**Vmax**	120 mph
Empty Wt.	518 lbs.	**VCr**	97 mph
Payload (full fuel)	195 lbs.	**V_{S1} (stall clean)**	46 mph
Fuel Capacity	12 gal.	**V_{S0} (lndg config)**	46 mph
Seats	1	**Climb Rate @msl**	650 fpm
Cabin Width	28 in.	**Service Ceiling**	13,000 ft.

Std. Engine/HP	Usable HP Range	Construction Mtrl.	Est. Bldg. Time
VW / 60 hp	N/A	Metal	1500 hrs.
Plans	**Kit**	**Info Packet**	**Video**
$300.00	N/A	$5.00	N/A

Detachable wings and distinctive "T" tail. A rugged, time tested, design. Won "Outstanding New Design" at Oshkosh 1972. Fully enclosed cockpit. All-aluminum construction. Plans only.

Kits/Plans Sold: 728 **Completed & Flying:** 50

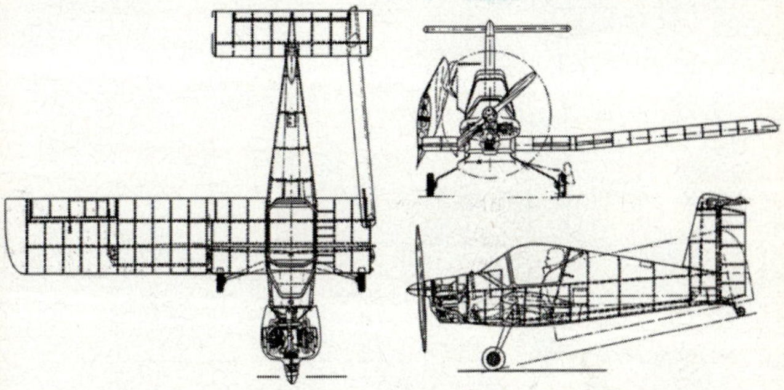

The PL-4A is intended to answer the need for a well designed, all-metal airplane which is easy to build, easy to fly, with low initial and operating costs, and folding wings so that it may be towed behind an automobile and stored at home.

It is VW-powered, with a closed or open cockpit large enough to accommodate very large persons, and a generous baggage compartment. A "T" tail configuration improves rudder effectiveness and prevents the stabilator from interfering with the folded wings.

Ed Nolan, Pilot/Builder, Bethlehem CT
"... I have "bumped" into several Internet pilots from New Zealand and Australia who have flown a PL-4A and to a man they say it is one of the most comfortable, pleasant, and safe planes they have ever flown. No vices, just great handling."

PAZMANY AIRCRAFT CORP.
P.O. Box 80051
San Diego CA 92138
Tel. 619-224-7330 • Fax 619-224-7358

Legend

Wing Span	28.50 ft.	**Range**	1500 sm
Wing Area	100.0 sq. ft.	**Takeoff Dist.**	800 ft.
Length	25.00 ft.	**Landing Dist.**	1000 ft.
Max Gross Wt.	3000 lbs.	**Vmax**	376 mph
Empty Wt.	2000 lbs.	**VCr**	362 mph
Payload (full fuel)	572 lbs.	**V_{S1} (stall clean)**	84 mph
Fuel Capacity	125 gal.	**V_{S0} (lndg config)**	72 mph
Seats	2	**Climb Rate @msl**	5500 fpm
Cabin Width	29 in.	**Service Ceiling**	25,000 ft.

Std. Engine/HP	Usable HP Range	Construction Mtrl.	Est. Bldg. Time
541 CID V-8 / 600 hp	400 - 1000	Composite/Carbon Fiber	2500 hrs.
Plans	**Kit**	**Info Packet**	**Video**
N/A	$79,850.00	$45.00	Yes

High performance, composite, two place aircraft. High speed, excellent maneuverability, the sound of the engine, and feel of the aircraft all evoke its fighter type lineage.

Kits/Plans Sold: 12 **Completed & Flying:** 1

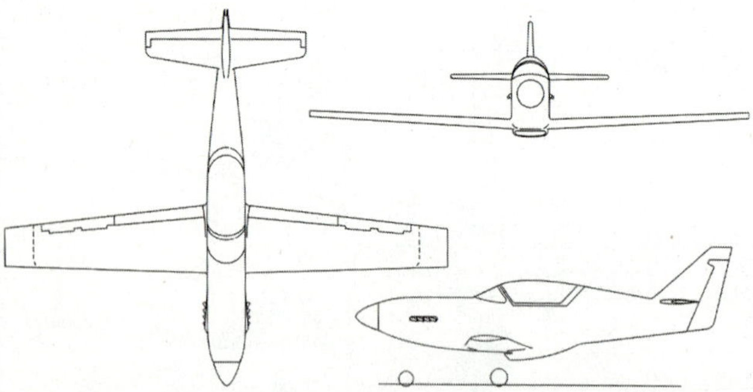

Bud Davisson, Sport Aviation Magazine
"I was impressed, very impressed, at how well it was put together both mechanically and aerodynamically."

PERFORMANCE AIRCRAFT
12901 West 151st St., Ste. C
Olathe KS 66062
Tel. 913-780-9140 • Fax 913-780-1774

JN-1

Wing Span	30.00 ft.	**Range**	110 sm
Wing Area	140.0 sq. ft.	**Takeoff Dist.**	250 ft.
Length	17.30 ft.	**Landing Dist.**	250 ft.
Max Gross Wt.	600 lbs.	**Vmax**	65 mph
Empty Wt.	320 lbs.	**VCr**	60 mph
Payload (full fuel)	N/A	**V_{S1} (stall clean)**	28 mph
Fuel Capacity	5 gal.	**V_{S0} (lndg config)**	N/A
Seats	1	**Climb Rate @msl**	600 fpm
Cabin Width	N/A	**Service Ceiling**	10,000 ft.

Std. Engine/HP	Usable HP Range	Construction Mtrl.	Est. Bldg. Time
Kaw. 440cc / 38 hp	N/A	Wood, Foam, Fiberglass	800 hrs.
Plans	**Kit**	**Info Packet**	**Video**
$68.00	N/A	$9.00	N/A

A plans only, detachable wing design. Originally designed by Jim Peris. Informational video currently in production. Build time is estimated at 500 to 800 hours.

Kits/Plans Sold: 625 **Completed & Flying:** 20

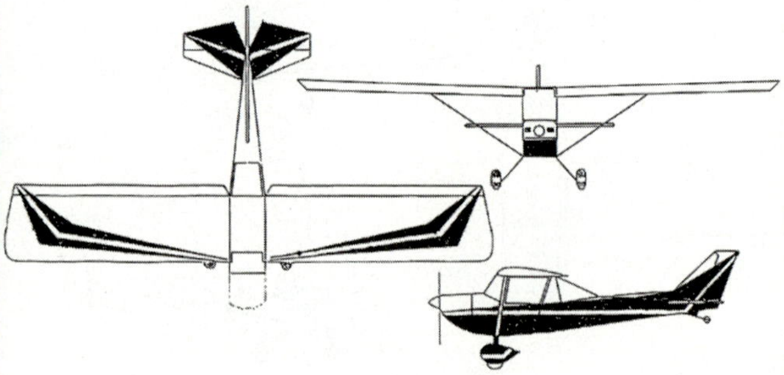

NANCY PERIS
149 S. Eastland Drive
Lancaster PA 17602
Tel. 717-393-5928

Phantom X1

Wing Span	28.50 ft.	**Range**	145 sm
Wing Area	142.0 sq. ft.	**Takeoff Dist.**	300 ft.
Length	16.90 ft.	**Landing Dist.**	450 ft.
Max Gross Wt.	510 lbs.	**Vmax**	61 mph
Empty Wt.	254 lbs.	**VCr**	55 mph
Payload (full fuel)	N/A	**V_{S1} (stall clean)**	26 mph
Fuel Capacity	5 gal.	**V_{S0} (lndg config)**	N/A
Seats	1	**Climb Rate @msl**	600 fpm
Cabin Width	N/A	**Service Ceiling**	14,500 ft.

Std. Engine/HP	Usable HP Range	Construction Mtrl.	Est. Bldg. Time
Rotax 447 / 40 hp	40 - 70	Aluminum Tube, Fabric	100 hrs.
Plans	**Kit**	**Info Packet**	**Video**
N/A	$8,998.00	$3.00	N/A

Amphibious version available. Options include brakes, chute, Rotax 503 and 582 engines. Excellent cross wind handling and good cross country performance. Rated to +9, -6 g's. Complete kit or in component groups. A two seat version, the Phantom X2 is also available for $12,995. **Completed & Flying:** 1000

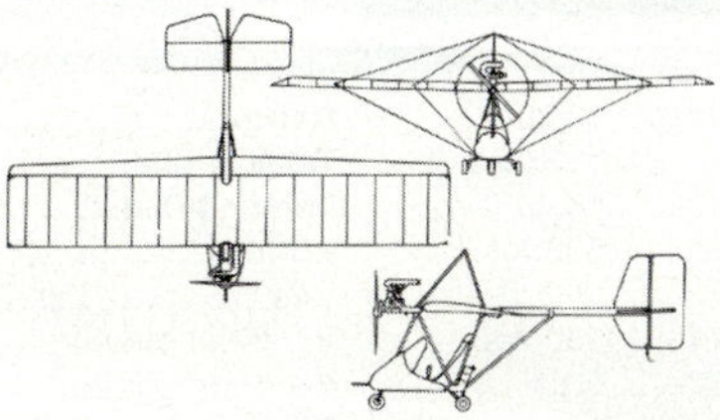

PHANTOM AIRCRAFT COMPANY
6154 West G Avenue • Kalamazoo MI 49009
Ph: 616-375-0505 • Fax: 616-375-0551
E-Mail: Phantom@complink.net

B4F Air Camper

Wing Span	29.00 ft.	**Range**	350 sm
Wing Area	145.0 sq. ft.	**Takeoff Dist.**	400 ft.
Length	17.67 ft.	**Landing Dist.**	400 ft.
Max Gross Wt.	1040 lbs.	**Vmax**	105 mph
Empty Wt.	620 lbs.	**VCr**	80 mph
Payload (full fuel)	N/A	**V_{S1} (stall clean)**	40 mph
Fuel Capacity	18 gal.	**V_{S0} (lndg config)**	N/A
Seats	2	**Climb Rate @msl**	500 fpm
Cabin Width	N/A	**Service Ceiling**	N/A

Std. Engine/HP	Usable HP Range	Construction Mtrl.	Est. Bldg. Time
Chev. Corvair / 60 hp	45 - 85	Wood, Fabric	1000 hrs.
Plans	**Kit**	**Info Packet**	**Video**
$75.00	N/A	$5.00	N/A

One of the Pietenpol family of aircraft evoking visions of the 1920's and 30's era. Engine options include "A" Ford, Continental, Lycoming, and Franklin from 40 to 85 hp.

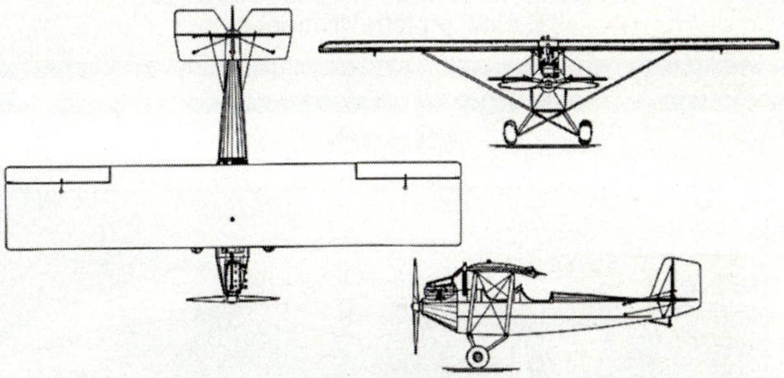

DON PIETENPOL
1604 Meadow Circle S.E. • Rochester MN 55904-5251
Tel. 507-289-2436

Sky Scout

DON PIETENPOL
1604 Meadow Circle S.E. • Rochester MN 55904-5251
Tel. 507-289-2436

Wing Span	27.00 ft.	**Range**	200 sm
Wing Area	135.0 sq. ft.	**Takeoff Dist.**	150 ft.
Length	16.25 ft.	**Landing Dist.**	250 ft.
Max Gross Wt.	750 lbs.	**Vmax**	62 mph
Empty Wt.	500 lbs.	**VCr**	55 mph
Payload (full fuel)	N/A	**V_{S1} (stall clean)**	35 mph
Fuel Capacity	6 gal.	**V_{S0} (lndg config)**	N/A
Seats	1	**Climb Rate @msl**	250 fpm
Cabin Width	N/A	**Service Ceiling**	N/A

Std. Engine/HP	Usable HP Range	Construction Mtrl.	Est. Bldg. Time
Ford T / 25 hp	25 - 65	Wood, Fabric, Steel	1000 hrs.
Plans	**Kit**	**Info Packet**	**Video**
$75.00	N/A	$5.00	N/A

A classic parasol, single seat Pietenpol aircraft. Motor options include the "A" Ford and others, 25 to 65 hp. Buckeye Pietenpol Association and International Pietenpol Association publications include much useful information from builders and flyers.

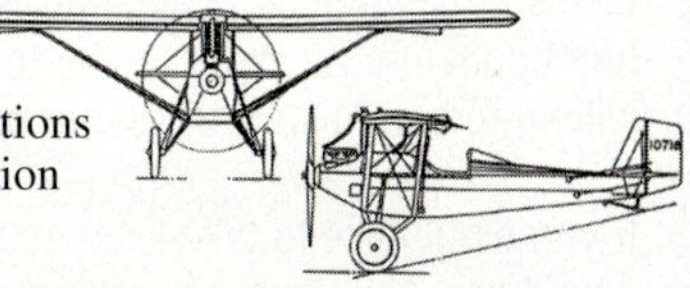

Spider

Wing Span	34.40 ft.	**Range**	310 sm
Wing Area	N/A	**Takeoff Dist.**	98 ft.
Length	N/A	**Landing Dist.**	98 ft.
Max Gross Wt.	990 lbs.	**Vmax**	111 mph
Empty Wt.	286 lbs.	**VCr**	75 mph
Payload (full fuel)	N/A	**V_{S1} (stall clean)**	34 mph
Fuel Capacity	11 gal.	**V_{S0} (Indg config)**	N/A
Seats	2	**Climb Rate @msl**	984 fpm
Cabin Width	N/A	**Service Ceiling**	22,000 ft.

Std. Engine/HP	Usable HP Range	Construction Mtrl.	Est. Bldg. Time
Rotax 503 / 53 hp	N/A	N/A	150 hrs.
Plans	**Kit**	**Info Packet**	**Video**
N/A	Yes	N/A	N/A

Trike available in eight different models utilizing a variety of Rotax engines.

Kits/Plans Sold: 250 **Completed & Flying:** 250

PIPISTREL
Strancarjeva 11 • Ajdovscina 5270
SLOVINIA
Tel. 00386 65 63 873 • Fax 00386 65 61 263
E-Mail: pipistrel@eunet.si

Zing

Wing Span	25.40 ft.	**Range**	100 sm
Wing Area	118.7 sq. ft.	**Takeoff Dist.**	160 ft.
Length	17.00 ft.	**Landing Dist.**	200 ft.
Max Gross Wt.	500 lbs.	**Vmax**	70 mph
Empty Wt.	249 lbs.	**VCr**	60 mph
Payload (full fuel)	N/A	**V_{S1} (stall clean)**	27 mph
Fuel Capacity	5 gal.	**V_{S0} (Indg config)**	N/A
Seats	1	**Climb Rate @msl**	650 fpm
Cabin Width	22 in.	**Service Ceiling**	5,000 ft.

Std. Engine/HP	Usable HP Range	Construction Mtrl.	Est. Bldg. Time
Kawasaki / 30 hp	28 - 50	Wood, Fabric	450 hrs.
Plans	**Kit**	**Info Packet**	**Video**
$60.00	$2,950.00	$5.00	$20.00

Liquid cooled Kawasaki powered Ultralight. A tough all-wood parasol. Spring aluminum one piece gear and prefab spindles are standard kit items. Sub and complete kits are available. Many are currently under construction. See Ad on Page 55.

POP'S PROPS
R.R. #1 Box 98
Cooksville IL 61730
Tel. 309-829-8343 • Fax 309-828-2031

Cloudster

POP'S PROPS
R.R. #1 Box 98
Cooksville IL 61730
Tel. 309-829-8343 • Fax 309-828-2031

Wing Span	30.00 ft.	**Range**	60 sm
Wing Area	138.0 sq. ft.	**Takeoff Dist.**	175 ft.
Length	16.40 ft.	**Landing Dist.**	200 ft.
Max Gross Wt.	500 lbs.	**Vmax**	60 mph
Empty Wt.	250 lbs.	**VCr**	50 mph
Payload (full fuel)	230 lbs.	**V_{S1} (stall clean)**	22 mph
Fuel Capacity	3 gal.	**V_{S0} (Indg config)**	N/A
Seats	1	**Climb Rate @msl**	700 fpm
Cabin Width	10 in.	**Service Ceiling**	5,000 ft.

Std. Engine/HP	Usable HP Range	Construction Mtrl.	Est. Bldg. Time
Kaw.340 WC / 35 hp	30 - 40	Wood, Fabric	400 hrs.
Plans	**Kit**	**Info Packet**	**Video**
$65.00	$2,750.00	$5.00	$20.00

Ultralight designed for safe, low and slow pasture flying. Easy, simple construction for the first-time builder. Estimate 600 hours construction time from plans.

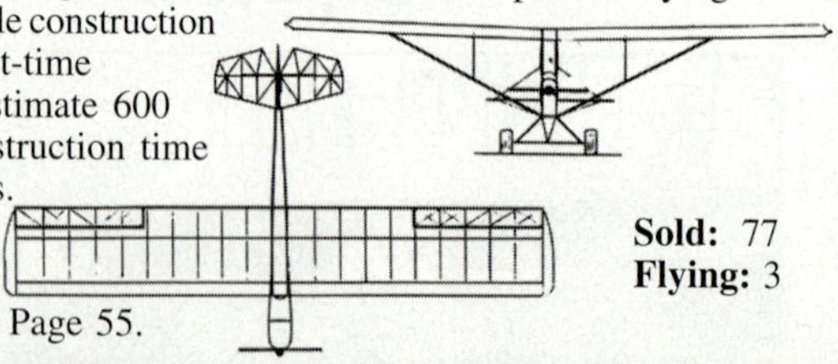

See Ad on Page 55.

Sold: 77
Flying: 3

HOMEBUILT "FAQ'S" *(See Page 68)*

Q: Can I take lessons/get my license in my homebuilt?

A. Like so many things in aviation, the answer is, "Yes, but..." There is no regulation to prevent your taking lessons or your flight test in a homebuilt aircraft. Your ability to do so will depend on finding an instructor willing to instruct in a homebuilt, as well as an Examiner willing to administer the test in the aircraft. Neither is automatic, and you'll have little recourse if they refuse.

If you would like to take lessons in your homebuilt, ask around at your local EAA chapter. Many EAA members are instructors, and would probably be more willing. Manyl chapters have CFIs who administer BFR's in members' homebuilts.

Pinocchio

Wing Span	25.80 ft.	**Range**	100 sm
Wing Area	115.8 sq. ft.	**Takeoff Dist.**	150 ft.
Length	16.50 ft.	**Landing Dist.**	225 ft.
Max Gross Wt.	500 lbs.	**Vmax**	70 mph
Empty Wt.	254 lbs.	**VCr**	55 mph
Payload (full fuel)	211 lbs.	**V_{S1} (stall clean)**	27 mph
Fuel Capacity	5 gal.	**V_{S0} (lndg config)**	N/A
Seats	1	**Climb Rate @msl**	750 fpm
Cabin Width	22 in.	**Service Ceiling**	5,000 ft.

Std. Engine/HP	Usable HP Range	Construction Mtrl.	Est. Bldg. Time
1/2 VW / 45 hp	40 - 50	Wood, Fabric	500 hrs.
Plans	**Kit**	**Info Packet**	**Video**
$120.00	$3,995.00	$5.00	$20.00

Offering antique appearance, nostalgia, and barnstorming with a romantic appeal. Affordable and simple with all-wood construction. The entire airframe requires no welding or fiberglass. See Ad on Page 55.

Kits/Plans Sold: 46 **Completed & Flying:** 3

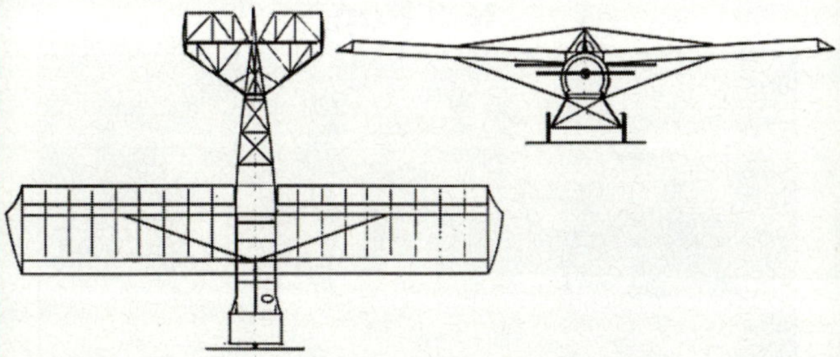

POP'S PROPS
R.R. #1 Box 98
Cooksville IL 61730
Tel. 309-829-8343 • Fax 309-828-2031

Isaacs Fury II

Wing Span	21.00 ft.	**Range**	300 sm
Wing Area	124.0 sq. ft.	**Takeoff Dist.**	450 ft.
Length	19.25 ft.	**Landing Dist.**	180 ft.
Max Gross Wt.	1000 lbs.	**Vmax**	115 mph
Empty Wt.	730 lbs.	**VCr**	100 mph
Payload (full fuel)	N/A	**V_{S1} (stall clean)**	45 mph
Fuel Capacity	18 gal.	**V_{S0} (lndg config)**	45 mph
Seats	1	**Climb Rate @msl**	1600 fpm
Cabin Width	24 in.	**Service Ceiling**	N/A

Std. Engine/HP	Usable HP Range	Construction Mtrl.	Est. Bldg. Time
Lyc 0-290 / 125 hp	65-125	Wood, Fabric	N/A
Plans	**Kit**	**Info Packet**	**Video**
$120.00	N/A	$10.00	N/A

70% scale version of the Hawker Fury used by the RAF in the early 1930's. First flown Sept. 1963. Aerobatic and stressed to +9, -4 G's (fully factored). 85% of plans were redrawn and improved during 1993-94. Plans incorporate alternative American hardware specifications. **Plans Sold:** 200 **Flying:** 24

The Fury II is a single seat sports biplane. Its external shape being based on a 7/10 scale version of the Hawker Fury which was used by the RAF in the early 1930's. The Fury was noted for its purity of line, exceptionally fine flying characteristics, and top speed of over 200 mph.

The original Isaacs Fury, G-ASCM, was first flown in 1963 fitted with a 65 hp Walter Mikron engine. Since then it has been developed to accept flat-four type engines of between 65 and 125 hp and stressed to +9, -4.5G's. The prototype was brought up to Mk II standards by the installation of a 125 hp Lycoming 0-290 in 1967. Amateur built Furies have flown successfully in the UK, Canada, New Zealand, and the US.

Capt. Mike Clark, UK
"The aircraft is absolutely beautiful in the air with superb handling."

Barry Thompson, Australia
"She flies like a dream and has a most incredible short-field performance."

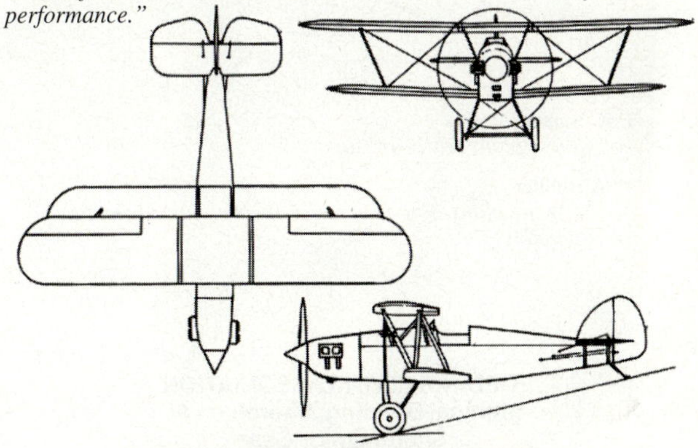

POPULAR FLYING ASSOCIATION
Terminal Building, Shoreham Ai • Shoreham-by-Sea
West Sussex BN43 5FF
UNITED KINGDOM
Tel. 1273-461616 • Fax 1273-463390

Isaacs Spitfire

Wing Span	22.10 ft.	Range	250 sm
Wing Area	87.00 sq. ft.	Takeoff Dist.	600 ft.
Length	19.30 ft.	Landing Dist.	450 ft.
Max Gross Wt.	1100 lbs.	Vmax	150 mph
Empty Wt.	805 lbs.	VCr	134 mph
Payload (full fuel)	N/A	V_{S1} (stall clean)	52 mph
Fuel Capacity	12 gal.	V_{S0} (lndg config)	47 mph
Seats	1	Climb Rate @msl	1100 fpm
Cabin Width	24 in.	Service Ceiling	15,000 ft.

Std. Engine/HP	Usable HP Range	Construction Mtrl.	Est. Bldg. Time
Cont 0-200 / 100 hp	N/A	Wood	N/A
Plans	**Kit**	**Info Packet**	**Video**
$230.00	N/A	$10.00	N/A

Designed as an affordable 60% scale aircraft using the 100 hp Continental O-200. Total construction costs approximately $4,000. First flew May 1975 and shown at Oshkosh in 1976. Aerobatic and stressed to +9, -4 G's.

Completed & Flying: 1

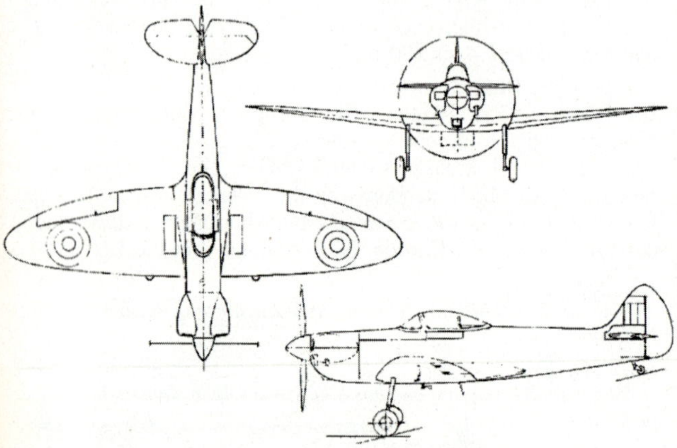

U.K. Builder
"...these plans really are superb."

USA Builder
"As to the plans, I've never seen their equal! The detail is astounding."

POPULAR FLYING ASSOCIATION
Terminal Building, Shoreham Ai
Shoreham-by-Sea
West Sussex BN43 5FF
UNITED KINGDOM
Tel. 1273-461616 • Fax 1273-463390

Stinger

Wing Span	30.50 ft.	Range	N/A
Wing Area	122.0 sq. ft.	Takeoff Dist.	100 ft.
Length	17.00 ft.	Landing Dist.	150 ft.
Max Gross Wt.	550 lbs.	Vmax	N/A
Empty Wt.	N/A	VCr	N/A
Payload (full fuel)	N/A	V_{S1} (stall clean)	N/A
Fuel Capacity	10 gal.	V_{S0} (lndg config)	N/A
Seats	1	Climb Rate @msl	1000 fpm
Cabin Width	N/A	Service Ceiling	N/A

Std. Engine/HP	Usable HP Range	Construction Mtrl.	Est. Bldg. Time
50 hp	N/A	N/A	N/A
Plans	**Kit**	**Info Packet**	**Video**
Yes	$8,995.00	N/A	Yes

Parasol modified basic N-3 Pup fuselage with open cockpit and larger 50 hp engine for higher performance. Kit price does not include engine, propeller or instruments.

Kits/Plans Sold: 3 **Completed & Flying:** 1

PRECEPTOR AIRCRAFT CORP.
4250 US 64/74 A Hwy
Rutherfordton NC 28139
Tel. 704-286-1926 • Fax 704-286-1936

N-3 Pup

Wing Span	30.50 ft.	Range	150 sm
Wing Area	122.0 sq. ft.	Takeoff Dist.	150 ft.
Length	16.50 ft.	Landing Dist.	100 ft.
Max Gross Wt.	535 lbs.	Vmax	N/A
Empty Wt.	254 lbs.	VCr	N/A
Payload (full fuel)	281 lbs.	V_{S1} (stall clean)	27 mph
Fuel Capacity	5 gal.	V_{S0} (lndg config)	N/A
Seats	1	Climb Rate @msl	600 fpm
Cabin Width	23 in.	Service Ceiling	10,000 ft.

Std. Engine/HP	Usable HP Range	Construction Mtrl.	Est. Bldg. Time
TEC 4-stroke / 40 hp	N/A	Tube, Fabric	300 hrs.
Plans	**Kit**	**Info Packet**	**Video**
$225.00	$8,500.00	$5.00	$15.00

Single place classic Cub design. 75% scale model of the Super Cub. Good choice for the new pilot. Easy to build and fly. Excellent overall performance. Kit price does not include engine, propeller, or instruments. **Sold:** 700 **Flying:** 500

PRECEPTOR AIRCRAFT CORP.
4250 US 64/74 A Hwy
Rutherfordton NC 28139
Tel. 704-286-1926 • Fax 704-286-1936

Super Pup

Wing Span	26.00 ft.	Range	300 sm
Wing Area	122.0 sq. ft.	Takeoff Dist.	150 ft.
Length	16.50 ft.	Landing Dist.	100 ft.
Max Gross Wt.	630 lbs.	Vmax	N/A
Empty Wt.	330 lbs.	VCr	N/A
Payload (full fuel)	300 lbs.	V_{S1} (stall clean)	N/A
Fuel Capacity	12 gal.	V_{S0} (lndg config)	N/A
Seats	1	Climb Rate @msl	800 fpm
Cabin Width	23 in.	Service Ceiling	N/A

Std. Engine/HP	Usable HP Range	Construction Mtrl.	Est. Bldg. Time
50 hp	N/A	Tube, Fabric	300 hrs.
Plans	**Kit**	**Info Packet**	**Video**
$250.00	$12,000.00	$5.00	$15.00

Higher powered version of the N-3 Pup. Excellent overall performance. Kit price does not include engine, propeller, or instruments. **Kits/Plans Sold:** 45 **Completed & Flying:** 25

PRECEPTOR AIRCRAFT CORP.
4250 US 64/74 A Hwy
Rutherfordton NC 28139
Tel. 704-286-1926 • Fax 704-286-1936

Ultra Pup

Wing Span	30.50 ft.	Range	350 sm
Wing Area	122.0 sq. ft.	Takeoff Dist.	250 ft.
Length	17.25 ft.	Landing Dist.	150 ft.
Max Gross Wt.	1000 lbs.	Vmax	N/A
Empty Wt.	450 lbs.	VCr	N/A
Payload (full fuel)	N/A	V_{S1} (stall clean)	N/A
Fuel Capacity	14 gal.	V_{S0} (lndg config)	N/A
Seats	2	Climb Rate @msl	1500 fpm
Cabin Width	25 in.	Service Ceiling	12,000 ft.

Std. Engine/HP	Usable HP Range	Construction Mtrl.	Est. Bldg. Time
60 hp	60-75	Tube, Fabric	450 hrs.
Plans	**Kit**	**Info Packet**	**Video**
$250.00	$13,500.00	$5.00	$15.00

Two place version of the N-3 Pup. Can be flown from front or rear seat. Design based on the J-3 Cub. Kit price does not include engine, propeller, or instruments.
Kits/Plans Sold: 75 **Completed & Flying:** 45

PRECEPTOR AIRCRAFT CORP.
4250 US 64/74 A Hwy
Rutherfordton NC 28139
Tel. 704-286-1926 • Fax 704-286-1936

SeaRey

Wing Span	30.83 ft.	Range	300 sm
Wing Area	157.0 sq. ft.	Takeoff Dist.	300 ft.
Length	22.40 ft.	Landing Dist.	350 ft.
Max Gross Wt.	1250 lbs.	Vmax	110 mph
Empty Wt.	750 lbs.	VCr	85 mph
Payload (full fuel)	N/A	V_{S1} (stall clean)	40 mph
Fuel Capacity	18 gal.	V_{S0} (lndg config)	N/A
Seats	2	Climb Rate @msl	750 fpm
Cabin Width	N/A	Service Ceiling	10,000 ft.

Std. Engine/HP	Usable HP Range	Construction Mtrl.	Est. Bldg. Time
Rotax 912 / 80 hp	65-100	Comp., Metal, Tube, Fabric	400 hrs.
Plans	**Kit**	**Info Packet**	**Video**
N/A	$19,000.00	$2.00	$15.00

Amphibian with large cabin and side-by-side seating. The hull is composed of lightweight reinforced fiberglass. Can also use Rotax 618, 74 hp engine. First flown November 1992. Offers stable handling even in gusty wind conditions. See Ad on Page 13.

Completed & Flying: 95

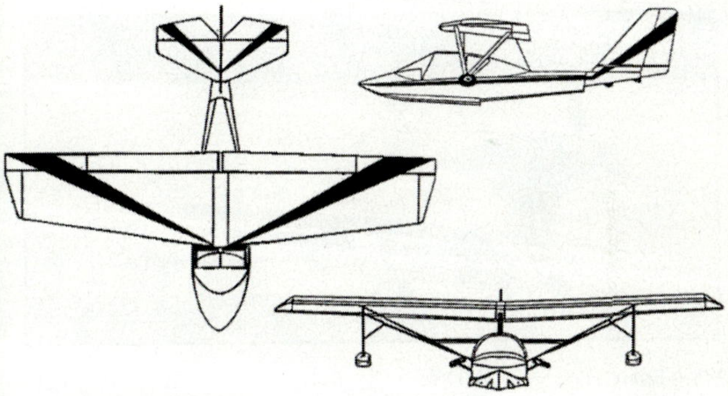

PROGRESSIVE AERODYNE, INC.
520 Clifton Street
Orlando FL 32808
Tel. 407-292-3700 • Fax 407-292-5555

Stingray

Wing Span	30.83 ft.	**Range**	100 sm
Wing Area	150.0 sq. ft.	**Takeoff Dist.**	190 ft.
Length	25.42 ft.	**Landing Dist.**	220 ft.
Max Gross Wt.	800 lbs.	**Vmax**	85 mph
Empty Wt.	475 lbs.	**VCr**	70 mph
Payload (full fuel)	N/A	**V_{S1} (stall clean)**	27 mph
Fuel Capacity	6 gal.	**V_{S0} (lndg config)**	N/A
Seats	1	**Climb Rate @msl**	600 fpm
Cabin Width	N/A	**Service Ceiling**	9,000 ft.

Std. Engine/HP	Usable HP Range	Construction Mtrl.	Est. Bldg. Time
Rotax 447 / 40 hp	40-65	Composite, Metal, Tube, Fabric	350 hrs.
Plans	**Kit**	**Info Packet**	**Video**
N/A	$14,000.00	$2.00	$15.00

Single place, high wing pusher amphibian. The hull is composed of lightweight reinforced fiberglass. Features stable handling in gusty wind conditions. See Ad on Page 13.

Completed & Flying: 5

PROGRESSIVE AERODYNE, INC.
520 Clifton Street
Orlando FL 32808
Tel. 407-292-3700 • Fax 407-292-5555

Jaguar

Wing Span	24.30 ft.	**Range**	1200 sm
Wing Area	104.0 sq. ft.	**Takeoff Dist.**	1100 ft.
Length	21.00 ft.	**Landing Dist.**	1200 ft.
Max Gross Wt.	2500 lbs.	**Vmax**	300 mph
Empty Wt.	1620 lbs.	**VCr**	250 mph
Payload (full fuel)	880 lbs.	**V_{S1} (stall clean)**	65 mph
Fuel Capacity	72 gal.	**V_{S0} (lndg config)**	65 mph
Seats	2	**Climb Rate @msl**	2000 fpm
Cabin Width	28 in.	**Service Ceiling**	N/A

Std. Engine/HP	Usable HP Range	Construction Mtrl.	Est. Bldg. Time
Rodeck V8 / 350 hp	N/A	Metal	2000 hrs.
Plans	**Kit**	**Info Packet**	**Video**
N/A	$78,545.00	$12.50	$24.95

Look and feel of a classic WWII era warbird but smaller, lighter, and less expensive to operate. Available in several sub-kits. Kit price includes engine and prop. Fully aerobatic with optional inverted fuel and oil package. **Sold:** 14 **Flying:** 2

PROWLER AVIATION, INC.
3777 A Meadow View Drive #800
Redding CA 96002
Tel. 916-365-4524 • Fax 916-365-4570

CORA Allegro

Wing Span	34.80 ft.	**Range**	310 sm
Wing Area	N/A	**Takeoff Dist.**	480 ft.
Length	20.00 ft.	**Landing Dist.**	450 ft.
Max Gross Wt.	990 lbs.	**Vmax**	98 mph
Empty Wt.	123 lbs.	**VCr**	80 mph
Payload (full fuel)	204 lbs.	**V_{S1} (stall clean)**	N/A
Fuel Capacity	13 gal.	**V_{S0} (lndg config)**	31 mph
Seats	2	**Climb Rate @msl**	590 fpm
Cabin Width	N/A	**Service Ceiling**	N/A

Std. Engine/HP	Usable HP Range	Construction Mtrl.	Est. Bldg. Time
Rotax 582 / 64 hp	49-64	Alum., Composite, Fabric	N/A
Plans	**Kit**	**Info Packet**	**Video**
Yes	Yes	Yes	N/A

Modern two seater with good aerodynamics, excellent flight qualities, and economic operation. Fuselage is fiberglass laminate. Wing and empenage are metal covered in part with fabric. Control stick is positioned between the seats. Flaps are actuated by hand or electrically as an option. CORA Legato version has longer (37.4 ft.) wingspan; 29 mph stall speed, and 690 fpm climb rate.

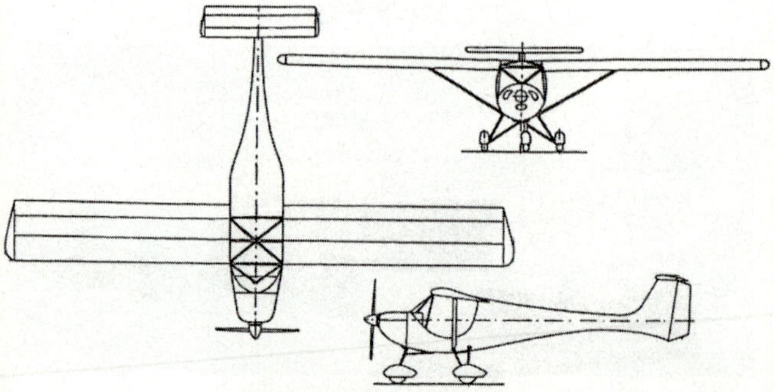

In North America contact:
A.K.S. INC.
PO Box 17161
Portland OR 97217
Ph: 503-289-1491
Fax: 503-247-9207
E-Mail:
trikes@teleport.com
See Ad on Pg. 47

PROXIMEX LTD.
Vinohradska 42 • 120 00 Praha 2
CZECH REPUBLIC
Tel. 00420-2-24253199 • Fax 00420-2-24247584
E-Mail: infipo@tradeinvest.cz

Challenger

Wing Span	31.50 ft.	Range	130 sm
Wing Area	142.0 sq. ft.	Takeoff Dist.	200 ft.
Length	18.50 ft.	Landing Dist.	200 ft.
Max Gross Wt.	500 lbs.	Vmax	63 mph
Empty Wt.	235 lbs.	VCr	55 mph
Payload (full fuel)	N/A	V_{S1} (stall clean)	25 mph
Fuel Capacity	5 gal.	V_{S0} (lndg config)	N/A
Seats	1	Climb Rate @msl	700 fpm
Cabin Width	N/A	Service Ceiling	12,500 ft.

Std. Engine/HP	Usable HP Range	Construction Mtrl.	Est. Bldg. Time
Rotax 277 / 28 hp	N/A	Aluminum Tube, Fabric	80 hrs.
Plans	**Kit**	**Info Packet**	**Video**
N/A	$6,246.00	$5.00	$10.00

Single seat, ultralight design with a 28 hp engine. Price does not include engine. Options include ten gallon fuel tank, Lexan doors, electric start, wheel pants and more.

QUAD CITY ULTRALIGHT AIRCRAFT
PO Box 370 • 3810 34th St. • Moline IL 61266-0370
Tel. 309-764-3515 • Fax 309-762-3920
E-Mail: daveatqcu@aol.com
Web: http://www.quadcitychallenger.com

Challenger Special

Wing Span	26.00 ft.	Range	200 sm
Wing Area	120.0 sq. ft.	Takeoff Dist.	150 ft.
Length	18.50 ft.	Landing Dist.	200 ft.
Max Gross Wt.	520 lbs.	Vmax	100 mph
Empty Wt.	260 lbs.	VCr	85 mph
Payload (full fuel)	N/A	V_{S1} (stall clean)	30 mph
Fuel Capacity	10 gal.	V_{S0} (lndg config)	N/A
Seats	1	Climb Rate @msl	1200 fpm
Cabin Width	N/A	Service Ceiling	12,000 ft.

Std. Engine/HP	Usable HP Range	Construction Mtrl.	Est. Bldg. Time
Rotax 447 / 42 hp	N/A	Aluminum Tube, Fabric	150 hrs.
Plans	**Kit**	**Info Packet**	**Video**
N/A	$7,029.00	$5.00	$10.00

"Clipped Wing" version of the Challenger. A single seat ultralight design. Good cross country cruiser. Price does not include engine. Several engine options. **Completed & Flying:** 70

QUAD CITY ULTRALIGHT AIRCRAFT
PO Box 370 • 3810 34th St. • Moline IL 61266-0370
Tel. 309-764-3515 • Fax 309-762-3920
E-Mail: daveatqcu@aol.com
Web: http://www.quadcitychallenger.com

Challenger II

Wing Span	31.50 ft.	Range	150 sm
Wing Area	173.0 sq. ft.	Takeoff Dist.	100 ft.
Length	20.00 ft.	Landing Dist.	200 ft.
Max Gross Wt.	820 lbs.	Vmax	90 mph
Empty Wt.	320 lbs.	VCr	75 mph
Payload (full fuel)	N/A	V_{S1} (stall clean)	22 mph
Fuel Capacity	5 gal.	V_{S0} (lndg config)	N/A
Seats	2	Climb Rate @msl	1200 fpm
Cabin Width	N/A	Service Ceiling	1,400 ft.

Std. Engine/HP	Usable HP Range	Construction Mtrl.	Est. Bldg. Time
Rotax 503 / 52 hp	N/A	Aluminum Tube, Fabric	100 hrs.
Plans	**Kit**	**Info Packet**	**Video**
N/A	$7,676.00	$5.00	$10.00

Superb, two seat trainer. Amphibious floats and ten gallon fuel tank available. Price does not include engine. Performance specs are for solo flight. **Completed & Flying:** 750

QUAD CITY ULTRALIGHT AIRCRAFT
PO Box 370 • 3810 34th St. • Moline IL 61266-0370
Tel. 309-764-3515 • Fax 309-762-3920
E-Mail: daveatqcu@aol.com
Web: http://www.quadcitychallenger.com

Challenger II Special

Wing Span	26.00 ft.	Range	170 sm
Wing Area	N/A	Takeoff Dist.	125 ft.
Length	20.00 ft.	Landing Dist.	200 ft.
Max Gross Wt.	850 lbs.	Vmax	100 mph
Empty Wt.	350 lbs.	VCr	85 mph
Payload (full fuel)	N/A	V_{S1} (stall clean)	30 mph
Fuel Capacity	10 gal.	V_{S0} (lndg config)	N/A
Seats	2	Climb Rate @msl	1200 fpm
Cabin Width	N/A	Service Ceiling	14,000 ft.

Std. Engine/HP	Usable HP Range	Construction Mtrl.	Est. Bldg. Time
Rotax 503 / 52 hp	N/A	Aluminum Tube, Fabric	200 hrs.
Plans	**Kit**	**Info Packet**	**Video**
N/A	$8,626.00	$5.00	$10.00

"Clipped Wing" version of the Challenger. Comes standard with Stits covering. Good cross-country cruiser. Price does not include engine. All performance specs are for solo flight.

QUAD CITY ULTRALIGHT AIRCRAFT
PO Box 370 • 3810 34th St. • Moline IL 61266-0370
Tel. 309-764-3515 • Fax 309-762-3920
E-Mail: daveatqcu@aol.com
Web: http://www.quadcitychallenger.com

GT

Wing Span	30.00 ft.	**Range**	98 sm
Wing Area	146.0 sq. ft.	**Takeoff Dist.**	80 ft.
Length	19.75 ft.	**Landing Dist.**	100 ft.
Max Gross Wt.	570 lbs.	**Vmax**	74 mph
Empty Wt.	306 lbs.	**VCr**	59 mph
Payload (full fuel)	234 lbs.	**V_{S1} (stall clean)**	29 mph
Fuel Capacity	5 gal.	**V_{S0} (lndg config)**	27 mph
Seats	1	**Climb Rate @msl**	940 fpm
Cabin Width	N/A	**Service Ceiling**	12,000 ft.

Std. Engine/HP	Usable HP Range	Construction Mtrl.	Est. Bldg. Time
Rotax 447 / 40 hp	40 - 50	Tube, Fabric	60 hrs.
Plans	**Kit**	**Info Packet**	**Video**
N/A	$11,360.00	$5.00	N/A

Single seat, tube and fabric, pusher design utilizing tricycle landing gear and a spring loaded, steerable nose wheel. Available with an open or fully enclosed cabin. Specs are with the optional enclosed cabin. Features include four position flaps, adjustable pilot seat and 15-minute setup and breakdown. Available with Rotax 447 or 503 engines.

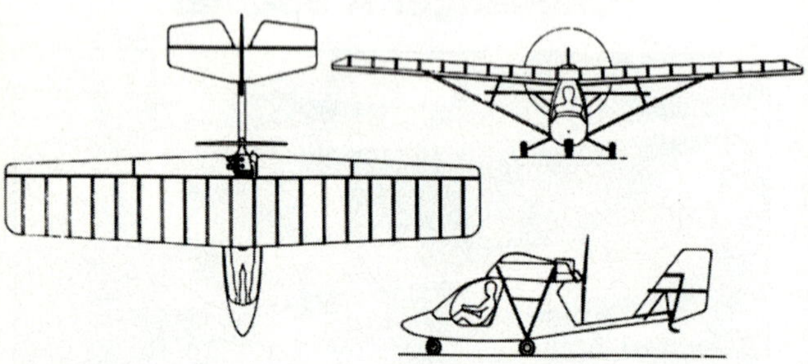

QUICKSILVER AIRCRAFT
27495 Diaz Rd.
Temecula CA 92590
Tel. 909-600-0818 • Fax 909-676-4883

GT 500

Wing Span	30.00 ft.	**Range**	133 sm
Wing Area	155.0 sq. ft.	**Takeoff Dist.**	220 ft.
Length	20.45 ft.	**Landing Dist.**	260 ft.
Max Gross Wt.	1000 lbs.	**Vmax**	103 mph
Empty Wt.	445 lbs.	**VCr**	72 mph
Payload (full fuel)	504 lbs.	**V_{S1} (stall clean)**	45 mph
Fuel Capacity	9 gal.	**V_{S0} (lndg config)**	39 mph
Seats	2	**Climb Rate @msl**	400 fpm
Cabin Width	N/A	**Service Ceiling**	10,000 ft.

Std. Engine/HP	Usable HP Range	Construction Mtrl.	Est. Bldg. Time
Rotax 582 / 62 hp	N/A	Tube, Fabric	190 hrs.
Plans	**Kit**	**Info Packet**	**Video**
N/A	$16,970.00	$5.00	N/A

A two place, tube and fabric pusher design with tricycle landing gear and a fully enclosed cabin. Removable doors. Standard features include a shock-mounted instrument panel, adjustable seats, steerable nose wheel with brake and a six month limited warranty. Available in kit form or fully assembled.

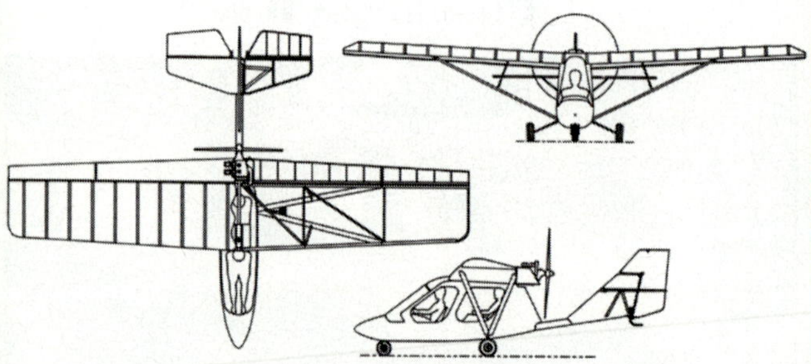

QUICKSILVER AIRCRAFT
27495 Diaz Rd.
Temecula CA 92590
Tel. 909-600-0818 • Fax 909-676-4883

MX II Sprint

Wing Span	32.75 ft.	**Range**	63 sm
Wing Area	180.0 sq. ft.	**Takeoff Dist.**	79 ft.
Length	17.50 ft.	**Landing Dist.**	75 ft.
Max Gross Wt.	720 lbs.	**Vmax**	75 mph
Empty Wt.	330 lbs.	**VCr**	50 mph
Payload (full fuel)	359 lbs.	**V_{S1} (stall clean)**	27 mph
Fuel Capacity	6 gal.	**V_{S0} (Indg config)**	N/A
Seats	2	**Climb Rate @msl**	1145 fpm
Cabin Width	N/A	**Service Ceiling**	10,000 ft.

Std. Engine/HP	Usable HP Range	Construction Mtrl.	Est. Bldg. Time
Rotax 582 / 64 hp	50 - 64	Tube, Fabric	60 hrs.
Plans	**Kit**	**Info Packet**	**Video**
N/A	$10,370.00	$5.00	N/A

A high wing pusher offering side-by-side seating and an open cockpit. Optional spray system available for agricultural applications. Can be configured for amphibious operation. Available with Rotax 503 or 582 engines.

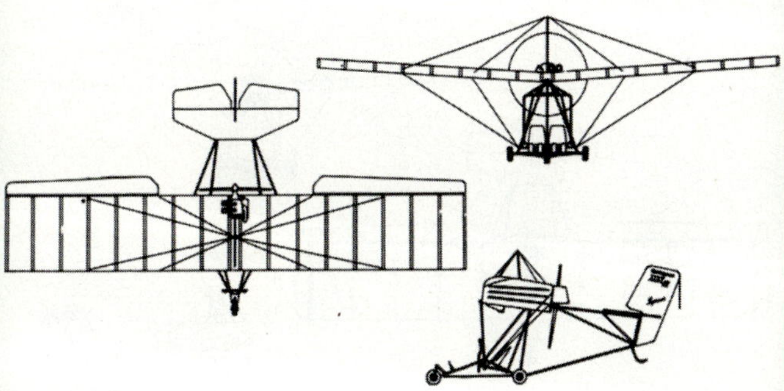

QUICKSILVER AIRCRAFT
27495 Diaz Rd.
Temecula CA 92590
Tel. 909-600-0818 • Fax 909-676-4883

MX Sport

Wing Span	28.00 ft.	**Range**	91 sm
Wing Area	156.0 sq. ft.	**Takeoff Dist.**	75 ft.
Length	18.08 ft.	**Landing Dist.**	70 ft.
Max Gross Wt.	525 lbs.	**Vmax**	74 mph
Empty Wt.	254 lbs.	**VCr**	49 mph
Payload (full fuel)	241 lbs.	**V_{S1} (stall clean)**	27 mph
Fuel Capacity	5 gal.	**V_{S0} (Indg config)**	N/A
Seats	1	**Climb Rate @msl**	850 fpm
Cabin Width	N/A	**Service Ceiling**	14,000 ft.

Std. Engine/HP	Usable HP Range	Construction Mtrl.	Est. Bldg. Time
Rotax 447 / 40 hp	N/A	Tube, Fabric	40 hrs.
Plans	**Kit**	**Info Packet**	**Video**
N/A	$9,995.00	$5.00	N/A

A higher performance version of the MX Sprint. Full-span ailerons and conventional three-axis controls. Very fast control response. Other features include a triangulated kingpost assembly and a tubular-braced tail.

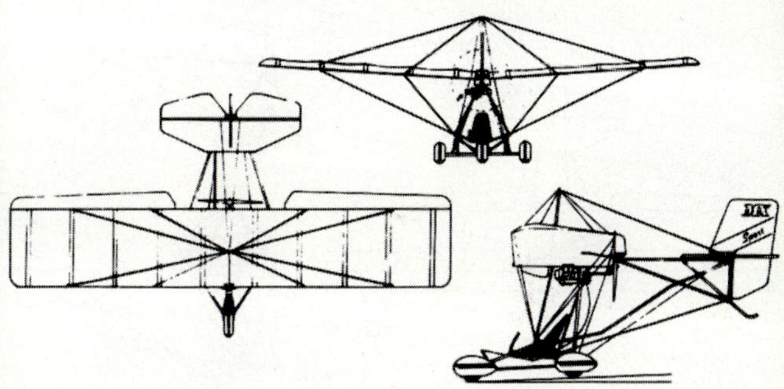

QUICKSILVER AIRCRAFT
27495 Diaz Rd.
Temecula CA 92590
Tel. 909-600-0818 • Fax 909-676-4883

Quicksilver Designs Distributed By:

SOUTH MISSISSIPPI LIGHT AIRCRAFT
1279 Marshall Smith Rd.
Lucedale MS 39452
Ph: 800-247-7652
Fax: 601-947-4959
See Ads On Pgs. 13 & 69

ULTRALIGHT AVIATION OF TEXAS
HCRS, Box 566F
Kerrville TX 78028
Ph: 210-896-2829
Fax: 210-896-7962

MX Sprint

Wing Span	28.00 ft.	**Range**	84 sm
Wing Area	156.0 sq. ft.	**Takeoff Dist.**	65 ft.
Length	18.08 ft.	**Landing Dist.**	60 ft.
Max Gross Wt.	525 lbs.	**Vmax**	65 mph
Empty Wt.	250 lbs.	**VCr**	47 mph
Payload (full fuel)	245 lbs.	**V_{S1} (stall clean)**	24 mph
Fuel Capacity	5 gal.	**V_{S0} (lndg config)**	N/A
Seats	1	**Climb Rate @msl**	900 fpm
Cabin Width	N/A	**Service Ceiling**	14,000 ft.

Std. Engine/HP	Usable HP Range	Construction Mtrl.	Est. Bldg. Time
Rotax 447 / 40 hp	N/A	Tube, Fabric	40 hrs.
Plans	**Kit**	**Info Packet**	**Video**
N/A	$9,695.00	$5.00	N/A

A high wing ultralight pusher utilizing a Rotax 447 engine. Features include conventional three-axis controls, flexible driveshaft coupler for reduced vibration, and a six month limited warranty.

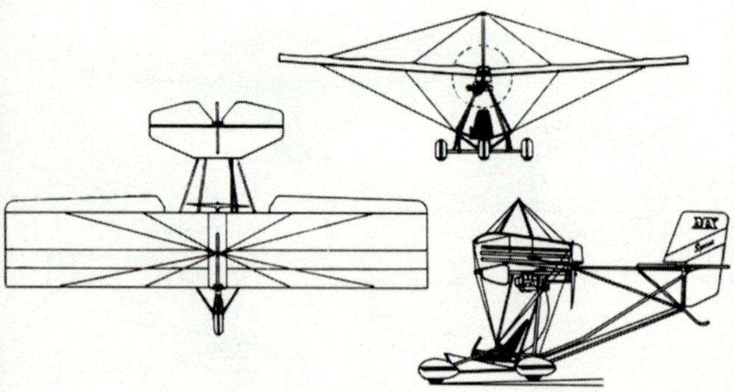

QUICKSILVER AIRCRAFT
27495 Diaz Rd.
Temecula CA 92590
Tel. 909-600-0818 • Fax 909-676-4883

MXL II Sport

Wing Span	32.75 ft.	**Range**	65 sm
Wing Area	180.0 sq. ft.	**Takeoff Dist.**	121 ft.
Length	17.50 ft.	**Landing Dist.**	75 ft.
Max Gross Wt.	720 lbs.	**Vmax**	75 mph
Empty Wt.	330 lbs.	**VCr**	52 mph
Payload (full fuel)	354 lbs.	**V_{S1} (stall clean)**	32 mph
Fuel Capacity	6 gal.	**V_{S0} (lndg config)**	N/A
Seats	2	**Climb Rate @msl**	1145 fpm
Cabin Width	N/A	**Service Ceiling**	10,000 ft.

Std. Engine/HP	Usable HP Range	Construction Mtrl.	Est. Bldg. Time
Rotax 582 / 64 hp	50 - 64	Tube, Fabric	60 hrs.
Plans	**Kit**	**Info Packet**	**Video**
N/A	$10,695.00	$5.00	N/A

A high wing pusher with side-by-side seating and tricycle landing gear. Optional spray system available for agricultural applications. Features include in-flight adjustable trim, conventional three-axis controls, steerable nose wheel and main wheel brakes. Available with Rotax 503 or 582 engines.

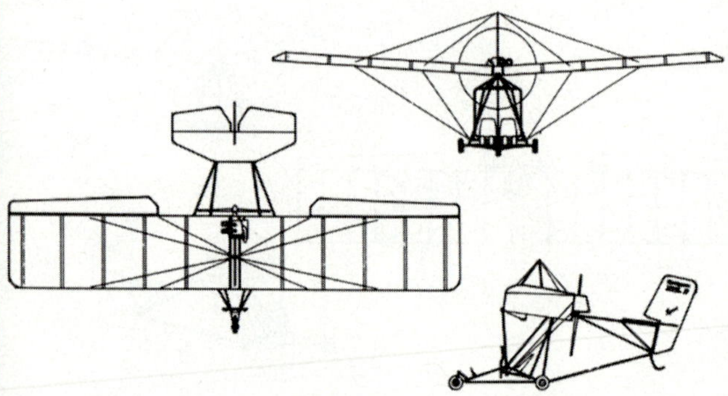

QUICKSILVER AIRCRAFT
27495 Diaz Rd.
Temecula CA 92590
Tel. 909-600-0818 • Fax 909-676-4883

See Ads On Pgs. 13 & 69

Glass Goose

Wing Span	27.00 ft.	**Range**	1100 sm
Wing Area	131.0 sq. ft.	**Takeoff Dist.**	900 ft.
Length	19.50 ft.	**Landing Dist.**	800 ft.
Max Gross Wt.	1800 lbs.	**Vmax**	160 mph
Empty Wt.	900 lbs.	**VCr**	140 mph
Payload (full fuel)	900 lbs.	**V_{S1} (stall clean)**	42 mph
Fuel Capacity	70 gal.	**V_{S0} (lndg config)**	N/A
Seats	2	**Climb Rate @msl**	1200 fpm
Cabin Width	42 in.	**Service Ceiling**	13,000 ft.

Std. Engine/HP	Usable HP Range	Construction Mtrl.	Est. Bldg. Time
Subaru/Lyc / 160 hp	150 - 185	Composite, Pre-Molded	900 hrs.
Plans	**Kit**	**Info Packet**	**Video**
$300.00	$25,000.00	$25.00	$25.00

Amphibian. Fast, long range, comfortable, responsive, cross country aircraft. Very stable. Large baggage area. All stainless steel weldments. Longest nonstop flight, Dallas to Oshkosh. Premolded parts make the Glass Goose a very fast build kit. Builder training included with kit. Engine, propeller, and instruments not included. Info pack includes a VHS Video. See Ad on Page 24. **Kits/Plans Sold:** 27

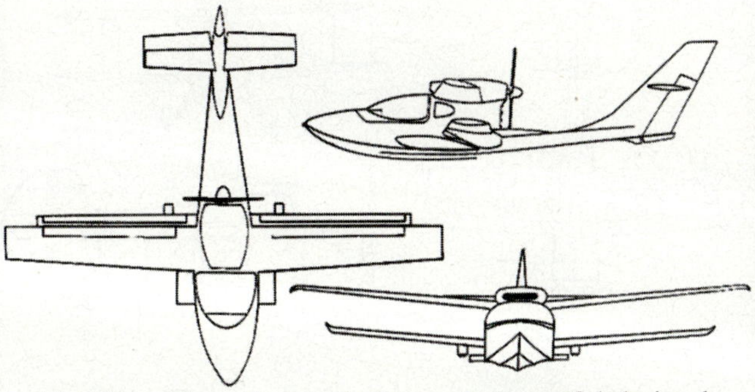

The Glass Goose is an exciting and successful design by Quikkit, Inc. Tom Scott purchased a Sea Hawker kit from AeroGare in October of 1984 and finished it, "by the plans", in March of 1986.

The performance and stability of that aircraft was not at all acceptable to Scott. Over a period of 5 years, he solicited the help of every source available to him in an effort to have an aircraft that would provide the qualities he wanted when he purchased that kit in 1984.

After an extensive redesign, the resulting Glass Goose included more wing area, more hull bottom area, and improved airflow around the pylon area plus numerous other improvements. Except for some visual similarity to the old Sea Hawker, there is no comparison between the two programs or products. The Glass Goose has been tested and proven under extreme conditions.

QUIKKIT CORPORATION
9002 Summer Glen
Dallas TX 75243-7445
Tel. 214-349-0462 • Fax 214-349-0462
E-Mail: quikkit@airmail.com
Web: http://webwrks.com/glassgoose/

Skylite

Wing Span	29.20 ft.	**Range**	120 sm
Wing Area	117.0 sq. ft.	**Takeoff Dist.**	350 ft.
Length	17.60 ft.	**Landing Dist.**	350 ft.
Max Gross Wt.	525 lbs.	**Vmax**	60 mph
Empty Wt.	245 lbs.	**VCr**	45 mph
Payload (full fuel)	250 lbs.	**V_{S1} (stall clean)**	27 mph
Fuel Capacity	5 gal.	**V_{S0} (lndg config)**	N/A
Seats	1	**Climb Rate @msl**	400 fpm
Cabin Width	24 in.	**Service Ceiling**	8,000 ft.

Std. Engine/HP	Usable HP Range	Construction Mtrl.	Est. Bldg. Time
Rotax 277 / 28 hp	25 - 48	Metal, Fabric	750 hrs.
Plans	**Kit**	**Info Packet**	**Video**
$75.00	N/A	$5.00	N/A

Plans only. Provides excellent visibility with large open sides and a parasol wing arrangement.The fuselage and rudder are welded 4130 steel tubing. The wings are aluminum tube and gusset construction. Various prefabricated component packages available to speed construction.

Kits/Plans Sold: 300 **Completed & Flying:** 26

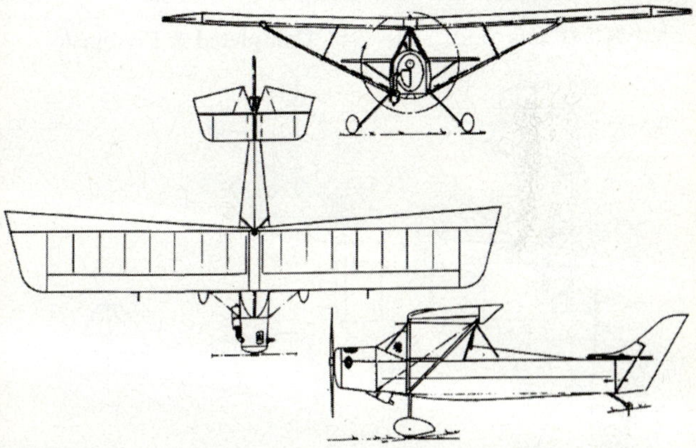

Ed Fisher, of Northeast Ohio, is an experienced aircraft designer and builder. His credentials include design of the Zippy Sport. In 1988, Ed and his wife Donna, dusted off some design sketches for a tube and rag ultralite aircraft with the classic parasol wing configuration. They wanted an inexpensive, open cockpit taildragger, easily constructed, that would appeal to the average handyman. After detail design, number crunching, mockups, static testing of components, along with a lot of help and hard work by friends, the Skylite was built and tested in time for Oshkosh, 1991, where it won the Grand Champion Ultralite award.

RACEAIR
21 Hemlock Dr.
Mayville NY 14757
Tel. 716-753-7110

Bearhawk

Wing Span	33.00 ft.	**Range**	800 sm
Wing Area	180.0 sq. ft.	**Takeoff Dist.**	450 ft.
Length	23.50 ft.	**Landing Dist.**	550 ft.
Max Gross Wt.	2500 lbs.	**Vmax**	148 mph
Empty Wt.	1150 lbs.	**VCr**	130 mph
Payload (full fuel)	N/A	**V_{S1} (stall clean)**	47 mph
Fuel Capacity	55 gal.	**V_{S0} (lndg config)**	40 mph
Seats	4	**Climb Rate @msl**	1400 fpm
Cabin Width	42 in.	**Service Ceiling**	18,000 ft.

Std. Engine/HP	Usable HP Range	Construction Mtrl.	Est. Bldg. Time
Lyc. O-360 / 170 hp	150 - 260	Alum., Tube, Fabric	2000 hrs.
Plans	**Kit**	**Info Packet**	**Video**
$230.00	N/A	$2.00	N/A

A four place, plans-only design with large flaps that provide excellent short field performance. The prototype, N6890R, is powered by a 170 hp Lycoming O-360 engine set up to burn auto fuel. The Bearhawk uses an all-metal wing with a fabric and steel tube fuselage. A Pitts style nosebowl, wheels, brakes, Cessna windshield, and modified lift struts are the only airframe components not built by the homebuilder. Plans price includes all current and back issues of the Newsletter

Kits/Plans Sold: 280 **Completed & Flying:** 1

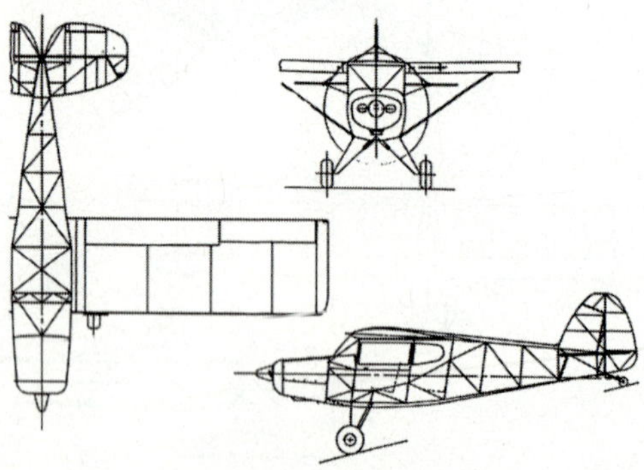

R & B AIRCRAFT CO.
7590 Breckinridge
Fincastle VA 24090
Tel. 540-473-3661

Special Bipe

Wing Span	18.00 ft.	**Range**	90 sm
Wing Area	120.0 sq. ft.	**Takeoff Dist.**	150 ft.
Length	14.60 ft.	**Landing Dist.**	250 ft.
Max Gross Wt.	525 lbs.	**Vmax**	63 mph
Empty Wt.	250 lbs.	**VCr**	60 mph
Payload (full fuel)	N/A	**V_{S1} (stall clean)**	26 mph
Fuel Capacity	5 gal.	**V_{S0} (lndg config)**	N/A
Seats	1	**Climb Rate @msl**	750 fpm
Cabin Width	N/A	**Service Ceiling**	10,000 ft.

Std. Engine/HP	Usable HP Range	Construction Mtrl.	Est. Bldg. Time
2SI / 45 hp	35 - 52	Wood, Fabric	350 hrs.
Plans	**Kit**	**Info Packet**	**Video**
$125.00	$2,500.00	$3.00	$15.00

Ultralight replica of the famous Pitts biplane. Detachable Wings.

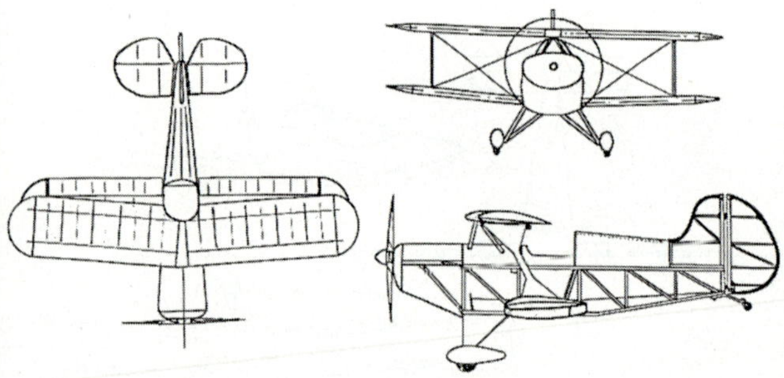

RAGWING AVIATION CO.
PO Box 39 • Tokeena Air Park • Townville SC 29689-0039
Tel. 864-225-3773 • Fax 864-972-5606

HOMEBUILT "FAQ'S" *(See Page 68)*

Q: I've made a tentative choice of an aircraft to build. Where can I see it and can I fly one?

A: The EAA Oshkosh fly-in in Wisconsin every summer is a great place to see virtually everything first hand, but it's not the best place to try for a ride because of traffic. If you do go to OSH, there will be builders' forums, dinners, and parties for type-specific gatherings. Make sure you take advantage of these. Same thing goes for the Sun 'n Fun fly-in in Florida in the spring.

Smaller fly-ins may be better for taking a ride **if** your favorite aircraft shows up. If your pet design is a little more obscure, it may take some effort.

Once you've expressed interest in a type (usually by buying the $10-$20 "info pack"), the seller may provide you with the addresses & phone numbers of customers who are willing to demo. This is not as unusual as one might think - builders often love to show off their toys!

For most popular kit aircraft, the companies usually keep a demonstrator around. You will probably have to go to them unless they are "touring" your area giving demos.

Midwing Sport

Wing Span	26.50 ft.	**Range**	160 sm
Wing Area	115.0 sq. ft.	**Takeoff Dist.**	125 ft.
Length	16.00 ft.	**Landing Dist.**	150 ft.
Max Gross Wt.	550 lbs.	**Vmax**	80 mph
Empty Wt.	235 lbs.	**VCr**	65 mph
Payload (full fuel)	270 lbs.	**V_{S1} (stall clean)**	26 mph
Fuel Capacity	5 gal.	**V_{S0} (lndg config)**	N/A
Seats	1	**Climb Rate @msl**	500 fpm
Cabin Width	24 in.	**Service Ceiling**	N/A

Std. Engine/HP	Usable HP Range	Construction Mtrl.	Est. Bldg. Time
2SI / 40 hp	N/A	Wood, Fabric	300 hrs.
Plans	**Kit**	**Info Packet**	**Video**
$52.00	N/A	$3.00	$15.00

Replica of the Church Midwing Sport Monoplane pylon racer. Bungee landing gear and optional chute. Wing gap allows seeing wheels or viewing ground the same as on the original '31 Model Church.

Kits/Plans Sold: 1000

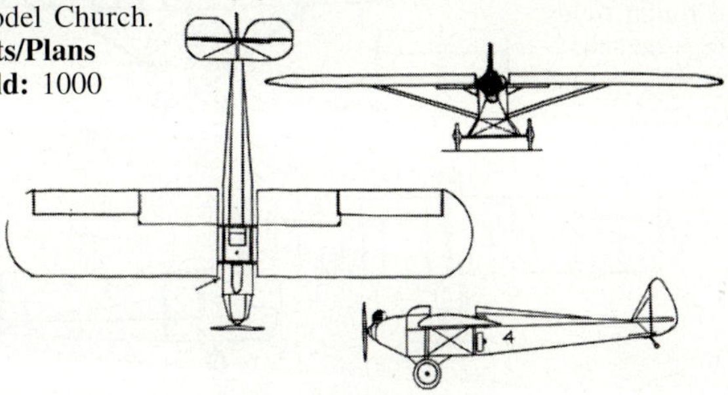

RAGWING AVIATION CO.
PO Box 39 • Tokeena Air Park • Townville SC 29689-0039
Tel. 864-225-3773 • Fax 864-972-5606

Motor Bipe

Wing Span	18.00 ft.	**Range**	125 sm
Wing Area	120.0 sq. ft.	**Takeoff Dist.**	125 ft.
Length	16.00 ft.	**Landing Dist.**	150 ft.
Max Gross Wt.	525 lbs.	**Vmax**	95 mph
Empty Wt.	225 lbs.	**VCr**	60 mph
Payload (full fuel)	270 lbs.	**V_{S1} (stall clean)**	26 mph
Fuel Capacity	5 gal.	**V_{S0} (lndg config)**	N/A
Seats	1	**Climb Rate @msl**	750 fpm
Cabin Width	N/A	**Service Ceiling**	N/A

Std. Engine/HP	Usable HP Range	Construction Mtrl.	Est. Bldg. Time
2SI / 35 hp	22-52	Wood, Fabric, Tube	175 hrs.
Plans	**Kit**	**Info Packet**	**Video**
$88.00	N/A	$5.00	$15.00

True "open air" flying with this ultralight. Easy construction and assembly.

Kits/Plans Sold: 1000

RAGWING AVIATION CO.
PO Box 39 • Tokeena Air Park • Townville SC 29689-0039
Tel. 864-225-3773 • Fax 864-972-5606

PT2S

RAGWING AVIATION CO.
PO Box 39 • Tokeena Air Park • Townville SC 29689-0039
Tel. 864-225-3773 • Fax 864-972-5606

Wing Span	30.00 ft.	**Range**	250 sm
Wing Area	147.5 sq. ft.	**Takeoff Dist.**	250 ft.
Length	17.00 ft.	**Landing Dist.**	400 ft.
Max Gross Wt.	900 lbs.	**Vmax**	105 mph
Empty Wt.	465 lbs.	**VCr**	75 mph
Payload (full fuel)	380 lbs.	**V_{S1} (stall clean)**	36 mph
Fuel Capacity	10 gal.	**V_{S0} (lndg config)**	N/A
Seats	2	**Climb Rate @msl**	600 fpm
Cabin Width	40 in.	**Service Ceiling**	10,000 ft.

Std. Engine/HP	Usable HP Range	Construction Mtrl.	Est. Bldg. Time
2SI / 70 hp	52-75	Wood, Fabric	500 hrs.
Plans	**Kit**	**Info Packet**	**Video**
$72.00	N/A	$3.00	N/A

Pilot trainer with dual side-by-side seating. Also available as single seat ALF version (Affordable Light Flyer). Tri-gear optional.

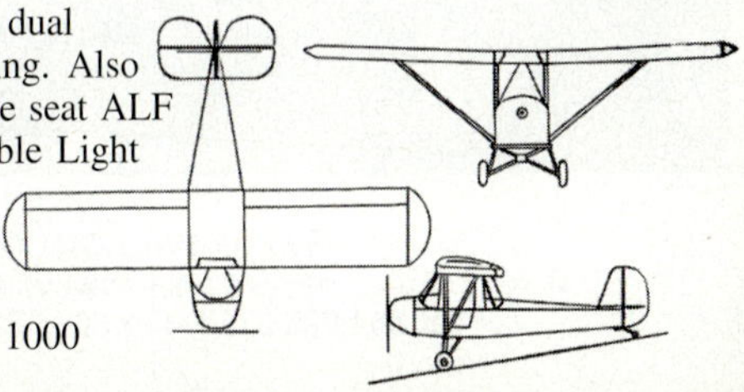

Plans Sold: 1000

RW5 Heath Replica

Wing Span	25.50 ft.	**Range**	165 sm
Wing Area	114.0 sq. ft.	**Takeoff Dist.**	125 ft.
Length	16.00 ft.	**Landing Dist.**	150 ft.
Max Gross Wt.	600 lbs.	**Vmax**	85 mph
Empty Wt.	240 lbs.	**VCr**	50 mph
Payload (full fuel)	280 lbs.	**V_{S1} (stall clean)**	26 mph
Fuel Capacity	5 gal.	**V_{S0} (lndg config)**	N/A
Seats	1	**Climb Rate @msl**	550 fpm
Cabin Width	24 in.	**Service Ceiling**	10,000 ft.

Std. Engine/HP	Usable HP Range	Construction Mtrl.	Est. Bldg. Time
2SI / 35 hp	20-50	Wood, Fabric	350 hrs.
Plans	**Kit**	**Info Packet**	**Video**
$52.00	N/A	$15.00	$15.00

Full scale ultralight replica of the famous Heath Parasol. Very low cost, long range. Pleasing classic lines. Spacious cockpit. Simple wooden construction. Short field capability. +4, -2 G's load factor. 14 sheet ply kits available for $225.

Plans Sold:
1000

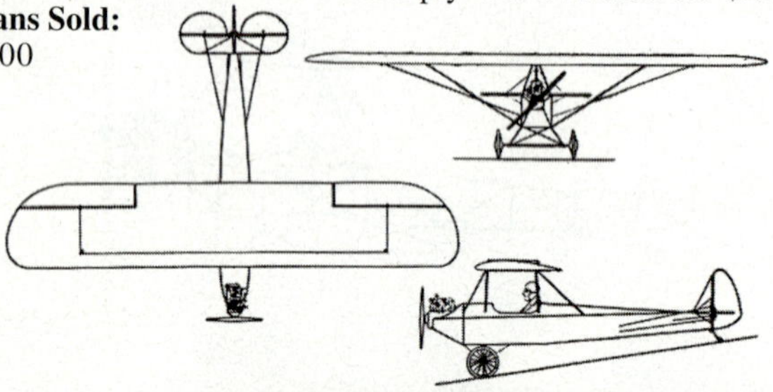

RAGWING AVIATION CO.
PO Box 39 • Tokeena Air Park • Townville SC 29689-0039
Tel. 864-225-3773 • Fax 864-972-5606

Rag-A-Muffin

Wing Span	25.50 ft.	**Range**	160 sm
Wing Area	114.0 sq. ft.	**Takeoff Dist.**	100 ft.
Length	16.50 ft.	**Landing Dist.**	150 ft.
Max Gross Wt.	600 lbs.	**Vmax**	85 mph
Empty Wt.	278 lbs.	**VCr**	65 mph
Payload (full fuel)	300 lbs.	**V_{S1} (stall clean)**	22 mph
Fuel Capacity	5 gal.	**V_{S0} (lndg config)**	N/A
Seats	1	**Climb Rate @msl**	750 fpm
Cabin Width	24 in.	**Service Ceiling**	10,000 ft.

Std. Engine/HP	Usable HP Range	Construction Mtrl.	Est. Bldg. Time
2SI / 45 hp	25 - 52	Wood, Fabric	350 hrs.
Plans	**Kit**	**Info Packet**	**Video**
$52.00	N/A	$15.00	$15.00

With large cargo areas for camping or cross country flying. OK for short and rough fields. Wings detach. Formerly named "Sport Parasol."

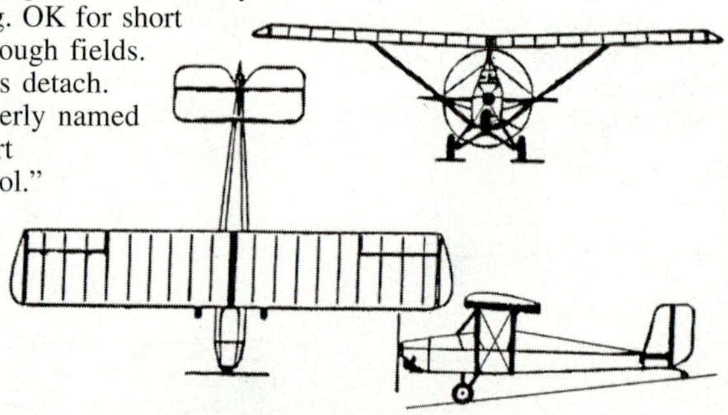

RAGWING AVIATION CO.
PO Box 39 • Tokeena Air Park • Townville SC 29689-0039
Tel. 864-225-3773 • Fax 864-972-5606

Ultra-Piet

RAGWING AVIATION CO.
PO Box 39 • Tokeena Air Park • Townville SC 29689-0039
Tel. 864-225-3773 • Fax 864-972-5606

Wing Span	26.50 ft.	**Range**	90 sm
Wing Area	114.0 sq. ft.	**Takeoff Dist.**	100 ft.
Length	15.00 ft.	**Landing Dist.**	150 ft.
Max Gross Wt.	525 lbs.	**Vmax**	62 mph
Empty Wt.	252 lbs.	**VCr**	55 mph
Payload (full fuel)	N/A	**V_{S1} (stall clean)**	26 mph
Fuel Capacity	5 gal.	**V_{S0} (lndg config)**	N/A
Seats	1	**Climb Rate @msl**	650 fpm
Cabin Width	N/A	**Service Ceiling**	10,000 ft.

Std. Engine/HP	Usable HP Range	Construction Mtrl.	Est. Bldg. Time
2SI / 35 hp	25 - 52	Wood, Fabric	350 hrs.
Plans	**Kit**	**Info Packet**	**Video**
$70.00	N/A	$3.00	$15.00

Ultralight replica of the famous Pietenpol Air Camper. 3/4 scale, folding wings, low cost. Will accept 1/2 VW engine 25 to 52 hp. Wings detach.

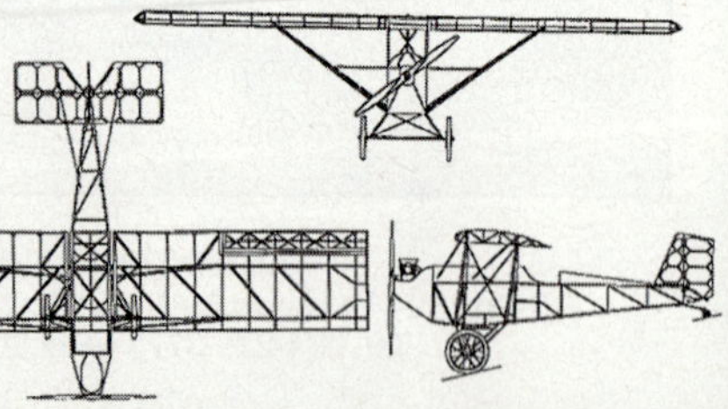

Ragabond

Wing Span	30.00 ft.	**Range**	250 sm
Wing Area	135.0 sq. ft.	**Takeoff Dist.**	300 ft.
Length	18.75 ft.	**Landing Dist.**	400 ft.
Max Gross Wt.	950 lbs.	**Vmax**	105 mph
Empty Wt.	450 lbs.	**VCr**	75 mph
Payload (full fuel)	500 lbs.	**V_{S1} (stall clean)**	36 mph
Fuel Capacity	20 gal.	**V_{S0} (lndg config)**	36 mph
Seats	2	**Climb Rate @msl**	650 fpm
Cabin Width	42 in.	**Service Ceiling**	10,000 ft.

Std. Engine/HP	Usable HP Range	Construction Mtrl.	Est. Bldg. Time
2SI / 70 hp	52 - 85	Wood, Fabric	500 hrs.
Plans	**Kit**	**Info Packet**	**Video**
$94.00	N/A	$5.00	$15.00

Vagabond replica. May also be built as a single seater to replicate the J-3 Cub. Drooped STOL wingtips are optional.

Kits/Plans Sold: 1000 **Completed & Flying:** 4

RAGWING AVIATION CO.
PO Box 39 • Tokeena Air Park • Townville SC 29689-0039
Tel. 864-225-3773 • Fax 864-972-5606

Graffiti

Wing Span	33.80 ft.	**Range**	200 sm
Wing Area	160.0 sq. ft.	**Takeoff Dist.**	100 ft.
Length	N/A	**Landing Dist.**	150 ft.
Max Gross Wt.	872 lbs.	**Vmax**	77 mph
Empty Wt.	341 lbs.	**VCr**	56 mph
Payload (full fuel)	N/A	**V_{S1} (stall clean)**	31 mph
Fuel Capacity	N/A	**V_{S0} (lndg config)**	N/A
Seats	2	**Climb Rate @msl**	N/A
Cabin Width	N/A	**Service Ceiling**	19,700 ft.

Std. Engine/HP	Usable HP Range	Construction Mtrl.	Est. Bldg. Time
Rotax 582 / 53 hp	41-64	N/A	N/A
Plans	**Kit**	**Info Packet**	**Video**
N/A	Yes	N/A	N/A

The result of an extensive design and development program aimed at optimizing safety, performance, and simplicity of construction. Available in a "Super Quiet" model.

RAISNER AIRCRAFT DEPOT, INC.
Meadowlake Airport • 8242 Cessna Dr.
Peyton CO 80831
Tel. 719-683-4357

Aerotrike SPIRIT

Wing Span	33.00 ft.	**Range**	N/A
Wing Area	160.0 sq. ft.	**Takeoff Dist.**	165 ft.
Length	N/A	**Landing Dist.**	130 ft.
Max Gross Wt.	880 lbs.	**Vmax**	80 mph
Empty Wt.	340 lbs.	**VCr**	60 mph
Payload (full fuel)	N/A	**V_{S1} (stall clean)**	35 mph
Fuel Capacity	14 gal.	**V_{S0} (lndg config)**	N/A
Seats	2	**Climb Rate @msl**	1000 fpm
Cabin Width	N/A	**Service Ceiling**	N/A

Std. Engine/HP	Usable HP Range	Construction Mtrl.	Est. Bldg. Time
Rotax 582 / 64 hp	50-80	Tube, Fabric	40 hrs.
Plans	**Kit**	**Info Packet**	**Video**
N/A	$11,495.00	Yes	N/A

Designed, developed, and flown by a trike world champion. Easy to assemble kit (approximately 30-40 hours) with all tubing cut, drilled, anodized or powder coated, with complete assembly manual and drawings. Kit is complete, including engine and prop and freight to U.S. (FOB Indiana).

U.S. Distributor:
ROLLISON AIRPLANE COMPANY, INC.
R#2, Box 186
Bloomfield IN 47424
Ph: 812-384-4972 • Fax: 812-384-0518
E-Mail: rollison@viaduct.custom.net

For four and a half months in 1995, Mike Blyth flew his standard Rotax powered AEROTRIKE from Johannesburg, South Africa all the way up through Africa, across the Mediterranean Sea, through Europe to Hammerfest, and around the North Cape at the northernmost tip of Europe. The trip was 14,375 miles long and is by far the longest unsupported trip ever taken in a trike-type microlight.

RAINBOW AIRCRAFT(PTY) LTD.
Private Bag X10021 • 1610 Edenvale
SOUTH AFRICA
Tel. 27 11 817 2298 • Fax 27 11 817 2297
E-Mail: flying@iafrica.com

KR-1

Wing Span	17.00 ft.	**Range**	1400 sm
Wing Area	62.00 sq. ft.	**Takeoff Dist.**	350 ft.
Length	12.75 ft.	**Landing Dist.**	850 ft.
Max Gross Wt.	750 lbs.	**Vmax**	200 mph
Empty Wt.	375 lbs.	**VCr**	180 mph
Payload (full fuel)	N/A	**V_{S1} (stall clean)**	52 mph
Fuel Capacity	35 gal.	**V_{S0} (lndg config)**	N/A
Seats	1	**Climb Rate @msl**	1200 fpm
Cabin Width	N/A	**Service Ceiling**	15,000 ft.

Std. Engine/HP	Usable HP Range	Construction Mtrl.	Est. Bldg. Time
VW / 80 hp	60-80	Composite, Wood	1000 hrs.
Plans	**Kit**	**Info Packet**	**Video**
$60.00	$5,500.00	$8.00	$24.95

Single seat, retractable gear design. Powered by the smaller VW engines up to the VW 2100. The KR-1B version is the standard KR-1 with modified outer wing sections. With the wing conversion, the original KR-1 becomes an excellent self powered glider. Retractable taildragger or tricycle gear options. See Ad on Page 22. **Completed & Flying:** 700

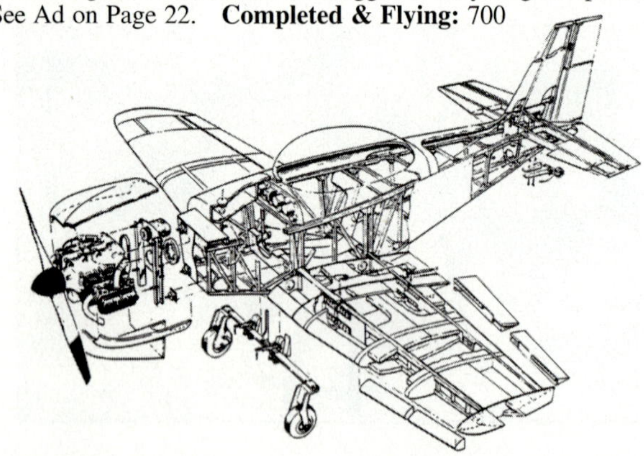

Kent Rand's design intent for the KR Series was to provide the cleanest airplanes and the best performance possible with the available, reliable and affordable Volkswagen engine. The cost and construction time for achieving this goal is minimized by using a unique construction system which combines wood, polyurethane foam, glass cloth, and epoxy resin. The result is structurally strong, clean, hard surfaced and exceptionally fast airplanes.

Kent pioneered this construction system, now 'state of the art' for homebuilts, in 1972 after designing his KR1. The KR1 is the single seat, low wing, retractable which is powered by smaller VW engines up to 2100.

The KR2 is a high speed, two place, side-by-side aircraft. As with all the KR's, the KR2 is equipped with removable wings and a steerable tail wheel.

RAND ROBINSON ENGINEERING
15641 Product Lane, A-5
Huntington Beach CA 92649
Tel. 714-898-3811 • Fax 714-890-1658
E-Mail: pilot@beegroup.com
Web: http://www.fly-kr.com

KR-2

Wing Span	20.66 ft.	**Range**	1600 sm
Wing Area	80.00 sq. ft.	**Takeoff Dist.**	350 ft.
Length	14.50 ft.	**Landing Dist.**	900 ft.
Max Gross Wt.	900 lbs.	**Vmax**	200 mph
Empty Wt.	480 lbs.	**VCr**	180 mph
Payload (full fuel)	N/A	**V_{S1} (stall clean)**	52 mph
Fuel Capacity	35 gal.	**V_{S0} (lndg config)**	N/A
Seats	2	**Climb Rate @msl**	1200 fpm
Cabin Width	N/A	**Service Ceiling**	15,000 ft.

Std. Engine/HP	Usable HP Range	Construction Mtrl.	Est. Bldg. Time
VW 2100 / 80 hp	76-80	Wood, Fiberglass	800 hrs.
Plans	**Kit**	**Info Packet**	**Video**
$165.00	$6,200.00	$8.00	$24.95

A well-proven, two seat, side-by-side design. Premolded parts - no machining or welding is required in building the airframe. Removable wings. Taildragger or tricycle gear options. Can use VW engines from 1834 to 2200 cc. See Ad on Page 22.

Completed & Flying: 1200

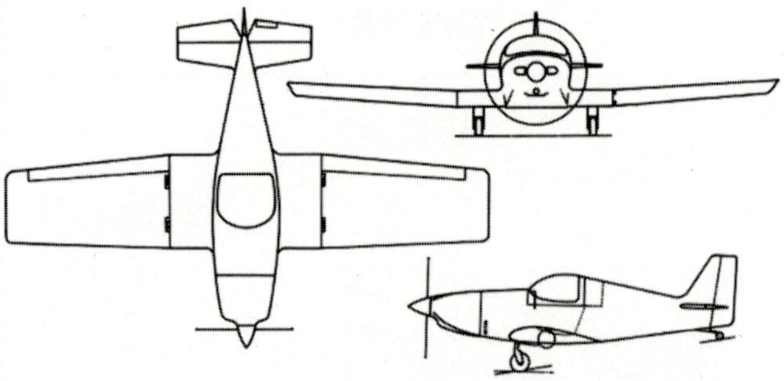

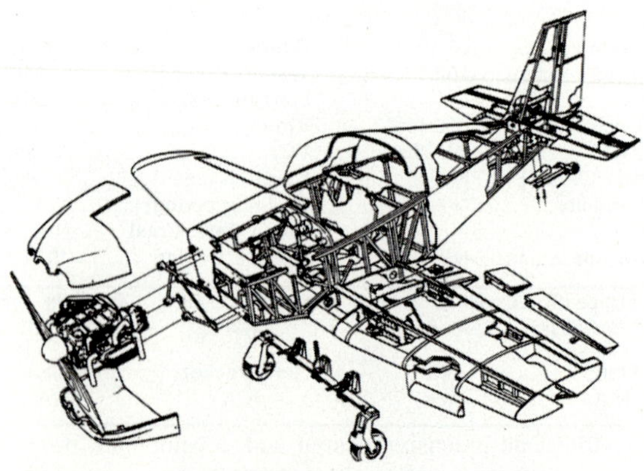

RAND ROBINSON ENGINEERING
15641 Product Lane, A-5
Huntington Beach CA 92649
Tel. 714-898-3811 • Fax 714-890-1658
E-Mail: pilot@beegroup.com
Web: http://www.fly-kr.com

KR-2S

Wing Span	23.00 ft.	**Range**	1080 sm
Wing Area	82.00 sq. ft.	**Takeoff Dist.**	350 ft.
Length	16.00 ft.	**Landing Dist.**	600 ft.
Max Gross Wt.	980 lbs.	**Vmax**	200 mph
Empty Wt.	520 lbs.	**VCr**	180 mph
Payload (full fuel)	N/A	**V_{S1} (stall clean)**	52 mph
Fuel Capacity	24 gal.	**V_{S0} (lndg config)**	N/A
Seats	2	**Climb Rate @msl**	1200 fpm
Cabin Width	N/A	**Service Ceiling**	15,000 ft.

Std. Engine/HP	Usable HP Range	Construction Mtrl.	Est. Bldg. Time
VW 2180 / 80 hp	76-100	Wood, Fiberglass	800 hrs.
Plans	**Kit**	**Info Packet**	**Video**
$240.00	$7,800.00	$8.00	$24.95

Stretched version of the KR-2:16" longer and 3" more headroom in the cockpit. A sleek, two seat, side-by-side design utilizing wood, foam and fiberglass composite construction. Premolded parts made of high temperature epoxy based prepregs and all high temperature, core materials. Taildragger or tricycle gear options. New 14 piece premolded wing skin grouping available. See Ad on Page 22.

Completed & Flying: 4

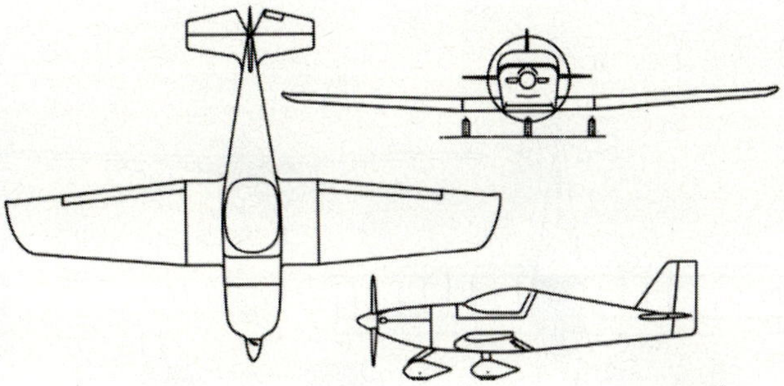

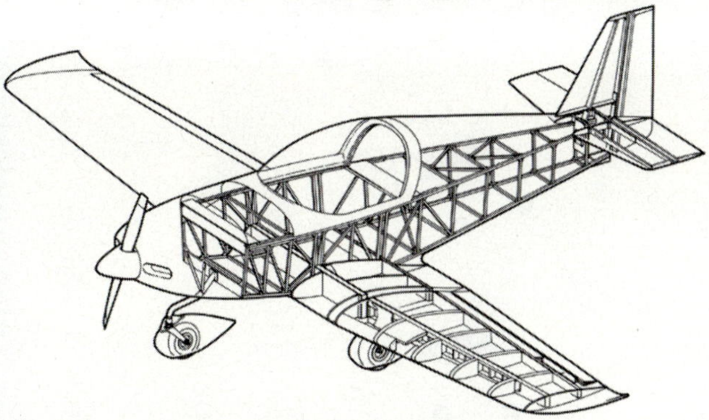

RAND ROBINSON ENGINEERING
15641 Product Lane, A-5
Huntington Beach CA 92649
Tel. 714-898-3811 • Fax 714-890-1658
E-Mail: pilot@beegroup.com
Web: http://www.fly-kr.com

S-4/5 Coyote I

Wing Span	29.50 ft.	**Range**	150 sm
Wing Area	127.0 sq. ft.	**Takeoff Dist.**	60 ft.
Length	17.00 ft.	**Landing Dist.**	60 ft.
Max Gross Wt.	587 lbs.	**Vmax**	95 mph
Empty Wt.	290 lbs.	**VCr**	55 mph
Payload (full fuel)	297 lbs.	**V_{S1} (stall clean)**	29 mph
Fuel Capacity	5 gal.	**V_{S0} (lndg config)**	27 mph
Seats	1	**Climb Rate @msl**	800 fpm
Cabin Width	25 in.	**Service Ceiling**	15,000 ft.

Std. Engine/HP	Usable HP Range	Construction Mtrl.	Est. Bldg. Time
Rotax 447 / 42 hp	42 - 47	Tube, Fabric	130 hrs.
Plans	**Kit**	**Info Packet**	**Video**
N/A	$9,980.00	Yes	N/A

First flight in March 1983. Very short build time due to the high level of prefinished parts, pre-sewn skins, and ease of assembly. Can be built in an average garage using normal shop tools and skills. Trike and taildragger options. Foldable/detachable wings. Kit prices include engine, prop, and instruments. Also available with Rotax 503 engine. Video includes all models.

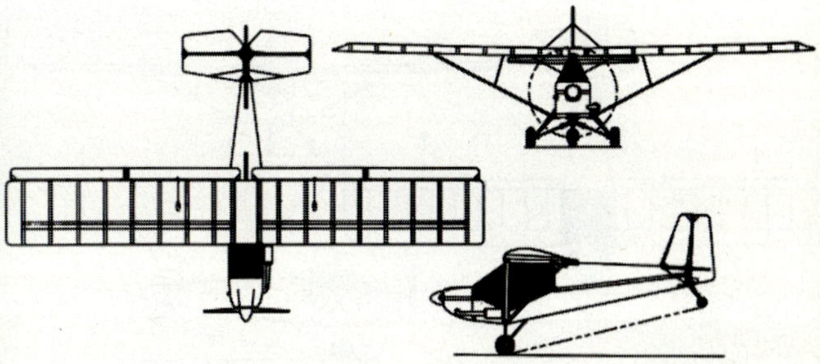

RANS INC.
4600 Highway 183 Alternate
Hays KS 67601
Tel. 785-625-6346 • Fax 785-625-2795

S-6ES Coyote II

Wing Span	34.50 ft.	**Range**	460 sm
Wing Area	155.3 sq. ft.	**Takeoff Dist.**	160 ft.
Length	20.00 ft.	**Landing Dist.**	200 ft.
Max Gross Wt.	930 lbs.	**Vmax**	120 mph
Empty Wt.	410 lbs.	**VCr**	65 mph
Payload (full fuel)	520 lbs.	**V_{S1} (stall clean)**	39 mph
Fuel Capacity	18 gal.	**V_{S0} (lndg config)**	34 mph
Seats	2	**Climb Rate @msl**	500 fpm
Cabin Width	41 in.	**Service Ceiling**	15,000 ft.

Std. Engine/HP	Usable HP Range	Construction Mtrl.	Est. Bldg. Time
Rotax 503 / 47 hp	47 - 80	Tube, Fabric	300 hrs.
Plans	**Kit**	**Info Packet**	**Video**
N/A	$16,300.00	Yes	$20.00

Second generation of the original S-6 design with "Extended Span" wings for increased performance. In production since April 1990. Can be built in a normal garage using normal shop tools. Three different wing options. Kit prices include engine, prop, and instruments. Video includes all models.

Kits/Plans Sold: 948 **Completed & Flying:** 853

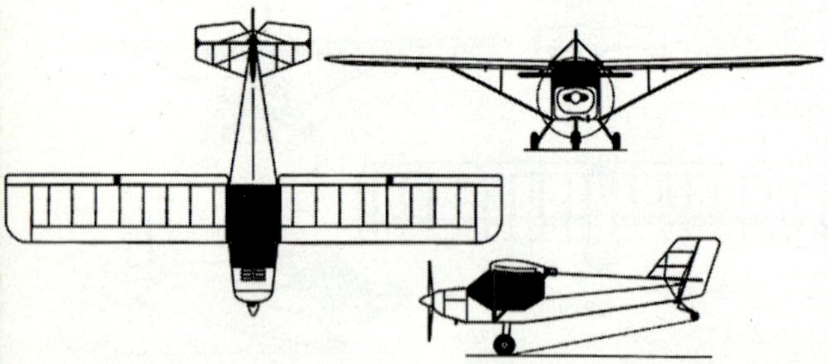

RANS INC.
4600 Highway 183 Alternate
Hays KS 67601
Tel. 785-625-6346 • Fax 785-625-2795

S-6S Coyote II

Wing Span	34.50 ft.	**Range**	460 sm
Wing Area	155.3 sq. ft.	**Takeoff Dist.**	160 ft.
Length	20.00 ft.	**Landing Dist.**	200 ft.
Max Gross Wt.	1050 lbs.	**Vmax**	120 mph
Empty Wt.	550 lbs.	**VCr**	75 mph
Payload (full fuel)	500 lbs.	**V_{S1} (stall clean)**	38 mph
Fuel Capacity	18 gal.	**V_{S0} (lndg config)**	33 mph
Seats	2	**Climb Rate @msl**	500 fpm
Cabin Width	41 in.	**Service Ceiling**	15,000 ft.

Std. Engine/HP	Usable HP Range	Construction Mtrl.	Est. Bldg. Time
Rotax 503 / 47 hp	47 - 80	Tube, Fabric	500 hrs.
Plans	**Kit**	**Info Packet**	**Video**
N/A	$17,635.00	Yes	$20.00

Take the very successful S-6ES Coyote II, add standard aircraft fabric and drag reducing fairings and you have the "Super Six." Notable among fabric planes for the high level of pre-finish. Number of kits sold includes the entire S-5 and S-6 Series.

Kits/Plans Sold: 948 **Completed & Flying:** 853

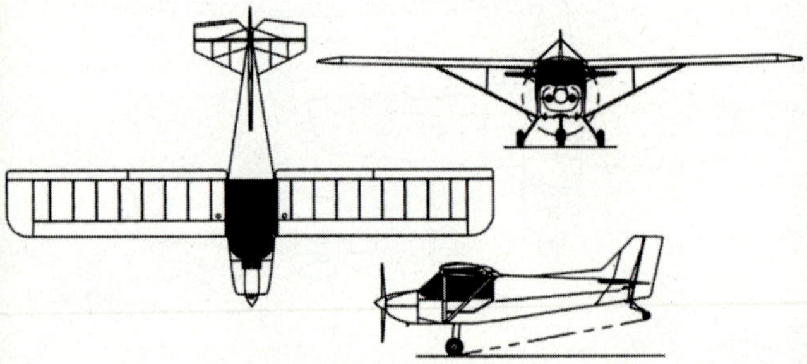

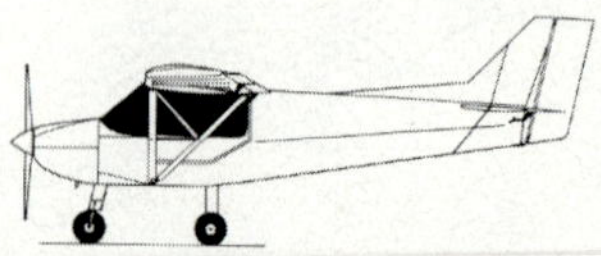

RANS INC.
4600 Highway 183 Alternate
Hays KS 67601
Tel. 785-625-6346 • Fax 785-625-2795

S-7 Courier

Wing Span	29.30 ft.	**Range**	450 sm
Wing Area	151.0 sq. ft.	**Takeoff Dist.**	175 ft.
Length	21.00 ft.	**Landing Dist.**	303 ft.
Max Gross Wt.	1025 lbs.	**Vmax**	130 mph
Empty Wt.	575 lbs.	**VCr**	80 mph
Payload (full fuel)	505 lbs.	**V_{S1} (stall clean)**	35 mph
Fuel Capacity	18 gal.	**V_{S0} (lndg config)**	30 mph
Seats	2	**Climb Rate @msl**	800 fpm
Cabin Width	26 in.	**Service Ceiling**	13,500 ft.

Std. Engine/HP	Usable HP Range	Construction Mtrl.	Est. Bldg. Time
Rotax 582 / 65 hp	65 - 80	Tube, Fabric	700 hrs.
Plans	**Kit**	**Info Packet**	**Video**
N/A	$16,300.00	Yes	N/A

First flight in November 1985. Originally designed as a trainer for the Coyote I. Tandem seating with a sweeping view from the rear seat makes this plane ideal for photo work. Can be built in an average two-car garage, usually in 500 to 700 hours. Kit prices include engine, prop, and instruments. Also available with Rotax 912 engine. Video includes all models.

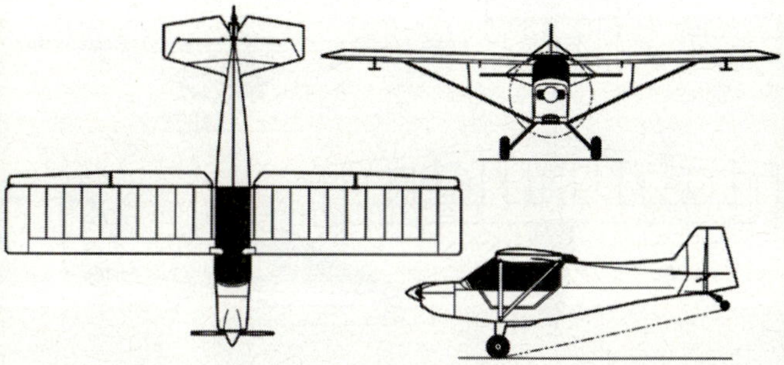

Jim Schilling's S-7 on Full Lotus Floats

RANS INC.
4600 Highway 183 Alternate
Hays KS 67601
Tel. 785-625-6346 • Fax 785-625-2795

S-9 Chaos

Wing Span	22.00 ft.	**Range**	210 sm
Wing Area	91.50 sq. ft.	**Takeoff Dist.**	200 ft.
Length	15.67 ft.	**Landing Dist.**	400 ft.
Max Gross Wt.	670 lbs.	**Vmax**	130 mph
Empty Wt.	320 lbs.	**VCr**	85 mph
Payload (full fuel)	350 lbs.	**V_{S1} (stall clean)**	33 mph
Fuel Capacity	9 gal.	**V_{S0} (lndg config)**	N/A
Seats	1	**Climb Rate @msl**	800 fpm
Cabin Width	25 in.	**Service Ceiling**	15,500 ft.

Std. Engine/HP	Usable HP Range	Construction Mtrl.	Est. Bldg. Time
Rotax 503 / 47 hp	65-100	Tube, Fabric	500 hrs.
Plans	**Kit**	**Info Packet**	**Video**
N/A	$10,625.00	Yes	$20.00

A mid wing monoplane designed for full aerobatic flight. "Full Inverted" capable. Accepts engines up to 100 hp. In production since 1986. All welding completed at the factory. Can be assembled in an average two-car garage in about 350 to 500 hours. Kits include engine, prop, and basic instruments. Also available with Rotax 582 engine.

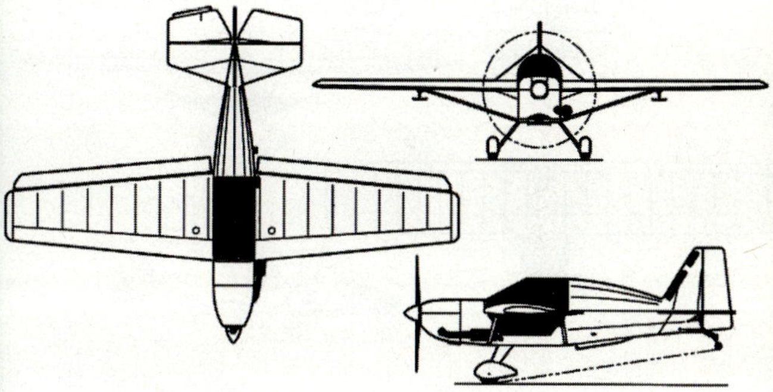

RANS INC.
4600 Highway 183 Alternate
Hays KS 67601
Tel. 785-625-6346 • Fax 785-625-2795

S-12 XL Airaile

Wing Span	31.00 ft.	Range	275 sm
Wing Area	152.0 sq. ft.	Takeoff Dist.	275 ft.
Length	20.50 ft.	Landing Dist.	130 ft.
Max Gross Wt.	920 lbs.	Vmax	120 mph
Empty Wt.	425 lbs.	VCr	65 mph
Payload (full fuel)	495 lbs.	V_{S1} (stall clean)	31 mph
Fuel Capacity	9 gal.	V_{S0} (lndg config)	27 mph
Seats	2	Climb Rate @msl	500 fpm
Cabin Width	41 in.	Service Ceiling	13,500 ft.

Std. Engine/HP	Usable HP Range	Construction Mtrl.	Est. Bldg. Time
Rotax 503 / 47 hp	47 - 80	Tube, Fabric	175 hrs.
Plans	**Kit**	**Info Packet**	**Video**
N/A	$13,500.00	Yes	$20.00

Very popular, economical, high wing, two place pusher with a 6061-T6 aluminum tail boom. Wings fold or detach. Stock, partial and full enclosures available. In production since March 1990. Can be built in an average garage using normal shop tools and skills. Kits include engine, prop, and basic instruments. Engine options include Rotax 582 and Rotax 912. Video includes all models.

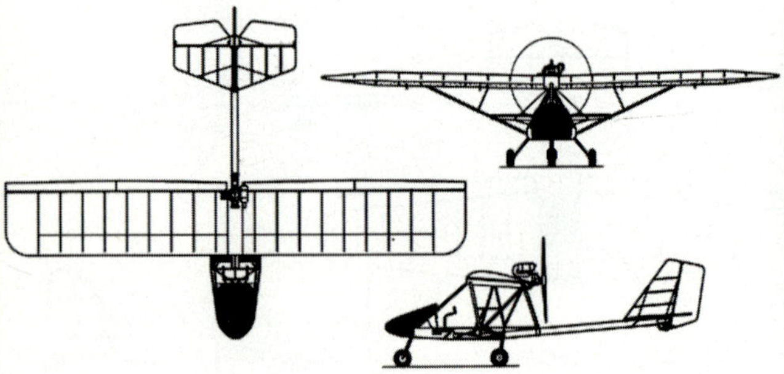

RANS INC.
4600 Highway 183 Alternate
Hays KS 67601
Tel. 785-625-6346 • Fax 785-625-2795

S-14 Airaile

Wing Span	26.38 ft.	Range	180 sm
Wing Area	100.0 sq. ft.	Takeoff Dist.	300 ft.
Length	18.83 ft.	Landing Dist.	130 ft.
Max Gross Wt.	675 lbs.	Vmax	120 mph
Empty Wt.	360 lbs.	VCr	65 mph
Payload (full fuel)	315 lbs.	V_{S1} (stall clean)	42 mph
Fuel Capacity	9 gal.	V_{S0} (lndg config)	38 mph
Seats	1	Climb Rate @msl	700 fpm
Cabin Width	30 in.	Service Ceiling	12,000 ft.

Std. Engine/HP	Usable HP Range	Construction Mtrl.	Est. Bldg. Time
Rotax 447 / 42 hp	42 - 65	Tube, Fabric	250 hrs.
Plans	**Kit**	**Info Packet**	**Video**
N/A	$11,300.00	Yes	$20.00

Designed to provide a performance aircraft with fine handling and balance in a quick build package. Welded steel cage and an aluminum tube superstructure. Two wing sizes available. Optional second wing tank increases fuel to18 gallons. Accepts Rotax 447, 503 and 582 engines. In production since 1991. Kit price includes basic instruments and engine. Video includes all models.

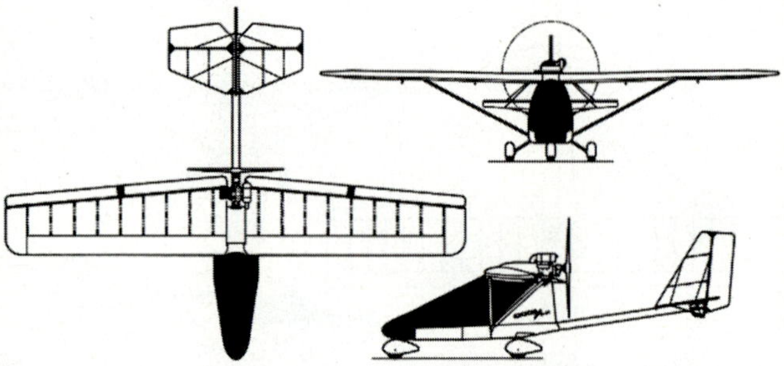

RANS INC.
4600 Highway 183 Alternate
Hays KS 67601
Tel. 785-625-6346 • Fax 785-625-2795

HOMEBUILT "FAQ'S" *(See Page 68)*

Q: How long will my aircraft take to finish?

A: Always longer than you think! A well-known writer on homebuilt topics is often quoted as saying "Firewall-forward is half the work" and, "The jobs you thought were simple take forever. The jobs you thought were tough turn out to be easy".

Designers often try to minimize their estimates of completion time for obvious reasons. And time varies significantly with builder skill and experience. So the best way to estimate this is to talk to other builders who have finished their projects.

For simple fixed gear kits, 500-1500 hours seems common. For complex planes (such as retractables) 1500-3000 is more realistic. For plans-built airplanes, anywhere from 1000-8000+ hours are involved. All of these numbers are highly dependent upon the type, and upon the builder's skills.

Don't be too dependant on "goal fixation." Take the time to do the job right. Above all, don't push it to try to get finished for a major fly-in. The risks aren't worth it. Besides, more attention you pay to small details the better you'll feel when you do fly to the big event.

S-10 Sakota

Wing Span	24.00 ft.	**Range**	260 sm
Wing Area	95.00 sq. ft.	**Takeoff Dist.**	300 ft.
Length	17.83 ft.	**Landing Dist.**	500 ft.
Max Gross Wt.	960 lbs.	**Vmax**	130 mph
Empty Wt.	460 lbs.	**VCr**	100 mph
Payload (full fuel)	500 lbs.	**V_{S1} (stall clean)**	48 mph
Fuel Capacity	18 gal.	**V_{S0} (lndg config)**	42 mph
Seats	2	**Climb Rate @msl**	1000 fpm
Cabin Width	37 in.	**Service Ceiling**	14,000 ft.

Std. Engine/HP	Usable HP Range	Construction Mtrl.	Est. Bldg. Time
Rotax 582 / 65 hp	65 - 80	Tube, Fabric	600 hrs.
Plans	**Kit**	**Info Packet**	**Video**
N/A	$14,800.00	Yes	$20.00

A sleek design. First flight in March 1988. Good cross country capabilities. No welding required for assembly. Low build time. Stressed to +6, -3 G's at gross weight. Comes standard with an inverted fuel system, shocks, roll over protection, and a high back tailcone. Kits include engine, prop, and basic instruments. Also available with Rotax 912 engine.

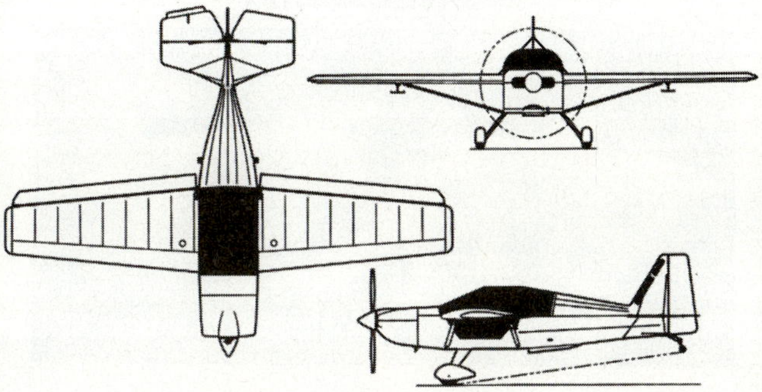

See Builder's Story on page 41.

RANS INC.
4600 Highway 183 Alternate
Hays KS 67601
Tel. 785-625-6346 • Fax 785-625-2795

S-16 Shekari

Wing Span	24.00 ft.	**Range**	850 sm
Wing Area	86.50 sq. ft.	**Takeoff Dist.**	400 ft.
Length	18.67 ft.	**Landing Dist.**	500 ft.
Max Gross Wt.	1300 lbs.	**Vmax**	248 mph
Empty Wt.	830 lbs.	**VCr**	170 mph
Payload (full fuel)	675 lbs.	**V_{S1} (stall clean)**	55 mph
Fuel Capacity	32 gal.	**V_{S0} (lndg config)**	50 mph
Seats	2	**Climb Rate @msl**	1200 fpm
Cabin Width	40 in.	**Service Ceiling**	14,000 ft.

Std. Engine/HP	Usable HP Range	Construction Mtrl.	Est. Bldg. Time
Cont. IO-240 / 130 hp	80 - 130	Composite, Tube	600 hrs.
Plans	**Kit**	**Info Packet**	**Video**
N/A	$39,000.00	Yes	N/A

Low wing metal and composite plane. First of Rans' new breed of QBHP (Quick-Build, High Performance) aircraft. Rated at +9, -9 G's. Kit price includes engine and basic instruments. Also available with a Rotax 912 engine at $30,000. Can be configured as a taildragger or with tricycle gear. Wings can be removed and installed by one person.

Kits/Plans Sold: 50

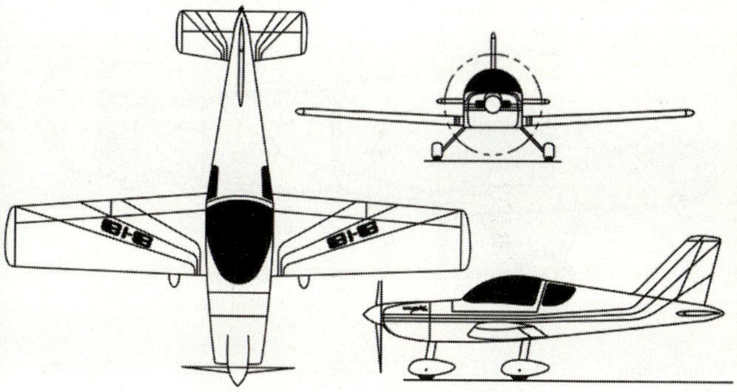

RANS INC.
4600 Highway 183 Alternate
Hays KS 67601
Tel. 785-625-6346 • Fax 785-625-2795

Fokker DR1

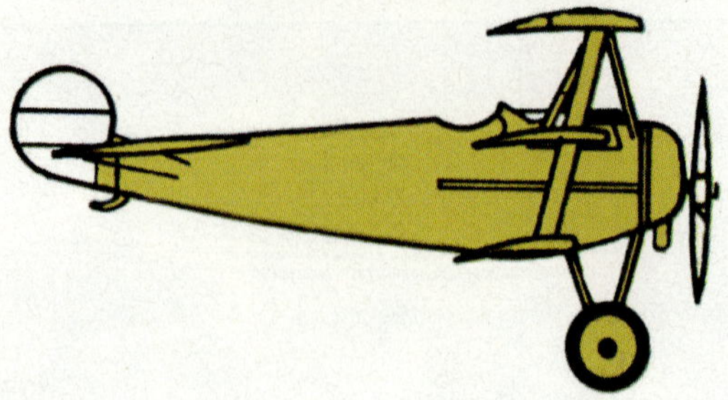

Wing Span	23.75 ft.	Range	250 sm
Wing Area	203.0 sq. ft.	Takeoff Dist.	150 ft.
Length	19.02 ft.	Landing Dist.	250 ft.
Max Gross Wt.	1450 lbs.	Vmax	125 mph
Empty Wt.	1100 lbs.	VCr	100 mph
Payload (full fuel)	N/A	V_{S1} (stall clean)	40 mph
Fuel Capacity	30 gal.	V_{S0} (lndg config)	N/A
Seats	1	Climb Rate @msl	1800 fpm
Cabin Width	N/A	Service Ceiling	18,000 ft.

Std. Engine/HP	Usable HP Range	Construction Mtrl.	Est. Bldg. Time
Warner / 145 hp	145-220	Wood, Tube, Fabric	2500 hrs.
Plans	**Kit**	**Info Packet**	**Video**
$60.00	N/A	$6.00	N/A

Replica of famous WWI aircraft. Very popular design. Typical wood, tube, and fabric construction. Add $25 for overseas delivery of plans. **Plans Sold:** 5000 **Completed & Flying:** 400

WALTER REDFERN COMPANY
S-211 Spencer
Post Falls ID 83854
Tel. 208-773-8280

Nieuport 17/24

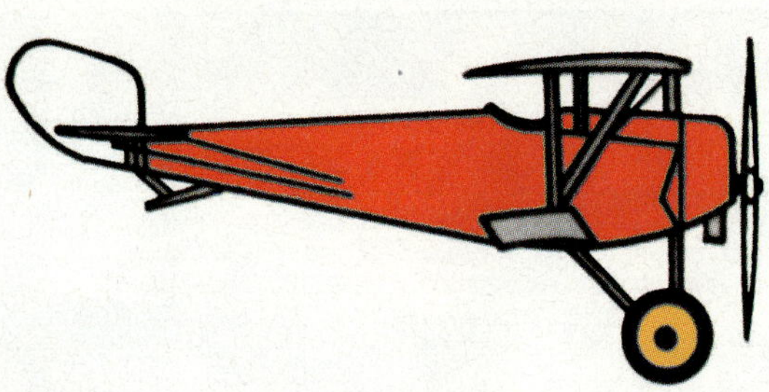

Wing Span	26.92 ft.	Range	225 sm
Wing Area	162.0 sq. ft.	Takeoff Dist.	125 ft.
Length	18.75 ft.	Landing Dist.	300 ft.
Max Gross Wt.	1280 lbs.	Vmax	125 mph
Empty Wt.	1000 lbs.	VCr	100 mph
Payload (full fuel)	N/A	V_{S1} (stall clean)	45 mph
Fuel Capacity	25 gal.	V_{S0} (lndg config)	N/A
Seats	1	Climb Rate @msl	1200 fpm
Cabin Width	N/A	Service Ceiling	12,500 ft.

Std. Engine/HP	Usable HP Range	Construction Mtrl.	Est. Bldg. Time
Warner / 145 hp	145-180	Wood, Tube, Fabric	2000 hrs.
Plans	**Kit**	**Info Packet**	**Video**
$100.00	N/A	$6.00	N/A

Famous WWI fighter aircraft replica. Generic design covers both the Model 17 and the Model 24. Brakes and tail wheel have been added to the original French design. Add $25 for overseas delivery of plans. **Completed & Flying:** 100

WALTER REDFERN COMPANY
S-211 Spencer
Post Falls ID 83854
Tel. 208-773-8280

DH-2

Wing Span	28.25 ft.	Range	250 sm
Wing Area	265.0 sq. ft.	Takeoff Dist.	100 ft.
Length	25.17 ft.	Landing Dist.	300 ft.
Max Gross Wt.	1450 lbs.	Vmax	95 mph
Empty Wt.	1100 lbs.	VCr	75 mph
Payload (full fuel)	N/A	V_{S1} (stall clean)	45 mph
Fuel Capacity	30 gal.	V_{S0} (lndg config)	N/A
Seats	1	Climb Rate @msl	650 fpm
Cabin Width	N/A	Service Ceiling	12,000 ft.

Std. Engine/HP	Usable HP Range	Construction Mtrl.	Est. Bldg. Time
Kinner / 125 hp	125-150	Wood, Tube, Fabric	2500 hrs.
Plans	**Kit**	**Info Packet**	**Video**
$150.00	N/A	$6.00	N/A

WWI replica. Metal tube has replaced much of the wood fuselage used in the original designs. Add $25 for overseas delivery of plans. **Completed & Flying:** 10

WALTER REDFERN COMPANY
S-211 Spencer
Post Falls ID 83854
Tel. 208-773-8280

Lightning Bug

Wing Span	17.83 ft.	Range	880 sm
Wing Area	40.00 sq. ft.	Takeoff Dist.	800 ft.
Length	17.46 ft.	Landing Dist.	1000 ft.
Max Gross Wt.	800 lbs.	Vmax	250 mph
Empty Wt.	475 lbs.	VCr	225 mph
Payload (full fuel)	N/A	V_{S1} (stall clean)	62 mph
Fuel Capacity	23 gal.	V_{S0} (lndg config)	N/A
Seats	1	Climb Rate @msl	1200 fpm
Cabin Width	25 in.	Service Ceiling	20,000 ft.

Std. Engine/HP	Usable HP Range	Construction Mtrl.	Est. Bldg. Time
AMW 808 / 100 hp	N/A	Fiberglass	300 hrs.
Plans	**Kit**	**Info Packet**	**Video**
N/A	$25,000.00	Yes	N/A

Utilizes fiberglass and epoxy molded parts and stainless steel. Kit includes engine, propeller and engine instruments. Retractable nose gear with swiveling 280/250-4 wheel and tire. Designed to FAR part 23 aerobatic category at 750 lbs. gross.

REFLEX FIBERGLASS WORKS
PO Box 497 • 179 Aviation Ave.
Walterboro SC 29488
Tel. 803-549-1800 • Fax 803-549-5139

White Lightning

Wing Span	27.67 ft.	**Range**	1638 sm
Wing Area	89.20 sq. ft.	**Takeoff Dist.**	1300 ft.
Length	23.33 ft.	**Landing Dist.**	1300 ft.
Max Gross Wt.	2700 lbs.	**Vmax**	280 mph
Empty Wt.	1400 lbs.	**VCr**	265 mph
Payload (full fuel)	1300 lbs.	**V_{S1} (stall clean)**	69 mph
Fuel Capacity	70 gal.	**V_{S0} (lndg config)**	N/A
Seats	4	**Climb Rate @msl**	1450 fpm
Cabin Width	43 in.	**Service Ceiling**	20,000 ft.

Std. Engine/HP	Usable HP Range	Construction Mtrl.	Est. Bldg. Time
Cont. IO-360 / 210 hp	N/A	Composite	1500 hrs.
Plans	**Kit**	**Info Packet**	**Video**
N/A	$45,000.00	$10.00	N/A

A high performance, four seater with retractable landing gear. Kit is available in four separate units which may be purchased in stages. Forty hours of builder instruction and consulting included with kit. Engine not included in kit.

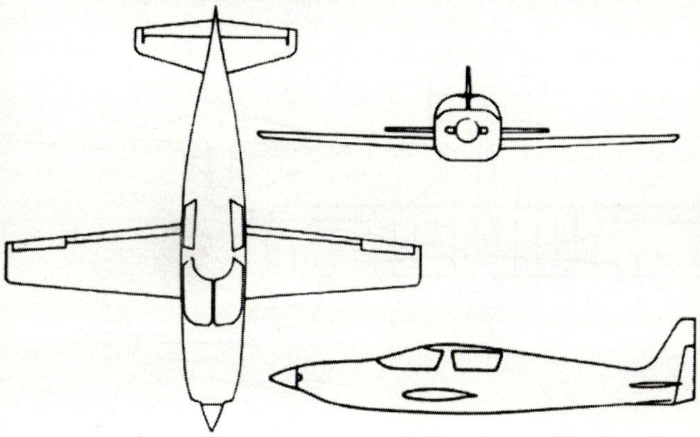

REFLEX FIBERGLASS WORKS
PO Box 497 • 179 Aviation Ave.
Walterboro SC 29488
Tel. 803-549-1800 • Fax 803-549-5139

Berkut

Wing Span	26.67 ft.	**Range**	1514 sm
Wing Area	110.0 sq. ft.	**Takeoff Dist.**	1000 ft.
Length	18.50 ft.	**Landing Dist.**	1000 ft.
Max Gross Wt.	2000 lbs.	**Vmax**	280 mph
Empty Wt.	1035 lbs.	**VCr**	162 mph
Payload (full fuel)	965 lbs.	**V_{S1} (stall clean)**	50 mph
Fuel Capacity	58 gal.	**V_{S0} (lndg config)**	N/A
Seats	2	**Climb Rate @msl**	2000 fpm
Cabin Width	24 in.	**Service Ceiling**	30,000 ft.

Std. Engine/HP	Usable HP Range	Construction Mtrl.	Est. Bldg. Time
Lyc. IO-540 / 260 hp	180-260	Composite	2000 hrs.
Plans	**Kit**	**Info Packet**	**Video**
N/A	$34,490.00	$20.00	$24.95

Canard pusher with tandem seating. Features high performance, modest cost, with ease and simplicity of construction. Engine, avionics, propeller, paint, and upholstery not included. Fully retractable landing gear and detachable molded wings. Aerobatic. See Ad on Page 21. **Kits Sold:** 57 **Flying:** 7

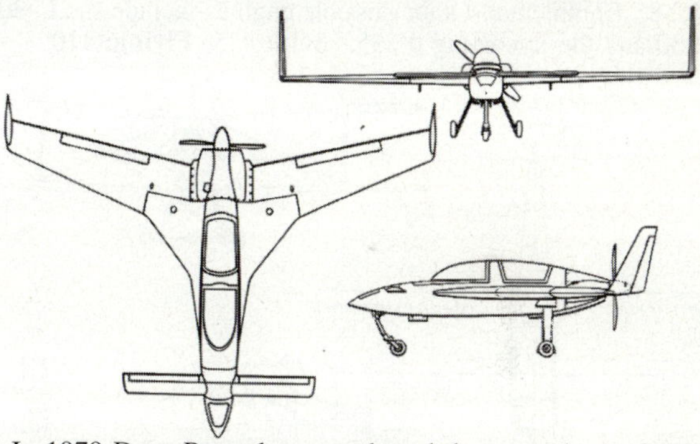

In 1979 Dave Ronneberg purchased the second set of Long-EZ plans ever sold and started a 10 year stint as a professional Long-EZ builder. In 1985 he began sketching a variation on the Long-EZ — bigger, longer and with a larger engine and retractable landing gear.

Construction began on the prototype in the Fall of 1989. The plane flew for the first time in June 1991 and made its public debut at Oshkosh a month later. After extensive testing, the first kit was sold at Oshkosh 1992.

Dick Rutan
"The best kitplane in the world - bar none."

Norm Howell, Berkut #012
"I love flying this plane! It's the next best thing to the F-16 I fly during my day job...and the best part of the Berkut is that mere mortals can afford to fly it."

Bob Higgenson, Berkut #035
"I'm on the large side - 6'8" and 290 lbs. I never thought that a kitplane would fit me, but the Berkut does. It has both the head room I need, and the performance to let me enjoy it."

RENAISSANCE COMPOSITES, INC.
3025 Airport Avenue • Santa Monica CA 90405
Tel. 310-391-1943 • Fax 310-391-8645
E-Mail: berkut@loop.com • Web: http://www.berkut.com

SE-5-A Replica

Wing Span	23.40 ft.	**Range**	250 sm
Wing Area	146.0 sq. ft.	**Takeoff Dist.**	400 ft.
Length	18.20 ft.	**Landing Dist.**	350 ft.
Max Gross Wt.	1150 lbs.	**Vmax**	100 mph
Empty Wt.	740 lbs.	**VCr**	85 mph
Payload (full fuel)	N/A	**V_{S1} (stall clean)**	35 mph
Fuel Capacity	18 gal.	**V_{S0} (lndg config)**	N/A
Seats	1	**Climb Rate @msl**	600 fpm
Cabin Width	24 in.	**Service Ceiling**	10,000 ft.

Std. Engine/HP	Usable HP Range	Construction Mtrl.	Est. Bldg. Time
Continental C-85	85 - 115	Wood, Fabric	2500 hrs.
Plans	**Kit**	**Info Packet**	**Video**
$120.00	N/A	$7.50	N/A

Scaled down, sport biplane with look and feel of the WWI SE-5-A Scout. One of the most popular WWI replica aircraft being built today. Looks like the original design but uses far simpler construction with more modern and readily available materials. Affordable, easy to build, and easy to fly. Specs based on C-85 Continental. Other possible engines include the C-90, O-200 and the Lycoming 0-235. **Sold:** 475 **Flying:** 110

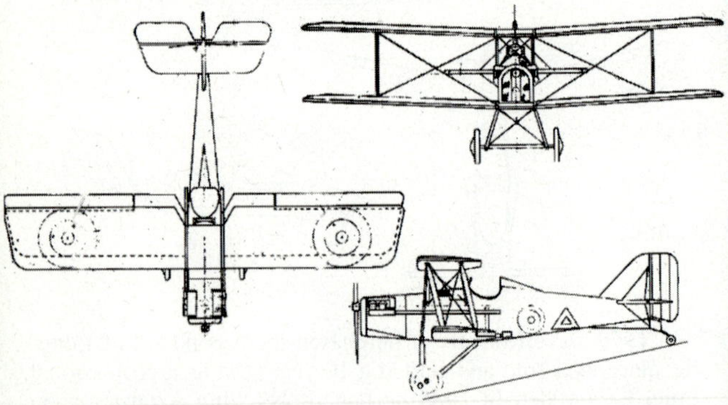

The fuselage is a 1/8 inch ply-skinned box structure with a fabric covered turtle deck, aluminum-covered forward top section and incorporates a standard cockpit layout with a left-hand quadrant throttle, standard panel, heel brakes, parking brake, small baggage compartment behind the seat, full shoulder harness, a headrest, and folding type windshield.

The original SE-5-A incorporated a tail skid only, but the replica is equipped with a steerable tail wheel enclosed in a dummy housing to make it appear as much as possible like a skid.

Sport Aviation 12/92
"The scaled down SE-5-A is one of the most popular types of WWI replica aircraft being built today."

Frank Ryder
"Can I build it? Can I afford to build it? Can I fly it? Yes. The SE-5-A is affordable, easy to build, and easy to fly."

REPLICA PLANS
4366 Eckert St. • Chilliwack, B.C. V2R 5J5
CANADA
Tel. 604-823-6428

Flaglor Scooter

Wing Span	28.00 ft.	**Range**	175 sm
Wing Area	115.0 sq. ft.	**Takeoff Dist.**	200 ft.
Length	15.00 ft.	**Landing Dist.**	250 ft.
Max Gross Wt.	625 lbs.	**Vmax**	90 mph
Empty Wt.	390 lbs.	**VCr**	65 mph
Payload (full fuel)	200 lbs.	**V_{S1} (stall clean)**	34 mph
Fuel Capacity	9 gal.	**V_{S0} (lndg config)**	N/A
Seats	1	**Climb Rate @msl**	325 fpm
Cabin Width	24 in.	**Service Ceiling**	N/A

Std. Engine/HP	Usable HP Range	Construction Mtrl.	Est. Bldg. Time
VW / 32 hp	25 - 43	Wood, Fabric	400 hrs.
Plans	**Kit**	**Info Packet**	**Video**
$100.00	N/A	$5.00	N/A

Winner of Outstanding Ultralight and Outstanding VW Powered Airplane awards in previous years. Plans are on 16 sheets of 18" x 24" black line, large format, Xerox paper. Plans includes construction manual. Total ready-to-fly cost approximately $7,000.

Kits/Plans Sold: 312 **Completed & Flying:** 101

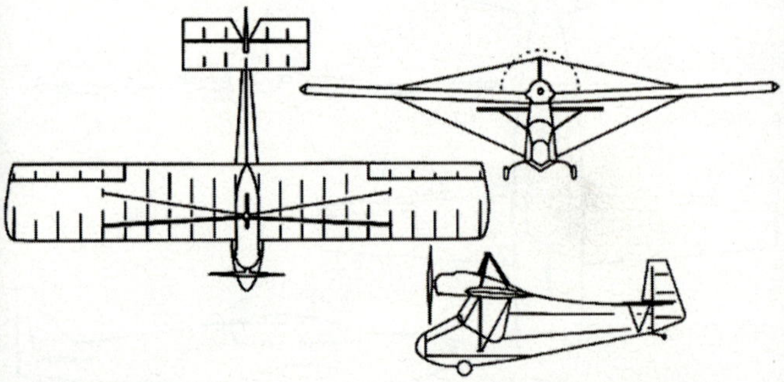

ROTOR-WINGS & FLYING MACHINES
10405 Button Quail Drive
Austin TX 78758-5032
Tel. 512-837-8041

Storm 300 Special

Wing Span	26.00 ft.	Range	650 sm
Wing Area	119.0 sq. ft.	Takeoff Dist.	450 ft.
Length	21.50 ft.	Landing Dist.	360 ft.
Max Gross Wt.	1155 lbs.	Vmax	200 mph
Empty Wt.	710 lbs.	VCr	171 mph
Payload (full fuel)	N/A	V_{S1} (stall clean)	42 mph
Fuel Capacity	38 gal.	V_{S0} (lndg config)	35 mph
Seats	2	Climb Rate @msl	1500 fpm
Cabin Width	44 in.	Service Ceiling	12,000 ft.

Std. Engine/HP	Usable HP Range	Construction Mtrl.	Est. Bldg. Time
Rotax / 115 hp	100-125	Aluminum	500 hrs.
Plans	**Kit**	**Info Packet**	**Video**
N/A	$23,000.00	$8.00	$12.50

Two seat, side-by-side sport configuration, high performance taildragger design. Kit price does not include freight charges. Performance specs based on gross weight. A Quick Build Kit is also available that reduces build time by 150 hours. See Ad on Page 60. **Kits/Plans Sold:** 20 **Completed & Flying:** 12

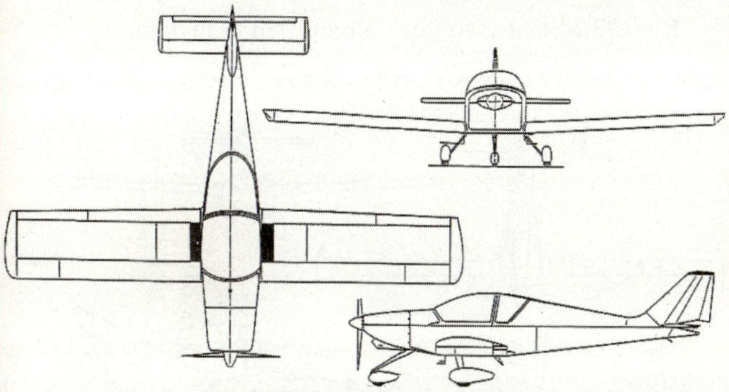

See list of distributors in next column.

S.G. AVIATION SRL
Zona Industriale via trav. Bia • Sabaudia (LT) 04016
ITALY
Ph: +39 (336) 765.541 • Fax: +39 (773) 511.407
E-Mail: sg@storm-sg.it
Web: http://www.storm-sg.it

Sea Storm Amphibian

Wing Span	30.30 ft.	Range	1040 sm
Wing Area	131.8 sq. ft.	Takeoff Dist.	400 ft.
Length	24.50 ft.	Landing Dist.	475 ft.
Max Gross Wt.	1340 lbs.	Vmax	202 mph
Empty Wt.	682 lbs.	VCr	148 mph
Payload (full fuel)	N/A	V_{S1} (stall clean)	44 mph
Fuel Capacity	26 gal.	V_{S0} (lndg config)	38 mph
Seats	2	Climb Rate @msl	1300 fpm
Cabin Width	49 in.	Service Ceiling	14,700 ft.

Std. Engine/HP	Usable HP Range	Construction Mtrl.	Est. Bldg. Time
Lyc. IO 540 / 115 hp	100-260	Composite, Aluminum	700 hrs.
Plans	**Kit**	**Info Packet**	**Video**
N/A	$24,700.00	$8.00	Yes

Two or four place, retractable gear, fast amphibian. Simple to construct composite fuselage. Aluminum wings with fowler flaps. Performance specs based on gross weight for two place model. Kit price does not include freight charges. Quick Build option reduces build time by 150 hours. Takeoff distance on water - 820 ft. (1100 ft. for four place). Landing distance on water - 620 ft. (820 ft. for four place). See Ad on Page 60.

Kits/Plans Sold: 25 **Completed & Flying:** 12

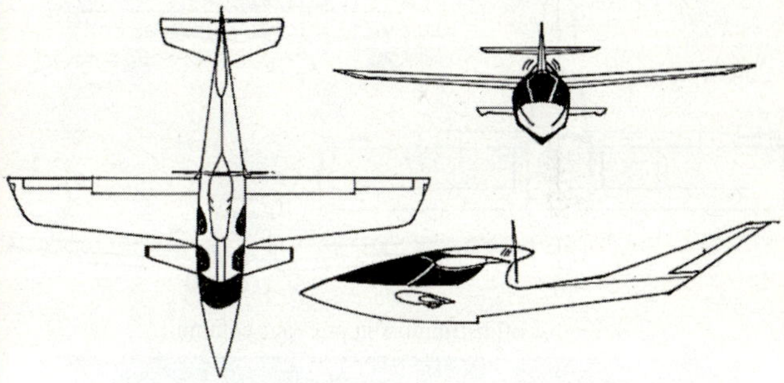

Some S.G. Aviation Distributors :

AVIO TECHNIK
Ingolstadter Str. 124
85077 Manching
GERMANY
Ph: 49-8459-6319
Fax: 49-8459-30971

CLASSIC AIRMOTIVE, INC.
7111-152 B Ave. Edmonton
Alberta T3C 2R9 CANADA
Ph: 403-473-1266 • Fax: 403-473-1266

J.C. DHOTEL
125 Chaussee de Spa
4910 Theux BELGIUM
Ph: 32-875 41454
Fax: 32-875 41454

AIRFLY
P.O. Box 12388
Thordon - Wellington
NEW ZEALAND
Ph: 64-4-4757906

S.G. AVIATION SRL
Zona Industriale via trav. Bia • Sabaudia (LT) 04016
ITALY
Ph: +39 (336) 765.541 • Fax: +39 (773) 511.407
E-Mail: sg@storm-sg.it
Web: http://www.storm-sg.it

Storm 280 - 320 E

Wing Span	28.20 ft.	**Range**	621 sm
Wing Area	119.1 sq. ft.	**Takeoff Dist.**	400 ft.
Length	21.49 ft.	**Landing Dist.**	500 ft.
Max Gross Wt.	992 lbs.	**Vmax**	182 mph
Empty Wt.	617 lbs.	**VCr**	151 mph
Payload (full fuel)	N/A	**V_{S1} (stall clean)**	40 mph
Fuel Capacity	16 gal.	**V_{S0} (lndg config)**	34 mph
Seats	2	**Climb Rate @msl**	1200 fpm
Cabin Width	44 in.	**Service Ceiling**	12,000 ft.

Std. Engine/HP	Usable HP Range	Construction Mtrl.	Est. Bldg. Time
Rotax 912 / 80 hp	80-115	Metal	400 hrs.
Plans	**Kit**	**Info Packet**	**Video**
N/A	$19,420.00	$8.00	$12.50

Two seat, side-by-side sport configuration. Kit price does not include freight charges. Performance specs based on gross weight. Also available in 280 version which has the same airframe but adds conventional gear. A Quick Build Kit is available that reduces build time by 150 hours. See Ad on Page 60.

Kits/Plans Sold: 100 **Completed & Flying:** 70

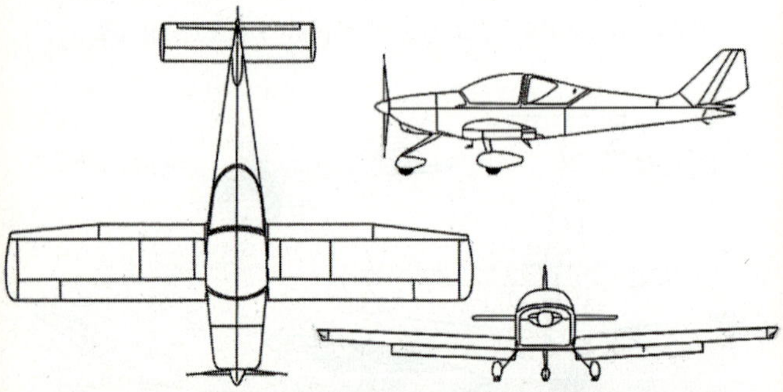

See list of distributors in previous column.

Storm 320 E

S.G. AVIATION SRL
Zona Industriale via trav. Bia • Sabaudia (LT) 04016
ITALY
Ph: +39 (336) 765.541 • Fax: +39 (773) 511.407
E-Mail: sg@storm-sg.it
Web: http://www.storm-sg.it

Seawind

Wing Span	35.00 ft.	**Range**	1460 sm
Wing Area	160.0 sq. ft.	**Takeoff Dist.**	820 ft.
Length	27.20 ft.	**Landing Dist.**	770 ft.
Max Gross Wt.	3400 lbs.	**Vmax**	200 mph
Empty Wt.	2300 lbs.	**VCr**	191 mph
Payload (full fuel)	N/A	**V_{S1} (stall clean)**	72 mph
Fuel Capacity	74 gal.	**V_{S0} (lndg config)**	59 mph
Seats	5	**Climb Rate @msl**	1250 fpm
Cabin Width	54 in.	**Service Ceiling**	20,000 ft.

Std. Engine/HP	Usable HP Range	Construction Mtrl.	Est. Bldg. Time
Lyc. IO-540 / 300 hp	260 - 315	Composite, Fiberglass	2000 hrs.
Plans	**Kit**	**Info Packet**	**Video**
N/A	$47,790.00	$10.00	$20.00

Sleek four place, fast amphibian. Fully retractable, tricycle landing gear. A variety of kit combinations are offered. They do not include engine, electronics, avionics, or upholstery. A new fast-build system reduces construction time to about 2,000 hours or 1,600 hours with a Kwick Kit. Approximately 130 being built in eleven countries.

Kits/Plans Sold: 120 **Completed & Flying:** 13

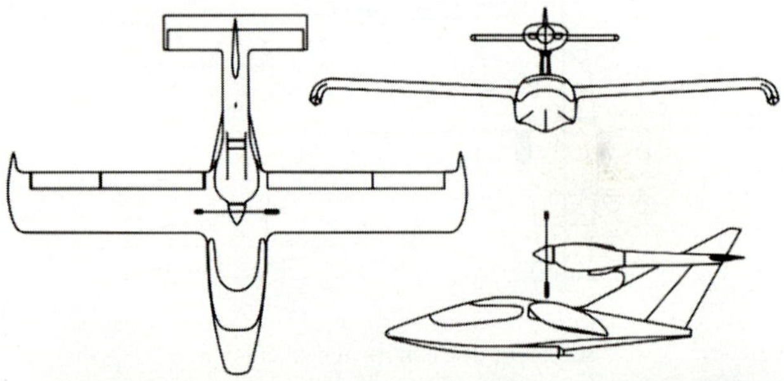

S.N.A., INC.
PO Box 607, Route 113
Kimberton PA 19442-0607
Tel. 610-983-3377 • Fax 610-933-3335

Sabre Trike

Wing Span	34.00 ft.	**Range**	130 sm
Wing Area	180.0 sq. ft.	**Takeoff Dist.**	100 ft.
Length	8.00 ft.	**Landing Dist.**	75 ft.
Max Gross Wt.	650 lbs.	**Vmax**	45 mph
Empty Wt.	205 lbs.	**VCr**	40 mph
Payload (full fuel)	N/A	**V_{S1} (stall clean)**	19 mph
Fuel Capacity	5 gal.	**V_{S0} (lndg config)**	N/A
Seats	2	**Climb Rate @msl**	1000 fpm
Cabin Width	N/A	**Service Ceiling**	10,000 ft.

Std. Engine/HP	Usable HP Range	Construction Mtrl.	Est. Bldg. Time
Kawasaki 340 / 32 hp	32 - 65	Aluminum, Fabric	2 hrs.
Plans	**Kit**	**Info Packet**	**Video**
N/A	$6,995.00	$15.00	$8.00

An entry level, one or two place trike. Complete with engine and prop, ready to fly with only minor assembly required. Very stable. Interchangable wings provide speeds from 16 to 87 mph. Durable frame, thoroughly tested. Also available with optional Rotax 447 and 503 engines at additional cost. See Ad on Page 22.

Completed & Flying: 150

SABRE AIRCRAFT, INC.
1300 S. Literfield Rd.
Goodyear AZ 85328
Tel. 602-925-6685 • Fax 602-925-6686

Excelsior

Wing Span	34.00 ft.	**Range**	125 sm
Wing Area	136.0 sq. ft.	**Takeoff Dist.**	200 ft.
Length	17.50 ft.	**Landing Dist.**	200 ft.
Max Gross Wt.	525 lbs.	**Vmax**	75 mph
Empty Wt.	251 lbs.	**VCr**	65 mph
Payload (full fuel)	N/A	**V_{S1} (stall clean)**	25 mph
Fuel Capacity	5 gal.	**V_{S0} (lndg config)**	N/A
Seats	1	**Climb Rate @msl**	600 fpm
Cabin Width	N/A	**Service Ceiling**	N/A

Std. Engine/HP	Usable HP Range	Construction Mtrl.	Est. Bldg. Time
Zenoah G-25 / 20 hp	20 - 30	Aluminum sheet,Tube	400 hrs.
Plans	**Kit**	**Info Packet**	**Video**
$79.00	N/A	$4.00	N/A

Sophisticated ultralight design with modest soaring capability. Tail pusher prop for outstanding stability. Some prefabricated components required from designer. Partial kits also available.

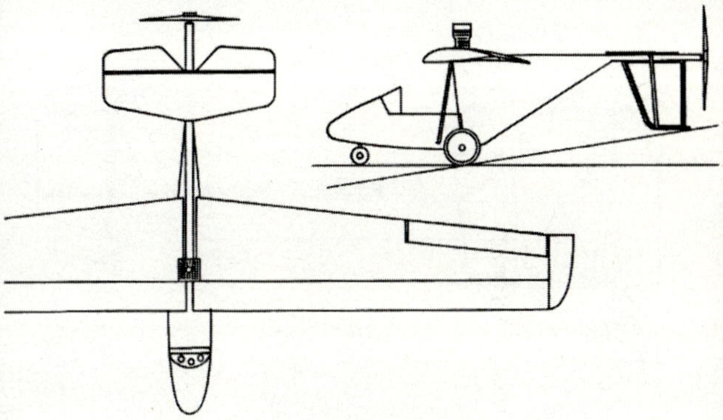

ST. CROIX AIRCRAFT
1139 State Highway #148
Corning IA 50841
Tel. 515-322-4041

Pietenpol Aerial

Wing Span	29.50 ft.	**Range**	150 sm
Wing Area	250.0 sq. ft.	**Takeoff Dist.**	200 ft.
Length	19.50 ft.	**Landing Dist.**	400 ft.
Max Gross Wt.	1300 lbs.	**Vmax**	105 mph
Empty Wt.	700 lbs.	**VCr**	85 mph
Payload (full fuel)	N/A	**V_{S1} (stall clean)**	36 mph
Fuel Capacity	25 gal.	**V_{S0} (lndg config)**	N/A
Seats	2	**Climb Rate @msl**	1000 fpm
Cabin Width	24 in.	**Service Ceiling**	N/A

Std. Engine/HP	Usable HP Range	Construction Mtrl.	Est. Bldg. Time
Continental / 125 hp	85 - 150	Wood, Tube, Fabric	1200 hrs.
Plans	**Kit**	**Info Packet**	**Video**
$105.00	Yes	$7.00	N/A

A delightful adaptation of the Pietenpol Aircamper to a biplane configuration. Very stable. Responsive four ailerons. For 1920's style pasture field fun. Truly "convertible" back and forth, to the monoplane Aircamper. Partial kits are available.

Kits/Plans Sold: 400

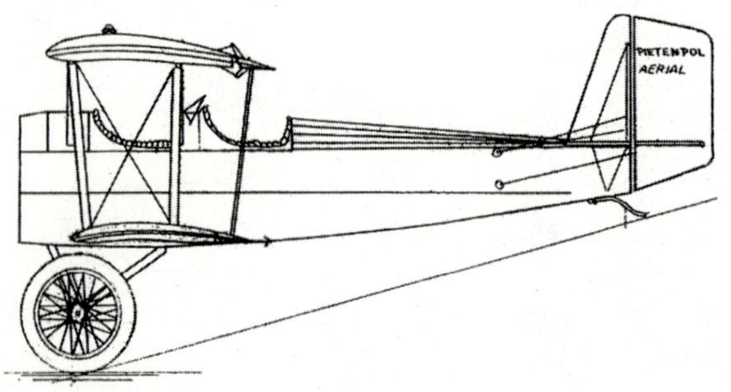

This aircraft was designed by Chad and Charles Wille in 1974. They had built several Pietenpol Aircampers since 1941. The biplane was a successful experiment to see what could be done to the existing Aircamper. The first flight for the biplane was in 1977. Plans are offered as supplements to the Aircamper drawings. This aircraft offers shorter takeoff and landing distances and a lower stall speed along with greater stability in wind and rough air than the Aircamper. One of the easier biplanes to build.

ST. CROIX AIRCRAFT
1139 State Highway #148
Corning IA 50841
Tel. 515-322-4041

Pietenpol Aircamper

Wing Span	29.50 ft.	**Range**	N/A
Wing Area	148.0 sq. ft.	**Takeoff Dist.**	750 ft.
Length	19.50 ft.	**Landing Dist.**	750 ft.
Max Gross Wt.	1100 lbs.	**Vmax**	105 mph
Empty Wt.	650 lbs.	**VCr**	70 mph
Payload (full fuel)	N/A	**V_{S1} (stall clean)**	40 mph
Fuel Capacity	12 gal.	**V_{S0} (lndg config)**	N/A
Seats	2	**Climb Rate @msl**	500 fpm
Cabin Width	24 in.	**Service Ceiling**	N/A

Std. Engine/HP	Usable HP Range	Construction Mtrl.	Est. Bldg. Time
Ford "A" / 40 hp	65 - 125	Wood, Metal, Fabric	1000 hrs.
Plans	**Kit**	**Info Packet**	**Video**
$60.00	Yes	$7.00	N/A

A classic aircraft designed in 1929. Many hundreds have been built. Straight forward, easy, low cost wood construction. Can use many different engines. Partial kits are available.

Kits/Plans Sold: 1000

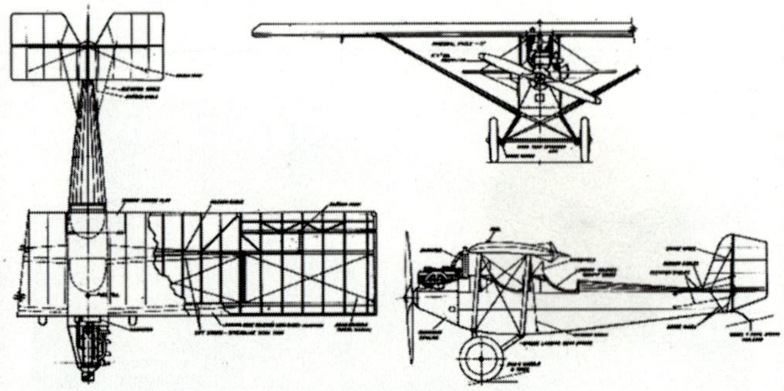

This design evolved during the 1920's to become the best two seater of the depression era homebuilts. Mr. Pietenpol's no nonsense approach has been carried on by thousands of builders over the years and used as an engine test bed for hundreds of different powerplants.

The Aircamper enjoyed a resurgence of interest starting in the 1960's as one of the few designs that could carry two men on low power in a full-sized airplane. The Aircamper is still being built in good numbers, many of them powered by the Model "A" Ford engine for authenticity and antique flavor. Many builders enjoy the ship so much they build a second one after finishing their first. A very honest airplane.

ST. CROIX AIRCRAFT
1139 State Highway #148
Corning IA 50841
Tel. 515-322-4041

Sopwith Triplane

Wing Span	26.50 ft.	**Range**	250 sm
Wing Area	231.0 sq. ft.	**Takeoff Dist.**	250 ft.
Length	19.50 ft.	**Landing Dist.**	750 ft.
Max Gross Wt.	2000 lbs.	**Vmax**	117 mph
Empty Wt.	1475 lbs.	**VCr**	97 mph
Payload (full fuel)	500 lbs.	**V_{S1} (stall clean)**	40 mph
Fuel Capacity	35 gal.	**V_{S0} (lndg config)**	N/A
Seats	2	**Climb Rate @msl**	1000 fpm
Cabin Width	27 in.	**Service Ceiling**	N/A

Std. Engine/HP	Usable HP Range	Construction Mtrl.	Est. Bldg. Time
Lyc 0-435 / 190 hp	150 - 250	Wood, Tube, Fabric	4000 hrs.
Plans	**Kit**	**Info Packet**	**Video**
$85.00	$26,000.00	$12.00	N/A

Replica of successful WWI Scout/Fighter. One or two seats. Excellent ground handling and flying characteristics. Fast cruise, slow landing speed. Relatively easy to operate compared to other WWI designs. Larger and taller than Fokker Tripe. Warm cockpit.

Kits/Plans Sold: 25 **Completed & Flying:** 155

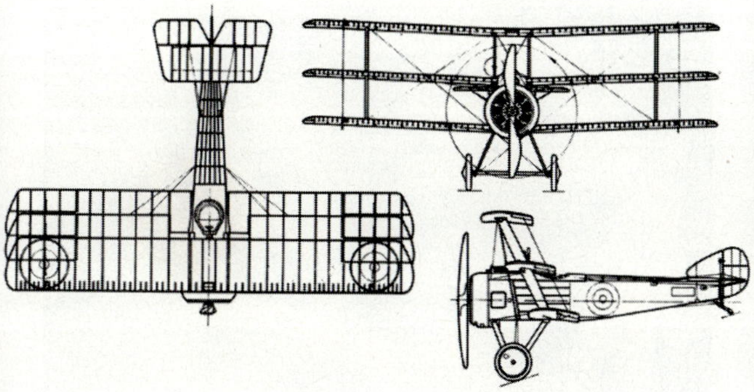

The original Sopwith was designed in 1916 by Herbert Smith at the Kingston-on-Thames Sopwith Works, this famous fighter design was to replace the Sopwith Pup. However, due to rivalry between the services, it ended up being produced for the Royal Naval Air Service instead. Famous Aces like Ray Collishaw scored many victories in this triplane. Its superb qualities led German pilots to demand similar airplanes for themselves. Tony Fokker's little triplane was the only one produced as a result of this demand but they came to the Western Front a year later - too late to have a significant impact on the war.

And now, St. Croix plans show the homebuilder how to create a modern replica of the Sopwith. They include everything necessary to make the conversion to steel tube fuselage, modern engine installation and other details that make this historic aircraft suitable for homebuilding.

ST. CROIX AIRCRAFT
1139 State Highway #148
Corning IA 50841
Tel. 515-322-4041

Fokker DR-1 Triplane

Wing Span	23.50 ft.	**Range**	250 sm
Wing Area	N/A	**Takeoff Dist.**	300 ft.
Length	19.20 ft.	**Landing Dist.**	200 ft.
Max Gross Wt.	1600 lbs.	**Vmax**	120 mph
Empty Wt.	1150 lbs.	**VCr**	110 mph
Payload (full fuel)	N/A	**V_{S1} (stall clean)**	42 mph
Fuel Capacity	23 gal.	**V_{S0} (lndg config)**	N/A
Seats	1	**Climb Rate @msl**	1800 fpm
Cabin Width	28 in.	**Service Ceiling**	15,000 ft.

Std. Engine/HP	Usable HP Range	Construction Mtrl.	Est. Bldg. Time
Lyc 4-cyl. / 185 hp	110 - 185	Wood, Tube, Fabric	3000 hrs.
Plans	**Kit**	**Info Packet**	**Video**
Yes	$32,000.00	$15.00	N/A

100% scale size replica of famous WWI German fighter. Authentic detailing. Engine, instruments, covering not included in kit. Airframe ready to bolt together - all parts pre-fit. Kits and parts available from Wickes Aircraft Supply, 800-221-9425. Uses Lycoming 0-320 to 0-360 engines. Information packet price overseas is $29.00. Call for plans price.

Kits/Plans Sold: 100 **Completed & Flying:** 15

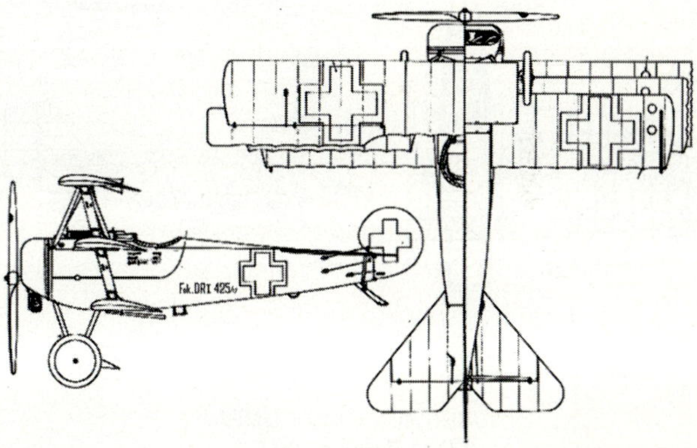

Ross T. Walton, Cedaredge CO
"Mr. Sands' plans are 26 pages of exquisite blueprints that address every detail of constrcution. He covers fundamentals such as how to cut plywood sheets to get the most for your money."

Fred Murrin, Greenville PA
"(Ron Sands') drawings are so thoroughly detailed and easy to understand that anybody could build a plane from them. Ron is very supportive of builders using his plans. He will make some of the more difficult parts for you if you do not have the skills or tooling required."

RON SANDS CO.
89 Forrest Drive
Mertztown PA 19539
Tel. 610-682-6788 • Fax 610-682-6788

P6E Replica

Wing Span	25.83 ft.	**Range**	425 sm
Wing Area	170.0 sq. ft.	**Takeoff Dist.**	N/A
Length	18.91 ft.	**Landing Dist.**	N/A
Max Gross Wt.	2040 lbs.	**Vmax**	185 mph
Empty Wt.	1425 lbs.	**VCr**	120 mph
Payload (full fuel)	475 lbs.	**V_{S1} (stall clean)**	60 mph
Fuel Capacity	27 gal.	**V_{S0} (lndg config)**	N/A
Seats	2	**Climb Rate @msl**	N/A
Cabin Width	N/A	**Service Ceiling**	N/A

Std. Engine/HP	Usable HP Range	Construction Mtrl.	Est. Bldg. Time
Chevy V8 / 212 hp	180 - 260	Tube, Wood, Fabric	N/A
Plans	**Kit**	**Info Packet**	**Video**
$285.00	N/A	$12.00	N/A

82% scale replica of the Curtiss Hawk P6E. Quick change to 2-place seating. Designed to use small block Chevy engines from 305 to 406 cu. in.

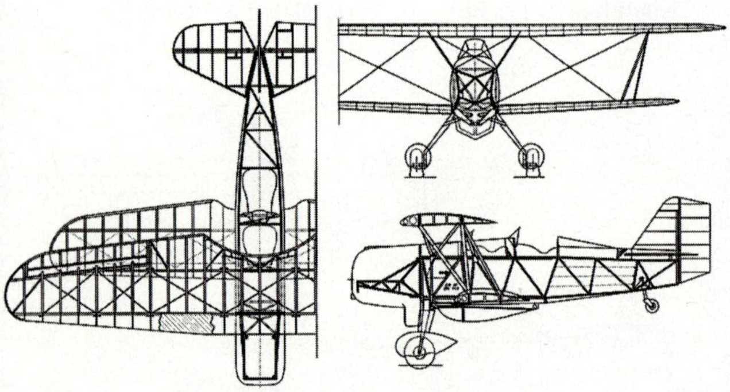

SAUSER AIRCRAFT COMPANY
13101 Ranchwood Rd.
Tustin CA 92782
Tel. 714-544-9015

Rowley P-40F

Wing Span	28.00 ft.	**Range**	403 sm
Wing Area	128.0 sq. ft.	**Takeoff Dist.**	N/A
Length	23.00 ft.	**Landing Dist.**	N/A
Max Gross Wt.	1200 lbs.	**Vmax**	120 mph
Empty Wt.	750 lbs.	**VCr**	110 mph
Payload (full fuel)	320 lbs.	**V_{S1} (stall clean)**	48 mph
Fuel Capacity	22 gal.	**V_{S0} (lndg config)**	N/A
Seats	1	**Climb Rate @msl**	1600 fpm
Cabin Width	N/A	**Service Ceiling**	N/A

Std. Engine/HP	Usable HP Range	Construction Mtrl.	Est. Bldg. Time
AMW Cuyuna / 100 hp	N/A	Aluminum, Wood	N/A
Plans	**Kit**	**Info Packet**	**Video**
$165.00	N/A	Yes	N/A

A plans-only, 3/4-scale Curtis P-40 Warhawk replica. Designed to be quick and easy to build using common shop tools and non-toxic materials that are insensitive to temperature and humidity. Fuselage construction consists of 4130 steel tube with 2024 T3 aluminum skin. Wings utilize aircraft Spruce box spar with aluminum front shear. Plans include a complete supplier list.

76TH FIGHTER SQUADRON, INC.
Box 12 • 13789 Judge Orr Road
Meadow Lake Airport CO 80831
Tel. 719-338-9348

HOMEBUILT "FAQ'S" *(See Page 68)*

Q: What tools and facilities will I need?

A: Many designers will specify the tools needed for construction of their particular aircraft. Again, other builders are a good source of guidance.

Every builder will need a set of basic mechanic's tools. Nothing fancy, but cheap tools often cost you money and time too. Another shop accessory almost everyone needs is a small (or big!) portable air compressor. From painting, to cleaning parts, to driving rivets, to general shop cleanup, compressed air is a welcome asset. A small variable speed power drill is another virtual necessity.

A small grinder and a drill press are other useful items. For wood, a small table saw and bandsaw are indispensable. An orbital sander is needed for wood and composite finishing. Wood builders simply "never" have enough clamps, or so it seems. There are many other items which are often nice to have, but are not necessarily mandatory.

Falco

Wing Span	26.25 ft.	**Range**	1000 sm
Wing Area	107.5 sq. ft.	**Takeoff Dist.**	570 ft.
Length	21.75 ft.	**Landing Dist.**	750 ft.
Max Gross Wt.	1880 lbs.	**Vmax**	212 mph
Empty Wt.	1212 lbs.	**VCr**	190 mph
Payload (full fuel)	N/A	**V_{S1} (stall clean)**	62 mph
Fuel Capacity	40 gal.	**V_{S0} (lndg config)**	N/A
Seats	2	**Climb Rate @msl**	1140 fpm
Cabin Width	N/A	**Service Ceiling**	19,000 ft.

Std. Engine/HP	Usable HP Range	Construction Mtrl.	Est. Bldg. Time
Lycoming / 180 hp	160 - 180	Wood	2000 hrs.
Plans	**Kit**	**Info Packet**	**Video**
$400.00	$70,000.00	$15.00	N/A

Fully aerobatic, all-wood construction. Uses a Hartzell two-blade, constant speed propeller. Designed for Lycoming 160 and 180 hp engines. Cruise speed is at 75%, 6000 feet. Info pack $20 overseas.

SEQUOIA AIRCRAFT CORP.
2000 Tomlynn Street
Richmond VA 23230
Tel. 804-353-1713 • Fax 804-359-2618

Sherwood Ranger LW

Wing Span	26.00 ft.	**Range**	100 sm
Wing Area	154.0 sq. ft.	**Takeoff Dist.**	100 ft.
Length	20.00 ft.	**Landing Dist.**	200 ft.
Max Gross Wt.	860 lbs.	**Vmax**	120 mph
Empty Wt.	400 lbs.	**VCr**	75 mph
Payload (full fuel)	388 lbs.	**V_{S1} (stall clean)**	36 mph
Fuel Capacity	6 gal.	**V_{S0} (lndg config)**	N/A
Seats	2	**Climb Rate @msl**	1200 fpm
Cabin Width	24 in.	**Service Ceiling**	14,000 ft.

Std. Engine/HP	Usable HP Range	Construction Mtrl.	Est. Bldg. Time
Hirth / 65 hp	50-80	Tube, Wood, Fabric	700 hrs.
Plans	**Kit**	**Info Packet**	**Video**
N/A	$15,125.00	$8.00	$10.00

Classic, lightweight, open cockpit, two seat biplane. Designed and manufactured by TCD Ltd. of England and distributed in North America by Sherwood America. Features good performance with economical operation. Good structural integrity (+9, -6 G's) and folding wings. Easy fast construction. ST standard version has 500 lb empty weight and 1000 lb gross. LW lightweight version meets U.K. Section S "microlight" specs.

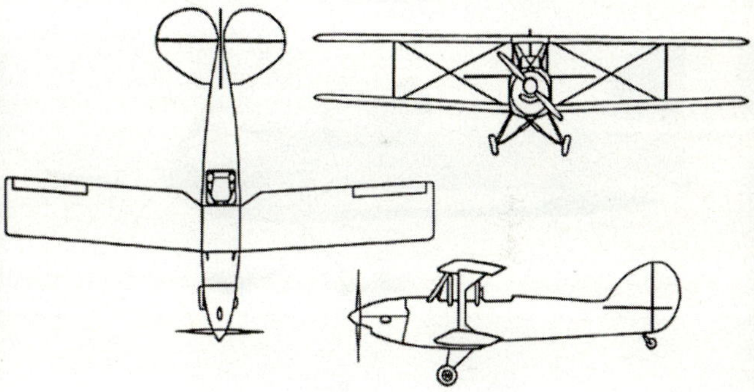

SHERWOOD AMERICA AVIATION
904 Silver Spur Rd., Suite 333
Rolling Hills Estates CA 90274
Tel. 310-325-3422 • Fax 310-378-7685
E-Mail: av8r@ix.netcom.com

Builders often get quite carried away with tools, and it's true that they can make a job easier and faster. But if you're in a squeeze for a particularly expensive tool, think about how often you will use it. It's entirely possible that you're better off renting or borrowing one.

It's amazing to hear of the places airplanes have been built. One grand champion airplane from the '92 season was built on the owner's back patio! But the favorite shop by far is the ubiquitous suburban garage. Most airplanes can have the majority of the work done on components in the garage, usually moving to the hangar at the final assembly stage.

Whatever space you use, make sure it's well lighted and ventilated. Composite aircraft may require winter heat in order for resins to cure properly. Above all, make sure you can get the assembly "out" of the shop before you start. More than one builder has had to "modify" his basement to extract a completed wing.

Sherwood Ranger XP

Wing Span	23.00 ft.	Range	200 sm
Wing Area	140.0 sq. ft.	Takeoff Dist.	100 ft.
Length	20.00 ft.	Landing Dist.	200 ft.
Max Gross Wt.	1000 lbs.	Vmax	120 mph
Empty Wt.	500 lbs.	VCr	85 mph
Payload (full fuel)	428 lbs.	V_{S1} (stall clean)	38 mph
Fuel Capacity	12 gal.	V_{S0} (lndg config)	N/A
Seats	2	Climb Rate @msl	1200 fpm
Cabin Width	24 in.	Service Ceiling	12,000 ft.

Std. Engine/HP	Usable HP Range	Construction Mtrl.	Est. Bldg. Time
Rotax 582 / 66 hp	66-100	Tube, Wood, Fabric	600 hrs.
Plans	**Kit**	**Info Packet**	**Video**
$450.00	$15,600.00	$8.00	$10.00

Clipped wing "Sport" model of the ST Ranger. Can go from flying condition to on the trailer for driving away in three minutes (one person). Engine options include full Rotax line, Jabiru, Ultratec, 2si. Quick Build option for additional $1200.00. Can be built from plans. Machined parts available.

SHERWOOD AMERICA AVIATION
904 Silver Spur Rd., Suite 333 • Rolling Hills Estates CA 90274
Tel. 310-325-3422 • Fax 310-378-7685 • E-Mail: av8r@ix.netcom.com

Sonex

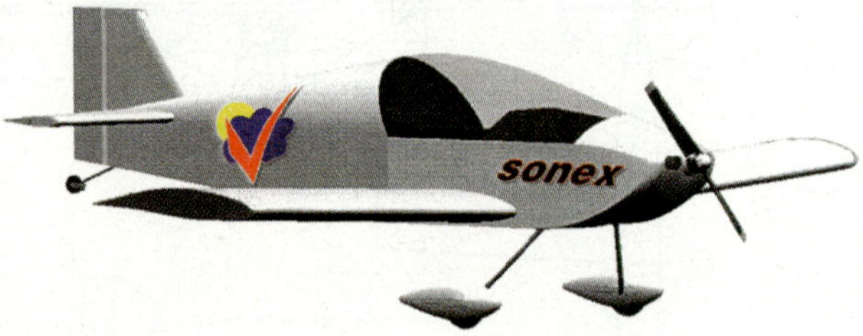

Wing Span	22.00 ft.	Range	320 sm
Wing Area	98.00 sq. ft.	Takeoff Dist.	650 ft.
Length	17.60 ft.	Landing Dist.	N/A
Max Gross Wt.	950 lbs.	Vmax	150 mph
Empty Wt.	525 lbs.	VCr	130 mph
Payload (full fuel)	N/A	V_{S1} (stall clean)	40 mph
Fuel Capacity	12 gal.	V_{S0} (lndg config)	N/A
Seats	N/A	Climb Rate @msl	N/A
Cabin Width	N/A	Service Ceiling	N/A

Std. Engine/HP	Usable HP Range	Construction Mtrl.	Est. Bldg. Time
Jabiru / 80 hp	N/A	Aluminum	N/A
Plans	**Kit**	**Info Packet**	**Video**
Yes	Yes	Yes	N/A

Designed to use inexpensive, easily worked 6061 aluminum alloys. Build from scratch or component kits for under $10,000. VW engine can also be used. Conventional or tri gear.

SKY STRUCK
167 North Oakwood Rd. • Oshkosh WI 54904
Tel. 800-759-7875 • Fax 414-231-8297
E-Mail: sonerai@vbe.com
Web: http://www.vbe.com~sonerai

SkyDancer SD-260

Wing Span	22.00 ft.	Range	320 sm
Wing Area	135.0 sq. ft.	Takeoff Dist.	600 ft.
Length	21.00 ft.	Landing Dist.	800 ft.
Max Gross Wt.	1850 lbs.	Vmax	185 mph
Empty Wt.	1250 lbs.	VCr	160 mph
Payload (full fuel)	N/A	V_{S1} (stall clean)	56 mph
Fuel Capacity	29 gal.	V_{S0} (lndg config)	56 mph
Seats	2	Climb Rate @msl	2200 fpm
Cabin Width	N/A	Service Ceiling	18,000 ft.

Std. Engine/HP	Usable HP Range	Construction Mtrl.	Est. Bldg. Time
Lyc. IO-540 / 260 hp	200 - 400	Tube, Fabric	1200 hrs.
Plans	**Kit**	**Info Packet**	**Video**
N/A	$35,000.00	Yes	N/A

Two seat, open cockpit biplane. Full sized. Will provide aerobatic capability and rugged sport flying utility for two large adults in comfort. Kit includes everything except engine, propeller, and radio.Three more SkyDancers, all with interchangeable components, are being developed. They include the SD-200 Basic, SD200C Classic, and the single seat SD-300S for advanced aerobatics.

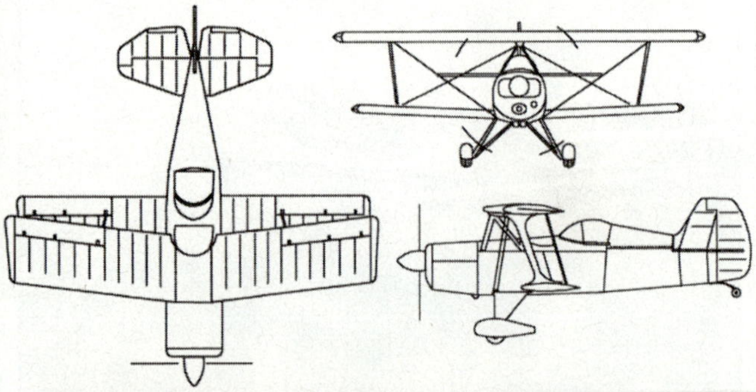

SKYDANCER AVIATION
1816 Overlook Terrace
Louisville KY 40205
Tel. 502-454-6295

Pulsar Series II

Wing Span	25.00 ft.	**Range**	900 sm
Wing Area	80.00 sq. ft.	**Takeoff Dist.**	250 ft.
Length	20.00 ft.	**Landing Dist.**	500 ft.
Max Gross Wt.	1200 lbs.	**Vmax**	190 mph
Empty Wt.	660 lbs.	**VCr**	180 mph
Payload (full fuel)	570 lbs.	**V_{S1} (stall clean)**	49 mph
Fuel Capacity	17 gal.	**V_{S0} (lndg config)**	N/A
Seats	2	**Climb Rate @msl**	2000 fpm
Cabin Width	39 in.	**Service Ceiling**	17,000 ft.

Std. Engine/HP	Usable HP Range	Construction Mtrl.	Est. Bldg. Time
Rot. 914 Turbo/115 hp	N/A	Composite	1000 hrs.
Plans	**Kit**	**Info Packet**	**Video**
N/A	$17,950.00	$10.00	N/A

Previously manufactured by Aero Designs of San Antonio, TX. A sleek composite aircraft with removable wings. Utilizes a light, simple design. Options include quick build kits, heater kit, and upholstery kit. Cargo bay capacity is 60 lbs. Engine options include the Rotax 912 and the Rotax 914 Turbo. Price is for the "Firewall Aft Airframe Kit." See Ad on Page 23.

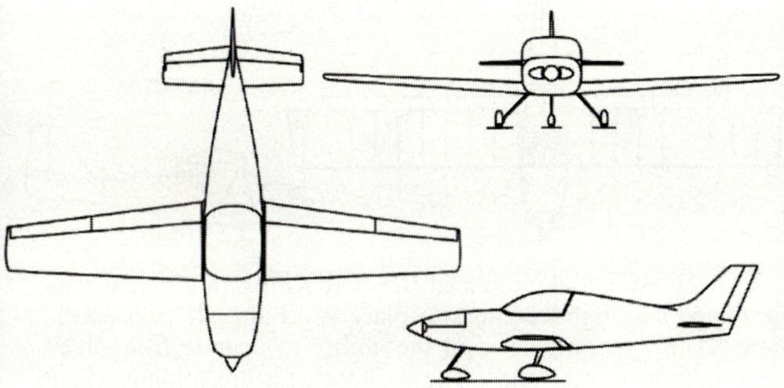

SKYSTAR AIRCRAFT CORP.
100 North Kings Rd.
Nampa ID 83651
Tel. 800-554-8369 • Fax 208-454-6464
Web: http://www.skystar.com

Kitfox Series 5 - Vixen

Wing Span	29.00 ft.	**Range**	630 sm
Wing Area	121.0 sq. ft.	**Takeoff Dist.**	350 ft.
Length	20.25 ft.	**Landing Dist.**	300 ft.
Max Gross Wt.	1550 lbs.	**Vmax**	140 mph
Empty Wt.	725 lbs.	**VCr**	115 mph
Payload (full fuel)	800 lbs.	**V_{S1} (stall clean)**	48 mph
Fuel Capacity	26 gal.	**V_{S0} (lndg config)**	43 mph
Seats	2	**Climb Rate @msl**	1000 fpm
Cabin Width	43 in.	**Service Ceiling**	13,000 ft.

Std. Engine/HP	Usable HP Range	Construction Mtrl.	Est. Bldg. Time
Rotax 912 / 80 hp	80 - 125	Tube, Fabric	750 hrs.
Plans	**Kit**	**Info Packet**	**Video**
N/A	$13,995.00	$15.00	$15.00

Popular tri-gear configuration with Series 5 comfort and cargo capacity (8.5 cubic feet, 150 pounds). Series 5 models now available with Rotax 912, Continental O-200 and IO-240, and Lycoming O-235 engines. Folding wings. Price does not include engine package. See Ad on Page 23.

Kits/Plans Sold: 2500 **Completed & Flying:** 1400

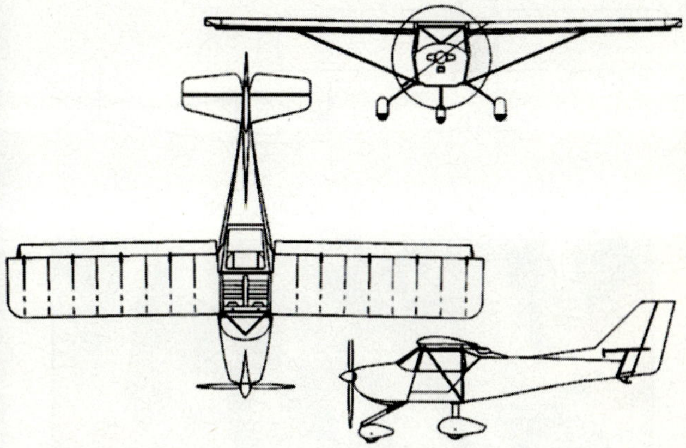

SKYSTAR AIRCRAFT CORP.
100 North Kings Rd.
Nampa ID 83651
Tel. 800-554-8369 • Fax 208-454-6464
Web: http://www.skystar.com

Kitfox Series 5 - Model 5

Wing Span	32.00 ft.	**Range**	605 sm
Wing Area	132.0 sq. ft.	**Takeoff Dist.**	380 ft.
Length	19.16 ft.	**Landing Dist.**	300 ft.
Max Gross Wt.	1550 lbs.	**Vmax**	115 mph
Empty Wt.	750 lbs.	**VCr**	110 mph
Payload (full fuel)	800 lbs.	**V_{S1} (stall clean)**	48 mph
Fuel Capacity	26 gal.	**V_{S0} (lndg config)**	43 mph
Seats	2	**Climb Rate @msl**	1000 fpm
Cabin Width	43 in.	**Service Ceiling**	19,000 ft.

Std. Engine/HP	Usable HP Range	Construction Mtrl.	Est. Bldg. Time
Rotax 912 / 80 hp	80 - 125	Tube, Fabric	750 hrs.
Plans	**Kit**	**Info Packet**	**Video**
N/A	$14,995.00	$10.00	$15.00

Series 5 models are available with Rotax 912, Rotax 914, Continental O-200 and IO-240, and Lycoming O-235 engines. Folding wings. Streamlined cowl and "bump" cowl designs are available. Available with standard and "Safari" options packages. Price does not include engine package. See Ad on Page 23.

Kits/Plans Sold: 2500 **Completed & Flying:** 1400

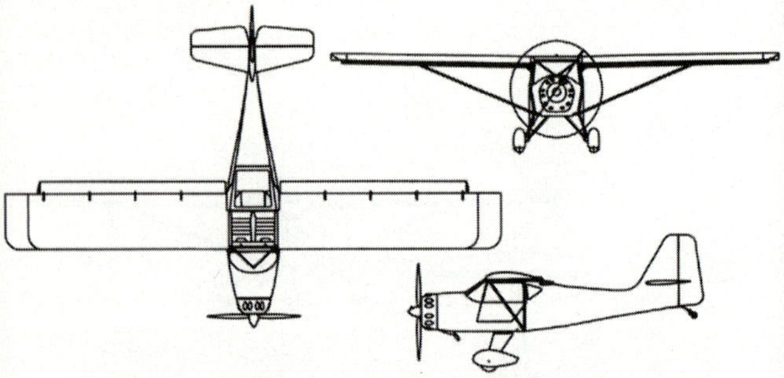

SKYSTAR AIRCRAFT CORP.
100 North Kings Rd.
Nampa ID 83651
Tel. 800-554-8369 • Fax 208-454-6464
Web: http://www.skystar.com

Kitfox Classic IV

Wing Span	32.00 ft.	**Range**	485 sm
Wing Area	132.0 sq. ft.	**Takeoff Dist.**	235 ft.
Length	18.10 ft.	**Landing Dist.**	250 ft.
Max Gross Wt.	1200 lbs.	**Vmax**	125 mph
Empty Wt.	560 lbs.	**VCr**	105 mph
Payload (full fuel)	450 lbs.	**V_{S1} (stall clean)**	44 mph
Fuel Capacity	26 gal.	**V_{S0} (lndg config)**	39 mph
Seats	2	**Climb Rate @msl**	840 fpm
Cabin Width	40 in.	**Service Ceiling**	12,000 ft.

Std. Engine/HP	Usable HP Range	Construction Mtrl.	Est. Bldg. Time
Rotax 912 / 80 hp	52 - 100	Tube, Fabric	650 hrs.
Plans	**Kit**	**Info Packet**	**Video**
N/A	$11,500.00	$15.00	$15.00

Popular, proven design available with Rotax 503, 582, 618, and 912 engines. All Kitfox models feature folding wings for easy car towing, hanger storage, or garage storage. Price does not include engine package. See Ad on Page 23.

Kits/Plans Sold: 2500 **Complete/Flying:** 1400

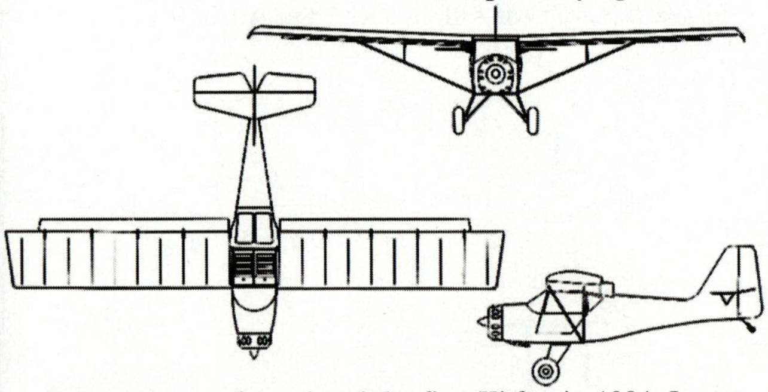

Denney Aerocraft produced the first Kitfox in 1984. It was designed as a lightweight, two place sport aircraft with excellent STOL performance and the ability to operate from short and unimproved airfields. These aircraft feature folding wings and are easily trailered, allowing owners to share hanger space or to keep their aircraft at home in a single car garage. Since 1984 over 2,500 Kitfoxes have been delivered to builders throughout the US and in over 40 foreign countries.

In 1992 Denney Aerocraft became SkyStar Aircraft Corporation. SkyStar immediately began development of anew Kitfox - the Series 5. This larger aircraft was designed to fit the needs of a growing segment of the marketplace that wanted a "Weekend Cruiser." It has become one of the most successful new introductions in the history of the kit aircraft industry.

Charles B. Holmes, Sr., St. Petersburg, FL

"...I can honestly say that building the Kitfox is reasonably easy and a most enjoyable pastime. ...I would like to commend all the personnel of Kitfox for providing a kit airplane that is professional in all respects and especially for the excellent and friendly handling of all needs and requirements."

SKYSTAR AIRCRAFT CORP.
100 North Kings Rd.
Nampa ID 83651
Tel. 800-554-8369 • Fax 208-454-6464
Web: http://www.skystar.com

Kitfox Series 5 - Safari

Wing Span	32.00 ft.	**Range**	N/A
Wing Area	132.0 sq. ft.	**Takeoff Dist.**	200 ft.
Length	19.20 ft.	**Landing Dist.**	300 ft.
Max Gross Wt.	1550 lbs.	**Vmax**	125 mph
Empty Wt.	N/A	**VCr**	120 mph
Payload (full fuel)	N/A	**V_{S1} (stall clean)**	N/A
Fuel Capacity	96 gal.	**V_{S0} (lndg config)**	43 mph
Seats	2	**Climb Rate @msl**	1400 fpm
Cabin Width	42.5 in.	**Service Ceiling**	N/A

Std. Engine/HP	Usable HP Range	Construction Mtrl.	Est. Bldg. Time
Cont. O-200/ 100 hp	80 - 125	Tube, Fabric	650 hrs.
Plans	**Kit**	**Info Packet**	**Video**
N/A	Yes	$15.00	$15.00

With a wider cabin than a Bonanza. Good back country aircraft. Options include Adjustable rudder pedals, Quick Build wings, cabin heater, tundra tires, floats, skis, and wheel pants. Engine options Rotax 912, Cont. O-200 and IO-240B, Lycoming O-235. See Ad on Page 23.

SKYSTAR AIRCRAFT CORP.
100 North Kings Rd. • Nampa ID 83651
Tel. 800-554-8369 • Fax 208-454-6464
Web: http://www.skystar.com

Slepcev Storch

Wing Span	32.17 ft.	**Range**	N/A
Wing Area	160.0 sq. ft.	**Takeoff Dist.**	30 ft.
Length	22.67 ft.	**Landing Dist.**	50 ft.
Max Gross Wt.	1100 lbs.	**Vmax**	N/A
Empty Wt.	600 lbs.	**VCr**	N/A
Payload (full fuel)	N/A	**V_{S1} (stall clean)**	N/A
Fuel Capacity	15 gal.	**V_{S0} (lndg config)**	14 mph
Seats	2	**Climb Rate @msl**	1200 fpm
Cabin Width	N/A	**Service Ceiling**	N/A

Std. Engine/HP	Usable HP Range	Construction Mtrl.	Est. Bldg. Time
83 hp	N/A	Tube, Fabric	800 hrs.
Plans	**Kit**	**Info Packet**	**Video**
$260.00	$16,950.00	$5.00	$15.00

Experimental 75% scale model of the well know WW II Fieseler Fi-156 "Storch." Design is approximately 90% true to the original in detail. Outstanding STOL performance.

Kits/Plans Sold: 45 **Completed & Flying:** 15

SLEPCEV AIRCRAFT INDUSTRIES
VRBAS Beechwood • N.S.W. 2446
AUSTRALIA
Tel. 61 65 856 458 • Fax 61 65 856 458

Speedtwin Mk. II

Wing Span	26.00 ft.	**Range**	1100 sm
Wing Area	120.0 sq. ft.	**Takeoff Dist.**	330 ft.
Length	22.80 ft.	**Landing Dist.**	510 ft.
Max Gross Wt.	2450 lbs.	**Vmax**	215 mph
Empty Wt.	1500 lbs.	**VCr**	200 mph
Payload (full fuel)	374 lbs.	**V_{S1} (stall clean)**	68 mph
Fuel Capacity	96 gal.	**V_{S0} (lndg config)**	60 mph
Seats	2	**Climb Rate @msl**	3050 fpm
Cabin Width	25 in.	**Service Ceiling**	25,000 ft.

Std. Engine/HP	Usable HP Range	Construction Mtrl.	Est. Bldg. Time
Twin / 160 hp	100 - 160	Metal, Carbon Fiber	1000 hrs.
Plans	**Kit**	**Info Packet**	**Video**
N/A	$55,000.00	Yes	N/A

Versatile, all-metal, twin engine aircraft. Well mannered, yet extremely capable. Exceptional over-the-nose visibility. Production model successor to the Mk I prototype which was under development for 10 years. Structure capable of +6, -3 G's aerobatics. Long range and high cruise speed. Usable on short unimproved, STOL airstrips. Kit price is approximate in US funds.

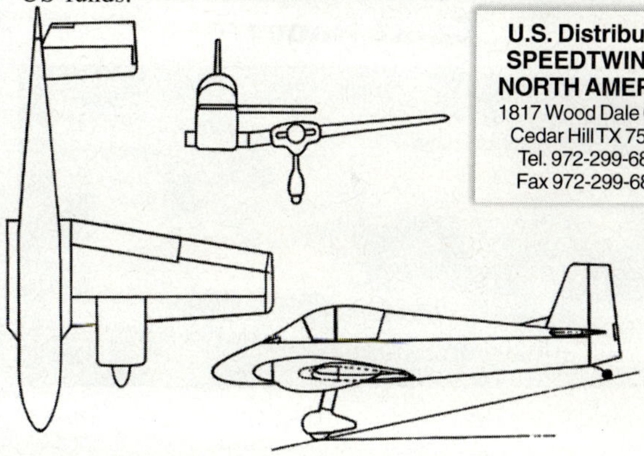

U.S. Distributor:
SPEEDTWIN OF NORTH AMERICA
1817 Wood Dale Circle
Cedar Hill TX 75104
Tel. 972-299-6804
Fax 972-299-6804

The Speedtwin line can trace its origin to the United Kingdom where in 1981 respected RAF Test and Aerobatic pilot, Peter J.C. Phillips, began to create a superior personal aircraft that possessed outstanding handling and stability characteristics combined with truly economical multi-engine safety. Nearly ten years were spent designing and building the proof-of-concept prototype Speedtwin Mk I that met the strictest international standards for handling and stability. Extensive additional development has produced the current model Speedtwin Mk II.

Bob Grimstead, Pilot, June '93

"Pure, delightful handling is the ultimate accolade. This is one of the few aircraft I have sampled meeting this criterion. For the connoisseur of fine handling, the Speedtwin is a singular aircraft from a remarkable man (Peter Phillips)."

SPEEDTWIN DEVELOPMENT, INC.
Upper Cae Garw Farm • Trelleck, Monmouth, Gwent
Wales NP5 4PJ
UNITED KINGDOM
Ph: 0600.860165 • Fax: 0600.860813

Spencer Air Car S-12-D

Wing Span	37.25 ft.	**Range**	700 sm
Wing Area	183.0 sq. ft.	**Takeoff Dist.**	750 ft.
Length	26.42 ft.	**Landing Dist.**	500 ft.
Max Gross Wt.	3150 lbs.	**Vmax**	152 mph
Empty Wt.	2100 lbs.	**VCr**	140 mph
Payload (full fuel)	N/A	**V_{S1} (stall clean)**	43 mph
Fuel Capacity	94 gal.	**V_{S0} (Indg config)**	N/A
Seats	4	**Climb Rate @msl**	900 fpm
Cabin Width	N/A	**Service Ceiling**	15,000 ft.

Std. Engine/HP	Usable HP Range	Construction Mtrl.	Est. Bldg. Time
Lyc. 0-540 / 260 hp	N/A	Wood, Tube, Fabric	3000 hrs.
Plans	**Kit**	**Info Packet**	**Video**
$350.00	$37,500.00	$10.00	N/A

Four place amphibian pusher. Roadable with fold up wings. Continental O-520 300 hp engine is optional. First flown May 25, 1970. Some prefab components are available.

Kits/Plans Sold: 140 **Completed & Flying:** 40

SPENCER AMPHIBIAN AIRCRAFT
P.O. Box 327
Kansas IL 61933
Tel. 217-948-5505

Talon Magnum

Wing Span	26.66 ft.	**Range**	N/A
Wing Area	134.0 sq. ft.	**Takeoff Dist.**	150 ft.
Length	21.75 ft.	**Landing Dist.**	N/A
Max Gross Wt.	700 lbs.	**Vmax**	80 mph
Empty Wt.	323 lbs.	**VCr**	78 mph
Payload (full fuel)	377 lbs.	**V_{S1} (stall clean)**	30 mph
Fuel Capacity	N/A	**V_{S0} (Indg config)**	N/A
Seats	1	**Climb Rate @msl**	1450 fpm
Cabin Width	N/A	**Service Ceiling**	N/A

Std. Engine/HP	Usable HP Range	Construction Mtrl.	Est. Bldg. Time
Rotax 582	N/A	Tube, Fabric	125 hrs.
Plans	**Kit**	**Info Packet**	**Video**
N/A	$9,500.00	Yes	$19.95

High wing tube and fabric design that can fall under Ultralight or Experimental category. Engine options include Rotax 447, 503, and 582. Also available in the Talon Typhoon aerobatic version for $9950. **Completed & Flying:** 82

SPORT FLIGHT AVIATION
20051 SE Tickle Creek Rd.
Boring OR 97009
Tel. 503-663-4949 • Fax 503-668-9321

Sport Racer

Wing Span	22.00 ft.	**Range**	525 sm
Wing Area	81.00 sq. ft.	**Takeoff Dist.**	1500 ft.
Length	21.15 ft.	**Landing Dist.**	1600 ft.
Max Gross Wt.	1850 lbs.	**Vmax**	225 mph
Empty Wt.	1175 lbs.	**VCr**	175 mph
Payload (full fuel)	N/A	**V_{S1} (stall clean)**	62 mph
Fuel Capacity	30 gal.	**V_{S0} (Indg config)**	N/A
Seats	2	**Climb Rate @msl**	900 fpm
Cabin Width	N/A	**Service Ceiling**	18,000 ft.

Std. Engine/HP	Usable HP Range	Construction Mtrl.	Est. Bldg. Time
Ford 230 V6 / 230 hp	N/A	Metal, Wood	1600 hrs.
Plans	**Kit**	**Info Packet**	**Video**
$375.00	N/A	$10.00	N/A

Mid wing, high performance monoplane designed to accept Ford conversions. Uses a steel and fabric fuselage with wood and fabric wings. Also see the Sport Racer Wichawk listing in the "Other Aircraft" section.

SPORT RACER, INC.
14 Hawthorne Road
Valley Center KS 67147
Tel. 316-755-0659

Talon XP

Wing Span	28.16 ft.	**Range**	N/A
Wing Area	142.0 sq. ft.	**Takeoff Dist.**	150 ft.
Length	21.75 ft.	**Landing Dist.**	N/A
Max Gross Wt.	850 lbs.	**Vmax**	85 mph
Empty Wt.	412 lbs.	**VCr**	80 mph
Payload (full fuel)	435 lbs.	**V_{S1} (stall clean)**	35 mph
Fuel Capacity	N/A	**V_{S0} (Indg config)**	N/A
Seats	2	**Climb Rate @msl**	1550 fpm
Cabin Width	N/A	**Service Ceiling**	N/A

Std. Engine/HP	Usable HP Range	Construction Mtrl.	Est. Bldg. Time
Rotax 582	N/A	Tube, Fabric	125 hrs.
Plans	**Kit**	**Info Packet**	**Video**
N/A	$11,500.00	Yes	N/A

Two seat version of the Talon. Engine options include Rotax 447, 503, and 582. Also available in the Talon Super Magnum version which removes the rear seat for additional cargo space. The Super Magnum is $10,500

SPORT FLIGHT AVIATION
20051 SE Tickle Creek Rd.
Boring OR 97009
Tel. 503-663-4949 • Fax 503-668-9321

Super Cyclone

Wing Span	38.00 ft.	**Range**	1000 sm
Wing Area	191.0 sq. ft.	**Takeoff Dist.**	250 ft.
Length	26.00 ft.	**Landing Dist.**	650 ft.
Max Gross Wt.	3500 lbs.	**Vmax**	175 mph
Empty Wt.	1800 lbs.	**VCr**	165 mph
Payload (full fuel)	N/A	**V_{S1} (stall clean)**	N/A
Fuel Capacity	90 gal.	**V_{S0} (lndg config)**	40 mph
Seats	4	**Climb Rate @msl**	1500 fpm
Cabin Width	N/A	**Service Ceiling**	N/A

Std. Engine/HP	Usable HP Range	Construction Mtrl.	Est. Bldg. Time
N/A	260-330	N/A	N/A
Plans	**Kit**	**Info Packet**	**Video**
Yes	Yes	$30.00	Yes

Higher gross weight/payload version of the Cyclone with 260-330 usable hp range.

ST. JUST AVIATION, INC.
13,361 Cure' Labelle
Mirabel, (St-janvier) QB J7J 1H1
CANADA
Tel. 514-437-6334 • Fax 514-437-5783

Firebolt

Wing Span	24.00 ft.	**Range**	500 sm
Wing Area	150.0 sq. ft.	**Takeoff Dist.**	400 ft.
Length	21.40 ft.	**Landing Dist.**	800 ft.
Max Gross Wt.	2000 lbs.	**Vmax**	235 mph
Empty Wt.	1325 lbs.	**VCr**	202 mph
Payload (full fuel)	675 lbs.	**V_{S1} (stall clean)**	62 mph
Fuel Capacity	39 gal.	**V_{S0} (lndg config)**	N/A
Seats	2	**Climb Rate @msl**	4000 fpm
Cabin Width	29 in.	**Service Ceiling**	18,000 ft.

Std. Engine/HP	Usable HP Range	Construction Mtrl.	Est. Bldg. Time
Lycoming / 300 hp	180 - 300	Tubing, Fabric, Wood, Metal	3000 hrs.
Plans	**Kit**	**Info Packet**	**Video**
$275.00	N/A	$10.00	N/A

Two place, fully aerobatic biplane based on the classic Skybolt design. Flush mounted, fighter-type canopy slides straight back. Overseas plans price is $350. Some prefab components are available. **Completed & Flying:** 6

STARFIRE AVIATION, INC.
907 S. Hohokam Drive
Tempe AZ 85281-5115
Tel. 602-731-9419 • Fax 602-731-9419

Cyclone

Wing Span	38.00 ft.	**Range**	870 sm
Wing Area	181.0 sq. ft.	**Takeoff Dist.**	250 ft.
Length	26.00 ft.	**Landing Dist.**	950 ft.
Max Gross Wt.	3000 lbs.	**Vmax**	170 mph
Empty Wt.	1700 lbs.	**VCr**	158 mph
Payload (full fuel)	N/A	**V_{S1} (stall clean)**	N/A
Fuel Capacity	N/A	**V_{S0} (lndg config)**	37 mph
Seats	4	**Climb Rate @msl**	1300 fpm
Cabin Width	N/A	**Service Ceiling**	18,000 ft.

Std. Engine/HP	Usable HP Range	Construction Mtrl.	Est. Bldg. Time
Cont. / 230 hp	200-250	Aluminum, Fiberglass, Steel	2000 hrs.
Plans	**Kit**	**Info Packet**	**Video**
Yes	$34,833.00	$30.00	Yes

A replication of the Cessna C-180 with improvements to enhance performance and utility. All skins are supplied already cut and have pick-up holes which eliminates complicated and expensive jigs. Price is for complete kit with all options.

ST. JUST AVIATION, INC.
13,361 Cure' Labelle
Mirabel, (St-janvier) QB J7J 1H1
CANADA
Tel. 514-437-6334 • Fax 514-437-5783

Warp 1-A

Wing Span	2.50 ft.	**Range**	N/A
Wing Area	87.50 sq. ft.	**Takeoff Dist.**	150 ft.
Length	N/A	**Landing Dist.**	150 ft.
Max Gross Wt.	575 lbs.	**Vmax**	110 mph
Empty Wt.	250 lbs.	**VCr**	95 mph
Payload (full fuel)	N/A	**V_{S1} (stall clean)**	40 mph
Fuel Capacity	5 gal.	**V_{S0} (lndg config)**	N/A
Seats	1	**Climb Rate @msl**	900 fpm
Cabin Width	N/A	**Service Ceiling**	12,000 ft.

Std. Engine/HP	Usable HP Range	Construction Mtrl.	Est. Bldg. Time
Rotax 503 / 50 hp	N/A	Composite	100 hrs.
Plans	**Kit**	**Info Packet**	**Video**
N/A	$26,995.00	$20.00	N/A

All-composite ultralight with low operating costs, detachable wings. Price includes engine, prop, and instruments. Wing utilizes alloy aluminum spar with streamlined S-Glass (TM) vinylester resin surface.

STAR-LITE ENGINEERING LTD.
11509 Diamond Mill Rd.
Englewood OH 45322
Tel. 513-836-9228
Web: http://www.star-lite.com

Knight Twister

Wing Span	15.50 ft.	**Range**	600 sm
Wing Area	55.00 sq. ft.	**Takeoff Dist.**	350 ft.
Length	13.50 ft.	**Landing Dist.**	1000 ft.
Max Gross Wt.	865 lbs.	**Vmax**	180 mph
Empty Wt.	517 lbs.	**VCr**	150 mph
Payload (full fuel)	N/A	**V_{S1} (stall clean)**	55 mph
Fuel Capacity	25 gal.	**V_{S0} (lndg config)**	N/A
Seats	1	**Climb Rate @msl**	1800 fpm
Cabin Width	N/A	**Service Ceiling**	19,000 ft.

Std. Engine/HP	Usable HP Range	Construction Mtrl.	Est. Bldg. Time
Lycoming / 100 hp	85 - 150	Steel Tube, Wood	2800 hrs.
Plans	**Kit**	**Info Packet**	**Video**
$200.00	N/A	$10.00	N/A

Other single place versions available with 17.5 and 19.5 foot wingspans. A two place, 22.5 foot version is also available. See Ad on Page 2.

Completed & Flying: 100

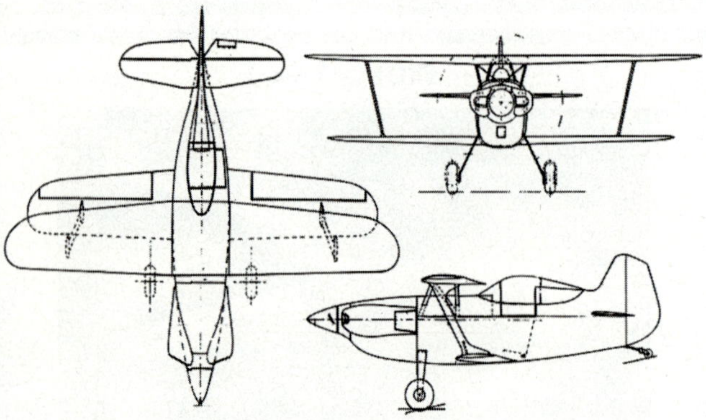

STEEN AERO LAB, INC.
1210 Airport Road
Marion NC 28752
Tel. 704-652-7382 • Fax 704-652-7382

Pitts S1

Wing Span	17.30 ft.	**Range**	275 sm
Wing Area	98.50 sq. ft.	**Takeoff Dist.**	300 ft.
Length	15.50 ft.	**Landing Dist.**	500 ft.
Max Gross Wt.	1150 lbs.	**Vmax**	203 mph
Empty Wt.	720 lbs.	**VCr**	143 mph
Payload (full fuel)	N/A	**V_{S1} (stall clean)**	64 mph
Fuel Capacity	20 gal.	**V_{S0} (lndg config)**	N/A
Seats	1	**Climb Rate @msl**	2600 fpm
Cabin Width	N/A	**Service Ceiling**	10,000 ft.

Std. Engine/HP	Usable HP Range	Construction Mtrl.	Est. Bldg. Time
Lyc IO-360 / 180 hp	125 - 200	Steel Tube, Wood	2800 hrs.
Plans	**Kit**	**Info Packet**	**Video**
$250.00	N/A	$10.00	N/A

The classic single place biplane with exceptional aerobatic capability yet within the ability of an average homebuilder to construct. A tried and proven design with hundreds flying all over the world. Available in the original version as the S1-C or in an advanced version with modern symmetrical airfoils as the S1-SS. Completely updated plans developed on Autocad are available. See Ad on Page 2. **Completed & Flying:** 500

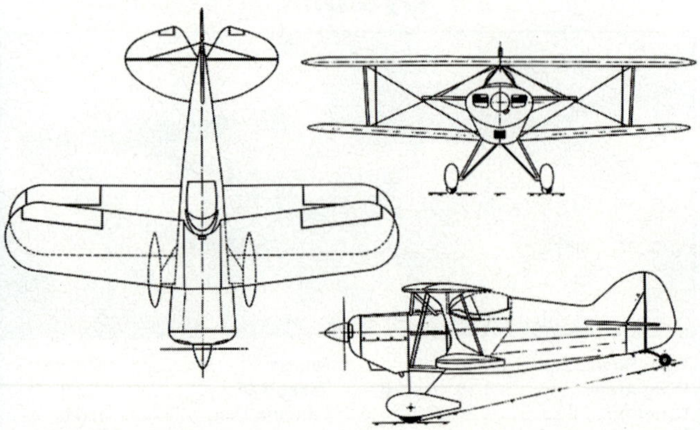

Randall Robinson, California

"I am very pleased with your information package and your sample drawings are the best I've seen. I've been looking at samples for about a year and have seen quite a few of them. The information in your package is complete and professional, my compliments."

Quin Martin, Louisiana

"Thanks for a great information packet. It's a real confidence builder knowing I can expect prompt, professional service at good prices."

STEEN AERO LAB, INC.
1210 Airport Road
Marion NC 28752
Tel. 704-652-7382 • Fax 704-652-7382

Skybolt

Wing Span	24.00 ft.	Range	450 sm
Wing Area	152.7 sq. ft.	Takeoff Dist.	300 ft.
Length	20.00 ft.	Landing Dist.	800 ft.
Max Gross Wt.	2000 lbs.	Vmax	200 mph
Empty Wt.	1250 lbs.	VCr	150 mph
Payload (full fuel)	N/A	V_{S1} (stall clean)	65 mph
Fuel Capacity	38 gal.	V_{S0} (lndg config)	N/A
Seats	2	Climb Rate @msl	3500 fpm
Cabin Width	29 in.	Service Ceiling	12,500 ft.

Std. Engine/HP	Usable HP Range	Construction Mtrl.	Est. Bldg. Time
Lyc. IO-540 / 260 hp	150 - 300	Tube, Wood, Fabric	3000 hrs.
Plans	**Kit**	**Info Packet**	**Video**
$165.00	Yes	$10.00	$25.00

Steen Skybolt is a strong, agile, proven aircraft that is relatively easy to build with only hand tools required.The airframe is stressed for aerobatic training and competition. Has been owned and flown by numerous well know pilots over the last 24 years. See Ad on Page 2. **Completed & Flying:** 425

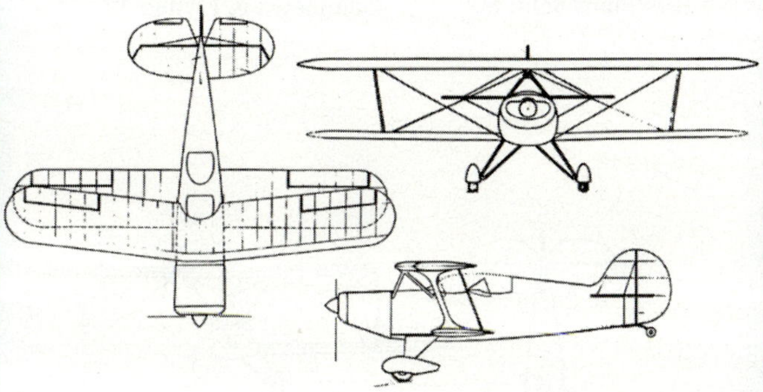

The Skybolt has enjoyed over twenty-four years of success with over 425 examples flying. Active building projects are underway in all 50 states and 33 other countries.

The Skybolt was designed as a fully aerobatic, honest two place biplane with simplicity of construction the prime goal. The prototype was powered by a 180 hp Lycoming and was capable of looping from level flight at Denver's 5,000 foot altitude. Static loading was performed as well as a professional stress analysis.The prototype was test flown to 225 mph with +8 and -6 G loads applied. Tens of thousands of hours of flying have shown the design to be strong and reliable.

Scotty Jergenson, Learjet test pilot & airshow performer

"The Skybolt is a well designed aircraft that exhibits excellent stability and control parameters for precision aerobatics. Ground handling is superb... I really like the lines. ... it presents itself very well esthetically on the ground as well as in the air and has been a good airshow performer for me."

STEEN AERO LAB, INC.
1210 Airport Road
Marion NC 28752
Tel. 704-652-7382 • Fax 704-652-7382

Mini-Imp

Wing Span	26.00 ft.	Range	500 sm
Wing Area	78.00 sq. ft.	Takeoff Dist.	600 ft.
Length	17.00 ft.	Landing Dist.	600 ft.
Max Gross Wt.	850 lbs.	Vmax	200 mph
Empty Wt.	600 lbs.	VCr	180 mph
Payload (full fuel)	N/A	V_{S1} (stall clean)	50 mph
Fuel Capacity	30 gal.	V_{S0} (lndg config)	N/A
Seats	1	Climb Rate @msl	1000 fpm
Cabin Width	N/A	Service Ceiling	20,000 ft.

Std. Engine/HP	Usable HP Range	Construction Mtrl.	Est. Bldg. Time
Cont.O-200 / 100 hp	60 - 110	Composite, Aluminum	2000 hrs.
Plans	**Kit**	**Info Packet**	**Video**
$200.00	N/A	$15.00	N/A

A high wing pusher design with inverted "V" tail. Comfortable, quiet cockpit big enough for large pilots. Retractable tricycle landing gear. Can use a variety of engines.

Completed & Flying: 12

RICHARD STEEVES
6958 Applewood Dr. • Madison WI 53719
Tel. 608-833-5586
E-Mail: rsteeves@facstaff.wisc.edu

Coot-A

Wing Span	36.00 ft.	Range	600 sm
Wing Area	180.0 sq. ft.	Takeoff Dist.	600 ft.
Length	22.00 ft.	Landing Dist.	500 ft.
Max Gross Wt.	2150 lbs.	Vmax	140 mph
Empty Wt.	1450 lbs.	VCr	130 mph
Payload (full fuel)	700 lbs.	V_{S1} (stall clean)	N/A
Fuel Capacity	45 gal.	V_{S0} (lndg config)	N/A
Seats	2	Climb Rate @msl	1000 fpm
Cabin Width	N/A	Service Ceiling	15,000 ft.

Std. Engine/HP	Usable HP Range	Construction Mtrl.	Est. Bldg. Time
Franklin / 180 hp	100 - 225	Wood, Metal	1000 hrs.
Plans	**Kit**	**Info Packet**	**Video**
$200.00	N/A	$15.00	N/A

Proven low wing, amphibian pusher design with retractable landing gear and folding wings. Towable. Can use a variety of suitably modified auto or standard aircraft engines.

Completed & Flying: 175

RICHARD STEEVES
6958 Applewood Dr. • Madison WI 53719
Tel. 608-833-5586
E-Mail: rsteeves@facstaff.wisc.edu

Stewart S-51D Mustang

Wing Span	26.00 ft.	**Range**	700 sm
Wing Area	123.0 sq. ft.	**Takeoff Dist.**	1080 ft.
Length	22.00 ft.	**Landing Dist.**	1800 ft.
Max Gross Wt.	3000 lbs.	**Vmax**	275 mph
Empty Wt.	2200 lbs.	**VCr**	260 mph
Payload (full fuel)	400 lbs.	**V_{S1} (stall clean)**	80 mph
Fuel Capacity	70 gal.	**V_{S0} (lndg config)**	70 mph
Seats	2	**Climb Rate @msl**	2780 fpm
Cabin Width	21 in.	**Service Ceiling**	24,000 ft.

Std. Engine/HP	Usable HP Range	Construction Mtrl.	Est. Bldg. Time
Liq. Cooled / 400 hp	300 - 600	Aluminum	2000 hrs.
Plans	**Kit**	**Info Packet**	**Video**
N/A	$64,000.00	$10.00	$20.00

Fully aerobatic, high performance, accurate reproduction of the North American P-51D Mustang. Structural limits are +9, -4.5 G's. Will accommodate two full size people and has dual controls. Performance approaches that of the P-51 except for top speed. Liquid cooled engines from 300 to 600 hp can be utilized. Price is for standard kit. Also available in a "fast build" kit for $89,000.

Kits/Plans Sold: 72 **Completed & Flying:** 2

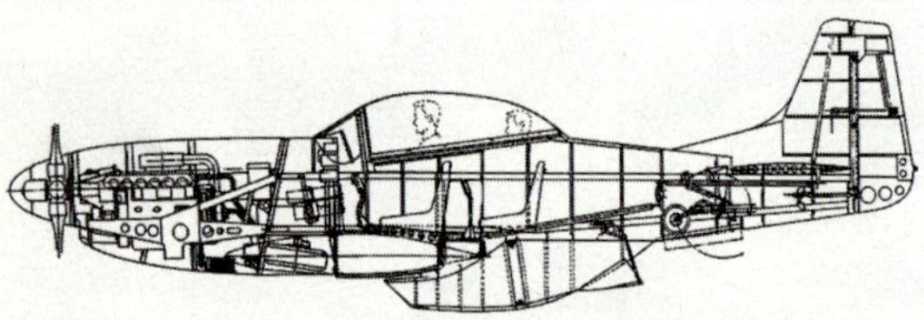

In 1989, Jim took a leave of absence from Pratt & Whitney to complete the design and tooling for the S-51 kit. The prototype S-51 made its initial flight on March 30, 1994. The required 40 hours of flight testing were completed without incident by late June 1994.

STEWART 51, INC.
3120 Airport W. Drive
Vero Beach FL 32960
Tel. 561-778-0051 • Fax 561-778-0051

Great Lakes

Wing Span	26.60 ft.	**Range**	300 sm
Wing Area	187.6 sq. ft.	**Takeoff Dist.**	350 ft.
Length	20.50 ft.	**Landing Dist.**	400 ft.
Max Gross Wt.	1800 lbs.	**Vmax**	135 mph
Empty Wt.	1200 lbs.	**VCr**	118 mph
Payload (full fuel)	600 lbs.	**V_{S1} (stall clean)**	50 mph
Fuel Capacity	26 gal.	**V_{S0} (lndg config)**	50 mph
Seats	2	**Climb Rate @msl**	1500 fpm
Cabin Width	26 in.	**Service Ceiling**	1,800 ft.

Std. Engine/HP	Usable HP Range	Construction Mtrl.	Est. Bldg. Time
Warner / 160 hp	90 - 360	Metal, Fabric	2000 hrs.
Plans	**Kit**	**Info Packet**	**Video**
$300.00	$25,000.00	N/A	$15.00

A classic biplane available with Warner, Ranger or 180 hp Lycoming engines. Completed sub-assemblies available: welded fuselage, all welded parts. Stamped ribs, all other parts available. "Great Lakes Brute" also available utilizing the Russian M-14 engine. See Ad on Page 42.

Kits/Plans Sold: 80 **Completed & Flying:** 40

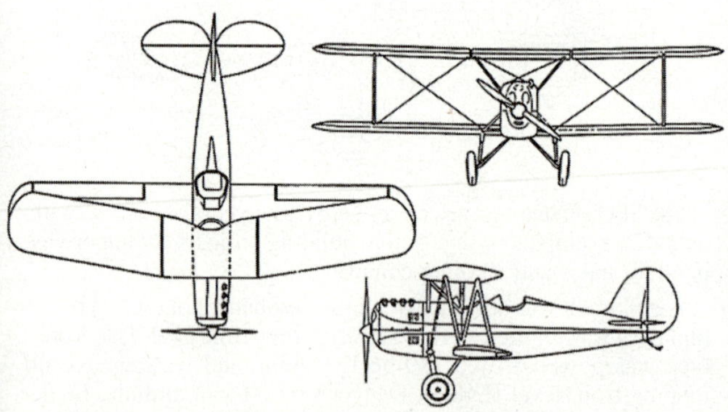

STICK & RUDDER CLUB
12518 Eveningside Drive
Sun City West AZ 95375
Tel. 800-633-2303 • Fax 520-684-3554
E-Mail: srcclub@aol.com
Web: http://members.aol.com/srcclub

Glasair Super II-S

Wing Span	22.30 ft.	**Range**	1080 sm
Wing Area	81.30 sq. ft.	**Takeoff Dist.**	N/A
Length	20.70 ft.	**Landing Dist.**	N/A
Max Gross Wt.	2100 lbs.	**Vmax**	238 mph
Empty Wt.	1325 lbs.	**VCr**	221 mph
Payload (full fuel)	100 lbs.	**V_{S1} (stall clean)**	65 mph
Fuel Capacity	48 gal.	**V_{S0} (lndg config)**	N/A
Seats	2	**Climb Rate @msl**	1700 fpm
Cabin Width	42 in.	**Service Ceiling**	19,000 ft.

Std. Engine/HP	Usable HP Range	Construction Mtrl.	Est. Bldg. Time
Lyc. O-360 / 180 hp	150 - 200	Fiberglass Composite	N/A
Plans	**Kit**	**Info Packet**	**Video**
N/A	$28,280.00	$20.00	$20.00

The Super II-S line consists of the FG (fixed gear), TD (tail-dragger) and RG (retractable gear) models. Specs are for the Super II-S RG. FG and TD models have slightly reduced performance. Available in a variety of assembly packages. Wing-tip tanks available for extended range. Price does not include engine, propeller, avionics, instruments, upholstery and paint.

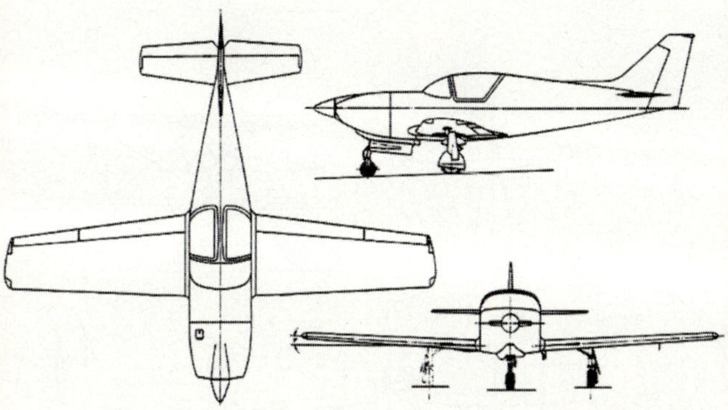

David Schultz, WA

"I think I have the best possible kit-built aircraft available. A lot of my feelings are because of the company I bought it from."

STODDARD-HAMILTON AIRCRAFT CO.
18701 58th Avenue N.E.
Arlington WA 98223
Tel. 360-435-8533 • Fax 360-435-9525
Web: http://www.stoddard-hamilton.com

Glasair III

Wing Span	23.30 ft.	**Range**	1100 sm
Wing Area	81.30 sq. ft.	**Takeoff Dist.**	700 ft.
Length	21.30 ft.	**Landing Dist.**	900 ft.
Max Gross Wt.	2400 lbs.	**Vmax**	270 mph
Empty Wt.	1625 lbs.	**VCr**	248 mph
Payload (full fuel)	100 lbs.	**V_{S1} (stall clean)**	73 mph
Fuel Capacity	61 gal.	**V_{S0} (lndg config)**	N/A
Seats	2	**Climb Rate @msl**	3750 fpm
Cabin Width	42 in.	**Service Ceiling**	24,000 ft.

Std. Engine/HP	Usable HP Range	Construction Mtrl.	Est. Bldg. Time
Lyc IO-540 / 300 hp	N/A	Fiberglass Composite	N/A
Plans	**Kit**	**Info Packet**	**Video**
N/A	$39,980.00	$20.00	$20.00

Weight and range based on standard fuel tanks. Also available with wing-tip tanks that increase range to 1,300 miles. Has gull-wing doors and a one piece, carry-through spar wing. Other versions include the Glasair III Turbo and the Turbine 250/III utilizing an Allison turboprop engine.

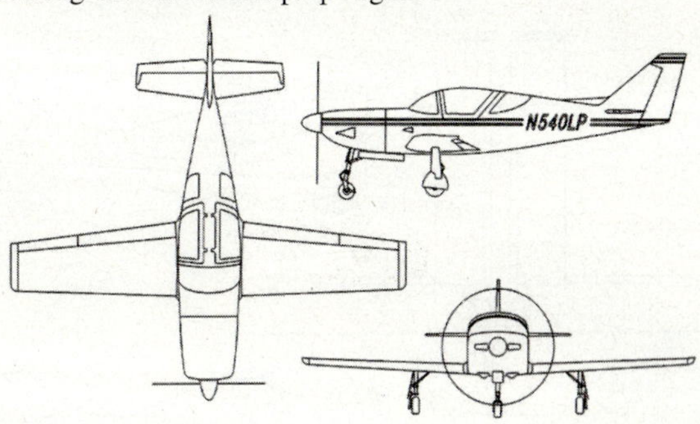

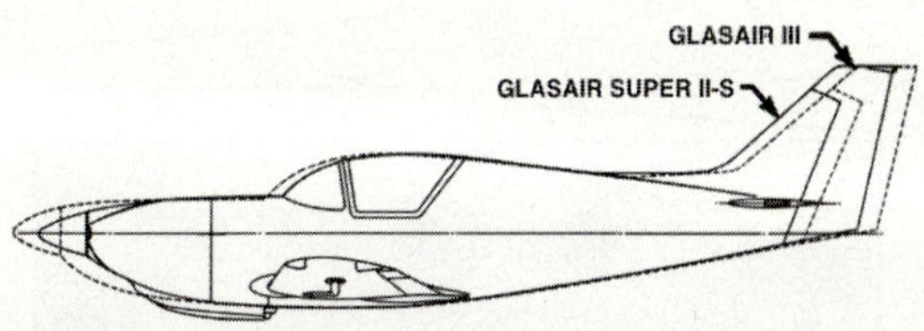

There are both dimensional and structural differences in the Glasair III airframe compared to the Glasair Super II-S. The changes are designed to accommodate the larger, six cylinder engine and to withstand the stresses and flight loads encountered at the faster speeds and higher gross weight.

Dimensionally, the fuselage is 10 inches longer tip to tail than the Glasair Super II-S models. The fuselage of both the Glasair III and II-S models incorporates the enlarged cockpit, giving more head and legroom. The overall fuselage length improves the longitudinal and directional stability making the Glasair III a slightly better cross country and IFR platform.

STODDARD-HAMILTON AIRCRAFT CO.
18701 58th Avenue N.E.
Arlington WA 98223
Tel. 360-435-8533 • Fax 360-435-9525
Web: http://www.stoddard-hamilton.com

Glastar

Wing Span	35.00 ft.	**Range**	778 sm
Wing Area	128.0 sq. ft.	**Takeoff Dist.**	N/A
Length	22.30 ft.	**Landing Dist.**	N/A
Max Gross Wt.	1900 lbs.	**Vmax**	170 mph
Empty Wt.	1100 lbs.	**VCr**	167 mph
Payload (full fuel)	250 lbs.	**V_{S1} (stall clean)**	54 mph
Fuel Capacity	33 gal.	**V_{S0} (lndg config)**	45 mph
Seats	2	**Climb Rate @msl**	1390 fpm
Cabin Width	44 in.	**Service Ceiling**	20,000 ft.

Std. Engine/HP	Usable HP Range	Construction Mtrl.	Est. Bldg. Time
Lyc. O-320 / 160 hp	125-180	Fiberglass, Composite, Tube	N/A
Plans	**Kit**	**Info Packet**	**Video**
N/A	$23,590.00	$20.00	$20.00

Available with either tricycle or taildragger landing gear. Various engine options. Folding wings and side-by-side seating in a 44" cabin. Seats adjustable for pilots from 5' to 6'6". Over 200 lbs. baggage capacity. Price does not include engine, propeller, avionics, instruments, upholstery and paint.

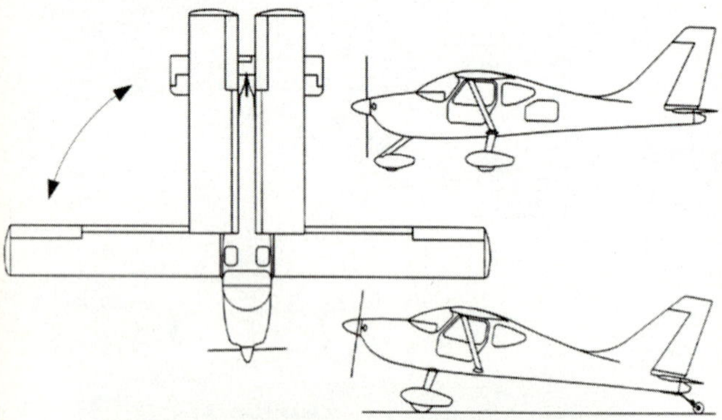

Mike D., Olympia, WA

"...it pleased me very much to know that a kitplane manufacturer as large as your's would take the time out to talk to a "small guy" like me."

STODDARD-HAMILTON AIRCRAFT CO.
18701 58th Avenue N.E.
Arlington WA 98223
Tel. 360-435-8533 • Fax 360-435-9525
Web: http://www.stoddard-hamilton.com

Acroduster Too SA750

Wing Span	21.40 ft.	**Range**	500 sm
Wing Area	130.0 sq. ft.	**Takeoff Dist.**	800 ft.
Length	18.50 ft.	**Landing Dist.**	1000 ft.
Max Gross Wt.	1700 lbs.	**Vmax**	185 mph
Empty Wt.	1100 lbs.	**VCr**	155 mph
Payload (full fuel)	N/A	**V_{S1} (stall clean)**	55 mph
Fuel Capacity	35 gal.	**V_{S0} (lndg config)**	N/A
Seats	2	**Climb Rate @msl**	2000 fpm
Cabin Width	N/A	**Service Ceiling**	16,000 ft.

Std. Engine/HP	Usable HP Range	Construction Mtrl.	Est. Bldg. Time
Lyc IO-360 / 200 hp	180-300	Metal, Tube, Fabric	2000 hrs.
Plans	**Kit**	**Info Packet**	**Video**
$150.00	$13,500.00	$10.00	$15.00

A fully aerobatic, two place design, stressed to +9, -9 G's. It has modified symmetrical wings and is capable of matching aerobatic performance with any two seat airplane now flying. Construction follows traditional biplane practices with steel tube structure and spruce wing spars. Ribs are plywood with cap strips. See Ad on Page 39.

Completed & Flying: 100

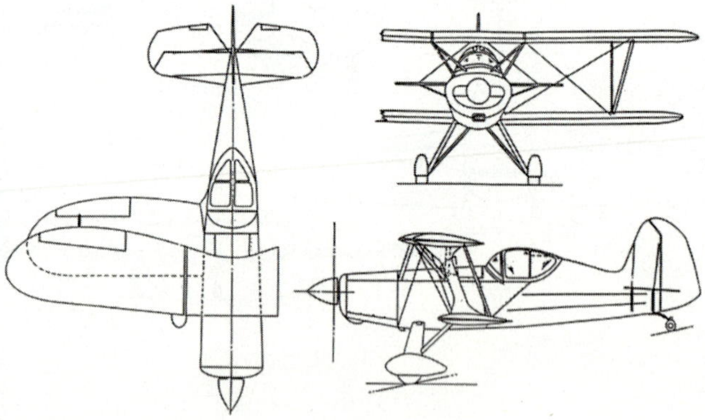

STOLP STARDUSTER CORP.
4301 Twining Street
Riverside CA 92509
Tel. 909-686-7943 • Fax 909-784-0072

Starduster Too

Wing Span	24.00 ft.	**Range**	600 sm
Wing Area	N/A	**Takeoff Dist.**	1200 ft.
Length	18.00 ft.	**Landing Dist.**	1800 ft.
Max Gross Wt.	1985 lbs.	**Vmax**	180 mph
Empty Wt.	1000 lbs.	**VCr**	140 mph
Payload (full fuel)	N/A	**V_{S1} (stall clean)**	55 mph
Fuel Capacity	44 gal.	**V_{S0} (lndg config)**	N/A
Seats	2	**Climb Rate @msl**	2000 fpm
Cabin Width	26 in.	**Service Ceiling**	12,500 ft.

Std. Engine/HP	Usable HP Range	Construction Mtrl.	Est. Bldg. Time
Lyc 0-360 / 180 hp	160 - 300	Tube, Wood, Fabric	2000 hrs.
Plans	**Kit**	**Info Packet**	**Video**
$150.00	$13,000.00	$10.00	$15.00

Designed as a reasonably sized, two place, open sport biplane. Not intended for aerobatic flight, although the structure is quite strong. The main structure is built of 4130 steel tubing and sheet stock and has no machined fittings or other complicated bends. The wings have spruce spars and plywood ribs. Design stressed to +6, -6 G's. All raw materials and certain prefabricated parts are available. See Ad on Page 39.

Completed & Flying: 600

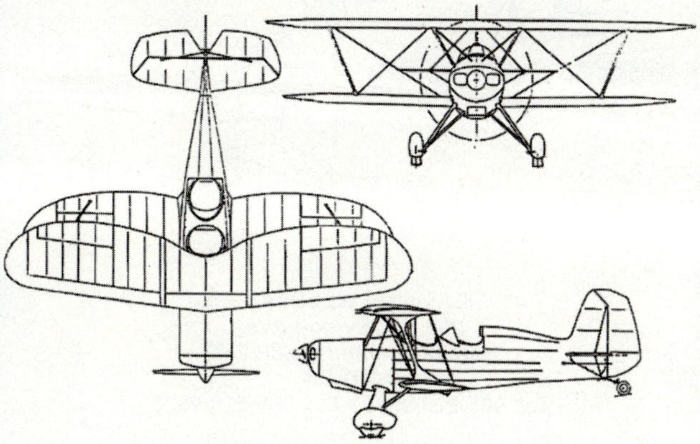

STOLP STARDUSTER CORP.
4301 Twining Street
Riverside CA 92509
Tel. 909-686-7943 • Fax 909-784-0072

Starlet SA500

Wing Span	25.00 ft.	**Range**	600 sm
Wing Area	83.00 sq. ft.	**Takeoff Dist.**	N/A
Length	17.00 ft.	**Landing Dist.**	N/A
Max Gross Wt.	1060 lbs.	**Vmax**	130 mph
Empty Wt.	700 lbs.	**VCr**	105 mph
Payload (full fuel)	N/A	**V_{S1} (stall clean)**	45 mph
Fuel Capacity	22 gal.	**V_{S0} (lndg config)**	N/A
Seats	1	**Climb Rate @msl**	N/A
Cabin Width	N/A	**Service Ceiling**	12,000 ft.

Std. Engine/HP	Usable HP Range	Construction Mtrl.	Est. Bldg. Time
VW 1500 cc / 65 hp	65-100	Metal, Wood, Fabric	1800 hrs.
Plans	**Kit**	**Info Packet**	**Video**
$125.00	$10,000.00	$5.00	$5.00

A stable, relatively fast, high wing, single seater with a parasol wing. All basic structure except the wing is made of 4130 steel tubing. The wing uses spruce spars and plywood ribs with capstrips. Drawings are complete and easy to follow with ribs printed to full size. See Ad on Page 39.

Completed & Flying: 35

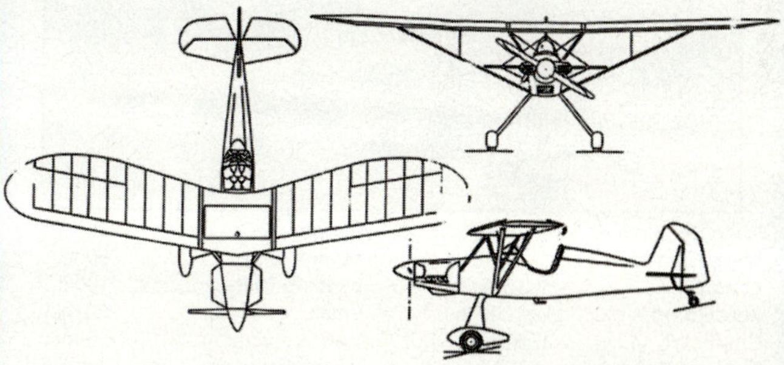

STOLP STARDUSTER CORP.
4301 Twining Street
Riverside CA 92509
Tel. 909-686-7943 • Fax 909-784-0072

V-Star SA900

Wing Span	23.00 ft.	**Range**	280 sm
Wing Area	141.0 sq. ft.	**Takeoff Dist.**	400 ft.
Length	17.00 ft.	**Landing Dist.**	600 ft.
Max Gross Wt.	1000 lbs.	**Vmax**	90 mph
Empty Wt.	700 lbs.	**VCr**	75 mph
Payload (full fuel)	N/A	**V_{S1} (stall clean)**	35 mph
Fuel Capacity	15 gal.	**V_{S0} (lndg config)**	N/A
Seats	1	**Climb Rate @msl**	600 fpm
Cabin Width	N/A	**Service Ceiling**	N/A

Std. Engine/HP	Usable HP Range	Construction Mtrl.	Est. Bldg. Time
Continental / 65 hp	N/A	Tube, Wood, Fabric	1800 hrs.
Plans	**Kit**	**Info Packet**	**Video**
$125.00	$10,000.00	$5.00	$5.00

A low cost, low hp, proven biplane that is easy to build. Light wing loading insures slow landing speed and short takeoff roll. Structure stressed to +9, -9 G's. Welded steel fuselage with spruce and plywood wings. See Ad on Page 39. **Flying:** 65

STOLP STARDUSTER CORP.
4301 Twining Street
Riverside CA 92509
Tel. 909-686-7943 • Fax 909-784-0072

Super Starduster

Wing Span	19.50 ft.	**Range**	450 sm
Wing Area	105.0 sq. ft.	**Takeoff Dist.**	250 ft.
Length	16.00 ft.	**Landing Dist.**	1100 ft.
Max Gross Wt.	1200 lbs.	**Vmax**	220 mph
Empty Wt.	825 lbs.	**VCr**	165 mph
Payload (full fuel)	N/A	**V_{S1} (stall clean)**	58 mph
Fuel Capacity	20 gal.	**V_{S0} (lndg config)**	N/A
Seats	1	**Climb Rate @msl**	3000 fpm
Cabin Width	N/A	**Service Ceiling**	16,000 ft.

Std. Engine/HP	Usable HP Range	Construction Mtrl.	Est. Bldg. Time
Lyc. IO-360 / 200 hp	180-200	Composite, Tube, Fabric	1800 hrs.
Plans	**Kit**	**Info Packet**	**Video**
$175.00	$12,000.00	$10.00	$15.00

Newer design, single place, high performance sport biplane. Raw materials and some prefabricated parts are available for scratch builders. See Ad on Page 39. **Completed & Flying:** 1

STOLP STARDUSTER CORP.
4301 Twining Street
Riverside CA 92509
Ph: 909-686-7943 • Fax: 909-784-0072

Ultimate 10-200

Wing Span	16.00 ft.	**Range**	500 sm
Wing Area	N/A	**Takeoff Dist.**	450 ft.
Length	16.50 ft.	**Landing Dist.**	500 ft.
Max Gross Wt.	1350 lbs.	**Vmax**	220 mph
Empty Wt.	925 lbs.	**VCr**	170 mph
Payload (full fuel)	N/A	**V_{S1} (stall clean)**	60 mph
Fuel Capacity	22 gal.	**V_{S0} (lndg config)**	N/A
Seats	1	**Climb Rate @msl**	2000 fpm
Cabin Width	N/A	**Service Ceiling**	N/A

Std. Engine/HP	Usable HP Range	Construction Mtrl.	Est. Bldg. Time
IO-360 Lyc. / 200 hp	N/A	Metal & Fabric	1200 hrs.
Plans	**Kit**	**Info Packet**	**Video**
$285.00	$22,363.00	$10.00	$20.00

Several related versions available including the 10-100, 10-180 and 10-300. F-84 type canopy. +7, -5 operating Gs. Inverted fuel system. 360 degrees/second roll rate.

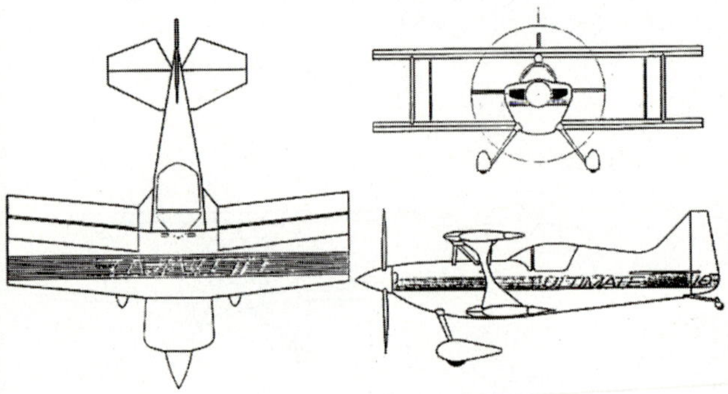

STREAMLINE WELDING, INC.
296 Homewood Ave.
Hamilton, Ontario L8P 2M9
CANADA
Tel. 905-526-9990 • Fax 905-526-9990

SNS-8 Hyperlite

Wing Span	22.00 ft.	**Range**	180 sm
Wing Area	140.0 sq. ft.	**Takeoff Dist.**	175 ft.
Length	15.50 ft.	**Landing Dist.**	175 ft.
Max Gross Wt.	500 lbs.	**Vmax**	98 mph
Empty Wt.	247 lbs.	**VCr**	60 mph
Payload (full fuel)	253 lbs.	**V_{S1} (stall clean)**	N/A
Fuel Capacity	5 gal.	**V_{S0} (lndg config)**	27 mph
Seats	1	**Climb Rate @msl**	700 fpm
Cabin Width	N/A	**Service Ceiling**	N/A

Std. Engine/HP	Usable HP Range	Construction Mtrl.	Est. Bldg. Time
Rotax 277 / 28 hp	N/A	Steel Tube, Aluminum, Fabric	150 hrs.
Plans	**Kit**	**Info Packet**	**Video**
N/A	$13,900.00	$15.00	N/A

Ultralight enclosed cabin, negative stagger biplane. Tractor configuration. Conventional rigid wings and tail, wire braced. Double surfaced airfoil section. All-aluminum flaperons. Heat shrunk dacron cover. Low operating cost. Welded aluminum aft section detaches in 10 minutes for trailering or storage. Also offered fully assembled and test flown for $19,900.

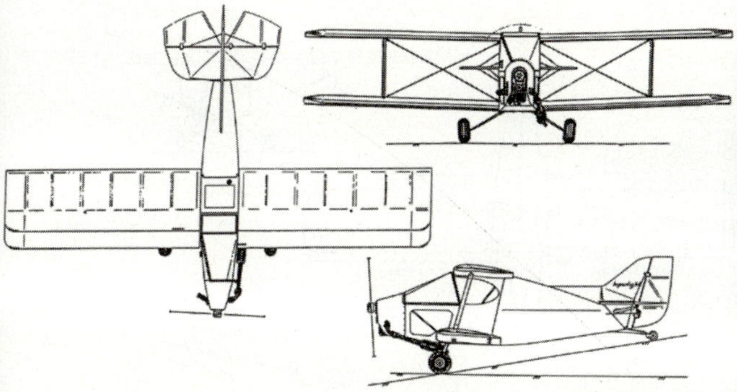

SUNRISE AIRCRAFT CORP.
21821 SW Rock Creek Rd.
Sheridan OR 97378
Tel. 503-843-3616 • Fax 503-843-4361

Jet Ranger

Wing Span	30.00 ft.	**Range**	200 sm
Wing Area	150.0 sq. ft.	**Takeoff Dist.**	330 ft.
Length	18.80 ft.	**Landing Dist.**	330 ft.
Max Gross Wt.	990 lbs.	**Vmax**	95 mph
Empty Wt.	407 lbs.	**VCr**	79 mph
Payload (full fuel)	583 lbs.	**V_{S1} (stall clean)**	32 mph
Fuel Capacity	16 gal.	**V_{S0} (lndg config)**	N/A
Seats	2	**Climb Rate @msl**	885 fpm
Cabin Width	37 in.	**Service Ceiling**	16,500 ft.

Std. Engine/HP	Usable HP Range	Construction Mtrl.	Est. Bldg. Time
Rotax	50 - 80	Tube, Fabric	100 hrs.
Plans	**Kit**	**Info Packet**	**Video**
N/A	$9,800.00	Yes	N/A

New ultralight with tandem seating. Good performance. Easy to build and fly. No welding tubes. Only straight tubes. Kit price is without engine. **Kits Sold:** 15 **Completed & Flying:** 12

SYNAIRGIE
BP 943 • 82009
Montauban, Cedex
FRANCE
Tel. 33-563-67-8080 • Fax 33-563-67-8080

Sky Ranger

Wing Span	31.00 ft.	**Range**	200 sm
Wing Area	161.0 sq. ft.	**Takeoff Dist.**	330 ft.
Length	18.20 ft.	**Landing Dist.**	330 ft.
Max Gross Wt.	996 lbs.	**Vmax**	95 mph
Empty Wt.	407 lbs.	**VCr**	79 mph
Payload (full fuel)	583 lbs.	**V_{S1} (stall clean)**	32 mph
Fuel Capacity	16 gal.	**V_{S0} (lndg config)**	N/A
Seats	2	**Climb Rate @msl**	985 fpm
Cabin Width	37 in.	**Service Ceiling**	16,500 ft.

Std. Engine/HP	Usable HP Range	Construction Mtrl.	Est. Bldg. Time
Rotax	50 - 80	Tube, Fabric	100 hrs.
Plans	**Kit**	**Info Packet**	**Video**
N/A	$10,000.00	Yes	N/A

Most popular (units sold) French ultralight. World Ultralight Champion '96/97. Spacious cockpit, good visibility, good performance. Easy to build and fly. No composites for main parts. Float version available. Kit price is without engine. **Flying:** 150

SYNAIRGIE
BP 943 • 82009
Montauban, Cedex
FRANCE
Tel. 33-563-67-8080 • Fax 33-563-67-8080

Sunny Sportster

Wing Span	23.00 ft.	**Range**	225 sm
Wing Area	185.0 sq. ft.	**Takeoff Dist.**	130 ft.
Length	12.00 ft.	**Landing Dist.**	130 ft.
Max Gross Wt.	820 lbs.	**Vmax**	80 mph
Empty Wt.	400 lbs.	**VCr**	65 mph
Payload (full fuel)	420 lbs.	**V_{S1} (stall clean)**	31 mph
Fuel Capacity	14 gal.	**V_{S0} (lndg config)**	N/A
Seats	2	**Climb Rate @msl**	1200 fpm
Cabin Width	N/A	**Service Ceiling**	N/A

Std. Engine/HP	Usable HP Range	Construction Mtrl.	Est. Bldg. Time
Rotax 582 / 65 hp	65 - 90	Tube, Fabric	N/A
Plans	**Kit**	**Info Packet**	**Video**
N/A	$8,500.00	$25.00	N/A

Unorthodox design but has docile flight characteristics. Will not stall or spin. Rudders located on the jointed wing tips. Elevons on lower wing. Supplied ready to bolt together. Assembled trainer version also available at additional cost. **Completed /Flying:** 145

TANDEM AIRCRAFT KG
Flugplotz D-88348 Saulgau
GERMANY
Tel. 0049.7581.8479 • Fax 0049.7581.81

TL-22 Duo

Wing Span	34.78 ft.	**Range**	200 sm
Wing Area	148.6 sq. ft.	**Takeoff Dist.**	N/A
Length	N/A	**Landing Dist.**	N/A
Max Gross Wt.	795 lbs.	**Vmax**	75 mph
Empty Wt.	353 lbs.	**VCr**	44 mph
Payload (full fuel)	N/A	**V_{S1} (stall clean)**	26 mph
Fuel Capacity	11 gal.	**V_{S0} (lndg config)**	N/A
Seats	2	**Climb Rate @msl**	N/A
Cabin Width	N/A	**Service Ceiling**	N/A

Std. Engine/HP	Usable HP Range	Construction Mtrl.	Est. Bldg. Time
Rotax	N/A	Tube, Fiberglass, Fabric	N/A
Plans	**Kit**	**Info Packet**	**Video**
N/A	Yes	Yes	N/A

Two seat, pusher, powered hang-glider. Very short takeoff and landing capabilities. Comfortable tandem seating.

Completed & Flying: 50

TL ULTRALIGHT CO.
Dobrovskeho 734
Hradec Kralove 500 02
CZECH REPUBLIC
Tel. (+42-49) 613378 • Fax (+42-49) 613378

Pegazair-100 STOL

Wing Span	29.50 ft.	**Range**	N/A
Wing Area	150.0 sq. ft.	**Takeoff Dist.**	250 ft.
Length	22.50 ft.	**Landing Dist.**	200 ft.
Max Gross Wt.	1350 lbs.	**Vmax**	122 mph
Empty Wt.	795 lbs.	**VCr**	105 mph
Payload (full fuel)	N/A	**V_{S1} (stall clean)**	18 mph
Fuel Capacity	15 gal.	**V_{S0} (lndg config)**	N/A
Seats	2	**Climb Rate @msl**	900 fpm
Cabin Width	N/A	**Service Ceiling**	N/A

Std. Engine/HP	Usable HP Range	Construction Mtrl.	Est. Bldg. Time
Cont. 0-200 / 100 hp	65 - 100	Tube, Fabric, Alum.	1400 hrs.
Plans	**Kit**	**Info Packet**	**Video**
$275.00	$15,000.00	$5.00	$20.00

A plans-only, STOL, bush plane equipped with full span retractable automatic, leading edge slats which provide good low speed handling. Tube and fabric fuselage with all-aluminum wings. At a gross weight of 1058 lbs., using a Rotax 912 engine, the Pegazair-80 meets Canadian ultralight regulation standards. Some prefabricated components available.

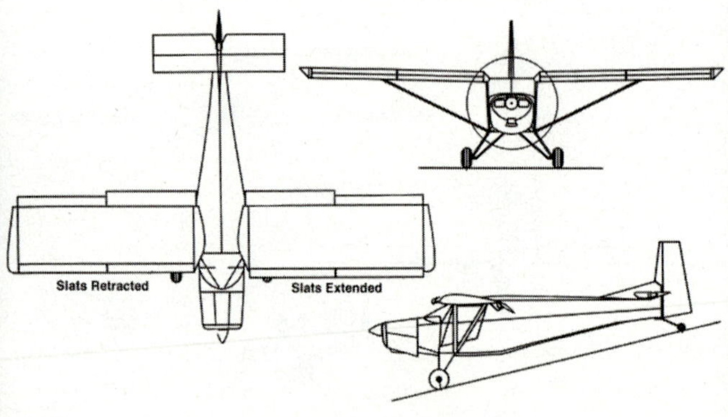

TAPANEE AVIATION, INC.
437 Route 309 Nord
Mont-St-Michel, Quebec J0W 1P0
CANADA
Tel. 819 586-2059 • Fax 819 586-2748

Taylor Monoplane

Wing Span	21.00 ft.	**Range**	290 sm
Wing Area	76.00 sq. ft.	**Takeoff Dist.**	350 ft.
Length	15.00 ft.	**Landing Dist.**	750 ft.
Max Gross Wt.	660 lbs.	**Vmax**	115 mph
Empty Wt.	430 lbs.	**VCr**	100 mph
Payload (full fuel)	180 lbs.	**V_{S1} (stall clean)**	40 mph
Fuel Capacity	8 gal.	**V_{S0} (lndg config)**	35 mph
Seats	1	**Climb Rate @msl**	1000 fpm
Cabin Width	N/A	**Service Ceiling**	13,500 ft.

Std. Engine/HP	Usable HP Range	Construction Mtrl.	Est. Bldg. Time
VW/1500 cc / 60 hp	40 - 60	Wood, Fabric	2200 hrs.
Plans	**Kit**	**Info Packet**	**Video**
$95.00	N/A	$5.00	N/A

This little craft was designed especially for amateur builders who don't have a comprehensive workshop available. The airplane offers excellent yet docile handling under all conditions. It will loop cleanly with an entry speed of 100 mph and perform most aerobatics. Prices are approximate in US funds.

Completed & Flying: 113+

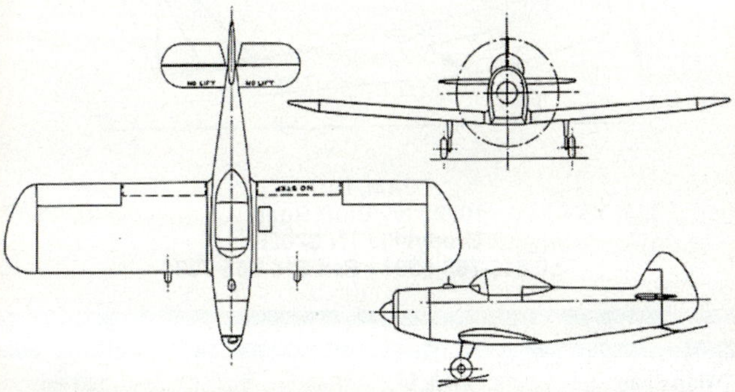

This aircraft was designed in 1956 and flew for the first time in 1959. At that time, it represented the first post war homebuilt design to come from England, was designed to be made in small spaces with a minimum of tools and material cost, requiring only average building skills from the constructor. The design is intended exclusively for use with lower power range engines such as the VW. These will provide acceptable cruise speeds with excellent economy. The Taylor Monoplane is semi-aerobatic. The airframe has been proof loaded to verify the stress calculations. No modification has been introduced since the prototype was approved. The total number flying to date is more than 113.

TERRY TAYLOR CO.
79 Spring Water Road • Leigh On Sea
Essex SS9 5BW
UNITED KINGDOM
Tel. 01702 521484

Taylor Titch

Wing Span	18.90 ft.	**Range**	380 sm
Wing Area	68.00 sq. ft.	**Takeoff Dist.**	350 ft.
Length	16.70 ft.	**Landing Dist.**	900 ft.
Max Gross Wt.	760 lbs.	**Vmax**	200 mph
Empty Wt.	505 lbs.	**VCr**	160 mph
Payload (full fuel)	180 lbs.	**V_{S1} (stall clean)**	53 mph
Fuel Capacity	12 gal.	**V_{S0} (lndg config)**	48 mph
Seats	1	**Climb Rate @msl**	2000 fpm
Cabin Width	N/A	**Service Ceiling**	18,000 ft.

Std. Engine/HP	Usable HP Range	Construction Mtrl.	Est. Bldg. Time
Cont. / 85 hp	60 - 105	Wood, Fabric	2000 hrs.
Plans	**Kit**	**Info Packet**	**Video**
$75.00	N/A	$5.00	N/A

High performance aircraft with pleasing appearance and different characteristics from the Monoplane. Available in racing form utilizing the lightest possible structure and without flaps. Requires a minimum of technical knowledge and skill to build. Prices are approximate in US funds.

Completed & Flying: 38

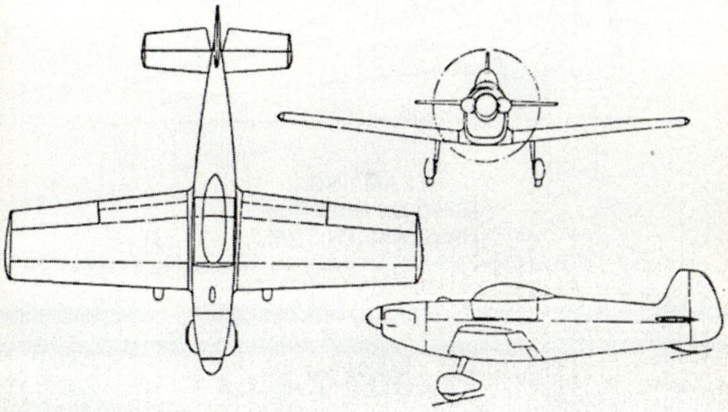

The Titch is the result of requests for an aircraft with higher performance than the Mono. It is designed for a powerplant range of from 60 to 105 hp. The aircraft first flew in 1967.

Titch construction is similar but actually utilizes fewer metal fittings than the Mono. Full size wing rib plans are supplied for the tapered wing panels. Total building cost of the Titch is slightly higher than the Mono because of the difference in engines.

With a cruise speed in the region of 160 mph, this aircraft is an effective cross country tourer. It is also fully aerobatic, having been proof loaded statically. This is an aircraft with very good overall performance, while still being safe for the amateur pilot. More than 35 have flown successfully to date.

TERRY TAYLOR CO.
79 Spring Water Road • Leigh On Sea
Essex SS9 5BW
UNITED KINGDOM
Tel. 01702 521484

1030R MAX 103 UL

Wing Span	25.00 ft.	**Range**	100 sm
Wing Area	112.5 sq. ft.	**Takeoff Dist.**	150 ft.
Length	16.00 ft.	**Landing Dist.**	180 ft.
Max Gross Wt.	500 lbs.	**Vmax**	90 mph
Empty Wt.	240 lbs.	**VCr**	55 mph
Payload (full fuel)	230 lbs.	**V_{S1} (stall clean)**	30 mph
Fuel Capacity	5 gal.	**V_{S0} (lndg config)**	27 mph
Seats	1	**Climb Rate @msl**	650 fpm
Cabin Width	22 in.	**Service Ceiling**	10,000 ft.

Std. Engine/HP	Usable HP Range	Construction Mtrl.	Est. Bldg. Time
Rotax 277 / 28 hp	N/A	Wood, Fabric	350 hrs.
Plans	**Kit**	**Info Packet**	**Video**
$125.00	$3,645.00	$15.00	$15.00

FAR 103 legal ultralight. Open cockpit with swing over canopy. Great looks and performance. Wheel pants optional. Engine not included in kit price. See Ad on Page 21. **Kits/Plans Sold:** 89 **Completed & Flying:** 51

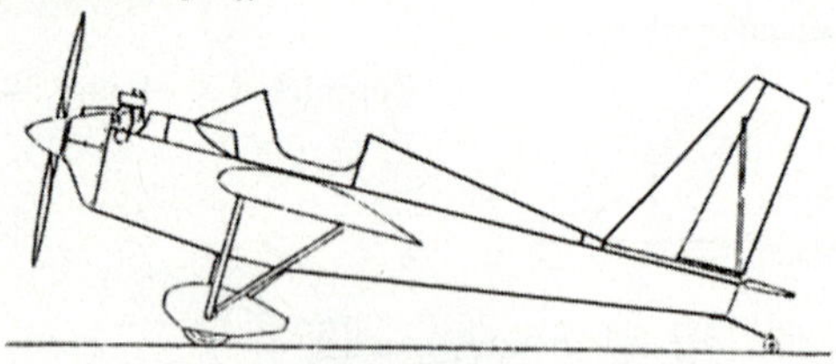

TEAM, INC.
10790 Ivy Bluff Road
Bradyville TN 37026
Tel. 615-765-5397 • Fax 615-765-7234

1100R Mini-MAX

Wing Span	25.00 ft.	**Range**	150 sm
Wing Area	112.5 sq. ft.	**Takeoff Dist.**	150 ft.
Length	15.50 ft.	**Landing Dist.**	180 ft.
Max Gross Wt.	560 lbs.	**Vmax**	100 mph
Empty Wt.	237 lbs.	**VCr**	55 mph
Payload (full fuel)	263 lbs.	**V_{S1} (stall clean)**	30 mph
Fuel Capacity	10 gal.	**V_{S0} (lndg config)**	28 mph
Seats	1	**Climb Rate @msl**	650 fpm
Cabin Width	24 in.	**Service Ceiling**	10,000 ft.

Std. Engine/HP	Usable HP Range	Construction Mtrl.	Est. Bldg. Time
Rotax 277 / 28 hp	28 - 42	Wood, Fabric	300 hrs.
Plans	**Kit**	**Info Packet**	**Video**
$100.00	$2,995.00	$10.00	$15.00

Original, open cockpit, MINI-MAX mid wing design. Can be upgraded to a model 1500R or 1600R. Legal ultralight with Rotax 277 engine and 5 gallon tank. Engine not included in kit price. See Ad on Page 21. **Kits/Plans Sold:** 510 **Completed & Flying:** 323

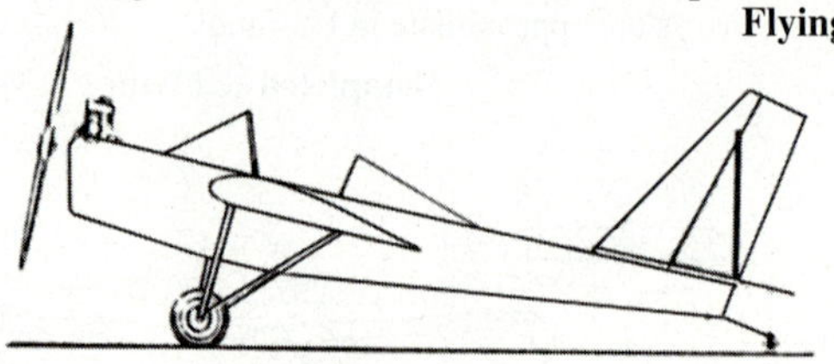

TEAM, INC.
10790 Ivy Bluff Road
Bradyville TN 37026
Tel. 615-765-5397 • Fax 615-765-7234

1200Z Z-Max

Jack Dalton, Jackson SC
"Building was straight forward and TEAM Aircraft answered the few questions I had. The plans were clear and easy to read."

TEAM, INC.
10790 Ivy Bluff Road
Bradyville TN 37026
Tel. 615-765-5397 • Fax 615-765-7234

Wing Span	25.00 ft.	**Range**	150 sm
Wing Area	112.5 sq. ft.	**Takeoff Dist.**	100 ft.
Length	16.00 ft.	**Landing Dist.**	210 ft.
Max Gross Wt.	560 lbs.	**Vmax**	100 mph
Empty Wt.	278 lbs.	**VCr**	65 mph
Payload (full fuel)	222 lbs.	**V_{S1} (stall clean)**	32 mph
Fuel Capacity	10 gal.	**V_{S0} (lndg config)**	29 mph
Seats	1	**Climb Rate @msl**	1000 fpm
Cabin Width	24 in.	**Service Ceiling**	10,000 ft.

Std. Engine/HP	Usable HP Range	Construction Mtrl.	Est. Bldg. Time
Zenoah G-50 / 45 hp	N/A	Wood, Fabric	350 hrs.
Plans	**Kit**	**Info Packet**	**Video**
$125.00	$3,645.00	$10.00	$15.00

The already proven MINI-MAX design, modified and improved to accommodate a larger, high performance 45 hp engine in a fully enclosed cowl. Engine not included in price. See Ad on Page 21. **Sold:** 75 **Flying:** 41

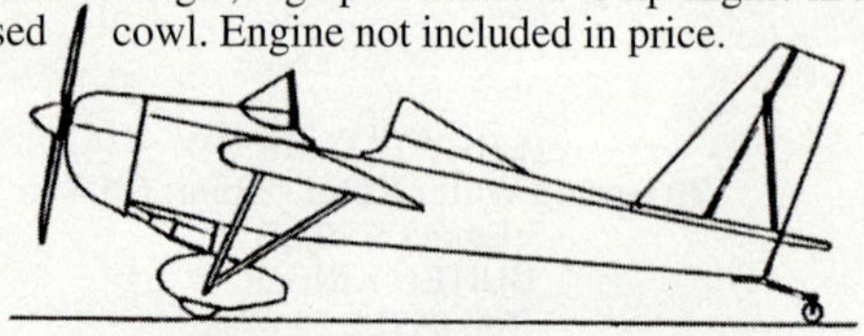

1300Z Z-Max

Wing Span	25.00 ft.	**Range**	150 sm
Wing Area	112.5 sq. ft.	**Takeoff Dist.**	100 ft.
Length	16.00 ft.	**Landing Dist.**	210 ft.
Max Gross Wt.	560 lbs.	**Vmax**	100 mph
Empty Wt.	310 lbs.	**VCr**	72 mph
Payload (full fuel)	190 lbs.	**V_{S1} (stall clean)**	33 mph
Fuel Capacity	10 gal.	**V_{S0} (lndg config)**	30 mph
Seats	1	**Climb Rate @msl**	1000 fpm
Cabin Width	24 in.	**Service Ceiling**	10,000 ft.

Std. Engine/HP	Usable HP Range	Construction Mtrl.	Est. Bldg. Time
Zenoah G-50 / 45 hp	N/A	Wood, Fabric	400 hrs.
Plans	**Kit**	**Info Packet**	**Video**
$150.00	$3,845.00	$15.00	$15.00

Enclosed cockpit with swing over canopy. Mid wing design. Vertical firewall engine mount provides rounded nose cowl. Wheel pants optional. Wood and fabric construction. Engine not included in kit price. Landing configuration spec. is with optional flaps. See Ad on Page 21. **Kits/Plans Sold:** 163 **Flying:** 104

TEAM, INC.
10790 Ivy Bluff Road
Bradyville TN 37026
Tel. 615-765-5397 • Fax 615-765-7234

1400Z Hi-MAX

Wing Span	25.00 ft.	**Range**	150 sm
Wing Area	112.5 sq. ft.	**Takeoff Dist.**	100 ft.
Length	16.00 ft.	**Landing Dist.**	210 ft.
Max Gross Wt.	560 lbs.	**Vmax**	100 mph
Empty Wt.	323 lbs.	**VCr**	70 mph
Payload (full fuel)	180 lbs.	**V_{S1} (stall clean)**	30 mph
Fuel Capacity	10 gal.	**V_{S0} (lndg config)**	N/A
Seats	1	**Climb Rate @msl**	1000 fpm
Cabin Width	24 in.	**Service Ceiling**	10,000 ft.

Std. Engine/HP	Usable HP Range	Construction Mtrl.	Est. Bldg. Time
Zenoah G-50 / 45 hp	N/A	Wood, Fabric	400 hrs.
Plans	**Kit**	**Info Packet**	**Video**
$150.00	$3,895.00	$15.00	$15.00

Strut-braced monoplane with an enclosed cabin. Vertical firewall engine mount provides rounded nose cowl. Wheel pants optional. Wood and fabric construction. Engine not included in kit price. See Ad on Page 21. **Kits/Plans Sold:** 92 **Flying:** 49

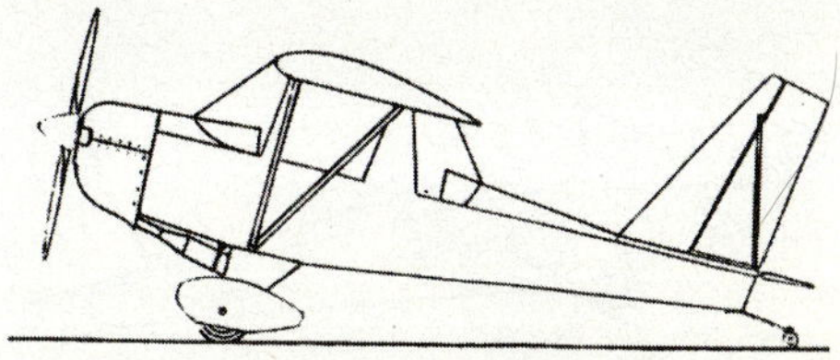

TEAM, INC.
10790 Ivy Bluff Road
Bradyville TN 37026
Tel. 615-765-5397 • Fax 615-765-7234

1500R Mini-MAX

TEAM, INC.
10790 Ivy Bluff Road
Bradyville TN 37026
Tel. 615-765-5397 • Fax 615-765-7234

Wing Span	25.00 ft.	**Range**	150 sm
Wing Area	112.5 sq. ft.	**Takeoff Dist.**	100 ft.
Length	16.00 ft.	**Landing Dist.**	210 ft.
Max Gross Wt.	560 lbs.	**Vmax**	100 mph
Empty Wt.	269 lbs.	**VCr**	65 mph
Payload (full fuel)	200 lbs.	**V_{S1} (stall clean)**	31 mph
Fuel Capacity	10 gal.	**V_{S0} (lndg config)**	28 mph
Seats	1	**Climb Rate @msl**	1000 fpm
Cabin Width	24 in.	**Service Ceiling**	10,000 ft.

Std. Engine/HP	Usable HP Range	Construction Mtrl.	Est. Bldg. Time
Rotax 447 / 42 hp	N/A	Wood, Fabric	350 hrs.
Plans	**Kit**	**Info Packet**	**Video**
$125.00	$3,645.00	$15.00	$15.00

Rotax powered mid wing, open cockpit monoplane with cowling and spinner. Wood and fabric construction. Engine not included. See Ad on Page 21.

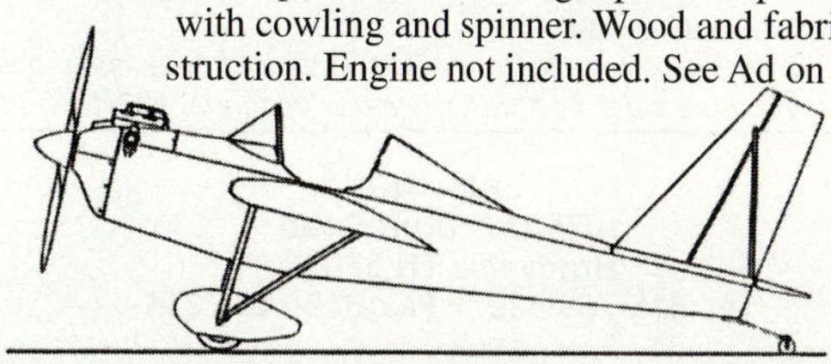

Sold: 153
Flying: 78

1550 V-MAX Sportplane

Wing Span	26.00 ft.	**Range**	150 sm
Wing Area	118.0 sq. ft.	**Takeoff Dist.**	150 ft.
Length	16.00 ft.	**Landing Dist.**	250 ft.
Max Gross Wt.	700 lbs.	**Vmax**	110 mph
Empty Wt.	400 lbs.	**VCr**	75 mph
Payload (full fuel)	240 lbs.	**V_{S1} (stall clean)**	35 mph
Fuel Capacity	10 gal.	**V_{S0} (lndg config)**	N/A
Seats	1	**Climb Rate @msl**	900 fpm
Cabin Width	24 in.	**Service Ceiling**	10,000 ft.

Std. Engine/HP	Usable HP Range	Construction Mtrl.	Est. Bldg. Time
Mosler / 50 hp	N/A	Wood, Fabric	350 hrs.
Plans	**Kit**	**Info Packet**	**Video**
$165.00	$4,175.00	$15.00	$10.00

The V-MAX is a redesigned MINI-MAX with a four cylinder 1600 cc Mosler engine. Wheel pants optional. Wood and fabric construction. Engine not included in kit price. See Ad on Page 21.

Kits/Plans Sold: 38 **Completed & Flying:** 24

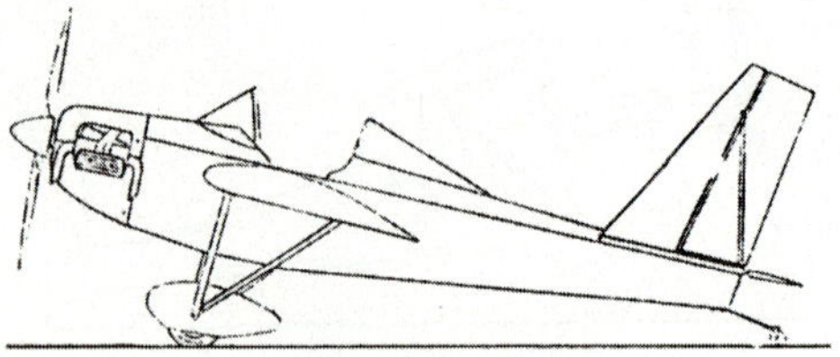

TEAM, INC.
10790 Ivy Bluff Road
Bradyville TN 37026
Tel. 615-765-5397 • Fax 615-765-7234

RX 40 Air Bike

Wing Span	26.00 ft.	**Range**	150 sm
Wing Area	118.0 sq. ft.	**Takeoff Dist.**	100 ft.
Length	16.00 ft.	**Landing Dist.**	200 ft.
Max Gross Wt.	540 lbs.	**Vmax**	80 mph
Empty Wt.	257 lbs.	**VCr**	63 mph
Payload (full fuel)	200 lbs.	**V_{S1} (stall clean)**	27 mph
Fuel Capacity	10 gal.	**V_{S0} (lndg config)**	N/A
Seats	1	**Climb Rate @msl**	1000 fpm
Cabin Width	N/A	**Service Ceiling**	10,000 ft.

Std. Engine/HP	Usable HP Range	Construction Mtrl.	Est. Bldg. Time
Rotax 447 / 42 hp	22 - 42	Wood, Metal, Fabric	200 hrs.
Plans	**Kit**	**Info Packet**	**Video**
N/A	$5,495.00	$10.00	$15.00

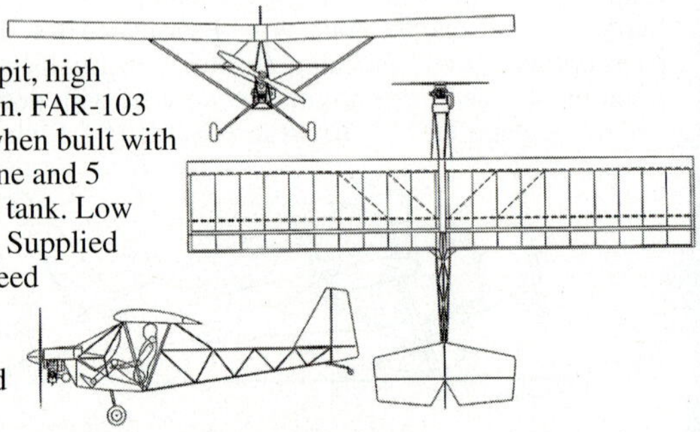

Open cockpit, high wing design. FAR-103 legal UL when built with 20 hp engine and 5 gallon fuel tank. Low build time. Supplied with air speed indicator, gas tank, throttle and covering. See Ad on Page 21.

Sold: 135
Flying: 51

TEAM, INC.
10790 Ivy Bluff Road
Bradyville TN 37026
Tel. 615-765-5397 • Fax 615-765-7234

1600R Encl. Cockpit Mini-MAX

Robert C. Ruhter, Bradyville, TN
"Well, we finally completed our MINI MAX and have completed all test flying. One word best describes this airplane, FANTASTIC!"

TEAM, INC.
10790 Ivy Bluff Road
Bradyville TN 37026
Tel. 615-765-5397 • Fax 615-765-7234

Wing Span	25.00 ft.	**Range**	150 sm
Wing Area	112.5 sq. ft.	**Takeoff Dist.**	100 ft.
Length	16.00 ft.	**Landing Dist.**	210 ft.
Max Gross Wt.	560 lbs.	**Vmax**	100 mph
Empty Wt.	301 lbs.	**VCr**	72 mph
Payload (full fuel)	200 lbs.	**V_{S1} (stall clean)**	31 mph
Fuel Capacity	10 gal.	**V_{S0} (lndg config)**	29 mph
Seats	1	**Climb Rate @msl**	1000 fpm
Cabin Width	24 in.	**Service Ceiling**	10,000 ft.

Std. Engine/HP	Usable HP Range	Construction Mtrl.	Est. Bldg. Time
Rotax 447 / 42 hp	N/A	Wood, Fabric	400 hrs.
Plans	**Kit**	**Info Packet**	**Video**
$150.00	$3,845.00	$15.00	$10.00

Enclosed cockpit with swing over canopy, cowling and spinner. Rotax powered. Experimental or Microlight category. Wheel pants optional. Wood and fabric construction. Engine is not included in kit price. See Ad on Page 21.

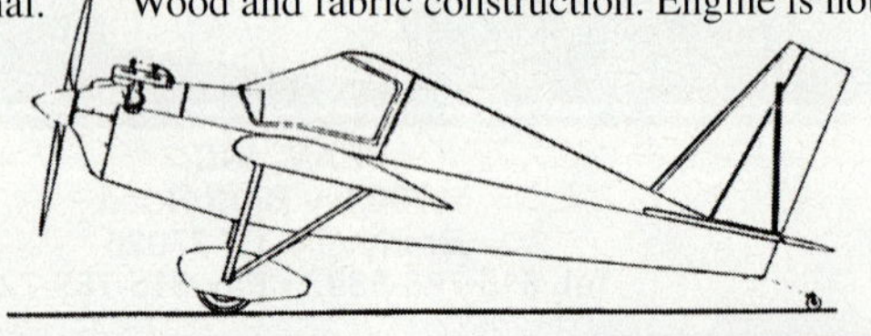

Sold: 138
Flying: 84

1650R Eros

Wing Span	26.40 ft.	**Range**	100 sm
Wing Area	118.0 sq. ft.	**Takeoff Dist.**	150 ft.
Length	16.00 ft.	**Landing Dist.**	250 ft.
Max Gross Wt.	625 lbs.	**Vmax**	110 mph
Empty Wt.	345 lbs.	**VCr**	85 mph
Payload (full fuel)	250 lbs.	**V_{S1} (stall clean)**	36 mph
Fuel Capacity	5 gal.	**V_{S0} (lndg config)**	33 mph
Seats	1	**Climb Rate @msl**	1200 fpm
Cabin Width	24 in.	**Service Ceiling**	10,000 ft.

Std. Engine/HP	Usable HP Range	Construction Mtrl.	Est. Bldg. Time
Rotax 503 / 50 hp	30 - 50	Metal, Wood, Fabric	400 hrs.
Plans	**Kit**	**Info Packet**	**Video**
$175.00	$4,395.00	$10.00	$15.00

Totally enclosed Rotax 503 DCDI (or optional Rotax 447, 40 hp) engine. Convertible swing over canopy available. Roll rate under two seconds from 45° to 45°. Plans, builder's manual and flight manual included in kit. Landing configuration spec. is with optional flaps. See Ad on Page 21.

Kits/Plans Sold: 49
Completed & Flying: 11

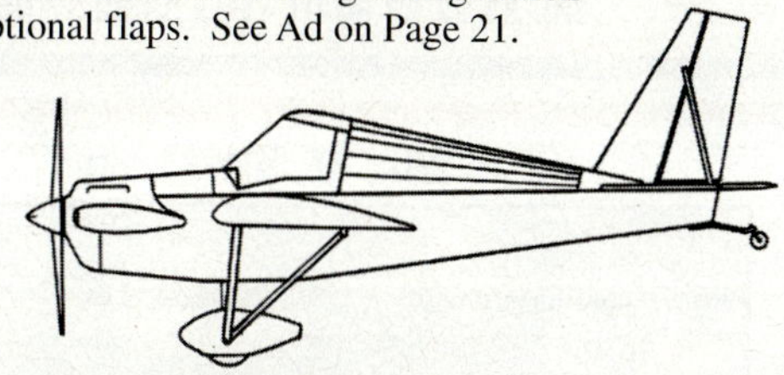

TEAM, INC.
10790 Ivy Bluff Road
Bradyville TN 37026
Tel. 615-765-5397 • Fax 615-765-7234

1700R Hi-Wing Hi-MAX

Wing Span	25.00 ft.	**Range**	150 sm
Wing Area	112.5 sq. ft.	**Takeoff Dist.**	100 ft.
Length	16.00 ft.	**Landing Dist.**	210 ft.
Max Gross Wt.	560 lbs.	**Vmax**	100 mph
Empty Wt.	319 lbs.	**VCr**	70 mph
Payload (full fuel)	180 lbs.	**V_{S1} (stall clean)**	30 mph
Fuel Capacity	10 gal.	**V_{S0} (lndg config)**	30 mph
Seats	1	**Climb Rate @msl**	1000 fpm
Cabin Width	24 in.	**Service Ceiling**	10,000 ft.

Std. Engine/HP	Usable HP Range	Construction Mtrl.	Est. Bldg. Time
Rotax 447 / 42 hp	N/A	Wood, Fabric	400 hrs.
Plans	**Kit**	**Info Packet**	**Video**
$150.00	$3,895.00	$10.00	$15.00

Strut-braced monoplane with an enclosed cabin. Rotax powered. Wheel pants optional. Wood and fabric construction. Engine not included in kit price. See Ad on Page 21. **Kits/Plans Sold:** 102
Flying: 49

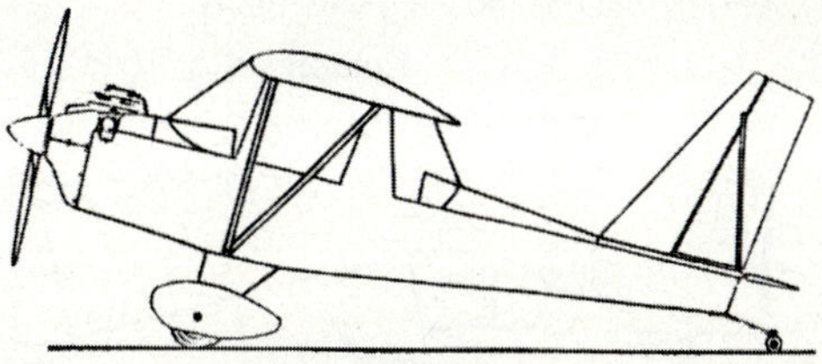

TEAM, INC.
10790 Ivy Bluff Road
Bradyville TN 37026
Tel. 615-765-5397 • Fax 615-765-7234

Tandem Air-Bike

TEAM, INC.
10790 Ivy Bluff Road
Bradyville TN 37026
Tel. 615-765-5397 • Fax 615-765-7234

Wing Span	31.40 ft.	**Range**	150 sm
Wing Area	141.00 sq. ft.	**Takeoff Dist.**	400 ft.
Length	18.30 ft.	**Landing Dist.**	200 ft.
Max Gross Wt.	900 lbs.	**Vmax**	90 mph
Empty Wt.	400 lbs.	**VCr**	70 mph
Payload (full fuel)	440 lbs.	**V_{S1} (stall clean)**	35 mph
Fuel Capacity	10 gal.	**V_{S0} (lndg config)**	28 mph
Seats	2	**Climb Rate @msl**	550 fpm
Cabin Width	N/A	**Service Ceiling**	10,000 ft.

Std. Engine/HP	Usable HP Range	Construction Mtrl.	Est. Bldg. Time
DC 503 / 52 hp	45 - 65	Metal, Wood, Fabric	300 hrs.
Plans	**Kit**	**Info Packet**	**Video**
N/A	$7,995.00	$10.00	$15.00

AirBike for two. Takes advantage of the waiver of Part 103 to provide a two place ultralight that meets the Part 103 exemption for training. All flight and engine controls are dual. Other standard features include two five gallon wing tanks and the spinner. Can be configured with optional electric trim, brakes, and electric flaps which are operable from both seats. Available as complete kit less engine or in four partial kits. See Ad on Page 21.

Kits/Plans Sold: 11 **Completed & Flying:** 3

Teenie Two

Wing Span	18.00 ft.	**Range**	400 sm
Wing Area	59.00 sq. ft.	**Takeoff Dist.**	600 ft.
Length	12.10 ft.	**Landing Dist.**	700 ft.
Max Gross Wt.	585 lbs.	**Vmax**	150 mph
Empty Wt.	310 lbs.	**VCr**	120 mph
Payload (full fuel)	N/A	**V_{S1} (stall clean)**	50 mph
Fuel Capacity	12 gal.	**V_{S0} (Indg config)**	N/A
Seats	1	**Climb Rate @msl**	1500 fpm
Cabin Width	N/A	**Service Ceiling**	15,000 ft.

Std. Engine/HP	Usable HP Range	Construction Mtrl.	Est. Bldg. Time
VW 1700cc / 42 hp	40-65	Aluminum	300 hrs.
Plans	**Kit**	**Info Packet**	**Video**
$150.00	$3,000.00	$15.00	Yes

First all-metal VW powered homebuilt. First flown in 1969. Wings detach in about five minutes for easy transportation. Uses non-retractable, tricycle type gear with a steerable nosewheel. Can utilize several different VW engines. Kit price is for raw materials only. A VHS video is included with plans.

Completed & Flying: 100

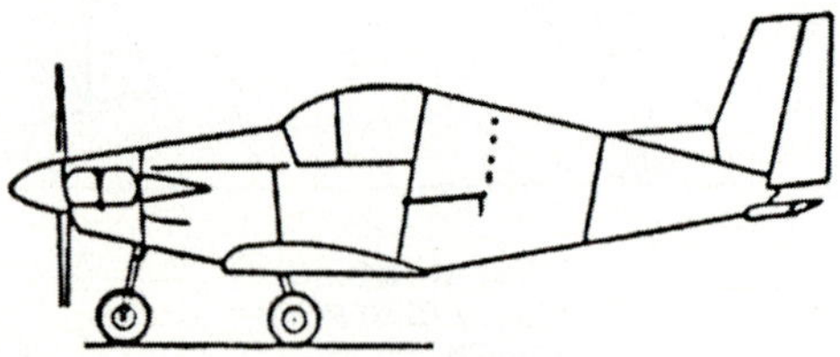

Mr. Parker's original aim in starting work on the Teenie Series was to build an aircraft which specifically utilized the Volkswagen automobile engine and, at the same time, to produce an all-metal design that would present few construction problems to homebuilders with virtually no metal-working experience. This was achieved, and no special tools or jigs are needed beyond a tool to close and form the cadmium plated steel pop rivets that are used for practically all assembly. One gauge of aluminum sheet and one size of light alloy angle section are used for almost all of the structure, except for CR-Mo steel tube and sheet which are required for construction of the landing gear and control actuation tubes respectively. For simplicity and economy, push pull tubes are used for all flying controls.

Teenie Two is considerably refined when compared to the original Teenie One, to produce a much "cleaner" airplane. Its structure is stressed for full aerobatics, but the fuel and oil systems are not suitable for inverted flight.

TEENIE COMPANY
P.O. Box 2092
Harker Heights TX 76543
Tel. 817-698-3283 • Fax 817-698-1207

TR 200

Wing Span	24.33 ft.	**Range**	590 sm
Wing Area	107.5 sq. ft.	**Takeoff Dist.**	1310 ft.
Length	21.02 ft.	**Landing Dist.**	1475 ft.
Max Gross Wt.	1917 lbs.	**Vmax**	202 mph
Empty Wt.	1256 lbs.	**VCr**	168 mph
Payload (full fuel)	407 lbs.	**V_{S1} (stall clean)**	59 mph
Fuel Capacity	42 gal.	**V_{S0} (Indg config)**	N/A
Seats	2	**Climb Rate @msl**	2200 fpm
Cabin Width	28 in.	**Service Ceiling**	18,000 ft.

Std. Engine/HP	Usable HP Range	Construction Mtrl.	Est. Bldg. Time
Lyc AEIO 360 / 200 hp	200-260	Wood	800 hrs.
Plans	**Kit**	**Info Packet**	**Video**
N/A	$68,000.00	N/A	N/A

High performance, two place tandem. Designed for aerobatic training and practice for "ultimate" category competition. Good also for cross country. First flown in August 1988. Price includes engine and three bladed propeller.

TECH AERO
Route De La Ferriere • Le Bois D'oissel
Glisolles 27190
FRANCE
Tel. 33 32 37 60 01 • Fax 33 32 37 00 14

Mk V Spitfire

Wing Span	23.00 ft.	**Range**	850 sm
Wing Area	81.66 sq. ft.	**Takeoff Dist.**	N/A
Length	18.50 ft.	**Landing Dist.**	N/A
Max Gross Wt.	1100 lbs.	**Vmax**	200 mph
Empty Wt.	600 lbs.	**VCr**	170 mph
Payload (full fuel)	N/A	**V_{S1} (stall clean)**	60 mph
Fuel Capacity	27 gal.	**V_{S0} (Indg config)**	45 mph
Seats	1	**Climb Rate @msl**	1500 fpm
Cabin Width	27 in.	**Service Ceiling**	N/A

Std. Engine/HP	Usable HP Range	Construction Mtrl.	Est. Bldg. Time
GeoStorm 1.6 l / 100 hp	N/A	Composite, Wood	600 hrs.
Plans	**Kit**	**Info Packet**	**Video**
N/A	$24,950.00	N/A	N/A

60% scale Spitfire replica. First shown at Sun 'n Fun '96. Fast build kit. Engine and instruments are not included. Blown canopy by Airplane Plastics. Performance figures are estimates.

Completed & Flying: 1

TIME WARP AIRCRAFT INC.
Lakeland Municipal Airport • 3905 Aero Place, Hangar Q
Lakeland FL 33811
Tel. 941-647-3487
Web: http://www.cyberspy.com/~tomr/spitfire

Speedster

Wing Span	17.50 ft.	**Range**	125 sm
Wing Area	77.50 sq. ft.	**Takeoff Dist.**	200 ft.
Length	13.67 ft.	**Landing Dist.**	200 ft.
Max Gross Wt.	460 lbs.	**Vmax**	75 mph
Empty Wt.	252 lbs.	**VCr**	62 mph
Payload (full fuel)	200 lbs.	**V_{S1} (stall clean)**	29 mph
Fuel Capacity	5 gal.	**V_{S0} (lndg config)**	N/A
Seats	1	**Climb Rate @msl**	700 fpm
Cabin Width	20 in.	**Service Ceiling**	N/A

Std. Engine/HP	Usable HP Range	Construction Mtrl.	Est. Bldg. Time
Kawasaki 440 / 40 hp	38 - 55	Alum., Wood, Steel, Foam	400 hrs.
Plans	**Kit**	**Info Packet**	**Video**
N/A	$4,994.00	Yes	$15.00

Golden age look alike in an ultralight kit. Reminiscent of staggerwing style biplane racer. Quick Build kit.

Completed & Flying: 2

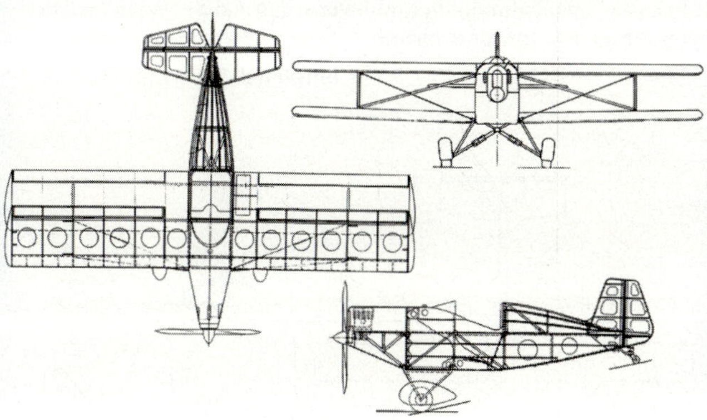

THEISS AVIATION
PO Box 1086
Salem OH 44460
Tel. 330-332-2031 • Fax 330-332-8273

Drifter SB

Wing Span	30.00 ft.	**Range**	300 sm
Wing Area	156.0 sq. ft.	**Takeoff Dist.**	350 ft.
Length	19.50 ft.	**Landing Dist.**	350 ft.
Max Gross Wt.	1100 lbs.	**Vmax**	N/A
Empty Wt.	550 lbs.	**VCr**	74 mph
Payload (full fuel)	N/A	**V_{S1} (stall clean)**	36 mph
Fuel Capacity	15 gal.	**V_{S0} (lndg config)**	N/A
Seats	2	**Climb Rate @msl**	N/A
Cabin Width	N/A	**Service Ceiling**	N/A

Std. Engine/HP	Usable HP Range	Construction Mtrl.	Est. Bldg. Time
Rotax 582 / 65 hp	65-80	Aluminum, Fabric	N/A
Plans	**Kit**	**Info Packet**	**Video**
N/A	$14,900.00	Yes	N/A

The Drifter SB is a strut braced, two place tandem aircraft with a large front pod. Engine option is the Jabiru 2200, 80hp, four cyl., four stroke. Designed and engineered by Austflight ULA Pty. Ltd., PO Box 84, Boonah, Queensland, Australia. Tel. 61 75 463 2755 Fax 61 75 463 2987.

Kits/Plans Sold: 500 **Completed & Flying:** 500

TIGER AVIATION, INC.
134 Cessna Drive
Trenton SC 29847
Tel. 803-663-6200 • Fax 803-663-6655
E-Mail: TREMFL@AOL.COM
Web: http://www.tigeraviation.com

HOMEBUILT "FAQ'S" *(See Page 68)*

Q: What design should I choose?

A: It's all up to the individual, but I'd suggest that you pay particular attention to the following two items:
(1) The desired mission for your airplane.
(2) Your available resources (money, space, ability, etc.)
Once you have a realistic and unemotional handle on these items, start checking out designs until you find two or three that fit your situation best. Then go ahead and start checking out differences between them. Don't dwell too much on factors like, "I've never done any welding" since you're going to have to learn to do lots of things you never considered before. Also, designers' claims for performance are often "stretched" a bit, so wait to talk to owners before making any final decisions.
When you're at this point, it's time to start checking out real aircraft. This is where the fun begins.

Tornado II

Wing Span	23.50 ft.	**Range**	500 sm
Wing Area	108.0 sq. ft.	**Takeoff Dist.**	300 ft.
Length	19.00 ft.	**Landing Dist.**	250 ft.
Max Gross Wt.	1000 lbs.	**Vmax**	150 mph
Empty Wt.	540 lbs.	**VCr**	120 mph
Payload (full fuel)	460 lbs.	**V_{S1} (stall clean)**	35 mph
Fuel Capacity	15 gal.	**V_{S0} (lndg config)**	35 mph
Seats	2	**Climb Rate @msl**	1800 fpm
Cabin Width	28 in.	**Service Ceiling**	14,500 ft.

Std. Engine/HP	Usable HP Range	Construction Mtrl.	Est. Bldg. Time
Rotax 912 / 80 hp	50-80	Composite, Metal, Fabric	350 hrs.
Plans	**Kit**	**Info Packet**	**Video**
N/A	$12,990.00	Yes	$10.00

High wing pusher. Single occupant climb rate is 1,200 to 1,400 fpm. Two occupants and full fuel on board yields climb rate in the 800 fpm range. The range of options include detachable wing tanks, electrical systems, lighting, external air vents, carpeting and more. Has exceptional flight characteristics and makes for a good training platform. Tornado II Trainer version costs $12,490.00.

Completed & Flying: 120

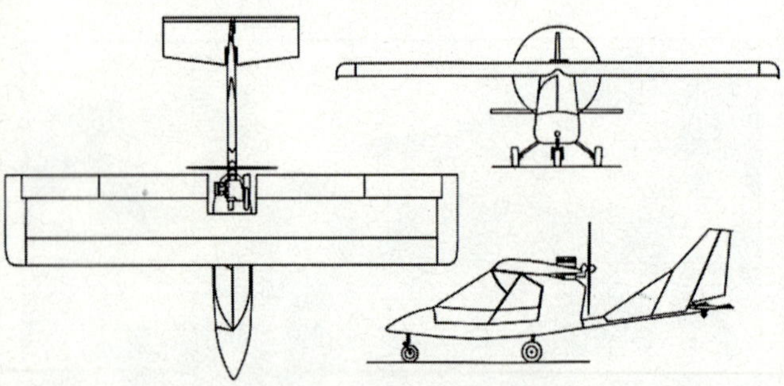

Developed in 1990, the aircraft is designed to be rugged, yet agile. Cr-Mo welded steel fuselage and all aluminum cantilever wing, it is capable of aerobatics, yet is still a friendly flyer. It is very stable in all flight regimes.

The Tornados have been static tested to their limit load with no permanent deformation. In the single place version, there is ample room for storage or an optional jump seat can be installed for carrying a passenger.

Bruce Borg, Eyota WI
"After owning both ultralight and general aviation airplanes, it is nice to have the wide speed range that provides both STOL performance and cross country capability in the same airplane."

Darryl Shade, West Palm FL
"The quality of this kit is excellent. Quality parts and materials. The parts are very well organized and easy to identify."

TITAN AIRCRAFT
2730 Walter Main Road • Geneva OH 44041
Tel. 216-466-0602 • Fax 216-466-7550
E-Mail: titan@interlaced.net
Web: http://webmill.com/titan

Tornado Sport

Wing Span	20.00 ft.	**Range**	250 sm
Wing Area	93.00 sq. ft.	**Takeoff Dist.**	250 ft.
Length	18.30 ft.	**Landing Dist.**	250 ft.
Max Gross Wt.	750 lbs.	**Vmax**	120 mph
Empty Wt.	320 lbs.	**VCr**	95 mph
Payload (full fuel)	360 lbs.	**V_{S1} (stall clean)**	35 mph
Fuel Capacity	10 gal.	**V_{S0} (lndg config)**	30 mph
Seats	1	**Climb Rate @msl**	1200 fpm
Cabin Width	26 in.	**Service Ceiling**	12,500 ft.

Std. Engine/HP	Usable HP Range	Construction Mtrl.	Est. Bldg. Time
Rotax 503 / 53 hp	N/A	Aluminum, Composite	300 hrs.
Plans	**Kit**	**Info Packet**	**Video**
N/A	$9,999.00	Yes	$10.00

High wing pusher. Good handling. Agile yet very stable and friendly. Can cruise in excess of 100 mph on 52 hp. Jump seat for carrying passengers is optional or area can be used for storage. Kit price does not include engine or options. The basic Tornado design was developed in 1990. It features a Cr-Mo welded steel fuselage, an all aluminum cantilever wing, and a specially developed high lift, low drag airfoil.

Completed & Flying: 70

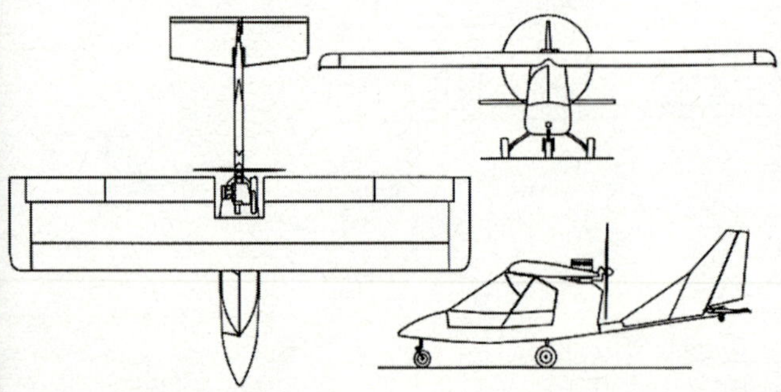

Tom Simko, Inkom ID
"I really did not plan to like it as much as I do, it is a very nice little airplane, light years away from the ultralights I flew in the 70's and 80's."

Roger Weinzirl, Equ Galle WI
"The more I fly this plane, the more I wish I never had to land. Very highly recommended."

TITAN AIRCRAFT
2730 Walter Main Road • Geneva OH 44041
Tel. 216-466-0602 • Fax 216-466-7550
E-Mail: titan@interlaced.net
Web: http://webmill.com/titan

Tornado U/L 103

Wing Span	20.00 ft.	**Range**	120 sm
Wing Area	93.00 sq. ft.	**Takeoff Dist.**	250 ft.
Length	18.30 ft.	**Landing Dist.**	150 ft.
Max Gross Wt.	550 lbs.	**Vmax**	90 mph
Empty Wt.	250 lbs.	**VCr**	60 mph
Payload (full fuel)	266 lbs.	**V_{s1} (stall clean)**	30 mph
Fuel Capacity	5 gal.	**V_{s0} (lndg config)**	28 mph
Seats	1	**Climb Rate @msl**	600 fpm
Cabin Width	26 in.	**Service Ceiling**	10,000 ft.

Std. Engine/HP	Usable HP Range	Construction Mtrl.	Est. Bldg. Time
Rotax 277 / 28 hp	N/A	Aluminum, Composite	250 hrs.
Plans	**Kit**	**Info Packet**	**Video**
N/A	$9,990.00	Yes	$10.00

Award-winning, state-of-the-art ultralight. Designed to meet FAR Part 103. Extremely agile flyer with very good handling in gusty conditions and still capable of mild aerobatics. Developed in 1990. The U/L 103 still retains the Cr-Mo welded steel fuselage, an all-aluminum cantilever wing, and a specially developed high lift, low drag airfoil. Kit price does not include engine or options.

Kits/Plans Sold: 15 **Completed & Flying:** 10

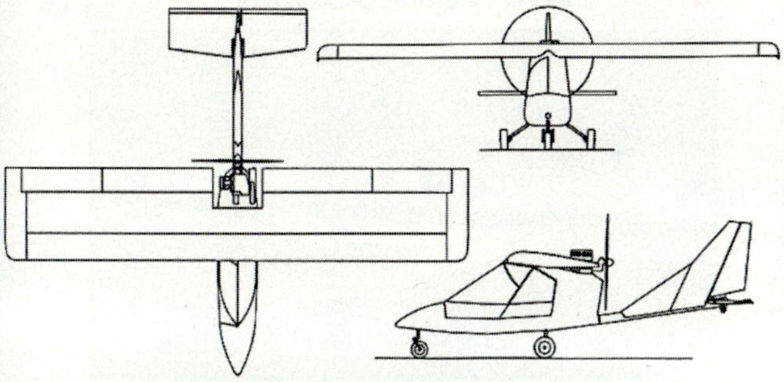

Don Cichocki, Huntington NY
"While not the rocket the Sport version is, this plane performs supurbly."

Paul Kepner, Garland TX
"Am enjoying working on the kit and pleasantly impressed by the planning, engineering and innovation shown."

TITAN AIRCRAFT
2730 Walter Main Road • Geneva OH 44041
Tel. 216-466-0602 • Fax 216-466-7550
E-Mail: titan@interlaced.net
Web: http://webmill.com/titan

Tornado II FP

Wing Span	23.50 ft.	**Range**	250 sm
Wing Area	108.0 sq. ft.	**Takeoff Dist.**	300 ft.
Length	19.00 ft.	**Landing Dist.**	250 ft.
Max Gross Wt.	1000 lbs.	**Vmax**	120 mph
Empty Wt.	580 lbs.	**VCr**	105 mph
Payload (full fuel)	400 lbs.	**V_{s1} (stall clean)**	38 mph
Fuel Capacity	10 gal.	**V_{s0} (lndg config)**	35 mph
Seats	2	**Climb Rate @msl**	1200 fpm
Cabin Width	28 in.	**Service Ceiling**	12,500 ft.

Std. Engine/HP	Usable HP Range	Construction Mtrl.	Est. Bldg. Time
Rotax 618 / 75 hp	N/A	Aluminum, Composite	350 hrs.
Plans	**Kit**	**Info Packet**	**Video**
N/A	$17,490.00	Yes	$10.00

High wing, pusher developed in 1990. Exceptional performance for a float plane. Float kit included in kit. It features a Cr-Mo welded steel fuselage, an all-aluminum cantilever wing, and a specially developed high lift, low drag airfoil. **Flying:** 6

TITAN AIRCRAFT
2730 Walter Main Road • Geneva OH 44041
Tel. 216-466-0602 • Fax 216-466-7550
E-Mail: titan@interlaced.net
Web: http://webmill.com/titan

Tornado MG

Wing Span	26.00 ft.	**Range**	375 sm
Wing Area	120.0 sq. ft.	**Takeoff Dist.**	250 ft.
Length	18.30 ft.	**Landing Dist.**	250 ft.
Max Gross Wt.	750 lbs.	**Vmax**	120 mph
Empty Wt.	330 lbs.	**VCr**	100 mph
Payload (full fuel)	N/A	**V_{s1} (stall clean)**	30 mph
Fuel Capacity	10 gal.	**V_{s0} (lndg config)**	N/A
Seats	1	**Climb Rate @msl**	1200 fpm
Cabin Width	N/A	**Service Ceiling**	12,500 ft.

Std. Engine/HP	Usable HP Range	Construction Mtrl.	Est. Bldg. Time
Rotax 447 / 40 hp	40-80	Composite, Metal, Fabric	350 hrs.
Plans	**Kit**	**Info Packet**	**Video**
N/A	$10,990.00	Yes	$10.00

An economical, easy to build design utilizing Titan's high lift, low drag airfoil. Engine options include the Rotax 503, 582, 618 and 912 engines. Prefab parts are available. Detachable wings.

TITAN AIRCRAFT
2730 Walter Main Road • Geneva OH 44041
Tel. 216-466-0602 • Fax 216-466-7550
E-Mail: titan@interlaced.net
Web: http://webmill.com/titan

Dolphin-3

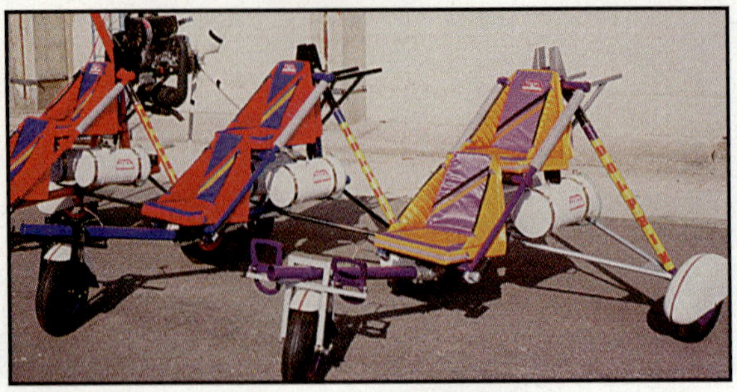

Wing Span	N/A	Range	N/A
Wing Area	N/A	Takeoff Dist.	N/A
Length	N/A	Landing Dist.	N/A
Max Gross Wt.	705 lbs.	Vmax	N/A
Empty Wt.	349 lbs.	VCr	N/A
Payload (full fuel)	N/A	V_{S1} (stall clean)	N/A
Fuel Capacity	10 gal.	V_{S0} (Indg config)	N/A
Seats	2	Climb Rate @msl	N/A
Cabin Width	N/A	Service Ceiling	N/A

Std. Engine/HP	Usable HP Range	Construction Mtrl.	Est. Bldg. Time
Rotax 503 / 50 hp	N/A	Alum., CR-MO	15 hrs.
Plans	**Kit**	**Info Packet**	**Video**
N/A	$11,700.00	Yes	N/A

Certified two seater trike. Features include folding mast, aerodynamic pod, fairings for all three wheels, toe operated front drum brake, and safety harness. Options include ballistic parachute and instruments. Engine options are SVS-1400 and Rotax 582

Eco-6

Wing Span	34.80 ft.	Range	N/A
Wing Area	167.0 sq. ft.	Takeoff Dist.	N/A
Length	N/A	Landing Dist.	N/A
Max Gross Wt.	995 lbs.	Vmax	70 mph
Empty Wt.	352 lbs.	VCr	40 mph
Payload (full fuel)	595 lbs.	V_{S1} (stall clean)	N/A
Fuel Capacity	10 gal.	V_{S0} (Indg config)	26 mph
Seats	1	Climb Rate @msl	N/A
Cabin Width	N/A	Service Ceiling	N/A

Std. Engine/HP	Usable HP Range	Construction Mtrl.	Est. Bldg. Time
Rotax 447 / 42 hp	N/A	Alum., CR-MO	15 hrs.
Plans	**Kit**	**Info Packet**	**Video**
N/A	$9,600.00	Yes	N/A

Ultralight single-seat trike. Standard features include folding mast, aerodynamic instrument panel, fairings for all three wheels, and a four point safety harness. Options include ballistic parachute and instruments. **Completed & Flying:** 30

TO-MI AVIATION
301 Recany n.
Labem
CZECH REPUBLIC

U.S. Distributor: A.K.S. INC.
PO Box 17161 • Portland OR 97217
Tel. 503-289-1491 Fax: 503-247-9207 • E-Mail: trikes@teleport.com
Web: http://www.teleport.com/~trikes
See See Ad On Page 47

TO-MI AVIATION
301 Recany n.
Labem
CZECH REPUBLIC

Cross-5

Wing Span	N/A	Range	N/A
Wing Area	N/A	Takeoff Dist.	N/A
Length	N/A	Landing Dist.	N/A
Max Gross Wt.	995 lbs.	Vmax	N/A
Empty Wt.	352 lbs.	VCr	N/A
Payload (full fuel)	N/A	V_{S1} (stall clean)	N/A
Fuel Capacity	10 gal.	V_{S0} (Indg config)	N/A
Seats	2	Climb Rate @msl	N/A
Cabin Width	N/A	Service Ceiling	N/A

Std. Engine/HP	Usable HP Range	Construction Mtrl.	Est. Bldg. Time
Rotax 503 / 50 hp	N/A	Alum., CR-MO	15 hrs.
Plans	**Kit**	**Info Packet**	**Video**
N/A	$11,600.00	Yes	N/A

Certified two seater trike.Features include dual folding mast, aerodynamic pod, fairings for wheels, toe operated front drum brake, and safety harness. Ballistic parachute optional. 6" longer Cross-5x version with wider and higher seats is available for $11,800.

TO-MI AVIATION
301 Recany n.
Labem
CZECH REPUBLIC

KIS TR-4 Cruiser

Wing Span	29.00 ft.	Range	800 sm
Wing Area	135.0 sq. ft.	Takeoff Dist.	1000 ft.
Length	25.50 ft.	Landing Dist.	1300 ft.
Max Gross Wt.	2300 lbs.	Vmax	190 mph
Empty Wt.	1200 lbs.	VCr	175 mph
Payload (full fuel)	N/A	V_{S1} (stall clean)	55 mph
Fuel Capacity	52 gal.	V_{S0} (Indg config)	N/A
Seats	4	Climb Rate @msl	1300 fpm
Cabin Width	44 in.	Service Ceiling	N/A

Std. Engine/HP	Usable HP Range	Construction Mtrl.	Est. Bldg. Time
N/A	160-180	Composite, Metal	1200 hrs.
Plans	**Kit**	**Info Packet**	**Video**
N/A	$24,900.00	$12.00	$15.00

A four-place fixed-gear design utilizing composite construction. All prefabricated metal parts are prewelded. Kit does not include engine, instruments, avionics, paint or upholstery.

TRI-R TECHNOLOGIES
1114 E. 5th Street
Oxnard CA 93030
Tel. 805-385-3680 • Fax 805-483-8366
E-Mail: flykis@aol.com

KIS TR-1

Wing Span	23.00 ft.	**Range**	750 sm
Wing Area	88.00 sq. ft.	**Takeoff Dist.**	600 ft.
Length	22.00 ft.	**Landing Dist.**	1200 ft.
Max Gross Wt.	1300 lbs.	**Vmax**	195 mph
Empty Wt.	750 lbs.	**VCr**	175 mph
Payload (full fuel)	N/A	**V_{S1} (stall clean)**	55 mph
Fuel Capacity	20 gal.	**V_{S0} (Indg config)**	N/A
Seats	2	**Climb Rate @msl**	1200 fpm
Cabin Width	N/A	**Service Ceiling**	17,000 ft.

Std. Engine/HP	Usable HP Range	Construction Mtrl.	Est. Bldg. Time
Continental / 125 hp	80-125	Composite	1000 hrs.
Plans	**Kit**	**Info Packet**	**Video**
N/A	$16,995.00	$15.00	$15.00

Tricycle version of the KIS design. Also available with Limbach 2000 engine. Options include engine mountings, wheel fairing kit, custom instrument panel and Sigtronics intercoms and headsets.

Completed & Flying: 25

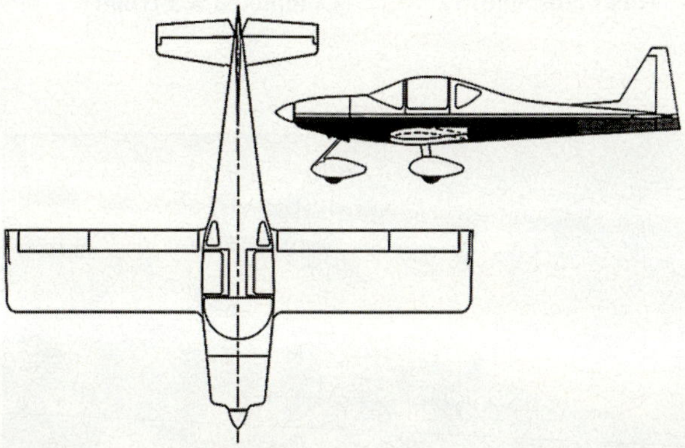

TRI-R TECHNOLOGIES
1114 E. 5th Street
Oxnard CA 93030
Tel. 805-385-3680 • Fax 805-483-8366
E-Mail: flykis@aol.com

KIS TR-1/TD

Wing Span	23.00 ft.	**Range**	750 sm
Wing Area	88.00 sq. ft.	**Takeoff Dist.**	600 ft.
Length	22.00 ft.	**Landing Dist.**	1200 ft.
Max Gross Wt.	1425 lbs.	**Vmax**	180 mph
Empty Wt.	800 lbs.	**VCr**	170 mph
Payload (full fuel)	N/A	**V_{S1} (stall clean)**	55 mph
Fuel Capacity	34 gal.	**V_{S0} (Indg config)**	N/A
Seats	2	**Climb Rate @msl**	1200 fpm
Cabin Width	N/A	**Service Ceiling**	17,000 ft.

Std. Engine/HP	Usable HP Range	Construction Mtrl.	Est. Bldg. Time
O-235-C1B / 108 hp	80-125	Composite	800 hrs.
Plans	**Kit**	**Info Packet**	**Video**
N/A	$16,900.00	$10.00	$15.00

Taildragger version of the KIS design. Also available with Limbach 2000 engine. Optional A/B building package allows purchase and construction of the plane in two parts. **Completed & Flying:** 8

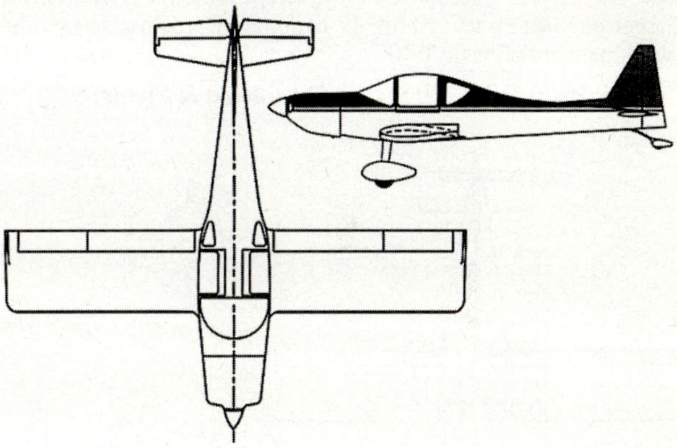

Rich Trickel is President of High Tech Composites in Oxnard, CA, the firm that makes molded airframe components for a number of homebuilt designs, most notably the Lancair 235, 320 and the four place Lancair IV.

While in Australia on company business, a client was dissatisfied with a rather radical design his company was attempting to develop, and wished a more conventional configuration had been chosen. Eager to help, Rich offered to do some 3-views. Rich made the drawings, which were well received but never paid for. So he brought them back home and eventually the plans were used as a starting point for the development of the KIS-1.

TRI-R TECHNOLOGIES
1114 E. 5th Street
Oxnard CA 93030
Tel. 805-385-3680 • Fax 805-483-8366
E-Mail: flykis@aol.com

Turner T-40A

Wing Span	23.80 ft.	**Range**	475 sm
Wing Area	96.00 sq. ft.	**Takeoff Dist.**	1250 ft.
Length	20.00 ft.	**Landing Dist.**	1200 ft.
Max Gross Wt.	1600 lbs.	**Vmax**	170 mph
Empty Wt.	1050 lbs.	**VCr**	147 mph
Payload (full fuel)	418 lbs.	**V_{S1} (stall clean)**	64 mph
Fuel Capacity	22 gal.	**V_{S0} (lndg config)**	59 mph
Seats	2	**Climb Rate @msl**	850 fpm
Cabin Width	40 in.	**Service Ceiling**	12,000 ft.

Std. Engine/HP	Usable HP Range	Construction Mtrl.	Est. Bldg. Time
Lycoming / 125 hp	100 - 125	Wood	2200 hrs.
Plans	**Kit**	**Info Packet**	**Video**
$200.00	N/A	$6.00	N/A

A plans only, non-aerobatic design with folding wings. +3.8, -2.0 G's limit. First flew July 29, 1966. Other variations include the single seat T-40 and the two seat Super T-40A. The Super T-40A retains the basic proven structure and folding wings of the T-40 and T-40A plus a bubble canopy, swept tail, and provision for larger engines up to 150 hp. 17 national awards for design and workmanship of basic T-40.

Kits/Plans Sold: 230 **Completed & Flying:** 9

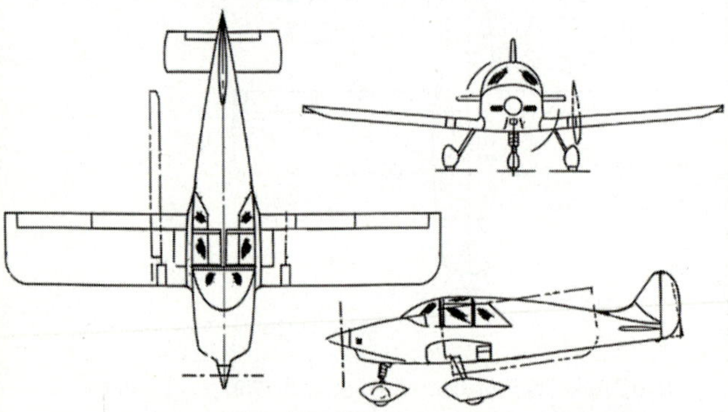

Turner Aircraft, Inc. (TAI) was founded by Gene Turner in 1962 when plans for the Turner T-40 single-engine folding-wing airplane were first offered for sale. Growth of the company was restricted for years as Gene was then working for the FAA and they only allowed him to sell plans for the T-40 and its variations, the T-40A and Super T-40A .

After retirement, Gene developed a new Ultralight, the Model T-100 Mariah I which is no longer on the market. He also produced the T-40B with a Tee tail, a GAW wing, full span flaps, spoilers, and slat. TAI has future plans for a two place amphibious airplane, a "primary" airplane, automobile engine power for the T-40, and an all new Tri-Plane with three tandem wings.

TURNER AIRCRAFT CO., INC.
Route 4, Box 115AB3 • Luscombe Acres Airport
Grandview TX 76050
Tel. 817-783-5350 • Fax 817-783-5350
E-Mail: turnert40@aol.com

Hornet

Wing Span	27.50 ft.	**Range**	400 sm
Wing Area	138.0 sq. ft.	**Takeoff Dist.**	200 ft.
Length	20.00 ft.	**Landing Dist.**	150 ft.
Max Gross Wt.	1000 lbs.	**Vmax**	95 mph
Empty Wt.	475 lbs.	**VCr**	85 mph
Payload (full fuel)	425 lbs.	**V_{S1} (stall clean)**	40 mph
Fuel Capacity	18 gal.	**V_{S0} (lndg config)**	32 mph
Seats	2	**Climb Rate @msl**	800 fpm
Cabin Width	27 in.	**Service Ceiling**	12,500 ft.

Std. Engine/HP	Usable HP Range	Construction Mtrl.	Est. Bldg. Time
Hirth / 55 hp	45 - 75	Tube, Fabric	300 hrs.
Plans	**Kit**	**Info Packet**	**Video**
N/A	$18,000.00	Yes	$10.00

Grand Champion Light Plane awards at both Sun 'n Fun '95 and Oshkosh '95. Design features custom hydraulic brakes, pneumatic suspension on all three wheels, electrically activated flaps and horizontal stabilizer, custom re-drive, and adjustable seats. Options include a custom fitted ballistic parachute, 65 hp Hirth engine, and dual 9 gallon tanks between spars. Kit price includes engine. $12,500 price without engine package.

Kits/Plans Sold: 12 **Completed & Flying:** 5

Mary Jones, EAA EXPERIMENTER, 11/95
"Attention to detail is usually the element that separates champion aircraft from "also rans" and Roger Chase's Hornet exemplified attention to detail throughout, from the incorporation of the soft-pack chute to its detailed upholstered covers."

U.S. LIGHT AIRCRAFT CORP.
27080 Rancho Ballena Lane
Ramona CA 92065
Tel. 619-789-8607 • Fax 619-789-8607

Calypso

Wing Span	29.00 ft.	**Range**	400 sm
Wing Area	125.0 sq. ft.	**Takeoff Dist.**	55 ft.
Length	18.00 ft.	**Landing Dist.**	50 ft.
Max Gross Wt.	630 lbs.	**Vmax**	90 mph
Empty Wt.	342 lbs.	**VCr**	62 mph
Payload (full fuel)	226 lbs.	**V_{S1} (stall clean)**	34 mph
Fuel Capacity	10 gal.	**V_{S0} (lndg config)**	N/A
Seats	1	**Climb Rate @msl**	600 fpm
Cabin Width	24 in.	**Service Ceiling**	N/A

Std. Engine/HP	Usable HP Range	Construction Mtrl.	Est. Bldg. Time
Rotax 447 / 40 hp	N/A	Wood, Fabric, Composite	250 hrs.
Plans	**Kit**	**Info Packet**	**Video**
N/A	$9,600.00	Yes	$25.00

Can be ready to fly from trailering in about 20 minutes. Large cockpit can accept pilots up to 6' 8" in height. Can also be equipped with the Rotax 503 single carburetor engine.

Kits/Plans Sold: 12 **Completed & Flying:** 5

ULTRACRAFT
St. Quirinusstraat 72 • Zolder B-3550
BELGIUM
Fax 011 43 66 78

Pelican PL 912/914

Wing Span	29.50 ft.	**Range**	700 sm
Wing Area	108.0 sq. ft.	**Takeoff Dist.**	350 ft.
Length	19.75 ft.	**Landing Dist.**	N/A
Max Gross Wt.	1400 lbs.	**Vmax**	133 mph
Empty Wt.	750 lbs.	**VCr**	121 mph
Payload (full fuel)	N/A	**V_{S1} (stall clean)**	37 mph
Fuel Capacity	26 gal.	**V_{S0} (lndg config)**	33 mph
Seats	2	**Climb Rate @msl**	1800 fpm
Cabin Width	46 in.	**Service Ceiling**	20,000 ft.

Std. Engine/HP	Usable HP Range	Construction Mtrl.	Est. Bldg. Time
Rotax 914 / 100 hp	80 - 100	Composite	800 hrs.
Plans	**Kit**	**Info Packet**	**Video**
N/A	$16,600.00	Yes	N/A

Stylish two place, fast build kit aircraft with composite fuselage and aluminum wings. Designed for speed and STOL performance. Rated +4, -2 G's. Fuel consumption at 75% power approximately 6.3 gph. Variety of quick build assembly options available. Performance specs based on 914 engine. Rotax 912 engine also available. Kit price is for basic airframe kit with tricycle gear.

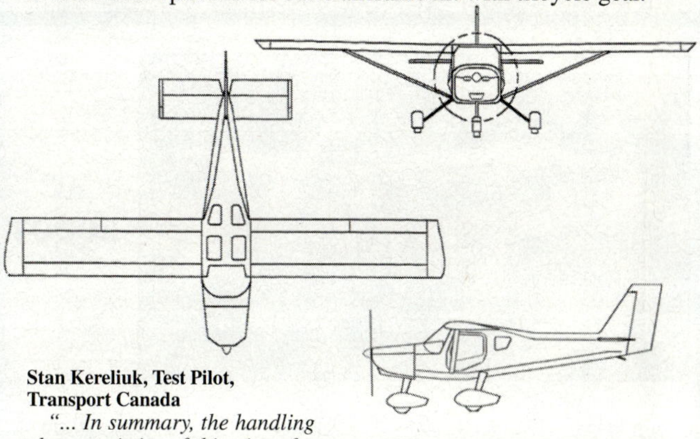

Stan Kereliuk, Test Pilot, Transport Canada
"... In summary, the handling characteristics of this aircraft make it a pleasure to fly."

ULTRAVIA AERO INT'L, INC.
300-D Airport Rd.
Mascouche, Montreal, Quebec J7K 3C1 • CANADA
Tel. 514-953-1491 • Fax 514-966-6299

HOMEBUILT "FAQ'S" *(See Page 68)*

Q: What separates a homebuilt from an ultralight?

A: Ultralights and Experimentals may both be built by the owner, but the ultralight may carry no passengers (except for instruction). It also has limits on weight, speed, and fuel, which the Experimental does not (outside of operating restrictions in FAR Part 91).

Often, licensed ultralight owners have chosen to obtain airworthiness certificates so that they can overcome one or more of these restrictions. This reclassifies the ultralight as an Experimental, and the pilot certificate, medical, and currency requirements become effective.

SR-1 Snoshoo

Wing Span	19.91 ft.	Range	N/A
Wing Area	66.00 sq. ft.	Takeoff Dist.	N/A
Length	17.91 ft.	Landing Dist.	N/A
Max Gross Wt.	746 lbs.	Vmax	280 mph
Empty Wt.	530 lbs.	VCr	220 mph
Payload (full fuel)	N/A	V_{S1} (stall clean)	65 mph
Fuel Capacity	6 gal.	V_{S0} (lndg config)	N/A
Seats	1	Climb Rate @msl	N/A
Cabin Width	22 in.	Service Ceiling	N/A

Std. Engine/HP	Usable HP Range	Construction Mtrl.	Est. Bldg. Time
Cont. 0-200 / 100 hp	95 - 100+	Tube, Wood, Composite	1500 hrs.
Plans	**Kit**	**Info Packet**	**Video**
$150.00	Yes	$5.00	N/A

A new, plans-built design offered to homebuilders as a Gold Race competitive Fomula One racer. First flight in 1997. Call regarding kits and components.

Kits/Plans Sold: 20

VSR
1834 White Oak
Wichita KS 67207
Tel. 316-684-2032 • Fax 316-687-9968

XC 2000T

Wing Span	32.00 ft.	Range	100 sm
Wing Area	176.0 sq. ft.	Takeoff Dist.	150 ft.
Length	20.00 ft.	Landing Dist.	150 ft.
Max Gross Wt.	890 lbs.	Vmax	90 mph
Empty Wt.	380 lbs.	VCr	40 mph
Payload (full fuel)	450 lbs.	V_{S1} (stall clean)	26 mph
Fuel Capacity	10 gal.	V_{S0} (lndg config)	20 mph
Seats	2	Climb Rate @msl	500 fpm
Cabin Width	43 in.	Service Ceiling	13,500 ft.

Std. Engine/HP	Usable HP Range	Construction Mtrl.	Est. Bldg. Time
2SI / 45 hp	40 - 75	Alum., Fabric	100 hrs.
Plans	**Kit**	**Info Packet**	**Video**
N/A	$9,995.00	$10.00	$15.00

Side-by-side seating. Use as ultralight trainer or experimental two seater. Introduced in the fall of 1997. Design evolved from the Starflight XC2000. Kit price does not include engine and prop. Options include enclosure and dual controls. **Sold:** 3 **Flying:** 2

VSTOL AIRCRAFT CORP.
PO Box 7534
Ft. Myers FL 33911
Tel. 941-936-1261 • Fax 941-936-1261

Super Solution 2000

Wing Span	32.00 ft.	Range	100 sm
Wing Area	165.0 sq. ft.	Takeoff Dist.	100 ft.
Length	20.00 ft.	Landing Dist.	150 ft.
Max Gross Wt.	890 lbs.	Vmax	90 mph
Empty Wt.	380 lbs.	VCr	40 mph
Payload (full fuel)	510 lbs.	V_{S1} (stall clean)	N/A
Fuel Capacity	10 gal.	V_{S0} (lndg config)	N/A
Seats	2	Climb Rate @msl	500 fpm
Cabin Width	34 in.	Service Ceiling	13,500 ft.

Std. Engine/HP	Usable HP Range	Construction Mtrl.	Est. Bldg. Time
2SI / 45 hp	40 - 75	Alum., Fabric	100 hrs.
Plans	**Kit**	**Info Packet**	**Video**
N/A	$7,495.00	$10.00	$15.00

Single engine, two-place ultralight trainer or two-place experimental airplane. Dacron sailcloth, and rugged fiberglass landing gear. Employs a high-lift, single-surface airfoil similar to those on ultralights. Performance specs are at full gross weight. Manufacturer claims aircraft can takeoff at 18 mph without passenger. Used for recreational flying and crop dusting. Kit price does not include engine and prop. Options include enclosure and dual controls. **Kits/Plans Sold:** 6 **Completed & Flying:** 2

VSTOL is a subsidiary of Venezuela STOL Aircraft C.A. Prototype work and small orders are produced in Ft. Myers, Florida. The company's main production plant is located in Caracas, Venezuela where it produces the XC Aglite which was the first specifically certified ultralight crop duster for developing countries. This aircraft uses only half the normal amount of chemicals with vegetable oil as the primary carrier.

Dick Turner, CEO of VSTOL, is one of the early ultralight pioneers. He founded Starflight Manufacturing, Inc. in 1980 and produced several different models including the TriStar and DBL which was the first tandem ultralight trainer. In 1985 Dick was the U.S.A. Ultralight Team manager and took the U.S. team to France for the first World Microlight Championships.

VSTOL AIRCRAFT CORP.
PO Box 7534
Ft. Myers FL 33911
Tel. 941-936-1261 • Fax 941-936-1261

RV-4

Wing Span	23.00 ft.	**Range**	650 sm
Wing Area	110.0 sq. ft.	**Takeoff Dist.**	450 ft.
Length	24.33 ft.	**Landing Dist.**	425 ft.
Max Gross Wt.	1500 lbs.	**Vmax**	204 mph
Empty Wt.	905 lbs.	**VCr**	192 mph
Payload (full fuel)	N/A	**V_{S1} (stall clean)**	54 mph
Fuel Capacity	32 gal.	**V_{S0} (lndg config)**	N/A
Seats	2	**Climb Rate @msl**	1650 fpm
Cabin Width	N/A	**Service Ceiling**	19,500 ft.

Std. Engine/HP	Usable HP Range	Construction Mtrl.	Est. Bldg. Time
Lyc. O-320 / 160 hp	N/A	Aluminum	1800 hrs.
Plans	**Kit**	**Info Packet**	**Video**
$165.00	$10,485.00	$8.00	$12.50

Very popular two seat, tandem aircraft. Aerobatic configuration. A variety of kit options are available. Can be fitted with Lycoming O-320 or O-360 engines. Kit price does not include freight charges. Performance specs based on gross weight.

Kits/Plans Sold: 4600 **Completed & Flying:** 550

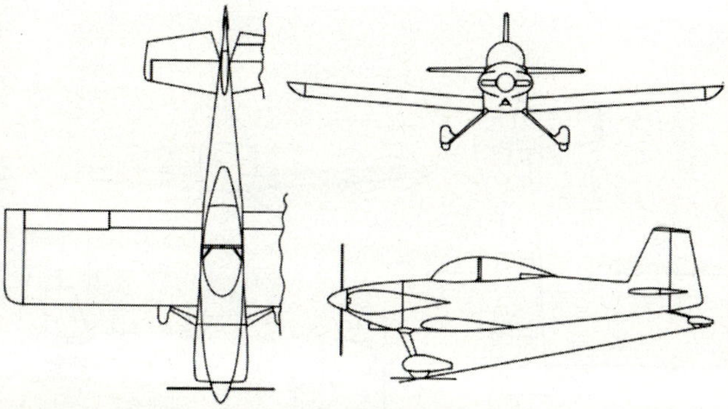

Brian Turner, Victoria, Australia

"Performs exactly as advertised, a delight to fly. Can't wait for our CAA to give aerobatic approval to the RVs!"

VAN'S AIRCRAFT
P.O. Box 160
North Plains OR 97133
Tel. 503-647-5117 • Fax 503-647-2206
Web: http://www.vansaircraft.com

RV-6

Wing Span	23.00 ft.	**Range**	775 sm
Wing Area	110.0 sq. ft.	**Takeoff Dist.**	525 ft.
Length	20.10 ft.	**Landing Dist.**	500 ft.
Max Gross Wt.	1600 lbs.	**Vmax**	201 mph
Empty Wt.	965 lbs.	**VCr**	171 mph
Payload (full fuel)	N/A	**V_{S1} (stall clean)**	49 mph
Fuel Capacity	38 gal.	**V_{S0} (lndg config)**	N/A
Seats	2	**Climb Rate @msl**	1500 fpm
Cabin Width	N/A	**Service Ceiling**	16,300 ft.

Std. Engine/HP	Usable HP Range	Construction Mtrl.	Est. Bldg. Time
Lyc. O-320 / 160 hp	N/A	Aluminum	2000 hrs.
Plans	**Kit**	**Info Packet**	**Video**
$195.00	$11,405.00	$8.00	$12.50

Two seat, side-by-side, taildragger configuration. Kit price does not include freight charges. Performance specs based on gross weight. Quick Build Kit, that reduces build time by about 50%, is available for $19,320. **Kits/Plans Sold:** 2000 **Flying:** 300

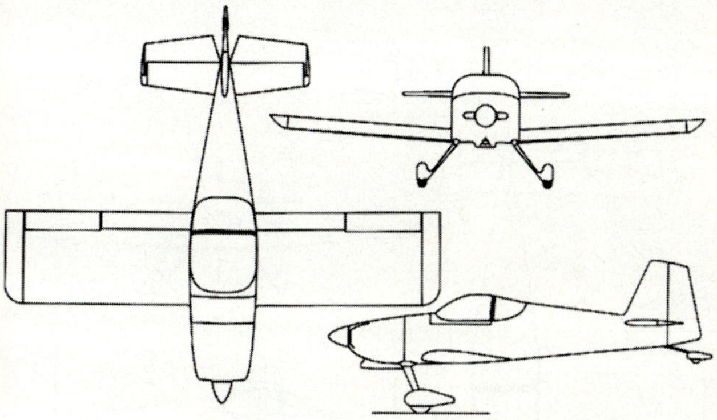

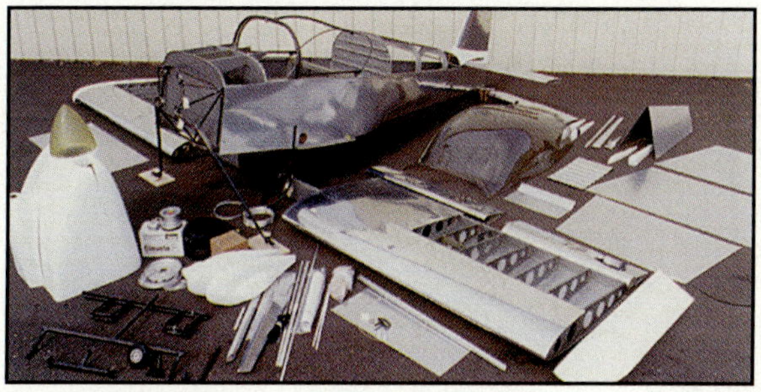

VAN'S AIRCRAFT
P.O. Box 160
North Plains OR 97133
Tel. 503-647-5117 • Fax 503-647-2206
Web: http://www.vansaircraft.com

RV-6A

Wing Span	23.00 ft.	**Range**	750 sm
Wing Area	110.0 sq. ft.	**Takeoff Dist.**	350 ft.
Length	19.75 ft.	**Landing Dist.**	400 ft.
Max Gross Wt.	1650 lbs.	**Vmax**	199 mph
Empty Wt.	985 lbs.	**VCr**	188 mph
Payload (full fuel)	N/A	**V_{S1} (stall clean)**	49 mph
Fuel Capacity	38 gal.	**V_{S0} (lndg config)**	N/A
Seats	2	**Climb Rate @msl**	1850 fpm
Cabin Width	N/A	**Service Ceiling**	16,300 ft.

Std. Engine/HP	Usable HP Range	Construction Mtrl.	Est. Bldg. Time
Lyc. O-320 / 160 hp	N/A	Aluminum	2000 hrs.
Plans	**Kit**	**Info Packet**	**Video**
$200.00	$12,085.00	$8.00	$12.50

Tricycle gear version of the RV-6. New Lycoming engines, propellers and avionics avialable at good prices to kit purchasers. A new Quick-Build Kit, that reduces build time by about 50%, is available for $19,995. Performance specs based on gross weight. Photo by George Orndorff.

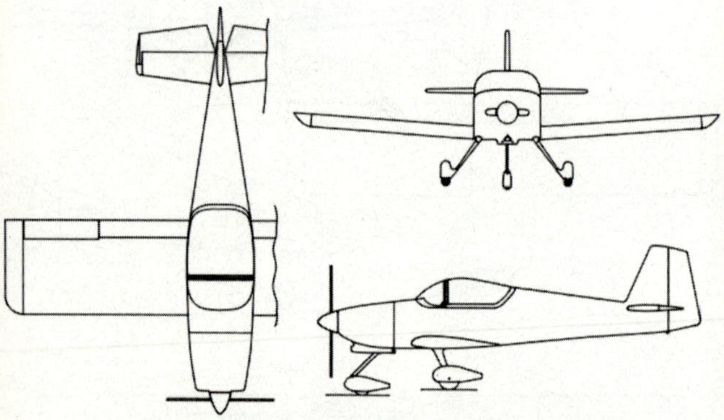

Jerry Dixon, East Wenatchee, WA

"I first flew my RV-6A on August 28, 1994 after four and a half years building time. What an experience, I'd never flown anything like it before. After flying a 172 for eight years, the RV-6A was a real pleasure."

VAN'S AIRCRAFT
P.O. Box 160
North Plains OR 97133
Tel. 503-647-5117 • Fax 503-647-2206
Web: http://www.vansaircraft.com

RV-8

Wing Span	23.00 ft.	**Range**	1000 sm
Wing Area	N/A	**Takeoff Dist.**	250 ft.
Length	21.00 ft.	**Landing Dist.**	300 ft.
Max Gross Wt.	1800 lbs.	**Vmax**	222 mph
Empty Wt.	1090 lbs.	**VCr**	210 mph
Payload (full fuel)	733 lbs.	**V_{S1} (stall clean)**	51 mph
Fuel Capacity	44 gal.	**V_{S0} (lndg config)**	N/A
Seats	2	**Climb Rate @msl**	2700 fpm
Cabin Width	N/A	**Service Ceiling**	25,000 ft.

Std. Engine/HP	Usable HP Range	Construction Mtrl.	Est. Bldg. Time
Lyc. IO-360 / 200 hp	150-200	Aluminum	2200 hrs.
Plans	**Kit**	**Info Packet**	**Video**
N/A	Yes	N/A	N/A

New tandem model with wider, longer fuselage, rounded engine cowl, and larger sliding canopy. Improved performance for cross country travel. Now available: Empennage Kit at $1,1190, Wing Kit at $4,350, and Fuselage Kit at $4,900; all pre-punched. The Finishing Kit is scheduled to be available late 1997.

Kits/Plans Sold: 400 **Completed & Flying:** 1

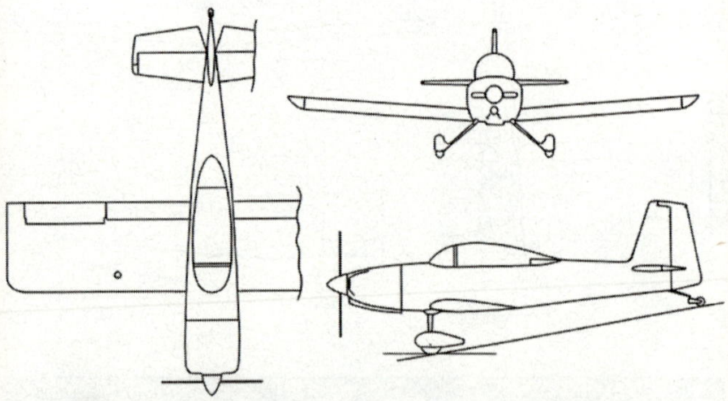

VAN'S AIRCRAFT
P.O. Box 160
North Plains OR 97133
Tel. 503-647-5117 • Fax 503-647-2206
Web: http://www.vansaircraft.com

Velocity LW

Wing Span	31.00 ft.	**Range**	911 sm
Wing Area	145.0 sq. ft.	**Takeoff Dist.**	1250 ft.
Length	19.16 ft.	**Landing Dist.**	1300 ft.
Max Gross Wt.	2400 lbs.	**Vmax**	149 mph
Empty Wt.	1350 lbs.	**VCr**	141 mph
Payload (full fuel)	1050 lbs.	**V_{S1} (stall clean)**	52 mph
Fuel Capacity	65 gal.	**V_{S0} (lndg config)**	N/A
Seats	4	**Climb Rate @msl**	1000 fpm
Cabin Width	42 in.	**Service Ceiling**	20,000 ft.

Std. Engine/HP	Usable HP Range	Construction Mtrl.	Est. Bldg. Time
Lyc. IO-360 / 200 hp	N/A	Composite	1500 hrs.
Plans	**Kit**	**Info Packet**	**Video**
N/A	$29,000.00	$12.00	Yes

The "Long Wing" version of the Velocity has a larger wing and 20% longer canard span than the Velocity Elite, increasing lifting surface by 23%. This provides lower landing speeds and more neutral stability. Specs are for the fixed-gear version. Also available with retractable gear. Retractable landing gear adds significantly to performance. Gull-wing doors are standard. See Ad on Page 20.

Kits/Plans Sold: 350 **Completed & Flying:** 70

Also Distributed By:

FUTURE AVIATION EUROPE
PO Box 75171
1117 ZS Schiphol Airport
The Netherlands
Ph: 31 23-5614881
Fax: 31 23-5631891

VELOCITY INC. WEST
1410-B Flightline Dr.
Lincoln CA 95648
Ph: 916-645-6866
Fax: 916-645-6944

VELOCITY AIRCRAFT
200 West Airport Road
Sebastian FL 32958
Tel. 561-589-1860 • Fax 561-589-1893
E-Mail: WebMaster@VelocityAircraft.com
Web: http://www.velocityaircraft.com

Velocity Elite

Wing Span	29.33 ft.	**Range**	911 sm
Wing Area	121.8 sq. ft.	**Takeoff Dist.**	1400 ft.
Length	19.00 ft.	**Landing Dist.**	1500 ft.
Max Gross Wt.	2250 lbs.	**Vmax**	156 mph
Empty Wt.	1230 lbs.	**VCr**	147 mph
Payload (full fuel)	1020 lbs.	**V_{S1} (stall clean)**	56 mph
Fuel Capacity	60 gal.	**V_{S0} (lndg config)**	N/A
Seats	4	**Climb Rate @msl**	1000 fpm
Cabin Width	44 in.	**Service Ceiling**	20,000 ft.

Std. Engine/HP	Usable HP Range	Construction Mtrl.	Est. Bldg. Time
Lyc IO-360 / 200 hp	N/A	Composite	1500 hrs.
Plans	**Kit**	**Info Packet**	**Video**
N/A	$25,000.00	$12.00	Yes

High performance, four place canard pusher available with fixed and retractable landing gear options. Specs are for fixed gear version. Gull-wing doors are standard. Other options include upholstry and pre-built wings. For overseas customers, an information packet including video is available for $44.50. Kit price does not include engine or avionics. See Ad on Page 20.

Kits/Plans Sold: 350 **Completed & Flying:** 70

Douglas Karlsen, Turbine Design Inc.
"I built and flew a Velocity RG for a year. I can tell you that the plane is easy to build and the factory support was second to none. You will enjoy this plane."

Richard Westwater, Merrimack NH
"For a first-time builder, the fuselage and wings were the easy part. Installing the the Lycoming engine and the avionics took a year."

VELOCITY AIRCRAFT
200 West Airport Road
Sebastian FL 32958
Tel. 561-589-1860 • Fax 561-589-1893
E-Mail: WebMaster@VelocityAircraft.com
Web: http://www.velocityaircraft.com

Velocity XL

Wing Span	31.00 ft.	**Range**	911 sm
Wing Area	122.5 sq. ft.	**Takeoff Dist.**	1400 ft.
Length	20.00 ft.	**Landing Dist.**	1500 ft.
Max Gross Wt.	2700 lbs.	**Vmax**	171 mph
Empty Wt.	1640 lbs.	**VCr**	163 mph
Payload (full fuel)	1060 lbs.	**V_{S1} (stall clean)**	56 mph
Fuel Capacity	70 gal.	**V_{S0} (lndg config)**	N/A
Seats	4	**Climb Rate @msl**	1200 fpm
Cabin Width	48 in.	**Service Ceiling**	20,000 ft.

Std. Engine/HP	Usable HP Range	Construction Mtrl.	Est. Bldg. Time
Lyc. IO-540 / 260 hp	N/A	Composite	1500 hrs.
Plans	**Kit**	**Info Packet**	**Video**
N/A	$34,000.00	$12.00	Yes

Velocity's newest "extra large" aircraft features a larger engine and larger fuel tanks for increased range and a more spaceous cabin. Available with fixed or retractable gear. Performance specs with retractable gear are slightly higher. Gull wing doors are standard. 100 gallon fuel tank is optional. See Ad on Page 20.

Also Distributed By:

FUTURE AVIATION EUROPE
PO Box 75171
1117 ZS Schiphol Airport
The Netherlands
Ph: 31 23-5614881
Fax: 31 23-5631891

VELOCITY INC. WEST
1410-B Flightline Dr.
Lincoln CA 95648
Ph: 916-645-6866
Fax: 916-645-6944

VELOCITY AIRCRAFT
200 West Airport Road
Sebastian FL 32958
Tel. 561-589-1860 • Fax 561-589-1893
E-Mail: WebMaster@VelocityAircraft.com
Web: http://www.velocityaircraft.com

Dragonfly Mk I

Wing Span	22.00 ft.	**Range**	500 sm
Wing Area	92.20 sq. ft.	**Takeoff Dist.**	1150 ft.
Length	19.00 ft.	**Landing Dist.**	2000 ft.
Max Gross Wt.	1150 lbs.	**Vmax**	168 mph
Empty Wt.	605 lbs.	**VCr**	165 mph
Payload (full fuel)	545 lbs.	**V_{S1} (stall clean)**	48 mph
Fuel Capacity	15 gal.	**V_{S0} (lndg config)**	N/A
Seats	2	**Climb Rate @msl**	850 fpm
Cabin Width	43 in.	**Service Ceiling**	18,500 ft.

Std. Engine/HP	Usable HP Range	Construction Mtrl.	Est. Bldg. Time
VW 1835cc / 60 hp	N/A	Composite	700 hrs.
Plans	**Kit**	**Info Packet**	**Video**
$259.00	$8,500.00	$10.00	$59.50

Equal span canard sportplane. Original version now called the Mk I. Wheels are mounted in wing tip housings. Best suited to paved runways and wide taxiways because of its wide wheel spacing. Designed for optimum performance with a VW engine. Best performance of any aircraft in its horsepower class. Both canard and wing provide lift. Side-by-side seating with dual controls. Prefabricated components available. Plans include construction video. Building from plans will take 1,200+ hours. More than 65 Mk I's now flying.

Kits/Plans Sold: 2000 **Completed & Flying:** 500

VIKING AIRCRAFT, INC.
P.O. Box 646
Elkhorn WI 53121-0646
Tel. 414-723-1048 • Fax 414-723-1049
E-Mail: viking@pensys.com

Dragonfly Mk II

Wing Span	22.00 ft.	**Range**	500 sm
Wing Area	92.20 sq. ft.	**Takeoff Dist.**	1150 ft.
Length	19.00 ft.	**Landing Dist.**	2000 ft.
Max Gross Wt.	1150 lbs.	**Vmax**	168 mph
Empty Wt.	605 lbs.	**VCr**	165 mph
Payload (full fuel)	545 lbs.	**V_{S1} (stall clean)**	48 mph
Fuel Capacity	15 gal.	**V_{S0} (lndg config)**	N/A
Seats	2	**Climb Rate @msl**	850 fpm
Cabin Width	43 in.	**Service Ceiling**	18,500 ft.

Std. Engine/HP	Usable HP Range	Construction Mtrl.	Est. Bldg. Time
VW 1835cc / 60 hp	N/A	Composite	700 hrs.
Plans	**Kit**	**Info Packet**	**Video**
$259.00	$8,500.00	N/A	$59.50

Mark II version with conventional tail dragger landing gear. Equal span canard sportplane. Designed for optimum performance with a VW engine. Both canard and wing provide lift. Comfortable cockpit with side- by-side seating and dual controls. Prefabricated components available. Plans include construction video. "Number of units sold/flying" includes all models. Estimated building time is for kit-built.

Kits/Plans Sold: 2000 **Completed & Flying:** 500

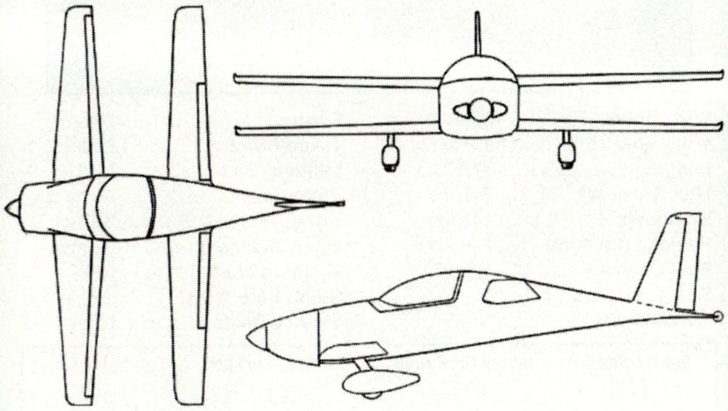

VIKING AIRCRAFT, INC.
P.O. Box 646
Elkhorn WI 53121-0646
Tel. 414-723-1048 • Fax 414-723-1049
E-Mail: viking@pensys.com

Dragonfly Mk III

Wing Span	22.00 ft.	**Range**	500 sm
Wing Area	92.20 sq. ft.	**Takeoff Dist.**	1150 ft.
Length	19.00 ft.	**Landing Dist.**	2000 ft.
Max Gross Wt.	1150 lbs.	**Vmax**	168 mph
Empty Wt.	605 lbs.	**VCr**	165 mph
Payload (full fuel)	545 lbs.	**V_{S1} (stall clean)**	48 mph
Fuel Capacity	15 gal.	**V_{S0} (lndg config)**	N/A
Seats	2	**Climb Rate @msl**	850 fpm
Cabin Width	43 in.	**Service Ceiling**	18,500 ft.

Std. Engine/HP	Usable HP Range	Construction Mtrl.	Est. Bldg. Time
VW 1835cc / 60 hp	N/A	Composite	700 hrs.
Plans	**Kit**	**Info Packet**	**Video**
$300.00	$8,500.00	N/A	$59.50

Mark III version with tricycle landing gear. Equal span canard sportplane. Optimum performance with VW engine. Side-by-side seating with dual controls. Construction video included. Prefab components available.

VIKING AIRCRAFT, INC.
P.O. Box 646
Elkhorn WI 53121-0646
Tel. 414-723-1048 • Fax 414-723-1049
E-Mail: viking@pensys.com

Cygnet Model SF-2A

Wing Span	30.00 ft.	**Range**	400 sm
Wing Area	N/A	**Takeoff Dist.**	700 ft.
Length	19.00 ft.	**Landing Dist.**	700 ft.
Max Gross Wt.	1100 lbs.	**Vmax**	108 mph
Empty Wt.	585 lbs.	**VCr**	100 mph
Payload (full fuel)	N/A	**V_{S1} (stall clean)**	48 mph
Fuel Capacity	15 gal.	**V_{S0} (lndg config)**	N/A
Seats	2	**Climb Rate @msl**	580 fpm
Cabin Width	39 in.	**Service Ceiling**	N/A

Std. Engine/HP	Usable HP Range	Construction Mtrl.	Est. Bldg. Time
VW 1835cc / 82 hp	60 - 82	Wood, Tube, Fabric	1800 hrs.
Plans	**Kit**	**Info Packet**	**Video**
$150.00	N/A	$10.00	N/A

VW powered utility aircraft. STOL performance on low horsepower. Side-by-side seating and 70 pounds of baggage storage. Design award at Oshkosh '73. The Cygnet is plans built from raw materials only. See Wicks Aircraft catalog. **Flying:** 100

VIKING AIRCRAFT, INC.
P.O. Box 646
Elkhorn WI 53121-0646
Tel. 414-723-1048 • Fax 414-723-1049
E-Mail: viking@pensys.com

Gypsy

Wing Span	32.00 ft.	Range	75 sm
Wing Area	144.0 sq. ft.	Takeoff Dist.	100 ft.
Length	17.00 ft.	Landing Dist.	250 ft.
Max Gross Wt.	415 lbs.	Vmax	55 mph
Empty Wt.	165 lbs.	VCr	35 mph
Payload (full fuel)	250 lbs.	V_{S1} (stall clean)	22 mph
Fuel Capacity	3 gal.	V_{S0} (Indg config)	N/A
Seats	1	Climb Rate @msl	500 fpm
Cabin Width	N/A	Service Ceiling	N/A

Std. Engine/HP	Usable HP Range	Construction Mtrl.	Est. Bldg. Time
25 hp	20 - 25	Tube, Fabric	250 hrs.
Plans	**Kit**	**Info Packet**	**Video**
$40.00	N/A	N/A	N/A

Designed by John Chotia in 1980 as an aluminum tube easier-to-build and stronger refinement of the "Woodhopper" ultralight. Aluminum tube spars with styrofoam ribs and doped fabric wings.

VINTAGE ULTRALIGHT ASSOC.
74 Brookwood Dr.
Marietta GA 30064
Tel. 770-973-3860

SR-1 Hornet

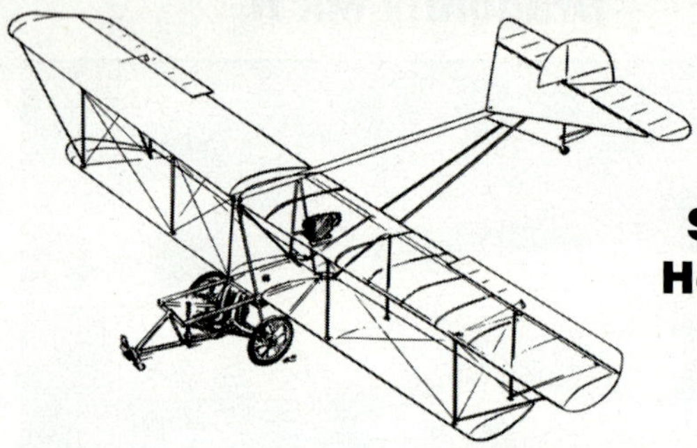

Wing Span	33.00 ft.	Range	100 sm
Wing Area	220.0 sq. ft.	Takeoff Dist.	75 ft.
Length	N/A	Landing Dist.	75 ft.
Max Gross Wt.	600.0 lbs.	Vmax	63 mph
Empty Wt.	245 lbs.	VCr	50 mph
Payload (full fuel)	320 lbs.	V_{S1} (stall clean)	16 mph
Fuel Capacity	5 gal.	V_{S0} (Indg config)	N/A
Seats	1	Climb Rate @msl	800 fpm
Cabin Width	N/A	Service Ceiling	N/A

Std. Engine/HP	Usable HP Range	Construction Mtrl.	Est. Bldg. Time
Cuyuna 430	30 - 35	Tube, Fabric	250 hrs.
Plans	**Kit**	**Info Packet**	**Video**
$75.00	N/A	N/A	N/A

Unique aluminum tube ultralight biplane widely sold in the 1980's. Largest wing area and lightest wing loading of any ultralight. Limited building assistance available from original designers.

VINTAGE ULTRALIGHT ASSOC.
74 Brookwood Dr.
Marietta GA 30064
Tel. 770-973-3860

Mr. Easy

Wing Span	24.00 ft.	Range	120 sm
Wing Area	145.0 sq. ft.	Takeoff Dist.	175 ft.
Length	N/A	Landing Dist.	175 ft.
Max Gross Wt.	485 lbs.	Vmax	63 mph
Empty Wt.	250 lbs.	VCr	50 mph
Payload (full fuel)	205 lbs.	V_{S1} (stall clean)	28 mph
Fuel Capacity	5 gal.	V_{S0} (Indg config)	N/A
Seats	1	Climb Rate @msl	800 fpm
Cabin Width	N/A	Service Ceiling	N/A

Std. Engine/HP	Usable HP Range	Construction Mtrl.	Est. Bldg. Time
Rotax 447 / 42 hp	N/A	Tube, Fabric	250 hrs.
Plans	**Kit**	**Info Packet**	**Video**
$25.00	N/A	N/A	N/A

Pusher biplane. Last of a series of six planes designed by Lyle Mathews and associates.

VINTAGE ULTRALIGHT ASSOC.
74 Brookwood Dr.
Marietta GA 30064
Tel. 770-973-3860

Woodhopper

Wing Span	32.00 ft.	Range	30 sm
Wing Area	157.0 sq. ft.	Takeoff Dist.	80 ft.
Length	17.50 ft.	Landing Dist.	60 ft.
Max Gross Wt.	345 lbs.	Vmax	45 mph
Empty Wt.	145 lbs.	VCr	35 mph
Payload (full fuel)	200 lbs.	V_{S1} (stall clean)	22 mph
Fuel Capacity	1 gal.	V_{S0} (Indg config)	N/A
Seats	1	Climb Rate @msl	250 fpm
Cabin Width	N/A	Service Ceiling	N/A

Std. Engine/HP	Usable HP Range	Construction Mtrl.	Est. Bldg. Time
25 hp	15 - 25	Wood, Fabric	250 hrs.
Plans	**Kit**	**Info Packet**	**Video**
$40.00	N/A	N/A	N/A

Original project commissioned by Popular Mechanics Magazine in 1978 in support of a "Build It At Home" economical wood airplane kit.

VINTAGE ULTRALIGHT ASSOC.
74 Brookwood Dr.
Marietta GA 30064
Tel. 770-973-3860

Whing Ding 71

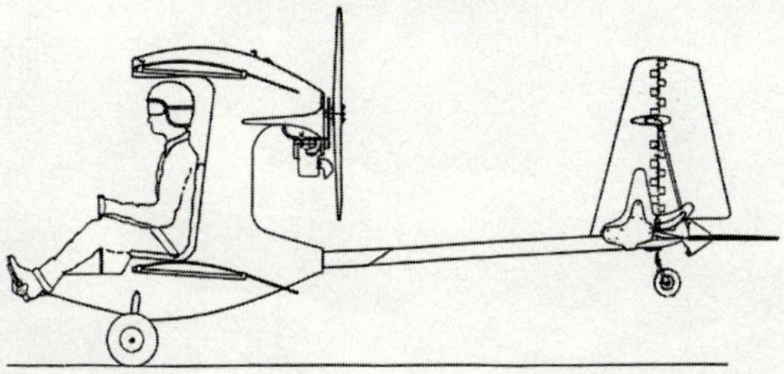

Wing Span	17.00 ft.	**Range**	20 sm
Wing Area	98.00 sq. ft.	**Takeoff Dist.**	250 ft.
Length	13.00 ft.	**Landing Dist.**	250 ft.
Max Gross Wt.	310 lbs.	**Vmax**	50 mph
Empty Wt.	122 lbs.	**VCr**	40 mph
Payload (full fuel)	180 lbs.	**V_{S1} (stall clean)**	26 mph
Fuel Capacity	1 gal.	**V_{S0} (lndg config)**	N/A
Seats	1	**Climb Rate @msl**	200 fpm
Cabin Width	N/A	**Service Ceiling**	3,000 ft.

Std. Engine/HP	Usable HP Range	Construction Mtrl.	Est. Bldg. Time
McCulloch / 12 hp	N/A	Wood, Fabric	250 hrs.
Plans	**Kit**	**Info Packet**	**Video**
$40.00	N/A	N/A	N/A

Wood biplane designed in 1971 using a go-kart engine with belt reduction drive.

VINTAGE ULTRALIGHT ASSOC.
74 Brookwood Dr.
Marietta GA 30064
Tel. 770-973-3860

San Francesco

Wing Span	31.80 ft.	**Range**	435 sm
Wing Area	N/A	**Takeoff Dist.**	656 ft.
Length	21.60 ft.	**Landing Dist.**	820 ft.
Max Gross Wt.	990 lbs.	**Vmax**	105 mph
Empty Wt.	616 lbs.	**VCr**	87 mph
Payload (full fuel)	N/A	**V_{S1} (stall clean)**	N/A
Fuel Capacity	N/A	**V_{S0} (lndg config)**	37 mph
Seats	2	**Climb Rate @msl**	590 fpm
Cabin Width	N/A	**Service Ceiling**	N/A

Std. Engine/HP	Usable HP Range	Construction Mtrl.	Est. Bldg. Time
Limbach 2000 / 80 hp	N/A	Tube, Fabric	N/A
Plans	**Kit**	**Info Packet**	**Video**
N/A	Yes	N/A	N/A

A practical design with simple, traditional construction methods.

VOLIAMO S.R.L.
Via Bologna
80143 - Napoli
ITALY
Tel. (39) 81 285133 • Fax (39) 81 5544864

VJ-22 Sportsman

Wing Span	36.50 ft.	**Range**	300 sm
Wing Area	180.0 sq. ft.	**Takeoff Dist.**	400 ft.
Length	24.00 ft.	**Landing Dist.**	500 ft.
Max Gross Wt.	1500 lbs.	**Vmax**	105 mph
Empty Wt.	1000 lbs.	**VCr**	85 mph
Payload (full fuel)	500 lbs.	**V_{S1} (stall clean)**	40 mph
Fuel Capacity	20 gal.	**V_{S0} (lndg config)**	N/A
Seats	2	**Climb Rate @msl**	500 fpm
Cabin Width	40 in.	**Service Ceiling**	13,000 ft.

Std. Engine/HP	Usable HP Range	Construction Mtrl.	Est. Bldg. Time
Cont C-85 / 100 hp	N/A	Wood, Fiberglass	3000 hrs.
Plans	**Kit**	**Info Packet**	**Video**
$250.00	N/A	$10.00	N/A

Veteran two place amphibian. Side-by-side seating in a closed cabin. Manually retractable main gear and tail wheel. Water takeoff and landing distance is about 1,000 feet. 100+ now flying.

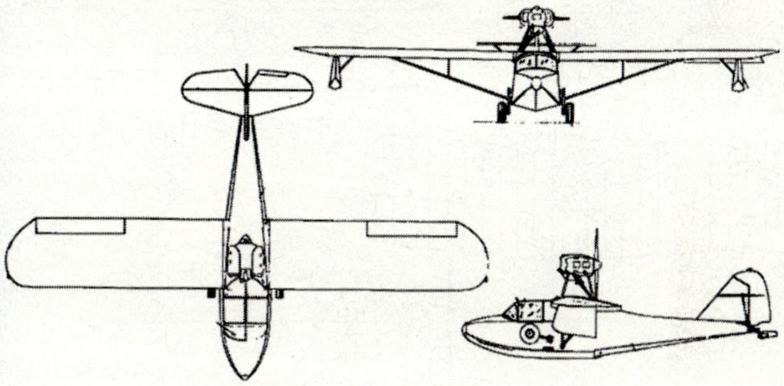

In 1925, at age 16, Volmer Jensen built his first glider inspired by a drawing in Boy Mechanic Magazine. A second glider was built the following year and another in 1927. In 1928, he built the first glider of his own design, the VJ-4, a 40 foot cantilever-winged utility glider. This was followed by the VJ-5 which proved to be the forerunner of one of his most successful gliders, the VJ-8.

In 1939 Volmer built the first two place, side-by-side sailplane in the US. Later, this ship was purchased by the Army for its Air Corps glider training program. As chief engineer of California Aero Glider, Volmer designed VJ-15 through VJ-20 for the Army but they were never constructed.

In 1949 he opened his own business, Production Model Shop, which produced the VJ-22. 800 plans have been sold and more than 100 are reported to be flying.

VOLMER AIRCRAFT
Box 5222
Glendale CA 91201
Tel. 818-247-8718

D-3 Sunwheel

Wing Span	22.96 ft.	Range	217 sm
Wing Area	N/A	Takeoff Dist.	N/A
Length	18.70 ft.	Landing Dist.	N/A
Max Gross Wt.	880 lbs.	Vmax	90 mph
Empty Wt.	484 lbs.	VCr	81 mph
Payload (full fuel)	N/A	V_{S1} (stall clean)	N/A
Fuel Capacity	N/A	V_{S0} (lndg config)	34 mph
Seats	2	Climb Rate @msl	N/A
Cabin Width	N/A	Service Ceiling	N/A

Std. Engine/HP	Usable HP Range	Construction Mtrl.	Est. Bldg. Time
Rotax 912 / 80 hp	65-80	Composite, Tube, Fabric	N/A

Plans	Kit	Info Packet	Video
N/A	Yes	N/A	N/A

Ultralight (European) traditional two place biplane utilizing modern construction methods.

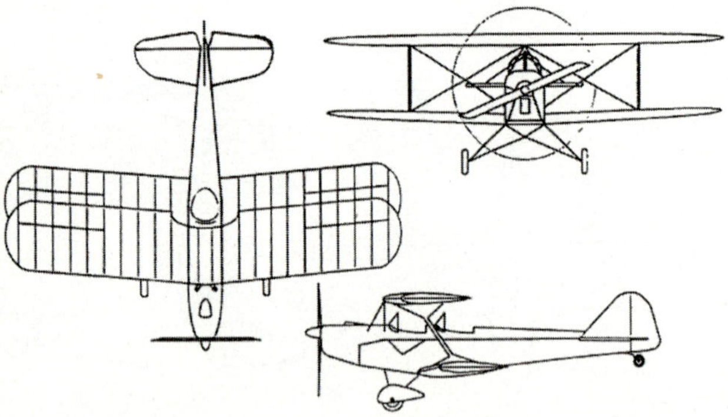

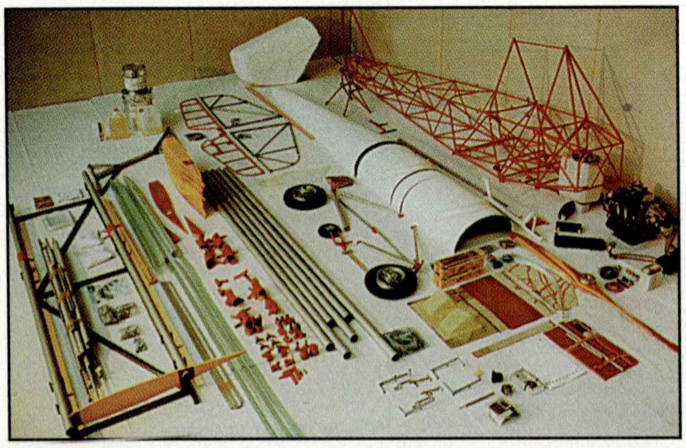

W. D. FLUGZEUGLEICHTBAU GMBH • FORSTRASSE 15
73529 Sch. Gmund Strassdorf • GERMANY
Tel. 071 71/48 04 • Fax 071 71/4 41 98

Dallach D-4 Fascination

Wing Span	29.50 ft.	Range	690 sm
Wing Area	113.9 sq. ft.	Takeoff Dist.	230 ft.
Length	20.24 ft.	Landing Dist.	230 ft.
Max Gross Wt.	1200 lbs.	Vmax	190 mph
Empty Wt.	550 lbs.	VCr	160 mph
Payload (full fuel)	425 lbs.	V_{S1} (stall clean)	N/A
Fuel Capacity	18 gal.	V_{S0} (lndg config)	40 mph
Seats	2	Climb Rate @msl	1600 fpm
Cabin Width	43 in.	Service Ceiling	13,000 ft.

Std. Engine/HP	Usable HP Range	Construction Mtrl.	Est. Bldg. Time
Rotax 912 / 80 hp	80-115	Aluminum, Fabric, Composite	N/A

Plans	Kit	Info Packet	Video
N/A	$24,500.00	Yes	N/A

German kit aircraft designed by Wolfgang Dallach. Fully retractable gear gives a high cruise speed while its fowler flaps allow a stall speed of 40 mph. Minimum sink rate 300 fpm. Dismantles in about five minutes for transport and stowage. The fuselage and all metal parts are pre-welded. Price is for fixed gear, firewall aft airframe kit.

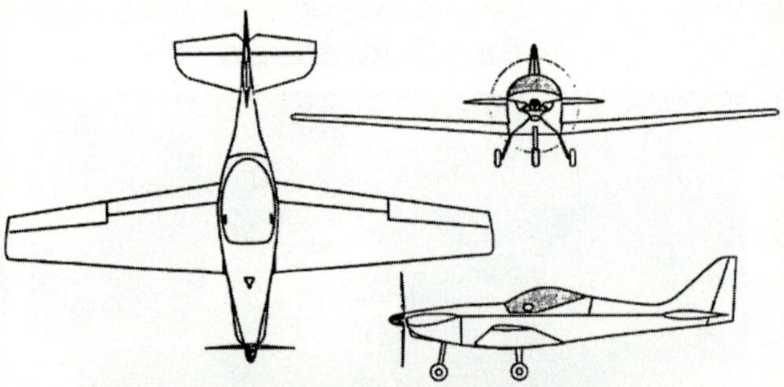

D-4 Fascination Fast-Build Kit

U.S. DISTRIBUTOR:
SIGGI'S AIRPLANE WORKS, INC.
372 Briarwood Road • Venice FL 34293
Fax: 941-493-0230 E-Mail:siggiplane@acun.com

W. D. FLUGZEUGLEICHTBAU GMBH • FORSTRASSE 15
73529 Sch. Gmund Strassdorf • GERMANY
Tel. 071 71/48 04 • Fax 071 71/4 41 98

ATO Taperwing

Wing Span	30.42 ft.	**Range**	690 sm
Wing Area	N/A	**Takeoff Dist.**	475 ft.
Length	22.42 ft.	**Landing Dist.**	355 ft.
Max Gross Wt.	2600 lbs.	**Vmax**	200 mph
Empty Wt.	1677 lbs.	**VCr**	138 mph
Payload (full fuel)	1015 lbs.	**V_{S1} (stall clean)**	57 mph
Fuel Capacity	60 gal.	**V_{S0} (lndg config)**	57 mph
Seats	3	**Climb Rate @msl**	1100 fpm
Cabin Width	36 in.	**Service Ceiling**	17,000 ft.

Std. Engine/HP	Usable HP Range	Construction Mtrl.	Est. Bldg. Time
Wright / 235 hp	N/A	Steel Tube, Wood, Fabric	1800 hrs.
Plans	**Kit**	**Info Packet**	**Video**
$350.00	$38,000.00	N/A	N/A

A classic, high performance biplane from the 1930's. All plans are redrawn from the original WACO plans. They include a book of drawings for the small fittings and an instruction manual. Kit price includes prefabricated components but does not include engine. Various parts and materials kits are available.

Kits/Plans Sold: 28

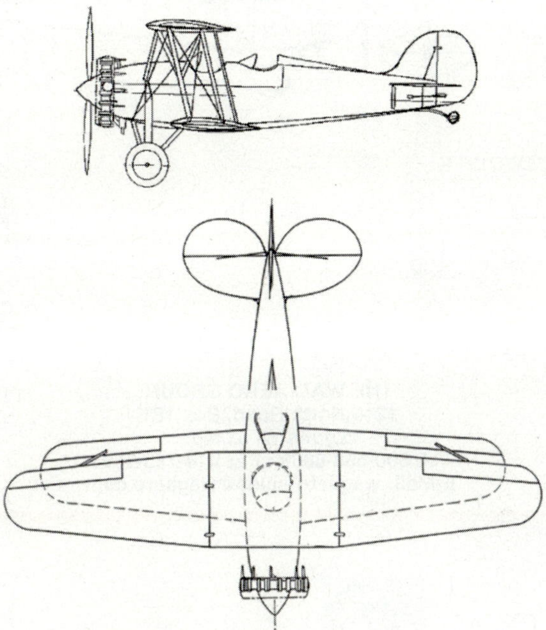

THE WACO AIRCRAFT CO., INC.
Quillayute Prairie Airport • 5144 Quillayute Rd.
Forks WA 98331
Tel. 360-374-6025 • Fax 360-374-6025
E-Mail: wacoii@olypen.com
Web: http://cyberxpress.com/waco

Sport Trainer

Wing Span	35.25 ft.	**Range**	220 sm
Wing Area	N/A	**Takeoff Dist.**	400 ft.
Length	22.38 ft.	**Landing Dist.**	425 ft.
Max Gross Wt.	1400 lbs.	**Vmax**	120 mph
Empty Wt.	720 lbs.	**VCr**	94 mph
Payload (full fuel)	680 lbs.	**V_{S1} (stall clean)**	39 mph
Fuel Capacity	12 gal.	**V_{S0} (lndg config)**	N/A
Seats	2	**Climb Rate @msl**	490 fpm
Cabin Width	N/A	**Service Ceiling**	11,200 ft.

Std. Engine/HP	Usable HP Range	Construction Mtrl.	Est. Bldg. Time
Cont C-85 / 100 hp	65 - 100	N/A	1230 hrs.
Plans	**Kit**	**Info Packet**	**Video**
$65.00	Yes	Yes	N/A

Replica of J-3 Cub. Proven performance and serviceability. Design has extensive baggage area. 26 gallon auxiliary fuel tank available; extends cruising range to 455 miles. See Ad on Page13.

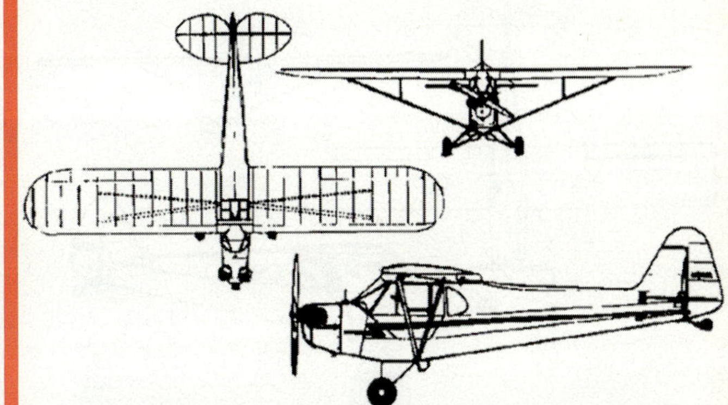

THE WAG-AERO GROUP
1216 North Road, Box 181
Lyons WI 53148
Tel. 800-558-6868 • Fax 414-763-7595
E-Mail: wagaero-sales@wagaero.com

Sportsman 2+2

Wing Span	35.80 ft.	**Range**	670 sm
Wing Area	174.1 sq. ft.	**Takeoff Dist.**	1470 ft.
Length	23.42 ft.	**Landing Dist.**	1261 ft.
Max Gross Wt.	2200 lbs.	**Vmax**	130 mph
Empty Wt.	1080 lbs.	**VCr**	125 mph
Payload (full fuel)	1120 lbs.	**V_{S1} (stall clean)**	50 mph
Fuel Capacity	39 gal.	**V_{S0} (lndg config)**	N/A
Seats	4	**Climb Rate @msl**	500 fpm
Cabin Width	39 in.	**Service Ceiling**	14,800 ft.

Std. Engine/HP	Usable HP Range	Construction Mtrl.	Est. Bldg. Time
Lyc. O-290 / 135 hp	N/A	Metal, Fabric	2000 hrs.
Plans	**Kit**	**Info Packet**	**Video**
$89.00	Yes	Yes	N/A

Replica of Piper PA-14 Cruiser with improvements. Room for four plus extra baggage and fuel. See Ad on Page 13.

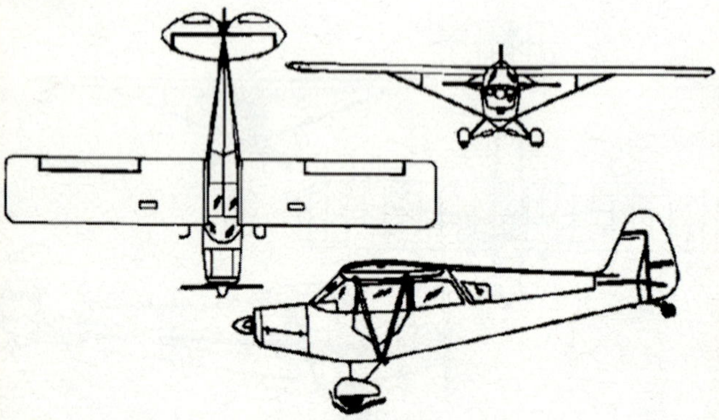

THE WAG-AERO GROUP
1216 North Road, Box 181
Lyons WI 53148
Tel. 800-558-6868 • Fax 414-763-7595
E-Mail: wagaero-sales@wagaero.com

Wag-A-Bond Traveler

Wing Span	29.30 ft.	**Range**	600 sm
Wing Area	148.0 sq. ft.	**Takeoff Dist.**	1470 ft.
Length	18.70 ft.	**Landing Dist.**	1261 ft.
Max Gross Wt.	1450 lbs.	**Vmax**	135 mph
Empty Wt.	725 lbs.	**VCr**	105 mph
Payload (full fuel)	725 lbs.	**V_{S1} (stall clean)**	55 mph
Fuel Capacity	12 gal.	**V_{S0} (lndg config)**	N/A
Seats	2	**Climb Rate @msl**	625 fpm
Cabin Width	40 in.	**Service Ceiling**	14,000 ft.

Std. Engine/HP	Usable HP Range	Construction Mtrl.	Est. Bldg. Time
Lyc. O-235 / 100 hp	N/A	Metal, Fabric	2000 hrs.
Plans	**Kit**	**Info Packet**	**Video**
$65.00	N/A	Yes	N/A

Replica of famous Piper Vagabond. Unique recreational aircraft, fast and maneuverable. Interesting to build and fun to fly. See Ad on Page 13.

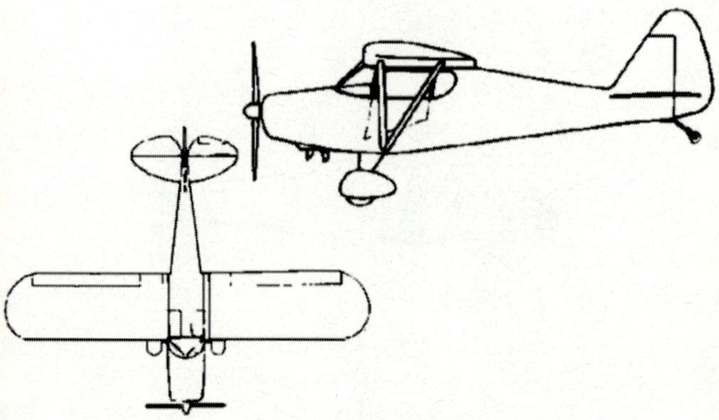

THE WAG-AERO GROUP
1216 North Road, Box 181
Lyons WI 53148
Tel. 800-558-6868 • Fax 414-763-7595
E-Mail: wagaero-sales@wagaero.com

F4U Corsair

Wing Span	20.00 ft.	**Range**	400 sm
Wing Area	N/A	**Takeoff Dist.**	1000 ft.
Length	16.00 ft.	**Landing Dist.**	1500 ft.
Max Gross Wt.	1200 lbs.	**Vmax**	170 mph
Empty Wt.	921 lbs.	**VCr**	140 mph
Payload (full fuel)	280 lbs.	**V_{S1} (stall clean)**	N/A
Fuel Capacity	14 gal.	**V_{S0} (Indg config)**	N/A
Seats	1	**Climb Rate @msl**	1400 fpm
Cabin Width	N/A	**Service Ceiling**	12,500 ft.

Std. Engine/HP	Usable HP Range	Construction Mtrl.	Est. Bldg. Time
Cont. 0-200 / 100 hp	90 - 125	Wood, Composite	1500 hrs.
Plans	**Kit**	**Info Packet**	**Video**
$245.00	N/A	$8.00	N/A

Half-size replica of the famous WWII U.S. Navy and Marine fighter. One of the series of WWII replicas designed and tested in California between 1974 and 1988. Wood aircraft structure requires only standard wood working tools. Some minor metal work also required. Plans for the P-51 Mustang at $295 are also available but not detailed here.

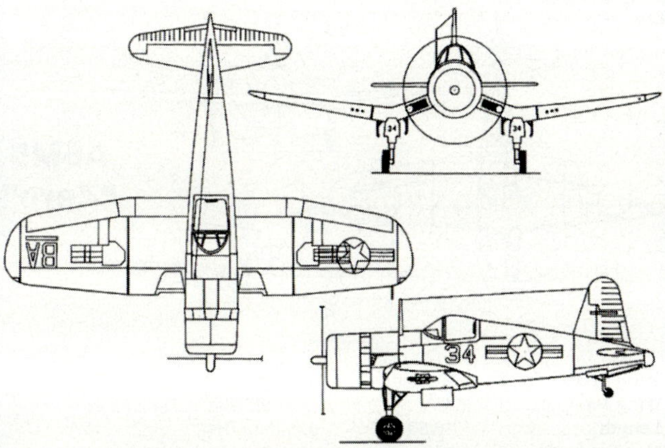

WAR AIRCRAFT REPLICAS OF FLORIDA
PO Box 79007
Tampa FL 33619
Tel. 813-620-0631

FW 190

Wing Span	20.00 ft.	**Range**	400 sm
Wing Area	100.0 sq. ft.	**Takeoff Dist.**	1000 ft.
Length	16.59 ft.	**Landing Dist.**	1800 ft.
Max Gross Wt.	900 lbs.	**Vmax**	195 mph
Empty Wt.	630 lbs.	**VCr**	145 mph
Payload (full fuel)	270 lbs.	**V_{S1} (stall clean)**	55 mph
Fuel Capacity	12 gal.	**V_{S0} (Indg config)**	N/A
Seats	1	**Climb Rate @msl**	1000 fpm
Cabin Width	N/A	**Service Ceiling**	12,500 ft.

Std. Engine/HP	Usable HP Range	Construction Mtrl.	Est. Bldg. Time
Cont. 0-200 / 100 hp	90 - 125	Wood, Composite	N/A
Plans	**Kit**	**Info Packet**	**Video**
$245.00	N/A	$8.00	N/A

Half-size replica of the famous WWII German Luftwaffe fighter. One of the series of WWII replicas designed and tested in California between 1974 and 1988. Wood aircraft structure requires only standard wood working tools. Some minor metal work also required. Plans for the P-51 Mustang at $295 are also available but not detailed here.

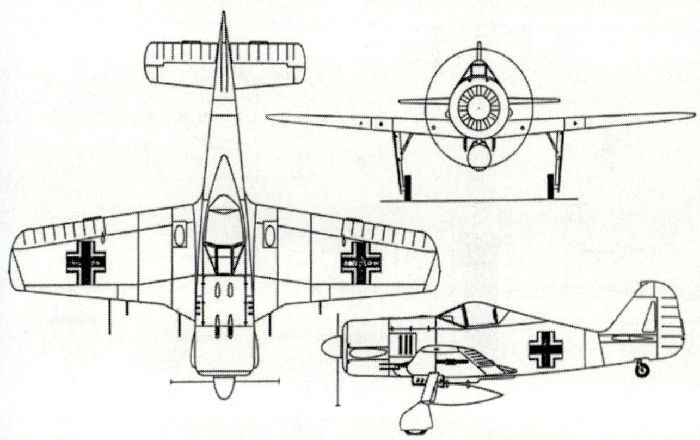

WAR AIRCRAFT REPLICAS OF FLORIDA
PO Box 79007
Tampa FL 33619
Tel. 813-620-0631

Captain Crumble often wondered if the chickens might seek revenge.

P-47 Thunderbolt

Wing Span	20.00 ft.	**Range**	400 sm
Wing Area	120.0 sq. ft.	**Takeoff Dist.**	1000 ft.
Length	17.00 ft.	**Landing Dist.**	1600 ft.
Max Gross Wt.	920 lbs.	**Vmax**	210 mph
Empty Wt.	620 lbs.	**VCr**	135 mph
Payload (full fuel)	300 lbs.	**V_{S1} (stall clean)**	55 mph
Fuel Capacity	14 gal.	**V_{S0} (lndg config)**	N/A
Seats	1	**Climb Rate @msl**	700 fpm
Cabin Width	N/A	**Service Ceiling**	12,500 ft.

Std. Engine/HP	Usable HP Range	Construction Mtrl.	Est. Bldg. Time
Cont. 0-200 / 100 hp	90 - 125	Wood, Composite	1500 hrs.
Plans	**Kit**	**Info Packet**	**Video**
$245.00	N/A	$8.00	N/A

Half-size replica of the famous WWII U.S. Air Force fighter. One of the series of WWII replicas designed and tested in California between 1974 and 1988. Wood aircraft structure requires only standard wood working tools. Some minor metal work also required. Plans for the P-51 Mustang at $295 are also available but not detailed here.

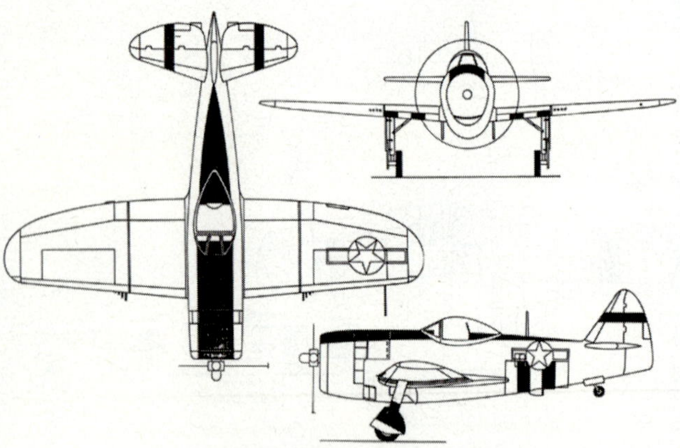

WAR AIRCRAFT REPLICAS OF FLORIDA
PO Box 79007
Tampa FL 33619
Tel. 813-620-0631

Sea Fury

Wing Span	20.00 ft.	**Range**	400 sm
Wing Area	N/A	**Takeoff Dist.**	1000 ft.
Length	16.75 ft.	**Landing Dist.**	1800 ft.
Max Gross Wt.	900 lbs.	**Vmax**	195 mph
Empty Wt.	630 lbs.	**VCr**	145 mph
Payload (full fuel)	270 lbs.	**V_{S1} (stall clean)**	55 mph
Fuel Capacity	14 gal.	**V_{S0} (lndg config)**	N/A
Seats	1	**Climb Rate @msl**	1000 fpm
Cabin Width	N/A	**Service Ceiling**	12,500 ft.

Std. Engine/HP	Usable HP Range	Construction Mtrl.	Est. Bldg. Time
Cont. 0-200 / 100 hp	90 - 125	Wood, Composite	1500 hrs.
Plans	**Kit**	**Info Packet**	**Video**
$245.00	N/A	$8.00	N/A

Half-size replica of the famous WWII British Navy fighter. Wood structure requires only standard wood working tools. Some minor metal work also required.

WAR AIRCRAFT REPLICAS OF FLORIDA
PO Box 79007
Tampa FL 33619
Tel. 813-620-0631

A6M5 "Zero"

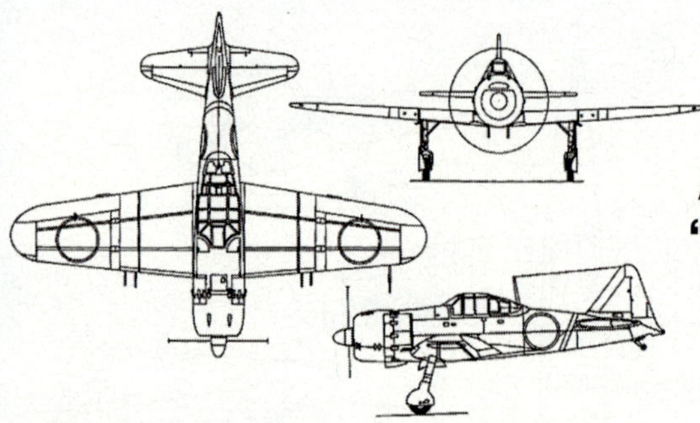

Wing Span	21.00 ft.	**Range**	N/A
Wing Area	N/A	**Takeoff Dist.**	N/A
Length	16.50 ft.	**Landing Dist.**	N/A
Max Gross Wt.	N/A	**Vmax**	N/A
Empty Wt.	N/A	**VCr**	N/A
Payload (full fuel)	N/A	**V_{S1} (stall clean)**	N/A
Fuel Capacity	N/A	**V_{S0} (lndg config)**	N/A
Seats	N/A	**Climb Rate @msl**	N/A
Cabin Width	N/A	**Service Ceiling**	N/A

Std. Engine/HP	Usable HP Range	Construction Mtrl.	Est. Bldg. Time
Cont. 0-200 / 100 hp	N/A	Wood, Metal	N/A
Plans	**Kit**	**Info Packet**	**Video**
$245.00	N/A	$8.00	N/A

Half-size replica of the famous WWII Japanese fighter. Wood structure requires only standard wood working tools. Some minor metal work also required.

WAR AIRCRAFT REPLICAS OF FLORIDA
PO Box 79007
Tampa FL 33619
Ph: 813-620-0631

Revolution II

Wing Span	28.00 ft.	**Range**	300 sm
Wing Area	126.0 sq. ft.	**Takeoff Dist.**	400 ft.
Length	19.75 ft.	**Landing Dist.**	450 ft.
Max Gross Wt.	1400 lbs.	**Vmax**	160 mph
Empty Wt.	800 lbs.	**VCr**	115 mph
Payload (full fuel)	N/A	**V_{S1} (stall clean)**	42 mph
Fuel Capacity	17 gal.	**V_{S0} (lndg config)**	N/A
Seats	2	**Climb Rate @msl**	1200 fpm
Cabin Width	N/A	**Service Ceiling**	14,000 ft.

Std. Engine/HP	Usable HP Range	Construction Mtrl.	Est. Bldg. Time
Cont./Lyc. / 125 hp	85 - 160	Tube, Wood, Fabric	N/A hrs.
Plans	**Kit**	**Info Packet**	**Video**
$495.00	$18,995.00	$9.95	$19.95

Two seat version of the Revolution, classic design reminiscent of a 1930's era low wing monoplane. Wood, tube and fabric construction. Variety of materials packages available. Kit price does not include engine but does include pre-welded Lycoming/Continental motor mount. Sold in Australia as the Space Walker II. See Amax Engineering.

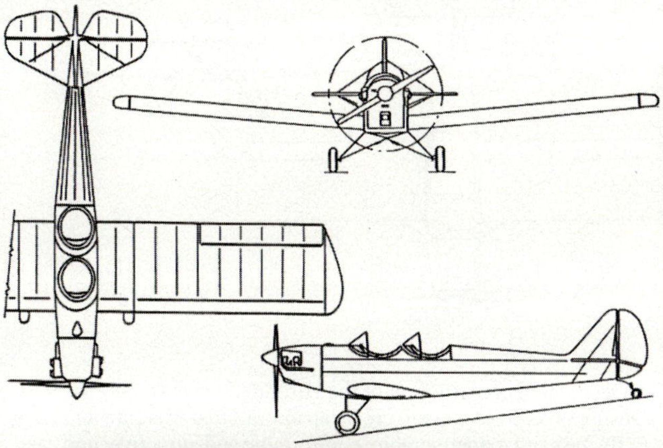

WARNER AEROCRAFT COMPANY
9415 Laura Ct. • Seminole FL 33776-1625
Ph: 813-595-2382
E-Mail: revolution@usa.net
Web: http://www.ij.net/revolution

Revolution I

Wing Span	26.00 ft.	**Range**	300 sm
Wing Area	N/A	**Takeoff Dist.**	N/A
Length	18.00 ft.	**Landing Dist.**	N/A
Max Gross Wt.	840 lbs.	**Vmax**	125 mph
Empty Wt.	540 lbs.	**VCr**	100 mph
Payload (full fuel)	N/A	**V_{S1} (stall clean)**	45 mph
Fuel Capacity	N/A	**V_{S0} (lndg config)**	N/A
Seats	1	**Climb Rate @msl**	900 fpm
Cabin Width	N/A	**Service Ceiling**	14,000 ft.

Std. Engine/HP	Usable HP Range	Construction Mtrl.	Est. Bldg. Time
Continental / 65 hp	N/A	Tube, Wood, Fabic	N/A hrs.
Plans	**Kit**	**Info Packet**	**Video**
$495.00	Yes	$9.95	$19.95

Classic low wing monoplane using wood, tube and fabric construction. Sold in Australia as the Space Walker I. See Amax Engineering.

WARNER AEROCRAFT COMPANY
9415 Laura Ct. • Seminole FL 33776-1625
Ph: 813-595-2382
E-Mail: revolution@usa.net
Web: http://www.ij.net/revolution

T-411 Wolverine

Wing Span	41.33 ft.	**Range**	510 sm
Wing Area	258.9 sq. ft.	**Takeoff Dist.**	345 ft.
Length	30.71 ft.	**Landing Dist.**	400 ft.
Max Gross Wt.	4190 lbs.	**Vmax**	150 mph
Empty Wt.	2425 lbs.	**VCr**	130 mph
Payload (full fuel)	1765 lbs.	**V_{S1} (stall clean)**	40 mph
Fuel Capacity	88 gal.	**V_{S0} (lndg config)**	N/A
Seats	5	**Climb Rate @msl**	990 fpm
Cabin Width	45 in.	**Service Ceiling**	10,000 ft.

Std. Engine/HP	Usable HP Range	Construction Mtrl.	Est. Bldg. Time
VMKB M-14P / 360 hp	300-400	Aluminum, Steel, Fabric	1000 hrs.
Plans	**Kit**	**Info Packet**	**Video**
$1,000.00	$50,000.00	Yes	$25.00

STOL aircraft for use on rough airfields with minimal or no maintenance facilities. Based on the proven Russian Aist-2 design. Tube and fabric strut-braced construction. Includes provision for skis and floats. Can be recowled to accept Continental or Lycoming 350 engines. **Completed & Flying:** 10

WASHINGTON AEROPROGRESS
8535 Perimeter Rd. S. • Seattle WA 98108
Tel. 206-763-4915 • Fax 206-768-0661

Weed-hopper Deluxe

Wing Span	28.00 ft.	**Range**	N/A
Wing Area	168.0 sq. ft.	**Takeoff Dist.**	N/A
Length	18.25 ft.	**Landing Dist.**	N/A
Max Gross Wt.	502 lbs.	**Vmax**	65 mph
Empty Wt.	252 lbs.	**VCr**	55 mph
Payload (full fuel)	N/A	**V_{S1} (stall clean)**	22 mph
Fuel Capacity	N/A	**V_{S0} (lndg config)**	N/A
Seats	1	**Climb Rate @msl**	1000 fpm
Cabin Width	N/A	**Service Ceiling**	10,000 ft.

Std. Engine/HP	Usable HP Range	Construction Mtrl.	Est. Bldg. Time
Rotax 447 / 40 hp	N/A	Tube, Fabric	12 hrs.
Plans	**Kit**	**Info Packet**	**Video**
N/A	$6,500.00	Yes	N/A

A tube and fabric trike also available in a two seat, side-by-side version utilizing a Rotax 503 engine. Price includes engine. Options include windshield, deluxe instruments, and parachute.

WEEDHOPPER, INC.
PO Box 1377
Clinton MS 39056
Ph: 601-924-0806

PGK-1 Hirondelle

Wing Span	26.00 ft.	**Range**	600 sm
Wing Area	117.0 sq. ft.	**Takeoff Dist.**	750 ft.
Length	20.58 ft.	**Landing Dist.**	900 ft.
Max Gross Wt.	1475 lbs.	**Vmax**	155 mph
Empty Wt.	933 lbs.	**VCr**	145 mph
Payload (full fuel)	N/A	**V_{S1} (stall clean)**	61 mph
Fuel Capacity	30 gal.	**V_{S0} (lndg config)**	N/A
Seats	2	**Climb Rate @msl**	950 fpm
Cabin Width	N/A	**Service Ceiling**	12,500 ft.

Std. Engine/HP	Usable HP Range	Construction Mtrl.	Est. Bldg. Time
Lyc O-235 / 100 hp	100-160	Wood, Foam/FG, Fabric	1400 hrs.
Plans	**Kit**	**Info Packet**	**Video**
$150.00	N/A	$5.00	N/A

Two place, cross country and recreational design. Prices are US currency. Set of prefab components, comprising more than half the kit, is available for $2,600 US.

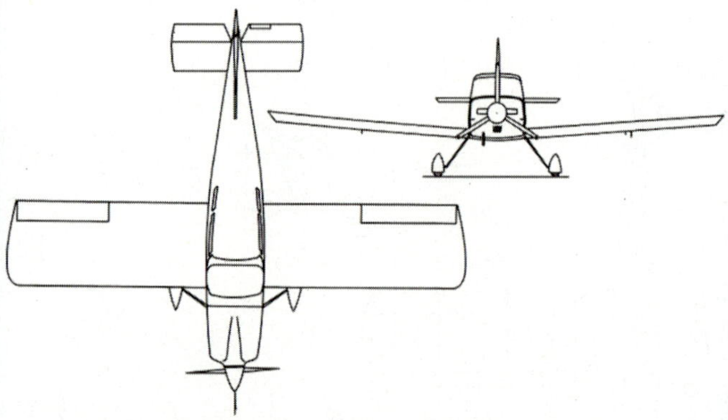

Design and construction of the Hirondelle was started in 1970. Design objectives were to create an aircraft that was safe and easy to fly, fast and maneuverable, comfortable and attractive, and easy and economical to build. The designers feel that all of these objectives were met in the prototype aircraft C-GWYL which was first flown in June, 1976.

Field testing was extensive and proved the aircraft to be very durable and reliable. For example, when a weak point showed up in the landing gear mount after approximately six hours of operation, the problem was quickly solved. Later, a bad landing caused the aircraft to bounce over ten feet into the air with no ill effect. All testing was completed by July, 1977.

The aircraft was designed by homebuilt veterans Jean Peters, manager of Western Aircraft Supplies, Glenn Gibb, and John Kopala.

WESTERN AIRCRAFT SUPPLIES
623 Markerville Road N.E.
Calgary Alberta, T2E 5X1
CANADA
Tel. 403-275-3513 • Fax 403-276-3087

Flitplane

Wing Span	27.00 ft.	**Range**	N/A
Wing Area	145.0 sq. ft.	**Takeoff Dist.**	300 ft.
Length	15.75 ft.	**Landing Dist.**	350 ft.
Max Gross Wt.	500 lbs.	**Vmax**	70 mph
Empty Wt.	130 lbs.	**VCr**	40 mph
Payload (full fuel)	270 lbs.	**V_{S1} (stall clean)**	27 mph
Fuel Capacity	5 gal.	**V_{S0} (lndg config)**	N/A
Seats	1	**Climb Rate @msl**	350 fpm
Cabin Width	30 in.	**Service Ceiling**	N/A

Std. Engine/HP	Usable HP Range	Construction Mtrl.	Est. Bldg. Time
2SI / 35 hp	N/A	Steel, Alum.Tube, Fabric	550 hrs.
Plans	**Kit**	**Info Packet**	**Video**
$100.00	$7,995.00	$5.00	N/A

Simple ultralight. Part 103 legal. Low cost, easy to build. Open, tube fuselage. Fabric covered wing and tail. Kit price includes engine.

Kits/Plans Sold: 35 **Completed & Flying:** 1

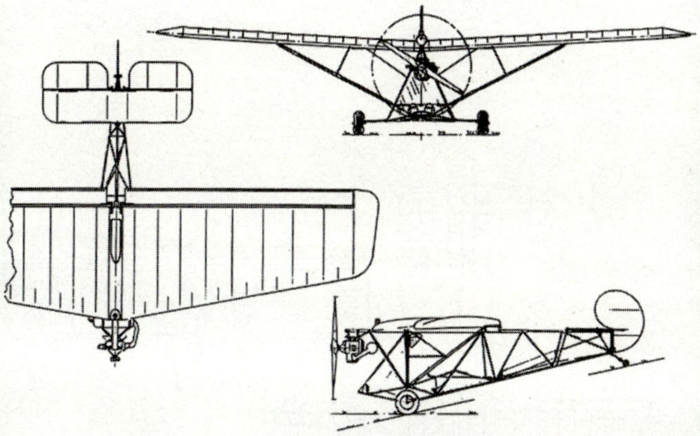

WINGS OF FREEDOM AVIATION, INC.
PO Box 835
Middlefield OH 44062
Tel. 216-834-3254 • Fax 216-693-4075

Anderson Kingfisher

Wing Span	36.08 ft.	**Range**	280 sm
Wing Area	185.0 sq. ft.	**Takeoff Dist.**	600 ft.
Length	23.58 ft.	**Landing Dist.**	600 ft.
Max Gross Wt.	1600 lbs.	**Vmax**	95 mph
Empty Wt.	1060 lbs.	**VCr**	85 mph
Payload (full fuel)	468 lbs.	**V_{S1} (stall clean)**	N/A
Fuel Capacity	20 gal.	**V_{S0} (lndg config)**	N/A
Seats	2	**Climb Rate @msl**	500 fpm
Cabin Width	44 in.	**Service Ceiling**	N/A

Std. Engine/HP	Usable HP Range	Construction Mtrl.	Est. Bldg. Time
Cont O-200 / 100 hp	100 - 125	Wood, Fiberglass	2000 hrs.
Plans	**Kit**	**Info Packet**	**Video**
$190.00	N/A	$10.00	N/A

A plans only, high wing amphibian with retractable gear. Conventional tractor mounted engine. Water takeoff distance is approximately 1,800 feet.

Kits/Plans Sold: 250 **Completed & Flying:** 100

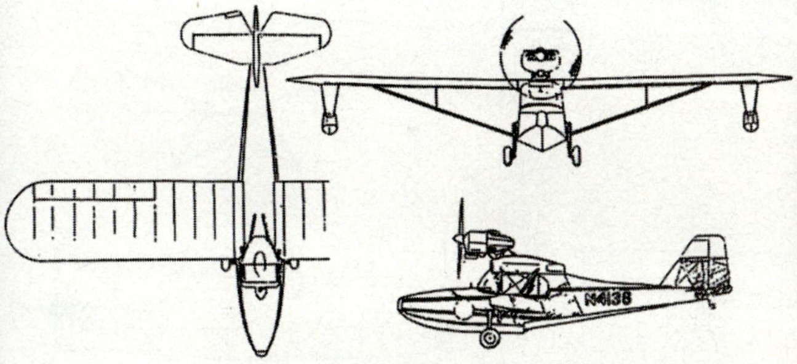

WINGS UNLIMITED
6230 Rock Island Rd.
Charlotte NC 28278
Tel. 704-588-9249 • Fax 704-588-9249

W-II Boredom Fighter

Wing Span	20.00 ft.	**Range**	450 sm
Wing Area	N/A	**Takeoff Dist.**	150 ft.
Length	15.70 ft.	**Landing Dist.**	350 ft.
Max Gross Wt.	770 lbs.	**Vmax**	120 mph
Empty Wt.	473 lbs.	**VCr**	110 mph
Payload (full fuel)	N/A	**V_{S1} (stall clean)**	42 mph
Fuel Capacity	15 gal.	**V_{S0} (lndg config)**	N/A
Seats	1	**Climb Rate @msl**	1400 fpm
Cabin Width	N/A	**Service Ceiling**	15,000 ft.

Std. Engine/HP	Usable HP Range	Construction Mtrl.	Est. Bldg. Time
Continental / 65 hp	N/A	Wood, Fabric	1500 hrs.
Plans	**Kit**	**Info Packet**	**Video**
$125.00	N/A	$3.00	N/A

A plans only biplane designed to appeal to the admirer of WWI vintage aircraft. It has the appearance of many of that era's aircraft but duplicates none of them in particular. Can accomodate pilots up to 6'1" and 200 lbs.

Completed & Flying: 40

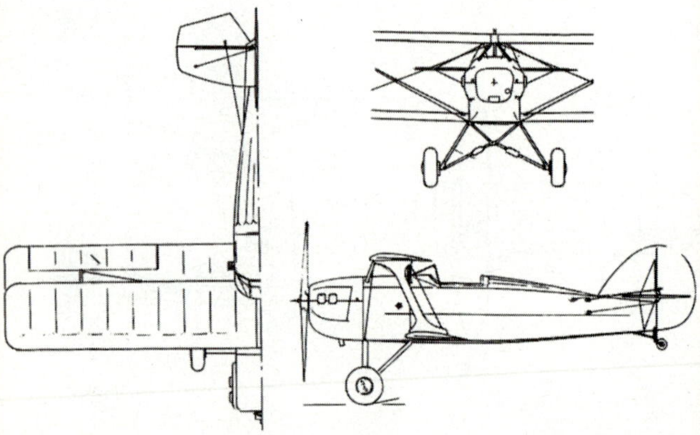

The W-11 is lively and responsive but normal in behavior and presents no problem to anyone who could land a Taylorcraft in a light crosswind solo. Although it flies like a fighter, it will go willingly where you point it. It is stable in all axes and presents a light, well proportioned, positive feel in all controls. Put the stick against the rear stop and a definite buffet is felt - whereupon the nose drops gently, wings unstall, and almost immediately the stall is reentered. This will continue until back pressure is eased off. It is quite easy to land and should be three pointed. Roll out is steady even with the tailwheel unlocked. Construction time varies. Actual time for the prototype, with no rush, was 1,500 hours.

DON S. WOLF CO.
53 Cedar Drive
Huntington NY 11743
Tel. 516-427-1992 • Fax 516-427-1992

Fokker D.VII

Wing Span	29.30 ft.	**Range**	N/A
Wing Area	N/A	**Takeoff Dist.**	N/A
Length	22.90 ft.	**Landing Dist.**	N/A
Max Gross Wt.	1870 lbs.	**Vmax**	116 mph
Empty Wt.	1540 lbs.	**VCr**	N/A
Payload (full fuel)	N/A	**V_{S1} (stall clean)**	N/A
Fuel Capacity	N/A	**V_{S0} (lndg config)**	N/A
Seats	1	**Climb Rate @msl**	865 fpm
Cabin Width	N/A	**Service Ceiling**	N/A

Std. Engine/HP	Usable HP Range	Construction Mtrl.	Est. Bldg. Time
Mercedes / 185 hp	160 - 185	Metal, Wood, Fabric	N/A
Plans	**Kit**	**Info Packet**	**Video**
$150.00	N/A	Yes	N/A

A classic WW1 biplane utilizing a steel tube fuselage, wood wings and no wire rigging. Drawings for full-scale reproduction developed by Herbert Kelley. Many other original or redrawn plans for 1900-1940 aircraft designs are offered.

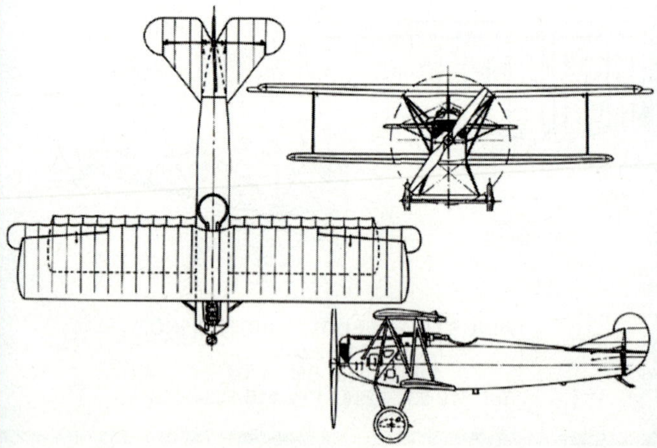

WORLD WAR I AEROPLANES, INC.
15 Crescent Road
Poughkeepsie NY 12601
Tel. 914-473-3679

Gemini CH 620

Wing Span	27.30 ft.	**Range**	650 sm
Wing Area	123.0 sq. ft.	**Takeoff Dist.**	450 ft.
Length	19.00 ft.	**Landing Dist.**	450 ft.
Max Gross Wt.	1450 lbs.	**Vmax**	155 mph
Empty Wt.	800 lbs.	**VCr**	145 mph
Payload (full fuel)	650 lbs.	**V_{S1} (stall clean)**	N/A
Fuel Capacity	34 gal.	**V_{S0} (lndg config)**	55 mph
Seats	2	**Climb Rate @msl**	1280 fpm
Cabin Width	44 in.	**Service Ceiling**	12,000 ft.

Std. Engine/HP	Usable HP Range	Construction Mtrl.	Est. Bldg. Time
Jabiru 2200 (x 2)/ 80 hp	N/A	All-Metal	N/A
Plans	**Kit**	**Info Packet**	**Video**
N/A	$19,950.00	Yes	N/A

Twin engine performance and reliability in a kit aircraft with low costs and easy construction. Based on the successful Zodiac CH 601 kit plane line. Proven easy-build, all-metal construction, full dual controls, outstanding 360 degree bubble canopy visibility, excellent short and rough field capability, and high useful loads. Airframe kit price is for conventional gear version. Airframe kit does not include engines and powerplant accessories, props, instruments, lights, paint, and upholstery. See Ad Inside Front Cover.

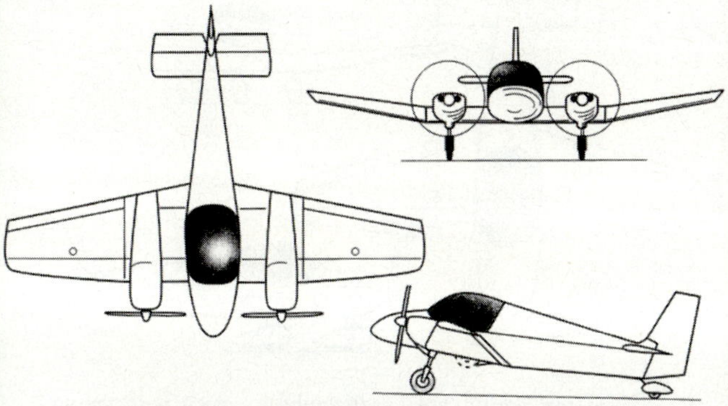

ZENITH AIRCRAFT CO.
Box 650 • Mexico Memorial Airport • Mexico MO 65265-0650
Tel. 573-581-9000 • Fax 573-581-0011
E-Mail: info@zenithair.com
Web: http://www.zenithair.com

STOL CH 701

Wing Span	27.00 ft.	**Range**	210 sm
Wing Area	122.0 sq. ft.	**Takeoff Dist.**	60 ft.
Length	20.00 ft.	**Landing Dist.**	80 ft.
Max Gross Wt.	960 lbs.	**Vmax**	85 mph
Empty Wt.	460 lbs.	**VCr**	75 mph
Payload (full fuel)	500 lbs.	**V_{S1} (stall clean)**	32 mph
Fuel Capacity	11 gal.	**V_{S0} (lndg config)**	27 mph
Seats	2	**Climb Rate @msl**	1400 fpm
Cabin Width	42 in.	**Service Ceiling**	15,000 ft.

Std. Engine/HP	Usable HP Range	Construction Mtrl.	Est. Bldg. Time
Rotax 582 / 65 hp	N/A	Aluminum, Metal	400 hrs.
Plans	**Kit**	**Info Packet**	**Video**
$320.00	$12,360.00	$15.00	$20.00

A STOL sportplane equipped with fixed leading-edge slats, steerable nosewheel, full-length flaperons, and all-flying rudder. Latest models feature a new "All-Terrain Gear" with tundra tires and hydraulic disk brakes for exceptional rough field capability. More than 400 now flying in 36 countries. Kit price does not include engine or instruments. Rotax 912 "firewall forward" package available. See Ad Inside Front Cover.

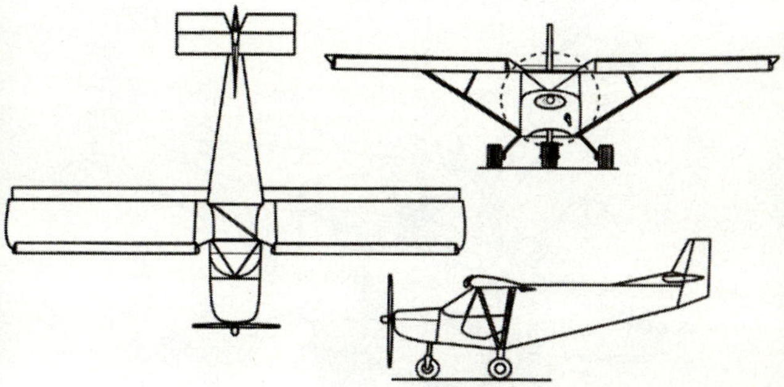

The STOL CH 701 was introduced in 1986 by designer Chris Heintz as an affordable off-airport short takeoff and landing aircraft. The design has been type-certificated in Israel and Mexico and is approved as an "advanced ultralight" in many countries where it is also successfully used as a trainer aircraft.

Plans-Builder Robert Guillet, Irrigon, OR
"Construction is very straight forward, with a very minimum of special tools, forms and jigs."

David & Karen Smith, Elgin, SC
"The building instructions are explicit and the company is knowledgeble and very professional. The kit is easier to build than many others, and the quality is excellent."

ZENITH AIRCRAFT CO.
Box 650 • Mexico Memorial Airport • Mexico MO 65265-0650
Tel. 573-581-9000 • Fax 573-581-0011
E-Mail: info@zenithair.com
Web: http://www.zenithair.com

STOL CH 701 Amphibian

Wing Span	27.00 ft.	**Range**	200 sm
Wing Area	122.0 sq. ft.	**Takeoff Dist.**	155 ft.
Length	20.00 ft.	**Landing Dist.**	225 ft.
Max Gross Wt.	1075 lbs.	**Vmax**	90 mph
Empty Wt.	575 lbs.	**VCr**	80 mph
Payload (full fuel)	500 lbs.	**V_{S1} (stall clean)**	32 mph
Fuel Capacity	11 gal.	**V_{S0} (lndg config)**	N/A
Seats	2	**Climb Rate @msl**	800 fpm
Cabin Width	42 in.	**Service Ceiling**	12,000 ft.

Std. Engine/HP	Usable HP Range	Construction Mtrl.	Est. Bldg. Time
Rotax 912 / 80 hp	64 - 100	Aluminum, Metal	520 hrs.
Plans	**Kit**	**Info Packet**	**Video**
$320.00	$11,670.00	$15.00	$20.00

An amphibious sportplane with real STOL performance from the water or ground. Features Zenair's exclusive all-metal amphibious floats. Available factory-finished as an "Advanced Ultralight" in Canada. Other engine options include Rotax 582, 618, C-65 and O-200. Floats, but not engine, included in kit price. See Ad Inside Front Cover.

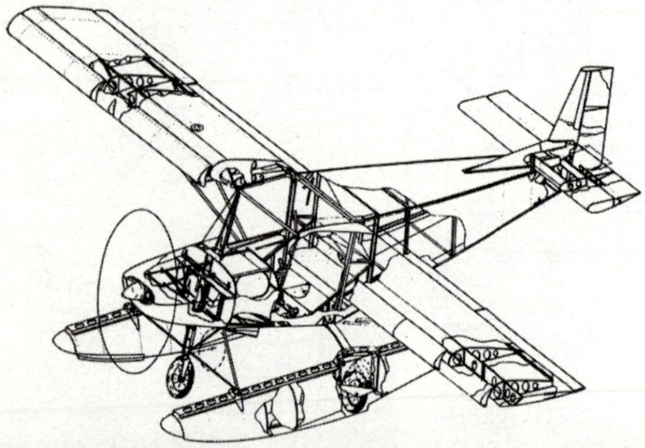

ZENITH AIRCRAFT CO.

Box 650 • Mexico Memorial Airport • Mexico MO 65265-0650
Tel. 573-581-9000 • Fax 573-581-0011
E-Mail: info@zenithair.com
Web: http://www.zenithair.com

Super Zodiac CH 601 HDS

Wing Span	23.00 ft.	**Range**	500 sm
Wing Area	98.00 sq. ft.	**Takeoff Dist.**	550 ft.
Length	19.00 ft.	**Landing Dist.**	450 ft.
Max Gross Wt.	1200 lbs.	**Vmax**	150 mph
Empty Wt.	580 lbs.	**VCr**	135 mph
Payload (full fuel)	516 lbs.	**V_{S1} (stall clean)**	62 mph
Fuel Capacity	16 gal.	**V_{S0} (lndg config)**	65 mph
Seats	2	**Climb Rate @msl**	1030 fpm
Cabin Width	44 in.	**Service Ceiling**	16,000 ft.

Std. Engine/HP	Usable HP Range	Construction Mtrl.	Est. Bldg. Time
Rotax 912 / 80 hp	N/A	Aluminum, Metal	400 hrs.
Plans	**Kit**	**Info Packet**	**Video**
$340.00	$13,640.00	$15.00	$20.00

A speedy, all-metal design featuring tapered 'speed' wings that are easily removable. Other standard features include electric trim tab, hydraulic disk brakes, and a large baggage compartment. 31 gallon fuel tank option available. "Sportster" canopy option available for open cockpit flying. Complete build kits available that require no special tools. Kit price does not include engine. See Ad Inside Front Cover. **Completed & Flying:** 100

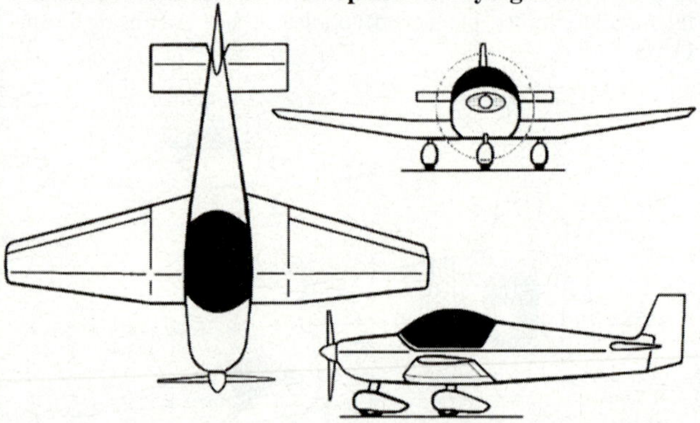

The "Sportster" option replaces the bubble canopy with a windshield, leaving the cabin top and sides open. "The open cockpit may not provide the full comfort of an enclosed cabin on cross country flights, but it certainly adds fun and excitement for recreational flights on hot summer days," said Nicholas Heintz, production manager and son of designer Chris Heintz.

John Desimone, Akron OH
"[Zenith's quality is] overall very impressive. Most all details are handled well. All parts seem very well [designed and contructed.]"

Ralph Morris, Florissant MO
"The quality is excellent. Cuts are extremely accurate as are the bends. Quality of materials and components are excellent."

ZENITH AIRCRAFT CO.
Box 650 • Mexico Memorial Airport • Mexico MO 65265-0650
Tel. 573-581-9000 • Fax 573-581-0011
E-Mail: info@zenithair.com
Web: http://www.zenithair.com

Zodiac CH 601 (UL)

Wing Span	27.00 ft.	**Range**	480 sm
Wing Area	130.0 sq. ft.	**Takeoff Dist.**	400 ft.
Length	19.00 ft.	**Landing Dist.**	380 ft.
Max Gross Wt.	1047 lbs.	**Vmax**	135 mph
Empty Wt.	537 lbs.	**VCr**	120 mph
Payload (full fuel)	510 lbs.	**V_{S1} (stall clean)**	44 mph
Fuel Capacity	16 gal.	**V_{S0} (lndg config)**	44 mph
Seats	2	**Climb Rate @msl**	1200 fpm
Cabin Width	44 in.	**Service Ceiling**	16,000 ft.

Std. Engine/HP	Usable HP Range	Construction Mtrl.	Est. Bldg. Time
Rotax 912 / 80 hp	N/A	Aluminum, Metal	400 hrs.
Plans	**Kit**	**Info Packet**	**Video**
$290.00	$12,380.00	$15.00	$20.00

The "Advanced Ultralight" version of the standard CH 601 HD. Available in tricycle or tailwheel configurations. It sports a bubble canopy that offers exceptional visibility. Features wing baggage lockers and quick-remove wings. Kit price does not include engine. See Ad Inside Front Cover. **Completed & Flying:** 90

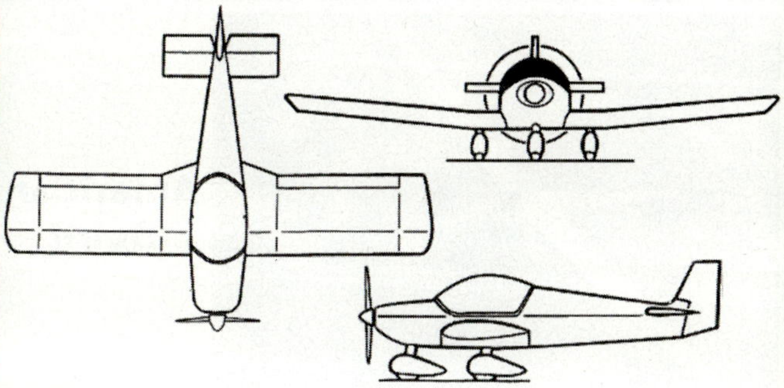

The Zodiac was introduced in 1984 by designer Chris Heintz as a primary low cost, all-purpose light aircraft, and has evolved into three models. Based on the standard CH 601 HD, the lightweight Zodiac CH 601 (UL) model is approved in Canada and European countries as an "Advanced Ultralight," where it is used successfully as a primary trainer aircraft. The Zodiac features docile characteristics, making it suitable for low time pilots, yet has the capabilities for economical cross country performance.

Loaded with standard features (electric trim tab, hydraulic disk brakes, new larger baggage compartment, and more...), this new Zodiac provides all-metal durability, first-class comfort for two large persons in a wide cabin, large baggage compartments, and much more.

ZENITH AIRCRAFT CO.
Box 650 • Mexico Memorial Airport • Mexico MO 65265-0650
Tel. 573-581-9000 • Fax 573-581-0011
E-Mail: info@zenithair.com
Web: http://www.zenithair.com

Zodiac CH 601 HD

Wing Span	27.00 ft.	**Range**	480 sm
Wing Area	130.0 sq. ft.	**Takeoff Dist.**	450 ft.
Length	19.00 ft.	**Landing Dist.**	450 ft.
Max Gross Wt.	1200 lbs.	**Vmax**	135 mph
Empty Wt.	640 lbs.	**VCr**	120 mph
Payload (full fuel)	464 lbs.	**V_{S1} (stall clean)**	49 mph
Fuel Capacity	16 gal.	**V_{S0} (lndg config)**	44 mph
Seats	2	**Climb Rate @msl**	1200 fpm
Cabin Width	44 in.	**Service Ceiling**	15,000 ft.

Std. Engine/HP	Usable HP Range	Construction Mtrl.	Est. Bldg. Time
Rotax 912 / 80 hp	N/A	Metal	400 hrs.
Plans	**Kit**	**Info Packet**	**Video**
$315.00	$12,620.00	$15.00	$20.00

Available in tricycle or tailwheel configurations. Features wing baggage lockers and quick-remove wings. Complete kits available that reduce building time. Kit price does not include engine. See Ad Inside Front Cover.

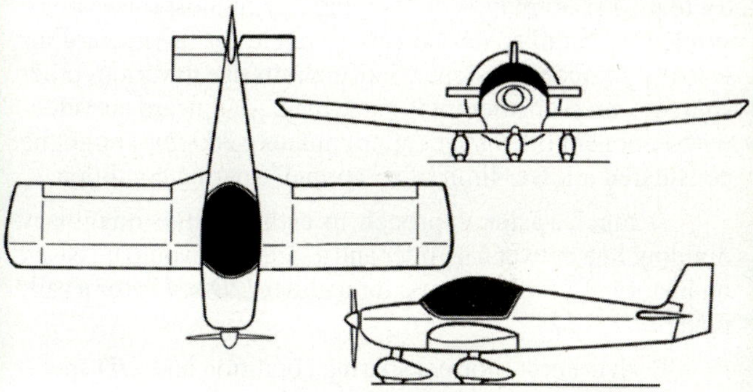

ZENITH AIRCRAFT CO.
Box 650 • Mexico Memorial Airport • Mexico MO 65265-0650
Tel. 573-581-9000 • Fax 573-581-0011
E-Mail: info@zenithair.com
Web: http://www.zenithair.com

SOARING GLOSSARY

Lately, the terms sailplane and glider are often used interchangeably (and we have placed *all* soaring aircraft into the AeroCrafter SAILPLANE Section). Also, the term "motor glider" has come into more common usage than "powered glider/sailplane."

The following definitions by S.O. Jenko, Dipl. Ing. ETH are from the 1979 NASA Symposium on Science and Technology of Low Speed and Motorless Flight and represent an attempt to standardize terminology among the engineering types.

Ed.

GLIDER (with or without auxiliary power): any manned flying device which is not capable of cross country soaring flight without any power, under "normal" soaring conditions.

SAILPLANE (with or without auxiliary power): any manned flying device which is fully capable of cross country flight without any power, under "normal" soaring conditions.

The above differentiation is based on technological progress from the inception of powerless flying. First it was gliding a few feet above the slope of a hill. Then, with substantial design improvements of gliders and discoveries of various atmospheric phenomena, foundations were laid for soaring, that is, flying without any use of power for a substantial length of time, gaining altitude, and flying a long distance in one or more flights.

The ability to reach these basic objectives of soaring depends on the skill of the pilot and on atmospheric conditions as well as on the performance of the sailplane. Thus the stipulation of "normal" soaring conditions may present a problem. What is a normal soaring day in one area of a country (e.g., Texas) may be a booming day in most other areas. While one could specify a certain range for the upward air velocity component (slope wind, updrafts due to various other sources), no such attempt is made here. An upward air velocity component of *1 m/sec* (approximately *200 fpm*) might be considered a lower limit of a "normal" soaring condition.

A much easier approach to establish the imaginary dividing line between a glider and a sailplane would be based on historical developments: for a glider $L/D < 17$; for a sailplane $L/D > 17$.

Both criteria (normal soaring condition and L/D specification) appear to be reasonably equivalent.

ULTRALIGHT GLIDER (includes Hang Glider) or **SAILPLANE**: A manned flying device having a wing loading $w < 10$ *kg/m2* (approximately *2 lb/ft*^{2}).

LIGHT GLIDER or **SAILPLANE**: A manned flying device having a wing loading $w = 10$ *to 25 kg/m*^{2} (approximately *2 to 5.1 lb/ft*^{2})

GLIDER or **SAILPLANE**: A manned flying device having a wing load $w = 25$ *to 40 kg/m*^{2} (approximately *5.1 to 8 lb/ft*^{2}).

AUXILIARY POWERED GLIDER or **SAILPLANE**: A manned flying device having an auxiliary engine to takeoff and to overfly with power any severe downdraft areas which would otherwise result in a landing; power loading $p > 9$ *kg/hp* (approximately *20 lb/hp*).

POWERED GLIDER or SAILPLANE: A glider or sailplane converted into a powered aircraft; the engine is essential for flight operation.

Sierra LS

Wing Span	42.65 ft.	**Range**	N/A
Wing Area	142.0 sq. ft.	**Takeoff Dist.**	N/A
Length	21.00 ft.	**Landing Dist.**	N/A
Max Gross Wt.	455 lbs.	**Vmax**	40 mph
Empty Wt.	205 lbs.	**VCr**	N/A
Payload (full fuel)	250 lbs.	V_{S1} **(stall clean)**	27 mph
Fuel Capacity	N/A	V_{S0} **(lndg config)**	27 mph
Seats	1	**Climb Rate @msl**	N/A
Cabin Width	26 in.	**Service Ceiling**	N/A

Std. Engine/HP	Usable HP Range	Construction Mtrl.	Est. Bldg. Time
N/A	N/A	Tube, Fabric, Composite	150 hrs.
Plans	**Kit**	**Info Packet**	**Video**
N/A	$9,995.00	$5.00	N/A

Lightweight, affordable sailplane with an L/D of 25:1. Good handling qualities and performance. Easy launching by ultralight air tow, auto tow, winch, or bungee.

ADVANCED AEROMARINE
152 East Eighth Street • Apopka Industrial Park
Apopka FL 32703
Tel. 407-877-7871 • Fax 407-877-7871

American Spirit

Wing Span	49.25 ft.	**Range**	N/A
Wing Area	106.3 sq. ft.	**Takeoff Dist.**	N/A
Length	21.60 ft.	**Landing Dist.**	N/A
Max Gross Wt.	1160 lbs.	**Vmax**	N/A
Empty Wt.	505 lbs.	**VCr**	N/A
Payload (full fuel)	N/A	V_{S1} **(stall clean)**	44 mph
Fuel Capacity	N/A	V_{S0} **(lndg config)**	N/A
Seats	1	**Climb Rate @msl**	N/A
Cabin Width	N/A	**Service Ceiling**	N/A

Std. Engine/HP	Usable HP Range	Construction Mtrl.	Est. Bldg. Time
Sailplane	N/A	Composite	N/A
Plans	**Kit**	**Info Packet**	**Video**
N/A	$17,980.00	N/A	N/A

Construction has a carbon-fiber spar with pre-molded fiberglass parts. Aramid-Fiber reinforced with a factory welded box frame. Aspect ratio is 24:1. L/D is 44:1. Kit includes basic instrumentation. Cockpit accepts 6'4" pilot. **Completed:** 18

ADVANCED SOARING CONCEPTS
4730 Calle Quetzal
Camarillo CA 93012
Tel. 805-389-3434 • Fax 805-482-3366

Swift

Wing Span	39.00 ft.	**Range**	N/A
Wing Area	135.0 sq. ft.	**Takeoff Dist.**	N/A
Length	N/A	**Landing Dist.**	N/A
Max Gross Wt.	310 lbs.	**Vmax**	75 mph
Empty Wt.	130 lbs.	**VCr**	N/A
Payload (full fuel)	N/A	**V_{S1} (stall clean)**	N/A
Fuel Capacity	N/A	**V_{S0} (lndg config)**	N/A
Seats	1	**Climb Rate @msl**	N/A
Cabin Width	N/A	**Service Ceiling**	N/A

Std. Engine/HP	Usable HP Range	Construction Mtrl.	Est. Bldg. Time
N/A	N/A	N/A	N/A
Plans	**Kit**	**Info Packet**	**Video**
N/A	Yes	N/A	$24.00

A foot or tow-launched sailplane/hang glider. SWIFT stands for "Swept Wing with Inboard Flap Trim." Has a maximum L/D of 25:1. Two new versions are now available; a 2-seat cage, adaptable on the standard wings, and a motorglider with 25 hp engine. Options include windshield and front fairing, full pilot fairing, tricycle gear for training, instrument panel, cartop transport box and towing hardware. A mandatory two day training program is included in purchase price.

Completed & Flying: 120

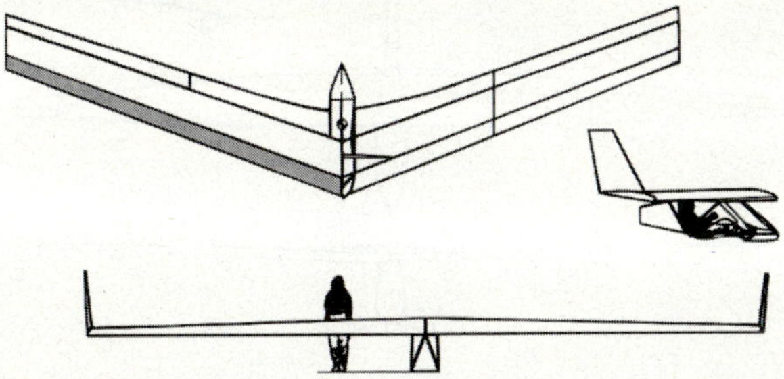

The SWIFT was developed at Stanford University along with Bright Star Gliders.

Assembling or disassembling the SWIFT can be done by one person in less than 30 minutes. The entire package fits into an optional, specially shaped cartop transport box. The SWIFT can be launched and landed either on foot or wheel. The wings are built from a foam core covered by an aerospace Kevlar skin and reinforced with carbon fiber. Repairs can be easily made as the main element of the wing (a carbon spar) is protected by the Kevlar skin and would rarely be damaged in a crash.

AERIANE
Rue de Poiriers, 7 • Gembloux B-5030
BELGIUM
Tel. 32 81 60 05 95 • Fax 32 81 60 05 96
E-Mail: Vincent.Piret@ping.be
Web: http://www.ping.be/~ping4026/aeriane.htm

Straton D-8

Wing Span	38.00 ft.	**Range**	240 sm
Wing Area	111.0 sq. ft.	**Takeoff Dist.**	180 ft.
Length	20.00 ft.	**Landing Dist.**	280 ft.
Max Gross Wt.	470 lbs.	**Vmax**	80 mph
Empty Wt.	250 lbs.	**VCr**	60 mph
Payload (full fuel)	220 lbs.	**V_{S1} (stall clean)**	18 mph
Fuel Capacity	5 gal.	**V_{S0} (lndg config)**	N/A
Seats	1	**Climb Rate @msl**	550 fpm
Cabin Width	24 in.	**Service Ceiling**	14,000 ft.

Std. Engine/HP	Usable HP Range	Construction Mtrl.	Est. Bldg. Time
Trabant/Rotax / 24 hp	20 - 30	Aluminum, Glass Fabric	400 hrs.
Plans	**Kit**	**Info Packet**	**Video**
N/A	$8,395.00	$5.00	N/A

Motorglider with a glide ratio of 18:1. American version of the Mini Straton D-7. Excellent for both soaring and cross country flight. All-aluminum frame. Choice of Trabant or Rotax engine. Total number completed and flying includes 23 in the US and 95 in Europe.

Completed & Flying: 118

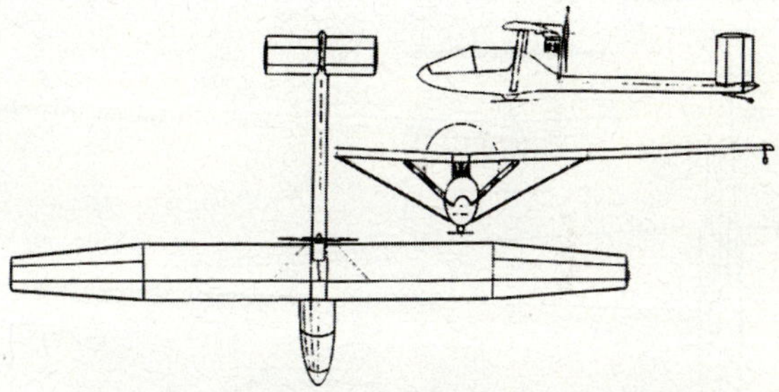

The D-8 is the successor to the Mini Straton D-7 which was designed in 1988 in the Czech Republic by top aeronautical engineers. The Mini Straton D-7 was first shown in the US at Oshkosh '92 and was very well received. The new D-8 offers numerous improvements in construction and performance.

AERO DOVRON, INC.
216 Hibiscus Street
Jupiter FL 33458
Tel. 407-747-6705 • Fax 407-747-6705

SZD 51-1 Junior

Wing Span	49.20 ft.	**Range**	N/A
Wing Area	134.0 sq. ft.	**Takeoff Dist.**	N/A
Length	21.90 ft.	**Landing Dist.**	N/A
Max Gross Wt.	836 lbs.	**Vmax**	136 mph
Empty Wt.	495 lbs.	**VCr**	N/A
Payload (full fuel)	N/A	**V_{S1} (stall clean)**	37 mph
Fuel Capacity	N/A	**V_{S0} (lndg config)**	N/A
Seats	1	**Climb Rate @msl**	N/A
Cabin Width	N/A	**Service Ceiling**	N/A

Std. Engine/HP	Usable HP Range	Construction Mtrl.	Est. Bldg. Time
N/A	N/A	Fiberglass, Composite	N/A
Plans	**Kit**	**Info Packet**	**Video**
N/A	$18,900.00	Yes	N/A

Designed as a club glider, the Junior features a large cockpit, fixed undercarriage with an oversize tire, and an unusually strong airframe. Has considerably more wing area than racing sailplanes and an airfoil optimized for safe handling and smooth stall characteristics.

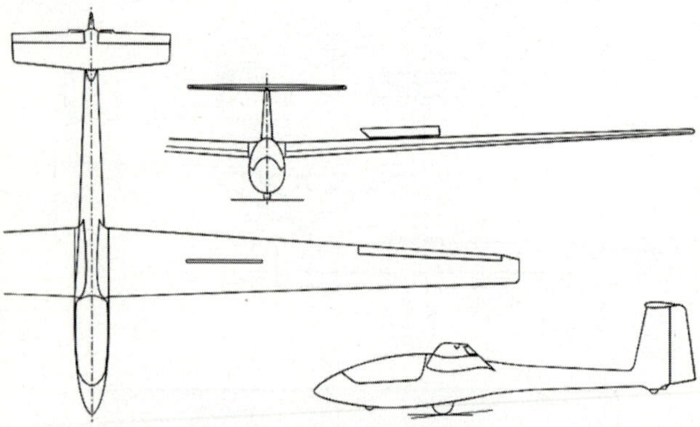

AEROUNION LTD.
901 Third Ave.
West Ashland WI 54806
Tel. 715-682-9339 • Fax 715-682-8364

Goeland

Wing Span	43.20 ft.	**Range**	N/A
Wing Area	125.0 sq. ft.	**Takeoff Dist.**	590 ft.
Length	N/A	**Landing Dist.**	265 ft.
Max Gross Wt.	660 lbs.	**Vmax**	95 mph
Empty Wt.	410 lbs.	**VCr**	65 mph
Payload (full fuel)	225 lbs.	**V_{S1} (stall clean)**	34 mph
Fuel Capacity	6 gal.	**V_{S0} (lndg config)**	N/A
Seats	1	**Climb Rate @msl**	680 fpm
Cabin Width	27 in.	**Service Ceiling**	N/A

Std. Engine/HP	Usable HP Range	Construction Mtrl.	Est. Bldg. Time
Konig SC 430 / 24 hp	N/A	Wood, Fabric, Fiberglass	300 hrs.
Plans	**Kit**	**Info Packet**	**Video**
$200.00	Yes	Yes	N/A

The Goeland includes a newly designd fuselage and wings from the German glider "Spatz." This configuration has an aspect ratio of 15:1 and an L/D ratio of 27:1. Other wings can be used to increase performance. Basic kit includes the two half-shells of the fuselage along with tail parts for $4,000 US. Cost of the finished fuselage kit without metal fittings or controls is $8,400 in US funds.

Kits/Plans Sold: 6 **Completed & Flying:** 2

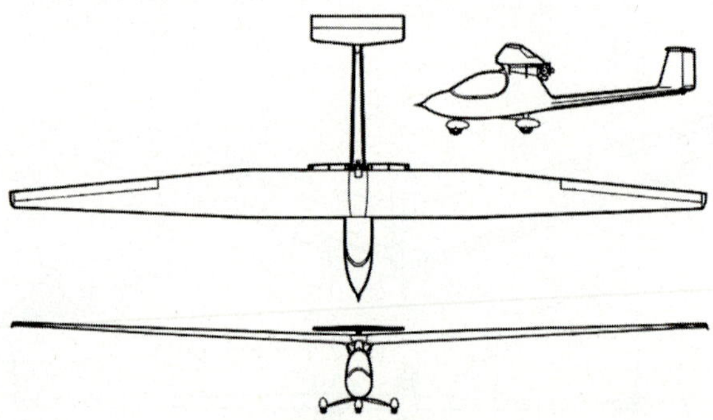

AIR EST SERVICES
37, rue St-Michel
Marly F-57155
FRANCE
Tel. 33 3 87 30 39 66 • Fax 33 3 87 63 91 48

HOMEBUILT "FAQ'S" *(See Page 68)*

Q: Why would someone want to build and fly a homebuilt when perfectly good certified aircraft are available?

A: In a sense, Cessnas & Pipers can be compared to older Chevys and Fords, in that they attempt to be "all things to all people." Such compromises may be acceptable for some, but not all pilots.

One of the remarkable things about homebuilts is the sheer diversity of designs and intended purposes. There are some compact aircraft that store in your garage and fly off of any short field.

Pelican

Wing Span	23.62 ft.	**Range**	220 sm
Wing Area	129.0 sq. ft.	**Takeoff Dist.**	590 ft.
Length	6.55 ft.	**Landing Dist.**	220 ft.
Max Gross Wt.	410 lbs.	**Vmax**	70 mph
Empty Wt.	185 lbs.	**VCr**	58 mph
Payload (full fuel)	205 lbs.	**V_{S1} (stall clean)**	28 mph
Fuel Capacity	3 gal.	**V_{S0} (lndg config)**	N/A
Seats	1	**Climb Rate @msl**	650 fpm
Cabin Width	25 in.	**Service Ceiling**	N/A

Std. Engine/HP	Usable HP Range	Construction Mtrl.	Est. Bldg. Time
Solo / 12 hp	12-24	Fiberglass, Fabric	300 hrs.
Plans	**Kit**	**Info Packet**	**Video**
N/A	$8,000.00	Yes	N/A

A unique twin-tail design. Can be fitted with a variety of engines. Price does not include engine, propeller or instruments. Available in finished form for $12,950.

Kits/Plans Sold: 1 **Completed & Flying:** 1

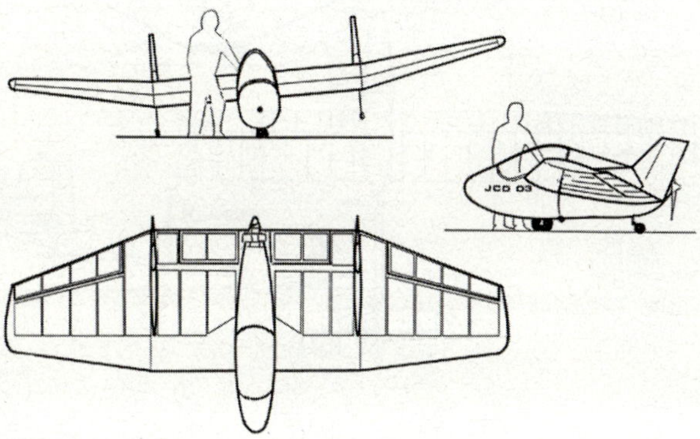

AIR EST SERVICES
37, rue St-Michel
Marly F-57155
FRANCE
Tel. 33 3 87 30 39 66 • Fax 33 3 87 63 91 48

Silent

Wing Span	39.36 ft.	**Range**	N/A
Wing Area	110.8 sq. ft.	**Takeoff Dist.**	N/A
Length	20.92 ft.	**Landing Dist.**	N/A
Max Gross Wt.	660 lbs.	**Vmax**	124 mph
Empty Wt.	341 lbs.	**VCr**	87 mph
Payload (full fuel)	242 lbs.	**V_{S1} (stall clean)**	32 mph
Fuel Capacity	N/A	**V_{S0} (lndg config)**	N/A
Seats	1	**Climb Rate @msl**	N/A
Cabin Width	N/A	**Service Ceiling**	N/A

Std. Engine/HP	Usable HP Range	Construction Mtrl.	Est. Bldg. Time
Sailplane	N/A	Composite	N/A
Plans	**Kit**	**Info Packet**	**Video**
N/A	Yes	N/A	N/A

Max L/D: 32 at 95 km/h. Minimum sink rate: 0.60 m/s at 65 km/h .The first ultralight (European) glider with a glide ratio of more than 30/1. Easy and enjoyable to fly. Can be landed in 150 ft. distance. Built entirely of composite materials including carbon fiber in both the wings and fuselage.

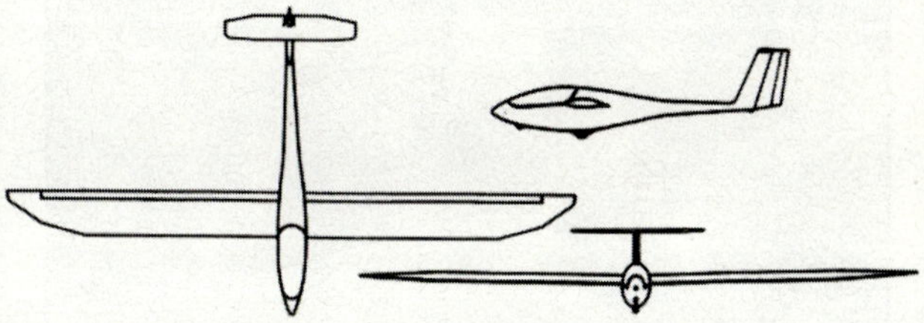

ALISPORT S.R.L.
Viale Verdi, 58
34074 Monfalcone (GO)
ITALY
Tel. 0481 778994

Another might carry four people 300 mph at 18,000 ft. while others could be flown in unlimited-class aerobatics.

Note that your garden variety Skyhawk can't do any of these things, but it's still a quite a useful & desirable airplane. It all depends on what you want your aircraft to do.

Silent op

Wing Span	39.40 ft.	**Range**	N/A
Wing Area	33.70 sq. ft.	**Takeoff Dist.**	N/A
Length	21.00 ft.	**Landing Dist.**	N/A
Max Gross Wt.	594 lbs.	**Vmax**	124 mph
Empty Wt.	341 lbs.	**VCr**	N/A
Payload (full fuel)	242 lbs.	**V_{S1} (stall clean)**	N/A
Fuel Capacity	5 gal.	**V_{S0} (lndg config)**	37 mph
Seats	1	**Climb Rate @msl**	N/A
Cabin Width	N/A	**Service Ceiling**	N/A

Std. Engine/HP	Usable HP Range	Construction Mtrl.	Est. Bldg. Time
Konig	N/A	Composite	N/A
Plans	**Kit**	**Info Packet**	**Video**
N/A	N/A	N/A	N/A

Max L/D: 26 at 84 km/h. Minimum sink rate: 0.80 m/s at 72 km/h. Motorized version of the Silent sailplane. With the propeller adjusted on the ground to fine pitch, a takeoff in 80 meters is possible. In coarse pitch the takeoff is 200 meters and a speed of 200 km/h in level flight can be attained.

ALISPORT S.R.L.
Viale Verdi, 58
34074 Monfalcone (GO)
ITALY
Tel. 0481 778994

Bagaliante

Wing Span	40.00 ft.	**Range**	N/A
Wing Area	140.0 sq. ft.	**Takeoff Dist.**	150 ft.
Length	21.00 ft.	**Landing Dist.**	150 ft.
Max Gross Wt.	407 lbs.	**Vmax**	54 mph
Empty Wt.	242 lbs.	**VCr**	37 mph
Payload (full fuel)	N/A	**V_{S1} (stall clean)**	26 mph
Fuel Capacity	3 gal.	**V_{S0} (lndg config)**	N/A
Seats	1	**Climb Rate @msl**	900 fpm
Cabin Width	N/A	**Service Ceiling**	N/A

Std. Engine/HP	Usable HP Range	Construction Mtrl.	Est. Bldg. Time
Rotax 277 / 26 hp	N/A	Wood, Metal	700 hrs.
Plans	**Kit**	**Info Packet**	**Video**
$250.00	N/A	$5.00	N/A

Wood and metal construction, fixed landing gear, and tapered wings.

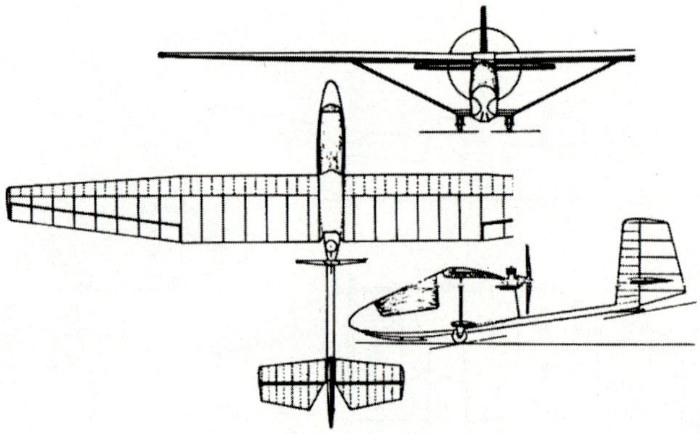

BAGALINI
via delle Rose 6
Milan 20147
ITALY
Tel. 39 2 4121784

TST-1-ALPIN

Wing Span	41.30 ft.	**Range**	220 sm
Wing Area	119.4 sq. ft.	**Takeoff Dist.**	500 ft.
Length	19.20 ft.	**Landing Dist.**	300 ft.
Max Gross Wt.	616 lbs.	**Vmax**	110 mph
Empty Wt.	330 lbs.	**VCr**	60 mph
Payload (full fuel)	242 lbs.	**V_{S1} (stall clean)**	37 mph
Fuel Capacity	4 gal.	**V_{S0} (lndg config)**	35 mph
Seats	1	**Climb Rate @msl**	N/A
Cabin Width	23 in.	**Service Ceiling**	8,000 ft.

Std. Engine/HP	Usable HP Range	Construction Mtrl.	Est. Bldg. Time
M-115 / 24 hp	N/A	Wood, Composite	600 hrs.
Plans	**Kit**	**Info Packet**	**Video**
$325.00	$9,840.00	$10.00	N/A

Economical, and easy to operate motorglider. Features outstanding soaring capability. Produced by TeST Aircraft in the Czech Republic. Fully enclosed cockpit. Special two stroke engine with flat configuration. Other engine options available. Certified and registered as a European Ultralight. Wings and tail fully detachable. Available as a Standard or 200 hour Express-Built Kit. Engine included in kit. Ready to fly aircraft offered for $15,250.

Kits/Plans Sold: 28 **Completed & Flying:** 22

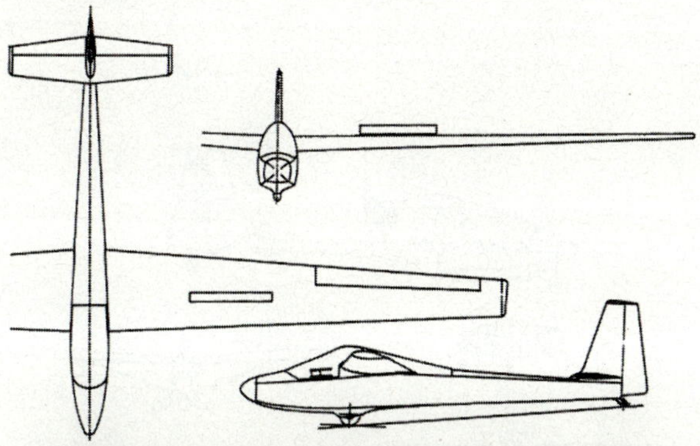

BOHEMICAN INTERNATIONAL
36 Buchanan Road
St. Catharines, Ontario L2M 4R6
CANADA
Tel. 905-935-4938 • Fax 905-935-4938
E-Mail: pdokter@niagara.com

TST-6 Duo

Wing Span	48.10 ft.	**Range**	220 sm
Wing Area	134.0 sq. ft.	**Takeoff Dist.**	750 ft.
Length	21.10 ft.	**Landing Dist.**	1200 ft.
Max Gross Wt.	990 lbs.	**Vmax**	110 mph
Empty Wt.	500 lbs.	**VCr**	65 mph
Payload (full fuel)	460 lbs.	**V_{S1} (stall clean)**	40 mph
Fuel Capacity	10 gal.	**V_{S0} (lndg config)**	38 mph
Seats	2	**Climb Rate @msl**	800 fpm
Cabin Width	42 in.	**Service Ceiling**	10,000 ft.

Std. Engine/HP	Usable HP Range	Construction Mtrl.	Est. Bldg. Time
M 215 / 50 hp	50 - 85	Wood, Composite	400 hrs.
Plans	**Kit**	**Info Packet**	**Video**
$325.00	$14,875.00	Yes	N/A

Two seat motorglider made by TeST Manufacturing of the Czech Republic. Fully enclosed, all wood and composite aircraft. Cantilever wing. Solid three wheel undercarriage, dual controls, and adjustable rudder pedals. Wings with flaps and spoilers; trim is standard. Wings and tail are fully detachable for transporting. Glide ratio is 20:1. Offered as Standard kit or 300 hour Express-Built kit. Ready to fly aircraft without engine is offered at $18,000. Instruments are included.

Kits/Plans Sold: 4 **Completed & Flying:** 6

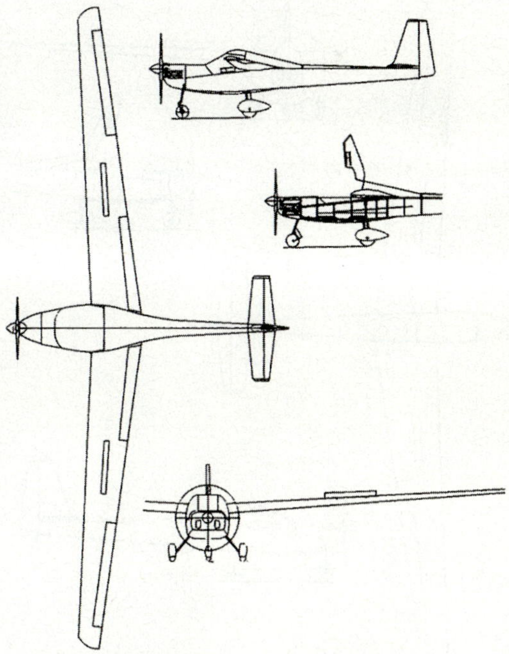

BOHEMICAN INTERNATIONAL
36 Buchanan Road
St. Catharines, Ontario L2M 4R6
CANADA
Tel. 905-935-4938 • Fax 905-935-4938
E-Mail: pdokter@niagara.com

TST-7 Junior

Wing Span	41.30 ft.	Range	230 sm
Wing Area	119.4 sq. ft.	Takeoff Dist.	500 ft.
Length	19.20 ft.	Landing Dist.	300 ft.
Max Gross Wt.	649 lbs.	Vmax	110 mph
Empty Wt.	395 lbs.	VCr	75 mph
Payload (full fuel)	225 lbs.	V_{S1} (stall clean)	39 mph
Fuel Capacity	5 gal.	V_{S0} (lndg config)	37 mph
Seats	1	Climb Rate @msl	N/A
Cabin Width	23 in.	Service Ceiling	12,000 ft.

Std. Engine/HP	Usable HP Range	Construction Mtrl.	Est. Bldg. Time
M 125 / 40 hp	40 - 50	Wood, Composite	400 hrs.
Plans	**Kit**	**Info Packet**	**Video**
N/A	N/A	$10.00	N/A

Single seat, all-wood, mid wing design. Junior is a modified version of the popular TST-1 Alpin motorglider. Glide ratio is 18:1. It is more of a sport cruiser than a glider. Three wheel undercarriage greatly improves ground handling and stability. Engines up to 55 hp are available for higher performance. Offered as a Standard or 250 hour Express-Built kit. Standard kit supplied without engine. Instruments are included. Ready to fly aircraft with engine are offered for $16,250.

Kits/Plans Sold: 8 **Completed & Flying:** 5

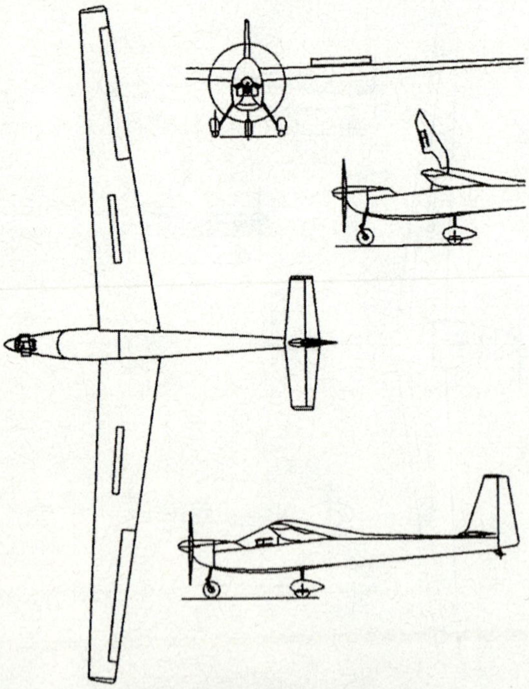

BOHEMICAN INTERNATIONAL
36 Buchanan Road
St. Catharines, Ontario L2M 4R6
CANADA
Tel. 905-935-4938 • Fax 905-935-4938
E-Mail: pdokter@niagara.com

Piuma Motorglider

Wing Span	39.00 ft.	Range	250 sm
Wing Area	125.0 sq. ft.	Takeoff Dist.	300 ft.
Length	19.70 ft.	Landing Dist.	300 ft.
Max Gross Wt.	530 lbs.	Vmax	62 mph
Empty Wt.	320 lbs.	VCr	50 mph
Payload (full fuel)	190 lbs.	V_{S1} (stall clean)	30 mph
Fuel Capacity	4 gal.	V_{S0} (lndg config)	N/A
Seats	1	Climb Rate @msl	500 fpm
Cabin Width	23 in.	Service Ceiling	10,000 ft.

Std. Engine/HP	Usable HP Range	Construction Mtrl.	Est. Bldg. Time
KFM 107 ER / 30 hp	20 - 30	Wood, Fabric	1000 hrs.
Plans	**Kit**	**Info Packet**	**Video**
$220.00	N/A	$10.00	N/A

An inexpensive, easy to build, ultralight motorglider. Plans only. Sure and easy to fly. Low fuel usage - about one gallon/hour. Long range with engine and much more when gliding. L/D max is 17:1 at 40 mph. Aspect ratio is 11:1. A 30 page construction manual is included with plans. An "E" model with an L/D of 20:1 at 44 mph and a number of other improvements is now in the prototype stage.

Kits/Plans Sold: 35 **Completed & Flying:** 4

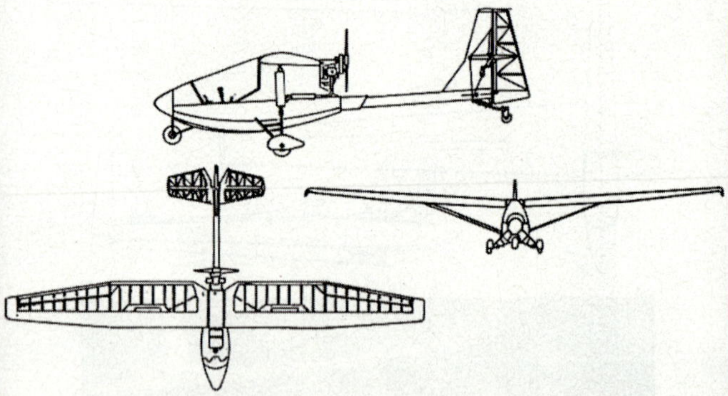

TIZIANO DANIELI
via dei Tamburini, 14
Schio (VI) 36015
ITALY
Tel. 39 445 527929

Soaring Gull

Wing Span	28.00 ft.	**Range**	180 sm
Wing Area	133.0 sq. ft.	**Takeoff Dist.**	115 ft.
Length	17.60 ft.	**Landing Dist.**	70 ft.
Max Gross Wt.	550 lbs.	**Vmax**	120 mph
Empty Wt.	254 lbs.	**VCr**	63 mph
Payload (full fuel)	266 lbs.	**V_{S1} (stall clean)**	24 mph
Fuel Capacity	5 gal.	**V_{S0} (lndg config)**	22 mph
Seats	1	**Climb Rate @msl**	850 fpm
Cabin Width	27 in.	**Service Ceiling**	N/A

Std. Engine/HP	Usable HP Range	Construction Mtrl.	Est. Bldg. Time
Rotax 277 / 27 hp	27 - 53	Steel tube, Composite	200 hrs.
Plans	**Kit**	**Info Packet**	**Video**
N/A	$10,580.00	$5.00	$25.00

A powered glider version of the Thunder Gull J. The 28 foot wing allows for a glide ratio of 16:1. As with all Thunder Gulls, the wing can be removed for ease in trailerability. Designed to meet FAR Part 103 regulations.

EARTHSTAR AIRCRAFT, INC.
Star Route 313
Santa Margarita CA 93453
Tel. 805-438-5235 • Fax 805-438-5235

Fauvel AV361

Wing Span	42.00 ft.	**Range**	N/A
Wing Area	157.0 sq. ft.	**Takeoff Dist.**	N/A
Length	10.00 ft.	**Landing Dist.**	50 ft.
Max Gross Wt.	570 lbs.	**Vmax**	137 mph
Empty Wt.	N/A	**VCr**	58 mph
Payload (full fuel)	304 lbs.	**V_{S1} (stall clean)**	30 mph
Fuel Capacity	N/A	**V_{S0} (lndg config)**	30 mph
Seats	1	**Climb Rate @msl**	N/A
Cabin Width	25 in.	**Service Ceiling**	N/A

Std. Engine/HP	Usable HP Range	Construction Mtrl.	Est. Bldg. Time
N/A	N/A	Wood, Fabric	1500 hrs.
Plans	**Kit**	**Info Packet**	**Video**
$300.00	$11,645.00	$10.00	N/A

Wood and fabric light sailplane designed for marginal soaring conditions. Less weight and materials provides a 30% savings in cost. Flying wing is spin-proof. Utilizes a single wheel landing gear. Setup about 10 minutes. **Sold:** 100 **Flying:** 10

FALCONAR AVIA, INC.
7739 - 81 Ave. • Edmonton, Alberta T6C 0V4
CANADA
Tel. 403-465-2024 • Fax 403-465-2029
E-Mail: falconar@tic.ab.ca

Genesis I

Wing Span	49.20 ft.	**Range**	N/A
Wing Area	120.5 sq. ft.	**Takeoff Dist.**	300 ft.
Length	16.00 ft.	**Landing Dist.**	300 ft.
Max Gross Wt.	1157 lbs.	**Vmax**	150 mph
Empty Wt.	510 lbs.	**VCr**	75 mph
Payload (full fuel)	N/A	**V_{S1} (stall clean)**	42 mph
Fuel Capacity	N/A	**V_{S0} (lndg config)**	N/A
Seats	1	**Climb Rate @msl**	N/A
Cabin Width	N/A	**Service Ceiling**	N/A

Std. Engine/HP	Usable HP Range	Construction Mtrl.	Est. Bldg. Time
N/A	N/A	Prepeg Carbon Fiber, Kevlar FG	300 hrs.
Plans	**Kit**	**Info Packet**	**Video**
N/A	$25,900.00	$10.00	Yes

A high performance Standard Class sailplane. Genesis I is a computer-optimized sailplane design. Designed to JAR 22 certification requirements. Kit completely prefabricated. Aspect ratio of 20:1, minimum sink of 114 fpm. L/D at 100 mph, 29:1. 400 lbs. water ballast. Emergency ballistic parachute included in kit.

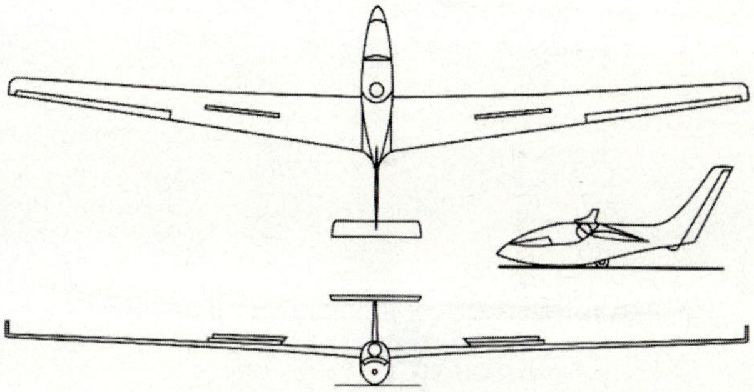

John Roncz and Jim Marske set out to create a reasonably priced kit sailplane with world-class performance. Their design was developed with emphasis on aerodynamic optimization through state-of-the-art computer modeling. The result is Genesis I, a high performance Standard Class sailplane constructed of high-temp prepreg composites.

The Genesis I is supplied in kit form with all major assemblies completed by the factory. This includes wing halves assembled in factory alignment jigs.

Collectively, the Group Genesis team has been involved in the design or construction of over 50 individual aircraft.The group has over a century of cumulative design and building experience.

GROUP GENESIS, INC.
1530 Pole Lane Road
Marion OH 43302
Tel. 614-387-9464 • Fax 614-387-0501

Mitchell Wing A-10

Wing Span	34.33 ft.	**Range**	200 sm
Wing Area	156.0 sq. ft.	**Takeoff Dist.**	200 ft.
Length	8.50 ft.	**Landing Dist.**	200 ft.
Max Gross Wt.	550 lbs.	**Vmax**	75 mph
Empty Wt.	270 lbs.	**VCr**	55 mph
Payload (full fuel)	N/A	**V_{S1} (stall clean)**	28 mph
Fuel Capacity	5 gal.	**V_{S0} (lndg config)**	N/A
Seats	1	**Climb Rate @msl**	800 fpm
Cabin Width	N/A	**Service Ceiling**	12,000 ft.

Std. Engine/HP	Usable HP Range	Construction Mtrl.	Est. Bldg. Time
Rotax 277 / 28 hp	20 - 35	Aluminum	100 hrs.
Plans	**Kit**	**Info Packet**	**Video**
N/A	$7,900.00	$10.00	N/A

A unique, ultralight motorglider for intermediate and advanced pilot skills. Utilizes an aluminum clad flying wing design. Soarable 16:1 glide ratio. 240 fpm minimum sink. Folds in 5 minutes. Trailerable. Price does not include engine.

Completed & Flying: 350

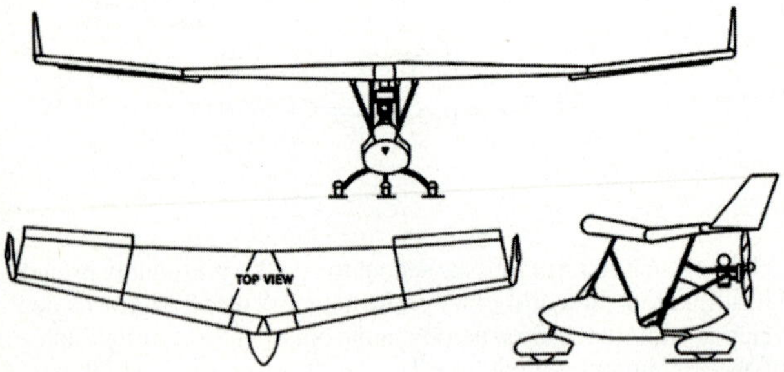

HIGHER PLANES, INC.
Box 4
Dover KS 66420
Tel. 913-256-6029

LO 120 S

Wing Span	49.00 ft.	**Range**	373 sm
Wing Area	N/A	**Takeoff Dist.**	N/A
Length	N/A	**Landing Dist.**	N/A
Max Gross Wt.	880 lbs.	**Vmax**	93 mph
Empty Wt.	440 lbs.	**VCr**	74 mph
Payload (full fuel)	N/A	**V_{S1} (stall clean)**	N/A
Fuel Capacity	N/A	**V_{S0} (lndg config)**	N/A
Seats	2	**Climb Rate @msl**	680 fpm
Cabin Width	N/A	**Service Ceiling**	N/A

Std. Engine/HP	Usable HP Range	Construction Mtrl.	Est. Bldg. Time
Hirth 2704 / 40 hp	N/A	Plywood, Metal, Composite	N/A
Plans	**Kit**	**Info Packet**	**Video**
N/A	Yes	N/A	N/A

Two seat, tandem Ultralight(European) characterized by twin booms and inverted vee tail. Can be built with short or long wings. Powered sailplane version shown.

LO-FLUGGERRATEBAU GMBH
Aspachstr 14
88267 Berg/Ettishofen
GERMANY
Tel. (49)0751/55 10 14 • Fax (49)0751/55 10 32

Lucas L-6A

Wing Span	46.00 ft.	**Range**	500 sm
Wing Area	172.0 sq. ft.	**Takeoff Dist.**	1000 ft.
Length	26.00 ft.	**Landing Dist.**	850 ft.
Max Gross Wt.	1380 lbs.	**Vmax**	90 mph
Empty Wt.	960 lbs.	**VCr**	75 mph
Payload (full fuel)	420 lbs.	**V_{S1} (stall clean)**	47 mph
Fuel Capacity	25 gal.	**V_{S0} (lndg config)**	N/A
Seats	2	**Climb Rate @msl**	N/A
Cabin Width	28 in.	**Service Ceiling**	19,000 ft.

Std. Engine/HP	Usable HP Range	Construction Mtrl.	Est. Bldg. Time
Lycoming / 80 hp	N/A	Metal	4000 hrs.
Plans	**Kit**	**Info Packet**	**Video**
$400.00	N/A	Yes	N/A

All-metal, folding wing, two place, powered sailplane version of the Lucas L6. Retractable landing gear, folding wings with removable wing extension, and variable pitch propeller. Plans only. Glide ratio is 20:1. Choice of Limbach or Lycoming 0-235.

Kits/Plans Sold: 5 **Completed & Flying:** 2

EMILE LUCAS
7 allee des Acacias • Lagny le Sec 60330 • FRANCE
Tel. 44 60 87 45 • Fax 16 44 60 04 81

Pioneer II-D

Wing Span	42.50 ft.	**Range**	N/A
Wing Area	144.0 sq. ft.	**Takeoff Dist.**	N/A
Length	12.50 ft.	**Landing Dist.**	200 ft.
Max Gross Wt.	600 lbs.	**Vmax**	130 mph
Empty Wt.	350 lbs.	**VCr**	95 mph
Payload (full fuel)	N/A	**V_{S1} (stall clean)**	36 mph
Fuel Capacity	N/A	**V_{S0} (lndg config)**	36 mph
Seats	1	**Climb Rate @msl**	N/A
Cabin Width	24 in.	**Service Ceiling**	N/A

Std. Engine/HP	Usable HP Range	Construction Mtrl.	Est. Bldg. Time
Sailplane	N/A	Wood, Fabric, Composite	600 hrs.
Plans	**Kit**	**Info Packet**	**Video**
$248.00	$6,550.00	$8.00	N/A

A forward sweep, mid wing, tailless sailplane with excellent stability and performance. Tri-axis control. Highly stall and spin resistant. Glide ratio is 35:1. Minimum sink rate is 138 fpm. Tow with auto or airplane.

Kits/Plans Sold: 146 **Completed & Flying:** 28

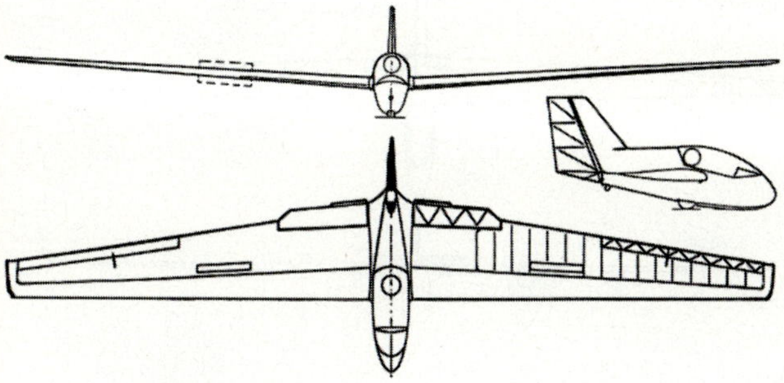

MARSKE AIRCRAFT CORP.
975 Loire Valley Dr.
Marion OH 43302
Tel. 614-389-3776

Monarch F

Wing Span	42.00 ft.	**Range**	N/A
Wing Area	186.0 sq. ft.	**Takeoff Dist.**	N/A
Length	12.50 ft.	**Landing Dist.**	100 ft.
Max Gross Wt.	450 lbs.	**Vmax**	70 mph
Empty Wt.	210 lbs.	**VCr**	60 mph
Payload (full fuel)	N/A	**V_{S1} (stall clean)**	26 mph
Fuel Capacity	N/A	**V_{S0} (lndg config)**	26 mph
Seats	1	**Climb Rate @msl**	N/A
Cabin Width	N/A	**Service Ceiling**	N/A

Std. Engine/HP	Usable HP Range	Construction Mtrl.	Est. Bldg. Time
Sailplane	N/A	Wood, Fabric, Composite	N/A
Plans	**Kit**	**Info Packet**	**Video**
$170.00	$7,040.00	$8.00	N/A

High, forward sweep wing, strutted tailless sailplane. Tri-axis control. Glide ratio is 20:1. Minimum sink rate is 160 fpm. Highly stall and spin resistant. Towable with auto or airplane.

Kits/Plans Sold: 170 **Completed & Flying:** 21

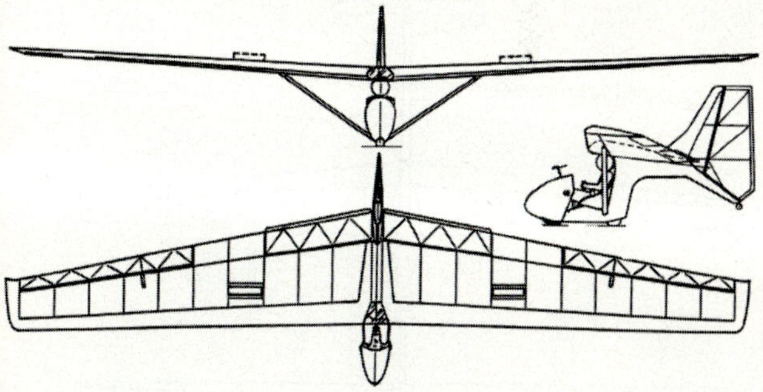

MARSKE AIRCRAFT CORP.
975 Loire Valley Dr.
Marion OH 43302
Tel. 614-389-3776

Carbon Dragon

Wing Span	44.00 ft.	**Range**	N/A
Wing Area	150.0 sq. ft.	**Takeoff Dist.**	N/A
Length	19.20 ft.	**Landing Dist.**	N/A
Max Gross Wt.	335 lbs.	**Vmax**	70 mph
Empty Wt.	145 lbs.	**VCr**	N/A
Payload (full fuel)	N/A	**V_{S1} (stall clean)**	19 mph
Fuel Capacity	N/A	**V_{S0} (lndg config)**	N/A
Seats	1	**Climb Rate @msl**	N/A
Cabin Width	25 in.	**Service Ceiling**	N/A

Std. Engine/HP	Usable HP Range	Construction Mtrl.	Est. Bldg. Time
Sailplane	N/A	Composite, Wood	1500 hrs.
Plans	**Kit**	**Info Packet**	**Video**
$160.00	N/A	$5.00	N/A

Best described as a "foot-launchable sailplane" although it meets the FAI definition of a hang-glider. Neither glider nor pilot need to be licensed. Can be launched by foot, auto tow, aero tow, or bungee. Basic wood construction with judicious use of carbon-epoxy to reduce weight and improve stiffness. Superb climbing ability. Manufacturer claims it soars in lighter conditions than any other soaring aircraft today. Has significantly exceeded world record for distance and out-and-back flights in its class. L/D is 25:1 at 32 mph. Min Sink 100 fpm at 26 mph. Stall speed is 19 knots with a 135 lb. pilot. While cabin width is noted at 17 inches, hips and shoulder width is 25 inches. Plans $160.00 in the U.S.A. and Canada; airmail overseas $180.00.

Kits/Plans Sold: 350 **Completed & Flying:** 35

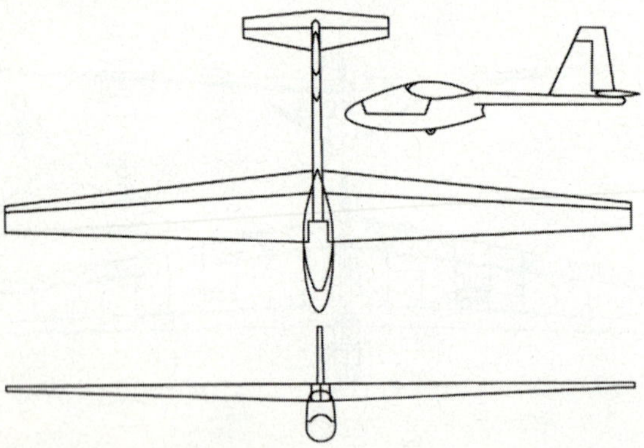

Mark Dunstone, Builder and Pilot

"It has been one of the most rewarding projects I have ever done - on a par with cruising the Pacific in a yacht."

Dan Armstrong, Pilot, Editor, SAILPLANE BUILDER

"Again and again the Dragon proved its ability to climb out in very light lifts that a hang glider or sailplane could not climb in. It did this with extremely gentle and easy handling characteristics. The Carbon Dragon is the easiest of all the gliders that I have flown, and yet, it is exceptionally fun to fly."

JIM MAUPIN, LTD.
24201 Rowel Court
Tehachapi CA 93561
Tel. 805-821-3450
E-Mail: maupin@mtxinu.com

Windrose

Wing Span	41.50 ft.	**Range**	N/A
Wing Area	96.00 sq. ft.	**Takeoff Dist.**	700 ft.
Length	21.60 ft.	**Landing Dist.**	N/A
Max Gross Wt.	740 lbs.	**Vmax**	132 mph
Empty Wt.	512 lbs.	**VCr**	N/A
Payload (full fuel)	220 lbs.	**V_{S1} (stall clean)**	N/A
Fuel Capacity	3 gal.	**V_{S0} (lndg config)**	N/A
Seats	1	**Climb Rate @msl**	700 fpm
Cabin Width	24 in.	**Service Ceiling**	N/A

Std. Engine/HP	Usable HP Range	Construction Mtrl.	Est. Bldg. Time
Rotax 503 / 33 hp	N/A	Composite, Wood	700 hrs.
Plans	**Kit**	**Info Packet**	**Video**
$175.00	N/A	$5.00	N/A

A simple, inexpensive, quick to build, self-launchable sailplane with an L/D of 29:1 at 52 mph. Min Sink is 2.3 fps at 46 mph. The wings consist of simple, primarily foam, construction. These have been called "six day wings." Plans $175.00 in the U.S.A. and Canada; airmail overseas $195.00. A higher performance version can be built with longer, 15 meter wings. Drawings and book cost is $50.

Kits/Plans Sold: 400 **Completed & Flying:** 40

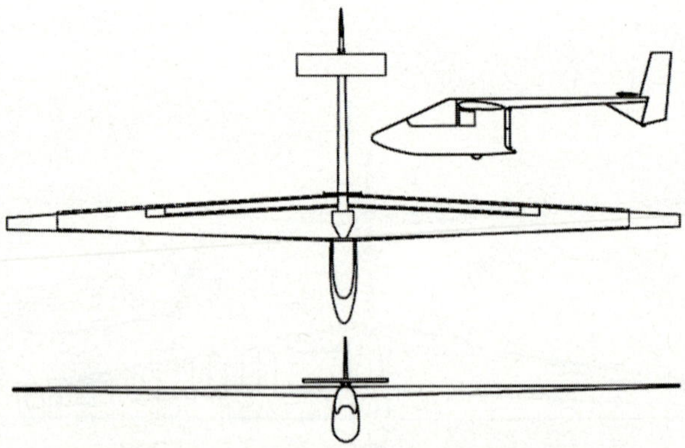

John Walkling, Builder & Pilot

"The Windrose was a joy to build and I'm thoroughly happy with it! Anyone who can build a model airplane can build a Windrose."

Bob Said, SOARING Magazine, Dec. '84

"The simplicity built into the design concept is visible everywhere. ... The purpose of all that simplicity was to make the Windrose both easy and inexpensive to build. ...I can say that the simplicity hasn't hurt the flying qualities a bit."

JIM MAUPIN, LTD.
24201 Rowel Court
Tehachapi CA 93561
Ph: 805-821-3450
E-Mail: maupin@mtxinu.com

Woodstock

Wing Span	39.00 ft.	Range	N/A
Wing Area	104.7 sq. ft.	Takeoff Dist.	N/A
Length	21.00 ft.	Landing Dist.	N/A
Max Gross Wt.	455 lbs.	Vmax	125 mph
Empty Wt.	245 lbs.	VCr	N/A
Payload (full fuel)	210 lbs.	V_{S1} (stall clean)	35 mph
Fuel Capacity	N/A	V_{S0} (lndg config)	N/A
Seats	1	Climb Rate @msl	N/A
Cabin Width	21 in.	Service Ceiling	N/A

Std. Engine/HP	Usable HP Range	Construction Mtrl.	Est. Bldg. Time
N/A	N/A	Wood, Fabric	1200 hrs.
Plans	**Kit**	**Info Packet**	**Video**
$125.00	N/A	$5.00	N/A

The Woodstock is a classic sailplane of all wood construction. Thoroughly test flown. It won the Sailplane Homebuilder's Design Competition in 1984. Smooth flying, handles nicely, and no bad habits. Specifications are for 12 meter version with an L/D max of 24:1 and a Min Sink of 2.6 fps. Plans $125.00 in the U.S.A. and Canada; airmail overseas $140.00. Higher performance 13 meter version available with an L/D max of 26.5:1 and a Min Sink of 2.3 fps. Longer wings, $35 additional.

Kits/Plans Sold: 650 **Completed & Flying:** 65

Design philosophy behind the Woodstock was to produce as small, light, simple, and inexpensive a sailplane as possible with an L/D in the mid-20's. The light weight aids in simplicity. Only 16 pieces of steel and 40 bolts are required. No special jigs or fixtures are needed. The whole sailplane is built on a simple assembly table 18 feet long made from a 1" x 10" and a 1" x 6". It can be built entirely with hand tools. The only welding required is for the control stick assembly, tow hook, aileron, and elevator.

Randy Beskow, Builder and Pilot

"My Woodstock made its first flight on March 19, 1994. It flew absolutely perfectly from the very first flight! ... an outstanding sailplane ... its handling is flawless, and it can outclimb almost anything else in the field."

Bill Ward, Builder and Pilot

"Everyone who sees the Woodstock says what a pretty little plane it is. I am proud of the workmanship and you can really be happy about the design and ideas. It really has been one of the most fun projects I have ever attempted."

JIM MAUPIN, LTD.
24201 Rowel Court
Tehachapi CA 93561
Tel. 805-821-3450
E-Mail: maupin@mtxinu.com

Choucas

Wing Span	46.80 ft.	Range	203 sm
Wing Area	230.5 sq. ft.	Takeoff Dist.	328 ft.
Length	17.40 ft.	Landing Dist.	493 ft.
Max Gross Wt.	1000 lbs.	Vmax	107 mph
Empty Wt.	578 lbs.	VCr	81 mph
Payload (full fuel)	422 lbs.	V_{S1} (stall clean)	40 mph
Fuel Capacity	10 gal.	V_{S0} (lndg config)	40 mph
Seats	2	Climb Rate @msl	N/A
Cabin Width	39 in.	Service Ceiling	N/A

Std. Engine/HP	Usable HP Range	Construction Mtrl.	Est. Bldg. Time
Rotax 503 / 50 hp	N/A	Fiberglass, Composite, Wood	N/A
Plans	**Kit**	**Info Packet**	**Video**
N/A	$9,046.00	$30.00	$60.00

Two seat "flying wing" motorized sailplane. Basic kit includes drawings, plans, and molded fuselage half shells. Also available in an advanced kit with all important and "tricky" parts prefabricated. Glide ratio is 22:1 at 52 mph.

Kits/Plans Sold: 2 **Completed & Flying:** 1

NOIN AERONAUTIQUES
RN85, Chateauvleux • 05130 Tallard • FRANCE
Tel. 33 4 92 54 15 04 • Fax 33 4 92 54 15 04

Tempest

Wing Span	42.00 ft.	Range	N/A
Wing Area	150.0 sq. ft.	Takeoff Dist.	N/A
Length	20.00 ft.	Landing Dist.	300 ft.
Max Gross Wt.	400 lbs.	Vmax	90 mph
Empty Wt.	200 lbs.	VCr	N/A
Payload (full fuel)	N/A	V_{S1} (stall clean)	28 mph
Fuel Capacity	N/A	V_{S0} (lndg config)	N/A
Seats	1	Climb Rate @msl	N/A
Cabin Width	22 in.	Service Ceiling	N/A

Std. Engine/HP	Usable HP Range	Construction Mtrl.	Est. Bldg. Time
Sailplane	N/A	Tube, Fabric	200 hrs.
Plans	**Kit**	**Info Packet**	**Video**
N/A	$10,000.00	Yes	N/A

Single wheel sailplane. Designed by Bob Bailey, Florida. An offspring of the Sierra design. Fiberglass monocockpit unit has canopy that hinges to one side. Completed price is $12,500.
Completed & Flying: 5

MOYES MICROLITES PTY. LTD.
2-4 Taylor St.
Waverley 2024 N.S.W.
AUSTRALIA
Tel. 02-9-387-5114 • Fax 02-9-369-3342

Sirius C

Wing Span	44.00 ft.	**Range**	150 sm
Wing Area	145.0 sq. ft.	**Takeoff Dist.**	328 ft.
Length	19.40 ft.	**Landing Dist.**	265 ft.
Max Gross Wt.	600 lbs.	**Vmax**	82 mph
Empty Wt.	335 lbs.	**VCr**	54 mph
Payload (full fuel)	265 lbs.	**V_{S1} (stall clean)**	30 mph
Fuel Capacity	5 gal.	**V_{S0} (lndg config)**	30 mph
Seats	1	**Climb Rate @msl**	N/A
Cabin Width	21 in.	**Service Ceiling**	N/A

Std. Engine/HP	Usable HP Range	Construction Mtrl.	Est. Bldg. Time
Konig / 24 hp	20 - 35	Composite	700 hrs.
Plans	**Kit**	**Info Packet**	**Video**
$220.00	$7,900.00	$30.00	$60.00

The Sirius-C is a glider in the European "ULM" catagory. The type of construction allows amateurs to work from "plans only" to several stages of completion. Glide ratio of 24:1.

Kits/Plans Sold: 25 **Completed & Flying:** 20

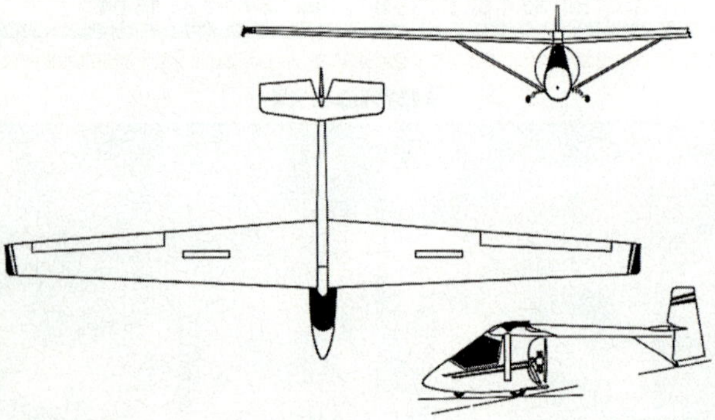

NOIN AERONAUTIQUES
RN85, Chateauvleux
05130 Tallard
FRANCE
Ph: 33 4 92 54 15 04 • Fax: 33 4 92 54 15 04

Sinus

Wing Span	49.00 ft.	**Range**	372 sm
Wing Area	N/A	**Takeoff Dist.**	396 ft.
Length	21.60 ft.	**Landing Dist.**	328 ft.
Max Gross Wt.	990 lbs.	**Vmax**	116 mph
Empty Wt.	495 lbs.	**VCr**	102 mph
Payload (full fuel)	N/A	**V_{S1} (stall clean)**	36 mph
Fuel Capacity	17 gal.	**V_{S0} (lndg config)**	32 mph
Seats	2	**Climb Rate @msl**	688 fpm
Cabin Width	44 in.	**Service Ceiling**	16,000 ft.

Std. Engine/HP	Usable HP Range	Construction Mtrl.	Est. Bldg. Time
Rotax 503 / 53 hp	53-8	Composite	300 hrs.
Plans	**Kit**	**Info Packet**	**Video**
N/A	Yes	Yes	N/A

Motorized sailplane built by Flysynthesis and Pipistrel (Slovenia). Uses the fuselage of the "Storch" and a 49 ft. wing derived from the "Silent" glider. 25:1 glide ratio.

Kits/Plans Sold: 2 **Completed & Flying:** 2

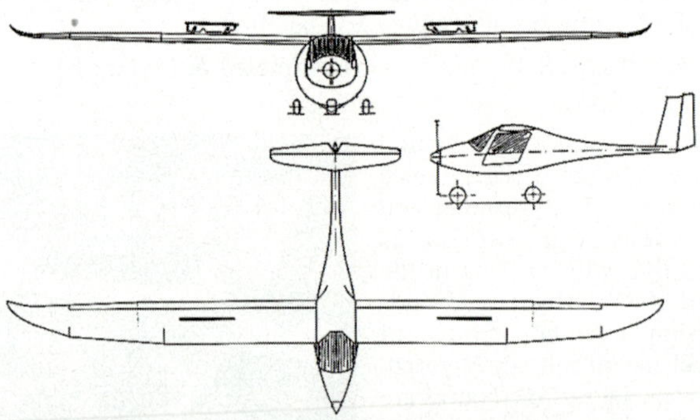

PIPISTREL
Strancarjeva 11
Ajdovscina 5270
SLOVINIA
Tel. 00386 65 63 873 • Fax 00386 65 61 263
E-Mail: pipistrel@eunet.si

D-8 Moby Dick

Wing Span	42.65 ft.	**Range**	N/A
Wing Area	150.7 sq. ft.	**Takeoff Dist.**	130 ft.
Length	23.29 ft.	**Landing Dist.**	N/A
Max Gross Wt.	851 lbs.	**Vmax**	82 mph
Empty Wt.	414 lbs.	**VCr**	56 mph
Payload (full fuel)	387 lbs.	**V_{S1} (stall clean)**	34 mph
Fuel Capacity	9 gal.	**V_{S0} (lndg config)**	N/A
Seats	2	**Climb Rate @msl**	800 fpm
Cabin Width	26 in.	**Service Ceiling**	N/A

Std. Engine/HP	Usable HP Range	Construction Mtrl.	Est. Bldg. Time
Rotax 447 / 41 hp	N/A	Composite, Wood	N/A
Plans	**Kit**	**Info Packet**	**Video**
N/A	$23,300.00	Yes	Yes

V-tailed motorglider. High wing, pusher design with completely enclosed cabin. Stress rated +4, -2 G's. Cruising fuel consumption is 1.85 gallons per hour. Can be certified as experimental or as an ultralight. U.S. price is dependant on import and shipping costs.

Kits/Plans Sold: 70 **Completed:** 37

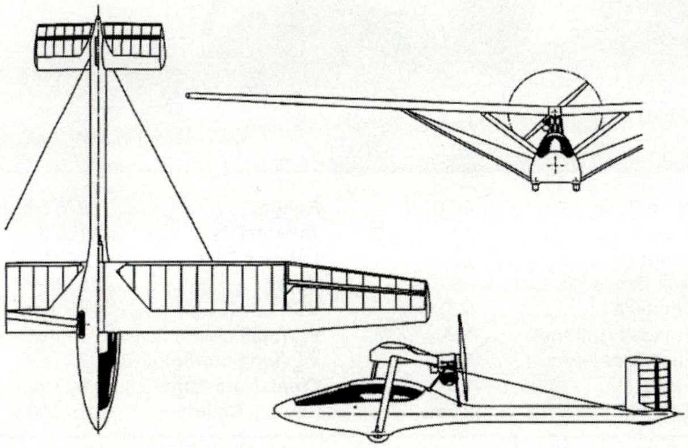

PRO FE
Lestinska 811
Nove Mesto nad Metuji 549 01
CZECH REPUBLIC
Tel. 00420 441-72353
Fax 00420 441-72353

D-10 Tukan

Wing Span	48.22 ft.	**Range**	N/A
Wing Area	160.0 sq. ft.	**Takeoff Dist.**	230 ft.
Length	23.95 ft.	**Landing Dist.**	N/A
Max Gross Wt.	903 lbs.	**Vmax**	82 mph
Empty Wt.	452 lbs.	**VCr**	N/A
Payload (full fuel)	400 lbs.	**V_{S1} (stall clean)**	37 mph
Fuel Capacity	9 gal.	**V_{S0} (lndg config)**	N/A
Seats	2	**Climb Rate @msl**	400 fpm
Cabin Width	29 in.	**Service Ceiling**	N/A

Std. Engine/HP	Usable HP Range	Construction Mtrl.	Est. Bldg. Time
Rotax 447 / 41 hp	N/A	Composite, Wood	N/A
Plans	**Kit**	**Info Packet**	**Video**
N/A	$25,000.00	Yes	Yes

A two-seat, T-tailed motorglider. High wing tractor design. Propeller retracts in less than 15 seconds for extended soaring. Enclosed cabin. Stress rated +3.8, -1.9 G's. Can be certified as experimental or as an ultralight. U.S. price is dependant on import and shipping costs.

Completed & Flying: 4

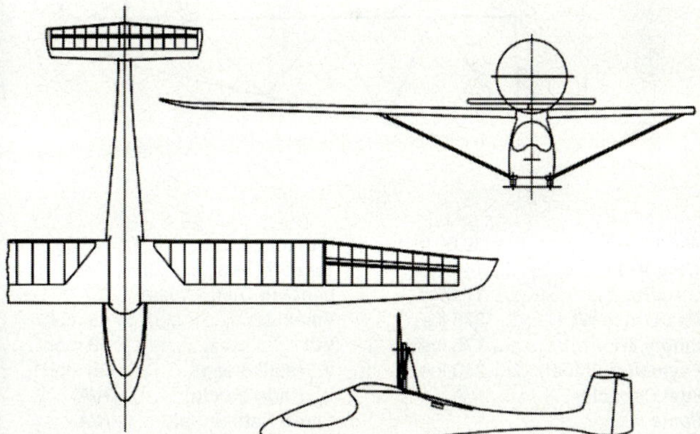

U.S. Distributor:
UPPER MIDWEST ULTRALIGHT
FREEDOM FIELD
1411 375TH AVE.
ESTHERVILLE IA 51334
PH: 712-362-4822
E-MAIL: MDJOHNSO@NCN.NET

PRO FE
Lestinska 811
Nove Mesto nad Metuji 549 01
CZECH REPUBLIC
Tel. 00420 441-72353
Fax 00420 441-72353

Banjo

Wing Span	43.60 ft.	**Range**	N/A
Wing Area	113.0 sq. ft.	**Takeoff Dist.**	N/A
Length	20.60 ft.	**Landing Dist.**	350 ft.
Max Gross Wt.	440 lbs.	**Vmax**	87 mph
Empty Wt.	220 lbs.	**VCr**	N/A
Payload (full fuel)	220 lbs.	**V_{S1} (stall clean)**	31 mph
Fuel Capacity	N/A	**V_{S0} (Indg config)**	N/A
Seats	1	**Climb Rate @msl**	N/A
Cabin Width	24 in.	**Service Ceiling**	N/A

Std. Engine/HP	Usable HP Range	Construction Mtrl.	Est. Bldg. Time
N/A	N/A	Composite, Wood	N/A
Plans	**Kit**	**Info Packet**	**Video**
N/A	$14,300.00	N/A	N/A

The Banjo has a glide ratio of 28:1 and a minimum sink rate of 134 fpm. U.S. Price is dependant on import and shipping costs.

Kits/Plans Sold: 8 **Completed:** 4

PRO FE
Lestinska 811
Nove Mesto nad Metuji 549 01
CZECH REPUBLIC
Tel. 00420 441-72353
Fax 00420 441-72353

Super Floater

Wing Span	38.00 ft.	**Range**	N/A
Wing Area	168.0 sq. ft.	**Takeoff Dist.**	50 ft.
Length	19.00 ft.	**Landing Dist.**	100 ft.
Max Gross Wt.	400 lbs.	**Vmax**	60 mph
Empty Wt.	179 lbs.	**VCr**	40 mph
Payload (full fuel)	N/A	**V_{S1} (stall clean)**	23 mph
Fuel Capacity	N/A	**V_{S0} (Indg config)**	23 mph
Seats	1	**Climb Rate @msl**	N/A
Cabin Width	24 in.	**Service Ceiling**	N/A

Std. Engine/HP	Usable HP Range	Construction Mtrl.	Est. Bldg. Time
Sailplane	N/A	Aluminum, Dacron	N/A
Plans	**Kit**	**Info Packet**	**Video**
N/A	$9,995.00	Yes	$9.95

Redesigned ultralight. Fully built and test flown. A modern day "Primary Glider." Sink Rate: 180 fpm @ 27 mph. Glide ratio is 15:1 @ 35 mph. Roll Rate: 3.5 sec 45 deg. to 45 deg. Supplied ready-to-fly.

US AVIATION
265 Echo Lane • South St. Paul MN 55075
Tel. 612-450-0930 • Fax 612-450-0930
E-Mail: CumulusMan@aol.com
Web: http://members.aol.com/CumulusMan

1929 Primary Glider

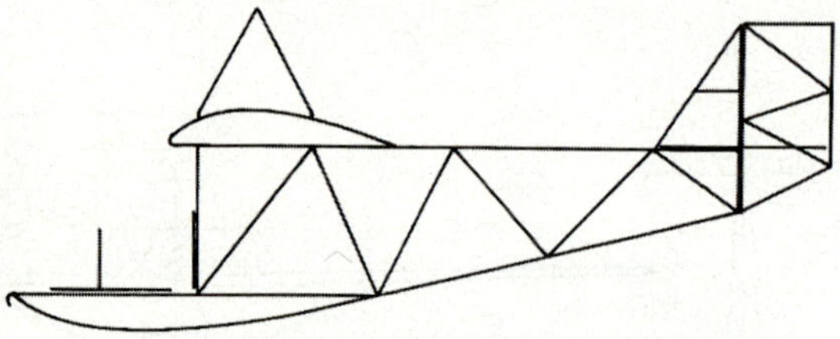

Wing Span	32.00 ft.	**Range**	N/A
Wing Area	160.0 sq. ft.	**Takeoff Dist.**	N/A
Length	17.67 ft.	**Landing Dist.**	N/A
Max Gross Wt.	375 lbs.	**Vmax**	45 mph
Empty Wt.	175 lbs.	**VCr**	38 mph
Payload (full fuel)	200 lbs.	**V_{S1} (stall clean)**	30 mph
Fuel Capacity	N/A	**V_{S0} (Indg config)**	N/A
Seats	1	**Climb Rate @msl**	N/A
Cabin Width	28 in.	**Service Ceiling**	N/A

Std. Engine/HP	Usable HP Range	Construction Mtrl.	Est. Bldg. Time
N/A	N/A	Wood, Tube, Fabric	200 hrs.
Plans	**Kit**	**Info Packet**	**Video**
$32.00	N/A	Yes	N/A

Drawing is of original 1929 Primary Glider. Simple wood, tube, and fabric construction. Plans include materials list and optional steel tube fuselage sketch. Materials available from Wickes Aircraft Supply, 800-221-9425. **Plans Sold:** 100 **Flying:** 55

RON SANDS CO.
89 Forrest Drive
Mertztown PA 19539
Tel. 610-682-6788 • Fax 610-682-6788

Cumulus

Wing Span	43.00 ft.	**Range**	275 sm
Wing Area	141.0 sq. ft.	**Takeoff Dist.**	170 ft.
Length	20.00 ft.	**Landing Dist.**	190 ft.
Max Gross Wt.	560 lbs.	**Vmax**	85 mph
Empty Wt.	320 lbs.	**VCr**	65 mph
Payload (full fuel)	N/A	**V_{S1} (stall clean)**	34 mph
Fuel Capacity	5 gal.	**V_{S0} (Indg config)**	32 mph
Seats	1	**Climb Rate @msl**	800 fpm
Cabin Width	24 in.	**Service Ceiling**	15,000 ft.

Std. Engine/HP	Usable HP Range	Construction Mtrl.	Est. Bldg. Time
Rotax 447 / 40 hp	N/A	Steel, Fabric	150 hrs.
Plans	**Kit**	**Info Packet**	**Video**
N/A	$9,995.00	Yes	$9.95

High performance design with detachable wings. One gph fuel consumption. Sink Rate: 195 fpm @ 34 mph. Glide ratio is 20:1 @ 43 mph. Roll Rate: 4.0 sec., 30 deg. to 30 deg.

US AVIATION
265 Echo Lane • South St. Paul MN 55075
Tel. 612-450-0930 • Fax 612-450-0930
E-Mail: CumulusMan@aol.com
Web: http://members.aol.com/CumulusMan

VJ-24W SunFun

Wing Span	36.50 ft.	**Range**	30 sm
Wing Area	163.0 sq. ft.	**Takeoff Dist.**	100 ft.
Length	18.50 ft.	**Landing Dist.**	100 ft.
Max Gross Wt.	345 lbs.	**Vmax**	42 mph
Empty Wt.	165 lbs.	**VCr**	28 mph
Payload (full fuel)	N/A	**V_{S1} (stall clean)**	20 mph
Fuel Capacity	1 gal.	**V_{S0} (lndg config)**	N/A
Seats	1	**Climb Rate @msl**	350 fpm
Cabin Width	N/A	**Service Ceiling**	N/A

Std. Engine/HP	Usable HP Range	Construction Mtrl.	Est. Bldg. Time
Yamaha / 15 hp	N/A	Metal, Fabric	250 hrs.
Plans	**Kit**	**Info Packet**	**Video**
$100.00	N/A	$10.00	N/A

Originally a foot launched glider. Constructed primarily of aluminum tubing plus fabric. Materials cost estimated at around $2400. Pilot weight limited to 200 lbs.

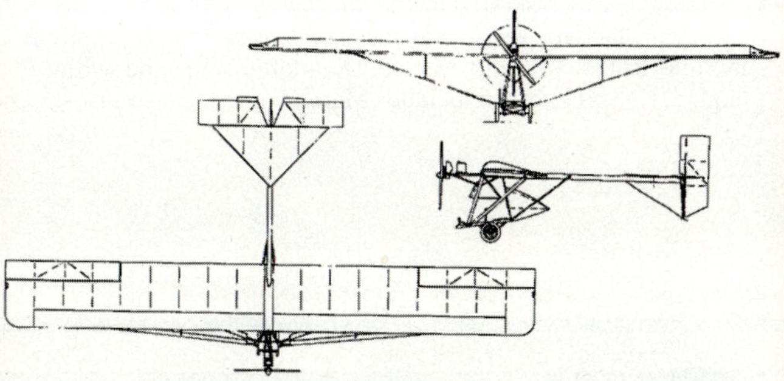

VOLMER AIRCRAFT
Box 5222
Glendale CA 91201
Tel. 818-247-8718

Windex 1200C

Wing Span	39.70 ft.	**Range**	320 sm
Wing Area	79.80 sq. ft.	**Takeoff Dist.**	800 ft.
Length	16.20 ft.	**Landing Dist.**	N/A
Max Gross Wt.	728 lbs.	**Vmax**	220 mph
Empty Wt.	386 lbs.	**VCr**	124 mph
Payload (full fuel)	342 lbs.	**V_{S1} (stall clean)**	50 mph
Fuel Capacity	6 gal.	**V_{S0} (lndg config)**	47 mph
Seats	1	**Climb Rate @msl**	570 fpm
Cabin Width	19 in.	**Service Ceiling**	13,000 ft.

Std. Engine/HP	Usable HP Range	Construction Mtrl.	Est. Bldg. Time
Konig 430cc / 24 hp	N/A	Fiberglass, Composite	1000 hrs.
Plans	**Kit**	**Info Packet**	**Video**
N/A	$39,000.00	Yes	Yes

Powered, high performance sailplane with max L/D of 38:1. Designed to JAR 22A with airframe stressed for aerobatic maneuvers. Also an efficient touring aircraft. A "pilot's airplane" with lively but pleasantly balanced control response. Stalling characteristics are docile. Kit includes engine, variable pitch propeller and pre-manufactured metal parts. Easy ground handling requires only one person.

Kits/Plans Sold: 14 **Completed & Flying:** 5

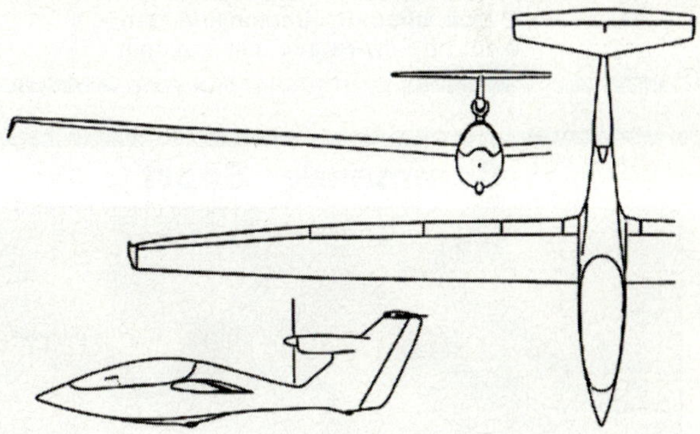

WINDEX AIRCRAFT CORP.
1121 Lewis Avenue • Sarasota FL 34237
Tel. 941-365-3860 • Fax 941-365-4431
E-Mail: brdsgn@interserv.com
Web: http://www.algonet.se/~windex/

Commander 147A

AIR COMMAND INTL., INC.
702 Cooper Drive • P.O. Box 1345
Wylie TX 75098
Tel. 972-442-6694 • Fax 972-442-9174
E-Mail: aircmd@aircommand.com
Web: http://www.aircommand.com

Rotor Span	31.00 ft.	**Range**	N/A
Rotor Area	N/A	**Takeoff Dist.**	N/A
Length	14.00 ft.	**Landing Dist.**	N/A
Max Gross Wt.	1500 lbs.	**Vmax**	120 mph
Empty Wt.	700 lbs.	**VCr**	85 mph
Payload (full fuel)	575 lbs.	**V_{S1} (stall clean)**	30 mph
Fuel Capacity	N/A	**V_{S0} (lndg config)**	N/A
Seats	2	**Climb Rate @msl**	N/A
Cabin Width	N/A	**Service Ceiling**	N/A

Std. Engine/HP	Usable HP Range	Construction Mtrl.	Est. Bldg. Time
Mazda 13B / 160 hp	N/A	N/A	N/A
Plans	**Kit**	**Info Packet**	**Video**
N/A	$30,000.00	$12.50	$25.00

A two seat, side-by-side gyro utilizing a 31' Skywheels main rotor and a 68" Warp Drive propeller. Kit price include engine. Available options include extended range tank, pre-rotator, electric trim and more.

Commander Tandem

AIR COMMAND INTL., INC.
702 Cooper Drive • P.O. Box 1345
Wylie TX 75098
Ph: 972-442-6694 • Fax: 972-442-9174
E-Mail: aircmd@aircommand.com
Web: http://www.aircommand.com

Rotor Span	25.00 ft.	**Range**	N/A
Rotor Area	N/A	**Takeoff Dist.**	N/A
Length	15.41 ft.	**Landing Dist.**	N/A
Max Gross Wt.	790 lbs.	**Vmax**	90 mph
Empty Wt.	330 lbs.	**VCr**	75 mph
Payload (full fuel)	460 lbs.	**V_{S1} (stall clean)**	20 mph
Fuel Capacity	N/A	**V_{S0} (lndg config)**	N/A
Seats	2	**Climb Rate @msl**	N/A
Cabin Width	N/A	**Service Ceiling**	N/A

Std. Engine/HP	Usable HP Range	Construction Mtrl.	Est. Bldg. Time
Rotax 582 / 65 hp	N/A	N/A	N/A
Plans	**Kit**	**Info Packet**	**Video**
N/A	$13,895.00	$12.50	$25.00

Two seat, tandem version of the Commander. Engine options include Rotax 582 and Mazda. Available with and without engine. Kit price shown includes Mazda engine.

Commander Sport

AIR COMMAND INTL., INC.
702 Cooper Drive • P.O. Box 1345
Wylie TX 75098
Tel. 972-442-6694 • Fax 972-442-9174
E-Mail: aircmd@aircommand.com
Web: http://www.aircommand.com

Rotor Span	23.00 ft.	**Range**	N/A
Rotor Area	N/A	**Takeoff Dist.**	100 ft.
Length	10.66 ft.	**Landing Dist.**	5 ft.
Max Gross Wt.	750 lbs.	**Vmax**	95 mph
Empty Wt.	275 lbs.	**VCr**	65 mph
Payload (full fuel)	475 lbs.	**V_{S1} (stall clean)**	8 mph
Fuel Capacity	9 gal.	**V_{S0} (lndg config)**	N/A
Seats	1	**Climb Rate @msl**	1200 fpm
Cabin Width	N/A	**Service Ceiling**	15,000 ft.

Std. Engine/HP	Usable HP Range	Construction Mtrl.	Est. Bldg. Time
Rotax 582 / 67 hp	N/A	Metal	30 hrs.
Plans	**Kit**	**Info Packet**	**Video**
N/A	$10,670.76	$12.50	$25.00

Single seat, all-aluminum triangulated frame gyro with black acrylic enamel finish. Steerable nose wheel. Engine options include Rotax 447, 503 and 582. Complete kit estimated to be buildable in three to four days. Available with and without engine. Kit price shown includes engine. Options include pre-rotator, fiberglass enclosure, wheel pants, extended range tank and more.

Bumble Bee

Rotor Span	23.00 ft.	**Range**	60 sm
Rotor Area	N/A	**Takeoff Dist.**	230 ft.
Length	8.00 ft.	**Landing Dist.**	N/A
Max Gross Wt.	500 lbs.	**Vmax**	64 mph
Empty Wt.	234 lbs.	**VCr**	40 mph
Payload (full fuel)	N/A	**V_{S1} (stall clean)**	N/A
Fuel Capacity	6 gal.	**V_{S0} (lndg config)**	N/A
Seats	1	**Climb Rate @msl**	1500 fpm
Cabin Width	N/A	**Service Ceiling**	N/A

Std. Engine/HP	Usable HP Range	Construction Mtrl.	Est. Bldg. Time
Rotax / 42 hp	N/A	Aluminum	500 hrs.
Plans	**Kit**	**Info Packet**	**Video**
$200.00	N/A	$29.00	$25.00

Single seat, open-frame gyroplane using an aluminum and composite main rotor. Manufacturer claims that it's the world's first ultralight gyro.

Sold: 280 **Completed & Flying:** 15

AIRCRAFT DESIGNS, INC.
5 Harris Court, Bldg. S
Monterey CA 93940
Tel. 408-649-6212 • Fax 408-649-5738

Sportster Gyro

Rotor Span	29.00 ft.	**Range**	90 sm
Rotor Area	N/A	**Takeoff Dist.**	350 ft.
Length	12.00 ft.	**Landing Dist.**	10 ft.
Max Gross Wt.	1100 lbs.	**Vmax**	100 mph
Empty Wt.	630 lbs.	**VCr**	70 mph
Payload (full fuel)	N/A	**V_{S1} (stall clean)**	N/A
Fuel Capacity	N/A	**V_{S0} (lndg config)**	N/A
Seats	2	**Climb Rate @msl**	700 fpm
Cabin Width	N/A	**Service Ceiling**	7,000 ft.

Std. Engine/HP	Usable HP Range	Construction Mtrl.	Est. Bldg. Time
Lyc 0-320 / 150 hp	N/A	Aluminum	1000 hrs.
Plans	**Kit**	**Info Packet**	**Video**
$250.00	N/A	N/A	$25.00

Two seat gyro design with a partially enclosed cockpit. First two seater in the Experimental category. First flew in 1976. Composite rotor blades.

Completed & Flying: 65

AIRCRAFT DESIGNS, INC.
5 Harris Court, Bldg. S
Monterey CA 93940
Tel. 408-649-6212 • Fax 408-649-5738

Double Eagle TT

Rotor Span	30.00 ft.	**Range**	300 sm
Rotor Area	N/A	**Takeoff Dist.**	200 ft.
Length	13.00 ft.	**Landing Dist.**	10 ft.
Max Gross Wt.	1150 lbs.	**Vmax**	100 mph
Empty Wt.	575 lbs.	**VCr**	70 mph
Payload (full fuel)	N/A	**V_{S1} (stall clean)**	N/A
Fuel Capacity	12 gal.	**V_{S0} (lndg config)**	N/A
Seats	2	**Climb Rate @msl**	1000 fpm
Cabin Width	N/A	**Service Ceiling**	14,000 ft.

Std. Engine/HP	Usable HP Range	Construction Mtrl.	Est. Bldg. Time
Subaru EA82 / 150 hp	N/A	Metal, Aluminum	100 hrs.
Plans	**Kit**	**Info Packet**	**Video**
$200.00	$16,900.00	$5.00	Yes

A two place gyroplane in a tandem configuration. The Subaru EA82 is a four cycle, liquid cooled engine that has 4.0 gph fuel consumption at 75%. Utilizes Amax Engineering's new "Tall Tail" technology.

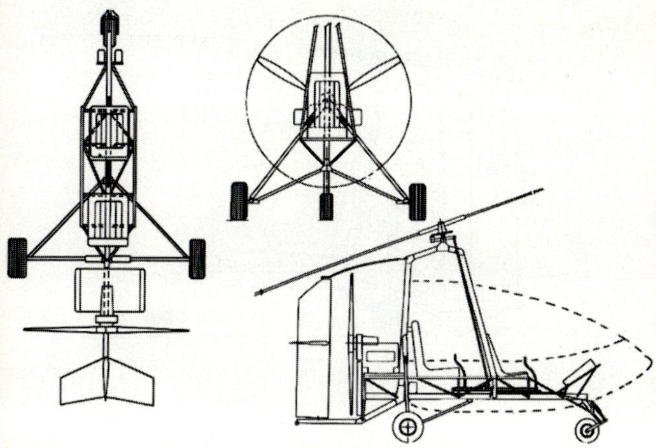

AMAX ENGINEERING
5 Utrecht Court
Donvale, Victoria, 3111
AUSTRALIA
Tel. 61-3-8423132 • Fax 61-3-8418177

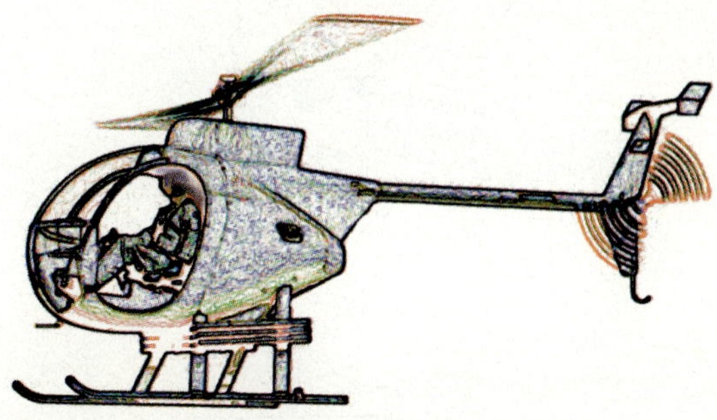

Eagle TT

Rotor Span	26.00 ft.	Range	300 sm
Rotor Area	N/A	Takeoff Dist.	100 ft.
Length	10.00 ft.	Landing Dist.	10 ft.
Max Gross Wt.	850 lbs.	Vmax	100 mph
Empty Wt.	410 lbs.	VCr	70 mph
Payload (full fuel)	N/A	V_{S1} (stall clean)	N/A
Fuel Capacity	12 gal.	V_{S0} (lndg config)	N/A
Seats	1	Climb Rate @msl	1500 fpm
Cabin Width	N/A	Service Ceiling	14,000 ft.

Std. Engine/HP	Usable HP Range	Construction Mtrl.	Est. Bldg. Time
Subaru EA81 / 110 hp	N/A	Metal, Aluminum	100 hrs.
Plans	**Kit**	**Info Packet**	**Video**
$200.00	$11,900.00	$5.00	N/A

Single place gyroplane utilizing Amax Engineering's new "Tall Tail" technology. Kit includes Subaru EA81 engine and basic instruments.

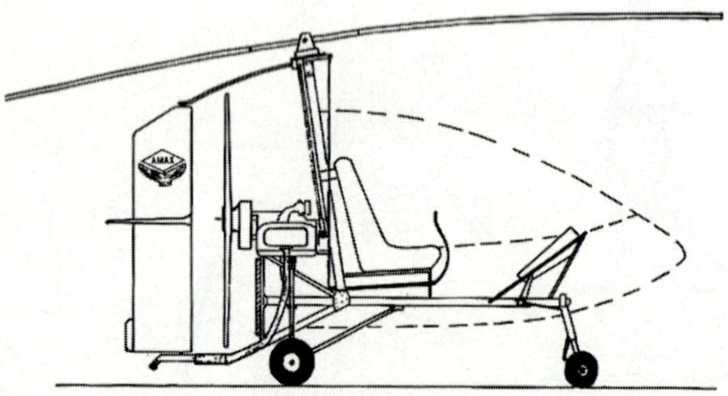

AMAX ENGINEERING
5 Utrecht Court
Donvale, Victoria, 3111
AUSTRALIA
Tel. 61-3-8423132 • Fax 61-3-8418177

Eagle

Rotor Span	26.00 ft.	Range	300 sm
Rotor Area	N/A	Takeoff Dist.	100 ft.
Length	10.00 ft.	Landing Dist.	10 ft.
Max Gross Wt.	850 lbs.	Vmax	90 mph
Empty Wt.	410 lbs.	VCr	70 mph
Payload (full fuel)	N/A	V_{S1} (stall clean)	N/A
Fuel Capacity	12 gal.	V_{S0} (lndg config)	N/A
Seats	1	Climb Rate @msl	1500 fpm
Cabin Width	N/A	Service Ceiling	14,000 ft.

Std. Engine/HP	Usable HP Range	Construction Mtrl.	Est. Bldg. Time
Subaru EA81 / 110 hp	N/A	Metal, Aluminum	100 hrs.
Plans	**Kit**	**Info Packet**	**Video**
$200.00	$11,500.00	$5.00	N/A

Single seat gyroplane. Kit includes Subaru EA81 engine and basic instruments. Most Outstanding Design award - 1992 P.R.A. Fly-In, Brookville, Ohio. Widely used in the Australian outback for herding live stock.

AMAX ENGINEERING
5 Utrecht Court
Donvale, Victoria, 3111
AUSTRALIA
Tel. 61-3-8423132 • Fax 61-3-8418177

UltraSport 331

Rotor Span	21.00 ft.	Range	150 sm
Rotor Area	N/A	Takeoff Dist.	N/A
Length	19.00 ft.	Landing Dist.	N/A
Max Gross Wt.	680 lbs.	Vmax	104 mph
Empty Wt.	330 lbs.	VCr	63 mph
Payload (full fuel)	320 lbs.	V_{S1} (stall clean)	N/A
Fuel Capacity	10 gal.	V_{S0} (lndg config)	N/A
Seats	1	Climb Rate @msl	1000 fpm
Cabin Width	N/A	Service Ceiling	12,000 ft.

Std. Engine/HP	Usable HP Range	Construction Mtrl.	Est. Bldg. Time
Hirth 2706 / 65 hp	N/A	Composite	60 hrs.
Plans	**Kit**	**Info Packet**	**Video**
N/A	$35,000.00	Yes	$15.00

Experimental single seat, fully enclosed helicopter. Overhead cyclic control operates with conventional control movements. Main rotor blades are 21 ft. diam. with 6.7" chord. Ring tail rotor blade is 2.6 ft. diam., 2" chord. Price includes doors and electric starter.

AMERICAN SPORTSCOPTER, INC.
875 Middle Ground Blvd.
Newport News VA 23606
Tel. 757-873-4914 • Fax 757-873-3711

UltraSport 254

Rotor Span	21.00 ft.	**Range**	90 sm
Rotor Area	N/A	**Takeoff Dist.**	N/A
Length	19.00 ft.	**Landing Dist.**	N/A
Max Gross Wt.	525 lbs.	**Vmax**	63 mph
Empty Wt.	252 lbs.	**VCr**	63 mph
Payload (full fuel)	273 lbs.	**V_{S1} (stall clean)**	N/A
Fuel Capacity	5 gal.	**V_{S0} (lndg config)**	N/A
Seats	1	**Climb Rate @msl**	1000 fpm
Cabin Width	N/A	**Service Ceiling**	12,000 ft.

Std. Engine/HP	Usable HP Range	Construction Mtrl.	Est. Bldg. Time
Hirth 2703 / 55 hp	N/A	Composite	60 hrs.
Plans	**Kit**	**Info Packet**	**Video**
N/A	$33,000.00	Yes	$15.00

True ultralight, single seat, partially enclosed helicopter. Meets FAA Part 103. Pilot's license not required. Composite body, main rotor and shrouded tail rotor. Overhead cyclic control operates with conventional control movements.

AMERICAN SPORTSCOPTER, INC.
875 Middle Ground Blvd.
Newport News VA 23606
Tel. 757-873-4914 • Fax 757-873-3711

Barnett J4B

Rotor Span	12.33 ft.	**Range**	110 sm
Rotor Area	N/A	**Takeoff Dist.**	350 ft.
Length	13.67 ft.	**Landing Dist.**	20 ft.
Max Gross Wt.	800 lbs.	**Vmax**	125 mph
Empty Wt.	512 lbs.	**VCr**	100 mph
Payload (full fuel)	N/A	**V_{S1} (stall clean)**	30 mph
Fuel Capacity	17 gal.	**V_{S0} (lndg config)**	N/A
Seats	1	**Climb Rate @msl**	1500 fpm
Cabin Width	N/A	**Service Ceiling**	12,000 ft.

Std. Engine/HP	Usable HP Range	Construction Mtrl.	Est. Bldg. Time
Cont O-200 / 100 hp	N/A	Metal, Tube, Fabric	350 hrs.
Plans	**Kit**	**Info Packet**	**Video**
$125.00	$12,000.00	$15.00	$25.00

A single seat craft with a fully or partially enclosed cockpit. Polycarbonate covered canopy. Streamlined fiberglass body shell. Subaru, Mazda, and Lycoming engines optional. Unfinished kit (without welding and machining completed) is $6,500.

BARNETT ROTORCRAFT
4307 Olivehurst Avenue
Olivehurst CA 95961
Tel. 916-742-7416 • Fax 916-743-6866

UltraSport 496

Rotor Span	23.00 ft.	**Range**	150 sm
Rotor Area	N/A	**Takeoff Dist.**	N/A
Length	20.00 ft.	**Landing Dist.**	N/A
Max Gross Wt.	1085 lbs.	**Vmax**	98 mph
Empty Wt.	495 lbs.	**VCr**	65 mph
Payload (full fuel)	590 lbs.	**V_{S1} (stall clean)**	N/A
Fuel Capacity	16 gal.	**V_{S0} (lndg config)**	N/A
Seats	2	**Climb Rate @msl**	1000 fpm
Cabin Width	N/A	**Service Ceiling**	12,000 ft.

Std. Engine/HP	Usable HP Range	Construction Mtrl.	Est. Bldg. Time
Hirth F-30 / 95 hp	N/A	Composite	60 hrs.
Plans	**Kit**	**Info Packet**	**Video**
N/A	$49,000.00	Yes	$15.00

Experimental or Ultralight Trainer. Two seat partial or fully enclosed helicopter. Composite body, main rotor and shrouded tail rotor. Overhead cyclic control. Full dual controls standard. Doors and ballistic parachute optional.

AMERICAN SPORTSCOPTER, INC.
875 Middle Ground Blvd.
Newport News VA 23606
Tel. 757-873-4914 • Fax 757-873-3711

Barnett J4B2

Rotor Span	25.33 ft.	**Range**	200 sm
Rotor Area	N/A	**Takeoff Dist.**	600 ft.
Length	13.70 ft.	**Landing Dist.**	25 ft.
Max Gross Wt.	1085 lbs.	**Vmax**	115 mph
Empty Wt.	577 lbs.	**VCr**	93 mph
Payload (full fuel)	N/A	**V_{S1} (stall clean)**	35 mph
Fuel Capacity	17 gal.	**V_{S0} (lndg config)**	N/A
Seats	2	**Climb Rate @msl**	500 fpm
Cabin Width	N/A	**Service Ceiling**	8,000 ft.

Std. Engine/HP	Usable HP Range	Construction Mtrl.	Est. Bldg. Time
Cont O-200 / 100 hp	N/A	Metal, Tube, Fabric	350 hrs.
Plans	**Kit**	**Info Packet**	**Video**
$175.00	$14,500.00	$15.00	$25.00

Enclosed tandem design. Kit price includes engine and instruments. Polycarbonate covered canopy. Streamlined fiberglass body shell. Subaru, Mazda, and Lyc. engines optional. Unfinished kit (without welding and machining completed) is $7500.

BARNETT ROTORCRAFT
4307 Olivehurst Avenue
Olivehurst CA 95961
Tel. 916-742-7416 • Fax 916-743-6866

KB-3 Gyroplane

Rotor Span	22.00 ft.	**Range**	75 sm
Rotor Area	N/A	**Takeoff Dist.**	300 ft.
Length	11.25 ft.	**Landing Dist.**	15 ft.
Max Gross Wt.	600 lbs.	**Vmax**	90 mph
Empty Wt.	250 lbs.	**VCr**	65 mph
Payload (full fuel)	N/A	**V_{S1} (stall clean)**	N/A
Fuel Capacity	5 gal.	**V_{S0} (lndg config)**	18 mph
Seats	1	**Climb Rate @msl**	1000 fpm
Cabin Width	N/A	**Service Ceiling**	10,000 ft.

Std. Engine/HP	Usable HP Range	Construction Mtrl.	Est. Bldg. Time
Rotax 582 / 65 hp	N/A	Metal	60 hrs.
Plans	**Kit**	**Info Packet**	**Video**
$125.00	$12,000.00	$7.00	$30.00

Completely unassembled kit. Design meets Part 103 ultralight requirements.

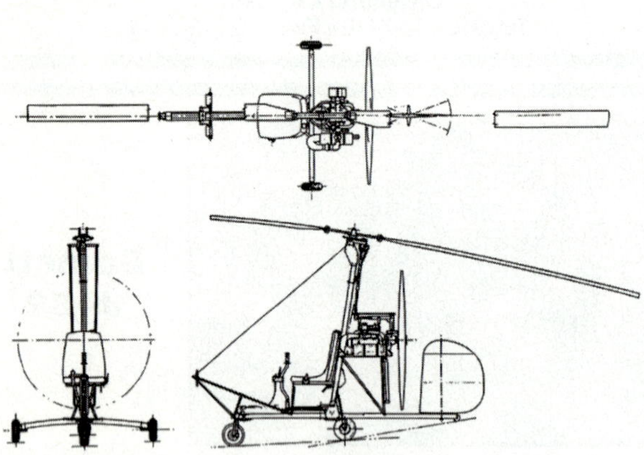

KEN BROCK MANUFACTURING, INC.
11852 Western Avenue
Stanton CA 90680
Tel. 714-898-4366 • Fax 714-894-0811

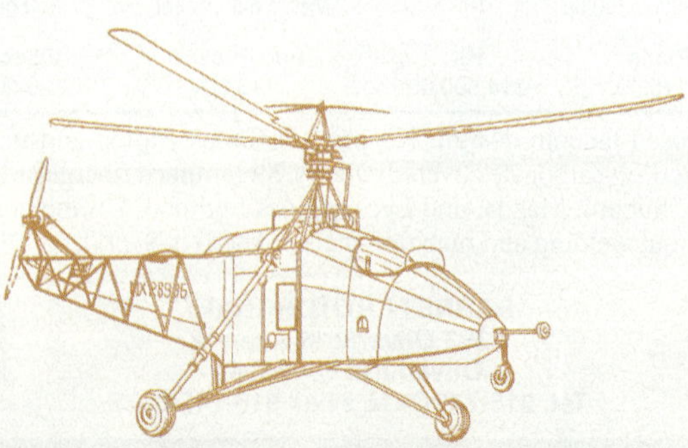

KB-2 Gyroplane

Rotor Span	22.00 ft.	**Range**	150 sm
Rotor Area	N/A	**Takeoff Dist.**	300 ft.
Length	12.00 ft.	**Landing Dist.**	10 ft.
Max Gross Wt.	600 lbs.	**Vmax**	95 mph
Empty Wt.	230 lbs.	**VCr**	75 mph
Payload (full fuel)	N/A	**V_{S1} (stall clean)**	N/A
Fuel Capacity	10 gal.	**V_{S0} (lndg config)**	18 mph
Seats	1	**Climb Rate @msl**	1200 fpm
Cabin Width	N/A	**Service Ceiling**	10,000 ft.

Std. Engine/HP	Usable HP Range	Construction Mtrl.	Est. Bldg. Time
McCulloch VW / 85 hp	72-90	Metal	60 hrs.
Plans	**Kit**	**Info Packet**	**Video**
$125.00	$12,000.00	$7.00	$30.00

A single seat open frame gyroplane. Nine different sub-kits are available. Uses a riveted aluminum rotor blade with an adjustable hub.

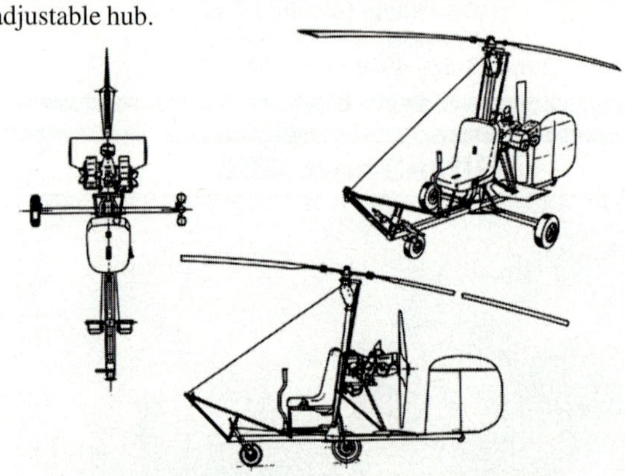

Ken Brock manufacturing has been a major supplier of Gyroplanes and homebuilt aircraft components since 1960. Originally, the company was the West Coast sales representative for Bensen Aircraft. Later, they developed their own designs, the KB-2 and KB-3. Contributions to the Gyroplane industry include a widely used, patented seat fuel tank, an aircraft type control system, and a quadrant throttle system.

Kas Thomas

"There may be a more exciting way to spend a Saturday than to buzz the back country at 75 mph, a mere 50 feet above the ground ... but if there is, I can't imagine it. When it comes to sheer out-and-out 'gotta do this some more' flying fun, the KB-2 Gyroplane is without equal."

KEN BROCK MANUFACTURING, INC.
11852 Western Avenue
Stanton CA 90680
Tel. 714-898-4366 • Fax 714-894-0811

Snobird Adventurer

Rotor Span	N/A	Range	205 sm
Rotor Area	N/A	Takeoff Dist.	150 ft.
Length	14.33 ft.	Landing Dist.	20 ft.
Max Gross Wt.	1187 lbs.	Vmax	90 mph
Empty Wt.	597 lbs.	VCr	65 mph
Payload (full fuel)	500 lbs.	V_{S1} (stall clean)	N/A
Fuel Capacity	15 gal.	V_{S0} (lndg config)	18 mph
Seats	2	Climb Rate @msl	1350 fpm
Cabin Width	N/A	Service Ceiling	11,000 ft.

Std. Engine/HP	Usable HP Range	Construction Mtrl.	Est. Bldg. Time
CAM 100	115	Composite, Aluminum	150 hrs.
Plans	**Kit**	**Info Packet**	**Video**
N/A	$19,995.00	$10.00	$20.00

Two place version. Can be used as a trainer for hire by an appropriately rated flight instructor when operated under the Popular Rotorcraft Assn's FAA exemption 5209B. Rotor is McCutchen composite or Dragon Wings aluminum, 26 feet diameter. Propeller is Warp Drive 3-blade. Airframe supplied factory jig drilled and powder coated. Unique main landing gear arrangement. Gyroplanes are generally easier to handle than a helicopter. Kit is FAA 51% approved. Kit includes instruments, night lights, and electric starter.

Kits Sold: 14 **Completed & Flying:** 9

CALUMET MOTORSPORTS INC.
PO Box 707
Lansing IL 60438
Tel. 708-895-0398 • Fax 708-895-0398
E-Mail: gyroplanes@aol.com
Web: http://members.aol.com/gyroplanes/index.html

Snobird Charger / Exciter

Rotor Span	23.00 ft.	Range	80 sm
Rotor Area	N/A	Takeoff Dist.	80 ft.
Length	11.92 ft.	Landing Dist.	20 ft.
Max Gross Wt.	670 lbs.	Vmax	90 mph
Empty Wt.	254 lbs.	VCr	70 mph
Payload (full fuel)	370 lbs.	V_{S1} (stall clean)	N/A
Fuel Capacity	8 gal.	V_{S0} (lndg config)	10 mph
Seats	1	Climb Rate @msl	1100 fpm
Cabin Width	N/A	Service Ceiling	12,000 ft.

Std. Engine/HP	Usable HP Range	Construction Mtrl.	Est. Bldg. Time
Rotax 582 / 67 hp	50 - 67	Composite, Aluminum	80 hrs.
Plans	**Kit**	**Info Packet**	**Video**
N/A	$12,950.00	$10.00	$20.00

Gyroplane design available either with a Rotax 503 engine as the Exciter 503 or with a Rotax 582 engine as the Charger. Specs are for the Charger 582. "Limited Time" pricing is $8995 for the Exciter and $9995 for the Charger. Contact manufacturer for current pricing. Airframe is supplied factory jig drilled and powder coated. Rotor is McCutchen composite or Dragon Wings aluminum, 23 feet diameter. Propeller is Warp Drive 3-blade. Unique main landing gear arrangement. Flight instruction by Gyroplane CFI.

Completed & Flying: 25

CALUMET MOTORSPORTS INC.
PO Box 707
Lansing IL 60438
Tel. 708-895-0398 • Fax 708-895-0398
E-Mail: gyroplanes@aol.com
Web: http://members.aol.com/gyroplanes/index.html

Baby Belle

Rotor Span	25.00 ft.	Range	200 sm
Rotor Area	N/A	Takeoff Dist.	N/A
Length	29.60 ft.	Landing Dist.	N/A
Max Gross Wt.	1450 lbs.	Vmax	100 mph
Empty Wt.	920 lbs.	VCr	85 mph
Payload (full fuel)	500 lbs.	V_{S1} (stall clean)	N/A
Fuel Capacity	28 gal.	V_{S0} (lndg config)	N/A
Seats	2	Climb Rate @msl	1000 fpm
Cabin Width	49 in.	Service Ceiling	10,000 ft.

Std. Engine/HP	Usable HP Range	Construction Mtrl.	Est. Bldg. Time
Lycoming / 160 hp	150 - 160	4130 Cr-Mo Tubing	1200 hrs.
Plans	**Kit**	**Info Packet**	**Video**
$150.00	$41,500.00	$20.00	$15.00

Engine not included in kit price. All gear driven, steel drive shafts with aluminum transmission housings. Oil bath lubrication with sight glasses. Design offers complete corrosion protection.

Stuart Fields, Inyokern CA

"Canadian Home Rotors Inc. has the best two seat kit helicopter and the best company ethics of any that I found during a seven month search involving both hands-on review flying and talking to other builders."

CANADIAN HOME ROTORS, INC.
P.O. Box 104
Ear Falls, Ontario P0V 1T0
CANADA
Tel. 807-222-2474 • Fax 807-222-2474

Helicycle

Rotor Span	20.00 ft.	**Range**	160 sm
Rotor Area	N/A	**Takeoff Dist.**	N/A
Length	18.50 ft.	**Landing Dist.**	N/A
Max Gross Wt.	672 lbs.	**Vmax**	105 mph
Empty Wt.	395 lbs.	**VCr**	90 mph
Payload (full fuel)	N/A	**V_{S1} (stall clean)**	N/A
Fuel Capacity	12 gal.	**V_{S0} (lndg config)**	N/A
Seats	1	**Climb Rate @msl**	900 fpm
Cabin Width	N/A	**Service Ceiling**	9,000 ft.

Std. Engine/HP	Usable HP Range	Construction Mtrl.	Est. Bldg. Time
Rotax 618 / 60 hp	55 - 70	Composite, Tube	N/A
Plans	**Kit**	**Info Packet**	**Video**
N/A	$14,750.00	$15.00	$18.00

A "compound helicopter" design utilizing optional stub wings and a McCullock 10 hp pusher engine on each wing. It can operate in three modes: conventional helicopter mode, compound mode using main rotor torque and pusher thrust, and autogyro mode using the pusher engines only. Specs are for helicopter mode. Wet wings provide two hours of fuel. Kit price does not include engines.

Completed & Flying: 1

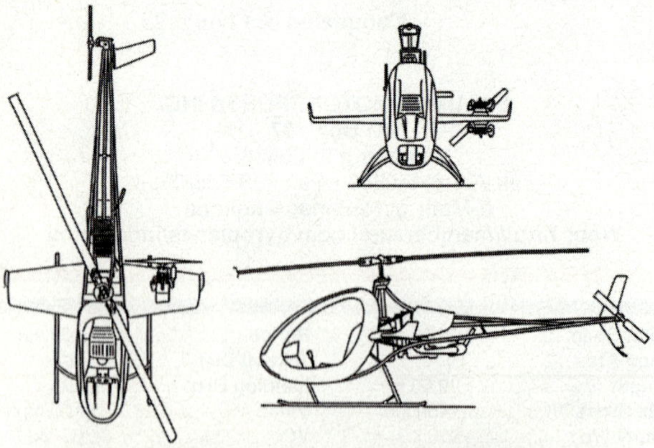

EAGLE RESEARCH AND DEVELOPMENT
2203 S 10th Ave.
Caldwell ID 83605
Tel. 208-466-6011

Eagle's Perch

Rotor Span	13.50 ft.	**Range**	100 sm
Rotor Area	N/A	**Takeoff Dist.**	N/A
Length	14.00 ft.	**Landing Dist.**	N/A
Max Gross Wt.	800 lbs.	**Vmax**	80 mph
Empty Wt.	480 lbs.	**VCr**	60 mph
Payload (full fuel)	N/A	**V_{S1} (stall clean)**	N/A
Fuel Capacity	10 gal.	**V_{S0} (lndg config)**	N/A
Seats	1	**Climb Rate @msl**	3000 fpm
Cabin Width	N/A	**Service Ceiling**	N/A

Std. Engine/HP	Usable HP Range	Construction Mtrl.	Est. Bldg. Time
Hirth 2706 / 50 hp	N/A	Prewelded Steel	240 hrs.
Plans	**Kit**	**Info Packet**	**Video**
N/A	$17,995.00	Yes	$35.00

Coaxial pendulum helicopter. Coaxial rotor blades. Doesn't have the complex components of a typical helicopter. Simple control system utilizes only stick, rudder, and throttle - similar to the controls of a fixed wing aircraft. No tail rotor, collective pitch or cyclic control are needed. Ballistic chute available. The Eagle's Perch has already won a number of different awards in the aviation community. It was judged "Grand Champion Helicopter" at the Popular Rotorcraft Association national convention in 1994.

This new type of rotary wing aircraft, the Coaxial Helicopter, was created in a backyard garage. The aircraft was developed by two brothers who had never been involved in aviation before they started developing their dream of flight. Because of their complete lack of aeronautical knowledge and training, they ended up creating an entirely new type of rotary wing aircraft. The difference between this and other helicopters is this helicopter doesn't have the complex components of a typical helicopter. It uses stick, rudder and throttle for control - similar to the controls of a fixed wing airplane. It is much easier to fly than a typical helicopter, and it performs all of the typical multi-dimensional rotary wing flight maneuvers. This aircraft can be used by anybody who requires vertical takeoff, landing, and hovering capabilities in their flight profile. It also has twin engines and, like a fixed wing airplane, will maintain flight on a single engine.

EAGLE'S PERCH, INC.
15268 Carrollton Blvd.
Carrollton VA 23314
Tel. 757-238-2820 • Fax 757-238-3268

18A Heliplane

Rotor Span	35.00 ft.	**Range**	200 sm
Rotor Area	N/A	**Takeoff Dist.**	150 ft.
Length	20.00 ft.	**Landing Dist.**	50 ft.
Max Gross Wt.	2000 lbs.	**Vmax**	110 mph
Empty Wt.	1350 lbs.	**VCr**	95 mph
Payload (full fuel)	482 lbs.	**V_{S1} (stall clean)**	N/A
Fuel Capacity	28 gal.	**V_{S0} (lndg config)**	N/A
Seats	2	**Climb Rate @msl**	800 fpm
Cabin Width	25 in.	**Service Ceiling**	10,000 ft.

Std. Engine/HP	Usable HP Range	Construction Mtrl.	Est. Bldg. Time
Lyc 0-360 / 180 hp	180-300	Tube, Alum., Wood, Glass	N/A
Plans	**Kit**	**Info Packet**	**Video**
N/A	$109,995.00	$20.00	N/A

Tandem seating gyroplane. Combines best attributes of a helicopter with the operating and maintenance costs of a fixed wing aircraft. Rotor is not powered in flight. Inflow of air in forward motion drives the rotor and provides lift. Full engine power is available to provide thrust. Sophisticated rotor and spin-up system enables takeoffs and landings with no ground roll. Enclosed cabin. An aerial platform unmatched for cost-effectiveness in patrol and short point-to-point missions where runways are not available for airplanes. An alternative to the light helicopter with VSTOL performance.

Kits/Plans Sold: 20 **Completed & Flying:** 20

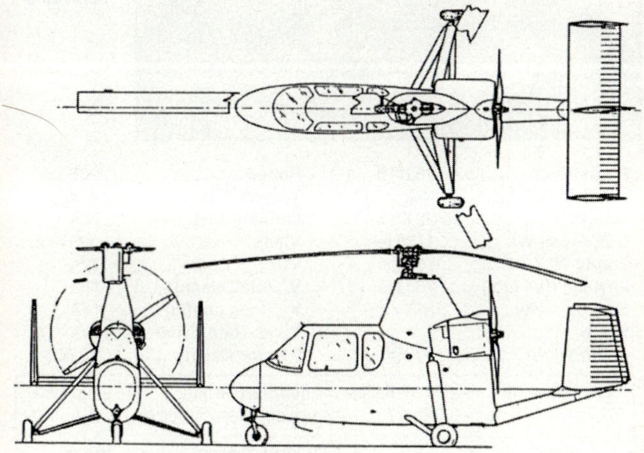

FARRINGTON AIRCRAFT
4460 Shemwell Lane
Paducah KY 42003
Tel. 502-898-2403 • Fax 502-898-8691

Twinstar Gyroplane

Rotor Span	30.00 ft.	**Range**	130 sm
Rotor Area	N/A	**Takeoff Dist.**	200 ft.
Length	13.00 ft.	**Landing Dist.**	50 ft.
Max Gross Wt.	1200 lbs.	**Vmax**	90 mph
Empty Wt.	700 lbs.	**VCr**	65 mph
Payload (full fuel)	480 lbs.	**V_{S1} (stall clean)**	N/A
Fuel Capacity	20 gal.	**V_{S0} (lndg config)**	N/A
Seats	2	**Climb Rate @msl**	1000 fpm
Cabin Width	24 in.	**Service Ceiling**	12,000 ft.

Std. Engine/HP	Usable HP Range	Construction Mtrl.	Est. Bldg. Time
Lyc O-320 / 150 hp	120 - 180	Tube, Alum., Fiberglass	200 hrs.
Plans	**Kit**	**Info Packet**	**Video**
N/A	$10,995.00	$25.00	$20.00

Tandem seating gyroplane. Classed as Experimental aircraft. Simplified version for recreational market. Same handling qualities and reliability as the 18A. Requires a short roll for takeoffs and landings. Total cost about $25,000. **Sold:** 25 **Flying:** 5

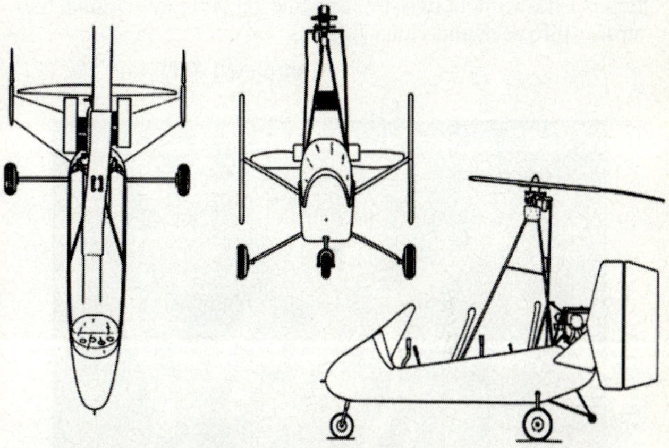

FARRINGTON AIRCRAFT
4460 Shemwell Lane
Paducah KY 42003
Tel. 502-898-2403 • Fax 502-898-8691

CH-7 Angel

Rotor Span	19.00 ft.	**Range**	170 sm
Rotor Area	N/A	**Takeoff Dist.**	N/A
Length	23.00 ft.	**Landing Dist.**	N/A
Max Gross Wt.	792 lbs.	**Vmax**	100 mph
Empty Wt.	451 lbs.	**VCr**	80 mph
Payload (full fuel)	301 lbs.	**V_{S1} (stall clean)**	N/A
Fuel Capacity	11 gal.	**V_{S0} (lndg config)**	N/A
Seats	1	**Climb Rate @msl**	1900 fpm
Cabin Width	32 in.	**Service Ceiling**	11,000 ft.

Std. Engine/HP	Usable HP Range	Construction Mtrl.	Est. Bldg. Time
Rotax 582 / 65 hp	65-80	Steel, Aluminum	200 hrs.
Plans	**Kit**	**Info Packet**	**Video**
N/A	Yes	Yes	Yes

Design utilizes a 19 ft. composite main rotor and a 40" metal tail rotor. Construction is steel trestle weld and bolted aluminum tubing. Instrument panel includes airspeed, altimeter, vertical speed, dual engine/rotor rpm and EGT/OT/EGT/WT. Kit price is 57,000 Italian Liras without two stroke engine; 63,900 Liras without four stroke. Info pack and video 70 Liras.

Completed & Flying: 50

U.S. Distributor:
ELISPORT USA, TEXAS
P.O. Box 152719
Dallas TX 75315
Ph: 972-222-2665

HELI-SPORT S.R.L.
Strada Traforo del Pino, 102
10132 Torino
ITALY
Tel. 39-11-899-6730 • Fax 39-11-899-5550

CH-7 Kompress

Rotor Span	20.30 ft.	**Range**	343 sm
Rotor Area	N/A	**Takeoff Dist.**	N/A
Length	24.60 ft.	**Landing Dist.**	N/A
Max Gross Wt.	992 lbs.	**Vmax**	129 mph
Empty Wt.	540 lbs.	**VCr**	100 mph
Payload (full fuel)	412 lbs.	**V_{S1} (stall clean)**	N/A
Fuel Capacity	11 gal.	**V_{S0} (lndg config)**	N/A
Seats	2	**Climb Rate @msl**	2000 fpm
Cabin Width	33 in.	**Service Ceiling**	16,000 ft.

Std. Engine/HP	Usable HP Range	Construction Mtrl.	Est. Bldg. Time
Rotax 914 T / 115 hp	N/A	Steel, Aluminum	250 hrs.
Plans	**Kit**	**Info Packet**	**Video**
N/A	Yes	Yes	Yes

Two place variant of the popular CH-7 Angel. Kit price, not including engine, is 79,900 Italian Liras.

HELI-SPORT S.R.L.
Strada Traforo del Pino, 102
10132 Torino
ITALY
Tel. 39-11-899-6730 • Fax 39-11-899-5550

Turbine Exec

Rotor Span	25.00 ft.	**Range**	225 sm
Rotor Area	N/A	**Takeoff Dist.**	N/A
Length	22.00 ft.	**Landing Dist.**	N/A
Max Gross Wt.	1485 lbs.	**Vmax**	125 mph
Empty Wt.	750 lbs.	**VCr**	N/A
Payload (full fuel)	735 lbs.	**V_{S1} (stall clean)**	N/A
Fuel Capacity	N/A	**V_{S0} (lndg config)**	N/A
Seats	2	**Climb Rate @msl**	N/A
Cabin Width	40 in.	**Service Ceiling**	13,500 ft.

Std. Engine/HP	Usable HP Range	Construction Mtrl.	Est. Bldg. Time
T62 Turbine/150 hp	150 - 170	Aluminum, Composite, Steel	N/A
Plans	**Kit**	**Info Packet**	**Video**
N/A	$20,000.00	Yes	N/A

Turbine retrofit for the Rotorway International Exec. Reduces empty weight and increases useful load. Shaft driven tail rotor with gear boxes. Includes 140 lbs. engine, gear reducer clutch, tail rotor shaft, main gear box, hangar bearings and fuel cell.

HILLBERG HELICOPTERS
PO Box 8974
Fountain Valley CA 92728
Tel. 909-279-5678

RotorMouse EH1-01

Rotor Span	20.00 ft.	**Range**	300 sm
Rotor Area	N/A	**Takeoff Dist.**	N/A
Length	20.10 ft.	**Landing Dist.**	N/A
Max Gross Wt.	1300 lbs.	**Vmax**	160 mph
Empty Wt.	635 lbs.	**VCr**	130 mph
Payload (full fuel)	665 lbs.	**V_{S1} (stall clean)**	N/A
Fuel Capacity	33 gal.	**V_{S0} (lndg config)**	N/A
Seats	1	**Climb Rate @msl**	2700 fpm
Cabin Width	24 in.	**Service Ceiling**	N/A

Std. Engine/HP	Usable HP Range	Construction Mtrl.	Est. Bldg. Time
Garret 36-55/145 hp	145 - 250	Aluminum	N/A
Plans	**Kit**	**Info Packet**	**Video**
$350.00	$58,000.00	$30.00	Yes

Turbine powered, single place helicopter. All-aluminum construction utilizing a semi-monocoque airframe using 2024-T3 aluminum skin, bulkheads and keel. Crash resistant fuel cell. Pro-Seal full service electrical system. Video included in info pack.

Kits/Plans Sold: 12 **Completed & Flying:** 1

HILLBERG HELICOPTERS
PO Box 8974
Fountain Valley CA 92728
Tel. 909-279-5678

LW-2

Rotor Span	26.00 ft.	**Range**	N/A
Rotor Area	N/A	**Takeoff Dist.**	200 ft.
Length	18.00 ft.	**Landing Dist.**	20 ft.
Max Gross Wt.	700 lbs.	**Vmax**	85 mph
Empty Wt.	350 lbs.	**VCr**	70 mph
Payload (full fuel)	300 lbs.	**V_{S1} (stall clean)**	N/A
Fuel Capacity	9 gal.	**V_{S0} (lndg config)**	N/A
Seats	1	**Climb Rate @msl**	1000 fpm
Cabin Width	24 in.	**Service Ceiling**	10,000 ft.

Std. Engine/HP	Usable HP Range	Construction Mtrl.	Est. Bldg. Time
2SI 540-L-70 / 70 hp	50 - 90	4130 Steel, Fabric	200 hrs.
Plans	**Kit**	**Info Packet**	**Video**
$175.00	N/A	$5.00	$10.00

Single seat, fully enclosed, tractor autogyro. Controlled by conventional elevator in pitch, and by rotorhead tilt in lateral axis. Ground adjustable pitch axis trim. A two-place version, LW-2+2, is also offered. **Kits/Plans Sold:** 13 **Flying:** 3

LITTLE WING AUTOGYRO, INC.
746 Hiway 89 N.
Mayflower AR 72106
Tel. 501-470-7444 • E-Mail: rotopup@aol.com
Web: http://www.pra.org

LW-3

Rotor Span	25.00 ft.	**Range**	200 sm
Rotor Area	N/A	**Takeoff Dist.**	200 ft.
Length	18.00 ft.	**Landing Dist.**	20 ft.
Max Gross Wt.	700 lbs.	**Vmax**	90 mph
Empty Wt.	300 lbs.	**VCr**	75 mph
Payload (full fuel)	350 lbs.	**V_{S1} (stall clean)**	20 mph
Fuel Capacity	9 gal.	**V_{S0} (lndg config)**	N/A
Seats	1	**Climb Rate @msl**	800 fpm
Cabin Width	24 in.	**Service Ceiling**	10,000 ft.

Std. Engine/HP	Usable HP Range	Construction Mtrl.	Est. Bldg. Time
2SI 540-L-70 / 70 hp	50 - 100	4130 chromoly steel & fabric	200 hrs.
Plans	**Kit**	**Info Packet**	**Video**
$175.00	N/A	$5.00	$10.00

Single seat, fully enclosed, tractor autogyro. Controlled by fully-tilting rotor head, dampened by large horizontal stabilizer. Pylon is adjustable and foldable. Shock absorbing gear is optional. Plans only. Fuselage available at $2,500. **Sold:** 2 **Flying:** 1

LITTLE WING AUTOGYRO, INC.
746 Hiway 89 N.
Mayflower AR 72106
Tel. 501-470-7444 • E-Mail: rotopup@aol.com
Web: http://www.pra.org

Roto-Pup U/L

Rotor Span	23.00 ft.	**Range**	50 sm
Rotor Area	N/A	**Takeoff Dist.**	200 ft.
Length	16.00 ft.	**Landing Dist.**	50 ft.
Max Gross Wt.	550 lbs.	**Vmax**	64 mph
Empty Wt.	250 lbs.	**VCr**	50 mph
Payload (full fuel)	250 lbs.	**V_{S1} (stall clean)**	N/A
Fuel Capacity	5 gal.	**V_{S0} (lndg config)**	N/A
Seats	1	**Climb Rate @msl**	800 fpm
Cabin Width	24 in.	**Service Ceiling**	10,000 ft.

Std. Engine/HP	Usable HP Range	Construction Mtrl.	Est. Bldg. Time
2SI 460 / 45 hp	45 - 70	4130 Steel, Fabric	200 hrs.
Plans	**Kit**	**Info Packet**	**Video**
N/A	$8,995.00	$5.00	$10.00

Ultralight single seat, partially enclosed, tractor autogyro. UL kit includes engine, prop, rotor system. Pre-welded fuselage and components available as options. Experimental model available with TEC 70hp engine. Cruise speed- 65 mph, range-100sm. Plans $175. Complete Kit $15,000.

Kits/Plans Sold: 13 **Completed & Flying:** 3

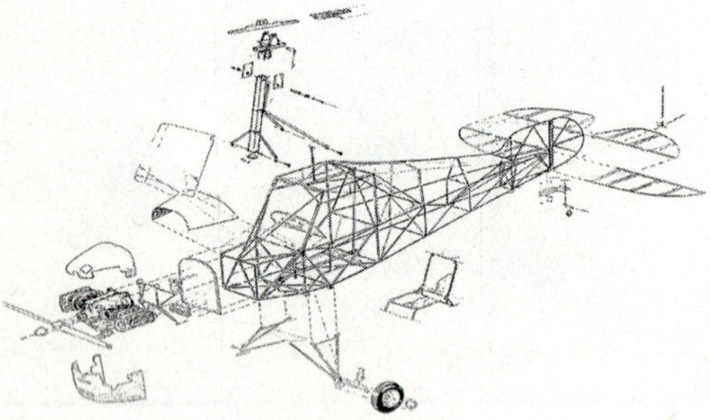

The Roto-Pups are second generation, tractor-configured, true autogyros. They are remniniscent of the old Cierva, Pitcairn design. The Roto Pup U/L was built using the time proven Preceptor Ultra-Pup airplane as a guideline.

The Pups utilize a unique control system originally used by David Kay in the early 1930's. This system controls airframe longitudinal attitude (pitch) by using a conventional horizontal stabilizer and elevator combination. Lateral control (roll) of the aircraft is accomplished by tilting the entire rotor head on the longitudinal axis. This method causes the rotor thrust vector to be offset from center and causes the craft to turn in the desired direction. Control is positive and not dependent on airspeed.

LITTLE WING AUTOGYRO, INC.
746 Hiway 89 N.
Mayflower AR 72106
Tel. 501-470-7444 • E-Mail: rotopup@aol.com
Web: http://www.pra.org

Zodiac Talon-Turbine

Rotor Span	25.00 ft.	**Range**	240 sm
Rotor Area	N/A	**Takeoff Dist.**	N/A
Length	31.00 ft.	**Landing Dist.**	N/A
Max Gross Wt.	1425 lbs.	**Vmax**	130 mph
Empty Wt.	600 lbs.	**VCr**	95 mph
Payload (full fuel)	N/A	**V_{S1} (stall clean)**	N/A
Fuel Capacity	48 gal.	**V_{S0} (lndg config)**	N/A
Seats	2	**Climb Rate @msl**	1100 fpm
Cabin Width	42 in.	**Service Ceiling**	15,000 ft.

Std. Engine/HP	Usable HP Range	Construction Mtrl.	Est. Bldg. Time
Turbine / 190 hp	95 - 225	Steel, Alum., Composite	1100 hrs.
Plans	**Kit**	**Info Packet**	**Video**
$225.00	N/A	$10.00	N/A

Proven airframe design utilizing state of the art high tech turbine engines. Turbines are preferred for small helicopter designs because they provide a steady power source to the rotors and have the ability to operate at high temperatures for extended periods. Specifications shown are with a Sunstrand 190 hp Turbine. Uses kerosene fuel. Rated at 2,000 hours TBO. This design can use custom main and rotor components or standard parts from Rotorway, Hughes, or Robinson. Current prototype has 1,200 hours flying time and is being utilized overseas in agricultural activities.

Kits/Plans Sold: 400 **Completed & Flying:** 15

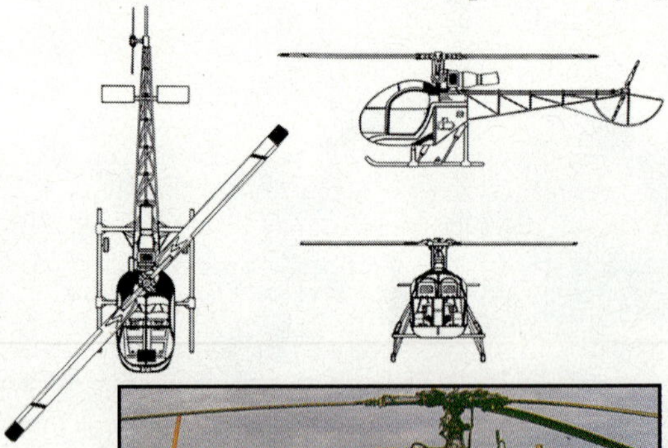

Victor Maxwell, AG Pilot

"The Zodiac Talon-Turbine has been a valuable asset to our agricultural research program. It has ample room for two pilots with photo or video equipment. With a seat removed, optional spray equipment can easily be installed in the cabin and between the skids."

MIDWEST ENGINEERING
PO Box 4706
Overland Park KS 66204-0706
Tel. 913-631-8447
E-Mail: mwengineering@juno.com

Pit Bull

Rotor Span	25.00 ft.	**Range**	85 sm
Rotor Area	N/A	**Takeoff Dist.**	100 ft.
Length	13.50 ft.	**Landing Dist.**	5 ft.
Max Gross Wt.	515 lbs.	**Vmax**	90 mph
Empty Wt.	254 lbs.	**VCr**	55 mph
Payload (full fuel)	N/A	**V_{S1} (stall clean)**	N/A
Fuel Capacity	5 gal.	**V_{S0} (lndg config)**	N/A
Seats	1	**Climb Rate @msl**	1050 fpm
Cabin Width	N/A	**Service Ceiling**	10,000 ft.

Std. Engine/HP	Usable HP Range	Construction Mtrl.	Est. Bldg. Time
Rotax 447 / 40 hp	N/A	Tube, Composite	30 hrs.
Plans	**Kit**	**Info Packet**	**Video**
N/A	$4,900.00	Yes	$15.00

Engine and propeller not included in kit price. Flying since 1993. Custom designed 6061-T6 cradle-bed engine mount. 4130 Cr-Mo down struts with elastomeric strut shock. Also available with a higher performance Subaru EA-81 engine.

Completed & Flying: 5

NORTH AMERICAN ROTORWERKS
PO Box 1084
Marysville WA 98270
Tel. 206-768-9826 • Fax 206-486-1694

Warrior

Rotor Span	21.00 ft.	**Range**	180 sm
Rotor Area	N/A	**Takeoff Dist.**	N/A
Length	15.00 ft.	**Landing Dist.**	N/A
Max Gross Wt.	850 lbs.	**Vmax**	100 mph
Empty Wt.	437 lbs.	**VCr**	80 mph
Payload (full fuel)	400 lbs.	**V_{S1} (stall clean)**	N/A
Fuel Capacity	14 gal.	**V_{S0} (lndg config)**	N/A
Seats	1	**Climb Rate @msl**	N/A
Cabin Width	N/A	**Service Ceiling**	10,000 ft.

Std. Engine/HP	Usable HP Range	Construction Mtrl.	Est. Bldg. Time
Hirth 2706 / 65 hp	N/A	Steel/Alum. Tube	80 hrs.
Plans	**Kit**	**Info Packet**	**Video**
N/A	$15,500.00	Yes	$15.00

Experimental helicopter. Kit price does not include engine. Main and tail rotors are one-piece aluminum extrusions. All major assemblies are preassembled. Simply bolt three major subassemblies together. All drive components are supplied ready to install. Detailed instruction video provided. Full cockpit enclosure with removable doors is optional. Two place version will be available in 1998.

Completed & Flying: 1

PAWNEE AVIATION INC.
505A Weaver Park Rd.
Longmont CO 80501
Tel. 303-682-5272 • Fax 303-682-9399

Raven Explorer I

Rotor Span	23.00 ft.	**Range**	100 sm
Rotor Area	N/A	**Takeoff Dist.**	250 ft.
Length	15.00 ft.	**Landing Dist.**	50 ft.
Max Gross Wt.	564 lbs.	**Vmax**	65 mph
Empty Wt.	254 lbs.	**VCr**	N/A
Payload (full fuel)	310 lbs.	**V_{S1} (stall clean)**	N/A
Fuel Capacity	5 gal.	**V_{S0} (lndg config)**	N/A
Seats	1	**Climb Rate @msl**	N/A
Cabin Width	28 in.	**Service Ceiling**	12,000 ft.

Std. Engine/HP	Usable HP Range	Construction Mtrl.	Est. Bldg. Time
AMW / 45 hp	45-65	Aluminum tube, Fiberglass	100 hrs.
Plans	**Kit**	**Info Packet**	**Video**
N/A	$9,995.00	Yes	$20.00

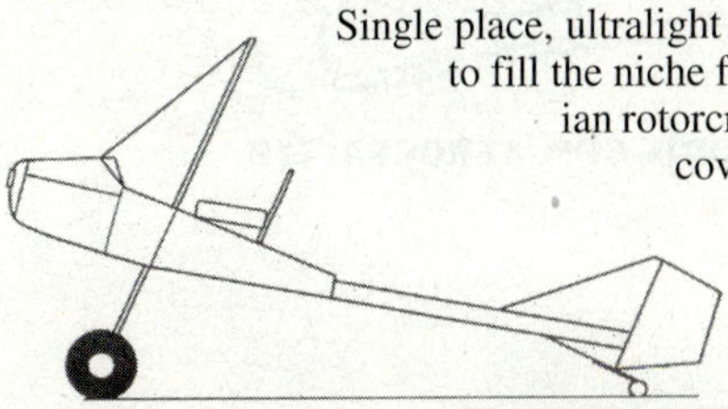

Single place, ultralight tractor gyro designed to fill the niche for a low cost, utilitarian rotorcraft. An enclosure will cover the front half of the craft, creating significant protection from prop draft and the elements.

RAVEN ROTORCRAFT, INC.
1025 Rosewood Ave., Suite 104
Boulder CO 80304
Tel. 303-440-6234 • Fax 303-440-4085
E-Mail: redrive@raven-rotor
Web: http://www.raven-rotor.com

Raven Explorer II

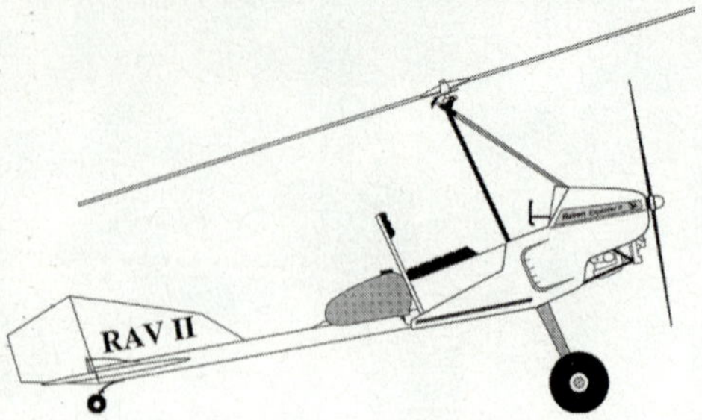

Rotor Span	27.00 ft.	Range	N/A
Rotor Area	N/A	Takeoff Dist.	400 ft.
Length	15.80 ft.	Landing Dist.	50 ft.
Max Gross Wt.	890 lbs.	Vmax	100 mph
Empty Wt.	365 lbs.	VCr	N/A
Payload (full fuel)	469 lbs.	V_{S1} (stall clean)	N/A
Fuel Capacity	8 gal.	V_{S0} (lndg config)	N/A
Seats	2	Climb Rate @msl	700 fpm
Cabin Width	28 in.	Service Ceiling	12,000 ft.

Std. Engine/HP	Usable HP Range	Construction Mtrl.	Est. Bldg. Time
Geo Metro / 90 hp	90-120	Aluminum tube, Fiberglass	100 hrs.
Plans	**Kit**	**Info Packet**	**Video**
N/A	$16,995.00	N/A	$20.00

Two place, tandem, tractor gyro with three cylinder or four cylinder supercharged Geo Metro/Suzuki engine.

RAVEN ROTORCRAFT, INC.
1025 Rosewood Ave., Suite 104
Boulder CO 80304
Tel. 303-440-6234 • Fax 303-440-4085
E-Mail: redrive@raven-rotor
Web: http://www.raven-rotor.com

Mini-500B

Rotor Span	19.17 ft.	Range	225 sm
Rotor Area	N/A	Takeoff Dist.	N/A
Length	22.50 ft.	Landing Dist.	N/A
Max Gross Wt.	840 lbs.	Vmax	120 mph
Empty Wt.	485 lbs.	VCr	75 mph
Payload (full fuel)	355 lbs.	V_{S1} (stall clean)	N/A
Fuel Capacity	15 gal.	V_{S0} (lndg config)	N/A
Seats	1	Climb Rate @msl	1100 fpm
Cabin Width	63 in.	Service Ceiling	10,000 ft.

Std. Engine/HP	Usable HP Range	Construction Mtrl.	Est. Bldg. Time
Rotax 582 / 67 hp	N/A	Composite, Tube	60 hrs.
Plans	**Kit**	**Info Packet**	**Video**
N/A	$24,500.00	$15.00	$30.00

Upgrade of the well-known Mini-500. Improved tail rotor tips, main transmission, rotorhead, airframe and more. Suitable for numerous civilian, law enforcement and military as well as recreational applications. Its patented "Triple Gimmel System" eliminates the complicated swashplate, allowing for more simplified assembly and maintenance. Mini-500s are now flying in South Africa, Spain, England, Canada, Australia, France, Italy, Brazil, Sri Lanka, New Zealand, and the Netherlands. Complete kit includes engine. **Kits/Plans Sold:** 415

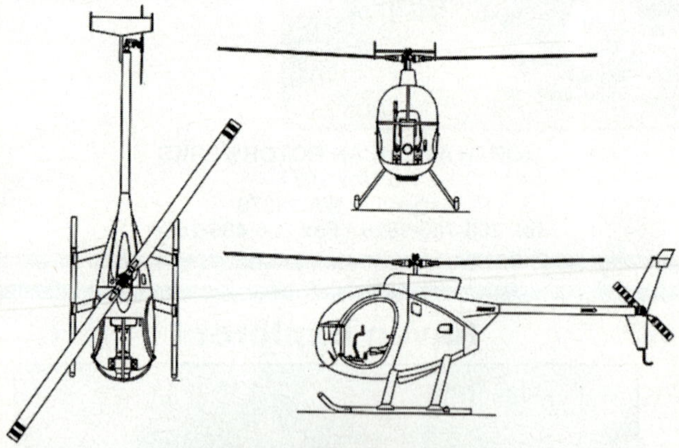

REVOLUTION HELICOPTER CORP., INC.
1905 W. Jesse James Road
Excelsior Springs MO 64024
Tel. 816-637-2800 • Fax 816-637-7936
E-Mail: sales@revolutionhelicopter.com
Web: http://www.revolutionhelicopter.com

Exec 162F

Rotor Span	25.00 ft.	Range	180 sm
Rotor Area	N/A	Takeoff Dist.	N/A
Length	22.00 ft.	Landing Dist.	N/A
Max Gross Wt.	1500 lbs.	Vmax	115 mph
Empty Wt.	975 lbs.	VCr	95 mph
Payload (full fuel)	425 lbs.	V_{S1} (stall clean)	N/A
Fuel Capacity	17 gal.	V_{S0} (lndg config)	N/A
Seats	2	Climb Rate @msl	1000 fpm
Cabin Width	40 in.	Service Ceiling	10,000 ft.

Std. Engine/HP	Usable HP Range	Construction Mtrl.	Est. Bldg. Time
RI 162F / 150 hp	N/A	Aluminum, Composite, Steel	300 hrs.
Plans	**Kit**	**Info Packet**	**Video**
Yes	$60,850.00	$15.00	$15.00

Mature proven design. Kit supplied complete. Powerplant utilizes electronic fuel injection and electronic ignition (FADEC). All have complete backup systems. Options include floats and an additional cargo holder that mounts between the landing skids. Info and video together is $25.00. See Ad on Page 8.

Kits/Plans Sold: 500 **Completed & Flying:** 300

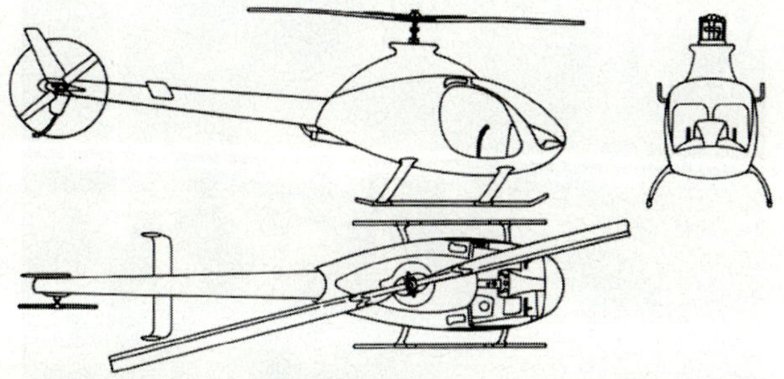

Ken Armstrong, Kit Planes Magazine
"The Exec Series has graduated ... to a real helicopter with excellent handling characteristics ..."

Mike Glair, Sport Pilot Magazine
"With sleek, great flying designs that win award after award, and get built in short order by their customers, there's no doubt who's number one in this segment of the industry."

Nigel Moll, Flying Magazine
"The Rotorway Exec 162F is a legitimate helicopter with good flying qualities and decent performance at a common man price."

ROTORWAY INTERNATIONAL
4140 W. Mercury Way
Chandler AZ 85226
Tel. 602-961-1001 • Fax 602-961-1514
E-Mail: rotorway@primenet.com

RAF 2000

Rotor Span	30.00 ft.	Range	420 sm
Rotor Area	N/A	Takeoff Dist.	75 ft.
Length	13.60 ft.	Landing Dist.	10 ft.
Max Gross Wt.	1125 lbs.	Vmax	100 mph
Empty Wt.	560 lbs.	VCr	75 mph
Payload (full fuel)	565 lbs.	V_{S1} (stall clean)	N/A
Fuel Capacity	23 gal.	V_{S0} (lndg config)	N/A
Seats	2	Climb Rate @msl	1200 fpm
Cabin Width	N/A	Service Ceiling	10,000 ft.

Std. Engine/HP	Usable HP Range	Construction Mtrl.	Est. Bldg. Time
EA 82 Subaru/100 hp	100 - 130	Metal	150 hrs.
Plans	**Kit**	**Info Packet**	**Video**
N/A	$19,500.00	$12.00	$20.00

Available in standard and GTX SE versions. Specs are for the standard version. The GTX SE includes a variety of performance enhancing options for $21,500. Engine, propeller, and instrument package included in kit. Glide ratio is 4:1. Endurance is 4.6 hours with a 0.5 hour reserve. See Ad on Page 3.

ROTARY AIR FORCE MKTG., INC.
Box 1236A • 1107 9th St. W. • Kindersley, Sask. S0L 1S0
CANADA
Tel. 306-463-6030 • Fax 306-463-6032
E-Mail: raf2000@sk.sympatico.ca
Web: http://www.raf2000.com

Dominator 1-Place

Rotor Span	24.00 ft.	**Range**	N/A
Rotor Area	N/A	**Takeoff Dist.**	50 ft.
Length	10.40 ft.	**Landing Dist.**	N/A
Max Gross Wt.	750 lbs.	**Vmax**	90 mph
Empty Wt.	375 lbs.	**VCr**	50 mph
Payload (full fuel)	275 lbs.	**V_{S1} (stall clean)**	N/A
Fuel Capacity	12 gal.	**V_{S0} (lndg config)**	N/A
Seats	1	**Climb Rate @msl**	1700 fpm
Cabin Width	20 in.	**Service Ceiling**	12,000 ft.

Std. Engine/HP	Usable HP Range	Construction Mtrl.	Est. Bldg. Time
Rotax 503 / 65 hp	65 - 74	Steel, Aluminum	500 hrs.
Plans	**Kit**	**Info Packet**	**Video**
$175.00	$16,000.00	Yes	$27.95

High performance gyroplane. Very stable even at high speed. Features centerline thrust, tall tail with horizontal stabilizer,and short and rough field capability. Full suspension landing gear. On-board computer with 2" x 3" LCD screen monitors all vital engine functions. Open frame made of bolted aluminum tubing. Performance Edition available. Build from plans or purchase kit subassemblies. For info pack send SASE.

Kits/Plans Sold: 200 **Completed & Flying:** 35

ROTOR FLIGHT DYNAMICS
19242 Grange Hall Loop
Lithia FL 33547
Tel. 813-634-3370 • Fax 813-634-3370

Dominator 2-Place

Rotor Span	27.00 ft.	**Range**	N/A
Rotor Area	N/A	**Takeoff Dist.**	400 ft.
Length	12.40 ft.	**Landing Dist.**	18 ft.
Max Gross Wt.	950 lbs.	**Vmax**	N/A
Empty Wt.	366 lbs.	**VCr**	N/A
Payload (full fuel)	525 lbs.	**V_{S1} (stall clean)**	N/A
Fuel Capacity	8 gal.	**V_{S0} (lndg config)**	N/A
Seats	2	**Climb Rate @msl**	750 fpm
Cabin Width	N/A	**Service Ceiling**	10,000 ft.

Std. Engine/HP	Usable HP Range	Construction Mtrl.	Est. Bldg. Time
Rotax / 74 hp	N/A	Steel, Aluminum	600 hrs.
Plans	**Kit**	**Info Packet**	**Video**
Yes	Yes	N/A	N/A

Two place, high performance gyroplane. Performs like a single place gyro while carrying two people. Approved and used by many CFI's for training. Can be flown single by students. Great for open air flying. On-board computer with 2" x 3" LCD screen monitors all vital engine functions. Open frame made of bolted aluminum tubing. Performance Edition available. Build from plans or purchase kit subassemblies.

Kits/Plans Sold: 10 **Completed & Flying:** 4

ROTOR FLIGHT DYNAMICS
19242 Grange Hall Loop
Lithia FL 33547
Tel. 813-634-3370 • Fax 813-634-3370

HOMEBUILT "FAQ'S" *(See Page 68)*

Q: Am I safe flying a homebuilt airplane?

A: You are as safe as you want to be. Little definitive data exists comparing homebuilts to certified aircraft. One set of opinions holds that accident rates are about the same for both certified and homebuilt aircraft, once the test period for the homebuilt is complete. Certified aircraft seem to have more unintentional IMC accidents, while homebuilts fare worse in accidents resulting from over-stress from aerobatics, forced landings, etc.

Safety, as in all aspects of aviation, is still a function of the pilot's level of training and ability to make intelligent decisions.

3D-RV Gyroplane

Rotor Span	25.00 ft.	**Range**	100 sm
Rotor Area	N/A	**Takeoff Dist.**	1000 ft.
Length	8.50 ft.	**Landing Dist.**	N/A
Max Gross Wt.	550 lbs.	**Vmax**	75 mph
Empty Wt.	250 lbs.	**VCr**	50 mph
Payload (full fuel)	300 lbs.	**V_{S1} (stall clean)**	N/A
Fuel Capacity	5 gal.	**V_{S0} (lndg config)**	10 mph
Seats	1	**Climb Rate @msl**	800 fpm
Cabin Width	N/A	**Service Ceiling**	8,000 ft.

Std. Engine/HP	Usable HP Range	Construction Mtrl.	Est. Bldg. Time
Rotax 503 / 50 hp	50 - 65	Aluminum	400 hrs.
Plans	**Kit**	**Info Packet**	**Video**
$100.00	N/A	$7.00	$15.00

Single seat, open-frame, ultralight gyroplane. Classed as FAR 103 Ultralight with Rotax 503 or 582; "Experimental" with other engines. Constructed from scratch with aluminum and 4130 steel. This gyro features hydraulic disk brakes and a light weight aluminum tall tail that provides excellent yaw control. It can be flown with aluminum manufactured rotor blades or with the 'Gyrotor' wood rotor blade design.

Kits/Plans Sold: 35 **Completed & Flying:** 3

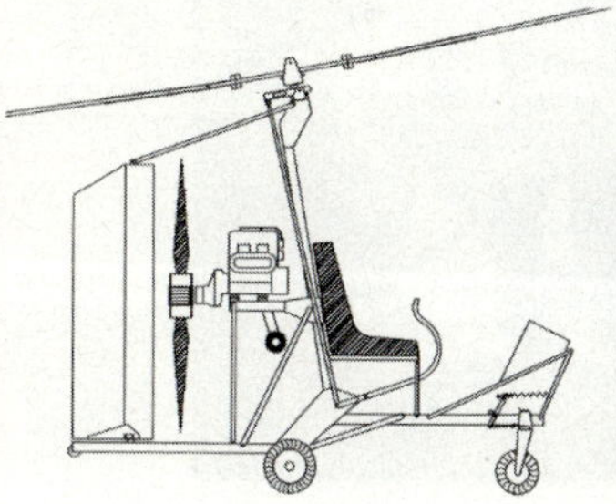

First prototype was built in 1989. It used a Max engine and flew with wood rotor blades which were only 8 feet in length with a 5 foot hub bar for a rotor diameter of 21 feet. Flight characteristics were excellent but it lacked lift on hot days.

Next, the keel was split near the mast to raise the front portion of the aircraft and pilots weight up so the center of gravity was in line with the prop thrust, providing stability in pitch. The tail design was chosen for its exceptional flight characteristics. The Rotax 503 is used because of its history of reliability. The current prototype performs very well with a 64 inch two blade prop but the plans built version will employ a 56 inch three blade prop.

Greg Hardy, Salt Lake City, UT
"... with occasional assistance from Monte (Hoskins), it's only taken me 7 months of spare time to build my gyro and I'll be ready for flying next month... if I wasn't out watching Monte fly his gyro, I could get more work done on mine..."

ROTORWING-AERO
PO Box 18614
Salt Lake City UT 84118
Tel. 801-969-1357 • Fax 801-969-1357
E-Mail: mhoskins@aros.net

Skytwister "Choppy"

Rotor Span	21.50 ft.	**Range**	N/A
Rotor Area	N/A	**Takeoff Dist.**	N/A
Length	15.75 ft.	**Landing Dist.**	N/A
Max Gross Wt.	670 lbs.	**Vmax**	85 mph
Empty Wt.	320 lbs.	**VCr**	65 mph
Payload (full fuel)	N/A	**V_{S1} (stall clean)**	N/A
Fuel Capacity	5 gal.	**V_{S0} (lndg config)**	N/A
Seats	1	**Climb Rate @msl**	1000 fpm
Cabin Width	N/A	**Service Ceiling**	N/A

Std. Engine/HP	Usable HP Range	Construction Mtrl.	Est. Bldg. Time
Rotax 582LC/65 hp	N/A	Aluminum, Steel	300 hrs.
Plans	**Kit**	**Info Packet**	**Video**
$60.00	N/A	$5.00	$20.00

Single seat, open frame, light helicopter. New design, tested and flown in 1995. Tube tail puts this aircraft into the 90's with a whole new look. Two bladed single main rotor. Anti-torque tail rotor. Bolted aluminum and steel airframe. A modified version of the Adams-Wilson Chopper. Designed with the amateur homebuilder in mind for scratch building from plans. Use W258 "B" gearbox.

Plans Sold: 200 **Completed & Flying:** 12

The Skytwister "Choppy" machine traces its origin back to the late fifties and two guys named Tom Adams and Paul Wilson. The original name for the craft was "Hobbycopter." It was a simplified, single place machine powered by a motorcycle engine available as plans-only.

In 1992, Ben Showers produced his offspring of the original Adams/Wilson machine. It is a modified, highly refined version of the A-W helicopter powered with a modern Rotax engine. He has added modern technology to an old design and breathed life into an almost forgotten machine.

SHOWERS-AERO
RD #3, Box 1200
Milton PA 17847
Tel. 717-742-4080 • Fax 717-473-1105

Lightning

Rotor Span	23.00 ft.	**Range**	240 sm
Rotor Area	N/A	**Takeoff Dist.**	300 ft.
Length	11.92 ft.	**Landing Dist.**	10 ft.
Max Gross Wt.	600 lbs.	**Vmax**	63 mph
Empty Wt.	252 lbs.	**VCr**	55 mph
Payload (full fuel)	300 lbs.	**V_{S1} (stall clean)**	N/A
Fuel Capacity	5 gal.	**V_{S0} (lndg config)**	N/A
Seats	1	**Climb Rate @msl**	1000 fpm
Cabin Width	71 in.	**Service Ceiling**	10,000 ft.

Std. Engine/HP	Usable HP Range	Construction Mtrl.	Est. Bldg. Time
Rotax 503 / 46 hp	N/A	Steel, Composite, Alum.	50 hrs.
Plans	**Kit**	**Info Packet**	**Video**
N/A	$10,495.00	$12.00	$15.00

Sport Copter's ultralight gyro. Field assembly in 10 minutes. Optional full body enclosure allows smooth flight up to 40 mph. Maneuver forward, backwards, sideways, hover, and descend vertically.

Kits/Plans Sold: 29 **Completed & Flying:** 13

Victor Gabas, Owner
"I'm really thrilled with its performance, and I wouldn't trade my (Lightning) Sport Copter for any other type of open-frame gyro!"

SPORT COPTER INTERNATIONAL, INC.
Scappoose Airport • 34012 N. Honeyman Road
Scappoose OR 97056
Tel. 503-543-7000 • Fax 503-543-7041
Web: http://www.sportcopter.com

Vortex

Rotor Span	25.00 ft.	**Range**	170 sm
Rotor Area	N/A	**Takeoff Dist.**	300 ft.
Length	12.00 ft.	**Landing Dist.**	10 ft.
Max Gross Wt.	760 lbs.	**Vmax**	100 mph
Empty Wt.	300 lbs.	**VCr**	80 mph
Payload (full fuel)	410 lbs.	**V_{S1} (stall clean)**	N/A
Fuel Capacity	10 gal.	**V_{S0} (lndg config)**	N/A
Seats	1	**Climb Rate @msl**	1200 fpm
Cabin Width	71 in.	**Service Ceiling**	13,000 ft.

Std. Engine/HP	Usable HP Range	Construction Mtrl.	Est. Bldg. Time
Rotax 582 / 67 hp	N/A	Metal, Alum.,Composite	50 hrs.
Plans	**Kit**	**Info Packet**	**Video**
N/A	$15,400.00	$12.00	$15.00

A high performance, single place gyroplane. Can be flown in the open or with a partial or full body enclosure. Body enclosure provides significant streamlining as well as protecting the engine from thermal shock. Options include a rough field package, retractable gear, and a shock-isolated folding mast.

Kits/Plans Sold: 42 **Completed & Flying:** 6

Chad Wille, Flight Instructor
"The new generation of Sport Copters have the best handling characteristics and safety of any gyro offered for sale."

SPORT COPTER INTERNATIONAL, INC.
Scappoose Airport • 34012 N. Honeyman Road
Scappoose OR 97056
Tel. 503-543-7000 • Fax 503-543-7041
Web: http://www.sportcopter.com

HOMEBUILT "FAQ'S" *(See Page 68)*

Q: If I sell my homebuilt, am I liable for it later?

A: Unfortunately, there is potential for a liability problem. Even though "free" legal advice is often available on the Internet, it is best to contact your attorney if you find this issue troubling.

Some people advocate having the buyer sign a "release form" which would be a promise not to sue if anything goes wrong. These are essentially worthless. The buyer can't sign over his or her "spouse's" rights.

Hummingbird

Rotor Span	33.00 ft.	**Range**	400 sm
Rotor Area	N/A	**Takeoff Dist.**	N/A
Length	30.60 ft.	**Landing Dist.**	N/A
Max Gross Wt.	2700 lbs.	**Vmax**	110 mph
Empty Wt.	1800 lbs.	**VCr**	90 mph
Payload (full fuel)	900 lbs.	**V_{S1} (stall clean)**	N/A
Fuel Capacity	57 gal.	**V_{S0} (lndg config)**	N/A
Seats	4	**Climb Rate @msl**	1200 fpm
Cabin Width	72 in.	**Service Ceiling**	11,000 ft.

Std. Engine/HP	Usable HP Range	Construction Mtrl.	Est. Bldg. Time
Alum. V-8 / 260 hp	N/A	Sheet Metal	750 hrs.
Plans	**Kit**	**Info Packet**	**Video**
$375.00	$140,500.00	$15.00	$15.00

A four place design based on the Sikorsky S-52. Kit includes engine, instruments, hardware, etc. Information packet $25.00 overseas. VHS Video $30.00 overseas.

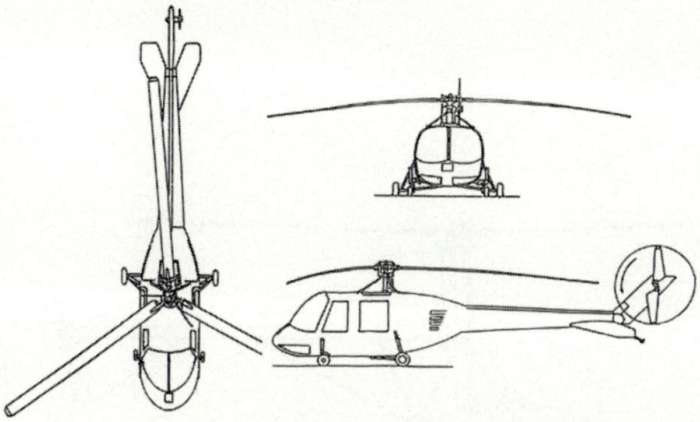

The Hummingbird provides an affordable kit with attention to safety and reliability. One innovative option in the Hummingbird is the aluminum block V-8 powerplant. Using the highest quality components, this engine with a dry oil sump gives full power with less fuel consumption.

VERTICAL AVIATION TECHNOLOGIES
P.O. Box 2527 • 1642 Hangar Road
Sanford FL 32772-2527
Tel. 407-322-9488 • Fax 407-330-2647

G-1 Ultralight Helicopter

Rotor Span	12.00 ft.	**Range**	120 sm
Rotor Area	N/A	**Takeoff Dist.**	N/A
Length	12.00 ft.	**Landing Dist.**	N/A
Max Gross Wt.	420 lbs.	**Vmax**	80 mph
Empty Wt.	150 lbs.	**VCr**	55 mph
Payload (full fuel)	N/A	**V_{S1} (stall clean)**	N/A
Fuel Capacity	5 gal.	**V_{S0} (lndg config)**	N/A
Seats	1	**Climb Rate @msl**	900 fpm
Cabin Width	30 in.	**Service Ceiling**	10,000 ft.

Std. Engine/HP	Usable HP Range	Construction Mtrl.	Est. Bldg. Time
McCulloch 101 / 42 hp	N/A	Aluminum	80 hrs.
Plans	**Kit**	**Info Packet**	**Video**
$27.95	$3,700.00	$10.00	N/A

An ultra-compact helicopter that meets ultralight aircraft criteria. Partial kits available consisting of airframe with seat, main and tail rotor blades, engine and bearings. Kit price is for minimum kit. Call for other kit prices. Overseas price of plans is $30.95 US. The three McCulloch go-cart engines can be replaced with a single engine such as the Rotax 447. Can be built from plans for under $3,000. See Ad on Page 21.

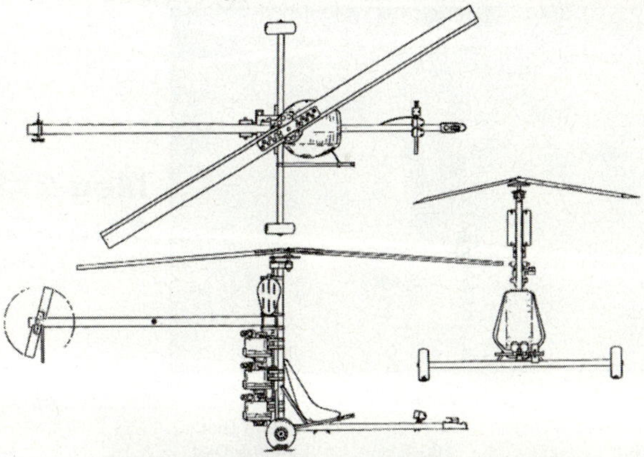

VORTECH, INC.
PO Box 511-KP
Fallston MD 21047
Tel. 410-692-2692 • Fax 410-692-5902
Web: http://www.webcom.com/neatstuf/helio

Kestrel

Rotor Span	25.00 ft.	Range	60 sm
Rotor Area	N/A	Takeoff Dist.	N/A
Length	12.00 ft.	Landing Dist.	N/A
Max Gross Wt.	500 lbs.	Vmax	95 mph
Empty Wt.	200 lbs.	VCr	63 mph
Payload (full fuel)	N/A	V_{S1} (stall clean)	N/A
Fuel Capacity	10 gal.	V_{S0} (lndg config)	N/A
Seats	1	Climb Rate @msl	950 fpm
Cabin Width	N/A	Service Ceiling	12,500 ft.

Std. Engine/HP	Usable HP Range	Construction Mtrl.	Est. Bldg. Time
2xG8-2-20 Jets/37 hp	N/A	Metal	40 hrs.
Plans	**Kit**	**Info Packet**	**Video**
$23.00	$9,995.00	$10.00	N/A

Complete kit; bolts together. Powered by dual 47 lb. thrust jets. Cabin option available. See Ad on Page 21.

Completed & Flying: 4

VORTECH, INC.
PO Box 511-KP
Fallston MD 21047
Tel. 410-692-2692 • Fax 410-692-5902
Web: http://www.webcom.com/neatstuf/helio

Meg-2XH Strap-On

Rotor Span	21.00 ft.	Range	35 sm
Rotor Area	N/A	Takeoff Dist.	N/A
Length	7.00 ft.	Landing Dist.	N/A
Max Gross Wt.	400 lbs.	Vmax	75 mph
Empty Wt.	120 lbs.	VCr	63 mph
Payload (full fuel)	250 lbs.	V_{S1} (stall clean)	N/A
Fuel Capacity	5 gal.	V_{S0} (lndg config)	N/A
Seats	1	Climb Rate @msl	1000 fpm
Cabin Width	N/A	Service Ceiling	12,500 ft.

Std. Engine/HP	Usable HP Range	Construction Mtrl.	Est. Bldg. Time
Jet / 37 hp	N/A	Metal	30 hrs.
Plans	**Kit**	**Info Packet**	**Video**
$136.00	$7,995.00	$10.00	N/A

Complete kit; bolts together. Design includes folding seat. Overseas plans price is $150 US. See Ad on Page 21.

Completed & Flying: 3

VORTECH, INC.
PO Box 511-KP
Fallston MD 21047
Tel. 410-692-2692 • Fax 410-692-5902
Web: http://www.webcom.com/neatstuf/helio

Skylark

Rotor Span	19.00 ft.	Range	80 sm
Rotor Area	N/A	Takeoff Dist.	N/A
Length	17.50 ft.	Landing Dist.	N/A
Max Gross Wt.	700 lbs.	Vmax	95 mph
Empty Wt.	325 lbs.	VCr	70 mph
Payload (full fuel)	350 lbs.	V_{S1} (stall clean)	N/A
Fuel Capacity	8 gal.	V_{S0} (lndg config)	N/A
Seats	1	Climb Rate @msl	1100 fpm
Cabin Width	N/A	Service Ceiling	12,500 ft.

Std. Engine/HP	Usable HP Range	Construction Mtrl.	Est. Bldg. Time
Rotax 582 / 65 hp	N/A	Metal	50 hrs.
Plans	**Kit**	**Info Packet**	**Video**
N/A	$19,995.00	$10.00	$28.00

Engine included in kit. Cabin option available. See Ad on Page 21.

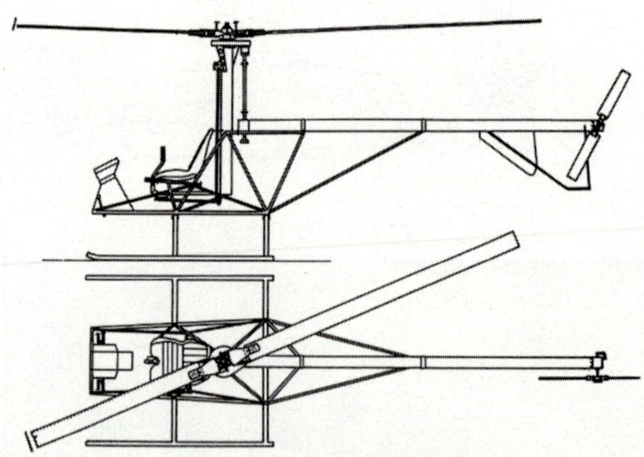

VORTECH, INC.
PO Box 511-KP
Fallston MD 21047
Tel. 410-692-2692 • Fax 410-692-5902
Web: http://www.webcom.com/neatstuf/helio

Aeros

Rotor Span	22.00 ft.	**Range**	65 sm
Rotor Area	N/A sq. ft.	**Takeoff Dist.**	N/A
Length	18.00 ft.	**Landing Dist.**	N/A
Max Gross Wt.	900 lbs.	**Vmax**	85 mph
Empty Wt.	540 lbs.	**VCr**	60 mph
Payload (full fuel)	360 lbs.	**V_{S1} (stall clean)**	N/A
Fuel Capacity	12 gal.	**V_{S0} (lndg config)**	N/A
Seats	1	**Climb Rate @msl**	950 fpm
Cabin Width	20 in.	**Service Ceiling**	12,000 ft.

Std. Engine/HP	Usable HP Range	Construction Mtrl.	Est. Bldg. Time
VW / 97 hp	N/A	Aluminum, Steel	300 hrs.
Plans	**Kit**	**Info Packet**	**Video**
$32.00	N/A	$2.00	N/A

Utilizes a bolt-together tubular frame with a monocoque tail boom. Flight characteristics similar to a Robinson R22.

Kits/Plans Sold: 40 **Completed & Flying:** 3

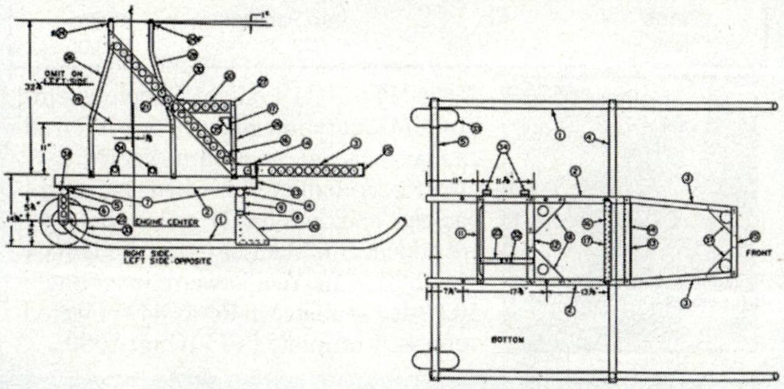

WINDSPIRE, INC.
PO Box 145-AX
Long Green MD 21092
Tel. 410-836-6824 • Fax 410-692-0595

H-1 Racer

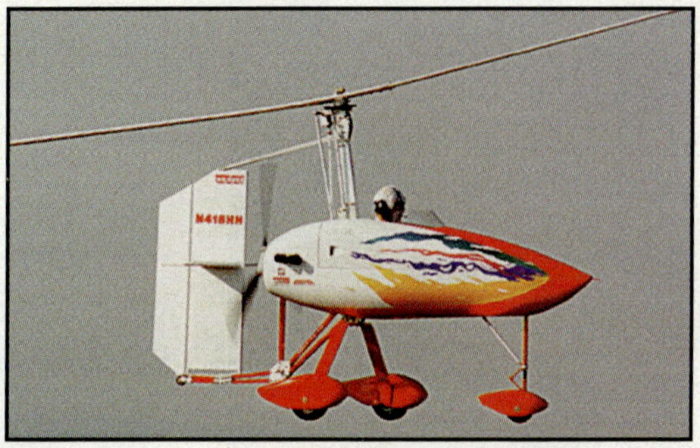

Rotor Span	25.00 ft.	**Range**	50 sm
Rotor Area	N/A sq. ft.	**Takeoff Dist.**	500 ft.
Length	12.59 ft.	**Landing Dist.**	10 ft.
Max Gross Wt.	630 lbs.	**Vmax**	90 mph
Empty Wt.	380 lbs.	**VCr**	65 mph
Payload (full fuel)	250 lbs.	**V_{S1} (stall clean)**	N/A
Fuel Capacity	5 gal.	**V_{S0} (lndg config)**	N/A
Seats	1	**Climb Rate @msl**	800 fpm
Cabin Width	22 in.	**Service Ceiling**	10,000 ft.

Std. Engine/HP	Usable HP Range	Construction Mtrl.	Est. Bldg. Time
Rotax DC 503 / 52 hp	47 - 120	Aluminum Tubing	150 hrs.
Plans	**Kit**	**Info Packet**	**Video**
$150.00	$13,000.00	$12.50	N/A

Inline C.G., prop thrust and horizontal stabilizer, hydraulic disc brakes, wheel pants, a forgiving suspension system, and self-aligning front wheel are standard. Options include a pre-rotator and a 12 lb. composite body kit. **Sold:** 100 **Flying:** 3

WINNERS CIRCLE ENGINEERING INC.
14 S. Water Street
Monrovia IN 46157
Tel. 317-996-3157 • Fax 317-996-3148

Wombat

Rotor Span	22.83 ft.	**Range**	190 sm
Rotor Area	N/A	**Takeoff Dist.**	300 ft.
Length	12.30 ft.	**Landing Dist.**	3 ft.
Max Gross Wt.	N/A	**Vmax**	87 mph
Empty Wt.	352 lbs.	**VCr**	60 mph
Payload (full fuel)	N/A	**V_{S1} (stall clean)**	N/A
Fuel Capacity	N/A	**V_{S0} (lndg config)**	N/A
Seats	1	**Climb Rate @msl**	1000 fpm
Cabin Width	N/A	**Service Ceiling**	1,000 ft.

Std. Engine/HP	Usable HP Range	Construction Mtrl.	Est. Bldg. Time
Rotax 532	N/A	Metal	N/A
Plans	**Kit**	**Info Packet**	**Video**
N/A	Yes	Yes	N/A

A lightweight gyroplane developed by Chris Julian. Endurance is about four hours at 60 mph. Available with a 15 or 17 gal. fuel tank, and a low profile seat allowing a full canopy to be fitted.

WOMBAT GYROCOPTERS
The Chalet, Higher Fraddon • Indian Queens
St. Columb, Cornwall TR9 6QN • UNITED KINGDOM
Tel. +44 01726 860662 • Fax +44 01726 860662
Web: http://www.demon.co.uk/amethyst/wombat.html

Dream Machine 582/503

Wing Span	N/A	**Range**	N/A
Wing Area	500.0 sq. ft.	**Takeoff Dist.**	500 ft.
Length	N/A	**Landing Dist.**	100 ft.
Max Gross Wt.	N/A	**Vmax**	30 mph
Empty Wt.	300 lbs.	**VCr**	30 mph
Payload (full fuel)	450 lbs.	**V_{S1} (stall clean)**	N/A
Fuel Capacity	8 gal.	**V_{S0} (lndg config)**	N/A
Seats	2	**Climb Rate @msl**	600 fpm
Cabin Width	N/A	**Service Ceiling**	10,500 ft.

Std. Engine/HP	Usable HP Range	Construction Mtrl.	Est. Bldg. Time
Rotax 582 / 66 hp	N/A	Aluminum, Steel	30 hrs.
Plans	**Kit**	**Info Packet**	**Video**
N/A	$12,789.60	$7.00	$10.00

Manufactured since 1985. Voted best in category at both Sun 'n Fun and Oshkosh. Also, Grand Champion of Powered Parachutes at the Northwest Fly-In at Arlington. Now also available in Rotax 503 powered version. **Completed & Flying:** 300

BUCKEYE INDUSTRIES, INC.
16095 Linden Road • Argos IN 46501
Tel. 219-892-5566 • Fax 219-892-6907
E-Mail: buckchute@aol.com
Web: http://buckchute.com

Eagle 503/447

Wing Span	N/A	**Range**	N/A
Wing Area	450.0 sq. ft.	**Takeoff Dist.**	300 ft.
Length	N/A	**Landing Dist.**	100 ft.
Max Gross Wt.	570 lbs.	**Vmax**	26 mph
Empty Wt.	230 lbs.	**VCr**	26 mph
Payload (full fuel)	310 lbs.	**V_{S1} (stall clean)**	N/A
Fuel Capacity	5 gal.	**V_{S0} (lndg config)**	N/A
Seats	1	**Climb Rate @msl**	700 fpm
Cabin Width	N/A	**Service Ceiling**	10,500 ft.

Std. Engine/HP	Usable HP Range	Construction Mtrl.	Est. Bldg. Time
Rotax 503 / 46 hp	N/A	Aluminum, Steel	30 hrs.
Plans	**Kit**	**Info Packet**	**Video**
N/A	$9,113.88	$7.00	$10.00

A legal Part 103 Ultralight. Simple assembly and operation make it a preferred choice for sport aviation newcomers. A great cross country flier when instrument package and optional 11 gallon fuel tank are added. Jim MacLeay won the USUA "Flight Of The Year" award in an Eagle 503. Also available in Rotax 447 powered version. **Completed & Flying:** 1000

BUCKEYE INDUSTRIES, INC.
16095 Linden Road • Argos IN 46501
Tel. 219-892-5566 • Fax 219-892-6907
E-Mail: buckchute@aol.com
Web: http://buckchute.com

Falcon-582

Wing Span	N/A	**Range**	N/A
Wing Area	450.0 sq. ft.	**Takeoff Dist.**	300 ft.
Length	N/A	**Landing Dist.**	100 ft.
Max Gross Wt.	N/A	**Vmax**	28 mph
Empty Wt.	240 lbs.	**VCr**	28 mph
Payload (full fuel)	350 lbs.	**V_{S1} (stall clean)**	N/A
Fuel Capacity	5 gal.	**V_{S0} (lndg config)**	N/A
Seats	1	**Climb Rate @msl**	850 fpm
Cabin Width	N/A	**Service Ceiling**	10,000 ft.

Std. Engine/HP	Usable HP Range	Construction Mtrl.	Est. Bldg. Time
Rotax 582 / 66 hp	N/A	Aluminum, Steel	30 hrs.
Plans	**Kit**	**Info Packet**	**Video**
N/A	$10,880.82	$7.00	$10.00

Canadian Distributor:
CANADIAN POWERED PARACHUTES LTD.
Box 1769 • Vegreville, Alberta T9C 1S8 CANADA
Ph: 403-658-2044
Fax: 403-658-2387
E-Mail: lindam@agt.net
Web: http://www.agt.net/public/lindam/cpp2.htm

The sports car of powered parachutes. Can lift a 350 lb. payload at 800 fpm. FAR Part 103 legal ultralight with the rotax 582 water cooled engine.

BUCKEYE INDUSTRIES, INC.
16095 Linden Road • Argos IN 46501
Tel. 219-892-5566 • Fax 219-892-6907
E-Mail: buckchute@aol.com
Web: http://buckchute.com

Glascat

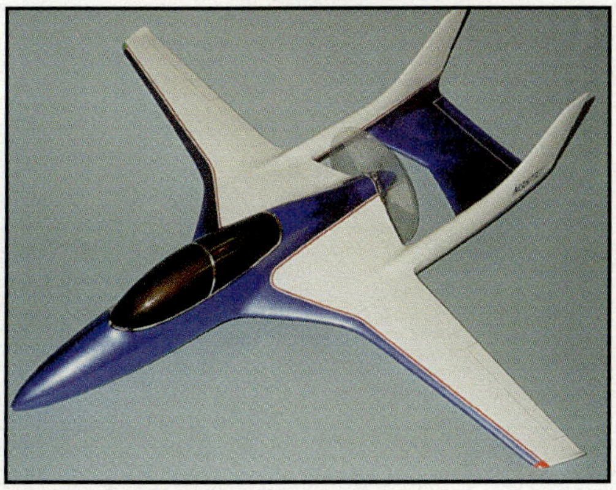

Wing Span	29.50 ft.	**Range**	1400 sm
Wing Area	120.0 sq. ft.	**Takeoff Dist.**	1200 ft.
Length	28.00 ft.	**Landing Dist.**	1450 ft.
Max Gross Wt.	3000 lbs.	**Vmax**	350 mph
Empty Wt.	1850 lbs.	**VCr**	300 mph
Payload (full fuel)	N/A	**V_{S1} (stall clean)**	72 mph
Fuel Capacity	N/A	**V_{S0} (lndg config)**	65 mph
Seats	2	**Climb Rate @msl**	2500 fpm
Cabin Width	35 in.	**Service Ceiling**	25,000 ft.

Std. Engine/HP	Usable HP Range	Construction Mtrl.	Est. Bldg. Time
Cont. TSIO-550/350 hp	280 - 420	Composite	2700 hrs.
Plans	**Kit**	**Info Packet**	**Video**
N/A	$49,000.00	N/A	N/A

High performance, two place pusher with excellent visibility and very quiet, roomy cockpit. Designed to simulate flight in a modern jet fighter. Ample room in nose-cone for weather radar. Ideal aircraft for Allison turboprop engine. Large baggage area behind rear seat. Fowler flaps for slow flight. Carbon fiber prepreg and honeycomb core construction.

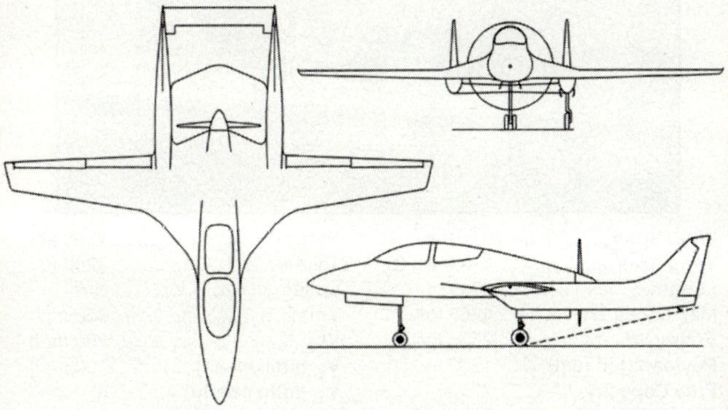

COMPOSITE AIRCRAFT TECHNOLOGIES
7902 E. Nora Ave.
Spokane WA 99212
Ph: 509-891-6704
E-Mail: comairtech@ior.com
Web: http://www.ior.com/~glascat/fullsize.html

MIFYTER

Wing Span	18.66 ft.	**Range**	300 sm
Wing Area	123.0 sq. ft.	**Takeoff Dist.**	75 ft.
Length	16.50 ft.	**Landing Dist.**	150 ft.
Max Gross Wt.	755 lbs.	**Vmax**	N/A
Empty Wt.	430 lbs.	**VCr**	75 mph
Payload (full fuel)	N/A	**V_{S1} (stall clean)**	37 mph
Fuel Capacity	15 gal.	**V_{S0} (lndg config)**	37 mph
Seats	N/A	**Climb Rate @msl**	N/A
Cabin Width	N/A	**Service Ceiling**	N/A

Std. Engine/HP	Usable HP Range	Construction Mtrl.	Est. Bldg. Time
Rotax 532 / 64 hp	N/A	Tube, Fabric	N/A
Plans	**Kit**	**Info Packet**	**Video**
N/A	Yes	Yes	N/A

Not a replica of a particular aircraft but representative of the era. Custom options that include variations of the cowlings, tail feathers, paint schemes, etc. can give standoff appearance of Spad/Nieuport/SE5, etc. Nostalgic appearance with modern construction methods.

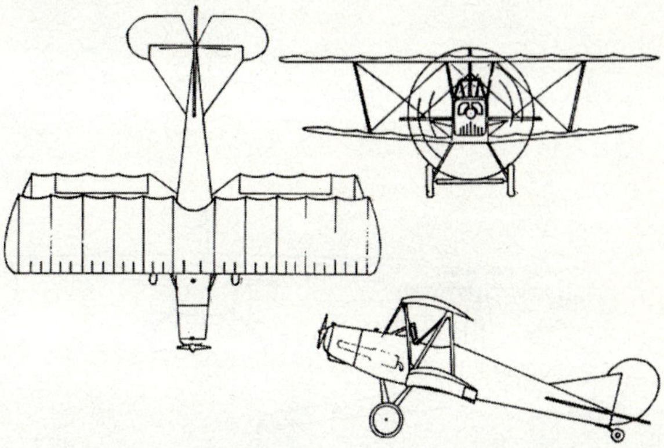

ROD COWGILL
1455 Crestway
Ontario OR 97914
Tel. 541-889-7536 • Fax 541-881-8129
E-Mail: maxaabjc@cmc.com

DJ2

Wing Span	26.33 ft.	Range	700 sm
Wing Area	105.0 sq. ft.	Takeoff Dist.	400 ft.
Length	16.00 ft.	Landing Dist.	700 ft.
Max Gross Wt.	1200 lbs.	Vmax	220 mph
Empty Wt.	600 lbs.	VCr	200 mph
Payload (full fuel)	600 lbs.	V_{S1} (stall clean)	61 mph
Fuel Capacity	30 gal.	V_{S0} (lndg config)	48 mph
Seats	2	Climb Rate @msl	1800 fpm
Cabin Width	43 in.	Service Ceiling	25,000 ft.

Std. Engine/HP	Usable HP Range	Construction Mtrl.	Est. Bldg. Time
Rotary / 150 hp	80-150	Composite	300 hrs.
Plans	**Kit**	**Info Packet**	**Video**
N/A	$24,000.00	$10.00	N/A

Advanced pusher design. Side-by-side seating. Central control. Very quiet and low vibration. Retractable/removable wings. Rotary engine runs on regular gasoline. Kit for U.S. is classed as Experimental. Kit for Canada is Amateur Build - Canadian Ultralight. Price is estimated in U.S. funds.

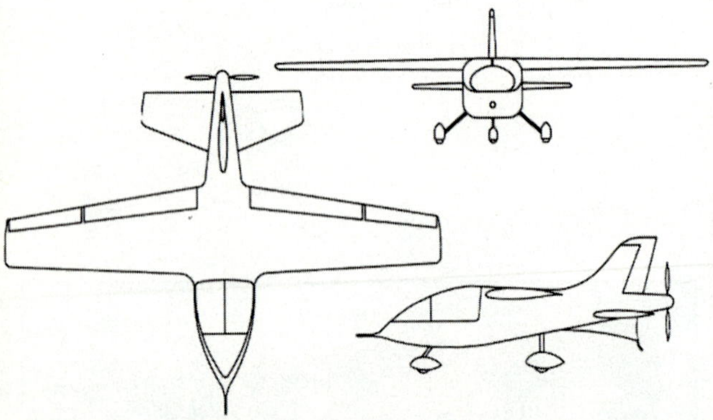

DJ AIRCRAFTS INTERNATIONAL
7594 Tessier Brossard
Ontario, Quebec J4W 2H5
CANADA
Tel. 514 466-1151

Pacifica Six

Wing Span	45.00 ft.	Range	3200 sm
Wing Area	205.0 sq. ft.	Takeoff Dist.	1500 ft.
Length	23.50 ft.	Landing Dist.	1200 ft.
Max Gross Wt.	3900 lbs.	Vmax	300 mph
Empty Wt.	1950 lbs.	VCr	300 mph
Payload (full fuel)	N/A	V_{S1} (stall clean)	N/A
Fuel Capacity	112 gal.	V_{S0} (lndg config)	N/A
Seats	6	Climb Rate @msl	1800 fpm
Cabin Width	N/A	Service Ceiling	17,500 ft.

Std. Engine/HP	Usable HP Range	Construction Mtrl.	Est. Bldg. Time
Chevy 502 / 500 hp	N/A	Composite	500 hrs.
Plans	**Kit**	**Info Packet**	**Video**
N/A	$178,000.00	Yes	N/A

Cabin class seats 6. Monolithic pressure molded hull. 9G wings. Fixed gear. Head room is 5 ft. Aisle slot can hold sports gear up to 8-1/2 ft. With pilot door and sliding passenger door. Estimate $200 fuel cost for cross country nonstop. Delivery in 1999.

DYNAMIC WING COMPANY
Hangar 13 • 1425 General Aviation Drive • Melbourne FL 32935
Tel. 407-253-1975 • Fax 407-253-3868 • E-Mail: wingco@iu.net
Web: http://www.wingco.com

Aurora

Wing Span	31.30 ft.	Range	1200 sm
Wing Area	110.0 sq. ft.	Takeoff Dist.	3200 ft.
Length	27.80 ft.	Landing Dist.	N/A
Max Gross Wt.	4266 lbs.	Vmax	330 mph
Empty Wt.	2550 lbs.	VCr	290 mph
Payload (full fuel)	1200 lbs.	V_{S1} (stall clean)	100 mph
Fuel Capacity	77 gal.	V_{S0} (lndg config)	83 mph
Seats	7	Climb Rate @msl	2100 fpm
Cabin Width	50 in.	Service Ceiling	25,000 ft.

Std. Engine/HP	Usable HP Range	Construction Mtrl.	Est. Bldg. Time
V-4 Diesels / 250 hp	500 - 1000	Composite	2500 hrs.
Plans	**Kit**	**Info Packet**	**Video**
N/A	$70,000.00	$25.00	N/A

Unique multi-engine homebuilt kit. Canard twin pusher design. Economical diesel power. Dual Univ. Tech Engines. Composite construction. Up to seven place. Cabin pressurization will be available.

EROSPACE TECHNOLOGIES INC.
288 Watercrest St. • Sebastian FL 32958
Tel. 407-388-0966 • Fax 407-388-0966
E-Mail: erospace@aol.com
Web: http://members.aol.com/erospace/

Finchbird

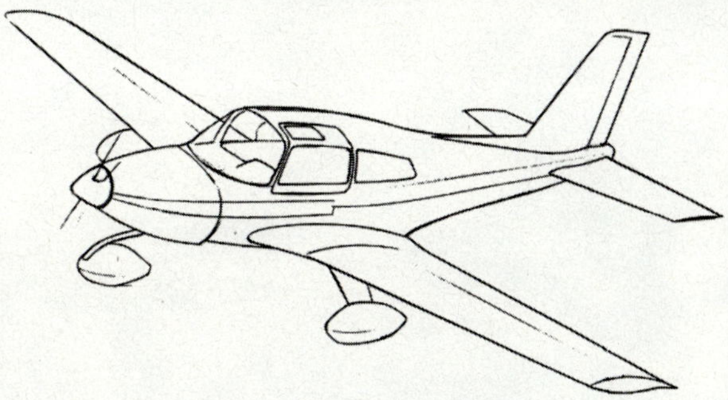

Wing Span	34.00 ft.	**Range**	1069 sm
Wing Area	136.0 sq. ft.	**Takeoff Dist.**	900 ft.
Length	23.40 ft.	**Landing Dist.**	650 ft.
Max Gross Wt.	2400 lbs.	**Vmax**	210 mph
Empty Wt.	1300 lbs.	**VCr**	175 mph
Payload (full fuel)	670 lbs.	**V_{S1} (stall clean)**	65 mph
Fuel Capacity	55 gal.	**V_{S0} (lndg config)**	60 mph
Seats	4	**Climb Rate @msl**	950 fpm
Cabin Width	46 in.	**Service Ceiling**	14,000 ft.

Std. Engine/HP	Usable HP Range	Construction Mtrl.	Est. Bldg. Time
Buick / 200 hp	150 - 350	Steel Tube, Alum.	2000 hrs.
Plans	**Kit**	**Info Packet**	**Video**
$450.00	N/A	$10.00	$30.00

A steel tube and sheet aluminum entered in the Aircraft Spruce & Specialty design contest. Can be scratch-built for under $10,000 not including the engine. Safety features include fire prevention and air bag crash protection for all occupants. Quick-Build design features are incorporated in the basic scratch-built plans. See Ad on Page 32. **Kits/Plans Sold:** 2

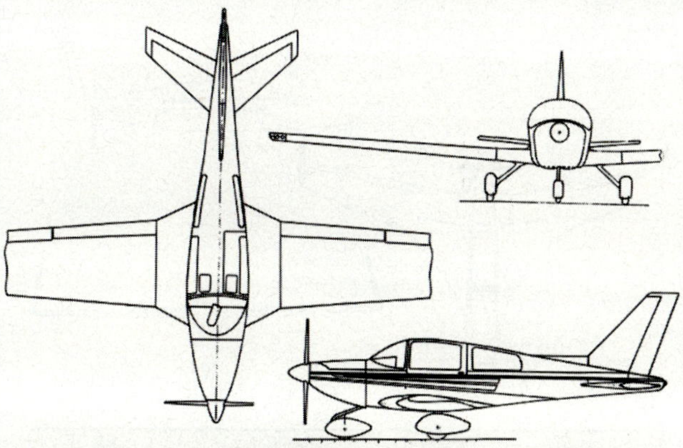

RICHARD FINCH
P.O. Box 3946
Santa Barbara CA 93130-3946
Tel. 805-683-5346 • Fax 805-683-8868

MF 200 Sea Bird

Wing Span	36.00 ft.	**Range**	1500 sm
Wing Area	N/A	**Takeoff Dist.**	N/A
Length	23.30 ft.	**Landing Dist.**	N/A
Max Gross Wt.	2420 lbs.	**Vmax**	177 mph
Empty Wt.	1276 lbs.	**VCr**	159 mph
Payload (full fuel)	N/A	**V_{S1} (stall clean)**	43 mph
Fuel Capacity	N/A	**V_{S0} (lndg config)**	N/A
Seats	4	**Climb Rate @msl**	N/A
Cabin Width	N/A	**Service Ceiling**	N/A

Std. Engine/HP	Usable HP Range	Construction Mtrl.	Est. Bldg. Time
100 hp	100x2	Composite	N/A
Plans	**Kit**	**Info Packet**	**Video**
N/A	Yes	Yes	N/A

Light, pusher twin engine amphibian to be built entirely of advanced composite materials. Retractable, tricycle gear.

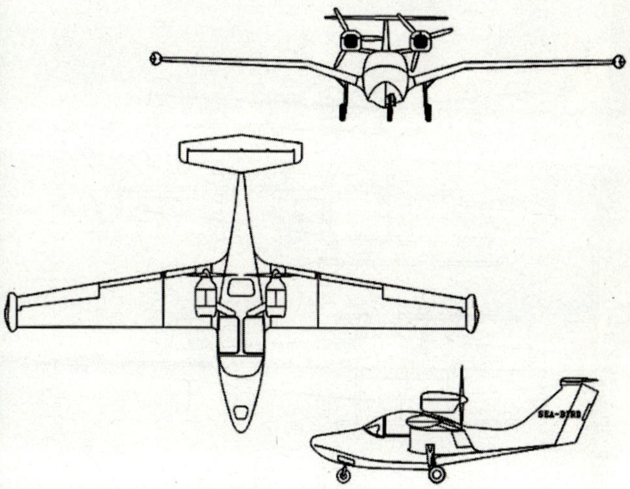

FLORIDA AERONAUTICS, INC.
2301 N.W. 33rd Ct.
Pompano Beach FL 33069
Tel. 954-970-7814 • Fax 954-970-5205

Two Place EH1-02

Wing Span	24.00 ft.	**Range**	390 sm
Wing Area	N/A	**Takeoff Dist.**	N/A
Length	28.00 ft.	**Landing Dist.**	N/A
Max Gross Wt.	1800 lbs.	**Vmax**	160 mph
Empty Wt.	850 lbs.	**VCr**	N/A
Payload (full fuel)	950 lbs.	**V_{S1} (stall clean)**	N/A
Fuel Capacity	45 gal.	**V_{S0} (lndg config)**	N/A
Seats	2	**Climb Rate @msl**	1500 fpm
Cabin Width	35 in.	**Service Ceiling**	N/A

Std. Engine/HP	Usable HP Range	Construction Mtrl.	Est. Bldg. Time
T62T32 / 150 hp	150 - 417	Aluminum	N/A
Plans	**Kit**	**Info Packet**	**Video**
$350.00	$68,000.00	$30.00	Yes

Two place version of the turbine powered RotorMouse EH1-01. Styling reminiscent of the Huey AH-1. Video included in info pack.

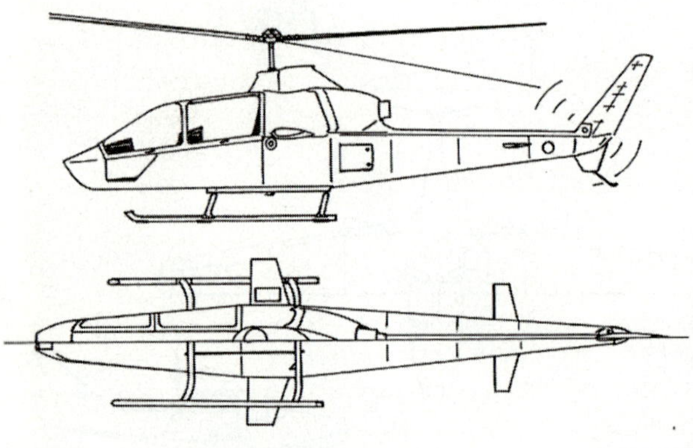

HILLBERG HELICOPTERS
PO Box 8974
Fountain Valley CA 92728
Tel. 909-279-5678

Infinity 1

Wing Span	26.83 ft.	**Range**	2239 sm
Wing Area	125.8 sq. ft.	**Takeoff Dist.**	950 ft.
Length	17.68 ft.	**Landing Dist.**	930 ft.
Max Gross Wt.	2000 lbs.	**Vmax**	320 mph
Empty Wt.	950 lbs.	**VCr**	290 mph
Payload (full fuel)	630 lbs.	**V_{S1} (stall clean)**	60 mph
Fuel Capacity	70 gal.	**V_{S0} (lndg config)**	N/A
Seats	2	**Climb Rate @msl**	3600 fpm
Cabin Width	27 in.	**Service Ceiling**	30,000 ft.

Std. Engine/HP	Usable HP Range	Construction Mtrl.	Est. Bldg. Time
Lyc. TIO-360 / 273 hp	273 - 324	Composite	1000 hrs.
Plans	**Kit**	**Info Packet**	**Video**
N/A	$25,000.00	$20.00	N/A

A prepreg, vacuum bag construction, tandem, canard design. All airframe parts included for quick build. Cockpit has room for two people 7 feet tall, or two smaller people in the rear seat. Kit includes OLEO strut retractable tricycle landing gear system with nose wheel steering & true HOTAS stick and throttle. Optional ballastic parachute recovery system. Canopy opens aft in a single piece ala F-16, but opening configuration can be varied.

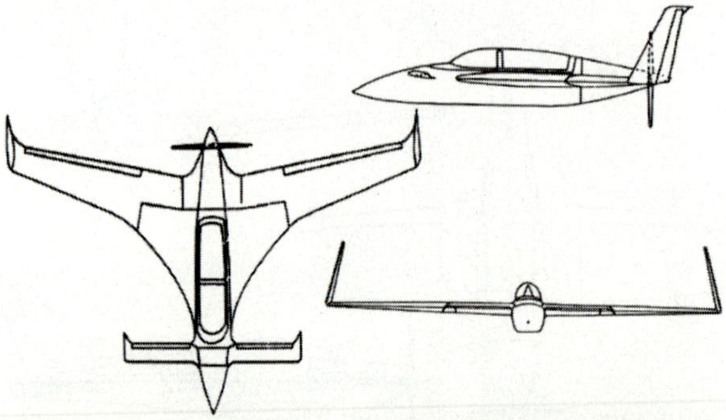

INFINITY AEROSPACE
P.O. Box 12275 Dept.KD
El Cajon CA 92022-2275
Tel. 619-448-5103 • Fax 619-448-5103
E-Mail: infaero@flash.net
Web: http://www.flash.net/~infaero

Starboy

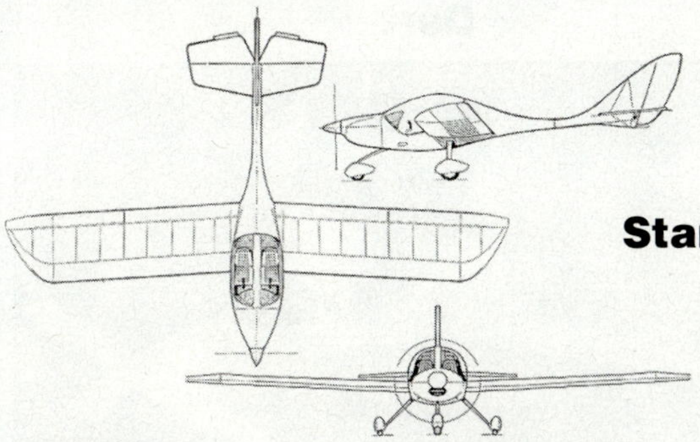

Wing Span	34.40 ft.	**Range**	N/A
Wing Area	133.6 sq. ft.	**Takeoff Dist.**	N/A
Length	24.80 ft.	**Landing Dist.**	N/A
Max Gross Wt.	992 lbs.	**Vmax**	137 mph
Empty Wt.	540 lbs.	**VCr**	93 mph
Payload (full fuel)	N/A	**V_{S1} (stall clean)**	N/A
Fuel Capacity	13 gal.	**V_{S0} (lndg config)**	34 mph
Seats	2	**Climb Rate @msl**	1386 fpm
Cabin Width	N/A	**Service Ceiling**	N/A

Std. Engine/HP	Usable HP Range	Construction Mtrl.	Est. Bldg. Time
Rotax 912 / 80 hp	N/A	Composite, aluminum	500 hrs.
Plans	**Kit**	**Info Packet**	**Video**
N/A	N/A	N/A	N/A

Sleek aluminum and composite two seater to be available factory completed or as a kit.

INTERPLANE SRO
84 05 Letiste Zbraslavice
CZECHOSLOVAKIA
Tel. 0327/92145 • Fax 0327/92145

U.S. Distributor:
INTERPLANE USA, INC.
International Center • 30 Lee Gate
Grosse Pointe Farms MI 48236
Ph: 313-882-3400
Fax: 313-882-3401
E-Mail: InPlaneUSA@aol.com

Vortex 230

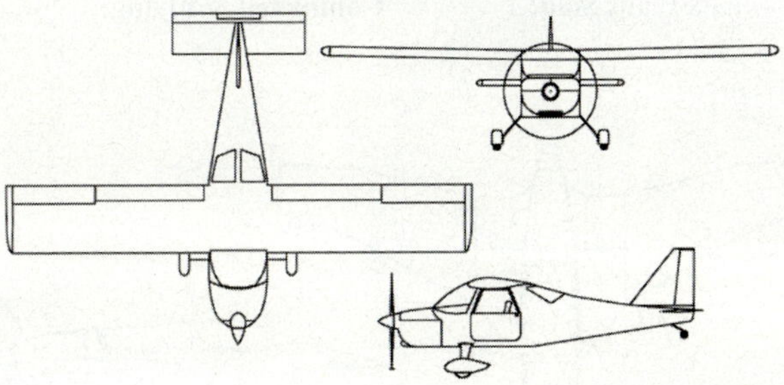

Wing Span	34.00 ft.	**Range**	1000 sm
Wing Area	170.0 sq. ft.	**Takeoff Dist.**	N/A
Length	24.25 ft.	**Landing Dist.**	1171 ft.
Max Gross Wt.	3000 lbs.	**Vmax**	204 mph
Empty Wt.	1500 lbs.	**VCr**	191 mph
Payload (full fuel)	1020 lbs.	**V_{S1} (stall clean)**	50 mph
Fuel Capacity	N/A	**V_{S0} (lndg config)**	N/A
Seats	4	**Climb Rate @msl**	986 fpm
Cabin Width	48 in.	**Service Ceiling**	22,000 ft.

Std. Engine/HP	Usable HP Range	Construction Mtrl.	Est. Bldg. Time
Ford V-6 / 230 hp	N/A	Wood, Composite	N/A
Plans	**Kit**	**Info Packet**	**Video**
Yes	N/A	$5.00	N/A

A high wing four seat design. Tri-gear or taildragger. Cockpit accepts pilots up to 6'3" tall. Wood and composite material used throughout for quick, strong and safe construction. Plans consist of construction booklet, drawings and AutoCad files.

JAY S. MCMULLAN
2451 E. Cinnabar Lane • Salt Lake City UT 84121
Tel. 801-944-0772
E-Mail: jaymc@sisna.com
Web: http://www.sisna.com/users/jaymc

Exel

Wing Span	41.00 ft.	**Range**	N/A
Wing Area	117.0 sq. ft.	**Takeoff Dist.**	310 ft.
Length	19.60 ft.	**Landing Dist.**	270 ft.
Max Gross Wt.	630 lbs.	**Vmax**	120 mph
Empty Wt.	366 lbs.	**VCr**	80 mph
Payload (full fuel)	264 lbs.	**V_{S1} (stall clean)**	37 mph
Fuel Capacity	5 gal.	**V_{S0} (lndg config)**	33 mph
Seats	1	**Climb Rate @msl**	N/A
Cabin Width	23 in.	**Service Ceiling**	N/A

Std. Engine/HP	Usable HP Range	Construction Mtrl.	Est. Bldg. Time
Zenoah / 22 hp	22-35	Fiberglass, Composite	200 hrs.
Plans	**Kit**	**Info Packet**	**Video**
Yes	$18,300.00	$30.00	$60.00

The newest motorized sailplane design from Claude Noin. Complete kit with all mold pieces will soon be available. 30:1 glide ratio. Assembly and dismantling in five minutes by one person.

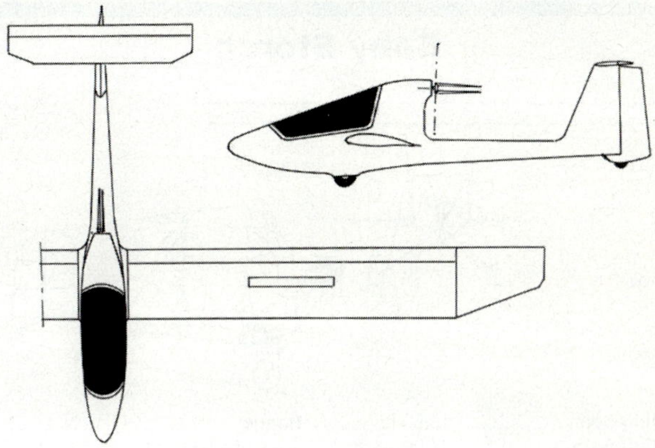

NOIN AERONAUTIQUES
RN85, Chateauvleux
05130 Tallard
FRANCE
Tel. 33 4 92 54 15 04 • Fax 33 4 92 54 15 04

TwinJet

Wing Span	N/A	Range	607 sm
Wing Area	N/A	Takeoff Dist.	1800 ft.
Length	N/A	Landing Dist.	1990 ft.
Max Gross Wt.	N/A	Vmax	440 mph
Empty Wt.	N/A	VCr	380 mph
Payload (full fuel)	N/A	V_{S1} (stall clean)	86 mph
Fuel Capacity	N/A	V_{S0} (lndg config)	N/A
Seats	4	Climb Rate @msl	N/A
Cabin Width	53 in.	Service Ceiling	25,000 ft.

Std. Engine/HP	Usable HP Range	Construction Mtrl.	Est. Bldg. Time
Twin GE T-58	N/A	N/A	N/A
Plans	**Kit**	**Info Packet**	**Video**
N/A	N/A	N/A	N/A

Spacious four place plus, twin turbojet design. Composite construction. Cruise of 380 mph, service ceiling of 25,000 feet. Utilizes twin GE T-58 powerplants derated to 650 lbs. of thrust.

MAVERICK AIR
32451 Walt Basett Ave.
Pueblo CO 81001
Ph: 719-251-3318
E-Mail: MavAir@aol.com

Baby Storch

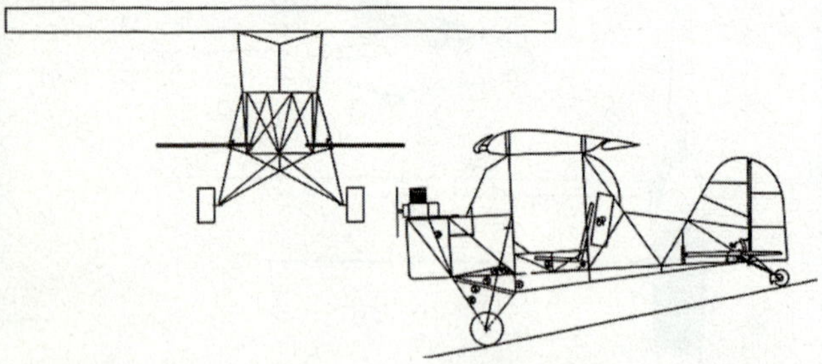

Wing Span	16.50 ft.	Range	N/A
Wing Area	84.00 sq. ft.	Takeoff Dist.	100 ft.
Length	12.00 ft.	Landing Dist.	50 ft.
Max Gross Wt.	500 lbs.	Vmax	N/A
Empty Wt.	254 lbs.	VCr	48 mph
Payload (full fuel)	N/A	V_{S1} (stall clean)	20 mph
Fuel Capacity	5 gal.	V_{S0} (lndg config)	N/A
Seats	1	Climb Rate @msl	500 fpm
Cabin Width	N/A	Service Ceiling	N/A

Std. Engine/HP	Usable HP Range	Construction Mtrl.	Est. Bldg. Time
Hirth F-33 / 25 hp	N/A	Metal, Tube, Fabric	400 hrs.
Plans	**Kit**	**Info Packet**	**Video**
$175.00	N/A	Yes	N/A

An ultralight based on the 1940's Fieseler Storch. Accepts pilots up to 6 feet tall and 190 lbs. Plans include 14 page construction manual, 13 page flight manual and 13 drawings. Building costs are estimated to be $4000 including engine.

GARY OLIVER
8301 SE Monterey Ave.
Portland OR 97266
Tel. 503-659-9090

Dart

Wing Span	23.00 ft.	Range	1000 sm
Wing Area	75.00 sq. ft.	Takeoff Dist.	N/A
Length	14.40 ft.	Landing Dist.	N/A
Max Gross Wt.	1200 lbs.	Vmax	200 mph
Empty Wt.	N/A	VCr	170 mph
Payload (full fuel)	N/A	V_{S1} (stall clean)	65 mph
Fuel Capacity	25 gal.	V_{S0} (lndg config)	65 mph
Seats	2	Climb Rate @msl	1250 fpm
Cabin Width	26 in.	Service Ceiling	20,000 ft.

Std. Engine/HP	Usable HP Range	Construction Mtrl.	Est. Bldg. Time
VW 2100 cc / 75 hp	75 - 150	Composite	1500 hrs.
Plans	**Kit**	**Info Packet**	**Video**
$300.00	N/A	$10.00	N/A

A plans-only canard design. Prototype first flown March 2, 1996. Specs are estimated. Acceptable hp range is 75 -150 hp with up to 225 lbs. engine weight. Uses only commercially available parts. Fuselage is glass sandwich and the wing/canard consist of solid core, hot wired airfoils of glass sandwich construction with glass spars. Cockpit layout features a console to the right with side stick controls in both cockpits. Rudder/brake pedals in foward cockpit only. Price for info pack outside the U.S. is $15.00.

Kits/Plans Sold: 1 **Completed & Flying:** 1

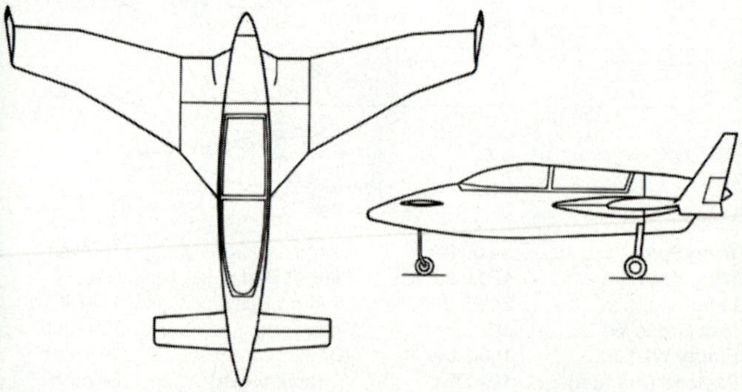

The design started in 1983 as a Delta wing pusher about the size of Jim Miller's Texas Gem Pusher. After doing more research on this design, some problems were discovered that led to the addition of a canard and winglets. The wings were stretched and narrowed. At this point it looked like a "sleeked up" EZ. Test flying of the prototype is scheduled to begin soon.

PARRISH AIRCRAFT XPERIMENTAL, INC.
1125 S.W. 49th Terrace
Plantation FL 33317-4420
Tel. 954-581-4477

PL-9 Stork

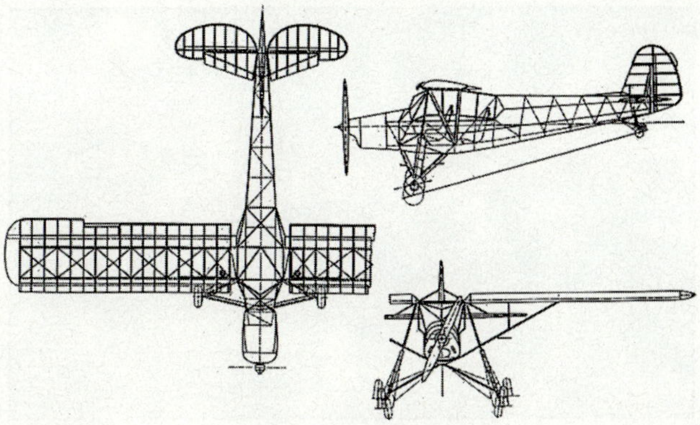

Wing Span	36.00 ft.	**Range**	450 sm
Wing Area	166.0 sq. ft.	**Takeoff Dist.**	246 ft.
Length	24.30 ft.	**Landing Dist.**	85 ft.
Max Gross Wt.	1739 lbs.	**Vmax**	116 mph
Empty Wt.	1132 lbs.	**VCr**	104 mph
Payload (full fuel)	380 lbs.	**V_{S1} (stall clean)**	40 mph
Fuel Capacity	41 gal.	**V_{S0} (lndg config)**	33 mph
Seats	2	**Climb Rate @msl**	1400 fpm
Cabin Width	39 in.	**Service Ceiling**	15,000 ft.

Std. Engine/HP	Usable HP Range	Construction Mtrl.	Est. Bldg. Time
Lyc. 0-320 / 150 hp	N/A	Tube, Metal, Fabric	800 hrs.
Plans	**Kit**	**Info Packet**	**Video**
$450.00	Yes	$5.00	N/A

75% scale model of famous WWII Luftwaffe Fieseler Fi-156 "Storch" known for its outstanding STOL performance. Under-powered prototype first flew 10/11/97. Performance data is estimated with larger engine. Plans are complete including 700+ square feet of drawings. A company in Indonesia will manufacture both kits and complete aircraft.

Kits/Plans Sold: 50

The Fieseler Fi-156 "Storch" was an outstanding WWII airplane, designed to takeoff and land on extremely short fields. The Storch is reported to have had a takeoff ground roll of 131 feet and a landing roll of 36 feet with 13 miles per hour head wind.

The aircraft was used by the Luftwaffe for reconnaissance, VIP transport, casualty evacuation, and other functions associated with a versatile utility aircraft. It was powered by a 240 hp, Argus As 10c engine (eight cylinder, air cooled, inverted "V"), turning a fixed pitch, wooden propeller.

Fifty-five years after the flight of the prototype Storch, a 3/4 scale replica of this aircraft, designed by L. Pazmany, is offered to the amateur builders. His designs include the PL-1, PL-2, PL-4, and Ryson ST-100. Maximum effort has been made to reproduce the performance of the Fi-156. The Pazmany PL-9 "Stork" has the same wing airfoil with fixed leading edge slats, Fowler flaps, drooped ailerons and folding wings. The PL-9 design includes only three major deviations from the original. Wing and empennage construction is aluminum sheet metal (fabric covered the Fi-156). The cockpit has been enlarged slightly from a true 3/4 scale. And, the PL-9 main tires and wheels are larger than a true 3/4 scale size.

PAZMANY AIRCRAFT CORP.
P.O. Box 80051
San Diego CA 92138
Tel. 619-224-7330 • Fax 619-224-7358

Formula GT

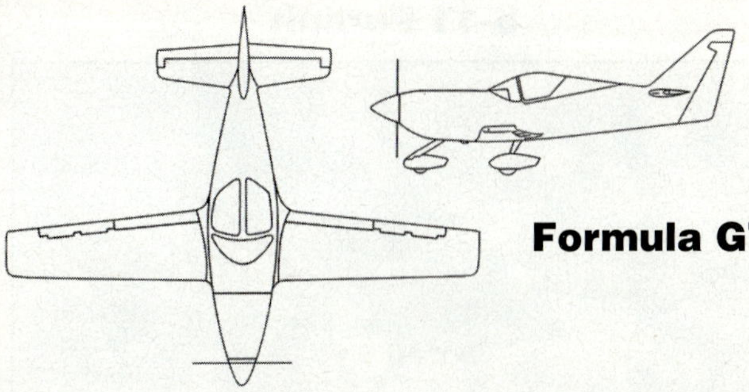

Wing Span	28.50 ft.	**Range**	1500 sm
Wing Area	100.0 sq. ft.	**Takeoff Dist.**	800 ft.
Length	22.00 ft.	**Landing Dist.**	600 ft.
Max Gross Wt.	2300 lbs.	**Vmax**	250 mph
Empty Wt.	1400 lbs.	**VCr**	230 mph
Payload (full fuel)	900 lbs.	**V_{S1} (stall clean)**	68 mph
Fuel Capacity	110 gal.	**V_{S0} (lndg config)**	61 mph
Seats	2	**Climb Rate @msl**	2000 fpm
Cabin Width	47 in.	**Service Ceiling**	23,000 ft.

Std. Engine/HP	Usable HP Range	Construction Mtrl.	Est. Bldg. Time
350 CID V-8 / 300 hp	250-350	Composite	1500 hrs.
Plans	**Kit**	**Info Packet**	**Video**
N/A	$25,500.00	Yes	N/A

Composite airframe with liquid cooled V-8 power. Large baggage compartment with external door. Allows for 225 pounds in each seat plus 100 pounds of baggage. Kit includes slotted electric flaps, electric elevator trim and dual side stick controls.

PERFORMANCE AIRCRAFT
12901 West 151st St., Ste. C
Olathe KS 66062
Tel. 913-780-9140 • Fax 913-780-1774

Ragwing Stork

Wing Span	32.00 ft.	**Range**	250 sm
Wing Area	180.0 sq. ft.	**Takeoff Dist.**	30 ft.
Length	22.00 ft.	**Landing Dist.**	30 ft.
Max Gross Wt.	1000 lbs.	**Vmax**	105 mph
Empty Wt.	497 lbs.	**VCr**	75 mph
Payload (full fuel)	503 lbs.	**V_{S1} (stall clean)**	22 mph
Fuel Capacity	20 gal.	**V_{S0} (lndg config)**	14 mph
Seats	2	**Climb Rate @msl**	1200 fpm
Cabin Width	26 in.	**Service Ceiling**	15,000 ft.

Std. Engine/HP	Usable HP Range	Construction Mtrl.	Est. Bldg. Time
2SI / 70 hp	N/A	Wood, Fabric	600 hrs.
Plans	**Kit**	**Info Packet**	**Video**
$225.00	$8,250.00	$5.00	$15.00

New design. Replica of Feiseler Storch. Plans and kits will be available early 1998. Flies at 15 mph with power on. Takeoff and landing in 30 feet. Video will verify performance figures.

Completed & Flying: 2

RAGWING AVIATION CO.
PO Box 39 • Tokeena Air Park
Townville SC 29689-0039
Tel. 864-225-3773 • Fax 864-972-5606

S-11 Pursuit

Wing Span	20.00 ft.	**Range**	N/A
Wing Area	N/A	**Takeoff Dist.**	375 ft.
Length	18.00 ft.	**Landing Dist.**	500 ft.
Max Gross Wt.	950 lbs.	**Vmax**	225 mph
Empty Wt.	575 lbs.	**VCr**	140 mph
Payload (full fuel)	375 lbs.	**V_{S1} (stall clean)**	48 mph
Fuel Capacity	20 gal.	**V_{S0} (lndg config)**	N/A
Seats	1	**Climb Rate @msl**	1500 fpm
Cabin Width	26 in.	**Service Ceiling**	18,000 ft.

Std. Engine/HP	Usable HP Range	Construction Mtrl.	Est. Bldg. Time
Rotax 912 / 80 hp	N/A	Metal, Composite	N/A
Plans	**Kit**	**Info Packet**	**Video**
N/A	Yes	$15.00	N/A

A unique 'Lifting Body' design - over 80% of lift is generated by the body. Rotax 912 engine standard, Rotax 582 engine optional.

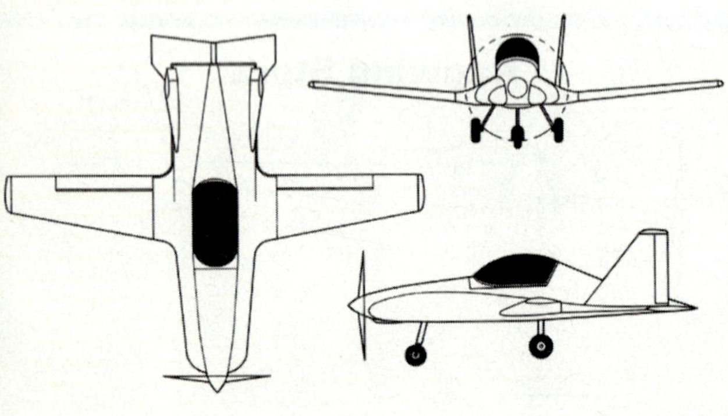

RANS INC.
4600 Highway 183 Alternate
Hays KS 67601
Tel. 785-625-6346 • Fax 785-625-2795

S-17 Stinger

Wing Span	23.50 ft.	**Range**	85 sm
Wing Area	102.5 sq. ft.	**Takeoff Dist.**	100 ft.
Length	17.25 ft.	**Landing Dist.**	175 ft.
Max Gross Wt.	570 lbs.	**Vmax**	63 mph
Empty Wt.	253 lbs.	**VCr**	55 mph
Payload (full fuel)	317 lbs.	**V_{S1} (stall clean)**	27 mph
Fuel Capacity	5 gal.	**V_{S0} (lndg config)**	N/A
Seats	1	**Climb Rate @msl**	800 fpm
Cabin Width	N/A	**Service Ceiling**	13,000 ft.

Std. Engine/HP	Usable HP Range	Construction Mtrl.	Est. Bldg. Time
Rotax 447 / 42 hp	N/A	N/A	200 hrs.
Plans	**Kit**	**Info Packet**	**Video**
N/A	$15,000.00	Yes	N/A

A Part-103 legal ultralight utilizing a one-piece metal fueslage and tube and fabric wings. All controls are internal. Foldable wings. Aircraft stores in three to five minutes.

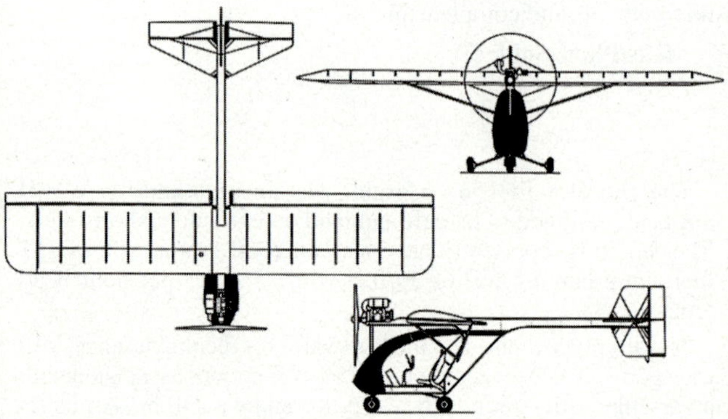

RANS INC.
4600 Highway 183 Alternate
Hays KS 67601
Tel. 785-625-6346 • Fax 785-625-2795

HOMEBUILT "FAQ'S" *(See Page 68)*

Q: How do I license my completed airplane, and what inspections are needed?

A: You *did* document the construction, didn't you?

Before you are issued an airworthiness certificate, an FAA inspector will require an inspection of the aircraft and all documentation of its construction. They used to require a "pre-cover" inspection of the internal structure, but no longer.

Stratonef

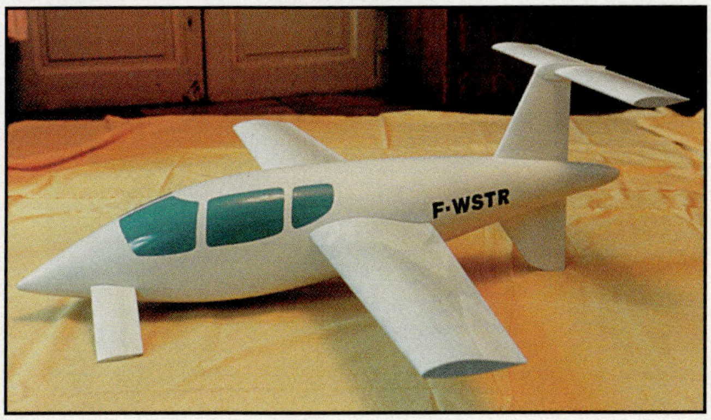

Wing Span	27.00 ft.	**Range**	1000 sm
Wing Area	108.0 sq. ft.	**Takeoff Dist.**	1400 ft.
Length	21.00 ft.	**Landing Dist.**	1500 ft.
Max Gross Wt.	2425 lbs.	**Vmax**	N/A
Empty Wt.	1268 lbs.	**VCr**	235 mph
Payload (full fuel)	835 lbs.	**V_{S1} (stall clean)**	80 mph
Fuel Capacity	53 gal.	**V_{S0} (lndg config)**	65 mph
Seats	5	**Climb Rate @msl**	1700 fpm
Cabin Width	46 in.	**Service Ceiling**	20,000 ft.

Std. Engine/HP	Usable HP Range	Construction Mtrl.	Est. Bldg. Time
Vortec / 230 hp	160 - 230	Composite, Steel Tube	1500 hrs.
Plans	**Kit**	**Info Packet**	**Video**
N/A	$50,000.00	$25.00	Yes

Unique anhedral canard pusher design with "T" tail. Will conform to FAR 23 specifications. Four or five place configuration. Designed for performance with affordable operating costs. Scheduled to be available in 1998. Price in U.S. funds is approximate and subject to change.

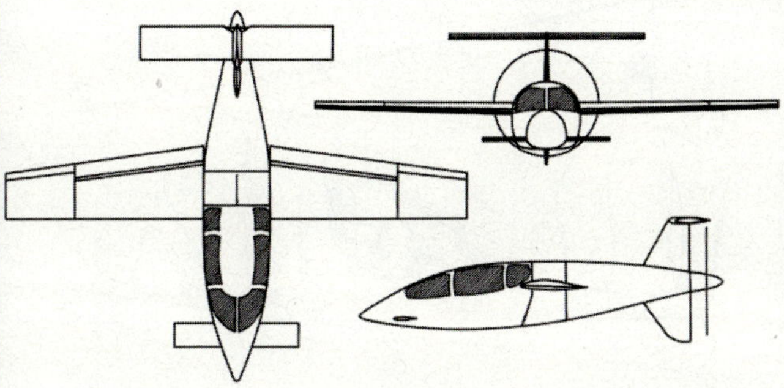

S.C.A.L.E.
15 rue Mirabeau
Angers 49000
FRANCE
Tel. 33-6-11-97-09-53 • Fax 33-2-11-87-72-87

SeAero 254

Wing Span	26.50 ft.	**Range**	90 sm
Wing Area	119.0 sq. ft.	**Takeoff Dist.**	101 ft.
Length	17.25 ft.	**Landing Dist.**	N/A
Max Gross Wt.	550 lbs.	**Vmax**	63 mph
Empty Wt.	289 lbs.	**VCr**	60 mph
Payload (full fuel)	N/A	**V_{S1} (stall clean)**	N/A
Fuel Capacity	5 gal.	**V_{S0} (lndg config)**	24 mph
Seats	1	**Climb Rate @msl**	N/A
Cabin Width	N/A	**Service Ceiling**	N/A

Std. Engine/HP	Usable HP Range	Construction Mtrl.	Est. Bldg. Time
Zenoah G50 / 45 hp	21-45	Fiberglass, Composite	N/A
Plans	**Kit**	**Info Packet**	**Video**
N/A	$14,990.00	N/A	N/A

The SeAero Design Group started with a clean sheet to create a totally new ultralight amphibian with superior open water performance and corrosion proofing as primary design goals. The kit features all molded components for rapid assembly. Wing attach system allows removal for trailering.

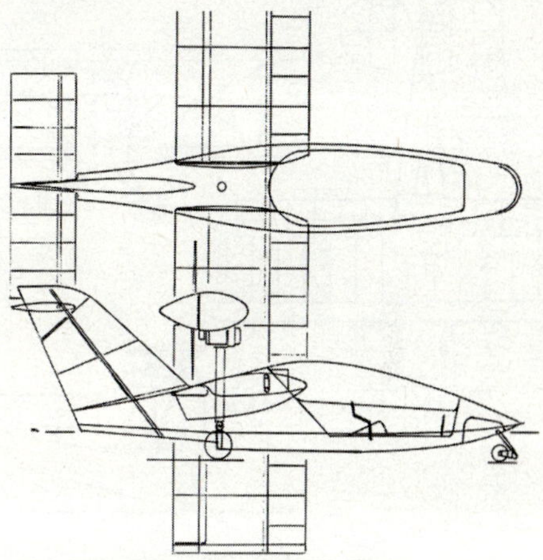

SEAERO CORPORATION
1304A Jamestown Rd.
Williamsburg VA 23185

Now they prefer that in-progress inspections are done by an EAA Technical Counselor, and the inspector will look for his/her comments in the construction log. When the FAA inspector arrives, they expect the aircraft to be ready for inspection (all covers removed), all taxi tests done and logged, and all documentation ready for review.

When the inspector is satisfied that your airplane is ready for in-flight testing, they will issue a restricted airworthiness certificate that describes a test period and testing requirements (or they may insist on changes or repairs if deficiencies

FAQ Continued on Page 358

Smith Sport Miniplane

Wing Span	17.00 ft.	**Range**	N/A
Wing Area	100.0 sq. ft.	**Takeoff Dist.**	400 ft.
Length	15.50 ft.	**Landing Dist.**	400 ft.
Max Gross Wt.	1000 lbs.	**Vmax**	135 mph
Empty Wt.	650 lbs.	**VCr**	125 mph
Payload (full fuel)	N/A	**V_{S1} (stall clean)**	N/A
Fuel Capacity	15 gal.	**V_{S0} (lndg config)**	N/A
Seats	1	**Climb Rate @msl**	N/A
Cabin Width	N/A	**Service Ceiling**	N/A

Std. Engine/HP	Usable HP Range	Construction Mtrl.	Est. Bldg. Time
115 hp	90 - 120	Wood, Tube, Fabric	1300 hrs.
Plans	**Kit**	**Info Packet**	**Video**
$140.00	Yes	$10.00	N/A

A new variation on the Smith Miniplane. Improvements include enlarged fuselage to accept larger pilots, new airfoil, redrawn plans and more. Variety of parts will be available.

SKY CLASSIC AIRCRAFT
108 Jefferson Avenue
Des Moines IA 50314
Tel. 515-967-6490 • Fax 515-243-3915

Sport Copter II

Rotor Span	30.00 ft.	**Range**	400 sm
Rotor Area	N/A	**Takeoff Dist.**	100 ft.
Length	17.00 ft.	**Landing Dist.**	10 ft.
Max Gross Wt.	1300 lbs.	**Vmax**	130 mph
Empty Wt.	650 lbs.	**VCr**	100 mph
Payload (full fuel)	N/A	**V_{S1} (stall clean)**	N/A
Fuel Capacity	25 gal.	**V_{S0} (lndg config)**	N/A
Seats	2	**Climb Rate @msl**	1500 fpm
Cabin Width	40 in.	**Service Ceiling**	15,000 ft.

Std. Engine/HP	Usable HP Range	Construction Mtrl.	Est. Bldg. Time
Subaru 2.2 / 160 hp	N/A	Alum, Fiberglass, Metal	100 hrs.
Plans	**Kit**	**Info Packet**	**Video**
N/A	$27,000.00	$12.00	N/A

New two seat design with stall-proof characteristics. Completion of prototype is expected by mid-1998. Video available September 1998.

Kits/Plans Sold: 3

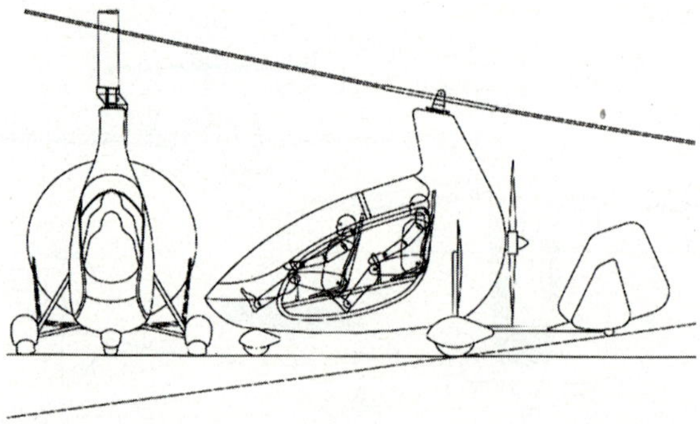

Jack Lenhardt, Hubbard, OR
"Love the Sport Copter...Lots of hours of fun."

SPORT COPTER INTERNATIONAL, INC.
Scappoose Airport • 34012 N. Honeyman Road
Scappoose OR 97056
Tel. 503-543-7000 • Fax 503-543-7041
Web: http://www.sportcopter.com

HOMEBUILT "FAQ" continued

are noted). The testing period is usually 40 flying hours (often 25 if you use a certified engine & prop), and limits you to a fixed testing area, normally a 25 mile radius from the home airport and over unpopulated areas. Passengers are not allowed during the testing period. While testing, keep a "detailed" log of all activities, repairs, and changes.

The inspector will evaluate your testing at the end of the test period, at which point he or she will issue a permanent airworthiness certificate. At that point, you are free to carry passengers and fly most anywhere you like.

FAQ Continued on Page 360

UltraAce

Wing Span	26.00 ft.	Range	N/A
Wing Area	117.0 sq. ft.	Takeoff Dist.	125 ft.
Length	17.50 ft.	Landing Dist.	150 ft.
Max Gross Wt.	525 lbs.	Vmax	63 mph
Empty Wt.	254 lbs.	VCr	60 mph
Payload (full fuel)	271 lbs.	V_{S1} (stall clean)	28 mph
Fuel Capacity	N/A	V_{S0} (Indg config)	N/A
Seats	N/A	Climb Rate @msl	1100 fpm
Cabin Width	N/A	Service Ceiling	N/A

Std. Engine/HP	Usable HP Range	Construction Mtrl.	Est. Bldg. Time
Christine Aero / 48 hp	N/A	Tube, Fabric, Alum.	400 hrs.
Plans	**Kit**	**Info Packet**	**Video**
Yes	Yes	Yes	N/A

A new high wing, open cockpit, taildragger, ultralight design reminiscent of the "thirties" era of aviation. Prototype testing to be finished September 1996. Performance specs are estimated.

SOUTHERN LIGHTS MFG.
2718 Mayo Rd.
Augusta GA 30907
Tel. 706-737-0090

Sprint WS202

Wing Span	27.00 ft.	Range	400 sm
Wing Area	130.0 sq. ft.	Takeoff Dist.	500 ft.
Length	19.00 ft.	Landing Dist.	500 ft.
Max Gross Wt.	1150 lbs.	Vmax	150 mph
Empty Wt.	670 lbs.	VCr	105 mph
Payload (full fuel)	480 lbs.	V_{S1} (stall clean)	41 mph
Fuel Capacity	14 gal.	V_{S0} (Indg config)	37 mph
Seats	2	Climb Rate @msl	500 fpm
Cabin Width	45 in.	Service Ceiling	10,000 ft.

Std. Engine/HP	Usable HP Range	Construction Mtrl.	Est. Bldg. Time
Subaru / 65 hp	65 - 100	Aluminum	1500 hrs.
Plans	**Kit**	**Info Packet**	**Video**
Yes	N/A	$5.00	N/A

"Scratch-built" from plans only for as little as $5,000 including engine. Plane is at testing stage. Designed for easy change between tri-gear and taildragger configurations. Several engine options. **Completed & Flying:** 1

W. SPRING
4493 Tremineer Avenue
Burlington, Ontario L7L 1H8
CANADA
Tel. 905-632-7693 • Fax 905-632-2269

Specter II

Wing Span	34.00 ft.	Range	N/A
Wing Area	147.0 sq. ft.	Takeoff Dist.	N/A
Length	16.50 ft.	Landing Dist.	N/A
Max Gross Wt.	1600 lbs.	Vmax	N/A
Empty Wt.	1000 lbs.	VCr	140 mph
Payload (full fuel)	450 lbs.	V_{S1} (stall clean)	54 mph
Fuel Capacity	23 gal.	V_{S0} (Indg config)	N/A
Seats	2	Climb Rate @msl	N/A
Cabin Width	22 in.	Service Ceiling	N/A

Std. Engine/HP	Usable HP Range	Construction Mtrl.	Est. Bldg. Time
Subaru / 118 hp	90 - 125	Composite	850 hrs.
Plans	**Kit**	**Info Packet**	**Video**
N/A	$29,495.00	N/A	N/A

Fast-build kit with all major structural items already bonded in place. Stall resistant canard design. Tandem seating for two.

Completed & Flying: 1

SPECTER AIRCRAFT
PO Box 87 • Bancroft ID 83217
Tel. 208-648-7668 • Fax 208-648-7742
E-Mail: specterair@aol.com

YT33

Wing Span	26.67 ft.	Range	1200 sm
Wing Area	110.0 sq. ft.	Takeoff Dist.	2000 ft.
Length	25.83 ft.	Landing Dist.	3000 ft.
Max Gross Wt.	2920 lbs.	Vmax	325 mph
Empty Wt.	2205 lbs.	VCr	300 mph
Payload (full fuel)	N/A	V_{S1} (stall clean)	75 mph
Fuel Capacity	200 gal.	V_{S0} (Indg config)	N/A
Seats	2	Climb Rate @msl	1500 fpm
Cabin Width	N/A	Service Ceiling	30,000 ft.

Std. Engine/HP	Usable HP Range	Construction Mtrl.	Est. Bldg. Time
Turbo Meca Turbojet	N/A	Composite	3000 hrs.
Plans	**Kit**	**Info Packet**	**Video**
N/A	N/A	$20.00	N/A

2/3 scale, composite replica of the T-33, America's first jet trainer. With 880 pound thrust engine.

Completed & Flying: 1

STARGATE
2932 NE Waggoner Dr.
McMinville OR 97128
Tel. 503-435-1524 • Fax 503-435-1524

Sportster

Wing Span	17.50 ft.	Range	N/A
Wing Area	82.00 sq. ft.	Takeoff Dist.	N/A
Length	14.00 ft.	Landing Dist.	N/A
Max Gross Wt.	460 lbs.	Vmax	N/A
Empty Wt.	252 lbs.	VCr	N/A
Payload (full fuel)	200 lbs.	V_{S1} (stall clean)	N/A
Fuel Capacity	5 gal.	V_{S0} (lndg config)	N/A
Seats	1	Climb Rate @msl	N/A
Cabin Width	20 in.	Service Ceiling	N/A

Std. Engine/HP	Usable HP Range	Construction Mtrl.	Est. Bldg. Time
N/A	N/A	Alum., Steel, Wood, Foam	400 hrs.
Plans	**Kit**	**Info Packet**	**Video**
N/A	$4,994.00	Yes	$15.00

A model to satisfy your nostalgic needs with many of the features of those Golden Thirties craft. Upper wing mounted on see-through pylon. Kits expected to be available in March, 1998.

Completed & Flying: 2

THEISS AVIATION
PO Box 1086
Salem OH 44460
Tel. 330-332-2031 • Fax 330-332-8273

Cheeky

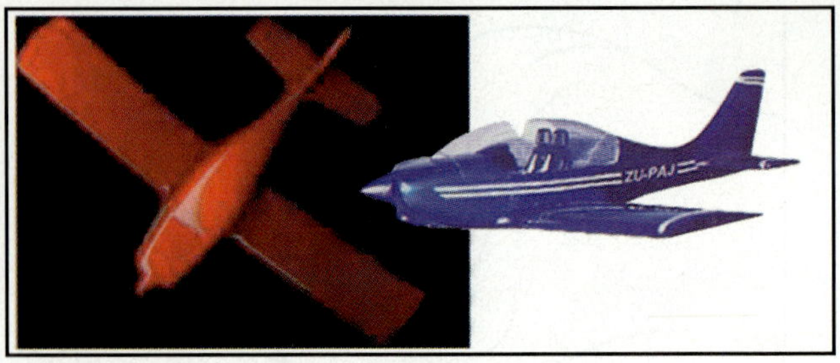

Wing Span	31.10 ft.	Range	N/A
Wing Area	133.8 sq. ft.	Takeoff Dist.	210 ft.
Length	19.50 ft.	Landing Dist.	197 ft.
Max Gross Wt.	N/A	Vmax	118 mph
Empty Wt.	N/A	VCr	101 mph
Payload (full fuel)	N/A	V_{S1} (stall clean)	34 mph
Fuel Capacity	N/A	V_{S0} (lndg config)	N/A
Seats	N/A	Climb Rate @msl	1570 fpm
Cabin Width	48 in.	Service Ceiling	N/A

Std. Engine/HP	Usable HP Range	Construction Mtrl.	Est. Bldg. Time
Rotax 912 / 80 hp	N/A	N/A	N/A
Plans	**Kit**	**Info Packet**	**Video**
Yes	N/A	N/A	N/A

A low wing design with a forward sliding canopy based in part on changing a Koala to a low wing configuration and installing a four stroke engine. Plans will be made available after the first prototype is built, flight tested and evaluated.

ULTRALIGHT MARKETING
PO Box 10 • Vereeniging 1930 • SOUTH AFRICA
Tel. 2716 42 42 21 2 • Fax 2716 42 42 21 2
E-Mail: cheeky@cyberserv.co.za
Web: http://www.geocities.com/Baja/9163/cheeky.html

HOMEBUILT "FAQ" continued

If any major modifications or repairs are done later, the airplane may need a re-inspection and retest. Call your local FSDO before doing this to find out what they want to see.

You can also be issued a Repairman's Certificate for *your* airplane only ("not" the type in general). This allows you to perform all repairs, inspections, annuals, etc. on the airframe, since they figure if you built it, you should be able to fix it. Note that in FAR Part 45, an "annual" is referred to as a "condition check," which is legally different from an "annual inspection," even if both actions are intended to accomplish the same goal.

Please note that the above information is valid only in the USA, other countries usually have very similar require ments, with some slight differences. Check with your local authorities before committing any large sums of money or time to a project.

Also note that this procedure is the general case. It is entirely possible that you may experience variations in the procedure. For instance, one builder commented that his inspector waived the second inspection and allowed standard experimental privileges immediately after the test time was flown off and logged. As FAA policy can vary between regions, expect some slight variations.

Pairadigm

Wing Span	32.00 ft.	Range	250 sm
Wing Area	160.0 sq. ft.	Takeoff Dist.	100 ft.
Length	22.00 ft.	Landing Dist.	200 ft.
Max Gross Wt.	1400 lbs.	Vmax	90 mph
Empty Wt.	700 lbs.	VCr	80 mph
Payload (full fuel)	700 lbs.	V_{S1} (stall clean)	26 mph
Fuel Capacity	24 gal.	V_{S0} (lndg config)	20 mph
Seats	2	Climb Rate @msl	1200 fpm
Cabin Width	43 in.	Service Ceiling	15,000 ft.

Std. Engine/HP	Usable HP Range	Construction Mtrl.	Est. Bldg. Time
Twin 2SI 45 hp / 90 hp	40 - 75	Alum., Fabric	250 hrs.
Plans	**Kit**	**Info Packet**	**Video**
N/A	$24,995.00	$10.00	$15.00

Twin engine, two place, high wing strut-braced aircraft. Engines are mounted on fuselage in a tractor-pusher configuration. With a unique high-lift double-surface wing. Available in July 1998. Engines and props not included. **Sold:** 10 **Flying:** 1

VSTOL AIRCRAFT CORP.
PO Box 7534
Ft. Myers FL 33911
Tel. 941-936-1261 • Fax 941-936-1261

Spitz S1

Wing Span	18.27 ft.	**Range**	N/A
Wing Area	116.0 sq. ft.	**Takeoff Dist.**	N/A
Length	15.23 ft.	**Landing Dist.**	N/A
Max Gross Wt.	475 lbs.	**Vmax**	90 mph
Empty Wt.	254 lbs.	**VCr**	63 mph
Payload (full fuel)	210 lbs.	**V_{S1} (stall clean)**	25 mph
Fuel Capacity	5 gal.	**V_{S0} (lndg config)**	N/A
Seats	1	**Climb Rate @msl**	N/A
Cabin Width	N/A	**Service Ceiling**	N/A

Std. Engine/HP	Usable HP Range	Construction Mtrl.	Est. Bldg. Time
Christine Aero / 48 hp	N/A	Tube, Wood, Fabric	N/A
Plans	**Kit**	**Info Packet**	**Video**
$150.00	$3,848.00	N/A	N/A

Specs are for Ultralight model. Experimental version has a maximum speed of 120 mph and a kit price of $4,154. Kit prices do not include engine or instruments. .

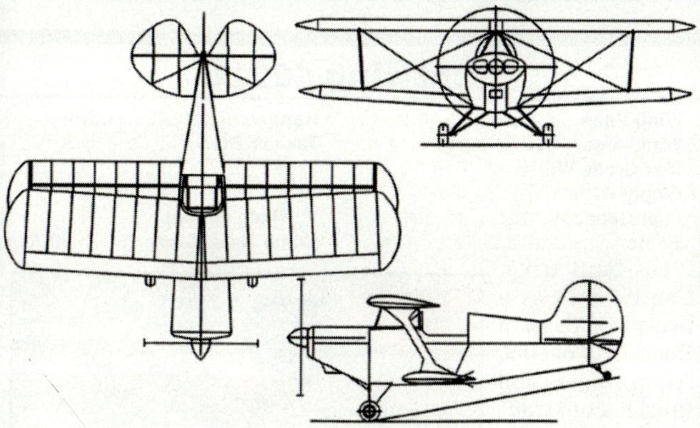

A. VANCIL
247 Murphy Lane
Belton SC 29627

Viperfan

Wing Span	24.50 ft.	**Range**	1500 sm
Wing Area	85.00 sq. ft.	**Takeoff Dist.**	1000 ft.
Length	26.50 ft.	**Landing Dist.**	1600 ft.
Max Gross Wt.	2500 lbs.	**Vmax**	500 mph
Empty Wt.	1450 lbs.	**VCr**	400 mph
Payload (full fuel)	600 lbs.	**V_{S1} (stall clean)**	74 mph
Fuel Capacity	100 gal.	**V_{S0} (lndg config)**	66 mph
Seats	2	**Climb Rate @msl**	4000 fpm
Cabin Width	31 in.	**Service Ceiling**	31,000 ft.

Std. Engine/HP	Usable HP Range	Construction Mtrl.	Est. Bldg. Time
TSIOL-520 / 350 hp	240 - 350	Composite, Tube	2000 hrs.
Plans	**Kit**	**Info Packet**	**Video**
N/A	$59,500.00	$20.00	N/A

Jet profile, two place, tandem seat, full dual control aircraft. Offered with a variety of engines including the IO-520, TSIO-520, and TSIOL-550. Designed to cruise from 295 mph with the VF-240 model to 412 mph with the VF-325T version. Rated +6, -6 G's with full aerobatic capability. **Completed & Flying:** 1

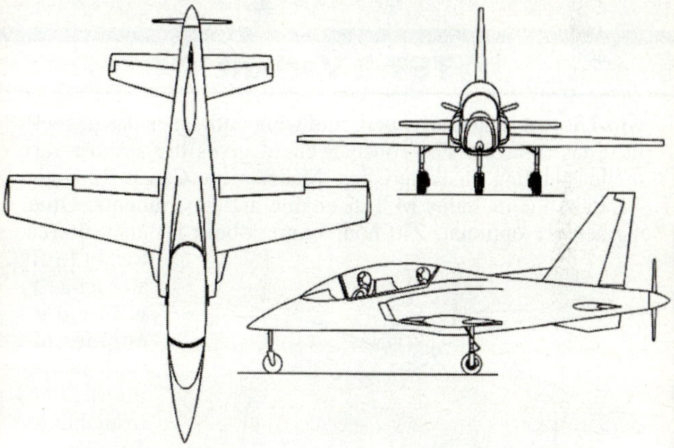

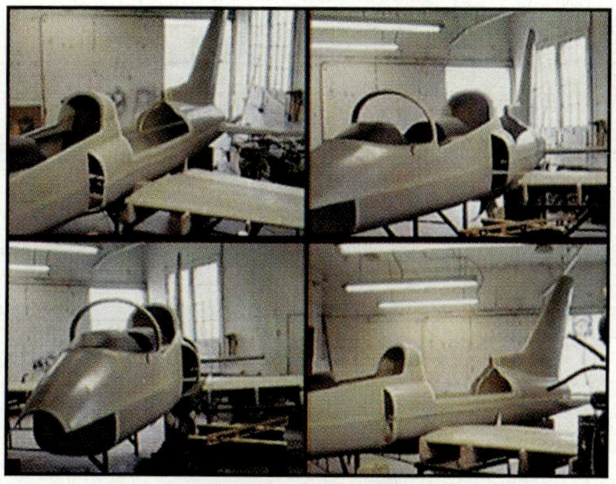

VIPER AIRCRAFT
2532 West 4th Place
Kennewick WA 99336
Ph: 509-543-3570 • Fax: 509-543-3570
E-Mail: shanchette@aol.com

Jet Fox 97

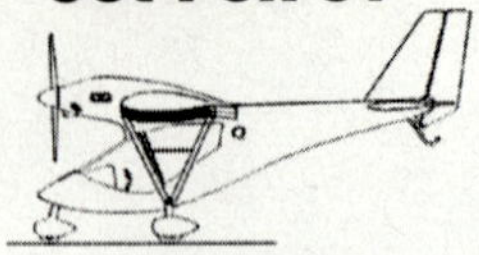

EUROALA
L.go Mosciano, 28
64046 Montorio Al Vomano (TE) ITALY
Tel: 0348/3395009 Fax: 0861/870979

Excalibur

Wing Span	31.50 ft.	Range	250 sm
Wing Area	173.0 sq. ft.	Takeoff Dist.	100 ft.
Length	20.00 ft.	Landing Dist.	200 ft.
Max Gross Wt.	1000 lbs.	Vmax	90 mph
Empty Wt.	450 lbs.	VCr	70 mph
Payload (full fuel)	N/A	V_{S1} (stall clean)	22 mph
Fuel Capacity	10 gal.	V_{S0} (lndg config)	N/A
Seats	2	Climb Rate @msl	1200 fpm
Cabin Width	N/A	Service Ceiling	14,000 ft.

Std. Engine/HP	Usable HP Range	Construction Mtrl.	Est. Bldg. Time
Rotax 503 / 52 hp	52 - 65	Alum. Tube, Fabric	350 hrs.

Plans	Kit	Info Packet	Video
N/A	Yes	$5.00	Yes

High wing amphibian pusher. Aluminum tube airframe and Superflite fabric construction. Some prefabricated components available. **Kits/Plans Sold:** 1

EXCALIBUR AIRCRAFT CO.
3812 Palazzo St.
Sebring FL 33872
Tel: 941-385-9486 • Fax: 941-385-9486

Z-21 Flitzer

Wing Span	18.00 ft.	Range	N/A
Fuel Capacity	9 gal.	V_{S0} (lndg config)	N/A
Seats	1	Climb Rate @msl	N/A

Std. Engine/HP	Usable HP Range	Construction Mtrl.	Est. Bldg. Time
VW 1834cc	N/A	Wood, Fabric	N/A hrs.

Plans	Kit	Info Packet	Video
Yes	Yes	N/A	N/A

Available as kit or plans-only. Kit does not include engine. Wood and fabric construction with minimal use of metal. Accepts 6'6" pilot with ease. 3+ hours endurance.

BELL AEROMARINE
153 Parker Drive • Leicester LE4 0JP
UNITED KINGDOM
Tel: +44 116 234 008 • Fax: +44 116 234 029

P-51 Mustang

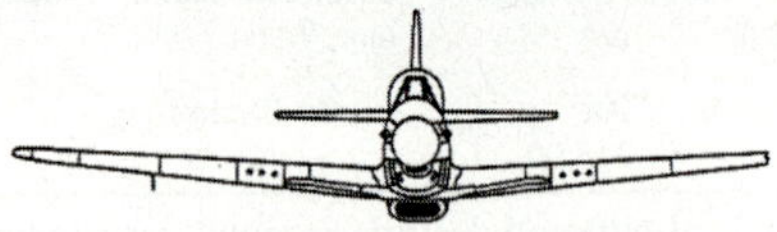

Information pack available.

EXCLUSIVE AVIATION
1705 19th Ave. North • Fargo ND 58102
Tel: 701-293-9331 • Fax 701-293-9331

TST-5 Variant

New UL two seater, all-wood, mid wing with an enclosed cockpit, dual controls. The constant chord gives the aircraft very docile handling characteristics. Made in the Czech Republic. Standard kit includes M 125 engine and instruments. Other engines are optional. 250 hour Express-built kit also offered. Ready to fly aircraft, without engine but with instruments, are available for $16,950.

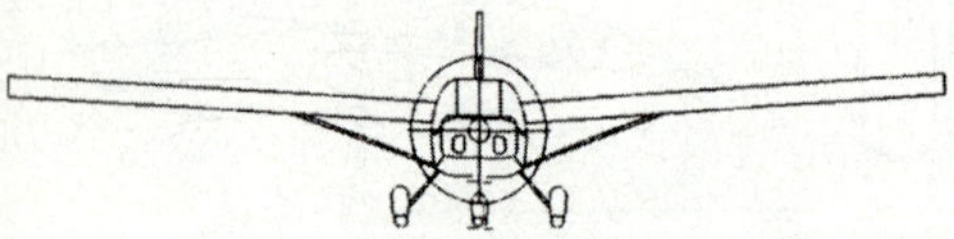

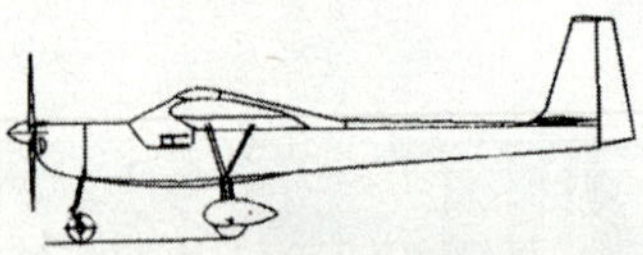

BOHEMICAN INTERNATIONAL
36 Buchanan Road • St. Catharines, Ontario L2M 4R6
CANADA
Tel: 905-935-4938 • Fax: 905-935-4938
E-Mail: pdokter@niagara.com

Huntwing 462LC

Wing Span	33.40 ft.	Range	N/A
Wing Area	161.0 sq. ft.	Takeoff Dist.	N/A
Max Gross Wt.	818 lbs.	Vmax	70 mph
Empty Wt.	N/A	VCr	60 mph
Fuel Capacity	13 gal.	V_{S0} (lndg config)	N/A
Seats	1	Climb Rate @msl	1200 fpm

Plans-built trike. Can be built as a two place. Uses a Rotax 462 or 582. Alum., tube, and fabric construction. Plans cost 125 UK pounds for 100 drawings and 100 descriptive pages.

JOHN HUNT
Old Railway Inn • Clydach Station Near Abergaven
Gwent NP7 0RD • UNITED KINGDOM
Tel: 44-873-830876

Cutting/Wilshire Spitfire

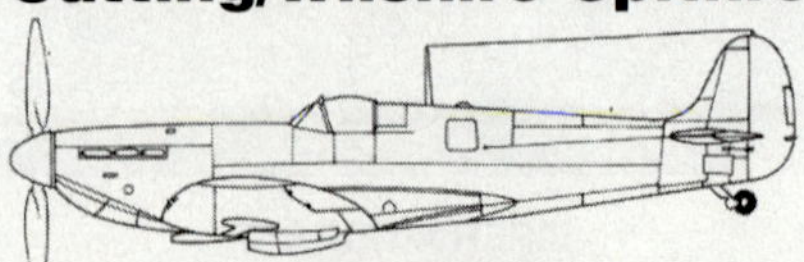

BOB CUTTING
4635 Britannia Dr. • Richmond, BC V7E 6B1
CANADA
Tel: 604-275-1603

Raptor

Functional, spacious, four place sport-utility. Taildragger, tri-gear, and float options.

INTER-TECH AIRCRAFT, INC.
P.O. Box 822
Powell WY 82435
Tel: 307-754-0274

Aerotek's Ultralite-Flea

Tandem wing pusher. FAR 103 ultralight. 18 ft. wing span. Stall proof. Plans only. $16.95.

POPULAR LINK
P.O. Box 1981
Sugar Land TX 77487
Web: http://www.popularlink.com/

Kestrel

Two place, wide cabin, high wing, taildragger. Tube and fabric construction.

INTER-TECH AIRCRAFT, INC.
P.O. Box 822
Powell WY 82435
Tel: 307-754-0274

FX Superlight

High wing single or two seater. Simple design. Traditional fabric covered wooden light airplane. Use Rotex 27 hp engine. FAR 103 ultralight. Plans only. $19.95.

POPULAR LINK
P.O. Box 1981
Sugar Land TX 77487
Web: http://www.popularlink.com/

Jeof Candiana

Wing Span	28.40 ft.	**Takeoff Dist.**	328 ft.
Wing Area	N/A	**Landing Dist.**	328 ft.
Length	21.90 ft.	**Vmax**	130 mph
Empty Wt.	622 lbs.	**VCr**	106 mph
Range	560 sm	**V_{S1} (stall clean)**	39 mph

Conceived as a testbed for the Sax 86 engine which is derived from the four cylinder Fiat "Fire" car engine. Steel, tube, and alum. construction. Tricycle and conventional gear versions. Available in Kits.

Completed: 10

JEOF S.R.L.
Via Valli Pontecasale, 36 • 35020 Candiana (PD)
ITALY
Tel: 39-49-5349320 • Fax 39-49-5349320

Keleher JK-1 Lark

Wing Span	23.00 ft.	**Range**	275 sm
Wing Area	81.00 sq. ft.	**Takeoff Dist.**	600 ft.
Length	17.00 ft.	**Landing Dist.**	700 ft.
Max Gross Wt.	810 lbs.	**Vmax**	130 mph
Empty Wt.	670 lbs.	**VCr**	120 mph
Payload (full fuel)	N/A	**V_{S1} (stall clean)**	58 mph
Fuel Capacity	14 gal.	**V_{S0} (lndg config)**	N/A
Seats	1	**Climb Rate @msl**	1000 fpm
Cabin Width	N/A	**Service Ceiling**	15,000 ft.

Std. Engine/HP	Usable HP Range	Construction Mtrl.	Est. Bldg. Time
Cont. CA-85 / 85 hp	N/A	Metal, Wood, Fabric	2500 hrs.

Plans	Kit	Info Packet	Video
$200.00	N/A	$15.00	N/A

Plans built, mid wing, single seater. VFR only. Some fabricated components available.

Kits/Plans Sold: 14

R&D AIRCRAFT
PO Box 2523
Castro Valley CA 94546
Tel: 510-581-8728 • Fax: 510-733-6082

Mong Sport

Small (14 ft. length, 16.8 ft. wing span) sport biplane. Plans $100. Some prewelded components available.

MIDGET BIPLANE ASSN.
21 Hemlock Dr.
Mayville NY 14757

X-Air

Two seat, 3-axis European ultralight. Tube and fabric construction. Assembles in about 40 hours.

RAND KAR
Sarl Canal de la Martiniere • Frossay 44320
FRANCE
Tel: 00 (33) 40 64 21 • Fax: 00 (33) 40 64 15

ACE Gyrocopter

70+ mph gyroplane. Very simple. Fly with stick, throttle, and rudder only. Use Rotex or VW engine. Can be assembled under FAR 103 ultralight. Plans only. $19.95.

POPULAR LINK
P.O. Box 1981
Sugar Land TX 77487
Web: http://www.popularlink.com/

Jodel D92

SOCITE AERONAUTIQUE DE BEAUNE (SAB)
37 route de Seurre • 21200 Beaune • FRANCE

Wichawk

Wing Span	24.00 ft.	**Range**	500 sm
Wing Area	180.0 sq. ft.	**Takeoff Dist.**	150 ft.
Length	19.00 ft.	**Landing Dist.**	300 ft.
Max Gross Wt.	2400 lbs.	**Vmax**	140 mph
Empty Wt.	1300 lbs.	**VCr**	127 mph
Fuel Capacity	40 gal.	**Service Ceiling**	18,000 ft.
Seats	2	**Climb Rate @msl**	1700 fpm

Std. Engine/HP	Usable HP Range	Construction Mtrl.	Est. Bldg. Time
Lyc. 0-360 / 180 hp	160 - 300	Metal Tube Wood Fabric Comp.	1800 hrs.

Side by side seating. Third seat optional. Variety of construction material choices. Plans cost $375, Info pack $20. **Flying:** 100

SPORT RACER, INC.
14 Hawthorne Road
Valley Center KS 67147
Tel: 316-755-0659

Breezy

Classic open frame and cockpit, fun flyer. Unger's prototype is in EAA's Air Adventure Museum. Wing drawings as the original was designed to use PA-12 Piper Super Cruiser wings. PA-14, PA-18, J-3/4/5 Cub wings can also be used.

CARL UNGER
8751 S. Kilboum
Hometown IL 60456

Ultimate 10-300

STREAMLINE WELDING, INC.
296 Homewood Ave. • Hamilton, Ontario L8P 2M9 CANADA
Tel: 905-526-9990 • Fax 905-526-9990

Stearman

Wing Span	24.30 ft.	**Max Gross Wt.**	1450 lbs.
Wing Area	170.0 sq. ft.	**Empty Wt.**	1000 lbs.
Length	18.80 ft.	**Fuel Capacity**	20 gal.

Stearman aircraft sub-assemblies. "You assemble." Certified Airworthy airframe $49,500.

VINTAGE AEROPLANE CO.
Tel: 601-467-7078

TL-132 Condor

Wing Span	34.78 ft.	**Vmax**	100 mph
Wing Area	145.3 sq. ft.	**VCr**	80 mph
Max Gross Wt.	N/A	V_{S1} **(stall clean)**	38 mph

Award winning ultralight design from the Czech Republic. Most popular two seater sold there. Engine is Rotax 503 52 hp. Certified in several European countries. Kit available in the U.S. for $20,000. Info packet available. **Flying:** 50+

TL ULTRALIGHT CO.
Dobrovskeho 734 • Hradec Kralove 500 02
CZECH REPUBLIC
Tel: (+42-49) 613378 • Fax (+42-49) 613378

Travel Air 2000

Plans and fabricated assemblies. Info kit $25.

THRUST, INC.
Tel: 407-324-9433

SUPPLIES AND SERVICES

- Supplies and services are listed alphabetically by company within the categories below. A Company Index is located on page 422.
- All information has been provided by the companies and from other sources. Listing of a company in AEROCRAFTER does not constitute the publisher's endorsement of the quality of the aircraft, plans, product or service offered, nor guarantee the accuracy of any claims or specifications.
- Company information is current at the time of publication but, as with all directories, addresses, phone numbers, etc. can change. We appreciate hearing about any such change.
- When contacting a company, be sure to mention you found them in AEROCRAFTER. If you know of a company that belongs in AEROCRAFTER, send us an e-mail at aerocftr@baicorp.com or drop us a note at the address on page 3.

SUPPLIES

SERVICES

• AUTOPILOTS, AVIONICS & ANTENNAS

Autopilots

Aircraft Spruce & Specialty
PO Box 4000, Corona CA 91718-9961
Ph: 800-824-1930 • Fax: 909-372-0555
E-Mail: info@aircraft-spruce.com
Web: http://www.aircraft-spruce.com
Literature/Catalog: $5.00
A complete selection of homebuilt and certified aircraft parts and materials. Shipping worldwide since 1965. Also see Aircraft Spruce East, Griffin GA, 800-831-2949.
See Display Ad On Page 24

American Avionics, Inc.
7675 Perimeter Road South
Seattle WA 98108
Ph: 206-763-8530 • Fax: 206-763-2036
E-Mail: sales@americanavionics.com
Literature/Catalog: Yes

Astronautics Corp. of America
5800 Cottage Grove Rd., Madison WI 53716
Ph: 608-221-9001 • Fax: 608-221-9104
Literature/Catalog: Yes
For helicopters.

Avcom Avionics & Instr., Inc.
3008 NW 82nd Ave., Miami FL 33122
Ph: 305-436-0300 • Fax: 305-436-0301
Literature/Catalog: Yes
Sales, service, and overhaul of all makes of avionics, instruments, radars and accessories.

Avionics Place
5257 Falcon Rd. • Rockford IL 61109
Ph: 815-229-5360 • Fax: 815-229-0308
Literature/Catalog: Yes
Sales, installation, and service of aircraft avionics.

Brittain Industries, Inc.
3266 N. Sheridian Road, Tulsa OK 74115
Ph: 918-836-7701 • Fax: 918-836-7703
E-Mail: brittainaps@juno.com
Literature/Catalog: Yes
Supplier of autopilots from basic stabilization to full 3 axis.

Century Flight Systems, Inc.
P.O. Box 610, Mineral Wells TX 76067
Ph: 800-433-5630 • Fax: 817-325-2546
E-Mail: century@centuryflight.com
Web: http://www.centuryflight.com
Literature/Catalog: Yes
Manufacturer of a wide range of systems, from single axis systems to 3 axis flight directors.

Empire Development Avionics
2107 N. 1st St. Suite 590, San Jose CA 95131-2027, E-Mail: chanik@aol.com
Web: http://members.aol.com/chanik/ed
The SE5B is a tri-axis, trim-tab autopilot implementing pitch and roll control with 1/4-scale servo motors. It uses a combination of GPS or Loran receivers for heading, bearing and XTE information.

Gator Aviation Co.
2826 NE 19th Drive, Gainesville FL 32609
Ph: 904-665-7930 Literature/Catalog: Yes

Gulf Coast Avionics Corp.
4243 N. Westshore Blvd., Tampa FL 33614
Ph: 813-879-9714 • Fax: 813-875-4514
E-Mail: gca@ix.netcom.com
Web: http://www.gulf-coast-avionics.com
Literature/Catalog: Yes
Avionics supplier. Custom instrument panel manufacturing.

Lafayette Avionics
Hangar 7, Purdue Airport
West Lafayette IN 47906
Ph: 317-743-3828 • Fax: 317-743-0000
E-Mail: ron@avionics-laf.com
Web: http://avionics-laf.com
Literature/Catalog: Yes
Avionics sales and service on most kinds of general aviation electronics.

Navaid Devices, Inc.
641 N. Market St., Chattanooga TN 37405
Ph: 423-267-3311 • Fax: 423-756-6154
Literature/Catalog: Yes
Manufacturers of autopilots and electronic ignitions for experimental category (kitbuilt) aircraft.

Otto Instrument Service, Inc.
Ontario Int'l. Airport • 1441 Valencia Place
Ontario CA 91761
Ph: 909-930-5800 • Fax: 909-930-5808
Instruments and autopilot overhauling. 50 years experience.

Porcine Associates
244 O'Connor Street, Menlo Park CA 94025
Ph: 415-326-2669 • Fax: 415-326-1071
E-Mail: jimham@porcine.com
Web: http://www.porcine.com
Literature/Catalog: Yes
Hand held GPS linked with autopilot. Will work with most hand held GPS receivers. $189.00 complete with manual and connector. Send self addressed stamped envelope for information.

Sky West Instruments
1749 W. 13th Street, Upland CA 91786
Ph: 909-985-5718 • Fax: 909-985-5718
Lexan instrument panel overlays with indirect lighting. Custom fabricated to your specifications. Instrument overhaul, sales and services.

Skycom Avionics, Inc.
9305 W. Appleton Avenue
Milwaukee WI 53225
Ph: 414-463-8180 • Fax: 414-463-8195
Literature/Catalog: Yes
Aircraft electronics sales, service and installation. Your source for new, rebuilt and recertified packaged instrumentation for every homebuilt aircraft. Free Skycom Avionics Newsletter on request. Branches: Kenosha, Phone 414-652-8120; Waukesha, Phone 414-521-8180; Madison, Phone 608-246-8180.

S-TEC Corp.
One S-TEC Way • Municipal Airport
Mineral Wells TX 76067-9236
Ph: 940-325-9406 • Fax: 940-325-3904
E-Mail: kenpaul@flash.net
Web: http://www.s-tec.com/kitplane
Literature/Catalog: Yes
Manufacturer of a complete line of flight control systems, autopilots and flight directors. TEC line avionics - NAV, COM, ADF, DME and transponder. H.S.I. and R.M.I. systems.

Val Avionics, Ltd.
P.O.Box 13025, Salem OR 97309-1025
Ph: 503-370-9429 • Fax: 503-370-9885
Literature/Catalog: Yes
Manufacture avionics including COM760TSO VHF transceivers, ILS400TSO instrument landing systems, and base station 7600 systems.

Western Avionics Inc.
19711 Campus Drive #A • John Wayne Airport, Santa Ana CA 92707-5203
Ph: 714-757-0290 • Fax: 714-757-0293
Literature/Catalog: Yes
Sales, service, installation of avionics, autopilots, radar systems, GPS, and moving maps.

Avionics & Antennas

II Morrow, Inc.
2345 Turner Road SE, Salem OR 97302
Ph: 800-525-6726 • Fax: 503-364-2138
E-Mail: IIM1JMB@iim.ups.com
Literature/Catalog: Yes
GPS, Loran, navigation management systems.

ACK Technologies, Inc.
440 W. Julian Street, San Jose CA 95110
Ph: 408-287-8021 • Fax: 408-971-6879
Literature/Catalog: Yes

Advanced Aircraft Electronics
PO Box 4111, Florissant MO 63032
Ph: 314-837-0106 • Fax: 314-837-0106
Literature/Catalog: Yes
Supplier of antennas for communication, navigation, transponders, and ELTs. Antennas are designed to work without a ground plane in composites, rag & tube airframes, and ultralights.

Advanced Composite Tech.
R.D. #5, Box 214C, Somerset PA 15501
Ph: 814-445-3802 • *Literature/Catalog: Yes*

Aero Tech Aviation Hardware
1860-B Joe Crosson Dr., El Cajon CA 92020
Ph: 800-448-4457 • Fax: 619-448-7479

AeroElectric Connection
6936 Bainbridge Road, Wichita KS
67226-1008 • Ph: 316-685-8617
E-Mail: nuckolls@aeroelectric.com
Web: http://www.aeroelectric.com
Literature/Catalog: Yes
The AeroElectric Connection is an electrical systems handbook for individuals who build and maintain their own airplanes. The AeroElectric Connection is adding a line of supplies and accessories. Brochure available for #10 SASE.

Aerosonic Corp.
1212 N. Hercules Avenue, Clearwater FL 34625
Ph: 813-461-3000 • Fax: 813-447-5926
Literature/Catalog: Yes
Altimeters, air speeds, vertical speeds, encoding altimeters, manifold pressure indicators, and test equipment. Founded in 1953.

Aircraft Spruce & Specialty
PO Box 4000, Corona CA 91718-9961
Ph: 800-824-1930 • Fax: 909-372-0555
E-Mail: info@aircraft-spruce.com
Web: http://www.aircraft-spruce.com
Literature/Catalog: $5.00
A complete selection of homebuilt and certified aircraft parts and materials. Shipping worldwide since 1965. Also see Aircraft Spruce East, Griffin GA, 800-831-2949.
See Display Ad On Page 24

Aire-Sciences, Inc.
19B Gardner Rd., Fairfield NJ 07004
Ph: 973-575-7725 • Fax: 973-575-4835
Literature/Catalog: Yes
Manufacturer of communication equipment for general aviation.

Airsport Corporation
1100 W. Cherokee, Sallisaw OK 74955
Ph: 918-775-4010 • Fax: 918-775-4000
E-Mail: Info@Airsport-Corp.com
Web: http://www. Airsport-Corp.com
Literature/Catalog: Yes
Altitude alerters, transponder data displays.

Allegro Avionics
1332 E. Hampton, Tucson AZ 85719
Ph: 520-327-3695 • Fax: 520-327-0426
Literature/Catalog: Yes
EGT/CHT engine analyzer, fuel flow totalizer and tachometer with unique features and many options. Affordable State-of-the-Art instrumentation.

Allied Signal Commercial Avionics
400 N. Rogers Rd., Olathe KS 66062-1212
Ph: 913-782-0400 • Fax: 913-791-1302
Web: http://www.alliedsignal.com/aerospace
Literature/Catalog: Yes
Bendix/King avionics. Full line of GPS and GPS/Comm units, Nav/Comms, ADF, DME, transponders, audio controls, weather radar, EFIS, and hand held radios.

American Avionics, Inc.
7675 Perimeter Road South, Seattle WA 98108
Ph: 206-763-8530 • Fax: 206-763-2036
E-Mail: sales@americanavionics.com
Literature/Catalog: Yes

Antenna Dynamics, Inc.
11100-8 Sepulveda Blvd., #565
Mission Hills CA 91345
Ph: 818-831-8640 • Fax: 818-831-8641
Literature/Catalog: Yes
Supplier of hidden antennas.

Antenna Specialist Co.
30500 Bruce Industrial Pkwy.
Cleveland OH 44139-3996
Ph: 216-349-8400 • Literature/Catalog: Yes

Archangel AVIONICS INC

Archangel Avionics Inc.
1500 Pumphrey Ave., Auburn AL 36832
Ph: 334-826-8008
See Display Ad On Page 20

Audio Flight Avionics

PO Box 28030
Scarborough, Ontario MIN 4E7 CANADA
Ph: 416-698-6928 • (800) 737-9185 Toll Free (USA & CANADA)
e-mail afa@user.rose.com
http://www.rose.com/~afa
Literature/Catalog: Yes
"First in Voice Alerting Engine Monitors" (You don't have to watch our gauges)
See Display Ad On Page 61

Avcom Avionics & Instr., Inc.
3008 NW 82nd Ave., Miami Fl 33122
Ph: 305-436-0300 • Fax: 305-436-0301
Litterature/Catalog: Yes
Sales, service, and overhaul of all makes of avionics, instruments, radars and accessories.

Avionics Place
5257 Falcon Rd., Rockford IL 61109
Ph: 815-229-5360 • Fax: 815-229-0308
Literature/Catalog: Yes
Sales, installation, and service of aircraft avionics.

Becker Avionics, Inc.
912 Airport Road, Roanoke TX 76262
Ph: 817-491-0080 • Fax: 817-491-0081
E-Mail: BeckerAv @ Flash.net
Literature/Catalog: Yes
Manufacturer of communication, VOR/ILS glide slope transponders, audio panels, ADF, moving map GPS systems, EICAS, fixed base and portable VHF ground stations.

C.A.A. Services
6-C Herman Drive, East Granby CT 06026
Ph: 203-658-8000 • Fax: 203-651-0885
Literature/Catalog: Yes
Avionics and accessories. Custom panels.

J.D. Chapdelaine Co.
5500 N.W. 21st Terrace
Ft. Lauderdale FL 33309
Ph: 305-772-4559 • Fax: 305-493-8960

chief AIRCRAFT

1301 Brookside Blvd., Grants Pass OR 97526
Ph: 800-447-3408 • Fax: 503-479-4431
E-Mail: chief@magick.net
Literature/Catalog: $2.00
See Display Ad On Page 59

Comant Industries, Inc.
12920 Park St., Santa Fe Springs CA 90670
Ph: 562-946-6694 X152 • Fax: 562-946-5150
Literature/Catalog: Yes
Antennas designed for VLA (metal or composite) and hand held GPS. Also multi-function antennas for GPS/COMM/FM/DGPS.

Crystal Shamrock, Inc.
6000 Douglas Drive N, Crystal MN 55429
Ph: 612-533-2214 • Fax: 612-533-5519
35 year old FBO. Certified repair station for upholstery, engines and service.

Davtron, Inc.
427 Hillcrest Way, Redwood City CA 94062
Ph: 415-369-1188 • Fax: 415-369-9988
Literature/Catalog: Yes

Eastern Avionics International
28000-A12 Airport Road, Punta Gorda FL 33982
Ph: 941-637-8585 • Fax: 941-637-0388
E-Mail: EAI Club@aol.com Web: http://avionix.com
Literature/Catalog: Yes
Supplier of all kinds of instruments and avionics by all manufacturers.

Flight Information Publication
P.O. Box 16616, St. Louis MO 63105
Ph: 314-469-1488 Literature/Catalog: Yes

Flightcraft, Inc.
7505 NE Airport Way, Portland OR 97218
Ph: 503-331-4200 • Fax: 503-331-4273
E-Mail: AAH07@AOL.COM
Custom panels for Glasair, Lancair. Factory authorized Lancair flight training.

Forbes Metal Stamping Co.
26555 Golden Valley Rd.
Saugus CA 91350
Ph: 805-250-1507 • Fax: 805-250-9406
E-Mail: fms76@aol.com
Literature/Catalog: Yes
Avionics hardware. Mic & phone jack mounting, instrument reducers, pilot & mechanics supplies. Wheel chock. Metal stamping of all kinds.

Garmin International
1200 151st Street, Olathe KS 66062-3426
Ph: 913-397-8200 • Fax: 913-397-8282
Manufacturer of a variety of navigation and communications electronics.

Grand Rapids Technologies
4526 Poinsettia SE, Kentwood MI 49508
Ph: 616-531-4893 • Fax: 616-249-0437
Literature/Catalog: Yes
Manufacturer of multifunction digital engine instruments. Four models provide complete instrumentation systems for nearly all types of aircraft engines including Rotax, Subaru, Lycoming, Continental, Hirth, etc.

Gulf Coast Avionics Corp.
4243 N. Westshore Blvd., Tampa FL 33614
Ph: 813-879-9714 • Fax: 813-875-4514
E-Mail: gca@ix.netcom.com
Web: http://www.gulf-coast-avionics.com
Literature/Catalog: Yes
Avionics supplier. Custom instrument panel manufacturing.

Heads Up Display Systems, Inc.
650 Normandy Rd.
Madeira Beach FL 33708-2371
Ph: 813-392-4228 • Fax: 813-538-0730
Literature/Catalog: Yes
Heads-Up display system: give airspeed, altitude, rate of climb. When interfaced with GPS displays ground speed, distance to waypoint, heading to fly.

ICE Corporation
240 Levee Dr., Manhattan KS 66502
Ph: 913-776-6423, E-Mail: sales@ice-ks.com
Web: http://www.ice-ks.com

ICOM America
2380 116th Avenue NE, Bellevue WA 98004
Ph: 425-454-8155 • Fax: 425-454-1509
Web: http://www.icomamerica.com
Literature/Catalog: Yes
Avionic hand helds, panel mounts, base stations, general coverage receivers, and VHF and UHF transceivers.

Instrument Tech Corp.
PO Box 1027, Addison TX 75001-1027
Ph: 800-229-9078 • Fax: 972-458-8789
Manufacturer, sales and service of aircraft instrumentation, accessories, lighting and related items.

J.A. Air Center/Joliet Avionic
3N060 Powis Road • DuPage Airport
West Chicago IL 60185
Ph: 630-584-3200 • Fax: 630-443-0169
Web: http:///aso.solid.com/seller/650/main.htm
Literature/Catalog: Yes
Complete FBO, avionics sales, service, installations, piston and turbine aircraft maintenance.

Kollsman Inc.
220 Daniel Webster Hwy., S.
Merrimack NH 03054
Ph: 603-889-2500 • Fax: 603-889-7966
Manufacturer of avionics products including flight and engine instruments, air data test equipment, and cabin pressurization systems.

Lafayette Avionics
Hangar 7, Purdue Airport
West Lafayette IN 47906
Ph: 317-743-3828 • Fax: 317-743-0000
E-Mail: ron@avionics-laf.com
Web: http://avionics-laf.com
Literature/Catalog: Yes
Sales, repair & installation of aircraft avionics.

Lone Star Aviation Corp.
804 N. Great S.W. Pkwy., Arlington TX 76011
Ph: 817-548-7768 • Fax: 817-633-6208
E-Mail: lsa@iamerica.net
Literature/Catalog: Yes
Noise filters for intercoms, stormscopes, etc. GPS power supplies for home use. Avionic cooling fans.

Magellan Systems Corp.
960 Overland Court, San Dimas CA 91773
Ph: 909-394-5000 • Fax: 909-394-7050
Manufacturer of global positioning systems receivers for aviation applications. Portable and panelmount.

McCoy Avionics Corp.
10761 Watkins Road
Marysville OH 43040-9544
Ph: 800-654-8124 • Fax: 937-642-0220
E-Mail: RPHartigan@juno.com
Web: http://www.McCoyAvionics.com
Literature/Catalog: Yes
Mac 1700C complete Nav/Com slide-in replacement. 760 channel comm. and 200 channel nav.

Narco Avionics, Inc.
270 Commerce Drive, Fort Washington PA 19034
Ph: 800-223-3636 • Fax: 215-643-0197
Literature/Catalog: Yes
Aircraft electronics.

Wil Neubert Aircraft Supply
1488 St. Rte. 162, Sprakers NY 12166
Ph: 518-922-5830 • Fax: 518-436-3632
Literature/Catalog: $5.00

Northstar Avionics
30 Sudbury Road, Acton MA 01720
Ph: 508-897-6600 • Fax: 508-897-7241
Web: http://www.northstarcmc.com
Northstar manufactures IFR and VFR GPS navigators and integrated GPS/VHF communication systems.

PS Engineering, Inc.
9800 Martel Road, Lenoir City TN 37772
Ph: 423-988-9800 • Fax: 423-988-6619
E-Mail: intercoms@ps-engineering.com
Web: http://www.ps-engineering.com
Literature/Catalog: Yes
Manufacturer of intercoms, audio selector panels, digital recorders, wire harnesses and modified headsets.

P.M. Research, Inc.
4110 Niles Hill Road, Wellsville NY 14895
Ph: 716-593-3169 • Fax: 716-593-5637

Pacific Coast-Avionics
1833 Auburn Way N., Auburn WA 98002
Ph: 800 353-0370 • Fax: 206-931-0584
E-Mail: PacAvionic@AOL.com
Full service avionics repair station with sales, service and installations. Specializing in panel package build-up for kit builders.

Porcine Associates
244 O'Connor Street, Menlo Park CA 94025
Ph: 415-326-2669 • Fax: 415-326-1071
E-Mail: jimham@porcine.com
Web: http://www.porcine.com
Hand held GPS linked with autopilot. Will work with most hand held GPS receivers. $189.00 complete with manual and connector.

Precision Dial Co.
7240 West KL Avenue, Kalamazoo MI 49009
Ph: 616-375-5601 • Fax: 616-375-1370
Dial and panel printing. Manufacture and refurbish instrument dials, instrument panels, radio/autopilot faces, backlit dials and panels. For homebuilt, warbird, and antique aircraft.

The GLASS GOOSE By Quikkit

Quikkit Corporation
9002 Summer Glen, Dallas TX 75243-7445
Ph: 214-349-0462 • Fax: 214-349-0462
E-Mail: quikkit@airmail.com
Web: http:////webwrks.com/glassgoose/
Literature/Catalog: Yes
See Display Ad On Page 24

RSE Inc.
1157 57th Drive SE, Auburn WA 98002-8720
Ph: 206-939-4105 • Fax: 206-939-4105
Literature/Catalog: Yes
Flight computers, weight and balance computers, aero design and performance software.

RST Engineering
13249 Grass Valley Avenue
Grass Valley CA 95945
Ph: 916-272-2203 • Fax: 916-272-2203
E-Mail: rst-engr@oro.net
Web: http://www.rst-engr.com
Literature/Catalog: Yes
Supplier of intercoms, headsets, and audio panel kits. Cable, connectors, antenna parts, etc.
See Display Ad On Page 14

Reliance Avionics, Inc.
PO Box 1528, Chandler AZ 85244-1528
Ph: 602-917-8034
Overvoltage protection module warns of high/low voltage condition. Solid state device. $99.00.

Rocky Mountain Instrument
PO Box 683 • 202 Fremont Street
Thermopolis WY 82443
Ph: 307-864-9300 • Fax: 307-864-9300
E-Mail: RMowRER@TRIB.com
Web: http://rkymtn.com
Literature/Catalog: Yes
Combination encoder, altimeter, airspeed, VSI, OAT and altitude alert system in 3-1/8" instrument. Complete air data output for all GPS systems included.

Ross Engineering Co.
12505-E Starkey Road, Largo FL 34643
Ph: 800-422-7676 • Fax: 813-535-4248
Literature/Catalog: Yes

Shadin Co., Inc.
6831 Oxford Street, St. Louis Park MN 55426
Ph: 800-328-0584 • Fax: 612-927-1111
E-Mail: sales@shadin.com
Web: http://www.shadin.com
Literature/Catalog: Yes
Fuel management systems, altitude encoders, altitude management systems, fuel/data computers, engine trend monitoring equipment.

Sigma-Tek, Inc.
1001 Industrial Rd., Augusta KS 67010
Ph: 316-775-6373 • Fax: 316-775-1416
E-Mail: info@sigmatek.com
Web: http://sigmatek.com
Literature/Catalog: Yes
Manufacturer of precision aviation instruments, avionics, accessories, control/display panels, and diagnostics systems. Capabilities are from product design through sales and service. Worldwide. Products for aircraft from general aviation trainers to commerical air carriers.

Sigtronics Corp.
949 N. Cataract Ave. #D, San Dimas CA 91773
Ph: 909-305-9399 • Fax: 909-305-9499
Headsets and intercoms for aviation, fire and law enforcement.

Skycom Avionics, Inc.
9305 W. Appleton Avenue
Milwaukee WI 53225
Ph: 414-463-8180 • Fax: 414-463-8195
Literature/Catalog: Yes
Aircraft electronics sales, service and installation. Your source for new, rebuilt and recertified packaged instrumentation for every homebuilt aircraft. Free Skycom Avionics Newsletter on request. Branches: Kenosha, Phone 414-652-8120; Waukesha, Phone 414-521-8180; Madison, Phone 608-246-8180.

SkySports
Hangar 1 • Linden-Price Airport
Linden MI 48451
Ph: 800-247-7883 • Fax: 810-735-1078
E-Mail: airstuff@aol.com
Literature/Catalog: Yes
Loran, GPS, transceivers.

Specialty Aircraft Co.
1440 NW 87th Ct., Redmond OR 97756
Ph: 800-500-6786 • Fax: 503-548-5336
Single engine Cessna salvage yard. Instruments and parts for homebuilt aircraft.

Sport Aircraft Supply
PO Box 13941, Scottsdale AZ 85267
Fax: 602-482-7882
Literature/Catalog: Yes
Complete antenna kit for composite aircraft.

S-TEC Corp.
One S-TEC Way • Municipal Airport
Mineral Wells TX 76067-9236
Ph: 940-325-9406 • Fax: 940-325-3904
E-Mail: kenpaul@flash.net
Web: http://www.s-tec.com/kitplane
Literature/Catalog: Yes
Manufacturer of a complete line of flight control systems, autopilots and flight directors. TEC line avionics - NAV, COM, ADF, DME and transponder. H.S.I. and R.M.I. systems.

Superflite
2149 E. Pratt Blvd., Elk Grove Village IL 60007
Ph: 800-323-0611 • Fax: 708-364-0358
Literature/Catalog: Yes
Portable GPS, Loran, comm radios, antennas. Supplier of a broad range of parts and accessories for personal aircraft.
See Display Ad On Page 38

TERF Inc.
11411 Culver, Brighton MI 48114
Ph: 810-229-2082 • Fax: 810-229-2082
E-Mail: Mtait@livgonline.com
Web: http://www.terf.com
Literature/Catalog: Yes
Solid state navlight systems. Wing tip lights, beacons, etc.,
See Display Ad On Page 52

Terra Avionics Corp.
3520 Pan-American Fwy NE
Albuquerque NM 87107
Ph: 505-884-2321 • Fax: 505-884-2384
Literature/Catalog: Yes
Full line of panel mounted avionics for both certificated and experimental aircraft. Fully digital line of equipment.

Trans-Cal Industries, Inc.
16141 Cohasset Street, Van Nuys CA 91406
Ph: 818-787-1221 • Fax: 818-787-8916
Literature/Catalog: Yes
Altitude encoders, mechanical & ss; altitude alert & monitor system RS 232 Interface.

Trans-Plata
3745 Ashwood Avenue, Los Angeles CA 90066
Ph: 310-397-0341 • Fax: 310-391-1998
Cockpit converter: reduces 24V to 12V for avionics, cellular phones, laptop computers. Inverter: 24V to 110V AC (250W) output. Flighttrack: pointing device for laptops that mounts on the control wheel. DigiVolt: digital voltmeter to plug into lighter socket and displays voltage.

Trimble Navigation, Ltd.
Avionics Div. • 2105 Donley Drive
Austin TX 78758
Ph: 800-767-8628 • Fax: 512-836-9413
Web: http://www.trimble.com
GPS Receivers, GPS/Loran & GPS/Omega

Tropic Aero
1090 NW 53rd Street, Ft. Lauderdale FL 33309
Ph: 305-491-6355 • Fax: 305-772-3327
Literature/Catalog: Yes
Discount portable G.P.S. New and used, buy, sell or trade all makes and models Loran or G.P.S.

Tru-Track Engineering
6824 Sandstone Court, Warrenton VA 22186
Ph: 504-341-7227
Track Wizard connects hand held GPS to autopilot. $119.00.

Val Avionics, Ltd.
P.O.Box 13025, Salem OR 97309-1025
Ph: 503-370-9429 • Fax: 503-370-9885
Literature/Catalog: Yes
Manufacture avionics including COM760TSO VHF transceivers, ILS400TSO instrument landing systems, and base station 7600 systems.

Valentine Aviation
14151 Montfort Dr., Suite 344
Dallas TX 75240-3067
Ph: 214-239-7911 • Fax: 214-385-1069
E-Mail: donaldv@airmail.net
Used avionics, specializing in storm scopes.

Vista Aviation, Inc.
12653 Osborne Street, Pacoima CA 91331
Ph: 818-896-6442 • Fax: 818-896-9541
Literature/Catalog: Yes
Avionics shop. Maintenance, parts and equipment. Flight school.

WAG-AERO
1216 North Road, Box 181, Lyons WI 53148
Ph: 800-558-6868 • Fax: 414-763-7595
E-Mail: wagaero-sales@wagaero.com
Literature/Catalog: Yes
NAV/COM radios, hand held antennas, decals.
See Display Ad On Page 13

Wentworth Aircraft Inc.
3015 Cedar Avenue, S., Minneapolis MN 55407
Ph: 612-722-0065 • Fax: 612-722-1024
Literature/Catalog: Yes
Engines, radios, instruments and miscellaneous single engine used parts.

Western Avionics Inc.
19711 Campus Drive #A • John Wayne Airport
Santa Ana CA 92707-5203
Ph: 714-757-0290 • Fax: 714-757-0293
Sales, service, installation of avionics, autopilots, radar systems, GPS, and moving maps.

Wings & Wheels
81 Jackson Avenue, W.E., Jamestown NY 14701
Ph: 716-664-6894 • Fax: 716-664-6895
Literature/Catalog: $4.00
Aircraft supplies, components, and avionics. Buy and sell, new and used. Becker, Icom, Terra, Trimble Magellan, Strong, National Parachutes, and much more.

Wire Specialties, Inc.
2515 Lucerne Drive, Janesville WI 53545
Ph: 608-758-1510 • Fax: 608-758-1701

Yingling Aircraft, Inc.
P.O. Box 9248, Wichita KS 67277
Ph: 316-943-3246 • Fax: 316-943-2484
All types of maintenance work and avionics supplies.

Batteries

Adcour, Inc.
PO Box 2649, Woburn MA 01888-1249
Ph: 617-784-8123 • Fax: 617-784-5563
Literature/Catalog: Yes

Aero Battery Sales & Service
3442 Church Street
Cincinnati OH 45244-3410
Ph: 513-271-9116 • Fax: 513-561-1950
Service and sales of aircraft batteries of all types - nickel cadmium and lead acid. Distributor for Concorde, Rapco, and Aerotech.

AIRCRAFT SPECIALTIES SERVICES

Aircraft Specialties Services
2860 North Sheridan Road • Tulsa International Airport, Tulsa OK 74115
Ph: 800-826-9252 • Fax: 918-836-4419
Literature/Catalog: Yes
FAA certified repair station D2R764K. Inspection, remachining of aircraft engine parts. Full service, new and used parts.
See Display Ad On Page 74

Aircraft Spruce & Specialty
PO Box 4000, Corona CA 91718-9961
Ph: 800-824-1930 • Fax: 909-372-0555
E-Mail: info@aircraft-spruce.com
Web: http://www.aircraft-spruce.com
Literature/Catalog: $5.00
A complete selection of homebuilt and certified aircraft parts and materials. Shipping worldwide since 1965. Also see Aircraft Spruce East, Griffin GA, 800-831-2949.
See Display Ad On Page 24

Approved Aircraft Accessories
29300 Goddard Road, Romulus MI 48174
Ph: 313-946-9000 • Fax: 313-946-5547
Literature/Catalog: Yes
Overhaul of various aircraft engines, landing gear, and accessories. Manufacture test equipment.

B & C Specialty Prod. Inc.
123 E. 4th Street • PO Box B
Newton KS 67114
Ph: 316-283-8000 • Fax: 316-283-7400
Literature/Catalog: Yes
Literature available, send large SASE.

Concorde Battery, Corp.
2009 San Bernardino Road
West Corvina CA 91790
Ph: 818-813-1234 • Fax: 818-338-3549
Literature/Catalog: Yes
Manufacturer of aircraft batteries.
See Display Ad On Page 28

Crystal Shamrock, Inc.
6000 Douglas Drive N, Crystal MN 55429
Ph: 612-533-2214 • Fax: 612-533-5519
35 year old FBO. Certified repair station for upholstery, engines and service.

Electro Marketing Battery Div.
7842 E. Gray, Scottsdale AZ 85260
Ph: 602-991-0110 • Literature/Catalog: Yes

ELTs Unlimited, Inc.
P.O. Box 1683 • 15 River Street
New Rochelle NY 10801
Ph: 800-382-0079 • Fax: 914-576-7075
Literature/Catalog: Yes • Replacement battery packs for Deft and Sharc ELTs.

Emergency Beacon Corp.
P.O. Box 178 • 15 River Street
New Rochelle NY 10802-0178
Ph: 914-235-9400 • Fax: 914-576-7075
Literature/Catalog: Yes
ELT replacement battery packs.

Falcon Ridge Industries
51 N. Hoagland Blvd., Kissimmee FL 34741
Ph: 407-396-7572 • Fax: 407-396-1696
Literature/Catalog: Yes • Battery boxes.

Goulet Aircraft Supply Ltd.
1137 Sanford St.
Winnepeg, MB R3E 3A1 CANADA
Ph: 204-786-1727 • Fax: 204-783-3302
General aviation parts distributor.

Keystone Battery Corp.
35 Holton Street, Winchester MA 01890
Ph: 617-729-8333 • Fax: 617-721-0127
Starting, deep-cycle, gel-cell, recombination, sealed valve regulated, marine, chargers, and associated battery products.

Light Power Engine Corp.
PO Box 3350, Morgantown WV 26503
Ph: 304-291-3843 • Fax: 304-292-1902
Literature/Catalog: $20.00

Linda Lou, Inc.
3514 Winhoma, Memphis TN 38118
Ph: 800-824-9912 • Fax: 901-365-9524

Wil Neubert Aircraft Supply
1488 St. Rte. 162, Sprakers NY 12166
Ph: 518-922-5830 • Fax: 518-436-3632
Literature/Catalog: $5.00

Seelye Equipment Specialist
1217 State Street, Charlevoix MI 49720
Ph: 616-547-9430 • Fax: 616-547-5522
Literature/Catalog: Yes
Solar electric (photovoltaic) panels for charging batteries. Power for remote hangers providing power for doors, lights, tools, outdoor security lights, runway lights. Portable, flexible panels for emergency battery charging and camping.

Superflite
2149 E. Pratt Blvd., Elk Grove Village IL 60007
Ph: 800-323-0611 • Fax: 708-364-0358
Literature/Catalog: Yes
Gill, Concorde, Merle and Artex ELT batteries. Supplier of a broad range of parts and accessories for personal aircraft.
See Display Ad On Page 38

Teledyne Aircraft Products Co.
Battery Products Unit • PO Box 7950
Redlands CA 92375 • Ph: 800-456-0070
Fax: 909-793-5818 • Literature/Catalog: Yes
Product brochure

Trans-Cal Industries, Inc.
16141 Cohasset Street, Van Nuys CA 91406
Ph: 818-787-1221 • Fax: 818-787-8916
Literature/Catalog: Yes
Battery warning system Hi/Lo Voltage Monitor-Audio/Visual.

True Air Service
6344 28th Avenue East, Palmetto FL 34221
Ph: 800-539-8842 • Fax: 941-729-8525
Literature/Catalog: Yes
Solargizer converter box emits a constant pulsating dc current that eliminates sulfates from a battery's lead plates.

Varga Enterprises, Inc.
2350 S. Airport Blvd., Chandler AZ 85249
Ph: 800-966-6936 • Fax: 602-899-0324
E-Mail: vargaair@vargaair.com
Web: http://www.vargaair.com
Literature/Catalog: Yes
Distributor of Concorde and Artex ELT batteries.

WAG-AERO
The Wag-Aero Group
1216 North Road, Box 181, Lyons WI 53148
Ph: 800-558-6868 • Fax: 414-763-7595
E-Mail: wagaero-sales@wagaero.com
Literature/Catalog: Yes
Battery accessories, aerobatic.
See Display Ad On Page 13

Wicks Aircraft Supply
410 Pine Street, Highland IL 62249
Ph: 800-221-9425 • Fax: 618-654-6253
Literature/Catalog: $5.00
Aircraft parts, components and supplies. Raw materials packages for 30+ aircraft designs such as the Aero Sport I & II, Barracuda, Corby Starlet, Cozy Mark IV, Defiant, E-Racer, Pober Pixie, Skybolt and many others. 350 page catalog with over 12,000 items.
See Display Ad On Page 46

Wilco, Inc.
3502 W. Harry, Witchita KS 67213
Ph: 800-767-7593 • Fax: 316-943-9664
E-Mail: wilco@feist.com
Web: http://www.globalair.com/wilco/
Lead acid, flooded type and maintenance free lead acid Concorde batteries.

Building Materials, Adhesives & Glues

A.C. Ultra Aviation, Ltd.
72 Wheatland Ave., Box 541
Smoky Lake, Alberta T0A 3C0 CANADA
Ph: 403-656-4207 • Fax: 403-656-4207
Literature/Catalog: Yes

AIN Plastics, Inc.
P.O. Box 151, Mt. Vernon NY 10550
Ph: 800-431-2451 • Fax: 914-668-8820
E-Mail: sales@ainplastics.com
Literature/Catalog: $5.00

Air Tech Coatings, Inc.
1 Paradise Park Road, Jacksonville AR 72076
Ph: 800-325-1650 • Fax: 501-985-1359
Literature/Catalog: Yes
Supplier of coating products for aviation.

Aircraft Spruce & Specialty
PO Box 4000, Corona CA 91718-9961
Ph: 800-824-1930 • Fax: 909-372-0555
E-Mail: info@aircraft-spruce.com
Web: http://www.aircraft-spruce.com
Literature/Catalog: $5.00
A complete selection of homebuilt and certified aircraft parts and materials. Shipping worldwide since 1965. Also see Aircraft Spruce East, Griffin GA, 800-831-2949.
See Display Ad On Page 24

Aircraft Supplies, Inc.
1745 Bathgate Avenue, Bronx NY 10457
Ph: 718-294-6370 • Fax: 718-716-8012
Aircraft engine spares, chemicals and tapes.

B & F Aircraft Supply, Inc.
22038 Howell Dr, Unit 6, New Lennox IL 60451
Ph: 815-462-8696 • Fax: 815-462-3697
Literature/Catalog: $1.00

Blue River Aircraft Supply
223 N. Clay, Harvard NE 68944
Ph: 800-653-7200 • Fax: 402-463-4476
Literature/Catalog: $2.00

Brown Aviation Tool Supply Co.
5700 North Rockwell • Building 2, Suite A
Bethany OK 73008
Ph: 800-587-3883 • Fax: 405-495-4992
E-Mail: browntool@aol.com
Literature/Catalog: Yes
Supplier of new, used and surplus aircraft assembly tools. Large selection of air, hand and cutting tools unique to the aviaiton industry.

DeMel Aircraft Corp.
Site 30 Comp. 27 RR2 • Pencticton Airport
Penticton BC V2A 6J7 CANADA
Ph: 250-490-9532 • Fax: 250-490-9538
Literature/Catalog: $3.00
Parts and supplies - AN hardware, paint, plywood and covering materials. 4130 steel-sheet and tube. Aluminum sheet and tube. Post or airmail delivery. Visa and MasterCard. Catalog in Canada $5.00.

Dymax Corporation
51 Greenwoods Rd., Torrington CT 06790
Ph: 860-482-1010 • Fax: 860-496-0608
E-Mail: dymax
J.Kay@postoffice.worldnet.att.net
Web: http://www.dymax.com
Literature/Catalog: Yes

Falconar Avia, Inc.
7739 - 81 Ave.
Edmonton, Alberta T6C 0V4 CANADA
Ph: 403-465-2024 • Fax: 403-465-2029
E-Mail: falconar@tic.ab.ca

Gougeon Brothers, Inc.
100 Patterson Avenue • PO Box 908
Bay City MI 48707
Ph: 517-684-7286 • Fax: 517-684-1374
E-Mail: westsys@concentric.net
Web: http://www.concentric.net/~westsys
Literature/Catalog: Yes

Hexcel Corp.-Resins Group
20701 Nordhoff St., Chatsworth CA 91311
Ph: 818-882-3022 • Fax: 818-709-0399
Literature/Catalog: Yes
Structural adhesives, epoxy laminating resins and wood laminating systems.

McBride Scientific Sales, Inc.
PO Box 1646, Rowlett TX 75030-1646
Ph: 214-494-5000 • Fax: 214-412-6922
Literature/Catalog: Yes
Uni-Wrap silicone rubber self-bonding tape.

AEROPOXY
PTM&W Industries, Inc.
10640 So. Painter Avenue
Santa Fe Springs CA 90670
Ph: 310-946-4511 • Fax: 310-941-4773
Literature/Catalog: Yes
Full line of High Performance Safe AEROPOXY Systems for adhesive bonding, coating, and structural components fabrication.
See Display Ad On Pages 14, 35

Superflite
2149 E. Pratt Blvd., Elk Grove Village IL 60007
Ph: 800-323-0611 • Fax: 708-364-0358
Literature/Catalog: Yes
Supplier of a broad range of parts and accessories for personal aircraft.
See Display Ad On Page 38

System Three Resins, Inc.
P.O. Box 70436, Seattle WA 98107
Ph: 800-333-5514 • Fax: 206-782-4426
E-Mail: support@system three.com
Web: http://system three.com
Literature/Catalog: Yes
Epoxy resins and adhesives. Water reduceable paints and coatings.

WAG-AERO
The Wag-Aero Group
1216 North Road, Box 181, Lyons WI 53148
Ph: 800-558-6868 • Fax: 414-763-7595
E-Mail: wagaero-sales@wagaero.com
Literature/Catalog: Yes
Paints, primers, fabrics, glues, and fillers.
See Display Ad On Page 13

Western Aircraft Supplies
623 Markerville Road N.E.
Calgary Alberta, T2E 5X1 CANADA
Ph: 403-275-3513 • Fax: 403-276-3087
Literature/Catalog: $5.00
Supplier of custom spruce kits, milled to specification, for wood design aircraft. Epoxy and adhesives for wood construction.

Wicks Aircraft Supply
410 Pine Street, Highland IL 62249
Ph: 800-221-9425 • Fax: 618-654-6253
Literature/Catalog: $5.00
Worldwide supplier for homebuilt projects specializing in raw material kits for 40+ different designs. 350 page catalog with over 12,000 items.
See Display Ad On Page 46

Building Materials, Carbon Fiber Composites

Aircraft Spruce & Specialty
PO Box 4000, Corona CA 91718-9961
Ph: 800-824-1930 • Fax: 909-372-0555
E-Mail: info@aircraft-spruce.com
Web: http://www.aircraft-spruce.com
Literature/Catalog: $5.00
A complete selection of homebuilt and certified aircraft parts and materials. Shipping worldwide since 1965. Also see Aircraft Spruce East, Griffin GA, 800-831-2949.
See Display Ad On Page 24

California Power Systems
790 139th Avenue, #4, San Leandro CA 94578
Ph: 800-247-9653 • Fax: 510-357-4429
Web: http://www.800-airwolf.com
Literature/Catalog: $6.95
Complete supplier of ultralight and sport aircraft parts and components. Rotax engines, parts and service. 500 page catalog, $6.95 US, $10.00 Canada and $20.00 foreign airmail.
See Display Ad On Page 69

Fitzgerald Composites, Inc.
16223 93rd Street, Bristol WI 53104-9617
Ph: 414-857-7500 • Fax: 414-857-7500
Web: http://fitzair@acronet.com
Literature/Catalog: $3.00
Equipped to supply any composite or metal part for any project. Specialists in custom built aircraft since 1968.

ICI Fiberite Corp.
4300 Jackson Street, Greenville TX 75401
Ph: 903-457-8500 • E-Mail: info@fiberite.com
Web: http://www.fiberite.com
Literature/Catalog: Yes

PCI - Newco
1745 Maize Road, Wichita KS 67209
Ph: 316-721-3606 • Fax: 316-721-3610
Literature/Catalog: Yes
Manufacturer of prepreg composite components. Ranging from flat panels to intricate ducts, fairings and radomes. Extensive capabilities in engineering and tooling. Call for possible applications for composite technology.

Reflex Fiberglass Works
PO Box 497 • 179 Aviation Ave.
Walterboro SC 29488
Ph: 803-549-1800 • Fax: 803-549-5139
Literature/Catalog: $10.00
Custom fiberglass mold making.

Skybolt Aeromotive Corp.
551 N. Park Avenue, Apopka FL 32712
Ph: 407-889-2613 • Fax: 407-889-8103
E-Mail: skybolt@mindspring
Web: http://skybolt.com
Composite manufacturing and fabrication. Aircraft hardware and accessories.

Textile Products, Inc.
2512 W. Woodland Drive
Anaheim CA 92801-2636
Ph: 714-761-0401 • Fax: 714-761-2928
Web: http://Textile.com • Literature/Catalog: Yes
Woven carbon kevlar and glass fabrics.

Wicks Aircraft Supply
410 Pine Street, Highland IL 62249
Ph: 800-221-9425 • Fax 618-654-6253
Literature/Catalog: $5.00
Worldwide supplier for homebuilt projects specializing in raw materialkits for 40+ different designs. 350 page catalog with over 12,000 items.
See Display Ad On Page 46

Building Materials, Fabrics

Aces High Light Aircraft Ltd.
RR 1, London, Ontario N6A 4B5 CANADA
Ph: 519-652-3020 • Fax: 519-652-3020
Literature/Catalog: $5.00
Aircraft kit & custom parts. Manufacturer able to supply most aircraft building materials and parts at competitive cost.

Action Air Parts, Inc.
106 Airport Drive, Port Huron MI 48074
Ph: 810-364-5885 • Fax: 810-364-5885
Literature/Catalog: Yes
Poly-Fiber Aircraft Covering System.

Air Tech Coatings, Inc.
1 Paradise Park Road • Jacksonville AR 72076
Ph: 800-325-1650 • Fax: 501-985-1359
Literature/Catalog: Yes
Supplier of coating products for aviation. Paints, coatings for leading edges, fabric covering materials.

Aircraft Finishings Systems
PO Box 1530, Ennis MT 59729
Ph: 800-653-7200 • E-Mail: AFS@3rivers.net
Literature/Catalog: Yes
One source for Ceconite fabrics and a full line of covering and coating products. Safe and EPA approved

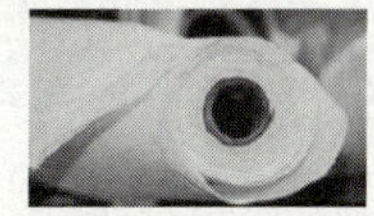

Aircraft Spruce & Specialty
PO Box 4000, Corona CA 91718-9961
Ph: 800-824-1930 • Fax: 909-372-0555
E-Mail: info@aircraft-spruce.com
Web: http://www.aircraft-spruce.com
Literature/Catalog: $5.00
A complete selection of homebuilt and certified aircraft parts and materials. Shipping worldwide since 1965. Also see Aircraft Spruce East, Griffin GA, 800-831-2949.
See Display Ad On Page 24

Airtex Products, Inc.
259 Lower Morrisville Road, Fallsington PA 19054
Ph: 215-295-4115 • Fax: 800-394-1247
Literature/Catalog: Yes

Anchor Reinforcements
5692 Buckingham Dr.
Huntington Beach CA 92649
Ph: 714-890-1432 • Fax: 714-890-1649
Literature/Catalog: Yes
Manufacturer of unidirectional, non-woven, non-stitched, non-impregnated carbon, E-glass, S-glass and Aramid fabrics. All biases and hybrids. Widths from 1 inch to 10 feet.

Blue River Aircraft Supply
223 N. Clay, Harvard NE 68944
Ph: 800-653-7200 • Fax: 402-463-4476
Literature/Catalog: $2.00

California Power Systems
790 139th Avenue, #4, San Leandro CA 94578
Ph: 800-247-9653 • Fax: 510-357-4429
Web: http://www.800-airwolf.com
Literature/Catalog: $6.95
Complete supplier of ultralight and sport aircraft parts and components. Rotax engines, parts and service. 500 page catalog, $6.95 US, $10.00 Canada and $20.00 foreign airmail.
See Display Ad On Page 69

Clark-Schwebel, Inc.
P.O. Box 2627 • 2200 South Murray Avenue, Anderson SC 29623-2627
Ph: 864-261-5174 • Fax: 864-260-6581
Literature/Catalog: Yes
Manufacturer of woven fabrics for advanced composites applications. Fibers include E Glass, S-Glass, and Kevlar. Certified to industry and airframe specifications.

DeMel Aircraft Corp.
Site 30 Comp. 27 RR2 • Pencticton Airport
Penticton BC V2A 6J7 CANADA
Ph: 250-490-9532 • Fax: 250-490-9538
Literature/Catalog: $3.00
Parts and supplies - AN hardware, paint, plywood and covering materials. 4130 steel-sheet and tube. Aluminum sheet and tube. Post or airmail delivery. Visa and MasterCard. Catalog in Canada $5.00.

Falconar Avia, Inc.
7739-81 Ave., Edmonton Alberta T6C 0V4 CANADA
Ph: 403-465-2024 • Fax: 403-465-2029
E-Mail: falconar@tic.ab.ca

Fisher Aero Corp.
7118 S.R. 335, Portsmouth OH 45662
Ph: 614-820-2219 • Fax: 614-820-2219

Goulet Aircraft Supply Ltd.
1137 Sanford St.
Winnepeg, MB R3E 3A1 CANADA
Ph: 204-786-1727 • Fax: 204-783-3302
General aviation parts distributor.

Wil Neubert Aircraft Supply
1488 St. Rte. 162, Sprakers NY 12166
Ph: 518-922-5830 • Fax: 518-436-3632
Literature/Catalog: $5.00

Poly-Fiber Aircraft Coatings
P.O. Box 3129, Riverside CA 92519
Ph: 909-684-4280 • Fax: 909-684-0518
Supplier of aircraft fabrics, coatings and paints.
See Display Ad On Page 51

R & R Aircraft Supply Co.
2716 E. US 12 • Niles MI 49120-5053
Ph: 616-683-2594 • Fax: 616-684-2369
Literature/Catalog: Yes
Distributor of Stits Polyfiber Covering system.

Redam Polyfix (Deer Park)
7242 Ohio Ave., Cincinnati OH 45326
Ph: 800-234-4583 • Fax: 606-727-2718
Literature/Catalog: Yes
Polyfix thermal fusion repair system for fiberglass and composites.

Rick's Aviation
Rt. 2, Box 319, Bristow OK 74010
Ph: 918-352-3083 • Fax: 918-352-2699
Literature/Catalog: Yes
Dacro-Shield UV Protective dressing for Dacron coverings. Shown to seal porosity, prevent blow-out and make older ultralight planes fly better. 1 quart covers 250 sq. ft. $21.95 plus $4.00 S/H per quart.

Superflite
2149 E. Pratt Blvd., Elk Grove Village IL 60007
Ph: 800-323-0611 • Fax: 708-364-0358
Literature/Catalog: Yes
Supplier of aircraft covering and finishing systems.
See Display Ad On Page 38

Textile Products, Inc.
2512 W. Woodland Drive
Anaheim CA 92801-2636
Ph: 714-761-0401 • Fax: 714-761-2928
Web: http://Textile.com
Literature/Catalog: Yes
Woven carbon kevlar and glass fabrics.

The Ultralight Store
36120 Pourroy Rd., Winchester CA 92596
Ph: 909-926-3196 • Fax: 909-926-3196
Literature/Catalog: Yes
Manufacturer of ultralight aircraft replacement sail sets and storage covers. Over 45 types of aircraft models available, many in stock. In business since 1983.

WAG-AERO
The Wag-Aero Group
1216 North Road, Box 181, Lyons WI 53148
Ph: 800-558-6868 • Fax: 414-763-7595
E-Mail: wagaero-sales@wagaero.com
Literature/Catalog: Yes Large selection.
See Display Ad On Page 13

Wicks Aircraft Supply
410 Pine Street, Highland IL 62249
Ph: 800-221-9425 • Fax 618-654-6253
Literature/Catalog: $5.00
Worldwide supplier for homebuilt projects specializing in raw materialkits for 40+ different designs. 350 page catalog with over 12,000 items.
See Display Ad On Page 46

Building Materials, Fiberglass

AIN Plastics, Inc.
P.O. Box 151, Mt. Vernon NY 10550
Ph: 800-431-2451 • Fax: 914-668-8820
E-Mail: sales@ainplastics.com
Literature/Catalog: $5.00

Aircraft Spruce & Specialty
PO Box 4000, Corona CA 91718-9961
Ph: 800-824-1930 • Fax: 909-372-0555
E-Mail: info@aircraft-spruce.com
Web: http://www.aircraft-spruce.com
Literature/Catalog: $5.00
A complete selection of homebuilt and certified aircraft parts and materials. Shipping worldwide since 1965. Also see Aircraft Spruce East, Griffin GA, 800-831-2949.
See Display Ad On Page 24

All Fab Corp.
1235 1/2 Lincoln Road, Allegan MI 49010
Ph: 616-673-6572 • Fax: 616-673-1644

Clark-Schwebel, Inc.
P.O. Box 2627 • 2200 South Murray Avenue
Anderson SC 29623-2627
Ph: 864-261-5174 • Fax: 864-260-6581
Manufacturer of woven fabrics for advanced composites applications. Fibers include E Glass, SZ Glass, and Keylar. Certified to industry and airframe specifications.

Gougeon Brothers, Inc.
100 Patterson Avenue • PO Box 908
Bay City MI 48707
Ph: 517-684-7286 • Fax: 517-684-1374
E-Mail: westsys@concentric.net
Web: http://www.concentric.net/~westsys
Literature/Catalog: Yes
We manufacture WEST SYSTEM Brand Epoxy and related construction and repair products. Free technical manual.

Maple Leaf Aviation Ltd.
Group 520 Box 16 RR#5
Brandon, MB R7A 5Y5 CANADA
Ph: 204-728-7618 • Fax: 204-725-1719
E-Mail: mla@access.tkm.mb.ca
Manufacturer of fiberglass replacement parts.

PCI - Newco
1745 Maize Road, Wichita KS 67209
Ph: 316-721-3606 • Fax: 316-721-3610
Literature/Catalog: Yes
Manufacturer of prepreg composite components. Ranging from flat panels to intricate ducts, fairings and radomes. Extensive capabilities in engineering and tooling. Call for possible applications for composite technology.

AEROPOXY
PTM&W Industries, Inc.
10640 So. Painter Avenue
Santa Fe Springs CA 90670
Ph: 310-946-4511 • Fax: 310-941-4773
Literature/Catalog: Yes
Full line of High Performance Safe AEROPOXY Systems for adhesive bonding, coating, and structural components fabrication.
See Display Ad On Pages 14, 35

Redam Polyfix (Deer Park)
7242 Ohio Ave., Cincinnati OH 45326
Ph: 800-234-4583 • Fax: 606-727-2718
Literature/Catalog: Yes
Polyfix thermal fusion repair system for fiberglass and composites.

Reflex Fiberglass Works
PO Box 497 • 179 Aviation Ave.
Walterboro SC 29488
Ph: 803-549-1800 • Fax: 803-549-5139
Literature/Catalog: $10.00
Custom fiberglass mold making.

Ron Sands Co.
89 Forrest Drive, Mertztown PA 19539
Ph: 610-682-6788 • Fax: 610-682-6788
Literature/Catalog: Yes
Cowls and fiberglass seats (German WWI). Send SASE for literature.

Skybolt Aeromotive Corp.
551 N. Park Avenue, Apopka FL 32712
Ph: 407-889-2613 • Fax: 407-889-8103
E-Mail: skybolt@mindspring
Web: http://skybolt.com
Composite manufacturing and fabrication. Aircraft hardware and accessories.

System Three Resins, Inc.
P.O. Box 70436, Seattle WA 98107
Ph: 800-333-5514 • Fax: 206-782-4426
E-Mail: support@system three.com
Web: http://system three.com
Literature/Catalog: Yes
Epoxy resins and adhesives. Water reduceable paints and coatings.

The Wag-Aero Group
1216 North Road, Box 181, Lyons WI 53148
Ph: 800-558-6868 • Fax: 414-763-7595
E-Mail: wagaero-sales@wagaero.com
Literature/Catalog: Yes
Replacement fenders, nose dowls, wingtips.
See Display Ad On Page 13

Wicks Aircraft Supply
410 Pine Street, Highland IL 62249
Ph: 800-221-9425 • Fax: 618-654-6253
Literature/Catalog: $5.00
Worldwide supplier for homebuilt projects specializing in raw material kits for 40+ different designs. 350 page catalog with over 12,000 items.
See Display Ad On Page 46

Building Materials, Finishes & Coatings

Action Air Parts, Inc.
106 Airport Drive, Port Huron MI 48074
Ph: 810-364-5885 • Fax: 810-364-5885
Literature/Catalog: Yes
Poly-Fiber Aircraft Covering System.

The Aeroplane Store
Kampel Airport, 8930 Carlisle
Wellsville PA 17365, Ph: 717-432-9688
Literature/Catalog: Yes

Air Tech Coatings, Inc.
1 Paradise Park Road, Jacksonville AR 72076
Ph: 800-325-1650 • Fax: 501-985-1359
Literature/Catalog: Yes
Supplier of coating products for aviation. Paints, coatings for leading edges, fabric covering materials.

Aircraft Finishings Systems
PO Box 1530, Ennis MT 59729
Ph: 800-653-7200 • E-Mail: AFS@3rivers.net
Literature/Catalog: Yes
One source for Ceconite fabrics and a full line of covering and coating products. Safe and EPA approved.

Aircraft Specialties Services
2860 North Sheridan Road • Tulsa International Airport, Tulsa OK 74115
Ph: 800-826-9252 • Fax: 918-836-4419
Literature/Catalog: Yes
FAA certified repair station D2R764K. Inspection, remachining of aircraft engine parts. Full service, new and used parts.
See Display Ad On Page 74

Aircraft Spruce & Specialty
PO Box 4000, Corona CA 91718-9961
Ph: 800-824-1930 • Fax: 909-372-0555
E-Mail: info@aircraft-spruce.com
Web: http://www.aircraft-spruce.com
Literature/Catalog: $5.00
A complete selection of homebuilt and certified aircraft parts and materials. Shipping worldwide since 1965. Also see Aircraft Spruce East, Griffin GA, 800-831-2949.
See Display Ad On Page 24

California Power Systems
790 139th Avenue, #4, San Leandro CA 94578
Ph: 800-247-9653 • Fax: 510-357-4429
Web: http://www.800-airwolf.com
Literature/Catalog: $6.95
Complete supplier of ultralight and sport aircraft parts and components. Rotax engines, parts and service. 500 page catalog, $6.95 US, $10.00 Canada and $20.00 foreign airmail.
See Display Ad On Page 69

Champions Choice, Inc.
1910 E. Via Burton Street, Anaheim CA 92806
Ph: 714-635-4491 • Fax: 714-635-9716

Chief Aircraft Inc.
1301 Brookside Blvd. • Grants Pass OR 97526
Ph: 800-447-3408 • Fax: 503-479-4431
E-Mail: chief@magick.net
Literature/Catalog: $2.00
See Display Ad On Page 59

Corrosion Technologies
P.O. Box 551625 • 2638 National Dr.
Dallas TX 75355-1625
Ph: 972-271-7361 • Fax: 972-278-9721
E-Mail: corrosnx@ix.netcom.com
Literature/Catalog: Yes
CorrosionX(tm)Aviation. New Fluid Thin Film Coating (FTFC) anti-corrosion formula kills existing corrosion and prevents it from starting. Qualified for Mil C-81309E, type II for application to airframes.

Dars Met-All Industries, Inc.
29-00 Review Avenue, Long Island City NY 11101
Ph: 718-937-3535 • Literature/Catalog: Yes

Falcon Air, Inc.
7332 N. Vincent • Portland OR 97217
Ph: 503-289-8287 • Literature/Catalog: Yes

Falconar Avia, Inc.
7739-81Ave., Edmonton, Alberta T6C 0V4 CANADA
Ph: 403-465-2024 • Fax: 403-465-2029
E-Mail: falconar@tic.ab.ca
Literature/Catalog: Yes
HIPEC flexible fine finishing and covering products.

Flitz International, Ltd.
821 Mohr Avenue, Waterford WI 53185
Ph: 414-534-5898 • Fax: 414-534-2991
Literature/Catalog: Yes
Manufacturer of Flitz metal polish products. Can be used on all types of metals as well as fiberglass and plexiglas.

Goulet Aircraft Supply Ltd.
1137 Sanford St., Winnepeg, MB R3E 3A1 CANADA
Ph: 204-786-1727 • Fax: 204-783-3302
General aviation parts distributor.

New Garden Aviation
PO Box 481 • 1235 Newark Road
Toughkenamon PA 19374
Ph: 610-268-2048 • Fax: 610-268-0342
Literature/Catalog: $2.00
Full service FBO specializing in antique and homebuilt aircraft. Complete aircraft finishing supplies.

P.M. Research, Inc.
4110 Niles Hill Road, Wellsville NY 14895
Ph: 716-593-3169 • Fax: 716-593-5637
P.M. Research manufactures and markets the best in radome erosion protection and leading edge tape.

Poly Fiber Aircraft Coatings
P.O. Box 3129 • Riverside CA 92519
Ph: 909-684-4280 • Fax: 909-684-0518
Supplier of aircraft fabrics, coatings and paints.
See Display Ad On Page 51

R & R Aircraft Supply Co.
2716 E. US 12 • Niles MI 49120-5053
Ph: 616-683-2594 • Fax: 616-684-2369
Literature/Catalog: Yes
Distributor of Stits Polyfiber Covering system.

Randolph Products Co.
P.O. Box 830 • Carlstadt NJ 07072-0830
Ph: 201-438-3700 • Fax: 201-438-4231
Web: http:/www.ourworld.compuserve.com/homepages/rpc • Aircraft paints and finishes.

Rick's Aviation
Rt. 2, Box 319, Bristow OK 74010
Ph: 918-352-3083 • Fax: 918-352-2699
Literature/Catalog: Yes
Dacro-Shield UV Protective dressing for Dacron coverings. Shown to seal porosity, prevent blow-out and make older ultralight planes fly better. 1 quart covers 250 sq. ft. $21.95 plus $4.00 S/H per quart.

Superflite
2149 E. Pratt Blvd., Elk Grove Village IL 60007
Ph: 800-323-0611 • Fax: 708-364-0358
Literature/Catalog: Yes
Supplier of aircraft covering and finishing systems. NEW instructional video on Superflite System II now available.
See Display Ad On Page 38

System Three Resins, Inc.
P.O. Box 70436, Seattle WA 98107
Ph: 800-333-5514 • Fax: 206-782-4426
E-Mail: support@system three.com
Web: http://system three.com
Literature/Catalog: Yes
Epoxy resins and adhesives. Water reduceable paints and coatings.

Wag-Aero
1216 North Road, Box 181, Lyons WI 53148
Ph: 800-558-6868 • Fax: 414-763-7595
E-Mail: wagaero-sales@wagaero.com
Literature/Catalog: Yes
Stits and Randolph products.
See Display Ad On Page 13

Wicks Aircraft Supply
410 Pine Street, Highland IL 62249
Ph: 800-221-9425 • Fax: 618-654-6253
Literature/Catalog: $5.00
Worldwide supplier for homebuilt projects specializing in raw material kits for 40+ different designs. 350 page catalog with over 12,000 items.
See Display Ad On Page 46

Building Materials, Foam

Aircraft Spruce & Specialty
PO Box 4000, Corona CA 91718-9961
Ph: 800-824-1930 • Fax: 909-372-0555
E-Mail: info@aircraft-spruce.com
Web: http://www.aircraft-spruce.com
Literature/Catalog: $5.00
A complete selection of homebuilt and certified aircraft parts and materials. Shipping worldwide since 1965. Also see Aircraft Spruce East, Griffin GA, 800-831-2949.
See Display Ad On Page 24

Garage Built Technologies
Box 271 Dept. AC, Wilmot OH 44689-0271
Literature/Catalog: $3.00
Plans for building a hot wire styrofoam cutter for composite aircraft parts, models, signs and molds, hand held or stationary.

Klegecell/Probalsa, Inc.
204 N. Dooley Street, Grapevine TX 76051
Ph: 817-481-3547 • Fax: 817-488-4816
Literature/Catalog: Yes
Klegecell structural PVC, foam, balsa, structural core, Estercore polyester co-polymer core.

Temperfoam-Sunmate
3710 Airpark Road, Lincoln NE 68524
Ph: 402-470-2346 Literature/Catalog: Yes
Comfort and impact absorbing seat cushion foam. Send SASE for information.

Wicks Aircraft Supply
410 Pine Street, Highland IL 62249
Ph: 800-221-9425 • Fax: 618-654-6253
Literature/Catalog: $5.00
Worldwide supplier for homebuilt projects specializing in raw material kits for 40+ different designs. 350 page catalog with over 12,000 items.
See Display Ad On Page 46

Building Materials, Prefab Components

Advanced Composite Tech.
R.D. #5, Box 214C, Somerset PA 15501
Ph: 814-445-3802 Literature/Catalog: Yes

Aero Specialty Co.
129 Chuck Yeager Way, Oroville CA 95965
Ph: 916-532-0919 • Fax: 916-532-0936
Literature/Catalog: Yes
Welded components such as exhaust systems, engine mounts and fuel tanks. FAA repair station ZSPR067L.

PREFAB COMPONENTS, SHEET METAL, BARS, TUBING

Aircraft Spruce & Specialty
PO Box 4000, Corona CA 91718-9961
Ph: 800-824-1930 • Fax: 909-372-0555
E-Mail: info@aircraft-spruce.com
Web: http://www.aircraft-spruce.com
Literature/Catalog: $5.00
A complete selection of homebuilt and certified aircraft parts and materials. Shipping worldwide since 1965. Also see Aircraft Spruce East, Griffin GA, 800-831-2949.
See Display Ad On Page 24

Airdrome Aeroplanes
929 NW Road 1571, HOLDEN MO 64040
Ph: 816-230-8585
Web: http://www.sunflower.org/~baron
Cowling wheel covers, spinners, firewalls. Send drawings for quotes.

All Fab Corp.
1235 1/2 Lincoln Road, Allegan MI 49010
Ph: 616-673-6572 • Fax: 616-673-1644

Alturair
1405 N. Johnson, El Cajon CA 92020
Ph: 619-449-1570 • Fax: 619-442-0481
Parts for BD5 aircraft. Also do sub-assemblies.

Alturair
1405 N. Johnson, El Cajon CA 92020
Ph: 619-449-1570 • Fax: 619-442-0481
Literature/Catalog: Yes
BD-5 parts and services including 100 hp rotary engine. Info kit $20.00

Bede Jet Corp.
18421 Edison Avenue, Chesterfield MO 63005
Ph: 314-537-2333 • Fax: 314-536-2822
Literature/Catalog: $20.00
Components for the BD-10 personal jet aircraft.

Boeve Fiberglass Components
37 W. 31 Street, Holland MI 49423
Ph: 616-392-1329
MJ-7 design, P51 Mustang components.

Bradley Aerospace
950 Muir Ave., Chico CA 95926
Ph: 916-899-7918 • Fax: 916-899-2942
Web: http://www.kgmsoftworks.com/bradley
See Display Ad On Page 40

Ken Brock Manufacturing, Inc.
11852 Western Avenue, Stanton CA 90680
Ph: 714-898-4366 • Fax: 714-894-0811
Literature/Catalog: $5.00
Gyroplane specialists. Stamped, machined, and welded sub-assemblies for most of the popular homebuilt models. Short-run stamping facilities.

C.S.N.
510 N.W. 46th Terrace, Plantation FL 33317-2044
Ph: 954-581-8835 • Literature/Catalog: Yes
Prefabricated components for the Corby Starlet CJ-1. Wing spar, control hinges, main landing gears, wing root fairings, leading edge fittings, pulley brackets, and "U" bolts. For info/price sheet, send self-addressed, stamped, #10 envelope.

Art Chard
PO Box 492, North Plains OR 97133
Ph: 503-647-9803
Newly constructed RV-4 wings include one-piece top skins, extended range fuel tanks (38 gallons) and foam-reinforced wing tips. Sub-assemblies for other RV models available.

Falcon Ridge Industries
51 N. Hoagland Blvd., Kissimmee FL 34741
Ph: 407-396-7572 • Fax: 407-396-1696
Literature/Catalog: Yes
Bolt on Cessna type axles; weld on axles -4130 steel. Spacer sand nuts for the above axles. Instrument panels. Fuel tank flanges. Battery boxes.

Featherlite Products, Inc.
13451 Airport Road • P.O. Box 781
Boonville CA 95415
Ph: 707-895-2718 • Fax: 707-895-2718
Literature/Catalog: Yes
Producers of prefabricated composite components for the home-built rutan designs as well as Cozy Mk III and Mk II aircraft. Also a source for wooden propellers with leading edge rain protection.

Fitzgerald Composites, Inc.
16223 93rd StreetBristol WI 53104-9617
Ph: 414-857-7500 • Fax: 414-857-7500
Web: http://fitzair@acronet.com
Literature/Catalog: $3.00
Equipped to supply any metal or composite part for any project. Specialists in custom built aircraft since 1968.

Harbor Products Co.
2390 Crenshaw Blvd., Ste. 164, Torrance CA 90501
Ph: 310-326-5609 • Fax: 310-560-7416
Literature/Catalog: Yes
Manufactures of high quality composite wheel pants, for antique classics, homebuilts and ultralight/microlight aircraft.

Hiett Aviation
PO Box 192, Superior AZ 85273
Ph: 520-689-5108 • Literature/Catalog: $5.00
Wing ribs. Modified Geottingen 387 airfoil type.

Homebuilders Gear
Claus Colling • Priel 5/a
Gammelsdorf 85408 GERMANY
Ph: 49 8766 844 • Fax: 49 8766 1351
Literature/Catalog: $5.00
Components for replica fighterplanes from 2/3 to full scale: retractable landing gear, spinners/backplates and flight control kits. Info packet available for US $5.

J.R. Metal Works, Inc.
P.O. Box 375, Monkton MD 21111
Ph: 410-357-4008 • Fax: 410-357-4008
Literature/Catalog: $5.00
Prefab parts and kits for BD-4.

Kosola & Associates, Inc.
5601 Newton Road • P.O. Box 3529
Albany GA 31707 • Ph: 800-456-7652
Fax: 912-888-5766 • Literature/Catalog: Yes
Manufactured A/C components.

Leading Edge Air Foils
Meadowlake Airport • 8242 Cessna Drive
Peyton CO 80831
Ph: 719-683-5323 • Fax: 719-683-5333
E-Mail: leafkp@kktv.com
Web: http://www.leadingedgeairfoils.com
Literature/Catalog: $6.00
Distributor for ultralight and experimental aircraft, parts, and engines.
See Display Ad On Page 34

PCI - Newco
1745 Maize Road, Wichita KS 67209
Ph: 316-721-3606 • Fax: 316-721-3610
Literature/Catalog: Yes
Manufacturer of prepreg composite components. Ranging from flat panels to intricate ducts, fairings and radomes. Extensive capabilities in engineering and tooling. Call for possible applications for composite technology.

The Glass Goose By Quikkit

Quikkit Corporation
9002 Summer Glen • Dallas TX 75243-7445
Ph: 214-349-0462 • Fax: 214-349-0462
E-Mail: quikkit@airmail.com
Web: http:////webwrks.com/glassgoose/
Literature/Catalog: Yes
See Display Ad On Page 24

Rattray Aero-Products
2357 Afton Road, Beloit WI 53511-8662
Ph: 608-362-4611 • Literature/Catalog: Yes
Manufacture of the world's largest variety of fiberglass aircraft parts.

Richfield Aero Partners
820 S. 300 E., Monroe UT 84754
Ph: 801-527-4294
Custom made aircraft parts.

ROMA Machine Tool & Die, Inc.
Manufacturer of Bellcrank Bearings since 1968

Roma Machine Tool & Die, Inc.
1990 East Maple Street, Kankakee IL 60901-5720
Ph: 815-937-1245 • Fax: 815-937-1285
Literature/Catalog: Yes
Manufacturersof bellcrank bearings, BC4W10 and BC5W11 propeller reductionspeed units, weld and bolt on axels, retractable landing gear, low wing, tubing fabrication, custom control panels, custom machine components for tail rotor on helicopters, variable pitch control and propeller hubs.
See Display Ad On Page 42

Ron Sands Co.
89 Forrest Drive, Mertztown PA 19539
Ph: 610-682-6788 • Fax: 610-682-6788
Literature/Catalog: $15.00
Fokker DR-1 components. Payment for info pack by US postal money order only.

Steelox Systems, Inc.
2412 Courseview Drive • P.O. Box 8181
Mason OH 45040-8181
Ph: 513-573-5200 • Fax: 513-573-5511

Systemware Labs
PO Box 492, North Plains OR 97133
Ph: 503-647-9803
RV-4 wings. Include one-piece top skins, 38-gallon extended range fuel tanks, and foam-reinforced wing tips. Sub-assemblies for other RV models are available.

The Ultralight Store
36120 Pourroy Rd., Winchester CA 92596
Ph: 909-926-3196 • Fax: 909-926-3196
Literature/Catalog: Yes
Manufacturer of ultralight aircraft replacement sail sets and storage covers. Over 45 types of aircraft models available, many in stock. In business since 1983.

The Wag-Aero Group
1216 North Road, Box 181, Lyons WI 53148
Ph: 800-558-6868 • Fax: 414-763-7595
E-Mail: wagaero-sales@wagaero.com
Literature/Catalog: Yes
Landing gear, cowlings, struts, etc.
See Display Ad On Page 13

Weldtech Aero
1925 Terminal Dr., Richland WA 99352-4924
Ph: 509-946-3269 • Pre-welded fuselages and other welded parts for the Kelly Hatz biplane.

Western Aircraft Supplies
623 Markerville Road N.E.
Calgary Alberta, T2E 5X1 CANADA
Ph: 403-275-3513 • Fax: 403-276-3087
Literature/Catalog: $5.00
We supply spruce kits to drawing's specification and manufacture and supply spars solid or laminated to customer's specification.

Zenair, Ltd.
Huronia Airport • Midland, Ont. L4R 4K8 CANADA
Ph: 705-526-2871 • Fax: 705-526-8022
E-Mail: zenair@barint.on.ca
ZODIAC CH 601 Series and STOL CH 701 individual factory made parts, kit assemblies, and components. Preassembled kits and factory-completed aircraft also available on custom order. Advanced Ultralight models.

Building Materials, Sheet Metal, Bars & Tubing

Aces High Light Aircraft Ltd.
RR 1, London, Ontario N6A 4B5 CANADA
Ph: 519-652-3020 • Fax: 519-652-3020
Literature/Catalog: $5.00
Aircraft kit & custom parts. Manufacturer able to supply most aircraft building materials and parts at competitive cost.

Action Air Parts, Inc.
106 Airport Drive, Port Huron MI 48074
Ph: 810-364-5885 • Fax: 810-364-5885
Literature/Catalog: Yes • Supplier of an extensive line of aircraft sheet metal tools.

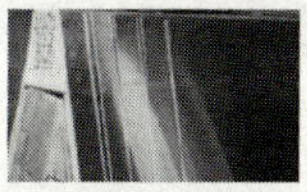

Aircraft Spruce & Specialty
PO Box 4000, Corona CA 91718-9961
Ph: 800-824-1930 • Fax: 909-372-0555
E-Mail: info@aircraft-spruce.com
Web: http://www.aircraft-spruce.com
Literature/Catalog: $5.00
A complete selection of homebuilt and certified aircraft parts and materials. Shipping worldwide since 1965. Also see Aircraft Spruce East, Griffin GA, 800-831-2949.
See Display Ad On Page 24

Aircraft Steel
923 W.C.R. 7, Erie CO 80516
Ph: 303-665-5817 • Literature/Catalog: Yes
4130 chromolly steel. Round, square, and streamline tubing. Sheet material. Material kits for many homebuilts. Aluminum tubing 1/4" to 1-1/2". Fax: 303-665-5817 press * twice and listen for fax tone.

Airparts, Inc.
2400 Merriam Lane, Kansas City KS 66106
Ph: 800-800-3229 • Fax: 913-831-1780
Literature/Catalog: Yes
Raw materials and hardware. Custom lengths of aluminum.

All Fab Corp.
1235 1/2 Lincoln Road, Allegan MI 49010
Ph: 616-673-6572 • Fax: 616-673-1644

Blue Ridge Machinery and Tools, Inc.
Your Complete Machine Shop Supplier!

Blue Ridge Machinery & Tools
P.O. Box 536 • 2905 Putnam Avenue
Hurricane WV 25526
Ph: 304-562-3538 • Fax: 304-562-5311
Literature/Catalog: Yes
Machine shop equipment and tooling for the home-shop machinist. Lathes, mills, drill rods and more. How-to videos and books.
See Display Ad On Page 22

Campbell Aircraft
P.O. Box 610518-A, San Jose CA 95161-0518
Ph: 408-251-3109 • Literature/Catalog: Yes

The Dillsburg Aeroplane Works
114 Sawmill Road, Dillsburg PA 17019
Ph: 717-432-4589 • Fax: 717-432-4589
4130 tubing, sheet, aluminum, bolts and more.

DJB Aluminum & Supply
3151 Lucas Drive, Bethlehem PA 18107
Ph: 610 866-6913 Literature/Catalog: $2.00
Aluminum, 6061-T6 and 2024-T4, rod, bar, angle, tubing, channel.

Eastside Ultralight Aircraft
4700 188th Street NE, Arlington WA 98223
Ph: 206-435-3737

Future Metals
P.O. Box 100639, Ft. Lauderdale FL 33310
Ph: 800-733-0960 • Fax: 305-731-7297
Literature/Catalog: Yes

Grass Roots Aviation
648 Adelaide Avenue W.
Oshawa, Ontario L1J 6S2 CANADA
Ph: 905-434-4651 • Fax: 905-728-2385
Literature/Catalog: $3.00
Raw material supplies; steel, aluminum, wood and hardware.

Industrial Metal Supply Co.
3303 N. San Fernando Blvd.
Burbank CA 91504
Ph: 213-849-3184 • Fax: 818-848-9551
Literature/Catalog: Yes
Steel, aluminum, stainless, brass, copper, lead, tool steel.

Industrial Metal Supply Co.
2052 Alton Parkway, Irvine CA 92714
Ph: 714-250-3343 • Fax: 714-250-4003
Literature/Catalog: Yes
Steel, aluminum, stainless, brass, copper, lead, tool steel.

Industrial Metal Supply Co.
7595 Raytheon Road, San Diego CA 92111
Ph: 619-277-8200 • Fax: 619-277-3865
E-Mail: ims@imsmetals.com
Web: http://imsmetal.com
Full line mill distributor and metal service center. Including hard to find sizes.

J. Lough
172 Boniface Avenue, Kitchener, Ontario
N2C 1L9 CANADA • Ph: 519-578-0968
Literature/Catalog: $5.00
Dandy dimple dies; hole flanging dies; bending brake drawings.

Rhino Aviation
607 Garber Street, Caldwell ID 83605
Ph: 800-701-9128 • Fax: 208-459-0819
Literature/Catalog: Yes
CNC cutting of sheet metal and metal plate to customer specifications. Items cut include instruments panels, stainless steel firewalls, metal brackets, tabs and plates. Custom prototype and production. For brochure, send self addressed stamped envelope.

ROMA *Machine Tool & Die, Inc.*
Manufacturer of Bellcrank Bearings since 1968

Roma Machine Tool & Die, Inc.
1990 East Maple Street • Kankakee IL
60901-5720 • Ph: 815-937-1245
Fax: 815-937-1285 • Literature/Catalog: Yes
Manufacturersof bellcrank bearings, BC4W10 and BC5W11 propeller reductionspeed units, weld and bolt on axels, retractable landing gear, low wing, tubing fabrication, custom control panels, custom machine components for tail rotor on helicopters, variable pitch control and propeller hubs.
See Display Ad On Page 42

Shapiro Supply
5617 Natural Bridge, St. Louis MO 63120
Ph: 314-382-7000 • Fax: 314-382-8557
Literature/Catalog: Yes
Aluminum: sheet, bar, tubing, extrusions. 4130 steel: sheet, tubing bar.

Superflite

2149 E. Pratt Blvd., Elk Grove Village IL 60007
Ph: 800-323-0611 • Fax: 708-364-0358
Literature/Catalog: Yes
Supplier of a broad range of parts and accessories for personal aircraft.
See Display Ad On Page 38

Tubular Dynamics
1207 NE Cookingham Drive
Kansas City MO 64155
Ph: 816-734-8211 • Fax: 816-734-8211
Literature/Catalog: Yes
Manufacturers of the original Jig-A-Joint and Jig-A-Joint2 tubing notchers.

WAG-AERO

The Wag-Aero Group
1216 North Road, Box 181 • Lyons WI 53148
Ph: 800-558-6868 • Fax: 414-763-7595
E-Mail: wagaero-sales@wagaero.com
Literature/Catalog: Yes
4130 chromolly seamless tubing, sheet steel and streamline tubing.
See Display Ad On Page 13

Wicks Aircraft Supply

410 Pine Street, Highland IL 62249
Ph: 800-221-9425 • Fax: 618-654-6253
Literature/Catalog: $5.00
Worldwide supplier for homebuilt projects specializing in raw material kits for 40+ different designs. 350 page catalog with over 12,000 items.
See Display Ad On Page 46

William Koucky
4656 Garden Court, Traverse City MI 49684
Ph: 616-947-6966
Fittings, 4130 steel plasma cut.

ZENITH
AIRCRAFT COMPANY

Box 650 • Mexico Memorial Airport
Mexico MO 65265-0650
Ph: 573-581-9000 • Fax: 573-581-0011
E-Mail: info@zenithair.com
Web: http://www.zenithair.com
Literature/Catalog: Yes
Building supplies for Zenair aircraft.
See Ad Inside Front Cover

Building Materials, Wood

The Aeroplane Store
Kampel Airport, 8930 Carlisle
Wellsville PA 17365 • Ph: 717-432-9688
Literature/Catalog: Yes

Aircraft Spruce & Specialty
PO Box 4000, Corona CA 91718-9961
Ph: 800-824-1930 • Fax: 909-372-0555
E-Mail: info@aircraft-spruce.com
Web: http://www.aircraft-spruce.com
Literature/Catalog: Yes
A complete selection of homebuilt and certified aircraft parts and materials. Shipping worldwide since 1965. Also see Aircraft Spruce East, Griffin GA, 800-831-2949.
See Display Ad On Page 24

B&D International, Inc.
1711 Center St., Tacoma WA 98409
Ph: 800-222-7853 • Fax: 206-272-6431
E-Mail: 72163.3671@compuserve.com
Literature/Catalog: Yes
Distributor of aircraft grade Finland birch plywood.

British Plywood Manufacturer
Alma Road, Ponders End,
Enfield, EN3 7BP UNITED KINGDOM
Ph: 44-181-804-2424 • Fax: 44-181-804-1293
British made aircraft plywood from mahogany or birch. CAA approved.

DeMel Aircraft Corp.
Site 30 Comp. 27 RR2 • Pencticton Airport
Penticton BC V2A 6J7 CANADA
Ph: 250-490-9532 • Fax: 250-490-9538
Literature/Catalog: $3.00
Parts and supplies - AN hardware, paint, plywood and covering materials. Post or airmail delivery. Visa and MasterCard. Catalog in Canada $5.00.

Fisher Aero Corp.
7118 S.R. 335, Portsmouth OH 45662
Ph: 614-820-2219 • Fax: 614-820-2219

Grass Roots Aviation
648 Adelaide Avenue W.
Oshawa, Ontario L1J 6S2 CANADA
Ph: 905-434-4651 • Fax: 905-728-2385
Literature/Catalog: $3.00
Raw material supplies; steel, aluminum, wood and hardware.

Harbor Sales Co., Inc. (The)
1401 Russell Street, Baltimore MD 21230-2089
Ph: 800-345-1712 • Fax: 410-752-0739
Literature/Catalog: Yes
Aircraft plywood, birch, mahogany, marine plywood, fir, and teak.

Klegecell/Probalsa, Inc.
204 N. Dooley Street, Grapevine TX 76051
Ph: 817-481-3547 • Fax: 817-488-4816
Literature/Catalog: Yes
Klegecell structural PVC, foam, balsa, structural core, Estercore polyester co-polymer core.

MISTER PLY&WOOD
PO Box 800, Baulkham Hills NSW 2153 Austrailia
Ph: 61-2-9898 9477 • Fax: 61-2-9838 7697
Literature/Catalog: Yes
Aircraft, Marine and Bending Plywood, Lite Ply (Poplar)

RagWing Aviation Co.
PO Box 39 • Tokeena Air Park
Townville SC 29689-0039
Ph: 864-225-3773 • Fax: 864-972-5606
Literature/Catalog: $3.00
Supplier of Northern White Pine for homebuilders.

RITECO Supply Inc.
12999 FM 529, Houston TX 77041
Ph: 713-896-6200 • Fax: 713-896-6100
Literature/Catalog: Yes
"Premium Grade" - Australian manufactured interior and exterior grade "bending" plywood. Flexible, no open hole knots, compression areas or voids, 100% Hoop Pine.

Sparcraft Aero-Marine Mfg. Co.
3001 So. Steele Street, Tacoma WA 98409
Ph: 206-474-2140 • Fax: 206-627-8395
E-Mail: sindlilj@pop3.puyallup.wednet.edu
Literature/Catalog: Yes
Wood aircraft parts. Specializing in wing kits for Pitts and Skybolt Biplanes.

WAG-AERO

The Wag-Aero Group
1216 North Road, Box 181, Lyons WI 53148
Ph: 800-558-6868 • Fax: 414-763-7595
E-Mail: wagaero-sales@wagaero.com
Literature/Catalog: Yes
Large selection.
See Display Ad On Page 13

Western Aircraft Supplies
623 Markerville Road N.E.
Calgary Alberta, T2E 5X1 CANADA
Ph: 403-275-3513 • Fax: 403-276-3087
Literature/Catalog: $5.00
Supplier of custom spruce kits, milled to specification for FALCO, EMERAUDE, OSPREY, KR-1, KR-2, CAVALIER, TAYLOR MONO-PLANE, TAYLOR TITCH, AIRCAMPER, SMITH MINI PLANE, BARRACUDA, CYGNET, HIRONDELLE. Epoxy and adhesives for wood construction.

Wicks Aircraft Supply

410 Pine Street, Highland IL 62249
Ph: 800-221-9425 • Fax: 618-654-6253
Literature/Catalog: $5.00
Worldwide supplier for homebuilt projects specializing in raw material kits for 40+ different designs. 350 page catalog with over 12,000 items.
See Display Ad On Page 46

Building Materials, Other

A&G Aviation Ltd.
Bldg 198, Bournemouth Intl. Christchurch,
Dorset BH23 6NE UNITED KINGDOM
Ph: 0202 578558 • Fax: 0202 570203
All Cessa & Piper spare parts.

AIN Plastics, Inc.
P.O. Box 151, Mt. Vernon NY 10550
Ph: 800-431-2451 • Fax: 914-668-8820
E-Mail: sales@ainplastics.com
Literature/Catalog: $5.00
Plastic sheet, rod, tube film, accessories.

Aircraft Lighting & Components
249 Brockway Road, Chehalis WA 98532
Ph: 360-748-0234 • Fax: 360-748-6455
Literature/Catalog: Yes
Custom fabrication of rotars swaged aircraft cable assemblies and vacuum molded plexiglass lens assemblies and other small plastic parts. OEM inquiries invited.

Aircraft Remanufacturing
1531 Avohill Dr. • Vista CA 92080-1325
Ph: 760-749-0239 • Fax: 760-749-6384
E-Mail: BJNash@connectnet.com
Literature/Catalog: Yes
Mill distributors for "Super Sound Proofing Mat". Water proof vinyl-nitryl foam available to 1" thick. Available from Aircraft Spruce and Wicks Aircraft Supply. FAA approved. Free "how-to" booklet on sound proofing material. Also available by e-mail infobot: SoundPRF@PDSIG.com.

Airparts, Inc.
2400 Merriam Lane, Kansas City KS 66106
Ph: 800-800-3229 • Fax: 913-831-1780
Literature/Catalog: Yes
Hydraulic hoses.

Alan Mfg. Co.
5017 Ritz Road, Marengo IL 60152
Ph: 815-568-6836 • Fax: 815 568-8696
Mounting Tabs for tube and fabric builders: .040" 4130, 5/8 x 13/16 with a 1/8 hole. $20.95 per 100 pcs. includes shipping

Apollo Safety & Industrial
1108 West 9th Street, Upland CA 91786-5703
Ph: 909-980-7071 • Fax: 909-985-6873
Full line of safety (head protection to foot protection) and industrial items. Includes "brushes for all industries". We also sell solid and blind rivets.

Evergreen Weigh, Inc.
19023 36th Avenue West, #E
Lynnwood WA 98036
Ph: 800-356-6060 • Fax: 206-775-4399
Manufacturer of Road Runner aircraft scales for large and small aircraft.

Fastening Systems Int'l.
P.O. Box 1372, Sonoma CA 95476
Ph: 800-344-2393 • Fax: 707-935-1828
Literature/Catalog: $1.00
Hydraulic, electric and battery operated blind riveters and nut setters. Up to 5,325 pounds of force. Complete inventory of blind fasteners by Cherry, Avdel, Gesipa/Olympic, Huck Intl., and Atlas Engineering.

Fisher Aero Corp.
7118 S.R. 335, Portsmouth OH 45662
Ph: 614-820-2219 • Fax: 614-820-2219
Literature/Catalog: Yes
Misc. building supplies.

Metal Masters
1027 Greenfield Dr. Ste. G, El Cajon CA 92021
Ph: 619-461-0168 • Literature/Catalog: $1.00
Aluminum landing gear made to your drawings. Send for kit $1.00.

Michael Engineering, Ltd.
4997 S. Crawford Road • Mt. Pleasant MI 48858
Ph: 517-772-4073 • Fax: 517-772-3500
Literature/Catalog: Yes
Epoxy ratio pumps & mixers, epoxy metering equipment.

Micro-AG Systems
Box 1932, Hwy 120 S. • Lake Geneva WI 53147
Ph: 414-248-2288 • Literature/Catalog: Yes
Ultra & hyper low volume crop spraying equipment.

National Products, Inc.
1017 S. Elmgrove St., Seattle WA 98108
Ph: 206-763-8361 • Fax: 206-763-9615
E-Mail: stuff@ram-mount.com
Web: http://www.ram-mount.com
Literature/Catalog: Yes
Manufacture Round-A-Mount ball and socket mounting system. Holds loads in high vibration conditions.

NORDAM
510 S. Lansing, Tulsa OK 74120
Ph: 918-587-4105 • Fax: 918-583-2604
Metalic & non-metalic composite bonding. Manufacture of flat & contoured composite components.

PCI - Newco
1745 Maize Road, Wichita KS 67209
Ph: 316-721-3606 • Fax: 316-721-3610
Literature/Catalog: Yes
Manufacturer of prepreg composite components. Ranging from flat panels to intricate ducts, fairings and radomes. Extensive capabilities in engineering and tooling. Call for possible applications for composite technology.

Perrone Leather Apparel, Inc.
48-50 South Main Street, Gloversville NY 12078
Ph: 518-725-9144 • Fax: 518-725-7044
E-Mail: interior@perroneleather
Web: http://www.perroneleather.com
Literature/Catalog: Yes
Fire retardant ulpholstery leather. "Townsend" interior leathers. FAR Part 25. Burn Certified.

AEROPOXY

PTM&W Industries, Inc.
10640 So. Painter Avenue
Santa Fe Springs CA 90670
Ph: 310-946-4511 • Fax: 310-941-4773
Literature/Catalog: Yes
Full line of High Performance Safe AEROPOXY Systems for adhesive bonding, coating, and structural components fabrication.
See Display Ad On Page 14, 35

Smithy Co.
170 Aprill Drive • PO Box 1517
Ann Arbor MI 48106
Ph: 800-345-6324 • Fax: 313-913-6663
Literature/Catalog: Yes
Manufacturer of multi-purpose power lathe, mill and drill. Three machines in one. Machine fits conveniently into home shop.

Topflight Design Group, Inc.
29 Sand Hill Road • P.O. Box 86
Hampton CT 06247-0086
Ph: 860-455-9298 • Fax: 860-455-9298
Compact cast aluminum pitot-static head; in development steerable/swiveling tailwheel assembly. Others pending.

Woofter-Saber Manufacturing
1951 NW 84th Terrace
Pembroke Pines FL 33024
Ph: 954 436-9496 • Fax: 954-436-0015
Literature/Catalog: Yes
Manufacturer of propeller extensions, our specialty for 12 years.

Canopies

Aircraft Lighting & Components
249 Brockway Road, Chehalis WA 98532
Ph: 360-748-0234 • Fax: 360-748-6455
Literature/Catalog: Yes

Aircraft Spruce & Specialty
PO Box 4000, Corona CA 91718-9961
Ph: 800-824-1930 • Fax: 909-372-0555
E-Mail: info@aircraft-spruce.com
Web: http://www.aircraft-spruce.com
Literature/Catalog: $5.00
A complete selection of homebuilt and certified aircraft parts and materials. Shipping worldwide since 1965. Also see Aircraft Spruce East, Griffin GA, 800-831-2949.
See Display Ad On Page 24

Aircraft Supply
Allegheny County Airport, West Mifflin PA 15122
Ph: 800-245-0690 • Fax: 412-469-8271
Literature/Catalog: Yes

Aircraft Windshield Co.
10871 Kyle Street, Los Alamitos CA 90720
Ph: 560-430-8108 • Fax: 560-598-0716
Custom and standard acrylic thermoforming.

Airplane Plastics Company
8300 Dayton Rd., Fairborn OH 45324
Ph: 937-864-5607 • Fax: 937-864-7010
E-Mail: Doug@Foxlite.com
Web: http://www.foxlite.com
Manufacturer of aircraft windshields, windows and canopies.

C.S.N.
510 N.W. 46th Terrace, Plantation FL 33317-2044
Ph: 954-581-8835 • Literature/Catalog: Yes
Prefabricated components for the Corby Starlet CJ-1. Canopy kits with frame, windscreen and side panels. High quality fiberglass parts. For info/price sheet, send self-addressed, stamped, #10 envelope.

Craftsman Steel Building
1845 Island Avenue, San Diego CA 92102
Ph: 619-233-4000 • Fax: 619-233-8373
Hangar buildings and canopies for airplanes and other hobbies.

Great Lakes Aero Products
915 Kearsley Park Blvd., Flint MI 48503
Ph: 800-532-2918 • Fax: 810-235-5260
E-Mail: www.tir.com~glap
Manufacturer of aircraft windshields and side windows.

Hummel Aviation
509 E. Butler Street, Bryan OH 43506-0880
Ph: 419-636-3390 • Literature/Catalog: $2.00
Hummel Bird and Team aircraft modification kits and canopy kits. Hummel Bird Canopies: $200.00. Mini Max Canopy Kits: $235.

Micro Surface Finishing Prod.
1217 West Third Street • PO Box 818
Wilton IA 52778
Ph: 800-225-3006 • Fax: 319-732-3390
E-Mail: msfpi@netins.net
Web: http://www.micro-surface.com
Literature/Catalog: Yes
Micro-Mesh abrasives. Fine finishing cushioned abrasive that removes crazing, scratches and other damage from acrylics. Used by the US military and all major airlines. Various repair kits available.

NORDAM
510 S. Lansing, Tulsa OK 74120
Ph: 918-587-4105 • Fax: 918-583-2604
Metalic & non-metalic composite bonding. Manufacture of flat & contoured composite components.

Thermo-Tec
14530 Lynshar Road
Grass Valley CA 95949-7507
Ph: 916-272-2556 • Fax: 916-272-2560
E-Mail: thermotc@inreach.com
Web: http://home.inreach.com/thermotc/
Plastics thermoforming - canopies, windshields and windows.

Wings & Wheels
81 Jackson Avenue, W.E.
Jamestown NY 14701
Ph: 716-664-6894 • Fax: 716-664-6895
Literature/Catalog: $4.00
Aircraft supplies, components, and avionics. Buy and sell, new and used. Becker, Icom, Terra, Trimble Magellan, Strong, National Parachutes, and much more.

Emergency Equipment

Aircraft Spruce & Specialty
PO Box 4000, Corona CA 91718-9961
Ph: 800-824-1930 • Fax: 909-372-0555
E-Mail: info@aircraft-spruce.com
Web: http://www.aircraft-spruce.com
Literature/Catalog: $5.00
A complete selection of homebuilt and certified aircraft parts and materials. Shipping worldwide since 1965. Also see Aircraft Spruce East, Griffin GA, 800-831-2949.
See Display Ad On Page 24

Microflight Products, Inc.
1356 Oakdale Drive, Suite 23
Griffin GA 30223-3929
Ph: 800-831-2949 • Fax: 770-229-2329
Literature/Catalog: $3.00
Complete line of instruments, gauges, fuel components, electrical parts, hardware, and accessories. Dealer and OEM programs available.

National Parachute Industries
47 E. Main Street, Box 1000
Flemington NJ 08822
Ph: 800-526-5946 • Fax: 908-782-5638
Literature/Catalog: Yes
Emergency parachutes available in three styles: back, chair and seat. Lightweight, comfortable designs. 19 years in business.

Strong Enterprises
11236 Satellite Blvd., Orlando FL 32837
Ph: 407-859-9317 • Fax: 407-850-6978
Literature/Catalog: Yes
Oldest and largest manufacturer of complete general aviation parachutes. Manufacturers of the Para-Cushion emergency parachute systems.

Val Avionics, Ltd.
P.O.Box 13025 • Salem OR 97309-1025
Ph: 503-370-9429 • Fax: 503-370-9885
Literature/Catalog: Yes
Manufacture avionics including COM760TSO VHF transceivers, ILS400TSO instrument landing systems, and base station 7600 systems.

Watkins Aviation Inc.
15770 Midway Road Hanger #6
Dallas TX 75244
Ph: 972-934-0033 • Fax: 972-960-2564
E-Mail: WATKINS@COMPUTEK.NET
Literature/Catalog: Yes
Military personal flight equipment. Helmets, flight suits, survival equipment and accessories. Catalog available; free in US, foreign countries $3.00.

Emergency Equipment, ELTs

ACK Technologies, Inc.
440 W. Julian Street, San Jose CA 95110
Ph: 408-287-8021 • Fax: 408-971-6879
Literature/Catalog: Yes

Aircraft Spruce & Specialty
PO Box 4000 • Corona CA 91718-9961
Ph: 800-824-1930 • Fax: 909-372-0555
E-Mail: info@aircraft-spruce.com
Web: http://www.aircraft-spruce.com
Literature/Catalog: $5.00
A complete selection of homebuilt and certified aircraft parts and materials. Shipping worldwide since 1965. Also see Aircraft Spruce East, Griffin GA, 800-831-2949.
See Display Ad On Page 24

Aircraft Super-Market
#10 Ellefson Dr. • PO Box 676, De Soto IA 50069
Ph: 515-834-2225 • Fax: 515-834-2152
E-Mail: bobdurly@netins.net
Web: http://www.netins.net/showcase/t-birds
Literature/Catalog: $5.00
Handles recreational aircraft of all kinds. New and used, ultralights, powered parachutes and U/L gliders. 2600' runway.
See Display Ad On Back Cover

B & F Aircraft Supply, Inc.
22038 Howell Dr., Unit 6, New Lennox IL 60451
Ph: 815-462-8696 • Fax: 815-462-3697
Literature/Catalog: $1.00

C.A.A. Services
6-C Herman Drive • East Granby CT 06026
Ph: 203-658-8000 • Fax: 203-651-0885
Literature/Catalog: Yes
Avionics and accessories. Custom panels.

J.D. Chapdelaine Co.
5500 N.W. 21st Terrace
Ft. Lauderdale FL 33309
Ph: 305-772-4559 • Fax: 305-493-8960

Eastern Avionics International
28000-A12 Airport Road, Punta Gorda FL 33982
Ph: 941-637-8585 • Fax: 941-637-0388
E-Mail: EAI Club@aol.com
Web: http://avionix.com • Literature/Catalog: Yes
Supplier of all kinds of instruments and avionics by all manufacturers.

ELTs Unlimited, Inc.
P.O. Box 1683 • 15 River Street
New Rochelle NY 10801
Ph: 800-382-0079 • Fax: 914-576-7075
Literature/Catalog: Yes
AP/AF tail mounted ELT.

Emergency Beacon Corp.
P.O. Box 178 • 15 River Street
New Rochelle NY 10802-0178
Ph: 914-235-9400 • Fax: 914-576-7075
Literature/Catalog: Yes
Manufacturer of TSO-C91A P/AP ELTs. Cabin or cockpit mounted with minimal installation.

Marv Golden Discount Sales
8690 Aero Drive, Ste. 102, San Diego CA 92123
Ph: 800-348-0014 • Fax: 619-569-4508
Web: http://www.MARVGOLDEN.COM
Literature/Catalog: Yes
Pointer, Ack, Ameriking ELTs. Blind encoders.

Intertech Aviation Services
3 Sunset Lane • Littleton CO 80121
Ph: 303-781-2577 • Fax: 303-781-2577
Literature/Catalog: Yes
Pointer ELT systems.

Microflight Products, Inc.
1356 Oakdale Drive, Suite 23
Griffin GA 30223-3929
Ph: 800-831-2949 • Fax: 770-229-2329
Literature/Catalog: $3.00
Complete line of instruments, gauges, fuel components, electrical parts, hardware, and accessories. Dealer and OEM programs available.

Narco Avionics, Inc.
270 Commerce Drive • Fort Washington PA 19034
Ph: 800-223-3636 • Fax: 215-643-0197
Literature/Catalog: Yes
Aircraft electronics.

Pacific Coast-Avionics
1833 Auburn Way N., Auburn WA 98002
Ph: 800 353-0370 • Fax: 206-931-0584
E-Mail: PacAvionic@AOL.com
Literature/Catalog: Yes
Full service avionics repair station with sales, service and installations. Complete line of pilot supplies. Specializing in panel package build-up for kit builders.

Pointer
1027 N. Stadem Drive, Tempe AZ 85281
Ph: 602-966-1674 • Fax: 602-968-8020
Literature/Catalog: Yes
Provider of ELT system meeting TSO C91A requirements. External and portable antenna, voice capability, 2 year battery pack, 1.9 pounds. Solid state, hermetically sealed transmitter encased in a military approved polycarbonate.
See Display Ad on Page 32

SkySports
Hangar 1 • Linden-Price Airport, Linden MI 48451
Ph: 800-247-7883 • Fax: 810-735-1078
E-Mail: airstuff@aol.com
Literature/Catalog: Yes

Superflite
2149 E. Pratt Blvd., Elk Grove Village IL 60007
Ph: 800-323-0611 • Fax: 708-364-0358
Literature/Catalog: Yes
Supplier of a broad range of parts and accessories for personal aircraft.
See Display Ad On Page 38

Tropic Aero
1090 NW 53rd Street • Ft. Lauderdale FL 33309
Ph: 305-491-6355 • Fax: 305-772-3327
Literature/Catalog: Yes
Discount portable G.P.S. New and used, buy, sell or trade all makes and models Loran or G.P.S.

Val Avionics, Ltd.
P.O.Box 13025, Salem OR 97309-1025
Ph: 503-370-9429 • Fax: 503-370-9885
Literature/Catalog: Yes
Manufacture avionics including COM760TSO VHF transceivers, ILS400TSO instrument landing systems, and base station 7600 systems.

Varga Enterprises, Inc.
2350 S. Airport Blvd., Chandler AZ 85249
Ph: 800-966-6936 • Fax: 602-899-0324
E-Mail: vargaair@vargaair.com
Web: http://www.vargaair.com
Pointer, Artex, Ack, and Ameri King.

The Wag-Aero Group
1216 North Road, Box 181, Lyons WI 53148
Ph: 800-558-6868 • Fax: 414-763-7595
E-Mail: wagaero-sales@wagaero.com
Literature/Catalog: Yes
ACK and AmeriKing. Replacement batteries for other models.
See Display Ad On Page 13

Western Avionics Inc.
19711 Campus Drive #A • John Wayne Airport
Santa Ana CA 92707-5203
Ph: 714-757-0290 • Fax: 714-757-0293
Literature/Catalog: Yes
Sales, service, installation of avionics, autopilots, radar systems, GPS, and moving maps.

Wicks Aircraft Supply
410 Pine Street, Highland IL 62249
Ph: 800-221-9425 • Fax: 618-654-6253
Literature/Catalog: $5.00
350 page catalog with over 12,000 items.

Wilco, Inc.
3502 W. Harry, Witchita KS 67213
Ph: 800-767-7593 • Fax: 316-943-9664
E-Mail: wilco@feist.com
Web: http://www.globalair.com/wilco/
Ameri-King

Yingling Aircraft, Inc.
P.O. Box 9248, Wichita KS 67277
Ph: 316-943-3246 • Fax: 316-943-2484

Emergency Equipment, Parachutes

A.K.S. Inc.
PO Box 17161, Portland OR 97217
Ph: 503-289-1491 • Fax: 503-247-9207
E-Mail: trikes@teleport.com
Web: http://www.teleport.com/~trikes
North American Distributor for GRS ballistic parachutes.
See Display Ad On Page 47

Aircraft Spruce & Specialty
PO Box 4000, Corona CA 91718-9961
Ph: 800-824-1930 • Fax: 909-372-0555
E-Mail: info@aircraft-spruce.com
Web: http://www.aircraft-spruce.com
Literature/Catalog: $5.00
A complete selection of homebuilt and certified aircraft parts and materials. Shipping worldwide since 1965. Also see Aircraft Spruce East, Griffin GA, 800-831-2949.
See Display Ad On Page 24

Aircraft Super-Market
#10 Ellefson Dr. • PO Box 676, De Soto IA 50069
Ph: 515-834-2225 • Fax: 515-834-2152
E-Mail: bobdurly@netins.net
Web: http://www.netins.net/showcase/t-birds
Literature/Catalog: $5.00
Handles recreational aircraft of all kinds. New and used, ultralights, powered parachutes and U/L gliders. 2600' runway.
See Display Ad On Back Cover

Ballistic Recovery Systems
1845 Henry Avenue, So. St. Paul MN 55075
Ph: 612-457-7491 • Fax: 612-457-8651
E-Mail: BRSchute@aol.com
Web: http://nsers.aol.com/BRSchute/BRS.HTML
Literature/Catalog: Yes
Ballistically deployed emergency parachutes to carry whole aircraft and their occupants to safety in the event of airborne disasters. For most Ultralights and some larger/faster aircraft.

Butler Parachute Systems, Inc.
1820 Loudon Avenue NW • PO Box 6098
Roanoke VA 24017-0098
Ph: 540-342-2501 • Fax: 540-342-4037
E-Mail: butlerchut@aol.com
Literature/Catalog: Yes
Design, test, and manufacture emergency parachute systems for pilots, aircraft (e.g. spin recovery systems), cargo, and munitions. Over 100 configurations of pilot rigs available, including custom built parachutes. 40+ page free information package.

California Power Systems
790 139th Avenue, #4, San Leandro CA 94578
Ph: 800-247-9653 • Fax: 510-357-4429
Web: http://www.800-airwolf.com
Literature/Catalog: $6.95
Complete supplier of ultralight and sport aircraft parts and components. Rotax engines, parts and service. 500 page catalog, $6.95 US, $10.00 Canada and $20.00 foreign airmail.
See Display Ad On Page 69

Colorado Unlimited Adventure
PO Box 100, Tumacacori AZ 85640-0100
Ph: 602-961-3580 • Fax: 602-961-3581
Second Chantz

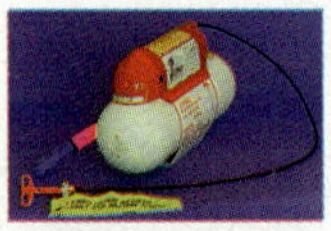

Galaxy s.r.o.
Trida 1. Maje 24A, 4600 Liberec
CZECH REPUBLIC
E-Mail: zlinak@teleport.com
Literature/Catalog: Yes
GRS ballistic parachutes.
See Display Ad On Page 47

Junkers Profly
Na Folimance 13,
Praha 2 120 00 CZECH REPUBLIC
Ph: 0042 2 522 376 • Fax: 0042 2 524 330
E-Mail: STRAKA@BIC.CVUT.CZ

National Parachute Industries
47 E. Main Street, Box 1000
Flemington NJ 08822
Ph: 800-526-5946 • Fax: 908-782-5638
Literature/Catalog: Yes
Emergency parachutes available in three styles: back, chair and seat. Lightweight, comfortable designs. 19 years in business.

Para-Gear Equipment Co., Inc.
3839 West Oakton, Stokie IL 60076
Ph: 708-679-5905
Literature/Catalog: $2.00
All types. Sales and repairs. 35 years in business. Also related products.

Para-Phernalia, Inc.
P.O. Box 3468, Arlington WA 98223-3468
Ph: 360-435-7220 • Fax: 360-435-7272
E-Mail: dant@pstbbs.com
Literature/Catalog: Yes
Softie emergency parachutes and container systems.

Parachute Associates, Inc.
69 Main Street, Vincentown NJ 08088
Ph: 800-526-2822 • Fax: 609-859-3974
Literature/Catalog: $2.00
Pilot emergency parachutes.

Parachute Shop
Pepperell Airport • 165 Nashua Rd.
Pepperell MA 01463
Ph: 800-872-2488 • Fax: 508-433-3264
Sales and service of personal emergency parachute systems. FAA certified loft/repair facility. Service BRS parachutes.

SkySports
Hangar 1 • Linden-Price Airport
Linden MI 48451
Ph: 800-247-7883 • Fax: 810-735-1078
E-Mail: airstuff@aol.com
Literature/Catalog: Yes

Sport Copter International, Inc.
Scappoose Airport • 34012 N. Honeyman Road
Scappoose OR 97056
Ph: 503-543-7000 • Fax: 503-543-7041
Web: http://www.sportcopter.com
JU 40 airplane and sailplane pilot's emergency parachute. Certified to standard: "TSO c 23C" category B. 440 sq. ft., 245 lb max load. Complete system weight 14.3 lb. Service life 15 years. Repacking - 6 month intervals.

Strong Enterprises
11236 Satellite Blvd., Orlando FL 32837
Ph: 407-859-9317 • Fax: 407-850-6978
Literature/Catalog: Yes
Oldest and largest manufacturer of complete general aviation parachutes. Manufacturers of the Para-Cushion emergency parachute systems.

Ultralight Safaris
23450 Vista Way, Quail Valley CA 92587
Ph: 800-RANSFLY • Fax: 909-765-5022
Literature/Catalog: $20.00
Dealer for RANS aircraft AIR CREATION TRIKE, SWIFT MOTORGLIDER and flight school. Brochure available at no cost.

Wings & Wheels
81 Jackson Avenue, W.E.
Jamestown NY 14701
Ph: 716-664-6894 • Fax: 716-664-6895
Literature/Catalog: $4.00
Aircraft supplies, components, and avionics. Buy and sell, new and used. Becker, Icom, Terra, Trimble Magellan, Strong, National Parachutes, and much more.

www.baicorp.com/aerocrafter

Emergency Equipment: Survival Gear

Aircraft Spruce & Specialty
PO Box 4000, Corona CA 91718-9961
Ph: 800-824-1930 • Fax: 909-372-0555
E-Mail: info@aircraft-spruce.com
Web: http://www.aircraft-spruce.com
Literature/Catalog: $5.00
A complete selection of homebuilt and certified aircraft parts and materials. Shipping worldwide since 1965. Also see Aircraft Spruce East, Griffin GA, 800-831-2949.
See Display Ad On Page 24

Apollo Safety & Industrial
1108 West 9th Street, Upland CA 91786-5703
Ph: 909-980-7071 • Fax: 909-985-6873

Asher Industrial
P.O. Box 238, Claremont CA 91711
Ph: 909-933-3168
Emergency medical kit for aircraft.

Aureus, Intl.
2860 S Circle Dr. # GL-22.
Colorado Springs CO 80906-4115
Ph: 719-540-9077 • Fax: 719-540-9078
AUREUS - an established supplier of NOMEX flightsuits, gloves and jackets. Also helmets, safety gear, parachutes, bags, boom mics, etc.

Butler Parachute Systems, Inc.
1820 Loudon Avenue NW • PO Box 6098
Roanoke VA 24017-0098
Ph: 540-342-2501 • Fax: 540-342-4037
E-Mail: butlerchut@aol.com
Design, test, and manufacture emergency parachute systems for pilots, aircraft (e.g. spin recovery systems), cargo, and munitions. Over 100 configurations of pilot rigs available, including custom built parachutes. 40+ page free information package.

J.D. Chapdelaine Co.
5500 N.W. 21st Terrace, Ft. Lauderdale FL 33309
Ph: 305-772-4559 • Fax: 305-493-8960

Intertech Aviation Services
3 Sunset Lane, Littleton CO 80121
Ph: 303-781-2577 • Fax: 303-781-2577
Literature/Catalog: Yes
"Res-Q-Pak" emergency preparedness pack.

MGS Enterprises
24 Bishopgate Dr. • Rochester NY 14624
Ph: 716-889-7756 • Fax: 716-889-5179
E-Mail: S/c@servtech.com
Literature/Catalog: Yes
Compact and lightweight first aid/survival kits for ultralights, powered parachutes, sailplanes, and rebuilts.

Superflite
2149 E. Pratt Blvd., Elk Grove Village IL 60007
Ph: 800-323-0611 • Fax: 708-364-0358
Literature/Catalog: Yes
Supplier of a broad range of parts and accessories for personal aircraft.
See Display Ad On Page 38

Watkins Aviation Inc.
15770 Midway Road Hanger #6
Dallas TX 75244
Ph: 972-934-0033 • Fax: 972-960-2564
E-Mail: WATKINS@COMPUTEK:NET
Literature/Catalog: Yes
Military personal flight equipment. Helmets, flightsuits, survival equipment and accessories Catalog available; free in US, foreign countries $3.00.

Engines

A.K.S. Inc.
PO Box 17161, Portland OR 97217
Ph: 503-289-1491 • Fax: 503-247-9207
E-Mail: trikes@teleport.com
Web: http://www.teleport.com/~trikes
North American distributor of Verner SVS-1400 engines.
See Display Ad On Page 47

AAA Small Car World
1500 Carson, Ft. Worth TX 76117
Ph: 817-831-0946 • Fax: 817-834-7453
Supplier of automotive engines.

Aberle Custom Aircraft
2141 S. Mission Road, Fallbrook CA 92028
Ph: 760-723-1731
Over 30 years assisting show quality and performance oriented builders.

Able Kit Planes
3 NW Maples Street
Fort Walton Beach FL 32548
Ph: 904-244-8482 • Fax: 904-244-8482
Literature/Catalog: Yes
VW based engines-all new parts-50 to 150 hp with electric start and alternator. Turbo systems available.

Aces High Light Aircraft Ltd.
RR 1, London, Ontario N6A 4B5 CANADA
Ph: 519-652-3020 • Fax: 519-652-3020
Literature/Catalog: $5.00
Aircraft kit & custom parts. Manufacturer able to supply most aircraft building materials and parts at competitive cost.

Aero-Bonner Co., Ltd.
Shoreham Airport
Shoreham, Sussex BN4 1FF UNITED KINGDOM

Air Salvage of Arkansas
2854 Hwy 8 East, Mena AR 71953
Ph: 501-394-1022 • Fax: 501-394-7475
Started 32 years ago. Dismantling mostly single engine aircraft, buying inventories, consignment selling of A/C and parts. Housing and storage.

Air-Tec, Inc.
1339 W. Washington Street
Orlando FL 32805-1756
Ph: 800-366-4746 • Fax: 407-468-7008
Literature/Catalog: Yes

Air-Ryder Manufacturing Inc.
PO Box 6896
Wetaskiwin, AB T9A 2G5 CANADA
Subaru EA81 $3900.00.

Aircraft Specialties Services
2860 North Sheridan Road • Tulsa International Airport, Tulsa OK 74115
Ph: 800-826-9252 • Fax: 918-836-4419
Literature/Catalog: Yes
FAA certified repair station D2R764K. Inspection, remachining of aircraft engine parts. Full service, new and used parts.
See Display Ad On Page 74

Aircraft Spruce & Specialty
PO Box 4000, Corona CA 91718-9961
Ph: 800-824-1930 • Fax: 909-372-0555
E-Mail: info@aircraft-spruce.com
Web: http://www.aircraft-spruce.com
Literature/Catalog: $5.00
A complete selection of homebuilt and certified aircraft parts and materials. Shipping worldwide since 1965. Also see Aircraft Spruce East, Griffin GA, 800-831-2949. Hirth engine authorized sales center.
See Display Ad On Page 24

Aircraft Super-Market
#10 Ellefson Dr. • PO Box 676
De Soto IA 50069
Ph: 515-834-2225 • Fax: 515-834-2152
E-Mail: bobdurly@netins.net
Web: http://www.netins.net/showcase/t-birds
Literature/Catalog: $5.00
Handles recreational aircraft of all kinds. New and used, ultralights, powered parachutes and U/L gliders. 2600' runway.
See Display Ad On Back Cover

Aircraft Supply
Allegheny County Airport, West Mifflin PA 15122
Ph: 800-245-0690 • Fax: 412-469-8271
Literature/Catalog: Yes

Aircraft Systems, Inc.
16606 W. Ryerson Road, New Berlin WI 53151
Ph: 800-322-2131 • Fax: 414-784-3099
Literature/Catalog: $15.00
Formula Power Dealer. Aircraft propulsion systems and accessories: Subaru engine conversions, EA81, Legacy and SVX, high performance heads and cams. Fuel injection, performance carbs, motor mounts, Subaru geared and belt reduction drives.

Airscrew Performance
7308 W. Griffin Avenue, Glendale AZ 85303
Ph: 602-931-6667
New, used and rebuilt engines, redrives, exhausts. Crankshaft rebuilding. CDI conversions.

Airstar Discount Sales
2251 Shady Willow Lane
Brentwood CA 94513
Ph: 800-247-7827 • Fax: 510-516-1188
Hirth engine authorized sales center.

Alturair
1405 N. Johnson, El Cajon CA 92020
Ph: 619-449-1570 • Fax: 619-442-0481
Literature/Catalog: Yes
Specializing in Mazda rotary engines. Firewall-forward installation kits. Manufacturer of A650-1 and A650-2 engines.

Amax Engineering
5 Utrecht Court
Donvale, Victoria, 3111 AUSTRALIA
Ph: 61-3-8423132 • Fax: 61-3-8418177
Literature/Catalog: $5.00
Supply of tested low milage and zero time rebuilt engines and aircraft type VW engines.

Ameratech Aviation
One Omni Dr. Ct. • Hangar 1, Cahokia IL 62206
Ph: 618-332-0082 • Fax: 618-332-0084
Sales and service for Limbach Flugmotoren. Firewall-forward installation kits. Manufacturer of Limbach L 1700 EA, Limbach L 1700 EA, Limbach L 2000AE/EO/EB/DA, Limbach L 2400 EB/EO/DE engines.

Arizona Light Aircraft
14473 Fountain Hills Blvd.
Fountain Hills AZ 85268
Ph: 602-837-5578 • Fax: 602-837-5578
E-Mail: titan@interramp.com
Web: http://webmill/titan
Literature/Catalog: Yes
Engine supplier for the Titan Tornado series of aircraft.

Arrow SNC
Via Badiaschi 25, Piacenza 29100 ITALY
Ph: 523-41932
Manufacturer of Arrow 250, Arrow 500 and Arrow 1000 engines.

Atkins Aviation, Inc.
16715 Meridian East, Bldg K • Unit A
Puyallup WA 98373
Ph: 253-848-7776 • Fax: 253-922-0118
E-Mail: Atkins@accessone.com
Web: http://Atkinsrotary.com
Manufacturer of Wankel Rotary 13B engines. Firewall-forward kits.

Atlas Motors Inc.
685 North Loudoun Street
Winchester VA 22601
Ph: 800-336-9878 • Fax: 540-665-2025
Literature/Catalog: Yes
Franklin PZL engines, accessories and parts for Certified and Homebuilt Aircraft.
See Display Ad On Page 73

Aviation Engines, Inc.
301 Airport Street, Centreville AL 35042
Ph: 800-344-7918 • Fax: 205-926-5186
E-Mail: aviationengines@msn.com
Repair station; engine/engine accessories overhaul and horsepower increases for experimentals/homebuilts.

Avtech of Montana Inc.
201 Cherry St., P.O. Box 138
Clancy MT 59634-0138 Ph: 406-933-5646
E-Mail: wa7bef@ixi.net

BAC-NPS
P.O. Box 1984 • 1050 Airport Road
W. Chester PA 19380-0075
Plans for converting VW engine to aircraft engine. 32 pages of instructions with drawings. $17.50 (USA).

Barrett Performance Aircraft, Inc.
2870-B N. Sheridan Road, Tulsa OK 74115
Ph: 918-835-1089 • Fax: 918-835-1754
Literature/Catalog: Yes

Thrust = Drag — Bates Engineering
2742 Swansboro Road, Placerville CA 95667
Ph: 530-622-1886 • Fax: 530-622-1886
E-Mail: bateseng@innercite.com
Web: http://www.innercite.com/~bateseng
Literature/Catalog: Yes
Airplane and propeller design services. Programs for aircraft design, performance, stability and control. Propeller design optimization software.
See Display Ad On Page 82

BD Micro Technologies, Inc.
1260 Wade Road, Siletz OR 97380
Ph: 541-444-1343
Literature/Catalog: $20.00
H-95 Turboprop and Quantum III Turbine-Prop Powerplant System. Engine weighs 75 lbs. and produces 100 SHP. Engine and propeller installed weigh less than 100 lbs. with custom gearbox.

Belted Air Power
1408 Western Avenue, Las Vegas NV 89102
Ph: 702-384-8006 • Fax: 702-384-8006
Manufacturer of Buick V-8, Vortec V-6 and Chevrolet 350 engines.

Better Half VW
PO Box 747, Brookshire TX 77423
Ph: 713-934-8450
E-Mail: kitvw@kcmo.com
Literature/Catalog: $20.00
Plans for 1/2 VW engine conversion for light aircraft. Propeller hubs and crank shafts available for VW engines.

Bradley Aerospace
950 Muir Ave., Chico CA 95926
Ph: 916-899-7918 • Fax: 916-899-2942
Web: http://www.kgmsoftworks.com/bradley
Literature/Catalog: $10.00
Bradley Aerospace VW conversion engines. 40-110 hp. Direct drive and Formula 1 series engines with PSRU. 1800 hour TBO.
See Display Ad On Page 40

California Power Systems
790 139th Avenue, #4, San Leandro CA 94578
Ph: 800-247-9653 • Fax: 510-357-4429
Web: http://www.800-airwolf.com
Literature/Catalog: $6.95
Complete supplier of ultralight and sport aircraft parts and components. Rotax engines, parts and service. 500 page catalog, $6.95 US, $10.00 Canada and $20.00 foreign airmail.
See Display Ad On Page 69

Cam-Fire Engineering Group Inc.
9755 W. Saanich Rd. • Sidney, B.C.
V8L 5T5 CANADA
Ph: 250-656-4774 • Fax: 250-656-8007
E-Mail: RLM@Camfire.BC.CA
Web: http://Camfire.BC.CA
Literature/Catalog: $10.00
Manufacturer of the Honda based CAM-100-175 liquid cooled aircraft engines. Propeller speed reduction units up to 500 hp.

Canadian Airmotive, Inc.
7400 Wilson Avenue
Delta, BC V4G 1E5 CANADA
Ph: 604-940-9378 • Fax: 604-940-9566
Literature/Catalog: $8.00
CAM 100 4 cycle 100 hp aircraft engine. Information packet US $8.00.

Carr Precision
Suite 303 • 6040 North Cutter Circle
Portland OR 97217
Ph: 503-735-9980 • Fax: 503-285-0553
E-Mail: carrprecision@worldnet.att.net
Literature/Catalog: $5.00
1/2 VW engines. 20 years experience. Four cycle, air cooled, lightweight aircraft engine. Instructions with detailed drawings: $60.00. Manufactures The Carr Twin engine.

Nick Carter Engines, Inc.
2116 West G Street, Elizabethton TN 37643
Ph: 423-542-2943 • Fax: 423-542-2633
E-Mail: ncarter@usit.net
Web: http://www.carterengines.com
Sales of Continental and Lycoming engines, cylinders and parts.

Certified Parts Corp.
1111 W. Racine • PO Box 8468
Janesville WI 53545
Ph: 608-752-9441 • Fax: 608-752-3528
Literature/Catalog: Yes
Ultralight engines and engine parts. Cuyuna and Zenoah engines.

Christine Aero Engines
PO Box 216 • Highway 178
Donalds SC 29638 • Ph: 803-379-8677
Manufactures C1200 engines.

Colorado Unlimited Adventure
PO Box 100, Tumacacori AZ 85640-0100
Ph: 602-961-3580 • Fax: 602-961-3581
Hirth and Rotax service center.

Cordray Research, Inc.
P.O. Box 720825
Oklahoma City OK 73172-0825

Crossflow Aero Corporation
RR#1, Palgrave ON L0N 1P0 CANADA
Ph: 519-942-9855 • Fax: 519-942-8503
Web: http://www.crossflow.com
Literature/Catalog: $5.00
Alternate power plant sources for recreational aircraft.

Jaramillo Crowe
Carrera 43F #17-45 • AP Aereo 791
Medellin COLUMBIA
Hirth engine authorized sales center.

Crystal Shamrock, Inc.
6000 Douglas Drive N, Crystal MN 55429
Ph: 612-533-2214 • Fax: 612-533-5519
35 year old FBO. Certified repair station for upholstery, engines and service.

Decker Aero Power, Inc.
P.O. Box 1524, Eagle River WI 54521
Ph: 715-479-3018 • Fax: 715-479-8286

Dumont Aircraft Engines, Inc.
1605 West State Road 64
Avon Park FL 33825
Ph: 941-453-2600 • Fax: 941-453-5864
Custom aircraft engines, overhaul and repair.

Dyna-Cam Aero Engine Corp.
23960 Madison St. • PO Box 1159
Torrance CA 90505-0159
Ph: 310-791-4642 • Fax: 310-791-4645
E-Mail: dyna-cam@pobox.com
Web: http://www.dynacam.com
Literature/Catalog: Yes
Cam-drive, free-piston 200 hp engine. Patented and FAA certified for aircraft and helicopters. Tech. Pak available for $20.00

Eastside Ultralight Aircraft
4700 188th Street NE, Arlington WA 98223
Ph: 206-435-3737

Eggenfellner Advanced Aircraft
427-3 Amherst St., Ste. 230, Nashua NH 03063
Ph: 800-840-4620 • Fax: 603-880-8076
Literature/Catalog: Yes
Subaru based aircraft engines, geared reduction drive units, and firewall forward packages.

EngineAir Inc.
209 Cessna Blvd., Daytona Beach FL 32124
Ph: 904-788-2852
Web: http:////home.aol.com/chevrolair
Manufactures GTSIV engines.

Engineering Systems Co., Inc.
5652-83 Sasaga • Matsumoto City
Nagano 399 JAPAN Ph: (81) 263-261212

Euravia Engineering
Largo Bacigalupo, 32 - Sc.G
Roma 00142 ITALY Ph: 011-396-519-3998
Literature/Catalog: Yes
Two stroke water and air cooled engines for ultralights and experimentals. From 22 to 92 hp.

Flight Dreams
RR1 Lot 27 Concession 14
Lakefield, Ontario K0L 2H0 CANADA
Ph: 705-657-3420
Hirth engine authorized sales center.

Flightcraft, Inc.
7505 NE Airport Way, Portland OR 97218
Ph: 503-331-4200 • Fax: 503-331-4273
E-Mail: AAH07@AOL.COM
Custom panels for Glasair, Lancair. Factory authorized Lancair flight training.

Formula Power
1060 Blady Park • PO Box 2168
Sandpoint ID 83864
Ph: 208-263-6637 • Fax: 208-263-6214
E-Mail: sybersys@rand.nidlink.com
Manufactures Subaru EA 81, Surbaru EA22 and SVX engines.

Formula Power East
4921 Fennell, Suffolk VA 23435
Ph: 804-484-33185
Literature/Catalog: Yes
Experimental aircraft engines. Best power to weight ratio. 110 to 375 hp. Multi-port fuel injected. Dual ignition option. Choice of reduction drives. F/W/F Kits for Murphy, Avid, Zenair, Kitfox and more.

G&N Aircraft, Inc.
1701 E. Main Street, Griffith IN 46319
Ph: 800-348-6504 • Fax: 219-924-1059
E-Mail: gnair@jorsm.com
Web: http://www.gnaircraft.com
Literature/Catalog: Yes
Custom overhaul, repairs and inspections for Continental and Lycoming piston engines. Overhaul and repair of all accessories: alternators, magnetos, carburetors, etc.
See Display Ad On Page 68

Geissinger Aero Services
8 Dogwood Dell Drive, Coatesville PA 19320
Ph: 215-383-4589

Don George, Inc.
1339 W. Washington St., A1, Orlando FL 32805
Ph: 407-422-0188 • Fax: 407-422-0836
Custom building and overhaul of aircraft engines. New and used parts available.

Great Plains Aircraft
P.O. Box 545, Boys Town NE 68010
Ph: 402-493-6507 • Fax: 402-333-7750
E-Mail: gpasc@earthlink.net
Web: http://www.greatplainsas.com
Literature/Catalog: $3.00
See Display Ad On Page 80

Green Sky Adventures, Inc.
2395 Cream Ridge Rd., Orwell OH 44076
Ph: 216-293-6624 • Fax: 216-293-6321
Web: http://www.greensky-adventures.com
Literature/Catalog: $2.00
All Rotax UL engines from 277-914. Parts for all engines.

HCI Aviation Inc.
3461 Dissen Rd., New Haven MO 63068
Ph: 573-237-3605 • Fax: 573-237-3605
E-Mail: jacki@fidnet
Web: http://users.aol.com/gmallaren/hci.html
Literature/Catalog: Yes
Five cylinder radial engine: 75 hp @ 2150 RPM, 122 lbs.

High Performance Aircraft Eng.
P.O. Box 1256, Mena AR 71953
Ph: 800-233-1099
Supplier of high performance aircraft engines and components.

Hirth Motor Corporation
12831 97th Street • Box 7068
Edmonton, Alberta T5E 5S9 CANADA
Ph: 403-448-0389 • Fax: 403-448-7270
Manufactures Hirth F-33, Hirth 263, F-23, Hirth 2702, Hirth 2704, Hirth 2703, Hirth 2706 and Hirth F-30 EP2 engines. U.S. distributors: Aircraft Spruce & Specialty; Airstar Discount Sales; Hoffman Flight Park; K & A Airport Services and Kitplanes; Midstate Planes & Cycle; Microflight Products. Canadian distributors: Flight Dreams; Recreation Aircraft Supplies. South American distributor: Jaramillo Crowe.

Hoffman Flight Park
Rt. 2 Box 164A, Poteau OK 74953
Ph: 918-647-3091
Hirth engine authorized sales center.

HPower
PO Box 690, Ellington CT 06029
Ph: 860-875-8185 • Fax: 860-870-5499
Web: http://www.hpower.ltd.com
Light aircraft engine.

Hummel Aviation
509 E. Butler Street
Bryan OH 43506-0880
Ph: 419-636-3390
Literature/Catalog: $2.00
Half-VW engines. 28 to 40 hp., 72 to 80 lbs. Plans $45. Kits $2195-$3195. US. Plans available. VHS video available at $42. Manufactures 1/2 VW engines.

J-Bird
Box 438 • 210 Main St., Kewaskum WI 53040
Ph. 414-626-2611
Rotax, Kawasaki, Cuyuna, Zenoah, Hirth, Solo and others.

Jabiru Aircraft Pty Ltd
PO Box 5186 • Bundaberg West
Queensland 4670 AUSTRALIA
Ph: 61-71-551-778 • Fax: 61-71-552-669
Manufacturer of the Jabiru 1600.

Jones Custom Aircraft Services
PO Box 157, Laurelton PA 17835
Ph: 717-922-4877
E-Mail: sjonesfly@AOL.com
Literature/Catalog: Yes
Chevy 350 V8 and 454/502 V-8 auto engine conversions. Belt prop drive units. Manufactures Chevrolet V-8 350 and Chevrolet V-8 502 engines.

K & A Airport Services & Kitplanes
3375 N. Hwy. 10, Huntington UT 84528
Ph: 801-687-2159
Hirth engine authorized sales center.

Kodiak Research Canada Ltd.
S22 C39 RR6 Stn Main • 6235 Okanagan Landing Road
Vernon British Columbia V2T 6Y5 CANADA
Ph: 604-542-4151 • Fax: 604-549-7111
US distributor for Rotax engines.

König

Konig Motorenbau
Friedrich Albricht Damm 72
Berlin D-13627 GERMANY
Ph: 0049 30 3443071 • Fax: 0049 30 3441595
Literature/Catalog: Yes
Three and four cycle radial engines, 6 kg/24 hp and 18.5 kg/28 hp, with or without reduction/electric starter, adjustable propellers. Manufacturer of Konig 430 and Konig 570 engines.
See Display Ad On Page 74

Krybus Aviation
822 Santa Maria Street • PO Box 14
Santa Paula CA 93061
Ph: 805-525-8764 • Fax: 805-525-1147
Four cylinder, inline, inverted, air cooled aircraft engines.

L.A.P.S. Inc.
PO Box 143, Burlington IL 60109
Ph: 847-683-4099 • Fax: 847-683-4099
Manufacturer of L.A.P.S. 430 engines. This is identical to the Cuyuna UL II-02 except that it is sold in kit form.

Leading Edge Air Foils
Meadowlake Airport • 8242 Cessna Drive
Peyton CO 80831
Ph: 719-683-5323 • Fax: 719-683-5333
E-Mail: leafkp@kktv.com
Web: http://www.leadingedgeairfoils.com
Literature/Catalog: $6.00
Rotax service center. Distributor for ultralight and experimental aircraft, parts, and engines.
See Display Ad On Page 34

Light Affordable Power Supply, Inc.
PO Box 184, Burlington IL 60109
Ph: 708-683-4099 • Fax: 708-683-4099
Manufactures Cuyuna UL II-02 (Laps 35) engines.

Light Aviation Research Corporation
391 Spry Road, Morrison TN 37357
Ph: 615-765-5511

Light Power Engine Corp.
PO Box 3350
Morgantown WV 26503
Ph: 304-291-3843 • Fax: 304-292-1902
Literature/Catalog: $20.00
Piston engines 125 - 700 hp, air cooled, liquid cooled ZM, AVIA, Franklin.

Lightning Engines
4099 Brick Church Pike, White Creek TN 37189
Ph: 615-865-1802 • Fax: 615-865-9164
Manufactures Chevrolet V-8 engines.

LIMBACH Flugmotoren
Aircraft-Engines the world over

Limbach Flugmotoren GmbH & Co
Kotthausener Strasse 5
Koenigswinter-Sassenberg D-53639 GERMANY
Ph: (01149) 224492010 • Fax: (01149) 224492013
Literature/Catalog: Yes
Four cylinder, four stroke engines with 1700, 1800, 2000 and 2400 ccm displacement from 60 to 100 hp for motorgliders and experimental sport aircraft. Mostly JAR certified.
See Display Ad On Page 71

Lockwood Aviation Supply, Inc.
280 Hendricks Way • Suite A
Sebring FL 33870
Ph: 941-655-5100 • Fax: 941-655-6225
E-Mail: lockwood@digital.net
Web: http://www.lockwood-aviation.com
Literature/Catalog: $9.95
Light aircraft engines, parts and accessories.

Gary McGill & Associates
P.O. Box 2566, Glendale AZ 85311-2566
E-Mail: gemcsr@juno.com
Literature/Catalog: $1.00
VW two cylinder engine conversion plans and parts, $15.00.

MDB Aerospace Corp.
13420 Reese Blvd. West, Huntersville NC 28078
Ph: 704-948-1234 • Fax: 704-875-0781
Manufactures FM 2600 engines.

Mid-West Engines
Hangar SE38 • Gloucestershire Airport
Staverton GL51 6SR UNITED KINGDOM
Ph: 452-857456 • Fax: 452-856519
E-Mail: midwestengines@globalnet.co.uk
Literature/Catalog: Yes
C.A.A. certified and non-certified 50 and 100 hp Petrol InjecteoAero Rotary Wankel type engines.

Midstate Planes & Cycle
3216 Lantanna Road, Lantanna FL 33462
Ph: 561-969-7719 • Fax: 561-969-9936
Hirth engine authorized sales center.

Midwest Engineering
PO Box 4706, Overland Park KS 66204-0706
Ph: 913-631-8447
E-Mail: mwengineering@juno.com
Literature/Catalog: $10.00
Turbine-Tech. Information, specifications, and plans for turbine pulse jet and ram jet engines, $10.00. Turbine engine demo video, $10.00. Both for $15.00.

Moravia, Inc.
P.O. Box 942
Thunder Bay, Ontario P7C 4X8 CANADA
Ph: 807-475-5113 • Fax: 807-475-9587
E-Mail: moravia@loon.norlink.net
Web: http://www.norlink.net/moravia
Literature/Catalog: Yes
Walter Mikron III AE and III B engines: four stroke, air cooled, inverted, direct drive, dual magnetos, dual oil pump, 65 - 75 hp. Walter-Lom engine: 120-210 hp, in-line, supercharger available.

Nelson Aircraft Corp.
8075 Penna Avenue • Irwin PA 15642

Nelson Engine of Florida, Inc.
420 Harbor Drive, Naples FL 33940
Ph: 813-263-1670 • Fax: 813-261-8345
Literature/Catalog: Yes
Manufacturer of the H-63-CP, FAA certified, 48 hp 2 stroke aircraft engine and the H-63 CP-2 twin engine, 90 hp single prop. aircraft engine. Complete engines and parts.

Northwest Aero Products
2801 Academy Dr. S.E., Auburn WA 98092
Ph: 206-735-5022 • Fax: 206-939-1578
E-Mail: nwaero@ix.netcom.com
Web: http://www.halcyon.com/wpowers/
Literature/Catalog: Yes
Complete "firewall forward" package including engine mount, engine, reduction drive and radiator using either Ford 3.8L V-6 or Chevrolet 4.3L V-6. Manufacturer of Chevrolet 4.3L V-6 and Ford 3.8L V-6 engines.

Northwest Formula Power
P.O. Box 2168, Sand Point ID 83864
Ph: 208-263-6637 • Literature/Catalog: Yes
Formula Power dealer.

19132 59th Dr. NE, Arlington WA 98223
Ph: 360-435-1055 • Fax: 360-435-4554
Manufacturer of EJ22-160-EFI, Turbo EJ22-EFI, EA81-98-TBI, EA81-118-TBI and Turbo EA81-130-EFI engines.

O'Neill Airplane Company
791 Livingston, Carlyle IL 62231
Ph: 618-594-2681 • Fax: 618-594-2681
E-Mail: troneill@accessus.net
Manufacturer of 351 W engines.

Olympic Ultralights
PO Box 388 • Port Angeles WA 98362
Ph: 360-452-0117 • Fax: 360-452-0117
Ultratec 80 four stroke, liquid cooled, fuel injected engine for light aircraft.
See Display Ad On Page 54

P & K Enterprises
11470 W. 800N • Colfax IN 46035-9636
Ph: 765-324-2424 • Fax: 765-324-2424
E-Mail: PKENTERPRISES@tctc.com
Web: http://pkenterprises/tctc/com/html
Rotax.

Pacific Continental Engines
16041 Arminta Street • Van Nuys CA 91406
Ph: 818-781-4947 • Fax: 818-781-0860
Web: http://www.pcenet.com
Literature/Catalog: Yes
Overhaul and sale of Textron, Lycoming, and Continental engines, components, and parts. Avionics shop sells new, repairs, and maintains all major brands. FBO offers aircraft maintenance.

Penn Yan Aero Service, Inc.
2499 Bath Road • Penn Yan NY 14527
Ph: 8007277230 • Fax: 315-536-2335
E-Mail: PnnYanAero@Aol.Com
Web: http://www.airtourist.com/Penn Yan
Literature/Catalog: Yes
Lycoming & Teledyne Continental Motors distributor. FAA certified engine repair staion NJ1R350K. Custom and exchange engine production, certificated engines, machine shop, cylinder and accessory repairs.

Piston Power, Inc.
136 Racquette Dr. • Fort Collins CO 80524
Ph: 970-472-5057 • Literature/Catalog: Yes
New and overhauled Lycoming engines. Overhauled Continental engines.

PowerKits-Technical Materials, Inc.
263 Quillayute Place • PO Box 80
La Conner WA 98257
Ph: 360-466-1071 • Fax: 360-466-1071
Literature/Catalog: Yes
100 hp turbo-charged Suzuki conversion for homebuilts and experimental aircraft. Manufacturer of Suzuki G10 Sprint engines.

Powersport, Inc.
12294 Hatch Lane • Aumsville OR 97325
Ph: 503-769-5469
Manufacturer of Mazda 13B engines.

Precision Aerial Services
207-2 Cessna Blvd.
Daytona Beach FL 32124
Ph: 904-761-4122 • Fax: 904-788-0150
Manufacturer of 164 CID/2.7 L engines.

Pro VW Parts & Machine
6614 Southeast Blvd. #1 • Derby KS 67037
Ph: 316-788-4625
Two cylinder and four cylinder VW engines built for light aircraft.

Propulsion Technologies, Inc.
7200 C.E. Dixon St. • Stockton CA 95206
Ph: 209-982-5558 • Fax: 209-982-5702
E-Mail: karlpti@msn.com
Literature/Catalog: Yes
Lom engines for experimental aircraft. Also a firewall-forward package of engine, prop, cowling, and accessories, for the Glastar.

The GLASS GOOSE By Quikkit

Quikkit Corporation
9002 Summer Glen • Dallas TX 75243-7445
Ph: 214-349-0462 • Fax: 214-349-0462
E-Mail: quikkit@airmail.com
Web: http:////webwrks.com/glassgoose/
Literature/Catalog: Yes
Rubello Racing (Formula Power) Subaru engines.
See Display Ad On Page 24

Raven Rotorcraft, Inc.
1025 Rosewood Ave., Suite 104
Boulder CO 80304
Ph: 303-440-6234 • Fax: 303-440-4085
E-Mail: redrive@raven-rotor
Web: http://www.raven-rotor.com
Literature/Catalog: Yes
Reduction drive kits for the 1.0 & 1.3 liter Geo Metro/Suzuki auto engines for kit aircraft. 65-120 horsepower.

Recreation Aircraft Supplies, Inc.
9825 33rd Avenue
Edmonton, Alberta T6N 1B6 CANADA
Ph: 403-448-9677 • Fax: 403-448-7370
Hirth engine authorized sales center.

Recreational Power Eng.
5479 E. Country Road 38 • Tiffin OH 44883
Ph: 419-585-7002 • Fax: 419-585-6004
Literature/Catalog: Yes
US distributor of Hirth two cycle aircraft engines. Sales, service, parts and engine overhauls for Hirth engines.

Reductions
Box 16, Group 15, RR 1
Dugald, MB R0E 0K0 CANADA
Ph: 204-853-7998
Web: http://www.mts.net/~davej
Manufacturer of Subaru EA81 engines.

Reg's Air Cooled Engines
9708 Princess Drive
Surrey, B.C. V3V 2T4 CANADA
Ph: 604-581-7414 • Fax: 604-581-7418
Literature/Catalog: Yes
Distributor and service center for Rotax ultralight aircraft engines.

Regale Limited, U.K.
27 Whistler Road, Eaton Ford
St. Neots, Cambridgeshire PE19-3RT
UNITED KINGDOM
Ph: 44-480-477975 • Fax: 44-480-477975
British & European military aircraft parts. Support & major items for BAC, Rolls Royce, Marine, BAE, and much more. $10,000,000 in stock.

Regale Limited, U.S.A.
RT. 1, Box 138A • Peculiar MO 64078
Ph: 816-779-6075 • Fax: 816-758-6054
British & European military aircraft parts. Support & major items for BAC, Rolls Royce, Marine, BAE, and much more. $10,000,000 in stock.

Revmaster Aviation
7146 Santa Fe Avenue E. • PO Box 402077
Hesperia CA 92345
Ph: 760-244-3074 • Fax: 760-244-4983
Literature/Catalog: $4.00
75 hp aircraft engines, complete.

RFI Powerplant Systems Group
2916 Tennessee Northeast
Albuquerque NM 87110
Ph: 505-298-9362 • Fax: 505-266-1789
Subaru and Mazda engine conversions, 60 to 200 hp. Conversion manuals.

Rick Giles
4287 East Winneshiek Road
Freeport IL 61032
Yamaha motorcyle engine into Nieuport 12. 1-1/2 hour video includes teardown of engine plus important features and modifications for installation. $29.50

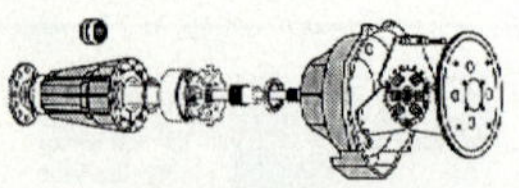

Ross Aero
3824 E. 37th Street • Tucson AZ 85713
Ph: 520-747-7877 • Literature/Catalog: Yes
Information pack for engines and engine accessories: $5.00.
See Display Ad On Page 4

Rotax
Bombardier-Rotax GmbH • Postfach 5
Gunskirchen A-4623 AUSTRIA
Ph: 072-462-710
Manufacturer of Rotax 185, Rotax 277, Rotax 377, Rotax 447, Rotax 503, Rotax 532, Rotax 582 and Rotax 618 engines. The distributor is Kodiak Canada, S22 C39 RR 6, Vernon BC V1T 6Y5, Phone (604) 545-4997.

Sacramento Sky Ranch, Inc.
6622 Freeport Blvd. • PO Box 22610
Sacramento CA 95822 • Ph: 916-421-7672
Literature/Catalog: Yes

Sadler Radial Engines, Inc.
PO Box 953 • Coupeville WA 98239-0953
Ph: 360-678-6994 • Fax: 360-678-2005
Literature/Catalog: $10.00
Engine kit: 4 cycle, 6 cylinder, dual ignition, 65 hp., compact, light-weight, radial engine. Kit can be purchased in two stages. Manual available $45.00. Literature/Catalog $15.00 overseas.

Scat Enterprises
1400 Kingsdale Avenue
Redondo Beach CA 90278
Ph: 310-370-5501 • Fax: 310-24-2285
Literature/Catalog: $6.00
Manufacturer of 2 and 4 cylinder water cooled and air cooled engines. 40 to 200 hp, 4 cycle. Horizontal and vertical configurations. Parts, kits and complete engines.

Recreational Power Eng.
5479 E. Country Road 38 • Tiffin OH 44883
Ph:419-585-7002 • Fax: 419-585-6004
Literature/Catalog: Yes
US distributor of Hirth two cycle aircraft engines. Sales, service, parts and engine overhauls for Hirth engines. Supplier of engine instrument systems and propellers.

T.W. Smith Engine Co., Inc.
4510 Airport Rd. • Hangar #1 Lunken Airport
Cincinnati OH 45226
Ph: 513-871-3500 • Fax: 513-533-244
Web: http://www/twsmith
Literature/Catalog: Yes
New and overhauled Continental and Lycoming engines and accessories. Certified Repair Station.

SOLO France
4 rue du Thal • Obernay 67210 FRANCE
Ph: 88 95 30 86

www.baicorp.com/aerocrafter

ROTAX

South Mississippi Light Air
1279 Marshall Smith Rd.
Lucedale MS 39452
Ph: 800-247-7652 • Fax: 601-947-4959
Literature/Catalog: Yes
Rotax engines in stock include the 447, 503, 582, 618, and 912. Rotax parts shipped UPS same day. Authorized Quicksilver Aircraft and Quicksilver aircraft parts distributor. Parts and sales, call 1-800-247-7652. For service, call 1-601-947-4953.
See Display Ad On Page 13, 69

Specialty Aircraft Co.
1440 NW 87th Ct. • Redmond OR 97756
Ph: 800-500-6786 • Fax: 503-548-5336
Single engine Cessna salvage yard. Instruments and parts for homebuilt aircraft.

Sport Copter International, Inc.
Scappoose Airport • 34012 N. Honeyman Road
Scappoose OR 97056
Ph: 503-543-7000 • Fax: 503-543-7041
Web: http://www.sportcopter.com
Literature/Catalog: $12.00
North American distributor for Verner SVS 1400 2 cylinder, 4 stroke engine from the Czech republic.

Stratus, Inc.
7750 12th Avenue NW • Seattle WA 98117
Ph: 206-783-3845 • Fax: 206-784-8511
E-Mail: Reinerh@aol.com
Literature/Catalog: Yes
Subaru EA 81 conversion, 100 hp, 182 lbs. dry. Subaru EJ 22 conversion, 160+ hp., 265 lbs. dry. Stratus EJ 22, 180 hp, fits directly to Lyc. O320/360 mounts.

Technical Materials, Inc.
PO Box 80 • LaConner WA 98257
Ph: 360-466-1071 • Fax: 360-466-1071
E-Mail: cline@fidalgo.net
Literature/Catalog: Yes
Kits to convert Suzuki G10 Sprint/Metro engines for use in experimental aircraft. Send SASE for information.

Tennessee Propellers, Inc.

125 Deerfield Drive • Normandy TN 37360
Ph: 615-455-4516 • Fax: 615-455-1309
Literature/Catalog: $5.00
World distributor for Zenoah, models G50 and G25B-1. The G50 is a twin cylinder opposed, 45 hp at 5800 rpm. Complete installed weight: 90 lbs. The G25B-1 is a single cylinder, 22 hp at 6600 rpm. Two cycle. Complete installed weight: 51 lbs.
See Display Ad On Page 75

Textron Lycoming
652 Oliver Street
Williamsport PA 17701
Ph: 717-327-7278 • Fax: 717-327-7022

The Cam-Fire Engineering Group, Ltd.
9755 W. Saanich Rd.
Sidney, BC V8L 5T5 CANADA
Ph: 604-656-4774 • Fax: 604-656-8007
Literature/Catalog: $10.00
Manufacture CAM 100, V-4, V-6 and V-8 liquid cooled engines. Up to 500 hp. Belt and helical cut PSRUs, CAM Puter redundant electronic ignitions, dyno testing, blueprinting.

Tiger Aviation
134 Cessna Drive • Trenton SC 29847
Ph: 803-663-6200 • Fax: 803-663-6655
E-Mail: TREMFL@AOL.COM
Web: http://www.tigeraviation.com
Distributor for Jabiru 2200 engines.

Toccoa Aviation, Inc.
Route 3 Airport • Toccoa GA 30577
Full maintenance for light airplanes.

Total Engine Concepts
2391 Old Dixie Hwy.
Riviera Beach FL 33404
Ph: 407-842-2813 • Fax: 407-845-3237
Literature/Catalog: Yes
Manufacturer of VW Aero Linblock and 1083 Defiant engines. Send self-addressed stamped envelope for literature.

2 Stroke International
Beaufort Industrial Park • 8 Schein Loop
Beaufort SC 29902
Ph: 803-846-2167 • Fax: 803-846-2169
E-Mail: engines@2si.com
Web: http://www.2si.com
Literature/Catalog: Yes
Lightweight high-power fan cooled and water cooled engines for aircraft. 20 to 120 hp.

Universal Engineering, Inc.
116 W. Main • PO Box 115
Winamac IN 46966
Ph: 219-946-6757 • Fax: 219-946-7471
Web: http://www.universalengineering.com
Literature/Catalog: $12.00
Manufacturer of Universal V-Tech 4 Turbo-Diesel engines.

Venture Aerodome
Rt. 1 Box 11631 • Roland OK 74954
Ph: 918-427-6490
Literature/Catalog: $4.00
Distributor of Hirth engines.

Verner Motors

Vikirovice 229
77813 Vikirovice CZECH REPUBLIC
Ph: 0042 649 211 576 • Fax: 0042 649 211 576
E-Mail: zlinak@teleport.com
Literature/Catalog: Yes
Verner SVS 1400 2 cylinder, 4 stroke engine
See Display Ad On Page 47

Dwayne Weber
5528 S. Nottingham • Chicago IL 60638
Ph: 312-889-3711
Sales of military gas turbine engines and other small gas turbines.

Wentworth Aircraft Inc.
3015 Cedar Avenue, S. • Minneapolis MN 55407
Ph: 612-722-0065 • Fax: 612-722-1024
Literature/Catalog: Yes
Engines, radios, instruments and miscellaneous single engine used parts.

Western Power Systems
420 30th Street • Bakersfield CA 93301
Ph: 805-323-3432
Literature/Catalog: Yes
Formula Power dealer.

Western Ultralight
4700 188th N.E. • Arlington WA 98223
Ph: 360-435-3737 • Fax: 360-435-3737
Literature/Catalog: Yes
Total Ultralight facility. Including flight instruction, sales, parts, and maintenance.

Windspire, Inc.
PO Box 145-AX • Long Green MD 21092
Ph: 410-836-6824 • Fax: 410-692-0595
Literature/Catalog: $2.00
Plans offered are for pulse jet engine construction. Manual includes plans for building two different pulsejet engines. A FREE bonus is included - plans for building two different jet helicopters. Jets may be used for one person helicopters & airplanes.

Michael Zoche Co.
Keferstrasse 13 • Muenchen
Munich 80802 GERMANY
Ph: +49 89 344591 • Fax: +49 89 342451
Literature/Catalog: Yes
150 and 300 hp lightweight diesel engines.

Engine Accessories

Aero Accessories
1240 Springwood Church Road
Gibsonville NC 27249
Ph: 800-822-3200 • Fax: 910-449-5461

Aero Battery Sales & Service
3442 Church Street
Cincinnati OH 45244-3410
Ph: 513-271-9116 • Fax: 513-561-1950

Aero Controls
17 Geneva Cr
St. Albert, Alta T8N 0Z3 CANADA
Ph: 403-460-2075 • Fax: 403-459-1928
Low friction throttle cable assemblies for two stroke engines.

Aero Fuel Transfer Pump
P.O. Box 78129 • Seattle WA 98178
Ph: 206-467-4170
Fuel pumps.

Aero Safe Corporation
Route 1 Box 289 • Millsap TX 76066
Ph: 800-433-5689 • Fax: 817-682-7662
Literature/Catalog: Yes
Standby vacuum/pressure systems.

Aero Specialty Co.
129 Chuck Yeager Way • Oroville CA 95965
Ph: 916-532-0919 • Fax: 916-532-0936
Literature/Catalog: Yes
Welded components such as exhaust systems, engine mounts and fuel tanks. FAA repair station ZSPR067L.

Air Drive Service Co.
12140 Washington Blvd. • Whittier CA 90606
Ph: 213-698-6321
Literature/Catalog: Yes

Air Support International, Inc.
1000 W. Washington, Sta.10 • PO Box 489
Marshfield MO 65706
Ph: 800-247-2738 • Fax: 417-329-5805
Literature/Catalog: Yes
Camshafts, crankshafts, followers, bearings.

Aircraft Accessories of Oklahoma
2740 N. Sheridan Road • Tulsa OK 74115
Ph: 918-835-9924 • Fax: 918-835-3681
E-Mail: airacc@busprod.com
Literature/Catalog: Yes

Aircraft Fuel Cell Repair
300 Airport Road • Eagle River WI 54521
Ph: 800-437-8732 • Fax: 715-479-6344
E-Mail: fuelcell@newnorth.net
Web: http://www.globalair.com/eagle.htm
Literature/Catalog: Yes
Custom built fuel cells for any aircraft.

Aircraft Projects
2269 Maywood Ave. • San Jose CA 95128
Ph: 408-277-0645 • Fax: 408-279-6535
E-Mail: BAFRANK@att.net
Literature/Catalog: $6.00
Engine mounts, fuel tanks for Ford V-6 engine for any homebuilt.

Aircraft Specialties Services
2860 North Sheridan Road • Tulsa International Airport • Tulsa OK 74115
Ph: 800-826-9252 • Fax: 918-836-4419
Literature/Catalog: Yes
FAA certified repair station D2R764K. Inspection, remachining of aircraft engine parts. Full service, new and used parts.
See Display Ad On Page 74

Aircraft Spruce & Specialty
PO Box 4000 • Corona CA 91718-9961
Ph: 800-824-1930 • Fax: 909-372-0555
E-Mail: info@aircraft-spruce.com
Web: http://www.aircraft-spruce.com
Literature/Catalog: $5.00
A complete selection of homebuilt and certified aircraft parts and materials. Shipping worldwide since 1965. Also see Aircraft Spruce East, Griffin GA, 800-831-2949.
See Display Ad On Page 24

Aircraft Super-Market
#10 Ellefson Dr., PO Box 676, De Soto IA 50069
Ph: 515-834-2225 • Fax 515-834-2152
E-Mail: bobdurfy@netins.net
Web: http://www.netins.net/showcase/t-birds
Literature/Catalog: $5.00
Handles recreational aircraft of all kinds. New and used, ultralights, powered parachutes and U/L gliders. 2600' runway.
See Display Ad On Back Cover

Aircraft Supplies, Inc.
1745 Bathgate Avenue • Bronx NY 10457
Ph: 718-294-6370 • Fax: 718-716-8012
Aircraft engine spares, chemicals and tapes.

Aircraft Systems, Inc.
16606 W. Ryerson Road • New Berlin WI 53151
Ph: 800-322-2131 • Fax: 414-784-3099
Literature/Catalog: $15.00
Formula Power dealer. Aircraft propulsion systems and accessories: Subaru engine conversions, EA81, Legacy and SVX, high performance heads and cams. Fuel injection, performance carbs, motor mounts, Subaru geared and belt reduction drives.

Airflow Performance, Inc.
P.O. Box 3364 • 101 Belton Dr.
Spartanburg SC 29304-3364
Ph: 864-576-4512 • Fax: 864-576-0201
Literature/Catalog: Yes
Manufacturer of fuel injection systems for experimental aircraft. Overhaul Bendix fuel injection system for experimental aircraft.

Airscrew Performance
7308 W. Griffin Avenue • Glendale AZ 85303
Ph: 602-931-6667
New, used and rebuilt engines, redrives, exhausts. Crankshaft rebuilding. CDI conversions.

Allegro Avionics
1332 E. Hampton • Tucson AZ 85719
Ph: 520-327-3695 • Fax: 520-327-0426
Literature/Catalog: Yes
EGT/CHT engine analyzer, fuel flow totalizer and tachometer with unique features and many options. Affordable 'State of the Art' instrumentation.

Amax Engineering
5 Utrecht Court
Donvale, Victoria, 3111 AUSTRALIA
Ph: 61-3-8423132 • Fax: 61-3-8418177
Literature/Catalog: $5.00
Subaru and VW reduction drives.

American Propeller Service
20208 Charlanne Dr. • Redding CA 96002
Ph: 800-292-7767 • Fax: 916-221-5201
Literature/Catalog: Yes
Distributors for all propeller manufacturers. Also custom designing of propellers and consultation.

Approved Aircraft Accessories
29300 Goddard Road • Romulus MI 48174
Ph: 313-946-9000 • Fax: 313-946-5547
Literature/Catalog: Yes
Overhaul of various aircraft engines, landing gear, and accessories. Manufacture test equipment.

Arctic Sparrow Aircraft
20748 Birchwood Spur Rd. • Hangar B
Chugiak AK 99567
Ph: 907-688-7001 • Fax: 907-688-7002
Literature/Catalog: Yes
ASA mixture control.

ARP Industries, Inc.
36 Bay Drive E. • Huntington NY 11743
Ph: 516-427-1585
Carburator ice detector. Manufactured and sold worldwide.

Atkins Aviation, Inc.
16715 Meridian East, Bldg K
Unit A Puyallup WA 98373
Ph: 253-848-7776 • Fax: 253-922-0118
E-Mail: Atkins@accessone.com
Web: http://Atkinsrotary.com
Manufacturer of Wankel Rotary 13B engines. Firewall-forward kits.

Atlas Motors Inc.
685 North Loudoun Street
Winchester VA 22601
Ph: 800-336-9878 • Fax 540-665-2025
Literature/Catalog: Yes
Franklin P2L engines, accessories and parts for Certified and Homebuilt Aircraft.
See Display Ad On Page 73

Aviation Engines, Inc.
301 Airport Street • Centreville AL 35042
Ph: 800-344-7918 • Fax: 205-926-5186
E-Mail: aviationengines@msn.com

B & C Specialty Prod. Inc.
123 E. 4th Street • PO Box B
Newton KS 67114
Ph: 316-283-8000 • Fax: 316-283-7400
Literature/Catalog: Yes
Voltage regulators, starters, alternators. For literature send large SASE.

BAC-NPS
P.O. Box 1984 • 1050 Airport Road
W. Chester PA 19380-0075
Vertex magneto plugs into distributor hole.

Barry Controls Aerospace
P.O. Box 7710 • 4510 Vanowen Street
Burbank CA 91505
Ph: 818-843-1000 • Fax: 818-845-6978
Literature/Catalog: Yes
Aircraft engine vibration isolators and engine mounts.

Belted Air Power
1408 Western Avenue • Las Vegas NV 89102
Ph: 702-384-8006 • Fax: 702-384-8006
Literature/Catalog: $12.50
Reduction drives for the Buick/Rover V-8 and the Chevy 350 small-block engines.

Bing Agency International
Box 1, South Main • Belvidere KS 67015
Ph: 316-862-5808 • Fax: 316-862-5699
Literature/Catalog: $6.95
Bing carburetors and Bing carburetor parts. Alcohol resistant fuel line, carburetor synchronizers. fuel pumps, and gas filters.

Brackett Aircraft Co., Inc.
7045 Flightline Drive • Kingman AZ 86401
Ph: 520-757-4005 • Fax: 520-757-1948
E-Mail: brackett@ctaz.com
Literature/Catalog: Yes
Engine air filters. Free "Filter Information Manual" for homebuilt aircraft engine air filter design.

Bradley Aerospace
950 Muir Ave. • Chico CA 95926
Ph: 916-899-7918 • Fax: 916-899-2942
Web: http://www.kgmsoftworks.com/bradley
See Display Ad On Page 40

California Power Systems, Inc.
790 139th Avenue, #4
San Leandro CA 94578
Ph: 800-247-9653 • Fax: 510-357-4429
Web: http://www.800-airwolf.com
Literature/Catalog: $6.95
Complete supplier of ultralight and sport aircraft parts and components. Rotax engines, parts and service. 500 page catalog, $6.95 US, $10.00 Canada and $20.00 foreign airmail.
See Display Ad On Page 69

Cam-Fire Engineering Group Inc.
9755 W. Saanich Rd. • Sidney, B.C.
V8L 5T5 CANADA
Ph: 250-656-4774 • Fax: 250-656-8007
E-Mail: RLM@Camfire.BC.CA
Web: http://Camfire.BC.CA
Literature/Catalog: $10.00
Manufacturer of the Honda based CAM-100-175 liquid cooled aircraft engines. Propeller speed reduction units up to 500 hp.

Nick Carter Engines, Inc.
2116 West G Street • Elizabethton TN 37643
Ph: 423-542-2943 • Fax: 423-542-2633
E-Mail: ncarter@usit.net
Web: http://www.carterengines.com
Continental and Lycoming engines, cylinders and parts.

Certified Parts Corp.
1111 W. Racine • PO Box 8468
Janesville WI 53545
Ph: 608-752-9441 • Fax: 608-752-3528
Fuel pumps and fuel pump kits.

Consolidated Fuel Systems
1400 East South Blvd. • Montgomery AL 36116
Ph: 334-286-8551 • Fax: 334-286-8018
Literature/Catalog: Yes
Carburetors and parts. Fuel pumps and parts. Fuel injectors.

Custom Airmotive, Inc.
2702-B North Sheridan Road • Tulsa OK 74115
Piston engine rebuild, repair and accessories. All models - Lycoming and Continental.

Diehl Aero-Nautical
1855 North Elm • Jenks OK 74037
Ph: 918-299-4445 • Fax: 918-299-4425
Literature/Catalog: Yes
VW engine conversion parts. Accessory case, electric starter, alternator and magneto drive.

Drake Air, Inc.
4085 Southwest Blvd. • Tulsa OK 74107
Ph: 800-542-6899 • Fax: 918-445-4155
E-Mail: 102117.171@compuserve.com
Literature/Catalog: Yes
Oil related equipment including coolers, pumps, and filter kits.

Dukes, Inc.
9060 Winetka • Northridge CA 91324
Ph: 818-998-9811 • Literature/Catalog: Yes
Fuel pumps.

Dumont Aircraft Engines, Inc.
1605 West State Road 64
Avon Park FL 33825
Ph: 941-453-2600 • Fax: 941-453-5864

Eagle Performance
2015 N. Elsea Smith Road
Independence MO 64056
Ph: 816-249-9608
Literature/Catalog: $7.00

EC-Elmore, Inc.
411 Hwy 301 Blvd. East, Bradenton FL 34208
Ph: 800-596-5566 • Fax: 941-737-0395
F.A.A. repair station with complete aircraft cylinder overhaul services available.

Eggenfellner Advanced Aircraft
427-3 Amherst St., Ste. 230, Nashua NH 03063
Ph: 800-840-4620 • Fax: 603-880-8076
Literature/Catalog: Yes
Subaru engine mount. CNC machined, nickle plated, 4130 steel Subaru Legacy engine mount.

Eklund Engineering, Inc.
19960 Elliott Road • PO Box 1510
Lockeford CA 95237-1510
Ph: 209-727-0318 • Fax: 209-727-0873
E-Mail: ThorpT18@compuserve.com
Prop ext.

Electrosystems, Inc.
P.O. Box 273 Airport Complex
Fort Deposit AL 36032
Ph: 334-227-8306 • Fax: 334-227-8596
Literature/Catalog: Yes
See Display Ad On Page 64

Elliott Crankshaft Corp.
250 N. Buena Vista Drive
Lake Alfred FL 33850 • Ph: 813-956-4452
E-Mail: EGC1967@AOL.com
Magnaflux, Grind and Nitride crankshafts for homebuilt aircraft.

Ellison Fluid Systems
350 Airport Way • Renton WA 98055
Ph: 206-271-3220 • Fax: 206-277-9333
Literature/Catalog: Yes
Throttle Body Injector replacement for carburetor.

Empire Development Avionics
2107 N. 1st St. Suite 590
San Jose CA 95131-2027
E-Mail: chanik@aol.com
Web: http://members.aol.com/chanik/ed
Electronic ignition system that is fully variable, Hall-effect based and distributorless. Optimized for use on 4 cylinder Lycoming, Continental and Franklin engines.

Falcon Ridge Industries
51 N. Hoagland Blvd. • Kissimmee FL 34741
Ph: 407-396-7572 • Fax: 407-396-1696
Literature/Catalog: Yes
Fuel tank flanges. Aluminum weld on and aluminum to be bonded into fiberglass structures.

Flame Engineering, Inc.
P.O. Box 577 • La Crosse KS 67548-0577
Ph: 800-255-2469 • Fax: 785-222-3619
E-Mail: Sales@flameeng.com
Web: http://www.flameeng.com
Literature/Catalog: Yes
Red Dragon portable propane burning engine preheaters. Available in 12, 24 and 115 volt versions.

Formula Power East
4921 Fennell • Suffolk VA 23435
Ph: 804-484-33185
Literature/Catalog: Yes
Reduction drives; gear and cog belt. Multi-port fuel injection - Subaru EA81/EJ22/EJ33. Dual ignition systems. Advanced electronic design. Performance parts, complete engine kits.

Fuel Manager
1903 NE Sawdust Hill Rd.
Poulsbo WA 98370 • Ph: 206-779-9373
The right Fuel Selector Valve for your engine-fuel tank configuration. 3-port, 6-port, and 10-port valves for optimum fuel management efficiency.

G&N Aircraft, Inc.

Master Engine Builders • Precision Aircraft Service Center

G&N Aircraft, Inc.

1701 E. Main Street • Griffith IN 46319
Ph: 800-348-6504 • Fax: 219-924-1059
E-Mail: gnair@jorsm.com
Web: http://www.gnaircraft.com
Literature/Catalog: Yes
Custom overhaul, repairs and inspections for Continental and Lycoming piston engines. Overhaul and repair of all accessories: alternators, magnetos, carburetors, etc.

See Display Ad On Page 68

Don George, Inc.

1339 W. Washington St., A1
Orlando FL 32805
Ph: 407-422-0188 • Fax: 407-422-0836
Custom building and overhaul of aircraft engines. New and used parts available.

Grand Rapids Technologies

4526 Poinsettia SE • Kentwood MI 49508
Ph: 616-531-4893 • Fax: 616-249-0437
Literature/Catalog: Yes
Manufacturer of multifunction digital engine instruments. Four models provide complete instrumentation systems for nearly all types of aircraft engines including Rotax, Subaru, Lycoming, Continental, Hirth, etc.

Great Plains Aircraft

P.O. Box 545 • Boys Town NE 68010
Ph: 402-493-6507 • Fax: 402-333-7750
E-Mail: gpasc@earthlink.net
Web: http://www.greatplainsas.com

See Display Ad On Page 80

Gyrodynamic Systems

P.O. Box 55 • Redlands CA 92373
Literature/Catalog: $6.00
R.P.M. reduction systems-50 hp to 350 hp. Planetary gear drive, V-belt drives, Cog-belt drives, chain drives, direct drive nose cones. Dual spark plug heads, Covair/Volkswagen.

Hal Hunt

6249 Longridge Avenue • Van Nuys CA 91401
Ph: 818-989-5534 • Literature/Catalog: Yes
Long-EZ, Cozy, new low drag pipes 321 stainless steel exhaust.

Insight Instrument Corp.

Box 194 Ellicott Station, Buffalo NY 14205-0194
Ph: 614-548-4122 • Fax: 905-871-5460
Manufacturer of engine temperature analyzers. The GEM 610 measures EGT, CHT and TIT with data-logging.

Intertech Aviation Services

3 Sunset Lane • Littleton CO 80121
Ph: 303-781-2577 • Fax: 303-781-2577
Literature/Catalog: Yes
Supplier of AVBLEND, the metal treatment for engines. FAA approved for all piston aircraft engines.

J-Bird

Box 438 • 210 Main St. • Kewaskum WI 53040
Ph: 414-626-2611
Parts and accessories for Rotax, Kawasaki, Cuyuna, Zenoah, Hirth, Solo and others.

J.P. Instruments, Inc.

P.O. Box 7033 • Huntington Beach CA 92615
Ph: 800-345-4574 • Fax: 714-557-9840
Literature/Catalog: Yes
Mfg. all types of engine gauges, also data recording systems.

Kearns Metal Works

8745 Highway 13 West • Carrier Mills IL 62917
Ph: 618-252-2044 • Literature/Catalog: Yes
Dimple dies, rivet sets, pneumatic rivet, metal brake plans, Nippondenso 35 amp alternators, and other metal working tools.

Kosola & Associates, Inc.

5601 Newton Road • P.O. Box 3529
Albany GA 31707
Ph: 800-456-7652 • Fax: 912-888-5766
Literature/Catalog: Yes
Engine mounts & isolators.

Mark Landoll Electrical Svc.

1205 Redbud Lane • Newcastle OK 73065
Ph: 405-392-3847 • Literature/Catalog: Yes
Lightweight starters, alternators and harmonic dampeners for Lycoming engines. Harmonic dampeners for use with wood propellers.

Light Power Engine Corp.

PO Box 3350 • Morgantown WV 26503
Ph: 304-291-3843 • Fax: 304-292-1902
Literature/Catalog: $20.00
Air conditioning systems, electric vacuum pumps, fuel pumps.

Light Speed Engineering

PO Box 549 • Hangar #15
Santa Paula CA 93060
Ph: 805-933-3299 • Fax: 805-525-0199
E-Mail: LSE@west.net
Literature/Catalog: Yes
"Plasma CDI" aircraft ignition systems for 4 and 6 cylinder engines.

Linda Lou, Inc.

3514 Winhoma • Memphis TN 38118
Ph: 800-824-9912 • Fax: 901-365-9524

Lockwood Aviation Supply, Inc.

280 Hendricks Way • Suite A
Sebring FL 33870
Ph: 941-655-5100 • Fax: 941-655-6225
E-Mail: lockwood@digital.net
Web: http://www.lockwood-aviation.com
Literature/Catalog: $9.95
Light aircraft engines, parts and accessories.

M. P. A. Company

P.O. Box 622 • Toccoa GA 30677-0622
Ph: 706-779-3446 • Fax: 707-779-2302
Vacuum pump drive system for VW engines.

Martin Induction Systems

Rt. 6, Box 611-B • Enid OK 73701
Ph: 800-848-6486 • Fax: 405-446-5411
Literature/Catalog: Yes
FAA repair station for overhauling and testing fuel systems, fuel pumps, turbochargers, waste gates, relief valves and controllers. Exchanges available. Also in-house machine shop. World-wide service.

McFarlane Aviation, Inc.

696 E. 1700 Road • Baldwin City KS 66006
Ph: 800-544-8594 • Fax: 913-594-3922
Custom swaged flight control cables. Custom push-pull controls. Cables and fitting supplies.

Monarch Air & Development, Inc

P.O. Box 419 • Oakland OR 97462
Ph: 541-459-2056 • Fax: 541-459-1765
E-Mail: wjbarton@telepat.com
Web: http://www.telepart.com/~wjbarton/
Literature/Catalog: Yes
Replacement fuel tanks, caps and air intake boots for single engine aircraft.

Moravia, Inc.

P.O. Box 942
Thunder Bay, Ontario P7C 4X8 CANADA
Ph: 807-475-5113 • Fax: 807-475-9587
E-Mail: moravia@loon.norlink.net
Web: http://www.norlink.net/moravia
Literature/Catalog: Yes

Nelson Engine of Florida, Inc.

420 Harbor Drive • Naples FL 33940
Ph: 813-263-1670 • Fax: 813-261-8345
Nelson H-63-CP 48 hp 2 stroke aircraft engine and the H-63 CP-2 twin engine, 90 hp single prop. aircraft engine. Complete engines and parts. FAA certified.

Northwest Aero Products

2801 Academy Dr. S.E. • Auburn WA 98092
Ph: 206-735-5022 • Fax: 206-939-1578
E-Mail: nwaero@ix.netcom.com
Web: http://www.halcyon.com/wpowers/
Literature/Catalog: Yes
Supplier of propeller shaft reduction drives for Ford V-6, Chev V-6, and Olds and Buick 215 V-8. Engine accessories include crank trigger ignition, crankshaft and accessory pulleys, alternators and brackets.

Northwest Formula Power

P.O. Box 2168 • Sand Point ID 83864
Ph: 208-263-6637 • Literature/Catalog: Yes
Formula Power dealer.

Oilamatic Inc.

P.O. Box 5284 • Englewood CO 80155
Ph: 303-720-0155 • Fax: 303-793-0493
E-Mail: grmccril@oilomatic.com
Web: http://oilamatic.com
Literature/Catalog: Yes
Aircraft piston engine pre-oiler. Electric-hydraulic pump to pre-lubricate engine before start.

Pacific Continental Engines

16041 Arminta Street • Van Nuys CA 91406
Ph: 818-781-4947 • Fax: 818-781-0860
Web: http://www.pcenet.com
Literature/Catalog: Yes
Overhaul and sale of Textron, Lycoming, and Continental engines, components, and parts. Avionics shop sells new, repairs, and maintains all major brands. FBO offers aircraft maintenance.

Penn Yan Aero Service, Inc.

2499 Bath Road • Penn Yan NY 14527
Ph: 8007277230 • Fax: 315-536-2335
E-Mail: PnnYanAero@Aol.Com
Web: http://www.airtourist.com/Penn Yan
Literature/Catalog: Yes
Lycoming & Teledyne Continental Motors distributor. FAA certified engine repair station NJ1R350K. Custom and exchange engine production, certificated engines, machine shop, cylinder and accessory repairs.

Phillips 66 Co.

723 Adams Building, 7th Floor
Bartlesville OK 74006
Ph: 800-766-0050 • Fax: 918-662-2062
Web: http://www.Phillips66.com
Literature/Catalog: Yes
Phillips 66 lubricants. X/C aviation oil for piston engines. Injex for two cycle engines.

PowerKits-Technical Materials, Inc.

263 Quillayute Place • PO Box 80
La Conner WA 98257
Ph: 360-466-1071 • Fax: 360-466-1071
Literature/Catalog: Yes
100 hp turbo-charged Suzuki conversion for homebuilts and experimental aircraft.

Prince Aircraft Co.

6774 Providence Street • PO Box 2669
Whitehouse OH 43571
Ph: 419-877-5557 • Fax: 419-877-5564
E-Mail: 75557.31555@Compuserve.Com
Literature/Catalog: Yes
Propeller extensions and crush plates 2" to 12" length and 8" diameter. Manufactured in aluminum or magnesium. Custom and prototype machining capabilities.

REAL Gasket Corp.

P.O. Box 2048 • Corvallis OR 97339-2048
Ph: 800-635-7325 • Fax: 800-635-7325
Literature/Catalog: Yes
Aircraft engine gaskets for Lycoming, Continental and Franklin engines. Pushrod tube seals for Continental 470,520,550 and 360. Pushrod tube kits for Continental A-65 to 0-300. Radial engine valve cover gaskets. Factory direct sales.

Real World Solutions

1936 Nugget Dr. • Clearwater FL 33755
Ph: 813-447-6956
E-Mail: 71175.606@compuserve.com
Literature/Catalog: Yes
Mazda 13B Rotary engine accessories including dual ignition controllers, EFI controllers, and planetary PSRU's.

Recreational Power

10888 Station Rd. • Columbia Station OH 44028
Ph: 800-583-3306 • Fax: 419-585-6004
Literature/Catalog: Yes
U.S. Hirth engine distributor.

Reg's Air Cooled Engines

9708 Princess Drive
Surrey, B.C. V3V 2T4 CANADA
Ph: 604-581-7414 • Fax: 604-581-7418
Literature/Catalog: Yes
Distributor and service center for Rotax ultralight aircraft engines.

Regale Limited, U.K.

27 Whistler Road, Eaton Ford
St. Neots, Cambridgeshire PE19-3RT
UNITED KINGDOM
Ph: 44-480-477975 • Fax: 44-480-477975
British & Europeon military aircraft parts. Support & major items for BAC, Roll Royce, Marine, BAE, and much more. $10,000,000 in stock.

Regale Limited, U.S.A.

RT. 1, Box 138A • Peculiar MO 64078
Ph: 816-779-6075 • Fax: 816-758-6054
British & Europeon military aircraft parts. Support & major items for BAC, Roll Royce, Marine, BAE, and much more. $10,000,000 in stock.

RFI Powerplant Systems Group

2916 Tennessee Northeast, Albuquerque NM 87110
Ph: 505-298-9362 • Fax: 505-266-1789
Reduction drive plans. Mazda 13B 170 hp powerplant with reduction drive system is pictured.

Rocky Mountain Instrument

PO Box 683 • 202 Fremont Street
Thermopolis WY 82443
Ph: 307-864-9300 • Fax: 307-864-9300
E-Mail: RMowRER@TRIB.com
Web: http://rkymtn.com
Literature/Catalog: Yes
Computer controlled 23 function engine instrument/monitor with fuel totalizer and multi-function clock in radio-stack size package.

ROMA Machine Tool & Die, Inc.
Manufacturer of Bellcrank Bearings since 1968

Roma Machine Tool & Die, Inc.

1990 East Maple Street
Kankakee IL 60901-5720
Ph: 815-937-1245 • Fax: 815-937-1285
Literature/Catalog: Yes
Manufacturersof bellcrank bearings, BC4W 10 and BC5W11 propeller reductionspeed units, weld and bolt on axels, retractable landing gear, low wing, tubing fabrication, custom control panels, custom machine components for tail rotor on helicopters, variable pitch control and propeller hubs.

See Display Ad On Page 42

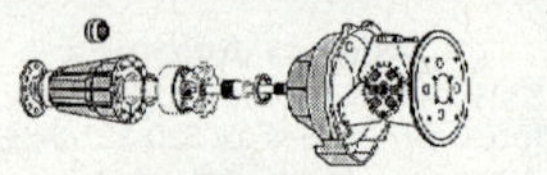

Ross Aero

3824 E. 37th Street • Tucson AZ 85713
Ph: 520-747-7877 • Literature/Catalog: Yes
Information pack for engines and engine accessories: $5.00.

See Display Ad On Page 4

Ron Sands Co.

89 Forrest Drive • Mertztown PA 19539
Ph: 610-682-6788 • Fax: 610-682-6788
Literature/Catalog: Yes • Radial engine mounting rings. Send SASE for literature.

Savage Magneto Service
1069 Langley St., PO Box 2401 • Airport Station
Oakland CA 94614 • Ph: 510-562-1593
Complete line of magnetos and parts. FAA approved repair station.

Scat Enterprises
1400 Kingsdale Avenue
Redondo Beach CA 90278
Ph: 310-370-5501 • Fax: 310-24-2285
Literature/Catalog: $6.00
Manufacturer of 2 and 4 cylinder water cooled and air cooled engines. 40 to 200 hp, 4 cycle. Horizontal and vertical configurations. Parts, kits and complete engines.

Sky-Tec
PO Box 1024 • Granbury TX 76048
Ph: 800-476-7896
Catalog/Brochure: Yes
12 and 24 volt starters utilizing an electromagnetic mechanical engagement system.
See Display Ad On Page 10

Recreational Power Eng.
5479 E. Country Road 38 • Tiffin OH 44883
Ph:419-585-7002 • Fax: 419-585-6004
Literature/Catalog: Yes
US distributor of Hirth two cycle aircraft engines. Sales, service, parts and engine overhauls for Hirth engines. Supplier of engine instrument systems and propellers.

Shannon Aircraft Motor Works
Bay 142 Industrial Estate • Shannon Co.
Clare IRELAND
Ph: 353-61-472466 • Fax: 353-61-472381
E-Mail: PSN@samw.ie
Web: http://www.iol.ie/~samw
Rewinding of aircraft electrical and generator parts.

Sigma-Tek, Inc.
1001 Industrial Rd. • Augusta KS 67010
Ph: 316-775-6373 • Fax: 316-775-1416
E-Mail: info@sigmatek.com
Web: http://sigmatek.com
Literature/Catalog: Yes
Manufacturer of precision aviation instruments, avionics, accessories, control/display panels, and diagnostics systems. Capabilities are from product design through sales and service. Worldwide. Products for aircraft from general aviation trainers to commerical air carriers.

T.W. Smith Engine Co., Inc.
4510 Airport Rd. • Hangar #1 Lunken Airport
Cincinnati OH 45226
Ph: 513-871-3500 • Fax: 513-533-244
Web: http://www/twsmith
Literature/Catalog: Yes
New and overhauled Continental and Lycoming engines and accessories. Certified Repair Station.

ROTAX

South Mississippi Light Air
1279 Marshall Smith Rd. • Lucedale MS 39452
Ph: 800-247-7652 • Fax: 601-947-4959
Literature/Catalog: Yes
Rotax engines in stock include the 447, 503, 582, 618, and 912. Rotax parts shipped UPS same day. Authorized Quicksilver Aircraft and Quicksilver aircraft parts distributor. Parts and sales, call 1-800-247-7652. For service, call 1-601-947-4953.
See Display Ad On Page 13, 69

Specialty Aircraft Co.
1440 NW 87th Ct. • Redmond OR 97756
Ph: 800-500-6786 • Fax: 503-548-5336
Single engine Cessna salvage yard. Instruments and parts for homebuilt aircraft.

Specialty Hose Aerospace
7802 Freedom Avenue NW
North Canton OH 44720-6978
Ph: 800-362-6533 • Fax: 330-497-0415
E-Mail: hose@cannet.com
Web: http://www.lmg.com/shc
Hose assemblies, fire sleeving, adapters, clamps.

Sun-Tech Innovations
P.O. Box 9154 • Canoga Park CA 91309
Ph: 800-786-8324 • Fax: 818-882-7859
E-Mail: Sun-tech@worldnet.att.net
Web: http://www.balancemasters.com/ultraschem1.html
Balance-Master add-on for propellers and Rotax engines. Reduces vibration on ultralights up to 50%.
See Display Ad On Page 55

Superflite
2149 E. Pratt Blvd. • Elk Grove Village IL 60007
Ph: 800-323-0611 • Fax: 708-364-0358
Literature/Catalog: Yes
Supplier of a broad range of parts and accessories for personal aircraft.
See Display Ad On Page 38

Technical Materials, Inc.
PO Box 80 • LaConner WA 98257
Ph: 360-466-1071 • Fax: 360-466-1071
E-Mail: cline@fidalgo.net
Literature/Catalog: Yes
Kits to convert Suzuki G10 Sprint/Metro engines for use in experimental aircraft. Send SASE for information.

The Cam-Fire Engineering Group, Ltd.
9755 W. Saanich Rd.
Sidney, BC V8L 5T5 CANADA
Ph: 604-656-4774 • Fax: 604-656-8007
Literature/Catalog: $10.00
Belt and helical cut PSRUs, CAM Puter redundant electronic ignitions, dyno testing, blueprinting.

Toccoa Aviation, Inc.
Route 3 Airport • Toccoa GA 30577
Full maintenance for light airplanes.

Total Engine Concepts
2391 Old Dixie Hwy. • Riviera Beach FL 33404
Ph: 407-842-2813 • Fax: 407-845-3237
Literature/Catalog: Yes
Total Engine Concepts manufactures new 2 cylinder, 4 cycle engines of 35 hp. and 4 cylinder V.W. based engines from 60 to 75 hp. Send self-addressed stamped envelope for literature.

Tuthill Corp.,Cablecraft Div.
P.O. Box 11372 • Tacoma WA 98411
Ph: 253-475-1080 • Fax: 253-474-1623
Design and manufacaturer mechanical push-pull control cables.

2 Stroke International
Beaufort Industrial Park • 8 Schein Loop
Beaufort SC 29902
Ph: 803-846-2167 • Fax: 803-846-2169
E-Mail: engines@2si.com
Web: http://www.2si.com
Literature/Catalog: Yes
Parts and accessories for lightweight high-power fan cooled and water cooled engines for aircraft. 20 to 120 hp.

Universal Engineering, Inc.
116 W. Main • PO Box 115 • Winamac IN 46966
Ph: 219-946-6757 • Fax: 219-946-7471
Web: http://www.universalengineering.com
Literature/Catalog: $12.00
Two reduction drives for 90 degree Chevrolet engines. One has 10" of propeller shaft rise above crankshaft, the other is a concentric planetary for up to 1200 horsepower.

Varga Enterprises, Inc.
2350 S. Airport Blvd. • Chandler AZ 85249
Ph: 800-966-6936 • Fax: 602-899-0324
E-Mail: vargaair@vargaair.com
Web: http://www.vargaair.com
Literature/Catalog: Yes Carburetors, alternators, starters, magnetos, fuel pumps, etc.

Vision Microsystems, Inc.
4071 Hannegan Rd. Ste. T
Bellingham WA 98226
Ph: 360-714-8203 • Fax: 360-714-8253
Literature/Catalog: Yes
Engine monitoring instruments and systems, fuel level systems, and electronic checklist and cautionary system. Since 1985.

Western Power Systems
420 30th Street • Bakersfield CA 93301
Ph: 805-323-3432 • Literature/Catalog: Yes
Formula Power dealer.

Western Ultralight
4700 188th N.E. • Arlington WA 98223
Ph: 360-435-3737 • Fax: 360-435-3737
Total Ultralight facility. Including flight instruction, sales, parts, and maintenance.

White Eagle Industries, Inc.
8282 Crane Road • Cranesville PA 16410
Ph: 814-734-7749 • Fax: 814-756-3577
Literature/Catalog: Yes
Manufacture reusable silicone valve cover gaskets.

Yingling Aircraft, Inc.
P.O. Box 9248 • Wichita KS 67277
Ph: 316-943-3246 • Fax: 316-943-2484

Zeftronics
1622 E. Whaley Street
Longview TX 75601-6830
Ph: 903-758-6661 • Fax: 903-236-9766
Literature/Catalog: Yes
Voltage regulators, alternator control units & electrical parts.
See Display Ad On Page 61

Floats & Skis

Aces High Light Aircraft Ltd.
RR 1, London, Ontario N6A 4B5 CANADA
Ph: 519-652-3020 • Fax: 519-652-3020
Literature/Catalog: $5.00
Aircraft kit & custom parts. Manufacturer able to supply most aircraft building materials and parts at competitive cost.

AEROCOMP, Inc.
2335 Newfound Harbor Drive
Merritt Island FL 32952
Ph: 407-453-6641 • Fax: 407-453-6641
E-Mail: ral@nexusprime.org
Web: http://aerocompinc.com
Literature/Catalog: Yes
SUPER FLOAT line of advanced composite construction floats that produce enough lift to fly themselves. Full line covers ten sizes from 500 to 2850 lbs. gross aircraft weight.
See Display Ad on Page 40

Aircraft Sales & Parts
PO Box 995 • Vernon, B.C. V1T 6N2 CANADA
Ph: 250-549-1102 • Fax: 250-549-3769
E-Mail: asap@junction.net
Web: http://www.ultralight.ca

Aircraft Super-Market
#10 Ellefson Dr. • PO Box 676
De Soto IA 50069
Ph: 515-834-2225 • Fax: 515-834-2152
E-Mail: bobdurly@netins.net
Web: http://www.netins.net/showcase/t-birds
Literature/Catalog: $5.00
Handles recreational aircraft of all kinds. New and used, ultralights, powered parachutes and U/L gliders. 2600' runway.
See Display Ad On Back Cover

Arctic Sparrow Aircraft
20748 Birchwood Spur Rd. • Hangar B
Chugiak AK 99567 • Ph: 907-688-7001
Fax: 907-688-7002 • Literature/Catalog: Yes
Snow skis for ultralights and experimentals.

Canadian Powered Parachutes Ltd.
Box 1769 • Vegreville, Alberta T9C 1S8 CANADA
Ph: 403-658-2044 • Fax: 403-658-2387
E-Mail: lindam@agt.net
Web: http://www.agt.net/public/lindam/cpp2.htm • Literature/Catalog: Yes

Full Lotus Mfg., Inc.
7400 Wilson Avenue
Delta, BC V4G 1E5 CANADA
Ph: 604-940-9378 • Fax: 604-940-9566
E-Mail: brenda@full-lotus.com
Literature/Catalog: $8.00
Inflatable floats in displacements up to 2,600 lbs.

T BIRD

Golden Circle Air Inc.
PO Box 676 • #10 Ellefson Drive
De Soto IA 50069-0676
Ph: 515-834-2225 • Fax: 515-834-2152
E-Mail: rdfoster@netins.net
Web: http://www.netins.net/showcase/sunflyer/tbird.htm
Light plane manufacturer offers a lightweight inflatable amphibious float system and fiberglass snow skis suitable for a variety of aircraft.
See Display Ad On Back Cover

Kenmore Air Harbor Inc.
Box 82064 • Kenmore WA 98028-0064
Ph: 206-486-1257
A full service charter Seaplane operation.

Lake & Air, Inc.
400 N. Airport Service Rd. • S. St. Paul MN 55075
Ph: 612-496-3870 • Fax: 612-496-4070
Seaplane supplies and accessories.

Parascender Technologies, Inc.
1040 E. Carroll St. Dept. KP
Kissimmee FL 34744
Ph: 407-935-0775 • Fax: 407-935-0778
E-Mail: para@iag.net
Web: http://www.iag.net/~para
Literature/Catalog: Yes
Snow Skis accessory for the Parascender, Buckeye & Aerochute.

Rotor Wing-Aero
PO Box 18614 • Salt Lake City UT 84118
Ph: 801-969-1357 • Fax: 801-969-1357
E-Mail: mhoskins@aros.net

Venture Aerodome
Rt. 1 Box 11631 • Roland OK 74954
Ph: 918-427-6490
Literature/Catalog: $4.00
Amphibious float systems.

WAG-AERO

The Wag-Aero Group
1216 North Road, Box 181 • Lyons WI 53148
Ph: 800-558-6868 • Fax: 414-763-7595
E-Mail: wagaero-sales@wagaero.com
Literature/Catalog: Yes
Spark plugs, engine mount bushings, vacuum pumps, carb boxes, oil tanks, oil filters, air filters, and preheaters.
See Display Ad On Page 13

Zenair, Ltd.
Huronia Airport
Midland, Ont. L4R 4K8 CANADA
Ph: 705-526-2871 • Fax: 705-526-8022
E-Mail: zenair@barint.on.ca
Literature/Catalog: $15.00
All-metal floats for kit aircraft, including amphibious floats. Kits or factory finished for Ultralights, Zenair STOL CH 701, ZODIAC Kitfox, RV6. Also, skies for STOL CH 701.

ZENITH AIRCRAFT COMPANY
Box 650 • Mexico Memorial Airport
Mexico MO 65265-0650
Ph: 573-581-9000 • Fax: 573-581-0011
E-Mail: info@zenithair.com
Web: http://www.zenithair.com
Literature/Catalog: $15.00
Amphibious floats for light aircraft.
See Ad Inside Front Cover

Hangars & Covers

Aircraft Components
PO Box 1003 • Sulphur Springs TX 75483
Ph: 903-885-2092
Literature/Catalog: $15.00
Hangar plans for up to four aircraft. Bolt construction, minimal welding required. Plans $295.00.

Aircraft Sales & Parts
PO Box 995 • Vernon, B.C. V1T 6N2 CANADA
Ph: 250-549-1102 • Fax: 250-549-3769
E-Mail: asap@junction.net
Web: http://www.ultralight.ca

Aircraft Spruce & Specialty
PO Box 4000 • Corona CA 91718-9961
Ph: 800-824-1930 • Fax: 909-372-0555
E-Mail: info@aircraft-spruce.com
Web: http://www.aircraft-spruce.com
Literature/Catalog: $5.00
A complete selection of homebuilt and certified aircraft parts and materials. Shipping worldwide since 1965. Also see Aircraft Spruce East, Griffin GA, 800-831-2949.
See Display Ad On Page 24

AIRCRAFT SUPER-MARKET
#10 Ellefson Dr. • PO Box 676
De Soto IA 50069
Ph: 515-834-2225 • Fax: 515-834-2152
E-Mail: bobdurly@netins.net
Web: http://www.netins.net/showcase/t-birds
Literature/Catalog: $5.00
Handles recreational aircraft of all kinds. New and used, ultralights, powered parachutes and U/L gliders. 2600' runway.
See Display Ad On Back Cover

Aviation North
1641 Airport Road • Alpena MI 49707
Ph: 517-356-6361 • Fax: 517-356-0952
Literature/Catalog: Yes
FBO, performing minor and major repairs to aircraft; hangar service; flight school; line service.

Butler Parachute Systems, Inc.
1820 Loudon Avenue NW • PO Box 6098
Roanoke VA 24017-0098
Ph: 540-342-2501 • Fax: 540-342-4037
E-Mail: butlerchut@aol.com

Craftsman Steel Building
1845 Island Avenue • San Diego CA 92102
Ph: 619-233-4000 • Fax: 619-233-8373
Hangar buildings and canopies for airplanes and other hobbies.

Delta-Therm Corp.
PO Box 345 • 398 W. Liberty Street
Wauconda IL 60084-0345
Ph: 800-526-7887 • Fax: 847-526-4456
Literature/Catalog: Yes
Hangar door rail de-icing systems. Manufacturers and Designers of electric and hydronic heat transfer systems.

ERECT-A-TUBE
PO Box 100 • Harvard IL 60033
Ph: 800-624-9219 • Fax: 815-943-4095
Web: http://www.erect-a-tube.com
Airplane hangar and hangar door systems.
See Display Ad On Page 38

Flexi-Dyne, Inc.
Rt. 1, Box 361 • S. Pittsburg TN 37380
Ph: 615-837-8537
Hangar door plans. No track or rollers. Modular design. Easy to build.

Hangar Door Co., Inc.
P.O. Box 296 • El Dorado Springs MO 64744
Ph: 417-876-6508 • Fax: 417-876-6506
Literature/Catalog: Yes
Manufacturer of electric operated bi-fold canopy hangar & industrial doors up to 130' wide x 34' high (30' clear).
See Display Ad On Page 13

Hi-Fold Door Corp.
N6170 1070th Street • River Falls WI 54022
Ph: 800-443-6536 • Fax: 715-262-3998
Literature/Catalog: Yes
High clearance bi-fold hangar doors.

Machine Tool & Fabrication
431 East North Street • Fostoria OH 44830
Ph: 419-435-7676 • Fax: 419-435-6999
Manufacturer of fabric hangar doors.

R & M Steel Company
PO Box 3198 • Boise ID 83703-0198
Ph: 208-853-1600 • Fax: 208-853-1853
Literature/Catalog: Yes
Pre-engineered steel structures for individual hangars, shade ports, maintenance hangars and FBO facilities.

RagWing Aviation Co.
PO Box 39 • Tokeena Air Park
Townville SC 29689-0039
Ph: 864-225-3773 • Fax: 864-972-5606
Portable arch hangar. Build from fence rail and tarp. Stands 40 + winds. Cost to build $400.00. Plans sold for $15.00. 20 pages.

Schweiss Distributing, Inc.
Box 48936A • Fairfax MN 55332
Ph: 507-426-8273 • Literature/Catalog: Yes
Bi-Fold Hangar Doors. Lose no headroom. Any size door.

Steelox Systems, Inc.
2412 Courseview Drive • P.O. Box 8181
Mason OH 45040-8181
Ph: 513-573-5200 • Fax: 513-573-5511

The Ultimate Door, Inc.
N8778 Sky Lane • East Troy WI 53120
Ph: 414-642-7788 • Fax: 414-642-5163
DIY hangar door kits. Hardware kit containing ready to assemble parts for constructing a hangar door. Other material purchased locally saving shipping costs.

The Ultralight Store
36120 Pourroy Rd. • Winchester CA 92596
Ph: 909-926-3196 • Fax: 909-926-3196
Literature/Catalog: Yes
Manufacturer of ultralight aircraft replacement sail sets and storage covers.

Venture Cinchers
PO Box 2163 • Ft. Smith AZ 72902
Ph: 501-648-4987
Literature/Catalog: $3.00

Wilson Industrial Doors Inc.
P.O. Box 348 • Elkhorn WI 53121
Ph: 800-558-5974 • Fax: 414-723-6433
Literature/Catalog: Yes
Aluminum aircraft hangar bi-fold doors.

Hardware

A. Wheels & Tie Rods
PO Box 174 • Ambler PA 19002
Ph: 215-843-7637 • Fax: 215-646-8122

Aberle Custom Aircraft
2141 S. Mission Road • Fallbrook CA 92028
Ph: 760-723-1731
Over 30 years assisting show quality and performance oriented builders.

Accurate Precision Fasteners
20 Honeck Street • Englewood NJ 07631
Ph: 201-567-9700 • Fax: 201-567-1965
Military standard hardware: AN-MS, NAS fasteners, pins, rivets, o-rings, Nylok, screws, washers, locknuts, bolts.

Achilles Aerospace Products
5435 Perkins Road • Bedford Hts. OH 44146
Ph: 216-963-5454 • Fax: 216-425-3633
Literature/Catalog: Yes
Mfg. of mechanical control components for aircraft.

The Aeroplane Store
Kampel Airport, 8930 Carlisle
Wellsville PA 17365
Ph: 717-432-9688
Literature/Catalog: Yes

Aerotech Aviation
1860 B Joe Crosson Drive
El Cajon CA 92020
Ph: 800-448-4457 • Fax: 619-448-7479
Literature/Catalog: Yes
Full line of aircraft fasteners - AN-MS-NAS specifications. Call for AN bolt price list.

Aerowood Aircraft Supply Co.
207 Schneider Road SE
North Canton OH 44720-3649
Ph: 216-494-1660 • Fax: 216-494-0418
Literature/Catalog: Yes

Air Salvage of Arkansas
2854 Hwy 8 East • Mena AR 71953
Ph: 501-394-1022 • Fax: 501-394-7475
Started 32 years ago. Dismantling mostly single engine aircraft, buying inventories, consignment selling of A/C and parts. Housing and storage.

Aircraft Lighting & Components
249 Brockway Road • Chehalis WA 98532
Ph: 360-748-0234 • Fax: 360-748-6455
Literature/Catalog: Yes
Aircraft cable assemblies.

Aircraft Spruce & Specialty
PO Box 4000 • Corona CA 91718-9961
Ph: 800-824-1930 • Fax: 909-372-0555
E-Mail: info@aircraft-spruce.com
Web: http://www.aircraft-spruce.com
Literature/Catalog: $5.00
A complete selection of homebuilt and certified aircraft parts and materials. Shipping worldwide since 1965. Also see Aircraft Spruce East, Griffin GA, 800-831-2949.
See Display Ad On Page 24

AIRCRAFT SUPER-MARKET
#10 Ellefson Dr. • PO Box 676
De Soto IA 50069
Ph: 515-834-2225 • Fax: 515-834-2152
E-Mail: bobdurly@netins.net
Web: http://www.netins.net/showcase/t-birds
Literature/Catalog: $5.00
Handles recreational aircraft of all kinds. New and used, ultralights, powered parachutes and U/L gliders. 2600' runway.
See Display Ad On Back Cover

Aircraft Supplies, Inc.
1745 Bathgate Avenue • Bronx NY 10457
Ph: 718-294-6370 • Fax: 718-716-8012
Aircraft engine spares, chemicals and tapes.

Airparts, Inc.
2400 Merriam Lane • Kansas City KS 66106
Ph: 800-800-3229 • Fax: 913-831-1780
Literature/Catalog: Yes
Raw materials and hardware.

All Fab Corp.
1235 1/2 Lincoln Road • Allegan MI 49010
Ph: 616-673-6572 • Fax: 616-673-1644

Armstrong Engineering
220 Mar Vista Dr. #69 • Aptos CA 95003
Ph: 800-919-7900 • Fax: 408-684-1007
E-Mail: armstrong@cruzio.com
Web: http://members.cruzio.com/~armstrong
Literature/Catalog: Yes
Mil-spec AN-MS-NAS hardware. Hard-to-find hardware.

Assoc. Aircraft & Marine Svcs.
POB 663 • Kent WA 98035
Ph: 206-631-3082 • Fax: 206-631-3082
Literature/Catalog: Yes
Aircraft hardware, stainless & cad screws, bolts, nuts, washers.

Aviation North
1641 Airport Road • Alpena MI 49707
Ph: 517-356-6361 • Fax: 517-356-0952
Literature/Catalog: Yes

Azusa Engineering Inc.
1542 Industrial Park St.
Covina CA 91722-3487
Ph: 818-967-4167 • Fax: 818-966-4071
E-Mail: azeng@earthlink.net

B & F Aircraft Supply, Inc.
22038 Howell Dr., Unit 6 • New Lennox IL 60451
Ph: 815-462-8696 • Fax: 815-462-3697
Literature/Catalog: $1.00

B & B Aircraft Supplies
Gardner Municipal Airport • 31905 W. 175th Street
Gardner KS 66030
Ph: 913-884-5930 • Fax: 913-884-6533

Beechhurst Industries Inc. Aviation
Aviation Division • 150-35 12 Avenue
Whitestone NY 11357
Ph: 718-746-3900 • Fax: 718-321-8436
Literature/Catalog: Yes
Control quadrants, cable assemblies, push-pull controls, custom hardware.

Ken Brock Manufacturing, Inc.
11852 Western Avenue • Stanton CA 90680
Ph: 714-898-4366 • Fax: 714-894-0811
Literature/Catalog: $5.00
Stamped, machined, and welded sub-assemblies. Short-run stamping facilities.

California Associated Products
4748-A Mission Blvd. • Ontario CA 91762
Ph: 909-591-3996 • Fax: 909-590-2973

California Power Systems
790 139th Avenue, #4 • San Leandro CA 94578
Ph: 800-247-9653 • Fax: 510-357-4429
Web: http://www.800-airwolf.com
Literature/Catalog: $6.95
Complete supplier of ultralight and sport aircraft parts and components. Rotax engines, parts and service. 500 page catalog, $6.95 US, $10.00 Canada and $20.00 foreign airmail.
See Display Ad On Page 69

The Originator of the "S/S Kit"

D & D Aircraft Supply
4 Stickney Terrace • PO Box 1200
Hampton NH 03842-1915
Ph: 800-468-8000 • Fax: 603-926-7855
E-Mail: doug@ddaircraft.com
Literature/Catalog: Yes
Stainless AN/MD fasteners, 1/4 turn Camloc and Southco studs, CRC aviation products.
See Display Ad On Page 28

DeMel Aircraft Corp.
Site 30 Comp. 27 RR2 • Penticton Airport
Penticton BC V2A 6J7 CANADA
Ph: 250-490-9532 • Fax: 250-490-9538
Literature/Catalog: $3.00
Parts and supplies - AN hardware, paint, plywood and covering materials. 4130 steel-sheet and tube. Aluminum sheet and tube. Post or airmail delivery. Visa and MasterCard. Catalog in Canada $5.00.

The Dillsburg Aeroplane Works
114 Sawmill Road • Dillsburg PA 17019
Ph: 717-432-4589 • Fax: 717-432-4589

DJB Aluminum & Supply
3151 Lucas Drive • Bethlehem PA 18107
Ph: 610 866-6913
Literature/Catalog: $2.00
AN bolts, clamps, shackles, rod ends. jig kits for the RV-4/6/6A. RV-4 Retro-Fit Kit converts RV-6 jig to RV-4. Hydraulic Hand Rivet Squeezer features: quick advance, auto retract, reversible handle for bench mount operation 360 degree swivel yoke allows for left or right handed operation.

Fastening Hardware Supply
PO Box 256 • Lancaster NY 14086
Ph: 716-681-1675 • Fax: 716-683-7315
Literature/Catalog: Yes
Stainless steel and plated aircraft hardware.

Fastening Systems Int'l.
P.O. Box 1372 • Sonoma CA 95476
Ph: 800-344-2393 • Fax: 707-935-1828
Literature/Catalog: $1.00
Hydraulic, electric and battery operated blind riveters and nutsetters. Up to 5,325 pounds of force. Complete inventory of blind fasteners by Cherry, Avdel, Gesipa/Olympic, Huck Intl., and Atlas Engineering.

Forbes Metal Stamping Co.
26555 Golden Valley Rd.
Saugus CA 91350
Ph: 805-250-1507 • Fax: 805-250-9406
E-Mail: fms76@aol.com
Literature/Catalog: Yes
Avionics hardware. Mic & phone jack mounting, instrument reducers, pilot & mechanics supplies. Wheel chock. Metal stamping of all kinds.

Gartmann Stainless, Inc.
206 Swedes Run Drive • Delran NJ 08075
Ph: 800-257-5855
Literature/Catalog: Yes

Goulet Aircraft Supply Ltd.
1137 Sanford St.
Winnepeg, MB R3E 3A1 CANADA
Ph: 204-786-1727 • Fax: 204-783-3302
General aviation parts distributor.

INFINITY Aerospace
P.O. Box 12275 Dept.KD
El Cajon CA 92022-2275
Ph: 619-448-5103 • Fax: 619-448-5103
E-Mail: infaero@flash.net
Web: http://www.flash.net/~infaero
The Infinity 1 Hotas Military-Style Stick Grip for the most important and frequently used cockpit functions. Left and right hand grips available. Send long self addressed stamped envelope for detailed information.

Instruments & Flight Research
2716 George Washington Blvd.
Wichita KS 67210-1585
Ph: 316-684-5177 • Fax: 316-684-0140
Literature/Catalog: Yes
Manufacture cabin lighting, panel lighting.

Kenmore Air Harbor Inc.
Box 82064 • Kenmore WA 98028-0064
Ph: 206-486-1257
A full service charter Seaplane opeartion.

Larsen Industries
2758 N. Las Posas Circle
Camarillo CA 93012
Ph: 805-491-2359 • Fax: 805-491-2559
Literature/Catalog: Yes
Surplus AN hardware. Large assortment of airframe bolts, stopnuts, platenuts, etc. Send SASE for complete list and pricing.

Leading Edge Air Foils
Meadowlake Airport • 8242 Cessna Drive
Peyton CO 80831
Ph: 719-683-5323 • Fax: 719-683-5333
E-Mail: leafkp@kktv.com
Web: http://www.leadingedgeairfoils.com
Literature/Catalog: $6.00
Distributor for ultralight and experimental aircraft, parts, and engines.
See Display Ad On Page 34

Linda Lou, Inc.
3514 Winhoma • Memphis TN 38118
Ph: 800-824-9912 • Fax: 901-365-9524

Lockwood Aviation Supply, Inc.
280 Hendricks Way • Suite A
Sebring FL 33870
Ph: 941-655-5100 • Fax: 941-655-6225
E-Mail: lockwood@digital.net
Web: http://www.lockwood-aviation.com
Literature/Catalog: $9.95
Light aircraft engines, parts and accessories.

M&M Aerospace Hardware, Inc.
1900 NW 89th Place • Miami FL 33172
Ph: 305-592-5155 • Fax: 305-592-7972
Literature/Catalog: Yes
Distributor of AN-MS-NAS hardware with a stock of over 100,000 items.

M&M Aerospace Hardware, Inc.
West Coast Division • 111 West Ocean Blvd. Ste. 560
Long Beach CA 90802
Ph: 310-901-1910 • Fax: 310-901-1915
Distributor of AN-MS-NAS hardware with a stock of over 100,000 items. Divisions in Miami FL and Long Beach CA.

Menzimer Aircraft Components
2525-8 Pioneer Avenue
Vista CA 92083
Ph: 760-598-0592 • Fax: 760-598-2967
Literature/Catalog: Yes
Electric trim systems and control stick grips. Small 14 volt servo motor for controlling elevator, aileron and rudder trim.

Mid-South Aircraft
2515 Alton Road
Birmingham AL 35210
Ph: 205-836-5101
Literature/Catalog: $1.00

Wil Neubert Aircraft Supply
1488 St. Rte. 162
Sprakers NY 12166
Ph: 518-922-5830 • Fax: 518-436-3632
Literature/Catalog: $5.00

Northwest Aero Products
2801 Academy Dr. S.E.
Auburn WA 98092
Ph: 206-735-5022 • Fax: 206-939-1578
E-Mail: nwaero@ix.netcom.com
Web: http://www.halcyon.com/wpowers/
Literature/Catalog: Yes
Miscellaneous hardware supplies including aircraft jacks, pulley assemblies, tow bars and more.

Northwest Custom Grips
4541 Harlan Drive N.E. • Salem OR 91305
Ph: 503-463-4433 • Fax: 503-463-4433
Custom handmade teakwood control grips. Provides comfort, dryness and coolness. Available with push-to-talk, 4-way trim, speed-brake and auto-pilot disconnect, any or all switch combination.

Pacific Continental Engines
16041 Arminta Street • Van Nuys CA 91406
Ph: 818-781-4947 • Fax: 818-781-0860
Web: http://www.pcenet.com
Literature/Catalog: Yes
Overhaul and sale of Textron, Lycoming, and Continental engines, components, and parts. Avionics shop sells new, repairs, and maintains all major brands. FBO offers aircraft maintenance.

Performance Propellers
P.O. Box 486, 63/27 Blue Haven • Nogales International Airport • Patagonia AZ 85624
Ph: 520-394-2059
Literature/Catalog: Yes
Propeller extensions, bolts & pressure plates.

The Glass Goose By Quikkit

Quikkit Corporation
9002 Summer Glen
Dallas TX 75243-7445
Ph: 214-349-0462 • Fax: 214-349-0462
E-Mail: quikkit@airmail.com
Web: http:////webwrks.com/glassgoose/
Literature/Catalog: Yes
Various weldments and hardware.
See Display Ad On Page 24

Regale Limited, U.K.
27 Whistler Road, Eaton Ford
St. Neots, Cambridgeshire PE19-3RT
UNITED KINGDOM
Ph: 44-480-477975 • Fax: 44-480-477975
British & Europeon military aircraft parts. Support & major items for BAC, Roll Royce, Marine, BAE, and much more. $10,000,000 in stock.

Regale Limited, U.S.A.
RT. 1, Box 138A
Peculiar MO 64078
Ph: 816-779-6075 • Fax: 816-758-6054
British & Europeon military aircraft parts. Support & major items for BAC, Roll Royce, Marine, BAE, and much more. $10,000,000 in stock.

Rich Industries, Inc.
P.O. Box 22396 • Bullhead City AZ 86439
Ph: 520-758-2777 • Fax: 520-758-7371
Literature/Catalog: Yes
AN, MS, NAS Fasteners Cherry Max Rivets, fittings, hilock pins, Jo-Bolts, electrical connectors, hose & fittings, stainless steel trim fasteners, quick release pins, rivnuts, AN426-AN470 solid rivets.

ROMA *Machine Tool & Die, Inc.*
Manufacturer of Bellcrank Bearings since 1968

Roma Machine Tool & Die, Inc.
1990 East Maple Street
Kankakee IL 60901-5720
Ph: 815-937-1245 • Fax: 815-937-1285
Literature/Catalog: Yes
Manufacturersof bellcrank bearings, BC4W 10 and BC5W11 propeller reductionspeed units, weld and bolt on axels, retractable landing gear, low wing, tubing fabrication, custom control panels, custom machine components for tail rotor on helicopters, variable pitch control and propeller hubs.
See Display Ad On Page 42

Shapiro Supply
5617 Natural Bridge • St. Louis MO 63120
Ph: 314-382-7000 • Fax: 314-382-8557
Literature/Catalog: Yes
AN, MS hardware.

Skybolt Aeromotive Corp.
551 N. Park Avenue • Apopka FL 32712
Ph: 407-889-2613 • Fax: 407-889-8103
E-Mail: skybolt@mindspring
Web: http://skybolt.com
Literature/Catalog: Yes
Composite manufacturing and fabrication. Aircraft hardware and accessories.

SkySports
Hangar 1 • Linden-Price Airport
Linden MI 48451
Ph: 800-247-7883 • Fax: 810-735-1078
E-Mail: airstuff@aol.com
Literature/Catalog: Yes

Smithy Co.
170 Aprill Drive • PO Box 1517
Ann Arbor MI 48106
Ph: 800-345-6324 • Fax: 313-913-6663
Literature/Catalog: Yes
Manufacturer of multi-purpose power lathe, mill and drill. Three machines in one. Machine fits conveniently into home shop.

Specialty Hose Aerospace
7802 Freedom Avenue NW
North Canton OH 44720-6978
Ph: 800-362-6533 • Fax: 330-497-0415
E-Mail: hose@cannet.com
Web: http://www.lmg.com/shc
Literature/Catalog: Yes

Spencer Aircraft
8308 12th Ave. South • Seattle WA 98108
Ph: 800-424-1160 • Fax: 206-763-4272
Literature/Catalog: Yes
N-M & NAS fasteners Aeroquip hoses and fittings.

Superflite
2149 E. Pratt Blvd. • Elk Grove Village IL 60007
Ph: 800-323-0611 • Fax: 708-364-0358
Literature/Catalog: Yes
Supplier of a broad range of parts and accessories for personal aircraft.
See Display Ad On Page 38

Toccoa Aviation, Inc.
Route 3 Airport • Toccoa GA 30577
Full maintenance for light airplanes.

Vista Aviation, Inc.
12653 Osborne Street • Pacoima CA 91331
Ph: 818-896-6442 • Fax: 818-896-9541
Literature/Catalog: Yes
Avionics shop. Maintenance, parts and equipment. Flight school.

WAG-AERO
The Wag-Aero Group
1216 North Road, Box 181 • Lyons WI 53148
Ph: 800-558-6868 • Fax: 414-763-7595
E-Mail: wagaero-sales@wagaero.com
Literature/Catalog: Yes
Nuts, bolts, washers, fittings, etc.
See Display Ad On Page 13

Wicks Aircraft Supply
410 Pine Street • Highland IL 62249
Ph: 800-221-9425 • Fax: 618-654-6253
Literature/Catalog: $5.00
Worldwide supplier for homebuilt projects specializing in raw material kits for 40+ different designs. 350 page catalog with over 12,000 items.
See Display Ad On Page 46

Yingling Aircraft, Inc.
P.O. Box 9248 • Wichita KS 67277
Ph: 316-943-3246 • Fax: 316-943-2484

Zenair, Ltd.
Huronia Airport • Midland, Ont. L4R 4K8 CANADA
Ph: 705-526-2871 • Fax: 705-526-8022
E-Mail: zenair@barint.on.ca
Literature/Catalog: $15.00
Complete hardware selection for ZODIAC CH 601 and STOL CH 701 aircraft. Including Avex blind rivets.

Instruments

AeroElectric Connection
6936 Bainbridge Road • Wichita KS 67226-1008
Ph: 316-685-8617
E-Mail: nuckolls@aeroelectric.com
Web: http://www.aeroelectric.com
Literature/Catalog: Yes
The AeroElectric Connection is an electrical systems handbook for individuals who build and maintain their own airplanes. The AeroElectric Connection is adding a line of supplies and accessories to aid builders in assembling their projects. Brochure available for #10 SASE.

Aerosonic Corp.
1212 N. Hercules Avenue
Clearwater FL 34625
Ph: 813-461-3000 • Fax: 813-447-5926
Literature/Catalog: Yes
Manufacturers of altimeters, encoding altimeters, blind encoders, air speeds, vertical speeds, manifold pressure indicator, and test equipment. Founded in 1953.

Air Salvage of Arkansas
2854 Hwy 8 East • Mena AR 71953
Ph: 501-394-1022 • Fax: 501-394-7475
Started 32 years ago. Dismantling mostly single engine aircraft, buying inventories, consignment selling of A/C and parts. Housing and storage.

Aircraft Quality Instruments
2661 West Esthner • Wichita KS 67213
Ph: 800-942-9938 • Fax: 316-942-9975
FAA cerfified repair station for aircraft instruments. We service, sell and exchange.

Aircraft Sales & Parts
PO Box 995 • Vernon, B.C. V1T 6N2 CANADA
Ph: 250-549-1102 • Fax: 250-549-3769
E-Mail: asap@junction.net
Web: http://www.ultralight.ca

Aircraft Spruce & Specialty — Since 1965
Aircraft Spruce & Specialty
PO Box 4000 • Corona CA 91718-9961
Ph: 800-824-1930 • Fax: 909-372-0555
E-Mail: info@aircraft-spruce.com
Web: http://www.aircraft-spruce.com
Literature/Catalog: $5.00
A complete selection of homebuilt and certified aircraft parts and materials. Shipping worldwide since 1965. Also see Aircraft Spruce East, Griffin GA, 800-831-2949.
See Display Ad On Page 24

Aircraft Super-Market
#10 Ellefson Dr. • PO Box 676
De Soto IA 50069
Ph: 515-834-2225 • Fax: 515-834-2152
E-Mail: bobdurly@netins.net
Web: http://www.netins.net/showcase/t-birds
Literature/Catalog: $5.00
Handles recreational aircraft of all kinds. New and used, ultralights, powered parachutes and U/L gliders. 2600' runway.
See Display Ad On Back Cover

Aircraft Supply
Allegheny County Airport
West Mifflin PA 15122
Ph: 800-245-0690 • Fax: 412-469-8271
Literature/Catalog: Yes

Airsport Corporation
1100 W. Cherokee • Sallisaw OK 74955
Ph: 918-775-4010 • Fax: 918-775-4000
E-Mail: Info@Airsport-Corp.com
Web: http://www. Airsport-Corp.com
Literature/Catalog: Yes
Altitude alerters, transponder data displays, mode T equipment.

Alcor, Inc.
12043 Colwick Street Ste 16
San Antonio TX 78216
Ph: 800-354-7233 • Fax: 210-308-8356
E-Mail: egt@alcorav.com
Web: http://alcorav.com • EGT indicators and thermocouples for piston aircraft.

Allegro Avionics
1332 E. Hampton • Tucson AZ 85719
Ph: 520-327-3695 • Fax: 520-327-0426
Literature/Catalog: Yes
EGT/CHT engine analyzer, fuel flow totalizer and tachometer with unique features and many options. Affordable 'State of the Art' instrumentation.

American Avionics, Inc.
7675 Perimeter Road South
Seattle WA 98108
Ph: 206-763-8530 • Fax: 206-763-2036
E-Mail: sales@americanavionics.com
Literature/Catalog: Yes

Archangel Avionics Inc
Archangel Avionics Inc.
1500 Pumphrey Ave. • Auburn AL 36832
Ph: 334-826-8008
See Display Ad On Page 20

ARP Industries, Inc.
36 Bay Drive E. • Huntington NY 11743
Ph: 516-427-1585
Carburator ice detector. Manufactured and sold worldwide.

Astronautics Corp. of America
5800 Cottage Grove Rd., Madison WI 53716
Ph: 608-221-9001 • Fax: 608-221-9104
Literature/Catalog: Yes
Aircraft flight/navigation instruments; multi-function CRT & AMLCD displays; mission/navigation computers; inertial navigation systems; synchros, resolvers/actuators, DC/AC motors.

Astrotech Corp.
10201 N. 21st Avenue • Phoenix AZ 85021
Ph: 602-870-3801 • Fax: 602-870-0358
Literature/Catalog: Yes
Supplier of digital clocks.

Avcom Avionics & Instr., Inc.
3008 NW 82nd Ave. • Miami FL 33122
Ph: 305-436-0300 • Fax: 305-436-0301
Literature/Catalog: Yes
Sales, service, and overhaul of all makes of avionics, instruments, radars and accessories.

Avionics Place
5257 Falcon Rd. • Rockford IL 61109
Ph: 815-229-5360 • Fax: 815-229-0308
Literature/Catalog: Yes
Sales, installation, and service of aircraft avionics.

B&F Technik V. Gmbh.
Am Neuen Rheinhafen 10
Speyer 67346 GERMANY
Ph: 06232/72076 • Fax: 06232/72078
Literature/Catalog: Yes
"All in one" UL-MIP Flight Log. Tracks engine speed, CHT, battery, OAT, EGT, fuel level and flow, time, date and flight log for last 20 flights.
See Display Ad On Page 8

Ball Variometers, Inc.
6595 Odell Place, Ste. C • Boulder CO 80301
Ph: 303-530-4940 • Fax: 303-530-4836

Becker Avionics, Inc.
912 Airport Road • Roanoke TX 76262
Ph: 817-491-0080 • Fax: 817-491-0081
E-Mail: BeckerAv @ Flash.net
Literature/Catalog: Yes
Manufacturer of communication, VOR/ILS glide slope transponders, audio panels, ADF, moving map GPS systems, EICAS, fixed base and portable VHF ground stations.

Bendix/King
400 N. Rogers Road • Olathe KS 66062-1212
Ph: 913-782-0400

California Power Systems
790 139th Avenue, #4 • San Leandro CA 94578
Ph: 800-247-9653 • Fax: 510-357-4429
Web: http://www.800-airwolf.com
Literature/Catalog: $6.95
Complete supplier of ultralight and sport aircraft parts and components. Rotax engines, parts and service. 500 page catalog, $6.95 US, $10.00 Canada and $20.00 foreign airmail.
See Display Ad On Page 69

Canadian Powered Parachutes Ltd.
Box 1769 • Vegreville, Alberta T9C 1S8 CANADA
Ph: 403-658-2044 • Fax: 403-658-2387
E-Mail: lindam@agt.net
Web: http://www.agt.net/public/lindam/cpp2.htm • Literature/Catalog: Yes

Cardinal Electronics
P.O. Box 70136 • Lansing MI 48907
Ph: 800-344-9469 • Fax: 517-321-4988
Literature/Catalog: Yes

Century Flight Systems, Inc.
P.O. Box 610 • Mineral Wells TX 76067
Ph: 800-433-5630 • Fax: 817-325-2546
E-Mail: century@centuryflight.com
Web: http://www.centureflight.com
Literature/Catalog: Yes
Manufacturer of HSI and turn & bank instruments.

Century Instruments Corp.
4440 Southeast Blvd. • Wichita KS 67210
Ph: 800-733-0116 • Fax: 316-522-7694
Literature/Catalog: Yes
Aircraft instrument repair station. Sell, buy, exchange and repair panel instruments.

J.D. Chapdelaine Co.
5500 N.W. 21st Terrace
Ft. Lauderdale FL 33309
Ph: 305-772-4559 • Fax: 305-493-8960

Chief Aircraft
1301 Brookside Blvd.
Grants Pass OR 97526
Ph: 800-447-3408 • Fax: 503-479-4431
E-Mail: chief@magick.net
Literature/Catalog: $2.00
See Display Ad On Page 59

Design Tech Systems
PO Box 338 • Lombard IL 60148
Ph: 708-971-9692 • Fax: 708-971-9742
Total communications packages including portable radios and intercoms.

DMA Instruments
14902 Forest Lodge Dr. • Houston TX 77070
Ph: 281-379-5430 • Fax: 281-379-5430
Literature/Catalog: Yes
Fuel Monitoring Systems, $645. Digital Tachometers, $325.

Eastern Avionics International
28000-A12 Airport Road • Punta Gorda FL 33982
Ph: 941-637-8585 • Fax: 941-637-0388
E-Mail: EAI Club@aol.com
Web: http://avionix.com • Literature/Catalog: Yes
Supplier of all kinds of instruments and avionics by all manufacturers.

Electronics International
12620 SW 231st Place
Hillsboro OR 97123-9329
Ph: 503-628-9113 • Fax: 503-628-3215
Literature/Catalog: Yes
See Display Ad On Page 42

Eventide Inc.
One Alsan Way • Little Ferry NJ 07643
Ph: 201-641-1200 • Fax: 201-641-1640
E-Mail: mike@eventide.com
Web: http://www.eventide.com
Literature/Catalog: Yes
Manufacturer of the ARGUS line of moving map displays. Includes flight planning option.

Falcon Ridge Industries
51 N. Hoagland Blvd. • Kissimmee FL 34741
Ph: 407-396-7572 • Fax: 407-396-1696
Instrument panels.

Flight Instrument Service, Inc
1619 Desoto Road • Sarasota FL 34234-3063
Ph: 941-355-2221 • Fax: 941-355-2221
FAA certified instrument service shop.

Flight Products International
P.O. Box 1558 • Kalispell MT 59901
Ph: 406-752-8783 • Fax: 406-257-7078
LED voltage monitor.

Flight Systems
81886 Sierra Avenue • Indio CA 92201
Ph: 800-824-1781 • Fax: 619-347-6795
Literature/Catalog: Yes
Digital Chronometer

Flightcraft, Inc.
7505 NE Airport Way • Portland OR 97218
Ph: 503-331-4200 • Fax: 503-331-4273
E-Mail: AAH07@AOL.COM
Custom panels for Glasair, Lancair. Factory authorized Lancair flight training.

Forbes Metal Stamping Co.
26555 Golden Valley Rd.
Saugus CA 91350
Ph: 805-250-1507 • Fax: 805-250-9406
E-Mail: fms76@aol.com
Literature/Catalog: Yes
Avionics hardware. Mic & phone jack mounting, instrument reducers, pilot & mechanics supplies. Wheel chock. Metal stamping of all kinds.

Grand Rapids Technologies
4526 Poinsettia SE • Kentwood MI 49508
Ph: 616-531-4893 • Fax: 616-249-0437
Literature/Catalog: Yes
Manufacturer of multifunction digital engine instruments. Four models provide complete instrumentation systems for nearly all types of aircraft engines including Rotax, Subaru, Lycoming, Continental, Hirth, etc.

Green Sky Adventures, Inc.
2395 Cream Ridge Rd. • Orwell OH 44076
Ph: 216-293-6624 • Fax: 216-293-6321
Web: http://www.greensky-adventures.com
Literature/Catalog: $2.00
EIS, VDO, Rotax.

Hamilton Instruments
106 Neubous • Houston TX 77061
Ph: 713-644-9008

Heads Up Display Systems, Inc.
650 Normandy Rd.
Madeira Beach FL 33708-2371
Ph: 813-392-4228 • Fax: 813-538-0730
Literature/Catalog: Yes
Heads-Up display system: give airspeed, altitude, rate of climb. When interfaced with GPS displays ground speed, distance to waypoint, heading to fly.

HORIZON INSTRUMENTS, INC.
556 S. State College Blvd.
Fullerton CA 92831
Ph: 800-541-8128 • Fax: 714-526-5937
E-Mail: horizoneml@aol.com
Web: http://members.aol.com/horizonins
Literature/Catalog: Yes
Manufactures electronic powerplant and electrical system indicators. Pilot friendly displays.
See Display Ad On Page 17

Insight Instrument Corp.
Box 194 Ellicott Station • Buffalo NY 14205-0194
Ph: 614-548-4122 • Fax: 905-871-5460
Manufacturer of engine temperature analyzers. The GEM 610 measures EGT, CHT and TIT with data-logging.

Instrument Tech Corp.
PO Box 1027 • Addison TX 75001-1027
Ph: 800-229-9078 • Fax: 972-458-8789
Manufacturer, sales and service of aircraft instrumentation, accessories, lighting and related items.

Instruments & Flight Research
2716 George Washington Blvd.
Wichita KS 67210-1585
Ph: 316-684-5177 • Fax: 316-684-0140
Literature/Catalog: Yes
Complete line of flight instruments. New vacuum gyros and other instruments in standard SAE sizes. Discounts available in quantity purchases.

J.P. Instruments, Inc.
P.O. Box 7033 • Huntington Beach CA 92615
Ph: 800-345-4574 • Fax: 714-557-9840
Literature/Catalog: Yes
Complete engine instrumentation in miniature gauges. Tachometer, manifold pressure, electrical, oil or fuel pressure, oil temperature, TIT, EGT. Mfg. all types of engine gauges, also data recording systems.

J.A. Air Center/Joliet Avionic
3N060 Powis Road • DuPage Airport
West Chicago IL 60185
Ph: 630-584-3200 • Fax: 630-443-0169
Web: http:////aso.solid.com/seller/650/main.htm
Literature/Catalog: Yes
J.A. is the midwest's most complete FBO, offering avionics sales, service, installations, piston and turbine aircraft maintenance.

J-AIR, INC.
24400 Highland Road, Ste. 62
Richmond Hts. OH 44143
Ph: 216-692-0010 • Fax: 216-692-0073
Literature/Catalog: Yes
Instruments/tools, gauges-press./vac. measure, test, certify, repair.

Kenmore Air Harbor Inc.
Box 82064 • Kenmore WA 98028-0064
Ph: 206-486-1257
A full service charter Seaplane operation.

Kollsman Inc.
220 Daniel Webster Hwy., S.
Merrimack NH 03054
Ph: 603-889-2500 • Fax: 603-889-7966
Manufacturer of avionics products including flight and engine instruments, air data test equipment, and cabin pressurization systems.

KS Avionics, Inc.
25216 Cypress Avenue • Hayward CA 94544
Ph: 800-346-4469 • Fax: 510-782-0329
Literature/Catalog: Yes
Manufacturer of exhaust gas & cylinder head temperture gauges. Available in both analog & digital formats.

Lockwood Aviation Supply, Inc.
280 Hendricks Way, Suite A
Sebring FL 33870
Ph: 941-655-5100 • Fax: 941-655-6225
E-Mail: lockwood@digital.net
Web: http://www.lockwood-aviation.com
Literature/Catalog: $9.95
Light aircraft engines, parts and accessories.

Microflight Products, Inc.
1356 Oakdale Drive, Suite 23
Griffin GA 30223-3929
Ph: 800-831-2949 • Fax: 770-229-2329
Literature/Catalog: $3.00
Complete line of instruments, gauges, fuel components, electrical parts, hardware, and accessories. Dealer and OEM programs available.

Mid-Continent West
16555 Sherman Way A-1
Van Nuys CA 91406
Ph: 818-786-0300 • Fax: 818-786-2734
Full service instrument repair and overhaul facility. Offering quality instrument sales and service.

On Final Aircraft Supplies
A Div. of Murphy Aircraft Mtg.
#2-8155 Aitken Road
Chilliwack, BC V2R 4H5 CANADA
Ph: 800-461-7767 • Fax: 604-792-7006
E-Mail: on-final@murphyair.com
Literature/Catalog: Yes

Otto Instrument Service, Inc.
Ontario Int'l. Airport • 1441 Valencia Place
Ontario CA 91761
Ph: 909-930-5800 • Fax: 909-930-5808
Instruments and autopilot overhauling. 50 years experience.

Pacific Coast-Avionics
1833 Auburn Way N. • Auburn WA 98002
Ph: 800 353-0370 • Fax: 206-931-0584
E-Mail: PacAvionic@AOL.com
Literature/Catalog: Yes
Full service avionics repair station with sales, service and installations. Complete line of pilot supplies. Specializing in panel package build-up for kit builders.

Panel Planner Software
P.O. Box 1042 • Boise ID 83701
Ph: 208 323-8724 • Fax: 208-939-8901
Instrument panel planning software.
See Display Ad On Page 57

Pflueger's
P.O. Box 187 • #23 Place Drive
Trinity Center CA 96091
Ph: 916 266-3236 • Fax: 916-266-3236
Literature/Catalog: Yes
Hardwood veneer aircraft panel used for replacement of plastic fascias or direct laminate of hardwood to aluminum panels. Lights can be built into overlays.

pointer
1027 N. Stadem Drive • Tempe AZ 85281
Ph: 602-966-1674 • Fax: 602-968-8020
Literature/Catalog: Yes
Provider of ELT system meeting TSO C91A requirements. External and portable antenna, voice capability, 2 year battery pack, 1.9 pounds. Solid state, hermetically sealed transmitter encased in a military approved polycarbonate.
See Display Ad On Page 32

Precision Dial Co.
7240 West KL Avenue
Kalamazoo MI 49009
Ph: 616-375-5601 • Fax: 616-375-1370
Dial and panel printing. Manufacture and refurbish instrument dials, instrument panels, radio/autopilot faces, backlit dials and panels. For homebuilt, warbird, and antique aircraft.

Rocky Mountain Instrument
PO Box 683 • 202 Fremont Street
Thermopolis WY 82443
Ph: 307-864-9300 • Fax: 307-864-9300
E-Mail: RMowRER@TRIB.com
Web: http://rkymtn.com
Literature/Catalog: Yes
Multi-function flight and engine instruments; encoder (see Avionics). Professional altitude alert with converge, hold and approach modes combined with altimeter, airspeed, VSI, OAT, and encoder in a 3-1/8" instrument. Complete air data output for all GPS systems included.

Ron Sands Co.
89 Forrest Drive • Mertztown PA 19539
Ph: 610-682-6788 • Fax: 610-682-6788
Literature/Catalog: Yes
P-51 fuel gauges. Send SASE for literature.

Sigma-Tek, Inc.
1001 Industrial Rd. • Augusta KS 67010
Ph: 316-775-6373 • Fax: 316-775-1416
E-Mail: info@sigmatek.com
Web: http://sigmatek.com
Literature/Catalog: Yes
Manufacturer of precision aviation instruments, avionics, accessories, control/display panels, and diagnostics systems. Capabilities are from product design through sales and service. Worldwide. Products for aircraft from general aviation trainers to commerical air carriers.

Sky West Instruments
1749 W. 13th Street • Upland CA 91786
Ph: 909-985-5718 • Fax: 909-985-5718
Lexan instrument panel overlays with indirect lighting. Custom fabricated to your specifications. Instrument overhaul, sales and service.

Skycom Avionics, Inc.
9305 W. Appleton Avenue
Milwaukee WI 53225
Ph: 414-463-8180 • Fax: 414-463-8195
Literature/Catalog: Yes
Aircraft electronics sales, service and installation. Your source for new, rebuilt and recertified packaged instrumentation for every homebuilt aircraft. Free Skycom Avionics Newsletter on request. Branches: Kenosha, Phone 414-652-8120; Waukesha, Phone 414-521-8180; Madison, Phone 608-246-8180

Skycraft Corp.
PO Box 984 • N. Hampton NH 03862
Ph: 603-964-1450 • Fax: 603-964-7604
Literature/Catalog: Yes
Supplier of spear type pitot-static head.

Skyflight
224 State St., Suite #5 • Portsmouth NH 03801
Ph: 603-431-4358
E-Mail: SkyFlyte@aol.com
Digital tachometers, EGT, temperature with preprogrammed alarm limits. Rugged design.

SkySports
Hangar 1 • Linden-Price Airport
Linden MI 48451
Ph: 800-247-7883 • Fax: 810-735-1078
E-Mail: airstuff@aol.com
Literature/Catalog: Yes
Flight and engine instruments.

Specialty Aircraft Co.
1440 NW 87th Ct.
Redmond OR 97756
Ph: 800-500-6786 • Fax: 503-548-5336
Single engine Cessna salvage yard. Instruments and parts for homebuilt aircraft.

S-TEC Corp.
One S-TEC Way • Municipal Airport
Mineral Wells TX 76067-9236
Ph: 940-325-9406 • Fax: 940-325-3904
E-Mail: kenpaul@flash.net
Web: http://www.s-tec.com/kitplane
Literature/Catalog: Yes
Manufacturer of a complete line of flight control systems, autopilots and flight directors. TEC line avionics - NAV, COM, ADF, DME and transponder. H.S.I. and R.M.I. systems.

Superflite
2149 E. Pratt Blvd. • Elk Grove Village IL 60007
Ph: 800-323-0611 • Fax: 708-364-0358
Literature/Catalog: Yes
Supplier of a broad range of parts and accessories for personal aircraft.
See Display Ad On Page 38

Taskem Corp.
12 Tickle's Road • Westport MA 02790
Ph: 508-679-9339 • Fax: 508-679-9339
E-Mail: TRiley@Meganet.net
Literature/Catalog: Yes
Model 2000 digital altimeter: to 37K feet, climb rate optional mode c, 3-1/8". Model J tach/timer: selectable redline, 3-1/8".

TEC Aviation

TEC Aviation-Aces Systems
10737 Lexington Drive • Knoxville TN 37932
Ph: 423-966-5856 • Fax: 423-675-1241
Literature/Catalog: Yes
Manufacturer of ground support test instruments—vibration analyzers, turbine engine balancers, propeller balancers, engine performance monitors and speed sensors.
See Display Ad On Page 70

Tennessee Propellers, Inc.

125 Deerfield Drive • Normandy TN 37360
Ph: 615-455-4516 • Fax: 615-455-1309
Literature/Catalog: $5.00
Distributor for Digatron digital instruments showing EGT, CHT in one degree increments, RPM in ten degree increments. Programmable to any temperature or RPM. Warning flash when set limit is reached. Complete with heavy duty probes.
See Display Ad On Page 75

Trans-Cal Industries, Inc.
16141 Cohasset Street • Van Nuys CA 91406
Ph: 818-787-1221 • Fax: 818-787-8916
Literature/Catalog: Yes
Altitude encoders, mechanical & ss; altitude alert & monitor system RS 232 Interface.

Universal Corp.
730 Independent Avenue
Grand Junction CO 81505
Ph: 970 242-5267 • Fax: 970 242-2572
Literature/Catalog: Yes
Mixture monitor, exhaust/gas temperature. Replacement probes.

Val Avionics, Ltd.
P.O.Box 13025 • Salem OR 97309-1025
Ph: 503-370-9429 • Fax: 503-370-9885
Literature/Catalog: Yes
Manufacture avionics including COM760TSO VHF transceivers, ILS400TSO instrument landing systems, and base station 7600 systems.

Vision Microsystems, Inc.
4071 Hannegan Rd. Ste. T
Bellingham WA 98226
Ph: 360-714-8203 • Fax: 360-714-8253
Literature/Catalog: Yes
Engine monitoring instruments and systems, fuel level systems, and electronic checklist and cautionary system. Since 1985.

WAG-AERO
The Wag-Aero Group
1216 North Road, Box 181
Lyons WI 53148
Ph: 800-558-6868 • Fax: 414-763-7595
E-Mail: wagaero-sales@wagaero.com
Literature/Catalog: Yes
Airspeeds, tachometers, altimeters, clocks, EGTs, CHTs, and accessories.
See Display Ad On Page 13

Wentworth Aircraft Inc.
3015 Cedar Avenue, S.
Minneapolis MN 55407
Ph: 612-722-0065 • Fax: 612-722-1024
Literature/Catalog: Yes
Engines, radios, instruments and miscellaneous single engine used parts.

Western Avionics Inc.
19711 Campus Drive #A • John Wayne Airport
Santa Ana CA 92707-5203
Ph: 714-757-0290 • Fax: 714-757-0293
Literature/Catalog: Yes
Sales, service, installation of avionics, autopilots, radar systems, GPS, and moving maps.

Western Ultralight
4700 188th N.E. • Arlington WA 98223
Ph: 360-435-3737 • Fax: 360-435-3737
Total Ultralight facility. Including flight instruction, sales, parts, and maintenance.

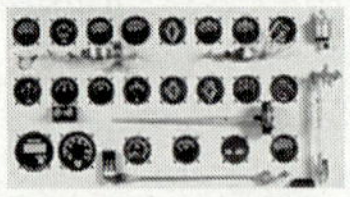

Wicks Aircraft Supply
410 Pine Street • Highland IL 62249
Ph: 800-221-9425 • Fax: 618-654-6253
Literature/Catalog: $5.00
Worldwide supplier for homebuilt projects specializing in raw material kits for 40+ different designs. 350 page catalog with over 12,000 items.
See Display Ad On Page 46

Wings & Wheels
81 Jackson Avenue, W.E.
Jamestown NY 14701
Ph: 716-664-6894 • Fax: 716-664-6895
Literature/Catalog: $4.00
Aircraft supplies, components, and avionics. Buy and sell, new and used. Becker, Icom, Terra, Trimble Magellan, Strong, National Parachutes, and much more.

Landing Gear

Grove Aircraft Company
1860 Joe Crosson Drive • El Cajon CA 92020
Ph: 619-562-1268 • Fax: 619-562-3274
Spring aluminum landing gear built to your specifications. Also in stock, one design of gun drilled landing gear.

Lighting

Adcour, Inc.
PO Box 2649 • Woburn MA 01888-1249
Ph: 617-784-8123 • Fax: 617-784-5563
Literature/Catalog: Yes

Aero Enhancements
14610 Chermoore • Chesterfield MO 63017
Ph: 314-527-2120 • Fax: 314-256-0370
Web: www.aeroenhance.com
Instrument panel and glare shield lighting with AV battery backup. Custom instrument panels.

Aircraft Specialties Services
2860 North Sheridan Road • Tulsa International Airport • Tulsa OK 74115
Ph: 800-826-9252 • Fax: 918-836-4419
Literature/Catalog: Yes
FAA certified repair station D2R764K. Inspection, remachining of aircraft engine parts. Full service, new and used parts.
See Display Ad On Page 74

Aircraft Spruce & Specialty
PO Box 4000 • Corona CA 91718-9961
Ph: 800-824-1930 • Fax: 909-372-0555
E-Mail: info@aircraft-spruce.com
Web: http://www.aircraft-spruce.com
Literature/Catalog: $5.00
A complete selection of homebuilt and certified aircraft parts and materials. Shipping worldwide since 1965. Also see Aircraft Spruce East, Griffin GA, 800-831-2949.
See Display Ad On Page 24

ARTEX Aircraft Supplies, Inc.
24368 S. Skylane Drive • Canby OR 97013
Ph: 800-547-8901

Avion Research Corp.
1022 W. Maude Ave. #102
Sunnyvale CA 94086
Ph: 408-738-1690 • Fax: 408-738-1678
Literature/Catalog: Yes
Replacements for all-metal, instrument panels, and lighted glare shields for Cessna, Piper, and Beechcraft.

AvLite Aviation Products, Inc.
3015 C Propeller Dr., C-7
Paso Robles CA 93446
Ph: 805-239-4037 • Fax: 805-239-2980
Grimes lighting fixtures and components. Interior, instrument, exterior, rotating beacons, strobes and landing lights.

Candlepower, Inc.
14626 Southlawn Lane • Rockville MD 20850
Ph: 800-368-2700 • Literature/Catalog: Yes

Great Plains Aircraft
P.O. Box 545 • Boys Town NE 68010
Ph: 402-493-6507 • Fax: 402-333-7750
E-Mail: gpasc@earthlink.net
Web: http://www.greatplainsas.com
Literature/Catalog: $3.00
See Display Ad On Page 80

Gulf Coast Avionics Corp.
4243 N. Westshore Blvd. • Tampa FL 33614
Ph: 813-879-9714 • Fax: 813-875-4514
E-Mail: gca@ix.netcom.com
Web: http://www.gulf-coast-avionics.com
Literature/Catalog: Yes
Panel build-up and pre-wiring information.

David Hoffman Products
HC 79, Box 102-H • Melba ID 83641
Ph: 208-495-2137 • Fax: 208-495-2137
Miniature, low-voltage lighting systems.

Illusion Aircraft Strobe Co.
1681 Cheryl Way • Aptos CA 95003
Ph: 408-688-9260 • Fax: 408-688-9260
Literature/Catalog: Yes
Strobe lights for aircraft with 2 cycle engines. 24 Joules operates on all electrical systems (with or without battery). OEM pricing. Send SASE

Instrument Tech Corp.
PO Box 1027 • Addison TX 75001-1027
Ph: 800-229-9078 • Fax: 972-458-8789
Manufacturer, sales and service of aircraft instrumentation, accessories, lighting and related items.

Instruments & Flight Research
2716 George Washington Blvd.
Wichita KS 67210-1585
Ph: 316-684-5177 • Fax: 316-684-0140
Literature/Catalog: Yes
Aircraft interior lighting systems and components. Custom building of straight or double-back bulbs for all special applications and dimensions.

Kenmore Air Harbor Inc.
Box 82064 • Kenmore WA 98028-0064
Ph: 206-486-1257
A full service charter Seaplane operation.

Kuntzleman Electronics
1660 South Hanover Street
Pottstown PA 19465
Ph: 610-326-9068 • Fax: 610-970-9078
Literature/Catalog: Yes
Anti-collision strobe lights and power supplies.

Linda Lou, Inc.
3514 Winhoma • Memphis TN 38118
Ph: 800-824-9912 • Fax: 901-365-9524

Manairco, Inc.
PO Box 111 • Mansfield OH 44901-0111
Ph: 419-524-2121 • Fax: 419-525-4790
Literature/Catalog: Yes
Manufacturers airport and heliport lighting equipment.

National Products, Inc.
1017 S. Elmgrove St. • Seattle WA 98108
Ph: 206-763-8361 • Fax: 206-763-9615
E-Mail: stuff@ram-mount.com
Web: http://www.ram-mount.com
Literature/Catalog: Yes

Pflueger's
P.O. Box 187 • #23 Place Drive
Trinity Center CA 96091
Ph: 916 266-3236 • Fax: 916-266-3236
Literature/Catalog: Yes
Hardwood veneer aircraft panel used for replacement of plastic fascias or direct laminate of hardwood to aluminum panels. Lights can be built into overlays.

Puritan-Bennett Aero Systems
10800 Pflumm Road • Lenexa KS 66215-4061
Ph: 913-338-9836 • Fax: 913-469-8419
Literature/Catalog: Yes
Manufacturer of oxygen systems, passenger service and convenience products, and closed circuit television systems.

R. V. Redmer Company
19N831 Big Timber Road • Hampshire IL 60140
Ph: 708-683-1319
Literature/Catalog: Yes
Cockpit spot light: adjustable, light weight, cool, bright red illumination for panel, map or interior.

Superflite

2149 E. Pratt Blvd. • Elk Grove Village IL 60007
Ph: 800-323-0611 • Fax: 708-364-0358
Literature/Catalog: Yes
Supplier of a broad range of parts and accessories for personal aircraft.
See Display Ad On Page 38

TERF Inc.

11411 Culver • Brighton MI 48114
Ph: 810-229-2082 • Fax: 810-229-2082
E-Mail: Mtait@livgonline.com
Web: http://www.terf.com
Literature/Catalog: Yes
Wing tip lights, beacons, etc.,
See Display Ad On Page 52

Universal Corp.
730 Independent Avenue
Grand Junction CO 81505
Ph: 970 242-5267 • Fax: 970 242-2572
Literature/Catalog: Yes
FS-4400 flight strobe.

Varga Enterprises, Inc.
2350 S. Airport Blvd. • Chandler AZ 85249
Ph: 800-966-6936 • Fax: 602-899-0324
E-Mail: vargaair@vargaair.com
Web: http://www.vargaair.com
Literature/Catalog: Yes
Distributor of Whelen lighting systems and GE lamps.

WAG-AERO
The Wag-Aero Group
1216 North Road, Box 181
Lyons WI 53148
Ph: 800-558-6868 • Fax: 414-763-7595
E-Mail: wagaero-sales@wagaero.com
Literature/Catalog: Yes
Whelen, Aeroflash Systems.
See Display Ad On Page 13

Whelen Engineering Co., Inc.
Route 145, Winthrop Road
Chester CT 06412
Ph: 203-526-9504 • Fax: 203-526-4784
Literature/Catalog: Yes
Aviation safety lighting. Strobe anticollision lights, landing lights, cockpit lights, recognition lights.

White Eagle Industries, Inc.
8282 Crane Road • Cranesville PA 16410
Ph: 814-734-7749 • Fax: 814-756-3577
Literature/Catalog: Yes
Manufacturer of runway lighting systems and pilot controlled switches.

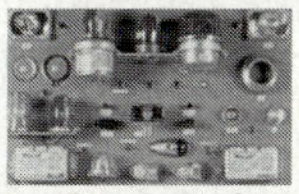

Wicks Aircraft Supply
410 Pine Street • Highland IL 62249
Ph: 800-221-9425 • Fax: 618-654-6253
Literature/Catalog: $5.00
Worldwide supplier for homebuilt projects specializing in raw material kits for 40+ different designs. 350 page catalog with over 12,000 items.
See Display Ad On Page 46

Navigation Equipment

Abbot Brown & Sons Ltd.
Barnfleet Works • Beaminster, Dorset DT8 3EJ UNITED KINGDOM
Ph: 01308 862404 • Fax: 01308 863402
Nameplates and labels, engraving and printing, rules and protractors.

Aero Products Research, Inc.
11201 Hindry Avenue
Los Angeles CA 90045
Ph: 310-641-7242 • Fax: 310-649-1703
Literature/Catalog: Yes
Supplier of hand held navigation equipment (E6-B computers, etc.), navigation plotters. Kneeboards, pilot flight desks, flight plan sheets and logs.

Aircraft Spruce & Specialty
PO Box 4000 • Corona CA 91718-9961
Ph: 800-824-1930 • Fax: 909-372-0555
E-Mail: info@aircraft-spruce.com
Web: http://www.aircraft-spruce.com
Literature/Catalog: $5.00
A complete selection of homebuilt and certified aircraft parts and materials. Shipping worldwide since 1965. Also see Aircraft Spruce East, Griffin GA, 800-831-2949.
See Display Ad On Page 24

American Avionics, Inc.
7675 Perimeter Road South
Seattle WA 98108
Ph: 206-763-8530 • Fax: 206-763-2036
E-Mail: sales@americanavionics.com
Literature/Catalog: Yes

Archangel Avionics Inc.
1500 Pumphrey Ave. • Auburn AL 36832
Ph: 334-826-8008
See Display Ad On Page 20

ARNAV Systems, Inc.
P.O. Box 73730 • Puyallup WA 98373-0730
Ph: 206-847-3550 • Fax: 206-848-3555
Web: http://arnav@aol.com
Literature/Catalog: Yes
GPS, Loran, Flight Management Systems TSO'd. Glass cockpit suites, primary flight displays, multi-function displays, engine instrumentation. Data link communications: GeoNet, ACARS, Mode S.

Avcom Avionics & Instr., Inc.
3008 NW 82nd Ave. • Miami FL 33122
Ph: 305-436-0300 • Fax: 305-436-0301
Literature/Catalog: Yes
Sales, service, and overhaul of all makes of avionics, instruments, radars and accessories.

Avionics Place
5257 Falcon Rd. • Rockford IL 61109
Ph: 815-229-5360 • Fax: 815-229-0308
Literature/Catalog: Yes
Sales, installation, and service of aircraft avionics.

Becker Avionics, Inc.
912 Airport Road • Roanoke TX 76262
Ph: 817-491-0080 • Fax: 817-491-0081
E-Mail: BeckerAv @ Flash.net
Literature/Catalog: Yes
Manufacturer of communication, VOR/ILS glide slope transponders, audio panels, ADF, moving map GPS systems, EICAS, fixed base and portable VHF ground stations.

C&R Aircraft
PO Box 281 • Quincy IL 62306-0281
Ph: 217-292-5967
Flight monitor system, complete and compact cross country flight planning forms, and complete aircraft monitoring forms.

C.A.A. Services
6-C Herman Drive • East Granby CT 06026
Ph: 203-658-8000 • Fax: 203-651-0885
Literature/Catalog: Yes
Avionics and accessories. Custom panels.

California Power Systems
790 139th Avenue, #4
San Leandro CA 94578
Ph: 800-247-9653 • Fax: 510-357-4429
Web: http://www.800-airwolf.com
Literature/Catalog: $6.95
Complete supplier of ultralight and sport aircraft parts and components. Rotax engines, parts and service. 500 page catalog, $6.95 US, $10.00 Canada and $20.00 foreign airmail.
See Display Ad On Page 69

Comant Industries, Inc.
12920 Park St. • Santa Fe Springs CA 90670
Ph: 562-946-6694 X152 • Fax: 562-946-5150
Literature/Catalog: Yes
Antennas designed for VLA (metal or composite) and hand held GPS. Also multi-function antennas for GPS/COMM/FM/DGPS.

Eastern Avionics International
28000-A12 Airport Road
Punta Gorda FL 33982
Ph: 941-637-8585 • Fax: 941-637-0388
E-Mail: EAI Club@aol.com
Web: http://avionix.com
Literature/Catalog: Yes
Supplier of all kinds of instruments and avionics by all manufacturers.

Emergency Beacon Corp.
P.O. Box 178 • 15 River Street
New Rochelle NY 10802-0178
Ph: 914-235-9400 • Fax: 914-576-7075
Literature/Catalog: Yes
Direction finders.

Eventide, Inc.
One Alsan Way • Little Ferry NJ 07643
Ph: 201-641-1200 • Fax: 201-641-1640
E-Mail: mike@eventide.com
Web: http://www.eventide.com
Literature/Catalog: Yes
Manufacturer of the ARGUS line of moving map displays. Includes flight planning option.

Flightcraft, Inc.
7505 NE Airport Way • Portland OR 97218
Ph: 503-331-4200 • Fax: 503-331-4273
E-Mail: AAH07@AOL.COM
Custom panels for Glasair, Lancair. Factory authorized Lancair flight training.

Garmin (Europe) Ltd.
Unit 5, The Quadrangle • Abbey Park
Romsey SO51 9AQ UNITED KINGDOM
Ph: 44-1794-519944 • Fax: 44-1794-519222
Literature/Catalog: Yes
Garmin supplies installed and portable GPS systems.

Garmin International
1200 151st Street • Olathe KS 66062-3426
Ph: 913-397-8200 • Fax: 913-397-8282
Manufacturer of a variety of navigation and communications electronics.

Marv Golden Discount Sales
8690 Aero Drive, Ste. 102
San Diego CA 92123
Ph: 800-348-0014 • Fax: 619-569-4508
Web: http://www.MARVGOLDEN.COM
Literature/Catalog: Yes
Portable GPS Receivers. Plotters, EGB computers. Magellan, Trimble, Garmin.

B.F. Goodrich FlightSystems
2001 Polaris Parkway
Columbus OH 43240-2001
Ph: 614-825-2001 • Fax: 614-825-2034
Literature/Catalog: Yes
Stormscope (R) Series II weather mapping systems. Detects electrical discharges from convective clouds and displays them as a moving map on a 3-inch CRT. Ranges to 100 NM. Six models with prices starting at less than $4,000.

Gulf Coast Avionics Corp.
4243 N. Westshore Blvd. • Tampa FL 33614
Ph: 813-879-9714 • Fax: 813-875-4514
E-Mail: gca@ix.netcom.com
Web: http://www.gulf-coast-avionics.com
Literature/Catalog: Yes
Avionics supplier. Custom instrument panel manufacturing.

Instrument Tech Corp.
PO Box 1027 • Addison TX 75001-1027
Ph: 800-229-9078 • Fax: 972-458-8789
Manufacturer, sales and service of aircraft instrumentation, accessories, lighting and related items.

Instruments & Flight Research
2716 George Washington Blvd.
Wichita KS 67210-1585
Ph: 316-684-5177 • Fax: 316-684-0140
Literature/Catalog: Yes
Manufacture navigation systems and gyros.

J.A. Air Center/Joliet Avionic
3N060 Powis Road • DuPage Airport
West Chicago IL 60185
Ph: 630-584-3200 • Fax: 630-443-0169
Web: http:///aso.solid.com/seller/650/main.htm • *Literature/Catalog: Yes*
J.A. is the midwest's most complete FBO, offering avionics sales, service, installations, piston and turbine aircraft maintenance.

Just Fly Aviation Gear
PO Box 450 • Kitty Hawk NC 27949
Ph: 919-480-3552 • Fax: 919-480-3552
Literature/Catalog: Yes
Garmin GPS units for ultralights.

Lafayette Avionics
Hangar 7, Purdue Airport
West Lafayette IN 47906
Ph: 317-743-3828 • Fax: 317-743-0000
E-Mail: ron@avionics-laf.com
Web: http://avionics-laf.com
Literature/Catalog: Yes
Sales, repair & installation of aircraft avionics.

Lone Star Aviation Corp.
804 N. Great S.W. Pkwy. • Arlington TX 76011
Ph: 817-548-7768 • Fax: 817-633-6208
E-Mail: lsa@iamerica.net
Literature/Catalog: Yes
Noise filters for intercoms, stormscopes, etc. GPS power supplies for home use. Avionic cooling fans.

McCoy Avionics Corp.
10761 Watkins Road
Marysville OH 43040-9544
Ph: 800-654-8124 • Fax: 937-642-0220
E-Mail: RPHartigan@juno.com
Web: http://www.McCoyAvionics.com
Literature/Catalog: Yes
Mac 1700C complete Nav/Com slide-in replacement. 760 channel comm. and 200 channel nav.

Micrologic, Inc.
9174 Deering Avenue
Chatsworth CA 91311-5801
Ph: 818-998-1216 • Fax: 818-709-3658
Literature/Catalog: Yes
Micrologic manufactures GPS and Loran consumer products.

Narco Avionics, Inc.
270 Commerce Drive
Fort Washington PA 19034
Ph: 800-223-3636 • Fax: 215-643-0197
Literature/Catalog: Yes
Aircraft electronics.

National Products, Inc.
1017 S. Elmgrove St. • Seattle WA 98108
Ph: 206-763-8361 • Fax: 206-763-9615
E-Mail: stuff@ram-mount.com
Web: http://www.ram-mount.com
Literature/Catalog: Yes

Northstar Avionics
30 Sudbury Road • Acton MA 01720
Ph: 508-897-6600 • Fax: 508-897-7241
Web: http://www.northstarcmc.com
Northstar manufactures IFR and VFR GPS navigators and integrated GPS/VHF communication systems.

Porcine Associates
244 O'Connor Street • Menlo Park CA 94025
Ph: 415-326-2669 • Fax: 415-326-1071
E-Mail: jimham@porcine.com
Web: http://www.porcine.com
Literature/Catalog: Yes
Hand held GPS linked with autopilot. Will work with most hand held GPS receivers. $189.00 complete with manual and connector. Send self addressed stamped envelope for information.

RMS Technology, Inc.
124 Berkley Avenue • PO Box 249
Molalla OR 97038
Ph: 503-829-6166 • Fax: 503-829-6568
E-Mail: 72163.3342@compuserve.com
Literature/Catalog: Yes
"Flitesoft" flight planning computer software. "Vista" moving map software.

Ross Engineering Co.
12505-E Starkey Road • Largo FL 34643
Ph: 800-422-7676 • Fax: 813-535-4248
Literature/Catalog: Yes

Skycom Avionics, Inc.
9305 W. Appleton Avenue • Milwaukee WI 53225
Ph: 414-463-8180 • Fax: 414-463-8195
Literature/Catalog: Yes
Aircraft electronics sales, service and installation. Your source for new, rebuilt and recertified packaged instrumentation for every homebuilt aircraft. Free Skycom Avionics Newsletter on request. Branches: Kenosha, Phone 414-652-8120; Waukesha, Phone 414-521-8180; Madison, Phone 608-246-8180.

S-TEC Corp.
One S-TEC Way • Municipal Airport
Mineral Wells TX 76067-9236
Ph: 940-325-9406 • Fax: 940-325-3904
E-Mail: kenpaul@flash.net
Web: http://www.s-tec.com/kitplane
Literature/Catalog: Yes
Manufacturer of a complete line of flight control systems, autopilots and flight directors. TEC line avionics - NAV, COM, ADF, DME and transponder. H.S.I. and R.M.I. systems.

Superflite
2149 E. Pratt Blvd. • Elk Grove Village IL 60007
Ph: 800-323-0611 • Fax: 708-364-0358
Literature/Catalog: Yes
Supplier of a broad range of parts and accessories for personal aircraft.
See Display Ad On Page 38

Trimble Navigation, Ltd.
Avionics Div. • 2105 Donley Drive
Austin TX 78758
Ph: 800-767-8628 • Fax: 512-836-9413
Web: http://www.trimble.com
GPS Receivers, GPS/Loran & GPS/Omega

Valentine Aviation
14151 Montfort Dr., Suite 344
Dallas TX 75240-3067
Ph: 214-239-7911 • Fax: 214-385-1069
E-Mail: donaldv@airmail.net
Used avionics, specializing in storm scopes.

Wentworth Aircraft Inc.
3015 Cedar Avenue, S.
Minneapolis MN 55407
Ph: 612-722-0065 • Fax: 612-722-1024
Literature/Catalog: Yes
Engines, radios, instruments and miscellaneous single engine used parts.

Western Avionics Inc.
19711 Campus Drive #A • John Wayne Airport
Santa Ana CA 92707-5203
Ph: 714-757-0290 • Fax: 714-757-0293
Literature/Catalog: Yes
Sales, service, installation of avionics, autopilots, radar systems, GPS, and moving maps.

Yingling Aircraft, Inc.
P.O. Box 9248 • Wichita KS 67277
Ph: 316-943-3246 • Fax: 316-943-2484

Oxygen Systems

Aerox Aviation Oxygen
200 Dillon Rd. • Hilton Head SC 29926
Ph: 800-237-6902 • Fax: 803-681-6828
E-Mail: aerox@digitel.net
Web: http://www.aerox.com
High-duration, lightweight, portable oxygen systems and retrofit kits for built-in systems or other portable systems.

Aircraft Spruce & Specialty Since 1965

Aircraft Spruce & Specialty
PO Box 4000 • Corona CA 91718-9961
Ph: 800-824-1930 • Fax: 909-372-0555
E-Mail: info@aircraft-spruce.com
Web: http://www.aircraft-spruce.com
Literature/Catalog: $5.00
A complete selection of homebuilt and certified aircraft parts and materials. Shipping worldwide since 1965. Also see Aircraft Spruce East, Griffin GA, 800-831-2949.
See Display Ad On Page 24

Aircraft Supply
Allegheny County Airport
West Mifflin PA 15122
Ph: 800-245-0690 • Fax: 412-469-8271
Literature/Catalog: Yes

Apollo Safety & Industrial
1108 West 9th Street
Upland CA 91786-5703
Ph: 909-980-7071 • Fax: 909-985-6873

Butler Parachute Systems, Inc.
1820 Loudon Avenue NW • PO Box 6098
Roanoke VA 24017-0098
Ph: 540-342-2501 • Fax: 540-342-4037
E-Mail: butlerchut@aol.com

Marv Golden Discount Sales
8690 Aero Drive, Ste. 102
San Diego CA 92123
Ph: 800-348-0014 • Fax: 619-569-4508
Web: http://www.MARVGOLDEN.COM
Literature/Catalog: Yes
Aerox systems for two or four people.

Mountain High Equipment & Supply Co.
516 12th Ave. • Salt Lake City UT 84103-3209
Ph: 800-468-8185
Web: http://www.mtn-high.com
Literature/Catalog: Yes
Carry-on and built-in oxygen systems. 5 to 115 cu. ft.

Nelson Aircraft Co.
21550 N.W. Nicholas Ct., #D
Hillsboro OR 97124
Ph: 503-629-5277 • Fax: 503-629-5817
E-Mail: info@nelsonair.com
Web: http://www.nelsonair.com
Literature/Catalog: Yes
Portable O2 systems. Flow control meters. Oxygen saving cannulas. Face masks (optional built-in mikes). O2 connectors and fittings. O2 cylinders and carrying cases.

Pacific Coast-Avionics
1833 Auburn Way N. • Auburn WA 98002
Ph: 800 353-0370 • Fax: 206-931-0584
E-Mail: PacAvionic@AOL.com
Literature/Catalog: Yes
Full service avionics repair station with sales, service and installations. Complete line of pilot supplies. Specializing in panel package build-up for kit builders.

Puritan-Bennett Aero Systems
10800 Pflumm Road • Lenexa KS 66215-4061
Ph: 913-338-9836 • Fax: 913-469-8419
Literature/Catalog: Yes
Manufacturer of oxygen systems, passenger service and convenience products, and closed circuit television systems.

Sky Ox

Sky-Ox, Ltd.
P.O. Box W • Edwardsburg MI 49112-0723
Ph: 800-253-0800 • Fax: 616-663-2579
Portable oxygen systems.
See Display Ad On Page 54

Superflite
2149 E. Pratt Blvd.
Elk Grove Village IL 60007
Ph: 800-323-0611 • Fax: 708-364-0358
Literature/Catalog: Yes
Supplier of a broad range of parts and accessories for personal aircraft.
See Display Ad On Page 38

WAG-AERO

The Wag-Aero Group
1216 North Road, Box 181
Lyons WI 53148
Ph: 800-558-6868 • Fax: 414-763-7595
E-Mail: wagaero-sales@wagaero.com
Literature/Catalog: Yes
Aerox and Sky-Ox systems.
See Display Ad On Page 13

Propellers

A.K.S. Inc.
PO Box 17161 • Portland OR 97217
Ph: 503-289-1491 • Fax: 503-247-9207
E-Mail: trikes@teleport.com
Web: http://www.teleport.com/~trikes
North American distributor for Sport Prop Propellers.
See Display Ad On Page 47

Aberle Custom Aircraft
2141 S. Mission Road • Fallbrook CA 92028
Ph: 760-723-1731
Over 30 years assisting show quality and performance oriented builders.

Aces High Light Aircraft Ltd.
RR 1 • London, Ontario N6A 4B5 CANADA
Ph: 519-652-3020 • Fax: 519-652-3020
Literature/Catalog: $5.00
Aircraft kit & custom parts. Manufacturer able to supply most aircraft building materials and parts at competitive cost.

Aero Marine Engineering
9727 Hagel Circle • Lorton VA 22079
Ph: 703-541-0766 • Fax: 703-550-1236
Literature/Catalog: Yes
Design, development, and engineering of low speed aircraft. Propeller design. CAD drafting. Licensed professional engineer.

Aerotec Propeller
1002 Chukar • Midland TX 79706
Ph: 915-685-0867 • Fax: 915-685-0867
Literature/Catalog: Yes

Air Salvage of Arkansas
2854 Hwy 8 East • Mena AR 71953
Ph: 501-394-1022 • Fax: 501-394-7475
Started 32 years ago. Dismantling mostly single engine aircraft, buying inventories, consignment selling of A/C and parts. Housing and storage.

Aircraft Accessories of Oklahoma
2740 N. Sheridan Road
Tulsa OK 74115
Ph: 918-835-9924 • Fax: 918-835-3681
E-Mail: airacc@busprod.com
Literature/Catalog: Yes

Aircraft Spruce & Specialty Since 1965

Aircraft Spruce & Specialty
PO Box 4000
Corona CA 91718-9961
Ph: 800-824-1930 • Fax: 909-372-0555
E-Mail: info@aircraft-spruce.com
Web: http://www.aircraft-spruce.com
Literature/Catalog: $5.00
A complete selection of homebuilt and certified aircraft parts and materials. Shipping worldwide since 1965. Also see Aircraft Spruce East, Griffin GA, 800-831-2949.
See Display Ad On Page 24

Aircraft Super-Market
#10 Ellefson Dr. • PO Box 676
De Soto IA 50069
Ph: 515-834-2225 • Fax: 515-834-2152
E-Mail: bobdurly@netins.net
Web: http://www.netins.net/showcase/t-birds
Literature/Catalog: $5.00
Handles recreational aircraft of all kinds. New and used, ultralights, powered parachutes and U/L gliders. 2600' runway.
See Display Ad On Back Cover

Aircraft Supply
Allegheny County Airport • West Mifflin PA 15122
Ph: 800-245-0690 • Fax: 412-469-8271
Literature/Catalog: Yes

Aircraft Systems, Inc.
16606 W. Ryerson Road • New Berlin WI 53151
Ph: 800-322-2131 • Fax: 414-784-3099
Literature/Catalog: $15.00
Midwest "Vari-Prop" distributor. Cockpit controllable pitch propeller. Pitch change over 90 degrees. Mounts to any engine.

American Propeller Service
20208 Charlanne Dr. • Redding CA 96002
Ph: 800-292-7767 • Fax: 916-221-5201
Literature/Catalog: Yes
Distributors for all propeller manufacturers. Also custom designing of propellers and consultation.

Arrowprop Co., Inc.
P.O. Box 610 • Meeker OK 74855
Ph: 405-279-3833 • Fax: 405-279-2377
Literature/Catalog: Yes
Manufacturers of Ultralight aircraft propellers, airboat propellers, Duster-Sprayer fans, test clubs, Hovercraft fans, decorator propellers, balloon inflator fans, experimental aircraft propellers, industrial wood fans.

Avia-Hamilton Std. Aviation
PO Box 22 • Hlavenecl 29
Stara Boleslav 250 02 CZECH REPUBLIC
Ph: 42-202-811114 • Fax: 42 202-811818
Literature/Catalog: Yes
Production, repairs, and overhauls of all metal constant speed, automatic and gound adjustable propellers in the range of 160-220 hp for Lycoming 0-320 and LOM M 332, 337 engines. 600-777 hp for Walter M 601 engines.

Aymar-DeMuth Propellers
8213 Elberta Drive • PO Box 853
Ellicott City MD 21041-0853
Ph: 410-461-4329 • Fax: 410-461-4329
Literature/Catalog: Yes
IFR props, composite props. Extensions and crush plates.

THRUST = DRAG
Bates Engineering
2742 Swansboro Road
Placerville CA 95667
Ph: 530-622-1886 • Fax: 530-622-1886
E-Mail: bateseng@innercite.com
Web: http://www.innercite.com/~bateseng
Literature/Catalog: Yes
Airplane and propeller design services. Programs for aircraft design, performance, stability and control. Propeller design optimization software.
See Display Ad On Page 82

Bernie Warnke Props
HRC 1, Box 1717
Tucson AZ 85736 • Ph: 602-822-1166

Bradley Aerospace
950 Muir Ave. • Chico CA 95926
Ph: 916-899-7918 • Fax: 916-899-2942
Web: http://www.kgmsoftworks.com/bradley
Ivoprop dealer.
See Display Ad On Page 40

California Power Systems
790 139th Avenue, #4 • San Leandro CA 94578
Ph: 800-247-9653 • Fax: 510-357-4429
Web: http://www.800-airwolf.com
Literature/Catalog: $6.95
Complete supplier of ultralight and sport aircraft parts and components. Rotax engines, parts and service. 500 page catalog, $6.95 US, $10.00 Canada and $20.00 foreign airmail.
See Display Ad On Page 69

Catto Aircraft Co.
P.O. Box 1104 • San Andreas CA 95249
Ph: 209-754-3553

Certified Wood Props
P.O. Box 61 • Zellwood FL 32798
Ph: 407-889-3451

Colorado Unlimited Adventure
PO Box 100
Tumacacori AZ 85640-0100
Ph: 602-961-3580 • Fax: 602-961-3581

Competition Aircraft, Inc.
11110 Gopher Mine Trail
Grass Valley CA 95949
Ph: 916-268-3048 • Fax: 916-268-2321
E-Mail: ultraprop@ord.net
Literature/Catalog: Yes
Ultra-Prop - composite propeller. Two, three and four blade adjustable-pitch composite propellers, up to 68" diameter for ultralight and light experimental aircraft.

Component Technology Corp.
R.R. 2, Box 336 • Colchester IL 62326
Ph: 800-237-9377
Literature/Catalog: Yes

Culver Props, Inc.
RD1, Box 176 • Galeton PA 16922
Ph: 814-435-1077 • Fax: 814-435-6523
E-Mail: rmiller@conterra.com
Literature/Catalog: Yes
Fixed pitch laminated wooden props for ultralights, experimentals, kitplanes, gyros, fans, and airboats.

Edward Sterba Aircraft Props.
412 S. 5th Street • Delavan WI 53115
Ph: 414-728-1367
Custom built propellers for all homebuilts. Hard maple laminates with urethane leading edges.

Executive Propellers, Inc.
16140 Covello Street • Van Nuys CA 91406
Ph: 818-779-7767 • Fax: 818-779-0848
Literature/Catalog: Yes
Sales and overhaul of propellers. Exchange and loaner props available.

Falconar Avia, Inc.
7739 - 81 Ave.
Edmonton, Alberta T6C 0V4 CANADA
Ph: 403-465-2024 • Fax: 403-465-2029
E-Mail: falconar@tic.ab.ca
Engine options include Hirth, Continental, VW, Lycoming, and Franklin. Basic structure in wood and plywood, Hipec coated synthetic fabric and Hipec finish. Utilizes a minimum of metal fittings.

Felix Propellers
W10508 Bell Rd.
Camp Douglas WI 54618
Ph: 800-776-7357 • Fax: 608-427-6544
Literature/Catalog: Yes
Custom designed wood, fixed pitch propellers for all experimental, replica, and antique aircraft.

GSC Systems Int'l., Ltd.
6255 Onanagan Landing Rd.
Vernon, B.C. V1T 6Y5 CANADA
Ph: 250-549-3772 • Fax: 250-549-3769
E-Mail: asap@junction.net
Tech series mechanically adjustable pitch propellers. Requires a hollow shaft to accommodate the pitch adjusting rod. Weight under 10 lbs. Pitch change mechanically controlled with a cam over an 18 degree pitch range. Tech V five bladed, ground adjustable propellers.

George J. Priester Aviation
Pal Waukee Airport • Wheeling IL 60090
Ph: 708-537-1200

Great Plains Aircraft
P.O. Box 545 • Boys Town NE 68010
Ph: 402-493-6507 • Fax: 402-333-7750
E-Mail: gpasc@earthlink.net
Web: http://www.greatplainsas.com
Literature/Catalog: $3.00
See Display Ad On Page 80

Green Sky Adventures, Inc.
2395 Cream Ridge Rd. • Orwell OH 44076
Ph: 216-293-6624 • Fax: 216-293-6321
Web: http://www.greensky-adventures.com
Literature/Catalog: $2.00
GSC, Culver, IVO

Hartzell Propeller, Inc.
One Propeller Place • FactoryServiceCenter
Piqua OH 45356-2634
Ph: 937-778-4201 • Fax: 937-778-4202
Hartzell propeller parts, service, repair and overhaul.

Hegy Propellers
407 W. Lincoln • Box 124 • Marfa TX 79843
Ph: 915-729-4249

G. Hess
881 Emory Court • Upland CA 91786
Ph: 909-982-5729
Carve your own propeller. For KR, Quickie II, Dragonfly, Sonerai, etc. Intended to avoid risking an expensive propeller during tests. $6.95 in the U.S., $10.00 foreign.

Hoffmann Propeller GmbH & Co. KG
Küpferlingstrasse 9
Rosenheim D-83022 GERMANY
Ph: 49-8031-1878-O • Fax: 49-8031-1878-78
Literature/Catalog: Yes
Manufacture & design of low-weight and noise-reduced propellers in wooden-composite constructions, and custom work. Advanced composite propellers for ultralight and light experimentals. US Distributor - Steen Aero Lab, Inc.

Hummel Aviation
509 E. Butler Street • Bryan OH 43506-0880
Ph: 419-636-3390 • Literature/Catalog: $2.00
All sizes of propellers in stock including Hummel Bird and The Right Prop, A Team Aircraft Dealer.

Paul Irlbeck Propellers
RR2 Box 21A • Wabasha MN 55981
Ph: 612-565-3653
Specialty-fixed pitch, wood (hard maple) propellers for RV-3, RV-4 and RV-6 aircraft.

IVOPROP CORP.
15903 Lakewood Blvd. • Bellflower CA 90706
Ph: 562-602-1451 • Fax: 562-602-1374
Literature/Catalog: Yes
Composite props, in-flight or ground adjustable. Carbon-graphite fiber blades with stainless steel leading edges. New high-horsepower magnum for engines up to 700 hp.
See Display Ad On Page 82

Junkers Profly
Na Folimance 13,
Praha 2 120 00 CZECH REPUBLIC
Ph: 0042 2 522 376 • Fax: 0042 2 524 330
E-Mail: STRAKA@BIC.CVUT.CZ
High performance, light weight "Sport Prop" composite propellers. Developed specifically for U/L and Experimental category aircraft.

Kinetics Inc.
438 Main Avenue North • Britt IA 50423
Ph: 515-843-3420

Mark Landoll Electrical Svc.
1205 Redbud Lane • Newcastle OK 73065
Ph: 405-392-3847
Literature/Catalog: Yes
Harmonic dampeners for use with wood propellers.

Light Power Engine Corp.
PO Box 3350 • Morgantown WV 26503
Ph: 304-291-3843 • Fax: 304-292-1902
Literature/Catalog: $20.00
Distributor for MT.

Light Speed Engineering
PO Box 549 • Hangar #15
Santa Paula CA 93060
Ph: 805-933-3299 • Fax: 805-525-0199
E-Mail: LSE@west.net
Literature/Catalog: Yes
High performance propellers, 100 hp to 210 hp. Composite spinners for pusher designs.

McCauley Accessory Division
Cessna Aircraft Company • 3535 McCauley Drive
Vandalia OH 45377-5053
Ph: 937-890-5246 • Fax: 937-890-6001

Gary McGill & Associates
P.O. Box 2566 • Glendale AZ 85311-2566
E-Mail: gemcsr@juno.com
Literature/Catalog: $1.00
Propeller carving machine plans; made from wood & common hardware, $15.00.

Midcoast Aviation
PO Box 10056 • St. Louis MO 63145
Ph: 314-731-7111

Moravia, Inc.
P.O. Box 942
Thunder Bay, Ontario P7C 4X8 CANADA
Ph: 807-475-5113 • Fax: 807-475-9587
E-Mail: moravia@loon.norlink.net
Web: http://www.norlink.net/moravia
Literature/Catalog: Yes

MT-Propeller Entwicklung GmbH
Flugplatz Straubing Wallmuehle
Atting D-94348 GERMANY
Ph: 49-9429-8433 • Fax: 49-9429-8432
Literature/Catalog: Yes
Fixed pitch wood composite, ground-adjustable or constant speed propellers with light weight wood-composite blades for acrobatic and high performance aircraft. FAA certified.
See Display Ad On Page 82

New England Propeller
PO Box 244 • East Haddam CT 06423
Ph: 203-873-1494 • Fax: 203-873-2388
Sell, service and build all makes of propellers and governors. McCauley, Hartzell, Sensenich, and others. Also build and sell experimental propellers including propellers for aircraft over 150 hp.

Ottosen Propeller/Accessories
105 South 28th Street • Phoenix AZ 85034
Ph: 602-275-8514

Pacesetter Propeller Works
P.O. Box 1245 • Hillsboro OR 97123
Ph: 503-628-2797 • Wood propellers.

Pacific Continental Engines
16041 Arminta Street • Van Nuys CA 91406
Ph: 818-781-4947 • Fax: 818-781-0860
Web: http://www.pcenet.com
Literature/Catalog: Yes
Overhaul and sale of Textron, Lycoming, and Continental engines, components, and parts. Avionics shop sells new, repairs, and maintains all major brands. FBO offers aircraft maintenance.

Performance Propellers
P.O. Box 486, 63/27 Blue Haven • Nogales International Airport • Patagonia AZ 85624
Ph: 520-394-2059 • Literature/Catalog: Yes
Two & three blade multi-laminated custom wood propellers.

Perry Experimental Propellers
Route 9, Box 1648 • Livingston TX 77351
Ph: 409-967-3467

Pop's Props
R.R. #1 Box 98 • Cooksville IL 61730
Ph: 309-829-8343 • Fax: 309-828-2031
Literature/Catalog: Yes
Specializing in two cylinder VW accessories, propellers, hubs, digital tachometers and hourmeters, and crankcase ventilation systems.
See Display Ad On Page 55

Powerfin, Inc.
4700 188th St. NE #G • Arlington WA 98223
Ph: 800-581-8207 • Fax: 360-403-0599
E-Mail: powerfin@supernal.net
Web: http://www.supernal.net/~powerfin
Literature/Catalog: Yes
Low inertia carbon fiber composite aircraft propellers for ultralights and small experimentals. Fully ground adjustable.

Precision Propellers, Inc.
PO Box 1392 • Vernal UT 84078
Ph: 801-789-3699 • Fax: 801-789-6059
Literature/Catalog: Yes
Infinitely ground adjustable propellers for ultralights and homebuilts. 2- and 3-bladed propellers in diameters from 20" to 74". Blades are 1/16" hard rock maple laminates. Hubs are aircraft aluminum and composites. Many designs - tested and custom made to match your application. Expert consultants. Established 1979.

Prince Aircraft Co.
6774 Providence Street • PO Box 2669
Whitehouse OH 43571
Ph: 419-877-5557 • Fax: 419-877-5564
E-Mail: 75557.31555@Compuserve.Com
Literature/Catalog: Yes
U.S. Prototype Manufacturer. Anti-Vortex P-TIP Propeller featuring high performance, low noise and automatic pitch control. Specializing in propeller prototype design applications. Special order aircraft production of aircraft, wind tunnel, UAV, airboat, and balloon inflators, computer designed for all engine/aircraft applications.

PROPS INC.
354 SE 2nd Street • Newport OR 97356
Ph: 541-265-3032 • Fax: 541-265-3032
Literature/Catalog: Yes
Ground adjustable and fixed pitch wood propellers for all kitbuilt, homebuilt, ultralights, and airboats. 2,3, and 4 blade configurations. Custom design services.
See Display Ad On Page 82

Propulsion Technologies, Inc.
7200 C.E. Dixon St. • Stockton CA 95206
Ph: 209-982-5558 • Fax: 209-982-5702
E-Mail: karlpti@msn.com
Literature/Catalog: Yes
Distributor for Avia-Hamilton Standard propellers. Also, a firewall-forward package of engine, prop, cowling, and accessories, for the Glastar.

The GLASS GOOSE By Quikkit

Quikkit Corporation
9002 Summer Glen • Dallas TX 75243-7445
Ph: 214-349-0462 • Fax: 214-349-0462
E-Mail: quikkit@airmail.com
Web: http:////webwrks.com/glassgoose/
Literature/Catalog: Yes
Warp drive propellers.
See Display Ad On Page 24

Recreational Power Eng.
10888 Station Rd.
Columbia Station OH 44028
Ph: 800-583-3306 • Fax: 419-585-6004
Literature/Catalog: Yes
US distributor of Hirth two cycle aircraft engines. Sales, service, parts and engine overhauls for Hirth engines. Supplier of engine instrument systems and propellers.

Regale Limited, U.K.
27 Whistler Road, Eaton Ford
St. Neots, Cambridgeshire PE19-3RT
UNITED KINGDOM
Ph: 44-480-477975 • Fax: 44-480-477975
British & European military aircraft parts. Support & major items for BAC, Roll Royce, Marine, BAE, and much more. $10,000,000 in stock.

Regale Limited, U.S.A.
RT. 1, Box 138A • Peculiar MO 64078
Ph: 816-779-6075 • Fax: 816-758-6054
British & European military aircraft parts. Support & major items for BAC, Roll Royce, Marine, BAE, and much more. $10,000,000 in stock.

RFI Powerplant Systems Group
2916 Tennessee Northeast
Albuquerque NM 87110
Ph: 505-298-9362 • Fax: 505-266-1789
Orion "P" tip propellers.

Ritz Propeller Co., Inc.
380 Shipmans Creek Road KD4
Wartrace TN 37183 • Ph: 615-857-3419

Saber Manufacturing
1951 NW 84th Terrace
Pembrook Pines FL 33024
Ph: 954-436-9496 • Fax: 954-436-0015
E-Mail: JSaber@iname.com
Literature/Catalog: Yes
Specializing in propeller extensions. Perfect safety record.

St. Croix Aircraft
1139 State Highway #148
Corning IA 50841 • Ph: 515-322-4041
Literature/Catalog: Yes
Custom wood propellers for antiques and replicas. Specialists at matching prop design to antique and modern engine installations for maximum performance. Two, three & four bladed props, diameters up to 120".

SensenichDirect

Sensenich Prop Manufacturing
14 Citation Lane • Lititz PA 17543
Ph: 717-569-0435 • Fax: 717-560-3725
Web: http://WWW.SENSENICH.COM
Manufacturer of Sensenich metal propellers.
See Display Ad On Page 85

Sensenich Propellers
14 Citation Lane • Lancaster PA 17543
Ph: 800-462-3412 • Fax: 717-560-3725
E-Mail: ez@sensenich.com
Web: http://www.sensenich.com

SENSENICH

Sensenich Wood Propellers Company
2008 Wood Ct. • Plant City FL 33567
Ph: 813-752-3711 • Fax: 813-752-2818
Manufacturer of Sensenich wood propellers.
See Display Ad On Page 85

Specialty Aircraft Co.
1440 NW 87th Ct. • Redmond OR 97756
Ph: 800-500-6786 • Fax: 503-548-5336
Single engine Cessna salvage yard. Instruments and parts for homebuilt aircraft.

Sport Copter International, Inc.
Scappoose Airport • 34012 N. Honeyman Road
Scappoose OR 97056
Ph: 503-543-7000 • Fax: 503-543-7041
Web: http://www.sportcopter.com
High performance, light weight "Sport Prop" composite propellers from the Czech Republic. Developed specifically for U/L and Experimental category aircraft.

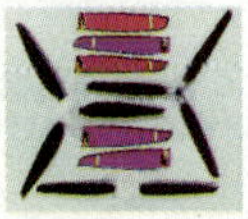

Sport Prop s.r.o.
PO Box 1 • 17100 posta
Praha 71 CZECH REPUBLIC
E-Mail: zlinak@teleport.com
Literature/Catalog: Yes
High performance composite propellers.
See Display Ad On Page 47

Sun Belt Aviation, Inc.
P.O. Box 193 • McGregor Municipal Airport
McGregor TX 76657
Ph: 817-848-5227 • Fax: 817-848-4798
E-Mail: 75176.3033@compuserve.com
FAA CRS #GPWR419E. Repairs and overhauls on McCauley, Hartzell, Sensenich and Hamilton standard propellers.

Sun-Tech Innovations

P.O. Box 9154 • Canoga Park CA 91309
Ph: 800-786-8324 • Fax: 818-882-7859
E-Mail: Sun-tech@worldnet.att.net
Web: http://www.balancemasters.com/ultraschem1.html
Balance-Master add-on for propellers and Rotax engines. Reduces vibration on ultralights up to 50%.
See Display Ad On Page 55

Ted's Custom Props
PO Box 824 • 600 Superior Street
Concrete WA 98237
Ph: 206-853-8947

Tennessee Propellers, Inc.

125 Deerfield Drive • Normandy TN 37360
Ph: 615-455-4516 • Fax: 615-455-1309
Literature/Catalog: $5.00
Since 1983, manufacturer of multi-laminated wooden propellers for most ultralights and gyro's. Worldwide.
See Display Ad On Page 75

Tiffin Aire, Inc.
1778 W. State Route 224 • Tiffin OH 44883
Ph: 800-553-7767 • Fax: 419-299-3244
E-Mail: docprop@juno.com
Web: http://www.aeroguide.com/tiffinaireprop/
Propeller sales and service for Hartzell, McCauley, Woodward, Sensenich, and Beechcraft.

Turbo West Inc.
10656 West 120th Avenue
Broomfiled CO 80021 • Ph: 303-469-6676

Vari-Prop International Inc.
P.O. Box 1856 • Cottonwood CA 96022
Ph: 916-347-6122
Literature/Catalog: Yes
In-flight adjustable propeller. Hydraulically controlled variable pitch propeller that is hydraulically controlled without needing a hollow crankshaft. Free brochure. 27-page Info Pack, $10.00.

VORTECH, Inc.

PO Box 511-KP • Fallston MD 21047
Ph: 410-692-2692 • Fax: 410-692-5902
Web: http://www.webcom.com/neatstuf/helio
Literature/Catalog: $10.00
All aluminum rotor blades for helicopters and gyroplanes. Manufactured as a single extrusion - no seams, rivets or bonding.
See Display Ad On Page 21

WAG-AERO

The Wag-Aero Group
1216 North Road, Box 181
Lyons WI 53148
Ph: 800-558-6868 • Fax: 414-763-7595
E-Mail: wagaero-sales@wagaero.com
Literature/Catalog: Yes
New and overhauling of existing Sensenich and McCauley propellers.
See Display Ad On Page 13

Warp Drive, Inc.
1207 Hwy 18 E. • PO Box 78
Ventura IA 50482
Ph: 800-833-9357 • Fax: 515-357-7592
Literature/Catalog: Yes
Solid carbon fiber propeller for ultralights, homebuilts, rotorcraft, etc. Free information on request. Lifetime guarantee of satisfaction.

Western Ultralight
4700 188th N.E. • Arlington WA 98223
Ph: 360-435-3737 • Fax: 360-435-3737
Total Ultralight facility. Including flight instruction, sales, parts, and maintenance.

Whirlwind Propellers Corporation
Hangar H • 1860 Joe Crosson Drive
El Cajon CA 92020 • Ph: 619-562-3725
E-Mail: wwpc@concentric.net
Composite constant speed propellers for Lycoming or Continental engines. Two and three blades for any high performance experimental.

Woofter-Saber Manufacturing
1951 NW 84th Terrace
Pembroke Pines FL 33024
Ph: 954 436-9496 • Fax: 954-436-0015
Literature/Catalog: Yes
Founded in 1984 by Lynn Woofter and Judith Saber. Prop extensions our specialty. Perfect safety record.

Radio, Intercoms and Headsets

Acousticom Corporation
28180 Clay Street • Elkhart IN 46517-1070
Ph: 800-664-0534 • Fax: 219-294-7250
E-Mail: acousticom@skyenet.net
Web: http://www.acousticom.com
Literature/Catalog: Yes
Headsets, intercom assemblies, and related spare parts for most domestic and imported headsets.

Aearo Peltor
90 Mechanic Street • Southbridge MA 01550
Ph: 800-327-6833 • Fax: 800-327-3291
Web: http://www.aearo.com
Peltor headsets.

Aerodeca, Inc.
P.O. Box 64AC • Monaca PA 15061
Ph: 800-727-0311
Literature/Catalog: Yes
Full-line authorized Hush-A-Com headset & intercom dealer.

AeroElectric Connection
6936 Bainbridge Road • Wichita KS 67226-1008
Ph: 316-685-8617
E-Mail: nuckolls@aeroelectric.com
Web: http://www.aeroelectric.com
Literature/Catalog: Yes
The AeroElectric Connection is an electrical systems handbook for individuals who build and maintain their own airplanes. The AeroElectric Connection is adding a line of supplies and accessories to aid builders in assembling their projects. Brochure available for #10 SASE.

Aircraft Sales & Parts
PO Box 995 • Vernon, B.C. V1T 6N2 CANADA
Ph: 250-549-1102 • Fax: 250-549-3769
E-Mail: asap@junction.net
Web: http://www.ultralight.ca

Aircraft Spruce & Specialty
PO Box 4000 • Corona CA 91718-9961
Ph: 800-824-1930 • Fax: 909-372-0555
E-Mail: info@aircraft-spruce.com
Web: http://www.aircraft-spruce.com
Literature/Catalog: $5.00
A complete selection of homebuilt and certified aircraft parts and materials. Shipping worldwide since 1965. Also see Aircraft Spruce East, Griffin GA, 800-831-2949.
See Display Ad On Page 24

American Avionics, Inc.
7675 Perimeter Road South
Seattle WA 98108
Ph: 206-763-8530 • Fax: 206-763-2036
E-Mail: sales@americanavionics.com
Literature/Catalog: Yes

Audio Com Electronics
1000 Harvard Way Apt.128
Reno NV 89502-2044

Aureus, Intl.
2860 S Circle Dr. # GL-22.
Colorado Springs CO 80906-4115
Ph: 719-540-9077 • Fax: 719-540-9078

Aviation Supplies & Academics
7005 132nd Pl. SE • Newcastle WA 98059
Ph: 206-235-1500 • Fax: 206-235-0128
E-Mail: 102105.223@compuserve.com
Manufacturer of pilot supplies.

Avionics Place
5257 Falcon Rd. • Rockford IL 61109
Ph: 815-229-5360 • Fax: 815-229-0308
Literature/Catalog: Yes
Sales, installation, and service of aircraft avionics.

Becker Avionics, Inc.
912 Airport Road • Roanoke TX 76262
Ph: 817-491-0080 • Fax: 817-491-0081
E-Mail: BeckerAv @ Flash.net
Literature/Catalog: Yes
Manufacturer of communication, VOR/ILS glide slope transponders, audio panels, ADF, moving map GPS systems, EICAS, fixed base and portable VHF ground stations.

C&R Aircraft
PO Box 281 • Quincy IL 62306-0281
Ph: 217-292-5967
Headset strain relief clip. Ease the strain on your headset cord and on your neck.

C.A.A. Services
6-C Herman Drive • East Granby CT 06026
Ph: 203-658-8000 • Fax: 203-651-0885
Literature/Catalog: Yes
Avionics and accessories. Custom panels.

California Power Systems
790 139th Avenue, #4
San Leandro CA 94578
Ph: 800-247-9653 • Fax: 510-357-4429
Web: http://www.800-airwolf.com
Literature/Catalog: $6.95
Complete supplier of ultralight and sport aircraft parts and components. Rotax engines, parts and service. 500 page catalog, $6.95 US, $10.00 Canada and $20.00 foreign airmail.
See Display Ad On Page 69

Canadian Powered Parachutes Ltd.
Box 1769
Vegreville, Alberta T9C 1S8 CANADA
Ph: 403-658-2044 • Fax: 403-658-2387
E-Mail: lindam@agt.net
Web: http://www.agt.net/public/lindam/cpp2.htm
Literature/Catalog: Yes

CEOTronics Inc.
2340 Trinity Mills Rd. #112, Carrollton TX 75006
Ph: 972-416-9500 • Fax: 972-416-9580
E-Mail: headsets@altinet.net
Web: http://www.ceotronics.com
International manufacturer of headsets and communication systems. Handcrafted cockpit headsets.

Chief Aircraft, Inc.
1301 Brookside Blvd. • Grants Pass OR 97526
Ph: 800-447-3408 • Fax: 503-479-4431
E-Mail: chief@magick.net
Literature/Catalog: $2.00
See Display Ad On Page 59

Classic Air Supply
801 S. Lincoln • Marshfield WI 54449
Ph: 715-387-3057 • Literature/Catalog: Yes
Econocom intercoms. System includes portable intercom and two headsets, all three for $175. Free brochure.

Colorado Unlimited Adventure
PO Box 100 • Tumacacori AZ 85640-0100
Ph: 602-961-3580 • Fax: 602-961-3581

Comant Industries, Inc.
12920 Park St. • Santa Fe Springs CA 90670
Ph: 562-946-6694 X152 • Fax: 562-946-5150
Literature/Catalog: Yes
Antennas designed for VLA (metal or composite) and hand held GPS. Also multi-function antennas for GPS/COMM/FM/DGPS.

Communications Specialists, Inc
426 W. Taft Avenue • Orange CA 92665-4296
Ph: 800-854-0547 • Fax: 800-850-0547
Literature/Catalog: Yes
TR-720-Hand held. Airband transceiver. Frequencies: NAV -108.0 to 118.0 MHZ, COM - 118.0 to 136.975 MHZ.

Comtronics Engineering, Ltd.
P.O. Box 146 • Almond WI 54909-0146
Ph: 715-366-7093 • Fax: 715-366-7091
Literature/Catalog: Yes
A manufacturing company in business making helmets, headsets, and intercoms. We have been in business for 15 years.

Crystal Shamrock, Inc.
6000 Douglas Drive N • Crystal MN 55429
Ph: 612-533-2214 • Fax: 612-533-5519
35 year old FBO. Certified repair station for upholstery, engines and service.

David Clark Company, Inc.
360 Franklin Street, Box 15054
Worcester MA 01615-0054
Ph: 508-751-5800 • Fax: 508-753-5827
E-Mail: Aviation@davidclark.com
Web: http://www.davidclark.com
Literature/Catalog: Yes
Manufacturers of noise attenuating headsets with noise cancelling microphones. Voice activated, portable, and panel mounted aircraft intercoms. Helmet kits and a complete line of comfort accessories.

DeMel Aircraft Corp.
Site 30 Comp. 27 RR2 • Pencticton Airport
Penticton BC V2A 6J7 CANADA
Ph: 250-490-9532 • Fax: 250-490-9538
Literature/Catalog: $3.00
Parts and supplies. Post or airmail delivery. Visa and MasterCard. Catalog in Canada $5.00.

Design Tech Systems
PO Box 338 • Lombard IL 60148
Ph: 708-971-9692 • Fax: 708-971-9742
Total communications packages including portable radios and intercoms.

Eastern Avionics International
28000-A12 Airport Road • Punta Gorda FL 33982
Ph: 941-637-8585 • Fax: 941-637-0388
E-Mail: EAI Club@aol.com
Web: http://avionix.com • Literature/Catalog: Yes
Supplier of all kinds of instruments and avionics by all manufacturers.

Electro-Voice, Inc.
600 Cecil Street • Buchanan MI 49107
Ph: 616-695-4714 • Fax: 616-695-7570
Literature/Catalog: Yes
Boom, mask and hand held microphones, handsets, and active noise reduction headsets.

Flightcom Corporation
7340 SW Durham Road • Portland OR 97224
Ph: 800-432-4342 • Fax: 503-620-2943
E-Mail: headset@teleport.com
Literature/Catalog: Yes

FlightTech Intercoms
PO Box 338 • Lombard IL 60148
Ph: 630-971-9692 • Fax: 630-971-9742
E-Mail: flighttech@2099.com
Web: http://www.flighttech.com
Portable intercoms for all types of aircraft and radios, including ICOM IC-A20, A21, A22, King KX99 and many poplular brands.

Marv Golden Discount Sales
8690 Aero Drive, Ste. 102 • San Diego CA 92123
Ph: 800-348-0014 • Fax: 619-569-4508
Web: http://www.MARVGOLDEN.COM
Literature/Catalog: Yes
David Clark, Telex, Peltor, Pilot and Marv's Brand, Icom, Delcom, Flightcom.

Warren Gregoire & Associates
229 El Pueblo Place • Clayton CA 94517
Ph: 800-634-0094 • Fax: 510-673-0538
E-Mail: info@warrengregoire.com
Literature/Catalog: Yes
Light weight noise-attenuating headsets and leather control yoke covers with built-in push-to-talk systems.

Gulf Coast Avionics Corp.
4243 N. Westshore Blvd. • Tampa FL 33614
Ph: 813-879-9714 • Fax: 813-875-4514
E-Mail: gca@ix.netcom.com
Web: http://www.gulf-coast-avionics.com
Literature/Catalog: Yes
Avionics supplier. Custom instrument panel manufacturing.

ICOM America
2380 116th Avenue NE • Bellevue WA 98004
Ph: 425-454-8155 • Fax: 425-454-1509
Web: http://www.icomamerica.com
Literature/Catalog: Yes
Avionic hand helds, panel mounts, base stations, general coverage receivers, and VHF and UHF transceivers.

Instrument Tech Corp.
PO Box 1027 • Addison TX 75001-1027
Ph: 800-229-9078 • Fax: 972-458-8789
Manufacturer, sales and service of aircraft instrumentation, accessories, lighting and related items.

J.A. Air Center/Joliet Avionic
3N060 Powis Road • DuPage Airport
West Chicago IL 60185
Ph: 630-584-3200 • Fax: 630-443-0169
Web: http:////aso.solid.com/seller/650/main.htm • Literature/Catalog: Yes
J.A. is the midwest's most complete FBO, offering avionics sales, service, installations, piston and turbine aircraft maintenance.

Lafayette Avionics
Hangar 7, Purdue Airport
West Lafayette IN 47906
Ph: 317-743-3828 • Fax: 317-743-0000
E-Mail: ron@avionics-laf.com
Web: http://avionics-laf.com
Literature/Catalog: Yes
Sales, repair & installation of aircraft avionics.

Lockwood Aviation Supply, Inc.
280 Hendricks Way • Suite A
Sebring FL 33870
Ph: 941-655-5100 • Fax: 941-655-6225
E-Mail: lockwood@digital.net
Web: http://www.lockwood-aviation.com
Literature/Catalog: $9.95
Light aircraft engines, parts and accessories.

Lone Star Aviation Corp.
804 N. Great S.W. Pkwy. • Arlington TX 76011
Ph: 817-548-7768 • Fax: 817-633-6208
E-Mail: lsa@iamerica.net
Literature/Catalog: Yes
Noise filters for intercoms, stormscopes, etc. GPS power supplies for home use. Avionic cooling fans.

Mason Electric Company, Inc.
605 8th Street • San Fernando CA 91340
Ph: 800-232-7700 X202 • Fax: 818-365-5505
E-Mail: Cogletree@Mason-electric.com
Web: http://www.Mason-electric.com
Literature/Catalog: Yes
Aviation style headsets, helmets, communications accessories and hardware.

McCoy Avionics Corp.
10761 Watkins Road
Marysville OH 43040-9544
Ph: 800-654-8124 • Fax: 937-642-0220
E-Mail: RPHartigan@juno.com
Web: http://www.McCoyAvionics.com
Literature/Catalog: Yes
Mac 1700C complete Nav/Com slide-in replacement. 760 channel comm. and 200 channel nav.

MicroCom
16141-6 Pine Ridge Road
Ft. Myers FL 33908
Ph: 813-454-6464 • Fax: 813-454-6652
Literature/Catalog: Yes

Microflight Products, Inc.
1356 Oakdale Drive, Suite 23
Griffin GA 30223-3929
Ph: 800-831-2949 • Fax: 770-229-2329
Literature/Catalog: $3.00
Complete line of instruments, gauges, fuel components, electrical parts, hardware, and accessories. Dealer and OEM programs available.

Moravia, Inc.
P.O. Box 942
Thunder Bay, Ontario P7C 4X8 CANADA
Ph: 807-475-5113 • Fax: 807-475-9587
E-Mail: moravia@loon.norlink.net
Web: http://www.norlink.net/moravia
Literature/Catalog: Yes

Narco Avionics, Inc.
270 Commerce Drive
Fort Washington PA 19034
Ph: 800-223-3636 • Fax: 215-643-0197
Literature/Catalog: Yes
Aircraft electronics.

On Final Aircraft Supplies
A Div. of Murphy Aircraft Mtg.
#2-8155 Aitken Road
Chilliwack, BC V2R 4H5 CANADA
Ph: 800-461-7767 • Fax: 604-792-7006
E-Mail: on-final@murphyair.com
Literature/Catalog: Yes

PS Engineering, Inc.
9800 Martel Road • Lenoir City TN 37772
Ph: 423-988-9800 • Fax: 423-988-6619
E-Mail: intercoms@ps-engineering.com
Web: http://www.ps-engineering.com
Literature/Catalog: Yes
Manufacturer of intercoms, audio selector panels, digital recorders, wire harnesses and modified headsets.

Pacific Coast-Avionics
1833 Auburn Way N. • Auburn WA 98002
Ph: 800 353-0370 • Fax: 206-931-0584
E-Mail: PacAvionic@AOL.com
Literature/Catalog: Yes
Full service avionics repair station with sales, service and installations. Complete line of pilot supplies. Specializing in panel package build-up for kit builders.

Peltor, Inc.
41 Commerical Way • East Providence RI 02914
Ph: 401-438-4800 • Fax: 410-434-1708
Literature/Catalog: Yes
Complete line of aviation headsets.

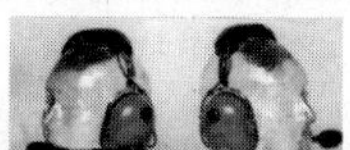

Pilot Avionics
10015 Muirlands Blvd. Unit G
Irvine CA 92618
Ph: 714-597-1012 • Fax: 714-474-4085
Web: http://www.pilot-avionics.com
Literature/Catalog: Yes
GEMINI Self-Contained Communication System. Two head sets with built-in VOX intercom. With superior squelch circuitry, noise cancelling electric microphones, super-cushion headpads, built-in push-to-talk switch. 5 year unconditional warranty. Power by 9 volt battery.
See Display Ad On Page 43

Plantronics
345 Encinal Street • Santa Cruz CA 95060
Ph: 800-544-4660 • Fax: 408-425-8654
Literature/Catalog: Yes
Inventor of the lightweight headset in 1961. Offers six models covering all applications.

Porcine Associates
244 O'Connor Street • Menlo Park CA 94025
Ph: 415-326-2669 • Fax: 415-326-1071
E-Mail: jimham@porcine.com
Web: http://www.porcine.com
Literature/Catalog: Yes
Handheld GPS linked with autopilot. Will work with most handheld GPS receivers. $189.00 complete with manual and connector. Send self addressed stamped envelope for information.

The GLASS GOOSE By Quikkit

Quikkit Corporation
9002 Summer Glen • Dallas TX 75243-7445
Ph: 214-349-0462 • Fax: 214-349-0462
E-Mail: quikkit@airmail.com
Web: http:////webwrks.com/glassgoose/
Literature/Catalog: Yes
Dealer for Wag-Aero radios.
See Display Ad On Page 24

RST Engineering
13249 Grass Valley Avenue
Grass Valley CA 95945
Ph: 916-272-2203 • Fax: 916-272-2203
E-Mail: rst-engr@oro.net
Web: http://www.rst-engr.com
Literature/Catalog: Yes
Supplier of intercoms, headsets, and audio panel kits. Cable, connectors, antenna parts, etc.
See Display Ad On Page 14

Ross Engineering Co.
12505-E Starkey Road • Largo FL 34643
Ph: 800-422-7676 • Fax: 813-535-4248
Literature/Catalog: Yes

Sennheiser Electronic Corp.
6 Vista Drive • P.O. Box 987
Old Lyme CT 06371
Ph: 860-434-9190 • Fax: 860-434-9022
E-Mail: mfeldbauer@sennheiserusa.com
Web: http://WWW.sennheiserusa.com
Literature/Catalog: Yes
Combination headsets and headphones with active noise cancellation.

Skycom Avionics, Inc.
9305 W. Appleton Avenue
Milwaukee WI 53225
Ph: 414-463-8180 • Fax: 414-463-8195
Literature/Catalog: Yes
Aircraft electronics sales, service and installation. Your source for new, rebuilt and recertified packaged instrumentation for every homebuilt aircraft. Free Skycom Avionics Newsletter on request. Branches: Kenosha, Phone 414-652-8120; Waukesha, Phone 414-521-8180; Madison, Phone 608-246-8180.

Softcomm Products, Inc.
2310 S. Airport Blvd. • Chandler AZ 85249-1725
Ph: 602-917-2328 • Fax: 602-917-3557
Web: http://.//gramercy.10s.com/~softcomm
Literature/Catalog: Yes
GA headsets and intercoms, active (ARR) and passive (NRR).

S-TEC Corp.
One S-TEC Way • Municipal Airport
Mineral Wells TX 76067-9236
Ph: 940-325-9406 • Fax: 940-325-3904
E-Mail: kenpaul@flash.net
Web: http://www.s-tec.com/kitplane
Literature/Catalog: Yes
Manufacturer of a complete line of flight control systems, autopilots and flight directors. TEC line avionics - NAV, COM, ADF, DME and transponder. H.S.I. and R.M.I. systems.

Superflite

2149 E. Pratt Blvd. Elk Grove Village IL 60007
Ph: 800-323-0611 • Fax: 708-364-0358
Literature/Catalog: Yes
Supplier of a broad range of parts and accessories for personal aircraft.
See Display Ad On Page 38

Telex Communications, Inc.
9600 Aldrich Avenue S.
Minneapolis MN 55420
Ph: 612-884-4051 • Fax: 612-884-0043
E-Mail: aircraft@telex.com
Web: http://www.telex.com
Literature/Catalog: Yes
Microphones, intercoms, headsets and headphones. Active noise reduction headphones and headsets.

Tropic Aero
1090 NW 53rd Street
Ft. Lauderdale FL 33309
Ph: 305-491-6355 • Fax: 305-772-3327
Literature/Catalog: Yes
Discount portable G.P.S. New and used, buy, sell or trade all makes and models, Loran or G.P.S.

Valentine Aviation
14151 Montfort Dr., Suite 344
Dallas TX 75240-3067
Ph: 214-239-7911 • Fax: 214-385-1069
E-Mail: donaldv@airmail.net
Used avionics, specializing in storm scopes.

Varga Enterprises, Inc.
2350 S. Airport Blvd.
Chandler AZ 85249
Ph: 800-966-6936 • Fax: 602-899-0324
E-Mail: vargaair@vargaair.com
Web: http://www.vargaair.com
Literature/Catalog: Yes
David Clark, Peltor, Softcomm, Sennheiser.

Venture Aerodome
Rt. 1 Box 11631 • Roland OK 74954
Ph: 918-427-6490
Literature/Catalog: $4.00

Vista Aviation, Inc.
12653 Osborne Street • Pacoima CA 91331
Ph: 818-896-6442 • Fax: 818-896-9541
Literature/Catalog: Yes
Avionics shop. Maintenance, parts and equipment. Flight school.

WAG-AERO

The Wag-Aero Group
1216 North Road, Box 181 • Lyons WI 53148
Ph: 800-558-6868 • Fax: 414-763-7595
E-Mail: wagaero-sales@wagaero.com
Literature/Catalog: Yes
David Clark, Flightcom, and Telex headsets.
See Display Ad On Page 13

Watkins Aviation Inc.
15770 Midway Road Hanger #6
Dallas TX 75244
Ph: 972-934-0033 • Fax: 972-960-2564
E-Mail: WATKINS@COMPUTEK:NET
Literature/Catalog: Yes
Military personal flight equipment. Catalog available; free in US, foreign countries $3.00.

Wentworth Aircraft Inc.
3015 Cedar Avenue, S. Minneapolis MN 55407
Ph: 612-722-0065 • Fax: 612-722-1024
Literature/Catalog: Yes
Engines, radios, instruments and miscellaneous single engine used parts.

Westwind
1850 Union St. • San Francisco CA 94123
Ph: 800-456-1665 • Fax: 415-285-3039
Manufacturer of Zephyr padded cordura headset cases and flight bags. Full inventory of headsets and intercoms.

Wings & Wheels
81 Jackson Avenue, W.E.
Jamestown NY 14701
Ph: 716-664-6894 • Fax: 716-664-6895
Literature/Catalog: $4.00
Aircraft supplies, components, and avionics. Buy and sell, new and used. Becker, Icom, Terra, Trimble Magellan, Strong, National Parachutes, and much more.

Rotor Blades

Rotor Wing-Aero
PO Box 18614 • Salt Lake City UT 84118
Ph: 801-969-1357 • Fax: 801-969-1357
E-Mail: mhoskins@aros.net
Construction plans and assembly manual for building a low cost, high quality, efficient rotor blade for a single place gyroplane. Auto cad drawings and sample of the actual rotor blade, $26.00 postage paid within USA, $32.00 outside USA.

Showers-Aero
RD #3, Box 1200 • Milton PA 17847
Ph: 717-742-4080 • Fax: 717-473-1105
Literature/Catalog: $2.00
Rotor blades, main and tail. All aluminum extruded sets include balance and mount plate extrusions. Including NACA 0012 for helicopters and 8-H-12 for gyros.

Sport Copter International, Inc.
Scappoose Airport • 34012 N. Honeyman Road
Scappoose OR 97056
Ph: 503-543-7000 • Fax: 503-543-7041
Web: http://www.sportcopter.com
Literature/Catalog: Yes
22' to 26' rotors with all-aluminum extrusion and skin and 7" chord. Bonded and riveted giving a greater peel strength for hot and humid conditions than is possible with other standard glues. Catalog available for $12.00.

VORTECH, Inc.

PO Box 511-KP • Fallston MD 21047
Ph: 410-692-2692 • Fax: 410-692-5902
Web: http://www.webcom.com/neatstuf/helio
Literature/Catalog: $10.00
Rotor blades for your homebuilt helicopter and gyroplane. With extruded aluminum construction for great strength and reliability. All sizes of both main and tail rotors.
See Display Ad On Page 21

Tires, Wheels and Brakes

Aero Battery Sales & Service
3442 Church Street • Cincinnati OH 45244-3410
Ph: 513-271-9116 • Fax: 513-561-1950

Air Salvage of Arkansas
2854 Hwy 8 East • Mena AR 71953
Ph: 501-394-1022 • Fax: 501-394-7475
Started 32 years ago. Dismantling mostly single engine aircraft, buying inventories, consignment selling of A/C and parts. Housing and storage.

Aircraft Sales & Parts
PO Box 995, Vernon, B.C. V1T 6N2 CANADA
Ph: 250-549-1102 • Fax: 250-549-3769
E-Mail: asap@junction.net
Web: http://www.ultralight.ca

AIRCRAFT SPECIALTIES SERVICES

Aircraft Specialties Services
2860 North Sheridan Road • Tulsa International Airport • Tulsa OK 74115
Ph: 800-826-9252 • Fax: 918-836-4419
Literature/Catalog: Yes
FAA certified repair station D2R764K. Inspection, remachining of aircraft engine parts. Full service, new and used parts.
See Display Ad On Page 74

Aircraft Spruce & Specialty
PO Box 4000 • Corona CA 91718-9961
Ph: 800-824-1930 • Fax: 909-372-0555
E-Mail: info@aircraft-spruce.com
Web: http://www.aircraft-spruce.com
Literature/Catalog: $5.00
A complete selection of homebuilt and certified aircraft parts and materials. Shipping worldwide since 1965. Also see Aircraft Spruce East, Griffin GA, 800-831-2949.
See Display Ad On Page 24

AIRCRAFT SUPER-MARKET

#10 Ellefson Dr. • PO Box 676
De Soto IA 50069
Ph: 515-834-2225 • Fax: 515-834-2152
E-Mail: bobdurly@netins.net
Web: http://www.netins.net/showcase/t-birds
Literature/Catalog: $5.00
Handles recreational aircraft of all kinds. New and used, ultralights, powered parachutes and U/L gliders. 2600' runway.
See Display Ad On Back Cover

Avpro
1400 East South Blvd. • Montgomery AL 36116
Ph: 334-286-4003 • Fax: 334-286-8018
Literature/Catalog: Yes
Manufacturer of new stainless steel and black steel brake discs and brake linings; replacement propeller parts and de-ice boots.

Azusa Engineering, Inc.
1542 Industrial Park St.
Covina CA 91722-3487 • Ph: 818-967-4167
Fax: 818-966-4071
E-Mail: azeng@earthlink.net

B.O.A.C.
P.O. Box 228 • Needham MA 02192
Literature/Catalog: Yes

Bradley Aerospace

950 Muir Ave. • Chico CA 95926
Ph: 916-899-7918 • Fax: 916-899-2942
Web: http://www.kgmsoftworks.com/bradley
See Display Ad On Page 40

Ken Brock Manufacturing, Inc.
11852 Western Avenue • Stanton CA 90680
Ph: 714-898-4366 • Fax: 714-894-0811
Literature/Catalog: $5.00
Stamped, machined, and welded sub-assemblies. Short-run stamping facilities.

Cleveland Wheels & Brakes
1160 Center Rd. • Avon OH 44011
Ph: 800-272-5464 • Fax: 216-937-5409
E-Mail: cleveland.tech@parker.h.com
Web: http://www.parker.com
Literature/Catalog: Yes
"Cleveland" brand wheels and brakes. In-house brake shop.

Condor Aircraft Tires
9700 Research Drive • Charlotte NC 28262
Ph: 704-548-2400 • Fax: 704-548-2599
Literature/Catalog: Yes
Aircraft tires and tubes.

Desser Tire & Rubber Co.
1230 Kansas Street • Memphis TN 38106
Ph: 800-247-8473 • Fax: 901-774-9043
Distributor of Goodyear, Michelin, Condor and McCreary aircraft tires and tubes.

Desser Tire & Rubber Co.
6900 Acco Street • P.O. Box 1028
Montebello CA 90640
Ph: 800-247-8473 • Fax: 213-721-7888
E-Mail: airtire@aol.com
Distributor of Goodyear, Michelin, Condor & McCreay aircraft tires and tubes.

B.F. Goodrich-General Aviation
250 N. Cleveland-Massillon Rd.
Akron OH 44313-0501

Great Airplanes, Inc.
P.O. Box 615
Needham Heights MA 02194-0005
Ph: 617-444-5480 • Fax: 617-444-5480
E-Mail: hswack@aol.com
Web: http:////aso.solid.com/greatair
Literature/Catalog: Yes
Cleveland TSO'd aircraft wheels and brakes. Send stamped addressed envelope for price and information.

Haigh Aviation Corp
4231 Sepulveda • San Bernardino CA 92404
Ph: 909-883-8488

Harbor Products Co.
2390 Crenshaw Blvd., Ste. 164
Torrance CA 90501
Ph: 310-326-5609 • Fax: 310-560-7416
Literature/Catalog: Yes
Manufacture of high quality composite wheel pants, for antique classics, homebuilts and ultralight/microlight aircraft.

Hawkins Aircraft Tires
PO Box 1725 • El Cajon CA 92022
Ph: 800-321-9717 • Fax: 619-449-2941
Literature/Catalog: Yes

Kenmore Air Harbor, Inc.
Box 82064 • Kenmore WA 98028-0064
Ph: 206-486-1257
A full service charter Seaplane operation.

Light Speed Engineering
PO Box 549 • Hangar #15
Santa Paula CA 93060
Ph: 805-933-3299 • Fax: 805-525-0199
E-Mail: LSE@west.net
Literature/Catalog: Yes
Fiberglass & epoxy wheel pants. Small, medium and large sizes.

Linda Lou, Inc.
3514 Winhoma • Memphis TN 38118
Ph: 800-824-9912 • Fax: 901-365-9524

Matco Mfg., Inc.
65 E. Kensington Avenue
Salt Lake City UT 84115
Ph: 801-486-7574 • Fax: 801-486-7581
E-Mail: tech@matcomfg.com
Web: http://www.matcomfg.com
Literature/Catalog: Yes
Wheel and brake assemblies, parking brake valves, master cylinders, hydraulic cylinders.

McCreary Tire & Rubber Co.
P.O. Box 749 • Indiana PA 15701
Ph: 412-349-9010
Literature/Catalog: Yes

Michelin Aircraft Tire Corp.
9700 Research Drive • Charlotte NC 28262
Ph: 704-548-2400 • Fax: 704-548-2599
Literature/Catalog: Yes
Aircraft Tires.

Microflight Products, Inc.
1356 Oakdale Drive, Suite 23
Griffin GA 30223-3929
Ph: 800-831-2949 • Fax: 770-229-2329
Complete line of instruments, gauges, fuel components, electrical parts, hardware, and accessories. Dealer and OEM programs available.

Motor Sports Supply
4742 E. Speedway • Tucson AZ 85712
Ph: 602-326-3103
Literature/Catalog: Yes

Tracy O'Brien Aircraft Supplies
156 Bunker Creek Road • Chehalis WA 98532
Ph: 360-748-4089 • Fax: 360-740-0504
E-Mail: tracy@localaccess.com
Literature/Catalog: Yes
5", 6", and 8" cast aluminum wheels with 5/8", 3/4", and 7/8" bearings. Hydraulic disk brakes and mechanical drum brakes. Bolt-on and weld-on axels in 5/8", 3/4", and 7/8" diameters. 5" tailwheel kits. 5", 6", and 8" tires.

Parker Hannifin Corp.
Cleveland Wheels & Brakes • 1160 Center Road
Avon OH 44011
Ph: 216-937-6211 • Fax: 216-937-5409
Literature/Catalog: Yes
Manufacturer of aircraft wheels, brakes, master cylinders, parking valves, and hydraulic systems and components.

The GLASS GOOSE By Quikkit

Quikkit Corporation
9002 Summer Glen • Dallas TX 75243-7445
Ph: 214-349-0462 • Fax: 214-349-0462
E-Mail: quikkit@airmail.com
Web: http:////webwrks.com/glassgoose/
Literature/Catalog: Yes
See Display Ad On Page 24

Topflight Design Group, Inc.
29 Sand Hill Road • P.O. Box 86
Hampton CT 06247-0086
Ph: 860-455-9298 • Fax: 860-455-9298
Compact cast aluminum pitot-static head; in development steerable/swiveling tail wheel assembly. Others pending.

WAG-AERO

The Wag-Aero Group
1216 North Road, Box 181 • Lyons WI 53148
Ph: 800-558-6868 • Fax: 414-763-7595
E-Mail: wagaero-sales@wagaero.com
Literature/Catalog: Yes
Complete line of specialty product tires; Goodyear, Cleveland wheels and brakes, brake cylinders, etc.
See Display Ad On Page 13

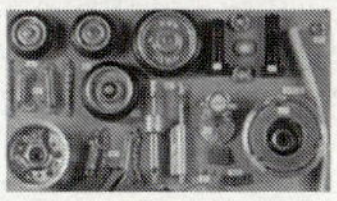

Wicks Aircraft Supply
410 Pine Street • Highland IL 62249
Ph: 800-221-9425 • Fax: 618-654-6253
Literature/Catalog: $5.00
Worldwide supplier for homebuilt projects specializing in raw material kits for 40+ different designs. 350 page catalog with over 12,000 items.
See Display Ad On Page 46

Wilco, Inc.
3502 W. Harry • Witchita KS 67213
Ph: 800-767-7593 • Fax: 316-943-9664
E-Mail: wilco@feist.com
Web: http://www.globalair.com/wilco/
Condor tires, wheel halves, wheel assemblies, brake assemblies and brake linings.

Tools

Aces Systems
10737 Lexington Drive
Knoxville TN 37932-3294
Ph: 423-966-5856 • Fax: 423-675-1241
E-Mail: Jerryaces@aol.com
Literature/Catalog: Yes
Full spectrum vibration analyzers, fan trim balance systems, dynamic digital propeller balancers, engine performance monitoring, trend analysis software, performance software.

Action Air Parts, Inc.
106 Airport Drive • Port Huron MI 48074
Ph: 810-364-5885 • Fax: 810-364-5885
Literature/Catalog: Yes
Supplier of an extensive line of aircraft sheet metal tools.

Aircraft Spruce & Specialty Since 1965

Aircraft Spruce & Specialty
PO Box 4000 • Corona CA 91718-9961
Ph: 800-824-1930 • Fax: 909-372-0555
E-Mail: info@aircraft-spruce.com
Web: http://www.aircraft-spruce.com
Literature/Catalog: $5.00
A complete selection of homebuilt and certified aircraft parts and materials. Shipping worldwide since 1965. Also see Aircraft Spruce East, Griffin GA, 800-831-2949.
See Display Ad On Page 24

Aircraft Tool & Supply Co.
P.O. Box 370 • 1000 Old US-23
Oscoda MI 48750
Ph: 517-739-1447 • Fax: 517-739-1448

Airparts, Inc.
2400 Merriam Lane • Kansas City KS 66106
Ph: 800-800-3229 • Fax: 913-831-1780
Literature/Catalog: Yes
Aluminum sheet, tubing, angle and bar. 4130 tubing and sheet.

All Fab Corp.
1235 1/2 Lincoln Road • Allegan MI 49010
Ph: 616-673-6572 • Fax: 616-673-1644

American Tool Companies
8400 Lakeview Parkway • Suite 400
Kenosha WI 53142
Ph: 800-866-5740 • Fax: 937-382-8199
Web: http://americantool.com
Hand tools & power tool accessories.

Apollo Safety & Industrial
1108 West 9th Street
Upland CA 91786-5703
Ph: 909-980-7071 • Fax: 909-985-6873

Avery Tools
2290 W. Hicks Rd. • Hangar 54-1 Unit #43
Ft. Worth TX 76131-3743
Ph: 800-652-8379 • Fax: 817-439-8402
E-Mail: averytools@why-net
Web: http://www.averytools.com
Literature/Catalog: Yes
Aircraft sheet metal tools: rivet, dimpling and hand tools. Aircraft maintenance publications and videos.

B & F Aircraft Supply, Inc.
22038 Howell Dr., Unit 6
New Lennox IL 60451
Ph: 815-462-8696 • Fax: 815-462-3697
Literature/Catalog: $1.00

BAKER'S School of Aeronautics
1645 Murfreesboro Road • Suite K
Nashville TN 37217
Ph: 615-361-6787 • Fax: 615-361-0010
E-Mail: KINGSSCH@Aol.com
Literature/Catalog: Yes
Graco/Croix HVLP spray equipment. Lightweight and portable. Call for free brochure.

BLUE RIDGE MACHINERY AND TOOLS, INC.
Your Complete Machine Shop Supplier!

Blue Ridge Machinery & Tools
P.O. Box 536 • 2905 Putnam Avenue
Hurricane WV 25526
Ph: 304-562-3538 • Fax: 304-562-5311
Literature/Catalog: Yes
Machine shop equipment and tooling for the home-shop machinist. Lathes, mills, drill rods and more. How-to videos and books.
See Display Ad On Page 22

Bondhus Corp.
P.O. Box 660 • 1400 E. Broadway
Monticello MN 55362
Ph: 800-328-8310 • Fax: 612-295-4440
Literature/Catalog: Yes

Bosch Power Tools
4300 W. Peterson • Chicago IL 60646
Ph: 773-286-7330 • Fax: 773-481-3654
Web: http://www.boschtools.com
Literature/Catalog: Yes
Power tools and accessories.

Brown Aviation Tool Supply Co.
5700 North Rockwell • Building 2, Suite A
Bethany OK 73008
Ph: 800-587-3883 • Fax: 405-495-4992
E-Mail: browntool@aol.com
Literature/Catalog: Yes
Supplier of new, used and surplus aircraft assembly tools. Large selection of air, hand and cutting tools unique to the aiviaton industry.

California Power Systems
790 139th Avenue, #4
San Leandro CA 94578
Ph: 800-247-9653 • Fax: 510-357-4429
Web: http://www.800-airwolf.com
Literature/Catalog: $6.95
Complete supplier of ultralight and sport aircraft parts and components. Rotax engines, parts and service. 500 page catalog, $6.95 US, $10.00 Canada and $20.00 foreign airmail.
See Display Ad On Page 69

Campbell Aircraft
P.O. Box 610518-A • San Jose CA 95161-0518
Ph: 408-251-3109
Literature/Catalog: Yes

Campbell Tools Co.
2100 Selma Road • Springfield OH 45505
Ph: 937-322-8562 • Fax: 937-322-8344
E-Mail: mackeen@islandnet.com
Literature/Catalog: $2.00
Lathes, Mills, Tools

Cleveland Tool and Supply
1804 First St., • Boone IA 50036
Ph: 800-368-1822 • Fax: 515-432-7804

CMI Tool Group
26290 W. Eight Mile Road • Southfield MI 48034

Dale Wilch Sales & Mfg.

PO Box 12031 • Kansas City KS 66112
Ph: 800-642-3288 • Fax: 913-788-9682
E-Mail: dale@dalewilch.com
Web: http://www.dalewilch.com
Literature/Catalog: Yes
"Ol Joint Jigger" tubing and pipe matching fixtures for use in fabricating with steel or aluminum tubing.
See Display Ad On Page 29

Daytona MIG, Inc.
1821 Holsonback Drive
Daytona Beach FL 32117
Ph: 800-331-9353 • Fax: 904-274-1237
E-Mail: daytonamig@aol.com
Web: http://www.daytonamig.com
Welders and Plasma Cutters.

DJB Aluminum & Supply
3151 Lucas Drive • Bethlehem PA 18107
Ph: 610 866-6913
Literature/Catalog: $2.00
Jig Kits, RV-4/6/6A. RV-4 Retro-Fit Kit converts RV-6 Jig to RV-4. Hydraulic Hand Rivet Squeezer features: quick advance, auto retract, reversible handle for bench mount operation 360 degree swivel yoke allows for left or right handed operation. An-bolts, clamps, shackles, rod ends.

Electronic Industries, Inc.
19 E. IrvingAvenue • P.O. Box 266
Oshkosh WI 54902
Ph: 414-235-8930 • Fax: 423-235-4233

Falcon Ridge Industries
51 N. Hoagland Blvd. • Kissimmee FL 34741
Ph: 407-396-7572 • Fax: 407-396-1696
Drill jig block - bolt, drill jig block - head, drill jig block - combination bolt and head. Bearing removal and installation tools. Sheet metal flanging dies. Air lines. Rivet squeezing equipment. English wheel kits.

Fastening Systems Int'l.
P.O. Box 1372 • Sonoma CA 95476
Ph: 800-344-2393 • Fax: 707-935-1828
Literature/Catalog: $1.00
Hydraulic, electric and battery operated blind riveters and nutsetters. Up to 5,325 pounds of force. Complete inventory of blind fasteners by Cherry, Avdel, Gesipa/Olympic, Huck Intl., and Atlas Engineering.

Forbes Metal Stamping Co.
26555 Golden Valley Rd.
Saugus CA 91350
Ph: 805-250-1507 • Fax: 805-250-9406
E-Mail: fms76@aol.com
Literature/Catalog: Yes
Avionics hardware. Mic & phone jack mounting, instrument reducers, pilot & mechanics supplies. Wheel chock. Metal stamping of all kinds.

Frey Engineering & Development
316 Turner Lane • West Chester PA 19380
Ph: 610-692-3553 • Fax: 610-692-9138
RV-3, RV-4 RV-6/6A precision steel jig systems. Options and accessories for the assembly and empennage of wing and fuselage kits.

Garage Built Technologies
Box 271 Dept. AC • Wilmot OH 44689-0271
Plans for building a hot wire styrofoam cutter for composite aircraft parts, models, signs and molds, hand held or stationary.

GLB Welding & Equipment
245 Oliver Rd. • Meansville GA 30256
Ph: 800-258-0126 • Fax: 770-567-5424
Literature/Catalog: Yes
Henrob oxyacetylene torch, Harris Calorific and Victor welding equipment. Abrasives, tubing notchers, aluminum flux, rod welding, tube bender, sheet metal press, regulators, supplies, books, etc. VHS video.

Heins Balancing Systems, Inc.
820 Railroad Avenue • Santa Paula CA 93060
Ph: 805-525-5445 • Fax: 805-525-7074
Manufacturer of balancing machines and tooling for overhauling a wide range of turbine and reciprocating engines and components.

Helsper Sewing Co., Inc.
994 Crosby Avenue • Sycamore IL 60178
Ph: 815-895-8700 • Fax: 815-895-8579
Manufacturer of canvas tool bags and wrench pouches.

Instruments & Flight Research
2716 George Washington Blvd.
Wichita KS 67210-1585
Ph: 316-684-5177 • Fax: 316-684-0140
Literature/Catalog: Yes
Manufacture flight and engine instruments and accessories.

J. Lough
172 Boniface Avenue
Kitchener, Ontario N2C 1L9 CANADA
Ph: 519-578-0968
Literature/Catalog: $5.00
Dandy Dimple Dies; hole flanging dies; bending brake drawings.

J-AIR, INC.
24400 Highland Road, Ste. 62
Richmond Hts. OH 44143
Ph: 216-692-0010 • Fax: 216-692-0073
Literature/Catalog: Yes
Instruments/tools, gauges-press./vac. measure, test, certify, repair.

The Jilson Corp.
20 Industrial Road • Lodi NJ 07644
Ph: 201-471-2400 • Fax: 201-471-3993
Literature/Catalog: Yes

Kearns Metal Works
8745 Highway 13 West • Carrier Mills IL 62917
Ph: 618-252-2044
Literature/Catalog: Yes
Dimple Dies, Rivet Sets, Pneumatic Rivet Dimplers, Metal Brake Plans. Nippondenso 35 amp alternators and other riveting tools.

Kett Tool Co.
5055 Madison Road
Cincinnati OH 45227-1494
Ph: 513-271-0333 • Fax: 513-271-5318
Literature/Catalog: Yes
Manufacturer of portable saws, shears and nibblers.

Leatherman Tool Group, Inc.
P.O. Box 20595 • Portland OR 97294
Ph: 503-253-7826 • Fax: 503-253-7830
Web: http://www.leatherman.com
Literature/Catalog: Yes
Multi-purpose folding hand tool.

LEM, Inc.
888 Home Avenue • Akron OH 44310
Ph: 216-535-6422 • Fax: 216-535-6422
Literature/Catalog: Yes
Original manufacturer of the BF Goodrich (R) hand tools. Rivet-Nut tools, spiral action, wrench type, and Keyway cutter for keyed fasteners.

Lex-Aire Spray Systems
34 Hutchinson Road • Arlington MA 02174
Ph: 800-537-2473 • Fax: 617-646-2827
Literature/Catalog: Yes
Supplier of spray painting equipment. Complete line of HVLP turbine systems. Full line of conversion guns usable with as small as 1.5 hp compressors. VHS video available for $7.50 U.S.

Williams Lowbuck Tools
4175 California Avenue • Norco CA 91760
Ph: 800-735-7844 • Fax: 909-735-1210
E-Mail: wlowbuck@aol.com
Literature/Catalog: $3.00
Metal fabrication tools for tubing and sheet metal.

Michael Engineering, Ltd.
4997 S. Crawford Road • Mt. Pleasant MI 48858
Ph: 517-772-4073 • Fax: 517-772-3500
Literature/Catalog: Yes
Epoxy ratio pumps, epoxy metering equipment.

The Milbar Company
530 E. Washington Street
Chagrin Falls OH 44022
Ph: 216-247-4600 • Fax: 216-247-3998
Literature/Catalog: Yes
Safety wire twisters, electrical stripping and crimping pliers, retaining ring pliers.

Motorsports Tools
Route 2 • Box 366 • MacClenny FL 32063
"English Wheel" metal shaping tools. 2 hr. VHS video on metal shaping techniques by John Glover - $39.95 plus $3.95 S&H.

NASSCO
3335 E. Deseret Dr. • PO Box 84771
St. George UT 84771
Ph: 800-8NASSCO • Fax: 435-652-1313
Plated bond diamond tools for abrasive machining composites

Wil Neubert Aircraft Supply
1488 St. Rte. 162 • Sprakers NY 12166
Ph: 518-922-5830 • Fax: 518-436-3632
Literature/Catalog: $5.00

Northern Michigan Enterprises
P.O. Box 6321 • Traverse City MI 49685-6321
Ph: 616-946-1071
Simple, easy to follow plans for building your own shop tools such as a tubing cutter/notcher, an english wheel for forming compound curves in sheet metal, a sand blast cabinet, a motorized ramp scooter ...and more. $7.50 ea., 3/$20, 5/$30.

Omni Manufacturing Co., Inc.
12961 Sunnyside Place
Santa Fe Springs CA 90670
Ph: 800-669-6213 • Fax: 562-944-3177
Literature/Catalog: Yes
Manufacturer of aircraft-quality cutting tools. Tight Fit Tools & Kits for close-quarter work. Allows repairs to be made quickly without having to disassemble other components.

Regale Limited, U.K.
27 Whistler Road, Eaton Ford
St. Neots, Cambridgeshire PE19-3RT
UNITED KINGDOM
Ph: 44-480-477975 • Fax: 44-480-477975
British & European military aircraft parts. Support & major items for BAC, Rolls Royce, Marine, BAE, and much more. $10,000,000 in stock.

Regale Limited, U.S.A.
RT. 1, Box 138A • Peculiar MO 64078
Ph: 816-779-6075 • Fax: 816-758-6054
British & European military aircraft parts. Support & major items for BAC, Rolls Royce, Marine, BAE, and much more. $10,000,000 in stock.

Right Angle Tool Co.
200 South Normal • Macomb IL 61455
Ph: 309-833-1311 • Fax: 309-833-1311
E-Mail: bob2@netins.net
Literature/Catalog: Yes
Machine tools and metal forming equipment.

Ralph C. Robinson Company
3449 Oceanview Blvd. • Glendale CA 91208
Ph: 800-900-9942 • Fax: 818-249-2084
Wire twisters, retaining ring pliers, electrical crimpers and strippers.

ROMA *Machine Tool & Die, Inc.*
Manufacturer of Bellcrank Bearings since 1968

Roma Machine Tool & Die, Inc.
1990 East Maple Street
Kankakee IL 60901-5720
Ph: 815-937-1245 • Fax: 815-937-1285
Literature/Catalog: Yes
Manufacturersof bellcrank bearings, BC4W10 and BC5W11 propeller reductionspeed units, weld and bolt on axels, retractable landing gear, low wing, tubing fabrication, custom control panels, custom machine components for tail rotor on helicopters, variable pitch control and propeller hubs.
See Display Ad On Page 42

Skybolt Aeromotive Corp.
551 N. Park Avenue • Apopka FL 32712
Ph: 407-889-2613 • Fax: 407-889-8103
E-Mail: skybolt@mindspring
Web: http://skybolt.com
Literature/Catalog: Yes
Composite manufacturing and fabrication. Aircraft hardware and accessories.

Smithy Co.
170 Aprill Drive • PO Box 1517
Ann Arbor MI 48106
Ph: 800-345-6324 • Fax: 313-913-6663
Literature/Catalog: Yes
Manufacturer of multi-purpose power lathe, mill and drill. Three machines in one. Machine fits conveniently into home shop.

Superflite
2149 E. Pratt Blvd. • Elk Grove Village IL 60007
Ph: 800-323-0611 • Fax: 708-364-0358
Literature/Catalog: Yes
Supplier of a broad range of parts and accessories for personal aircraft.
See Display Ad On Page 38

Tatco, Inc.
P.O. Box 277 • 4145 W. Marine Avenue
Lawndale CA 90260
Ph: 310-644-8721 • Fax: 310-644-0204
Aerospace and commercial aircraft tooling. Pneumatic tools, hand tools, riveting, drill and forming equipment. Sales, service and repair.

Tensitron, Inc.
1668 Valtec Lane • Boulder CO 80301
Ph: 303-444-5383 • Fax: 303-444-0204
Literature/Catalog: Yes
Aircraft cable tension meters. Five models available.

Tubular Dynamics
1207 NE Cookingham Drive
Kansas City MO 64155
Ph: 816-734-8211 • Fax: 816-734-8211
Literature/Catalog: Yes
Manufacturers of the original Jig-A-Joint and Jig-A-Joint2 tubing notchers.

U.S. Industrial Tool & Supply Company

U.S. Industrial Tool & Supply
15101 Cleat St. • Plymouth MI 48170-6098
Ph: 800-521-4800 • Fax: 313-455-3256
E-Mail: info@ustool.com
Web: http://www.ustool.com
Literature/Catalog: Yes
Manufacturers and distributors of sheet metal tools and equipment. Specializing in riveting, drilling, and assorted sheet metal tools.
See Display Ad On Page 1

USATCO / US Airtool Int'l.
60 Fleetwood Court • Dept. 106
Ronkonkoma NY 11779-6951
Ph: 516-471-3300 • Fax: 516-471-3308
Literature/Catalog: Yes
Manufacturers and distributors of sheet metal and composite tools for aircraft construction and maintenance.

USATCO International
16301 So. Broadway • Gardena CA 90248
Ph: 310-323-1953 • Fax: 310-329-8275
Web: http://www.usatco.com
Suppliers of sheet metal cutting and riveting tools.

V.R. Engineering
2149 East 5th St. #103 • Tempe AZ 85281
Ph: 602-903-3625 • Fax: 602-967-3391
"Center it" tube drilling V-block for drilling holes in tubes, pipes and round stock. When used in drill presses, holes are centered automatically without center-punching. Can also be used for cutting. $15.00.

WAG-AERO

The Wag-Aero Group
1216 North Road, Box 181
Lyons WI 53148
Ph: 800-558-6868 • Fax: 414-763-7595
E-Mail: wagaero-sales@wagaero.com
Literature/Catalog: Yes
Hand, powered, and shop tools.
See Display Ad On Page 13

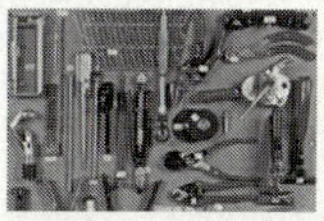

Wicks Aircraft Supply
410 Pine Street • Highland IL 62249
Ph: 800-221-9425 • Fax: 618-654-6253
Literature/Catalog: $5.00
Worldwide supplier for homebuilt projects specializing in raw material kits for 40+ different designs. 350 page catalog with over 12,000 items.
See Display Ad On Page 46

Wiha Tools
1348 Dundas Circle • Monticello MN 55362
Ph: 800 494-6104 • Fax: 800 494-7456
E-Mail: cservice@wihatools.com
Web: http://www.wihatools.com
Literature/Catalog: Yes
Precision screwdrivers, antistatic-conductive screwdrivers, multi-bit screwdriver sets.

Wilke Machinery Co.
3230 Susquehanna Trail
York PA 17404
Ph: 717-764-5000 • Fax: 717-764-3778
E-Mail: wilkemac@ptd.net
Web: http://www.wilkemach.com
Literature/Catalog: Yes
Woodworking and metalworking machinery and power tools. Home shop through industrial applications. Catalog $2.00.

Wood-Met Services, Inc.
3314 W. Shoff Circle Dept. KP
Peoria IL 61604-5464 • Ph: 309-637-9667
Literature/Catalog: $1.00
Plans to build shop tools and accessories. Plans from complete machines and attachments to hand tools for home and shop. Accessories for wood and metal cutting lathes, drills, saws and sanders.

Trailers

AIRCRAFT SUPER-MARKET
#10 Ellefson Dr. • PO Box 676
De Soto IA 50069
Ph: 515-834-2225 • Fax: 515-834-2152
E-Mail: bobdurly@netins.net
Web: http://www.netins.net/showcase/t-birds
Literature/Catalog: $5.00
Handles recreational aircraft of all kinds. New and used, ultralights, powered parachutes and U/L gliders. 2600' runway.
See Display Ad On Back Cover

ROMA Machine Tool & Die, Inc.
Manufacturer of Bellcrank Bearings since 1968

Roma Machine Tool & Die, Inc.
1990 East Maple Street
Kankakee IL 60901-5720
Ph: 815-937-1245 • Fax: 815-937-1285
Literature/Catalog: Yes
Manufacturersof bellcrank bearings, BC4W 10 and BC5W11 propeller reductionspeed units, weld and bolt on axels, retractable landing gear, low wing, tubing fabrication, custom control panels, custom machine components for tail rotor on helicopters, variable pitch control and propeller hubs.
See Display Ad On Page 42

Skyline Aviation
8298 Reid Road, Hangar #11
Swartz Creek MI 48473
Ph: 810-635-8724 • Fax: 810-635-2553
Literature/Catalog: Yes
Tail-Dragger Dragger (tm). An easy to use dolly for moving any tail wheel airplane. Works on most nose wheels also. Plans, kits, or complete.

TERF Inc.
11411 Culver • Brighton MI 48114
Ph: 810-229-2082 • Fax: 810-229-2082
E-Mail: Mtait@livgonline.com
Web: http://www.terf.com
Literature/Catalog: Yes
Airplane trailers
See Display Ad On Page 52

Wells Cargo, Inc.
P.O. Box 728 • Elkhart IN 46514
Ph: 219-264-9661 • Fax: 219-264-5938
Web: http://www.wellscargo.com
Literature/Catalog: Yes
Enclosed Industrial & Commercial Trailers.

Publications, Books

Charles Ackerman
1351 Cottontail Lane • La Jolla CA 92037
Ph: 619-454-4518
Literature/Catalog: $10.00
28 page brochure on VW conversions for aviation use.

Acro Sport, Inc.
PO Box 462 • Hales Corners WI 53130
Ph: 414-529-2609
"The Techniques of Aircraft Building" by Paul H. Poberezny. Builders manual for any custom built aircraft. $16.00 plus $2.50 for shipping and handling.

AeroElectric Connection
6936 Bainbridge Road
Wichita KS 67226-1008
Ph: 316-685-8617
E-Mail: nuckolls@aeroelectric.com
Web: http://www.aeroelectric.com
Literature/Catalog: Yes
The AeroElectric Connection is an electrical systems handbook for individuals who build and maintain their own airplanes. The AeroElectric Connection is adding a line of supplies and accessories to aid builders in assembling their projects. Brochure available for #10 SASE.

Aircraft Designs, Inc.
5 Harris Court, Bldg. S • Monterey CA 93940
Ph: 408-649-6212 • Fax: 408-649-5738
Literature/Catalog: $12.00
Aircraft design and engineering books. VHS video available for $25.00, PAL video for $58.

Aircraft Spruce & Specialty
PO Box 4000 • Corona CA 91718-9961
Ph: 800-824-1930 • Fax: 909-372-0555
E-Mail: info@aircraft-spruce.com
Web: http://www.aircraft-spruce.com
Literature/Catalog: $5.00
A complete selection of homebuilt and certified aircraft parts and materials. Shipping worldwide since 1965. Also see Aircraft Spruce East, Griffin GA, 800-831-2949.
See Display Ad On Page 24

AVIATION BOOK COMPANY
Aviation Provisioner Since 1964

Aviation Book Company
7201 Perimeter Rd. So., Ste. C • Boeing Field
Seattle WA 98108
Ph: 206-767-5232 • Fax: 206-763-3428
Literature/Catalog: Yes
Aviation books and videos, pilot supplies, and gifts
See Display Ad On Page 8

Aviation Publishers Company Limited
Box 1361 Station B
Ottawa Ontario K1P 5R4 CANADA
Ph: 613-745-2943 • Fax: 613-745-9851
E-Mail: aviationpub@igs.net
Web: http://www.aviationpublishers.com
Literature/Catalog: Yes
"From The Ground Up" 50th Anniversary edition. Latest technological and legislative aeronautical matters, $33.95. "From The Ground Up Workbook", $17.95. "Canadian Private Pilot Answer Guide" license exam review, $13.95.

Aviation Supplies & Academics
7005 132nd Pl. SE • Newcastle WA 98059
Ph: 206-235-1500 • Fax: 206-235-0128
E-Mail: 102105.223@compuserve.com
Literature/Catalog: Yes
Publishes aviation books for pilots and aviation maintenance technicians.

B & F Aircraft Supply, Inc.
22038 Howell Dr., Unit 6 • New Lennox IL 60451
Ph: 815-462-8696 • Fax: 815-462-3697
Literature/Catalog: $1.00

THRUST = DRAG
Bates Engineering
2742 Swansboro Road
Placerville CA 95667
Ph: 530-622-1886 • Fax: 530-622-1886
E-Mail: bateseng@innercite.com
Web: http://www.innercite.com/~bateseng
Literature/Catalog: Yes
Airplane and propeller design services. Programs for aircraft design, performance, stability and control. Propeller design optimization software.
See Display Ad On Page 82

BLUE RIDGE MACHINERY AND TOOLS, INC.
Your Complete Machine Shop Supplier!

Blue Ridge Machinery & Tools
P.O. Box 536 • 2905 Putnam Avenue
Hurricane WV 25526
Ph: 304-562-3538 • Fax: 304-562-5311
Literature/Catalog: Yes
Machine shop equipment and tooling for the home-shop machinist. Lathes, mills, drill rods and more. How-to videos and books.
See Display Ad On Page 22

C&R Aircraft
PO Box 281 • Quincy IL 62306-0281
Ph: 217-292-5967
Collector prints of 1930's Air Racers & "Heavenly Body" B-25.

C.R. Supply
14812 Washington St.
Omaha NE 68137-2552
Ph: 719-471-0110
Literature/Catalog: $0.50
6th Edition Modification/Maintenance Manual RV-3,4,6,6A, 102 pages. 6th Edition Flight Manual RV-3,4,6,6A, 46 pages. Not affilated with or endorsed by VAN'S Aircraft.

California Power Systems
790 139th Avenue, #4
San Leandro CA 94578
Ph: 800-247-9653 • Fax: 510-357-4429
Web: http://www.800-airwolf.com
Literature/Catalog: $6.95
Complete supplier of ultralight and sport aircraft parts and components. Rotax engines, parts and service. 500 page catalog, $6.95 US, $10.00 Canada and $20.00 foreign airmail.
See Display Ad On Page 69

Campbell Aircraft
P.O. Box 610518-A • San Jose CA 95161-0518
Ph: 408-251-3109
Literature/Catalog: Yes

Eric Clutton Co.
913 Cedar Lane • Tullahoma TN 37388
Ph: 615-455-2256
"Propeller Making for the Amateur" $8.00. Now retired from 40 years of propeller making, this "how-to" book is used by most commercial prop makers plus many hundreds of homebuilders.

Coot-Builder's Newsletter
6958 Applewood Drive • Madison WI 53719
Ph: 608-833-5586 • Fax: 608-263-9167
E-Mail: rsteeves@facstaff.wisc.edu
"Coot Story" $32.00.

Essco, Inc.
426 W. Turkeyfoot Lake Rd., Akron OH 44319
Ph: 330-644-7724 • Fax: 330-644-0886
Literature/Catalog: $1.00
Pilot, aircraft and avionics manuals. Civilian and military, new and used. Reproduction of most older aircraft, engine, propeller and avionics manuals.

Richard Finch Engineering
P.O. Box 3946 • Santa Barbara CA 93130-3946
Ph: 805-683-5346 • Fax: 805-683-8868
Literature/Catalog: Yes
Books on Turbocharging, Fire Safety, Welding, Auto Engine Conversions for Aircraft. Finchbird V-6 airplane plans.
See Display Ad On Page 32

Fisser Services
1127 SE 226th Avenue • Gresham OR 97030
Ph: 503-667-2414 • Fax: 503-667-2414
Literature/Catalog: Yes
Assembly manual to "Build Your Own Action and Motion Simulator". Use any computer, joy stick required. Add to the realism of your flight simulation. I.F.R. jet flighter combat or just for plane fun. Manual: $39.95. For info, send self addressed stamped envelope.

Flight Physics
P.O. Box 20919 • Billings MT 59104
Ph: 406-248-2606 • Fax: 406-245-9076
E-Mail: jlowry@mcn.net
Literature/Catalog: Yes
"The Bootstrap Approach" to propeller-airplane flight performance lets you fly the airplane for an hour or two, collecting data, and then calculate flight performance specifications.

Fly For Fun
71 South Shore Dr. • Cuba NY 14727
Ph: 716-968-3758 • Fax: 716-968-3758
Literature/Catalog: Yes
Books on aerobatics.

Flying Flea Archive USA
P.O. Box 892 • Wooster OH 44691-0892
Ph: 330-377-5013 • Fax: 330-377-5013
Literature/Catalog: Yes
Flying Flea Archival Material, Vol.1 $20. Technical Notes for the Amateur $10. Construction Manuals and Plans: HM-14 $45, HM-16 $125. Information Package: $5.

GA Airfoils
416 Riblett Lane • Wilmington DE 19808
Ph: 302-994-0479
Literature/Catalog: $1.00
"GA Airfoils", a catalog of new, safer airfoils with performance data." Available from EAA bookshop, 800-843-3612.

GLB Welding & Equipment
245 Oliver Rd. • Meansville GA 30256
Ph: 800-258-0126 • Fax: 770-567-5424
Literature/Catalog: Yes
Books on welding.

Marv Golden Discount Sales
8690 Aero Drive, Ste. 102
San Diego CA 92123
Ph: 800-348-0014 • Fax: 619-569-4508
Web: http://www.MARVGOLDEN.COM
Literature/Catalog: Yes
Test books. Various aviation titles.

Great Plains Aircraft
P.O. Box 545 • Boys Town NE 68010
Ph: 402-493-6507 • Fax: 402-333-7750
E-Mail: gpasc@earthlink.net
Web: http://www.greatplainsas.com
Literature/Catalog: $3.00
See Display Ad On Page 80

Historic Aviation Books
1401 Kingswood Road • Eagan MN 55122
Ph: 800-225-5575

Homebuilt Book Supply
PO Box 8351 • Scottsdale AZ 85252-8351
Ph: 602-990-3328 • Fax: 602-990-3328
Literature/Catalog: Yes
Over 60 books and videos devoted exclusively to experimental/sport aviation.

Hummel Aviation
509 E. Butler Street • Bryan OH 43506-0880
Ph: 419-636-3390 • Literature/Catalog: $2.00
Complete plans for Half-VW engines. 28 to 40 hp., 72 to 80 lbs. Plans $45. VHS video available on metal working and the 1/2-VW engine for $42.00 Complete kits available.

LesAir, Inc.
HCR 31 Box 441 • 817 Lindbergh Avenue
Sandy Valley NV 89019 8909
Ph: 702-723-1950 • Fax: 702-723-1950
"The Complete Gyroplane Information Manual." $23.50 including shipping.

Marshall Consulting
720 Appaloosa Drive • Walnut Creek CA 94596
Ph: 510-945-6051 • Fax: 510-945-1461
"Composite Basics" Fourth Edition covering all aspects of composite design and manufacturing. Book: $30.00.
See Display Ad On Page 38

MONI Motorglider
Monitor Newsletter • 2363 South Milton Ave.
Overland MO 63114-5531 • Ph: 314-423-5058
Monnett Moni Motorglider. 143 pages. Pub. 1991 and 1992. All information on mods, updates, etc. ever done to the Moni. $48.00 ppd. in US. $52.00 US in Canada. $54.00 US worldwide. Call M-F noon to 4 pm for more information.

On Final Aircraft Supplies
A Div. of Murphy Aircraft Mtg.
#2-8155 Aitken Road
Chilliwack, BC V2R 4H5 CANADA
Ph: 800-461-7767 • Fax: 604-792-7006
E-Mail: on-final@murphyair.com
Literature/Catalog: Yes

Paser Publications
5672 West Chestnut Aveue
Littleton CO 80123 • Ph: 303-904-3417
Aeronautical Engineer. Increase mph with less fuel consumption Mustangs, RV's, Glasairs, Lancairs, KR's and more. 176 large pages with photographs/illustrations. $24.95. Shipping and handling US: $3.50. Foreign shipping and handling: $10.00.

Real World Solutions
1936 Nugget Dr. • Clearwater FL 33755
Ph: 813-447-6956
E-Mail: 71175.606@compuserve.com
"Aviator's Guide To Mazda Rotary Conversion" $35.00 ppd.

RFI Powerplant Systems Group
2916 Tennessee Northeast
Albuquerque NM 87110
Ph: 505-298-9362 • Fax: 505-266-1789
Literature/Catalog: $5.00
Conversion manuals for Subaru engines. $24.00 U.S., $30.00 Canada/Mexico, $38.00 Foreign-Air Mail $50.00.

John Roby
3703 Nassau Dr. • San Diego CA 92115
Ph: 619-583-4264 • Fax: 619-286-2425
Literature/Catalog: $3.00
Aviation Technical Literature on any aspect. Out-of-Print and current. Send SASE for information on 47 Specialty lists. Most lists $3.00 each U.S. VHS video available.

Schneider Enterprises
PO Box 2267 • Gonzales LA 70707
Technical manual of ducted fan building and designing procedures.

Sport Aviation Publications
P.O. Box 101 • Mingoville PA 16856
Ph: 814-383-2569 • Fax: 814-383-2569
Literature/Catalog: Yes
Books and training manuals on hang gliding, paragliding and powered ultralight flying. Complete weather book for light aircraft pilots.

Swag Aeromotive
2521 North Fairview • Tucson AZ 85705
Ph: 520-622-6910 • E-Mail: swagaero@flash.net
Web: http://www.flash.net/~swagaero
Literature/Catalog: Yes
Books on converting automotive engines to electronic multi-port fuel-injected aircraft use. Plans for silent chain reductions.

TERF Inc.
11411 Culver • Brighton MI 48114
Ph: 810-229-2082 • Fax: 810-229-2082
E-Mail: Mtait@livgonline.com
Web: http://www.terf.com
The Rutan Aircraft Factory CD ROM Encyclopedia.
See Display Ad On Page 52

Thinc
PO Box 16611 • Greenville SC 29606
Ph: 864-292-5821 • Literature/Catalog: $5.00
HELICAR KIT. Manual and drawings show how to convert a car into a Helicopter.

Thunderbolt Publications
W1432 Highway E., Dept. B
Chilton WI 53014 • Ph: 414-853-3370
Literature/Catalog: Yes
Discount aviation books & supplies.

Toccoa Aviation, Inc.
Route 3 Airport • Toccoa GA 30577
Complete flight school and FBO.

United States Ultralight Assoc
P.O. Box 667 • Frederick MD 21705
Ph: 301-695-9100 • Fax: 301-695-0763
Literature/Catalog: Yes
Ultralight books and videos.

Vista Aviation, Inc.
12653 Osborne Street • Pacoima CA 91331
Ph: 818-896-6442 • Fax: 818-896-9541
Avionics shop. Maintenance, parts and equipment. Flight school.

Vogel Aviation
PO Box 370054 • Reseda CA 91335
"Elements of sport airplane design for the homebuilder." Design basics, recommended procedures, engineering data, tables, materials and construction data, design suggestions. 3rd edition, 105 pages, illustrated. $19.95 US.

Vortech, Inc.
PO Box 511-KP • Fallston MD 21047
Ph: 410-692-2692 • Fax: 410-692-5902
Web: http://www.webcom.com/neatstuf/helio
Literature/Catalog: $10.00
Rotor blades, all-aluminum for helicopters and gyroplanes.
See Display Ad On Page 21

WAG-AERO
The Wag-Aero Group
1216 North Road, Box 181 • Lyons WI 53148
Ph: 800-558-6868 • Fax: 414-763-7595
E-Mail: wagaero-sales@wagaero.com
Literature/Catalog: Yes
Informational books on several popular aircraft.
See Display Ad On Page 13

Waltz Publishing
PO Box 6088 • Fall River MA 02724
Ph: 508-673-1994 • Literature/Catalog: Yes
"The Paraflight Experience" John Richard Carr's experiences flying a powered parachute ultralight. 161 pages softcover, illustrated. $16.70 US by mail.

Wicks Aircraft Supply
410 Pine Street • Highland IL 62249
Ph: 800-221-9425 • Fax: 618-654-6253
Literature/Catalog: $5.00
Worldwide supplier for homebuilt projects specializing in raw material kits for 40+ different designs. 350 page catalog with over 12,000 items.
See Display Ad On Page 46

Wings Aviation
T-406 Kelly Avenue - Airport
Manchester NH 03103
Ph: 603-627-8663 • Fax: 603-627-3063
Literature/Catalog: Yes
Largest pilot supply store in the Northeast. 1,000 book titles in stock. Chart coverage worldwide plus flight training and aircraft rentals. Aircraft management program.

Wood-Met Services, Inc.
3314 W. Shoff Circle Dept. KP
Peoria IL 61604-5464 • Ph: 309-637-9667
Literature/Catalog: $1.00
Plans to build shop tools and accessories. Plans from complete machines and attachments to handtools for home and shop. Accessories for wood and metal cutting lathes, drills, saws and sanders.

Zenith Books
PO Box 1 • 729 Prospect Ave.
Osceola WI 54020
Ph: 715-294-3345 • Fax: 715-294-4448
E-Mail: mbibks@win.bright.net
Web: http://www.motorbooks.com
Literature/Catalog: $2.00
Wide selection of aviation, military, piloting, modeling books and videos. Thousands of titles. Many new listings added.

Publications, Periodicals

Aircraft Designs, Inc.
5 Harris Court, Bldg. S • Monterey CA 93940
Ph: 408-649-6212 • Fax: 408-649-5738
Literature/Catalog: $12.00
The Stallion; tips for the fabrication and design of composite material aircraft, and computer methods used in the design of high performance general aviation aircraft.

AOPA Pilot Magazine
Membership Dept. • 421 Aviation Way
Frederick MD 21701 • Ph: 301-695-2350
Magazine of the Aircraft Owners and Pilots Association. Membership is $39, including a subscription to AOPA PILOT.

Atlantic Flyer Aviation News
P.O. Box 668 • Litchfield CT 06759
Ph: 203-266-7947 • Fax: 203-266-7946
E-Mail: aflyer@tiac.net
Web: http://www.aflyer.com/

Aviation Book Company
7201 Perimeter Rd. So., Ste. C • Boeing Field
Seattle WA 98108
Ph: 206-767-5232 • Fax: 206-763-3428
Literature/Catalog: Yes
Aviation training manuals of all publishers.
See Display Ad On Page 8

BD Micro Technologies, Inc.
1260 Wade Road • Siletz OR 97380
Ph: 541-444-1343 • Literature/Catalog: $20.00
Builders service, kits & parts for the BD-5.

Contact! Magazine
2900 E. Weymouth Street • Tucson AZ 85716-1249
Ph: 520-881-2232 • Fax: 520-795-6776
E-Mail: contact1@flash.net
Web: http://flash.net/~contact1
Literature/Catalog: Yes
First hand reports on auto engine conversions. Bi-monthly.

Experimental Aircraft Assoc.
EAA Aviation Center • PO Box 3086
Oshkosh WI 54903-3086
Ph: 920-426-4858 • Fax: 920-426-6560
E-Mail: dnocenti@eaa.org
Web: http://www.eaa.org
Sport Aviation magazine. Subscription included with EAA membership.
See Display Ad On Page 33

Flyer Magazine
3, Kingsmead Square • Bath, Avon BA1 2AB ENGLAND • Ph: 01-225-481-40
Fax: 01-225-481-262
E-Mail: ians@seager.demon.co.uk
Web: http://www.flyer.co.uk
General aviation periodical published in the United Kingdom.

Flying Flea Archive USA
P.O. Box 892 • Wooster OH 44691-0892
Ph: 330-377-5013 • Fax: 330-377-5013
Literature/Catalog: Yes
Promotes and distributes information on Henri Mignet's designs. Publishes the POU REVIEW, a bi-monthly magazine plus books. Offers copies of original plans. Send SASE for more information.

Flying Magazine
PO Box 53647 • Boulder CO 80302
Ph: 303-447-9330
"World's most widely read aviation magazine."

General Aviation News & Flyer
P.O. Box 39099 • Tocoma WA 98439-0099
Ph: 253-471-9888 • Fax: 253-471-9911
E-Mail: bsclair@ganflyer.com
Web: http://www.ganflyer.com
An aviation publication printeted every other week, 26 issues a year.

Kit Aircraft Builder
1313 Paseo Alamos • San Dimas CA 91773
Ph: 800-827-5070 • Fax: 909-599-2691
E-Mail: KitAirBldr@aol.com
Literature/Catalog: Yes
Publication for all active builders. Covers all aspects of building over all types of kit aircraft. Send SASE for free info. Published by Coyne Publishing Company.

Kitplanes Magazine
Subscription Division • 1000 Quail St., Suite 190
Newport Beach CA 92660
Ph: 800-829-3340 • Web: http:Kitplanes.com
Monthly homebuilt aircraft magazine. $29.95 per year in the US. Outside the US call 904-446-6914.

Make It Light
PO Box 196 • Morrisville PA 19067-0196
Literature/Catalog: Yes
Improve aircraft performance by saving weight without sacrificing strength, durability or safety. Free information, send self addressed stamped envelope. 12 issues, $36.00 US, Canada, Mexico ($52.00 overseas).

Private Pilot Magazine
Subscription Service Dept. • PO Box 55064
Boulder CO 80322-5064
Ph: 303-786-7306 • "The pilot's guide to owning and flying private aircraft."

John Roby
3703 Nassau Dr. • San Diego CA 92115
Ph: 619-583-4264 • Fax: 619-286-2425
Literature/Catalog: $3.00
Aviation literature: any aspect. Including out-of-print and current. Specializing in technology. Most lists $3.00 each in U.S.

Sailplane Homebuilders Assoc.
21100 Angel Street • Tehachapi CA 93561
Ph: 805-822-8852
The Sailplane Homebuilder's Association provides a forum for exchanging information on building and operating soaring aircraft through its periodical "Sailplane Builder" and through workshops.

Sport Aviation Magazine
PO Box 3086 • Oshkosh WI 54903-3086
Ph: 800-843-3612 • Fax: 414-426-4837
The EAA's monthly membership magazine, for sport aviation enthusiasts in all catagories.

Sport Pilot Magazine
PO Box 16149 • North Hollywood CA 91606
Ph: 818-760-8983 • "Hot kits & homebuilts."

Dave Thomas
Box 309 • Monroe OR 97456
Literature/Catalog: Yes
Discussions of light STOL aircraft. Send SASE for literature.

ULTRA flight magazine
2167 14th Circle N. • St. Petersburg FL 33713
Ph: 813-894-4636
E-Mail: jbyers468@aol.com
Literature/Catalog: Yes
ULTRAFLIGHT covers all types of Ultralights. Monthly publication.
See Display Ad On Page 56

Ultralight Flying! Magazine
1085 Bailey Avenue • Chattanooga TN 37404
Ph: 423-629-5375 • Fax: 423-629-5379
Web: http://www.ulflyingmag.com
The magazine of the U.S. Ultralight Association for enthusiasts of Ultralight and Microlight Aviation. 12 issues $30 (U.S.); $34 (Canada and Mexico); $36 (outside North America).
See Display Ad On Page 52

Publications, Newsletters

Acro Sport, Inc.
PO Box 462 • Hales Corners WI 53130
Ph: 414-529-2609
Acro Sport Newsletter-Quarterly $12.00. Back issues available: $55.00 plus $4.50 for postage and handling.

Aircraft Newsletters
5630 South Washington Road
Lansing MI 48911-4999
Ph: 800-594-4634 • Fax: 517-882-8433
Literature/Catalog: Yes
Publisher of a variety of aircraft newsletters including the Varga. Free sample with business SASE or call.

Aircraft Projects
2269 Maywood Ave. • San Jose CA 95128
Ph: 408-277-0645 • Fax: 408-279-6535
E-Mail: BAFRANK@att.net
Literature/Catalog: $6.00
Newsletter for the Ford 3.8L V-6 engine aero conversion and the V-6 STOL homebuilt aircraft.

Aircraft Spruce & Specialty
PO Box 4000 • Corona CA 91718-9961
Ph: 800-824-1930 • Fax: 909-372-0555
E-Mail: info@aircraft-spruce.com
Web: http://www.aircraft-spruce.com
Literature/Catalog: Yes
A complete selection of homebuilt and certified aircraft parts and materials. Shipping worldwide since 1965. Also see Aircraft Spruce East, Griffin GA, 800-831-2949.
See Display Ad On Page 24

American Hatz Assn.
16225 143rd Avenue SE • Yelm WA 98597
Ph: 360-894-2862 • Fax: 360-894-2862
Literature/Catalog: Yes
International Association dedicated to building and flying the Hatz CB-1 Biplane. Quarterly newsletter. Forums at Oshkosh and Biplane Expo. Free sample issue.

Aviation Supplies & Academics
7005 132nd Pl. SE • Newcastle WA 98059
Ph: 206-235-1500 • Fax: 206-235-0128
E-Mail: 102105.223@compuserve.com
Literature/Catalog: Yes
Publishes aviation books for pilots and aviation maintenance technicians.

BD-4 Builders Newsletter
17605 SE 288 Pl.
Kent WA 98042
Ph: 253-631-5324
E-Mail: rmellema@who.net
Web: http://www.halcyon.com/www2/rmellema/BDindex.html
The BD-4 builder/flyer newsletter, containing sources of information, plans, hardware, and help for the BD-4.

The BD-5 Network
34311 Bodkin Terrace • Fremont CA 94555
Ph: 510-797-2804
E-Mail: BD5NET@AOL.COM
Literature/Catalog: Yes
The definitive BD-5 info source. Quarterly newsletter plus member/supplier list. $25/Year($29 US Overseas).

C.S.N.
510 N.W. 46th Terrace
Plantation FL 33317-2044
Ph: 954-581-8835
Corby Starlet newsletter. 14th year of publication. Construction methods, builder's reports, parts & suppliers. Subscription: $24.00 USA, $27.00 Canada airmail, $30.00 all others. Corby Starlet info pack: $8.00/$10.00 airmail.

Various Club Newsletters
PO Box 127 • Blakesburg IA 52536
Ph: 515-938-2773 • Fax: 515-938-2773
Newsletters-three issues of one title: $15.00. Choose from Corben, Culver, Fairchild, Great Lakes, Hatz, International Pietenpol, Parakeet, Interstate, Rearwin, Travel Air, Caribbean Breeze (6th Air Force). Future newsletters: Airmaster and Fly Baby.

Coot-Builder's Newsletter
6958 Applewood Drive • Madison WI 53719
Ph: 608-833-5586 • Fax: 608-263-9167
E-Mail: rsteeves@facstaff.wisc.edu
Coot - Builder's Newsletter: $15/year.

Corben Club Newsletter
PO Box 127 • Blakesburg IA 52536
Ph: 515-938-2773 • Fax: 515-938-2773
3 issues for $12.00.

Culver Club Newsletter
PO Box 127 • Blakesburg IA 52536
Ph: 515-938-2773 • Fax: 515-938-2773
3 issues for $12.00.

Fairchild Club Newsletter
PO Box 127 • Blakesburg IA 52536
Ph: 515-938-2773 • Fax: 515-938-2773
3 issues for $12.00.

Great Lakes Club Newsletter
PO Box 127 • Blakesburg IA 52536
Ph: 515-938-2773 • Fax: 515-938-2773
3 issues for $12.00.

Griffon Aerospace, Inc.
920 Yarbrough Rd. • Harvest AL 35749
Ph: 205-721-9055 • Fax: 205-837-1694
E-Mail: griffon@traveller.com
Web: http://www.griffon-aerospace.com
The LH Newsletter for Lionheart builders and admirers.

Hatz Club
PO Box 127 • Blakesburg IA 52536
Ph: 515-938-2773 • Fax: 515-938-2773
Hatz Herald. 16 to 24 page magazine published 1 to 2 times each year.

International Pietenpol News
PO Box 127 • Blakesburg IA 52536
Ph: 515-938-2773 • Fax: 515-938-2773
3 issues for $12.00.

Interstate Intercom Newsletter
PO Box 127 • Blakesburg IA 52536
Ph: 515-938-2773 • Fax: 515-938-2773
3 issues for $12.00.

KR Newsletter
624 W. UNiversity Dr., Ste. 19
Denton TX 76201 • Ph: 940-566-5871
Publisher of the "KR Newsletter" devoted to all aspects of KR aircraft. Not affiliated with Rand Robinson Engineering.

Mirage Aircraft Corp.
1091 W. Sahara Palms Dr.
Tucson AZ 85704
Ph: 520-797-2161
E-Mail: celerity@primemet.com
Bi-monthly newsletter for Celerity builders.

Moni Flyer
711 East Street • Ft. Atkinson WI 53538
Ph: 414-563-6502 • Fax: 414-563-9115
E-Mail: eisenbrg@intaccess.com
Literature/Catalog: $2.00
The Moni Flyer continues on where the Monitor Newsletter left off in 1992. 1994 and 1995 issues $16.00 US, $18.00 Canada in US funds and $21.00 US worldwide. 1996 issues same price.

Parakeet Club Newsletter
PO Box 127 • Blakesburg IA 52536
Ph: 515-938-2773 • Fax: 515-938-2773
3 issues $12.00.

Quicksilver Pilots Association
328 Luscombe Drive • Los Lunas NM 87031
Ph: 505-865-3466 • Fax: 505-866-7308
E-Mail: PFPA@aol.com
Publishers of Quicksilver Magazine.

Real World Solutions
1936 Nugget Dr. • Clearwater FL 33755
Ph: 813-447-6956
E-Mail: 71175.606@compuserve.com
Newsletter: "Rotary Aviation News"

Rearwin Newsletter
PO Box 127 • Blakesburg IA 52536
Ph: 515-938-2773 • Fax: 515-938-2773
3 issues $12.00.

Ronald Cook
PO Box 135 • Aguila AZ 85320
Ph: 520-685-2581

Rotax Owners Association
S22, C39, RR6 Stn Main • 6235 Ohanagan Landing Road • Vernon, B.C. V1T 6V5 CANADA
Fax: 604-549-7111
Newsletter providing service and technical information on ROTAX UL engines. Technical commentary on recent service and related information. $50.00 (US) per year in North America, $75.00 (US) per year international. Back issues available.

Skysmith International, Ltd.
2525 NW 71st Place • Ankeny IA 50021-9048
Ph: 800-743-1439 • Fax: 515-289-1235
E-Mail: SkySmith@aol.com
Skymaster, Pitts, and Smith Miniplane Newsletters and club management. Also offering group insurance, product plans, and pilot accessories.
See Display Ad On Page 60

Speed, Props & Pylons
PO Box 155 • 13045 Second Avenue S.
Zimmerman MN 55398
Ph: 612-783-8728 • Fax: 612-856-3138
Literature/Catalog: Yes
Newsletter for air racing since 1988. Nine issues per year. Complimentary issue available.

Star Review
1603 City View Dr. • Whichita Falls TX 76305-1101
Ph: 817-855-6977 • Fax: 817-855-6977
Newsletter of the StarLite Owners Association.

Super Prospector Notebook
1280 West 17th Street • Houston TX 77008-3440
Ph: 713-869-2128 • Literature/Catalog: Yes
Disgruntled purchasers of the incomplete Bounsall Super Prospector plans share their ideas for salvaging their investment through this newsletter. Free sample available.

Tri-R Technologies
1114 E. 5th Street • Oxnard CA 93030
Ph: 805-385-3680 • Fax: 805-483-8366
Literature/Catalog: $10.00
Newsletter for the KIS aircraft family. 4 issues per year, $15.00.

Ultralights
Phantom/Flightstar Pilots Assn • 328 Luscombe Drive • Los Lunas NM 87031
Ph: 505 865-3466 • Fax: 505 866-7308
E-Mail: PFPA@ Aol.com
Publishers of Phantom Driver.

Weedhopper Owners Club
74 Brookwood Dr. • Marietta, CA 30064
Ph: 770-973-3860 • 770-422-2597
Owners club and technical registry committed to the maintenance and enhancement of "Weedhopper" ultralight planes. Safety alerts, spare parts sources, buy/sell info etc. Shared via newsletter and member roster. Contact: Scott Perkins.

World War I Aeroplanes, Inc.
15 Crescent Road • Poughkeepsie NY 12601
Ph: 914-473-3679 • Literature/Catalog: $5.00
"WW1 Aero," The Journal of the Early Aeroplane (1900-1919) and "Skyways," The Journal of the Airplane (1920-1940). Founded in 1961. Devoted to builders, historians, modellers and restorers of historical aircraft. Contains information on current projects, photographs, scale modelling material, historical research and other current news.

ZENITH AIRCRAFT COMPANY
Box 650 • Mexico Memorial Airport
Mexico MO 65265-0650
Ph: 573-581-9000 • Fax: 573-581-0011
E-Mail: info@zenithair.com
Web: http://www.zenithair.com
Literature/Catalog: $15.00
Building newsletter that covers all Zenair designs.
See Ad Inside Front Cover

Publications, Training Materials

American Airman School
Gabreski Airport • Building 1
Westhampton Beach NY 11978
Ph: 800-377-1385 • Fax: 516-288-4632
Literature/Catalog: Yes
Preparation material for written, oral and practical A&P-IA exams. Material includes study guides, books, practice tests and videos.

Custom Aerial Photography
P.O. Box 771 • Lamar CO 81052
Literature/Catalog: Yes
Information to help pilots earn money by shooting aerial photos with a hand held camera.

Midwest Engineering
PO Box 4706 • Overland Park KS 66204-0706
Ph: 913-631-8447
E-Mail: mwengineering@juno.com
Operations manuals for the Robinson R-22, $15.00. Operations manuals for the Hughes OH6A, $15.00

Northern Michigan Enterprises
P.O. Box 6321 • Traverse City MI 49685-6321
Ph: 616-946-1071
Simple, easy to follow plans for building your own shop tools such as a tubing cutter/notcher, an english wheel for forming compound curves in sheet metal, a sand blast cabinet, a motorized ramp scooter ...and more. $7.50 ea., 3/$20, 5/$30.

Ron Sands Co.
89 Forrest Drive • Mertztown PA 19539
Ph: 610-682-6788 • Fax: 610-682-6788
Literature/Catalog: Yes
Manuals & parts catalogs (Copies). Send SASE for literature.

Sport Aviation Publications
P.O. Box 101 • Mingoville PA 16856
Ph: 814-383-2569 • Fax: 814-383-2569
Literature/Catalog: Yes
Books and training manuals on hang gliding, paragliding and powered ultralight flying. Complete weather book for light aircraft pilots.

Teledyne Aircraft Products Co.
Battery Products Unit • PO Box 7950
Redlands CA 92375 • Ph: 800-456-0070
Fax: 909-793-5818 • Literature/Catalog: Yes
Service manuals.

Westair International USA
Box 771 • Monument CO 80132
Ph: 719-481-2286 • Fax: 719-481-2203
E-Mail: Westair.Aerotecdata@world net.att.net
Military & civilian flight handbooks, maintenance manuals, structural repair, illustrated parts books, engine manuals, propeller manuals avionics, APU manuals. Engine overhaul and parts manuals.

Videos

Air-Ryder Manufacturing Inc.
PO Box 6896 • Wetaskiwin, AB T9A 2G5 CANADA
Subaru-Build Engine to O Time: $49.95. Engine mods & installation: $49.95. Both tapes $89.95.

Aircraft Spruce & Specialty
PO Box 4000 • Corona CA 91718-9961
Ph: 800-824-1930 • Fax: 909-372-0555
E-Mail: info@aircraft-spruce.com
Web: http://www.aircraft-spruce.com
Literature/Catalog: $5.00
A complete selection of homebuilt and certified aircraft parts and materials. Shipping worldwide since 1965. Also see Aircraft Spruce East, Griffin GA, 800-831-2949.
See Display Ad On Page 24

Aircraft Super-Market
#10 Ellefson Dr. • PO Box 676
De Soto IA 50069
Ph: 515-834-2225 • Fax: 515-834-2152
E-Mail: bobdurly@netins.net
Web: http://www.netins.net/showcase/t-birds
Literature/Catalog: $5.00
Handles recreational aircraft of all kinds. New and used, ultralights, powered parachutes and U/L gliders. 2600' runway.
See Display Ad On Back Cover

American Airman School
Gabreski Airport • Building 1
Westhampton Beach NY 11978
Ph: 800-377-1385 • Fax: 516-288-4632
Literature/Catalog: Yes
100% pass course, same as school, at home. Preparation material for written, oral and practical A&P-IA exams. Material includes study guides, books, practice tests and videos. A&P $245. IA $175.

Aviation Book Company
7201 Perimeter Rd. So., Ste. C • Boeing Field
Seattle WA 98108
Ph: 206-767-5232 • Fax: 206-763-3428
Literature/Catalog: Yes
Civil and military aviation videos.
See Display Ad On Page 8

Blue Ridge Machinery & Tools
P.O. Box 536 • 2905 Putnam Avenue
Hurricane WV 25526
Ph: 304-562-3538 • Fax: 304-562-5311
Literature/Catalog: Yes
Machine shop equipment and tooling for the home-shop machinist. Lathes, mills, drill rods and more. How-to videos and books.
See Display Ad On Page 22

Blue River Aircraft Supply
223 N. Clay • Harvard NE 68944
Ph: 800-653-7200 • Fax: 402-463-4476
Literature/Catalog: $2.00

C.R. Supply
14812 Washington St. • Omaha NE 68137-2552
Ph: 719-471-0110
Literature/Catalog: $0.75
Videos for light plane operators. "Let's Fly to Leadville and Rifle" and "Let's Fly Santa Fe, NM to Steamboat Springs, CO" Colorado mountain flying adventures.

California Power Systems
790 139th Avenue, #4
San Leandro CA 94578
Ph: 800-247-9653 • Fax: 510-357-4429
Web: http://www.800-airwolf.com
Literature/Catalog: $6.95
Complete supplier of ultralight and sport aircraft parts and components. Rotax engines, parts and service. 500 page catalog, $6.95 US, $10.00 Canada and $20.00 foreign airmail.
See Display Ad On Page 69

Experimental Aircraft Assoc.
EAA Aviation Center • PO Box 3086
Oshkosh WI 54903-3086
Ph: 920-426-4858 • Fax: 920-426-6560
E-Mail: dnocenti@eaa.org
Web: http://www.eaa.org
Literature/Catalog: Yes
The EAA is a membership organization geared towards sport flying. A variety of publications and instructional videos are offered.
See Display Ad On Page 33

Flying Flea Archive USA
P.O. Box 892 • Wooster OH 44691-0892
Ph: 330-377-5013 • Fax: 330-377-5013
http://www.netwizards.net/~mdaniels/flyingflea/
Literature/Catalog: Yes
Promotes and distributes information on Henri Mignet's designs. Publishes a bi-monthly magazine and offers a VHS video of various modern Flying Fleas, plus historical footage. Plans and construction manuals available for HM-14 and HM-16. Send SASE for more information.

France Invest
29 Av Gustave Nesureur
La Celle St. Cloud 78170 FRANCE
Ph: 30822929 • Fax: 30822720
Literature/Catalog: $20.00
Video with information package for an all metal STOL GT 2000 Trophy, a finished ultralight (without engine/instruments).

GBI - George & Becki Orndorff
2290 W. Hicks Rd., Unit 18 • Ft. Worth TX 76131
Ph: 817-439-3280 • Fax: 817-439-3280
E-Mail: orndorffg@aol.com
RV construction videos. Sheet metal working classes. Line of custom interior products.

Homebuilt Book Supply
PO Box 8351
Scottsdale AZ 85252-8351
Ph: 602-990-3328 • Fax: 602-990-3328
Literature/Catalog: Yes
Various videos that cover homebuilt aircraft construction techniques.

King Schools
3840 Calle Fortunada • San Diego CA 92123
Ph: 619-541-2200 • Fax: 619-541-2201
E-Mail: Acharney@kingschools.com
Web: http://www.kingschools.com
Literature/Catalog: Yes
Videos and software to train all levels of pilots.

Learn To Earn Video
540 California Avenue • Grants Pass OR 97526
Ph: 503-479-0241
Video on wet magnaflux welding.

Motorsports Tools
Route 2 • Box 366 • MacClenny FL 32063
"English Wheel" metal shaping tools. 2 hr. VHS video on metal shaping techniques by John Glover - $39.95 plus $3.95 S&H.

Northern Michigan Enterprises
P.O. Box 6321 • Traverse City MI 49685-6321
Ph: 616-946-1071
Simple, easy to follow plans for building your own shop tools such as a tubing cutter/notcher, an english wheel for forming compound curves in sheet metal, a sand blast cabinet, a motorized ramp scooter, plus many more.

Parascender Technologies, Inc.
1040 E. Carroll St. Dept. KP
Kissimmee FL 34744
Ph: 407-935-0775 • Fax: 407-935-0778
E-Mail: para@iag.net
Web: http://www.iag.net/~para
Powered parachute video tape show. Parascender Powered Parachute including single and two seater, new students learning, lots of flying footage, engine-out, AG spraying, and snow ski scenes. Video available for $20.00.

RFI Powerplant Systems Group
2916 Tennessee Northeast
Albuquerque NM 87110
Ph: 505-298-9362 • Fax: 505-266-1789
Subaru engine conversion video.

Rick Giles
4287 East Winneshiek Road • Freeport IL 61032
Yamaha motorcycle engine into Nieuport 12. 1-1/2 hour video includes teardown of engine plus important features and modifications for installation. $29.50

Ross Engineering Co.
12505-E Starkey Road • Largo FL 34643
Ph: 800-422-7676 • Fax: 813-535-4248
Literature/Catalog: Yes

Schweitzer Fisheries
Box 66 • Bedford IA 50833
Installation and maintenance of Ford 3.8L V-6 in RV6 and V6 Stol. Two hour video: $19.95.

Teledyne Aircraft Products Co.
Battery Products Unit • PO Box 7950
Redlands CA 92375
Ph: 800-456-0070 • Fax: 909-793-5818
Literature/Catalog: Yes
Service video.

TERF Inc.
11411 Culver • Brighton MI 48114
Ph: 810-229-2082 • Fax: 810-229-2082
E-Mail: Mtait@livgonline.com
Web: http://www.terf.com
See Display Ad On Page 52

Wicks Aircraft Supply
410 Pine Street • Highland IL 62249
Ph: 800-221-9425 • Fax: 618-654-6253
Worldwide supplier for homebuilt projects specializing in raw material kits for 40+ different designs. 350 page catalog with over 12,000 items.
See Display Ad On Page 46

Wings Aviation
T-406 Kelly Avenue - Airport
Manchester NH 03103
Ph: 603-627-8663 • Fax: 603-627-3063
Literature/Catalog: Yes
Largest pilot supply store in the Northeast. 1,000 book titles and videos in stock. Chart coverage worldwide plus flight training and aircraft rentals. Aircraft management program.

ZENITH AIRCRAFT COMPANY
Box 650 • Mexico Memorial Airport
Mexico MO 65265-0650
Ph: 573-581-9000 • Fax: 573-581-0011
E-Mail: info@zenithair.com
Web: http://www.zenithair.com
Literature/Catalog: $15.00
30 minute demo video of the popular Zenair designs.
See Ad Inside Front Cover

Zenith Books
PO Box 1 • 729 Prospect Ave.
Osceola WI 54020
Ph: 715-294-3345 • Fax: 715-294-4448
E-Mail: mbibks@win.bright.net
Web: http://www.motorbooks.com
Literature/Catalog: $2.00
Wide selection of aviation, military, piloting, modeling books and videos. Thousands of titles. Many new listings added.

Miscellaneous

Able Kit Planes
3 NW Maples Street
Fort Walton Beach FL 32548
Ph: 904-244-8482 • Fax: 904-244-8482
Literature/Catalog: Yes
Altitude compensating carburetor. Automatic and continuous altitude compensation of air/fuel mixture with Altimizer(tm) experimental carburetor. Standard kits for Rotax 447cc through 912 cc engines.

Aero Cosmetics
P.O. Box 460025 • San Antonio TX 78246
Ph: 800-927-4929 • Fax: 210-344-4967
Web: http://washwax.com
Literature/Catalog: Yes
Aero Cosmetics WashWaxALL, cleans and protects all surfaces and windows, dry or wet. Spray on and wipe dry. Meets aircraft manufacturer cleaning specifications. Full line of aircraft cleaning products.

Aero-K
Aero-K Aviation Products Manufactures 7 different styles of control stick grips, plain or with mil-spec quality push to talk switches. More than 9000 have been sold. Available from Wag-Aero, Leading Edge Airfoil, Wicks & Aircraft Spruce.

Aero Lift Corp.
1125 Morse Avenue • Schaumburg IL 60193
Ph: 847-891-7502 • Fax: 847-891-7548
Literature/Catalog: Yes
Self-jacking dollies for quick and easy repositioning of aircraft on wheels to maximize work and storage space.

Aero Products Research, Inc.
11201 Hindry Avenue • Los Angeles CA 90045
Ph: 310-641-7242 • Fax: 310-649-1703
Literature/Catalog: Yes
Pilot's personal equipment: flight logs, AM bolt gages, lighted and unlighted kneeboards and pilot's folding flight desk, etc.

Aero-K Aviation Products Co.
Redlands Municipal Airport • 1717 Sessums Dr.
Redlands CA 92374
Ph: 909-794-6631 • Fax: 909-794-9611
Literature/Catalog: Yes
Urethane plastic control stick grips.

A.I.R. Components, Inc.
P.O. Box 328 • Florence OR 97439
Ph: 503-997-2119 • Literature/Catalog: Yes
Fuel supply water contamination indicators.

Aircraft Belts, Inc.
2020 Anders Lane • PO Box 845
Kemah TX 77565-0845
Ph: 713-538-1284 • Fax: 713-538-2225
Literature/Catalog: Yes
Manufacturer & refurbishing of aircraft seat belts and harnesses.

Aircraft Designs, Inc.
5 Harris Court, Bldg. S • Monterey CA 93940
Ph: 408-649-6212 • Fax: 408-649-5738
Literature/Catalog: $12.00
Computer programs.

Aircraft Placards
6 Elm Road • N. Hampton NH 03862
Ph: 603-964-8905 • Fax: 603-926-7855
Literature/Catalog: Yes
Placards, checklists, anodized aluminum placards & panels.

Aircraft Spruce East
900 S. Pine Hill Rd. • Griffin GA 30223
Ph: 800-831-2949 • Fax: 770-229-2329
E-Mail: east@aircraft-spruce.com
Web: http://www.aircraft-spruce.com
East coast subsidiary of Aircraft Spruce and Specialty. See their listings and ad on back cover.

Airplane Plastics Company
8300 Dayton Rd. • Fairborn OH 45324
Ph: 937-864-5607 • Fax: 937-864-7010
E-Mail: Doug@Foxlite.com
Web: http://www.foxlite.com
Windshields-windows-canopies for all aircraft. Unmatched 1/2 price replacement warranty covers damage during installation and service for 6 months after purchase. Airplane Plastics Co., 8300 Dayton Rd., Fairborn, OH 45324. Phone 513-864-5607, FAX 513-864-7010.

AMSOIL, Inc.
AMSOIL Bldg. • Superior WI 54880
Ph: 715-392-7101 • Literature/Catalog: Yes

Antique Airplane Association
Antique Airfield • 22001 Bluegrass Road
Ottumwa IA 52501
Ph: 515-938-2773 • Fax: 515-938-2773
Literature/Catalog: Yes
Aircraft History Association for people interested in restoring and flying vintage aircraft. 24 active chapters and 11 active type clubs. Annual dues in AAA $36.00 includes Antique Airplane Association News and Antique Airplane Digest. Affiliated with Air Power Museum, Inc.

AVBlend
#900 Airport Drive • Springfield IL 62707
Engine lubricant.

Aviation Supplies & Academics
7005 132nd Pl. SE • Newcastle WA 98059
Ph: 206-235-1500 • Fax: 206-235-0128
E-Mail: 102105.223@compuserve.com
FAR/AIM flight computers, luggage, FAA reprints, pilot kits, FAA test books, kneeborads, headsets.

BD Micro Technologies, Inc.
1260 Wade Road • Siletz OR 97380
Ph: 541-444-1343
Literature/Catalog: $20.00
Builders service, kits & parts for the BD-5.

Bel-Ray Co., Inc.
Box 526 • Farmingdale NJ 07727

B-UR-O Company, Inc.
6986 El Camino Real, Dept. T71
Carlsbad CA 92009
Ph: 800-447-7818
Literature/Catalog: Yes
Lightweight, compact, electric power towbar for small aircraft-two models available.

C&R Aircraft
PO Box 281 • Quincy IL 62306-0281
Ph: 217-292-5967
Flight monitor system, headset strain relief clips, open cockpit safety lanyard, collector prints.

Candlepower, Inc.
14626 Southlawn Lane • Rockville MD 20850
Ph: 800-368-2700
Literature/Catalog: Yes
Tires & tubes.

Cardinal Helicopter Training
4184 Taylor Road, Suite P
Batavia OH 45103-9720
Ph: 513-732-5852 • Fax: 513-732-5853
Home builders center for airplanes and helicopters. Building is 7,000 square feet with heating and air conditioning. Power supply for both air and electric tools. Space rented by day, week, or month.

Century Series Designs Corp.
PO Box 987 • Hurley NM 88043
Ph: 505-538-5366
Literature/Catalog: $15.00
Design software.

Chadwick-Helmuth
4601 N. Arden Drive • El Monte CA 91731
Ph: 818-575-6161 • Fax: 818-350-4236
E-Mail: chadwick@chadwickhelmuth.com
The Vibrex 2000 is a high capability vibration analysis system.

Composite Aircraft Accessories
PO Box 21645 • Santa Barbara CA 93121-1645
Ph: 805-964-5453 • Fax: 805-683-8577
E-Mail: 103600.1401@compuserve.com
Literature/Catalog: $1.00
Electric cabin heaters and controllers for sport aviation aircraft.

Custom Products
19583 SW Sandra Lane • Aloha OR 97006
Ph: 503-649-8817 • Literature/Catalog: Yes
Custom bags & cases for radios, headsets, & helmets.

Datrim Corporation
PO Box 843 • Longmont CO 80502-9543
Ph: 303-684-9543 • Fax: 303-678-5235
Literature/Catalog: Yes
Air Plot - Airfoil plotting software for Microsoft Windows 3.1. Version 1.1 now available. Allows the display and printing of any airfoil to scale. Brochure available.

Davis Instruments
3465 Diablo Avenue • Hayward CA 94545
Ph: 510-732-9229 • Fax: 510-732-9188
E-Mail: sales@davisnet.com
Manufacturer of high precision weather stations.

DC Thermal, Inc.
12430 Hwy 3 • Suite E-20 • Webster TX 77598
Ph: 800-590-7500 • Fax: 281-486-0531
Literature/Catalog: Yes
Cabin heaters, DC electric. FAA approved for most aircraft. 12, 24, 36, and 48 nominal voltage models. Radiant (infra-red) and forced air (fan) heat. Compact, rugged, quartz heaters. Eliminates carbon monoxide hazard.
See Display Ad On Page 46

Delta-Therm Corp.
PO Box 345 • 398 W. Liberty Street
Wauconda IL 60084-0345
Ph: 800-526-7887 • Fax: 847-526-4456
Literature/Catalog: Yes
Hangar door rail de-icing systems. Manufacturers and Designers of electric and hydronic heat transfer systems.

Evergreen Weigh, Inc.
19023 36th Avenue West, #E
Lynnwood WA 98036
Ph: 800-356-6060 • Fax: 206-775-4399
Manufacturer of Road Runner aircraft scales for large and small aircraft.

Fastening Hardware Supply
PO Box 256 • Lancaster NY 14086
Ph: 716-681-1675 • Fax: 716-683-7315
Literature/Catalog: Yes
Stainless steel and plated airframe hardware and accessories (felt, chafe webbing, etc.)

Flight Physics
P.O. Box 20919 • Billings MT 59104
Ph: 406-248-2606 • Fax: 406-245-9076
E-Mail: jlowry@mcn.net
Literature/Catalog: Yes
Special engineering reports. Including detailed calculations for aviation litigation support.

Flight Products International
P.O. Box 1558 • Kalispell MT 59901
Ph: 406-752-8783 • Fax: 406-257-7078
Flight Workstation (Tripboard) (TM), Strapless Kneeboards, Swivelite (TM).

Flight Technology Int'l., Inc.
1571 Airport Road • Charlottesville VA 22911
Ph: 804-978-4359 • Fax: 804-978-7598
VOCALIST custom programmed audible checklist at your finger tips. 12 regular checklist positions and 12 emergency positions.

Flitz International, Ltd.
821 Mohr Avenue • Waterford WI 53185
Ph: 414-534-5898 • Fax: 414-534-2991
Literature/Catalog: Yes
Manufacturer of Flitz metal polish products. Can be used on all types of metals as well as fiberglass and plexiglass.

Forum Recordings
3410 St. Peters Road • Marion IA 52302
Ph: 319-377-4188
Literature/Catalog: Yes
Oshkosh forum audio cassettes.

Free Flite Enterprises
R.D. 2, Box 2121 • East Stroudsburg PA 18301
Ph: 717-421-8900
Manufacturers of Ultralite Disc Brakes.

Fullerton Air Parts
3029 E. Cardinal St. • Anaheim CA 92806-1002
Literature/Catalog: Yes

Great Airplanes Inc.
P.O. Box 615 • Needham Heights MA 02194-0005
Ph: 617-444-5480 • Fax: 617-444-5480
E-Mail: hswack@aol.com
Web: http:////aso.solid.com/greatair
Literature/Catalog: Yes
Macwhyte Flying and landing wires. Manufacturers of custom welded structures. Cleveland TSO'd aircraft wheels and brakes. Send stamped addressed #10 envelope for price and information.

Great Lakes Aero Products
915 Kearsley Park Blvd. • Flint MI 48503
Ph: 800-532-2918 • Fax: 810-235-5260
E-Mail: www.tir.com~glap
Manufacturer of aircraft windshields and side windows.

Hanley Innovations
PO Box 870 • Storrs CT 06268
Ph: 860-423-4060
E-Mail: hanley@hanleyinnovations.com
Web: http://www.hanleyinnovations.com
User friendly Windows 95 software to analyze the performance of airfoils. $99.00.

Hawk AirSports
P.O. Box 9056 • Knoxville TN 37940-0056
Ph: 423-212-4994 • Fax: 423-212-9854
Literature/Catalog: Yes
Portable, permanent, and replacement windsocks. All sizes, custom logo imprinting.

Heins Balancing Systems, Inc.
820 Railroad Avenue • Santa Paula CA 93060
Ph: 805-525-5445 • Fax: 805-525-7074
Manufacturer of balancing machines and tooling for overhauling a wide range of turbine and reciprocating engines and components.

Houde Enterprises
12573 U.S. Hwy. 26 • Riverton WY 82501
Ph: 307-856-9654
Instrument panel mockup stickers. Packet of 10 pages of full size photo reproductions of instruments, guages, switches, etc.

INFINITY Aerospace
P.O. Box 12275 Dept.KD
El Cajon CA 92022-2275
Ph: 619-448-5103 • Fax: 619-448-5103
E-Mail: infaero@flash.net
Web: http://www.flash.net/~infaero
Literature/Catalog: $15.00
Retractable main landing gear for Long-EZ, Cosy, Cozy Mk-IV and AeroCanard. Info pack and video available. US: $15.00 each or $25.00 both. Overseas: $20.00 each or $35.00 both.

INFINITY Aerospace
P.O. Box 12275 Dept.KD
El Cajon CA 92022-2275
Ph: 619-448-5103 • Fax: 619-448-5103
E-Mail: infaero@flash.net
Web: http://www.flash.net/~infaero
Literature/Catalog: Yes
The Infinity 1 Hotas Military-Style Stick Grip for the most important and frequeltly used cockpit functions. Left and right hand grips available. Send long self addressed stamped envelope for detailed information.

Insured Aircraft Title Service
PO Box 19527 • Oklahoma City OK 73144
Ph: 800-654-4882 • Fax: 405-681-9299
Literature/Catalog: Yes
Full service aircraft title company offering title search, escrow, closing service, owner/pilot listings, labels, diskettes, statistics, title insurance.

E.R. Jones Engineering
98 South Halifax Drive
Ormond Beach FL 32176-6539
Ph: 904 676-2344 • Literature/Catalog: Yes
Software. Propeller Design and Analysis (PDAP) for IBM PC and compatibles $49.95 ppd incl. manual. Add $4.00 for international.

Just Fly Aviation Gear
PO Box 450 • Kitty Hawk NC 27949
Ph: 919-480-3552 • Fax: 919-480-3552
Literature/Catalog: Yes
Just Fly supplies the aviation community with aviation related clothing and specializes in aviation screenprinting.

KC Graphics
1899 Orchard Lake Road, Ste. 1
Sylvan Lake MI 48320
Ph: 810-810-9119 • Fax: 810-332-9157
Literature/Catalog: Yes
Custom aircraft graphics, nose art and custom lettering like "Experimental" and "No Step".

King Schools
3840 Calle Fortunada • San Diego CA 92123
Ph: 619-541-2200 • Fax: 619-541-2201
E-Mail: Acharney@kingschools.com
Web: http://www.kingschools.com
Literature/Catalog: Yes
Videos and software to train all levels of pilots.

Kropp Enterprises
P.O. Box 120656 • Clermont FL 34711
Ph: 352-394-4727 • Fax: 352-394-1718
Literature/Catalog: Yes
Leather name plates for flight jackets, suits, etc. Civil and military insignia. Over 100 designs with prices starting at $9.00.

Langair Machining
33094 Church Road • Warren OR 97053
Ph: 503-397-1478 • Fax: 503-397-1498
Literature/Catalog: Yes
Wittman type landing gear legs for Homebuilt aircraft. Machined and ground from 6150 steel.

LEADING EDGE AIR FOILS
Meadowlake Airport
8242 Cessna Drive, Peyton CO 80831
Ph: 719-683-5323 • Fax: 719-683-5333
E-Mail: leafkp@kktv.com
Web: http://www.leadingedgeairfoils.com
Literature/Catalog: $6.00
Rotax engine info packets. Distributor for ultralight and experimental aircraft, parts, and engines.
See Display Ad On Page 34

LesAir, Inc.
HCR 31 Box 441 • 817 Lindbergh Avenue
Sandy Valley NV 89019-8909
Ph: 702-723-1950 • Fax: 702-723-1950
Folding mast plans: $75.00. "Thrustbuster" measures engine thrust up to 1000 lbs.

Log Book Organizer
2311 Swallow Road • Placerville CA 95667
Ph: 916-622-0684 • Fax: 916-622-0684
Literature/Catalog: Yes
Computerized maintenance log books.

McCutchen Sky Wheels
555 Alter Street #15 • Broomfield CO 80020
Ph: 303-465-5794 • Fax: 303-469-2801
Literature/Catalog: Yes
Composite rotor systems for autogyros up to 1500 lbs. gross weight and composite helicopter blades up to 13 feet long - 8" chord-symmetrical airfoils.

Menzimer Aircraft Components
2525-8 Pioneer Avenue • Vista CA 92083
Ph: 760-598-0592 • Fax: 760-598-2967
Literature/Catalog: Yes
Airframe accessories. Electric trim systems and control stick grips. Small 14 volt servo motor for controlling elevator, aileron and rudder trim.

Metal Masters
1027 Greenfield Dr. Stc. G • El Cajon CA 92021
Ph: 619-461-0168 • Literature/Catalog: $1.00
Aluminum landing gear made to your drawings. Send for kit $1.00.

Metalcraft, Inc.
Philadelphia Industrial Park • 9331-A
Philadelphia Road • Baltimore MD 21237
Ph: 410-687-5500 • Fax: 410-687-5503
Literature/Catalog: Yes
Supplier of fire extinguishers.

Micro Aerodynamics, Inc.
4000 Airport Road, Ste. D • Anacortes WA 98221
Ph: 800-677-2370 • Fax: 360-293-5499
E-Mail: micro@sos.com
Web: http://www.microaero.com/micro
Micro Vortex Generators for Long EZ and Murphy Rebel.

Micro-AG Systems
Box 1932, Hwy 120 S. • Lake Geneva WI 53147
Ph: 414-248-2288 • Literature/Catalog: Yes
Ultra & hyper low volume crop spraying equipment.

Micro Surface Finishing Prod.
1217 West Third Street • PO Box 818
Wilton IA 52778
Ph: 800-225-3006 • Fax: 319-732-3390
E-Mail: msfpi@netins.net
Web: http://www.micro-surface.com
Literature/Catalog: Yes
FAA Repair Station for aircraft acrylic windows. Fine finishing cushioned abrasive that removes crazing, scratches and other damage from acrylics. Various repair kits available.

Midwest Engineering
PO Box 4706 • Overland Park KS 66204-0706
Ph: 913-631-8447
E-Mail: mwengineering@juno.com
Literature/Catalog: $1.00
Midwest Letterface puts your favorite aircraft photo on a stamp. 72 stamps for $19.95. Sorry - no postal value but a great way to show-off your hobby. $1.00 kit includes samples.

National Products, Inc.
1017 S. Elmgrove St. • Seattle WA 98108
Ph: 206-763-8361 • Fax: 206-763-9615
E-Mail: stuff@ram-mount.com
Web: http://www.ram-mount.com
Literature/Catalog: Yes

Navaid Devices, Inc.
641 N. Market St. • Chattanooga TN 37405
Ph: 423-267-3311 • Fax: 423-756-6154
Literature/Catalog: Yes
Manufacturers of autopilots and electronic ignitions for experimental category (kitbuilt) aircraft.

Northwest Custom Grips
4541 Harlan Drive N.E. • Salem OR 91305
Ph: 503-463-4433 • Fax: 503-463-4433
Literature/Catalog: $2.00
Custom handmade teakwood control grips. Provides comfort, dryness and coolness. Available with push-to-talk, 4-way trim, speed-brake and auto-pilot disconnect, any or all switch combination.

Tracy O'Brien Aircraft Supplies
156 Bunker Creek Road • Chehalis WA 98532
Ph: 360-748-4089 • Fax: 360-740-0504
E-Mail: tracy@localaccess.com
Literature/Catalog: Yes
5", 6", and 8" cast aluminum wheels with 5/8", 3/4", and 7/8" bearings. Hydraulic disk brakes and mechanical drum brakes. Bolt-on and weld-on axels in 5/8", 3/4", and 7/8" diameters. 5" tailwheel kits. 5", 6", and 8" tires.

P.M. Research, Inc.
4110 Niles Hill Road • Wellsville NY 14895
Ph: 716-593-3169 • Fax: 716-593-5637
Erosion prevention leading edge tape. P.M. Research manufactures & markets the best in radome erosion protection & leading edge tape. Over 140 protective masks in stock.

Panel Planner Software
P.O. Box 1042 • Boise ID 83701
Ph: 208 323-8724 • Fax: 208-939-8901
Instrument panel planning software.
See Display Ad On Page 57

Parascender Technologies, Inc.
1040 E. Carroll St. Dept. KP
Kissimmee FL 34744
Ph: 407-935-0775 • Fax: 407-935-0778
E-Mail: para@iag.net
Web: http://www.iag.net/~para
Literature/Catalog: Yes
Agricultural spray boom accessory for the Parascender Powered Parachute. Electric rotary spray heads for application of oil and water based chemicals.

Parascender Technologies, Inc.
1040 E. Carroll St. Dept. KP
Kissimmee FL 34744
Ph: 407-935-0775 • Fax: 407-935-0778
E-Mail: para@iag.net
Web: http://www.iag.net/~para
Decks to hold instruments. Deluxe deck holds two 3" gauges, three 2" gauges, and ignition switches. Standard deck houses four 2" gauges and ignition. Black plastic with vibration dampening legs to attach to tubing.

Pflueger's
P.O. Box 187 • #23 Place Drive
Trinity Center CA 96091
Ph: 916 266-3236 • Fax: 916-266-3236
Hardwood veneer aircraft panel used for replacement of plastic fascias or direct laminate of hardwood to aluminum panels. Lights can be built into overlays.

Polycast
70 Carlisle Place • Stamford CT 06902
Ph: 800-245-9002
Quality cleaners

Porta-Bote International
1074 Independence Avenue
Mountain View CA 94043
Ph: 800-227-8882 • Fax: 650-961-3800
E-Mail: info@porta-bote.com
Web: http://www.porta-bote.com
Literature/Catalog: Yes
Folding boats for aviation use. Seat two to four. Folds to 4" flat.

Precise Flight, Inc.
63120 Powell Butte Road • PO Box 7168
Bend OR 97708
Ph: 800-547-2558 • Fax: 503-388-1105
Literature/Catalog: Yes
Speed brakes that provide rapid descent rate control without shock cooling the engine.

Precision Decal Co.
1729 Tilghman Street • Allentown PA 18104
Ph: 610-432-4911 • Fax: 610-432-9234
Literature/Catalog: Yes
Custom computerized signs, decals and logos. Specializing in aircraft applications.

Precision Dial Co.
7240 West KL Avenue • Kalamazoo MI 49009
Ph: 616-375-5601 • Fax: 616-375-1370
Dial and panel printing. Manufacture and refurbish instrument dials, instrument panels, radio/autopilot faces, backlit dials and panels. For homebuilt, warbird, and antique aircraft.

Protect All, Inc.
1910 East Via Burton Street
Anaheim CA 92806
Ph: 714-635-4491 • Fax: 714-635-9716
Literature/Catalog: Yes
Industrial lubrication oils, coolants, solvents & rust preventatives for the metal-working & machining industry. The Protect All product line includes a no rinse automotive wash & a wax, polish & protectant.

AEROPOXY

PTM&W Industries, Inc.
10640 So. Painter Avenue
Santa Fe Springs CA 90670
Ph: 310-946-4511 • Fax: 310-941-4773
Literature/Catalog: Yes
Full line of High Performance Safe AEROPOXY Systems for adhesive bonding, coating, and structural components fabrication.
See Display Ad On Page 14, 35

The GLASS GOOSE By Quikkit

Quikkit Corporation
9002 Summer Glen • Dallas TX 75243-7445
Ph: 214-349-0462 • Fax: 214-349-0462
E-Mail: quikkit@airmail.com
Web: http:////webwrkc.com/glassgoose/
Literature/Catalog: Yes
Manufacture of hydraulic cylinders for experimental aircraft.
See Display Ad On Page 24

RSE, Inc.
1157 57th Drive SE • Auburn WA 98002-8720
Ph: 206-939-4105 • Fax: 206-939-4105
Flight computers, weight and balance computers, aero design and performance software.

RST Engineering
13249 Grass Valley Avenue
Grass Valley CA 95945
Ph: 916-272-2203 • Fax: 916-272-2203
E-Mail: rst-engr@oro.net
Web: http://www.rst-engr.com
Literature/Catalog: Yes
Supplier of intercoms, headsets, and audio panel kits.
See Display Ad On Page 14

Rand Robinson Engineering

15641 Product Lane, A-5
Huntington Beach CA 92649
Ph: 714-898-3811 • Fax: 714-890-1658
E-Mail: pilot@beegroup.com
Web: http://www.fly-kr.com
Literature/Catalog: Yes • Brake systems.

See Display Ad On Page 22

Rick's Aviation

Rt. 2, Box 319 • Bristow OK 74010
Ph: 918-352-3083 • Fax: 918-352-2699
Literature/Catalog: Yes
Dacro-Shield UV Protective dressing for Dacron coverings. Shown to seal porosity, prevent blow-out and make older ultralight planes fly better. 1 quart covers 250 sq. ft. $21.95 plus $4.00 S/H per quart.

Rosen Product Development, Inc.

4678 Isabelle St. • Eugene OR 97402
Ph: 800 284-7677 • Fax: 541-485-8791
Literature/Catalog: Yes
Sun visors for commercial and private aviation. LCD monitor from 5.6" to 21".

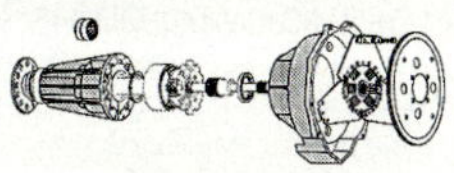

Ross Aero

3824 E. 37th Street • Tucson AZ 85713
Ph: 520-747-7877
P.S.R.V.'S (Planetary Reduction Drives).

See Display Ad On Page 4

Safety, Ltd.

31 W. 651 North Avenue
West Chicago IL 60185
Ph: 708-584-9366 • Fax: 708-584-9327
Seat Belts and restraint systems to your specifications and color.

Scott Engineering

P.O. Box 427 • Auburn IN 46706
Ph: 219-925-0316 • Brake systems.

Seahawker/Glass

GOOSE NEWSLETTER
9002 Summer Glen • Dallas TX 75243

Seelye Equipment Specialist

1217 State Street • Charlevoix MI 49720
Ph: 616-547-9430 • Fax: 616-547-5522
Literature/Catalog: Yes
Solar electric (photovoltaic) panels for charging batteries. Power for remote hangers providing power for doors, lights, tools, outdoor security lights, runway lights. Portable, flexible panels for emergency battery charging and camping.

Shapiro Supply

5617 Natural Bridge • St. Louis MO 63120
Ph: 314-382-7000 • Fax: 314-382-8557
Aluminium: sheet, bar, tubing, extrusions. 4130 steel: sheet, tubing, bar., AN, MS hardware.

Sky Dynamics Corporation

Rt. 1, Box 170 • Moneta VA 24121
Ph: 703-297-6754 • Fax: 703-297-2240
Literature/Catalog: Yes
Exhaust systems and exotic alloy fabrication.

Sky-Ox, Ltd.

P.O. Box W • Edwardsburg MI 49112-0723
Ph: 800-253-0800 • Fax: 616-663-2579
Windsocks, Windsock Frames

Skyline Aviation

8298 Reid Road, Hangar #11
Swartz Creek MI 48473
Ph: 810-635-8724 • Fax: 810-635-2553
Literature/Catalog: Yes
Tail-Dragger Dragger (tm). An easy to use dolly for moving any tailwheel airplane. Works on most nose wheels also. Plans, kits, or complete.

Skymax Aircraft Corporation

1921 NE 28th Street
Lighthouse Point FL 33064
Ph: 888-942-7611 • Fax: 954-942-5470
E-Mail: skymax@shadow.net
Literature/Catalog: Yes
CUSTOM: log pouches and bags, pitot tube covers, propeller banners, imprinted sportswear and screen printing.

Southwest Aero, Inc.

659 Strander Blvd., Bldg. F
Tukwila WA 98188
Ph: 206-575-8732 • Fax: 206-575-1201
Supplier of propeller governors.

Specialty Photo Products

PO Box 801241 • Houston TX 77080-1241
Ph: 713-939-0353 • Fax: 713-939-0352
Color reproductions from aircraft photo on a porcelain mug or ballcap. $13.95 + $2.50 S/H. Check or money order. Photos returned.

Steen Aero

Steen Aero Lab, Inc.

1210 Airport Road • Marion NC 28752
Ph: 704-652-7382 • Fax: 704-652-7382
Imported streamline wires & tie rods featuring stronger rolled threads. Stainless terminal assemblies and drag/anti-drag wires also available. Good prices, great delivery.

See Display Ad On Page 2

Stoffel Seals Corp.

P.O. Box 825' • 400 High Avenue
Nyack NY 10960
Ph: 914- 353-3800 • Fax: 914-353-3876
Manufacturer of Promotional Wings, tamper resistant security seals, labels and tags, name badges,key tags.

Tailwinds, Inc.

42 Digital Dr., #14 • Novato CA 94941
Ph: 415-883-7300 • Fax: 415-883-7599
Web: http://www.tailwinds.com
Unique gifts. Pilot watches, flight jackets, software simulation, computerized log books, flight planners.

Tana Wire Markers

P.O. Box 370 F • California MO 65018
Ph: 314-796-3812 • Fax: 314-796-3770
Literature/Catalog: Yes
Manufacturer of Tana Wire Markers. Mark power, communication and guy wires at airport or helicopter approach areas, river and canyon crossings, construction sites.

Tele-Optics

5514 Lawrence Avenue
Chicago IL 60630-3494
Ph: 773-283-7757 • Fax: 773-283-7757
Sales and repair of all makes of binoculars and telescopes. Literature available on request.

Texas Aeroplastics

Northwest Regional Airport • Rt. 10, Box 17
Roanoke TX 76262
Ph: 817-430-3651 • Fax: 817-491-4849
Literature/Catalog: Yes
FAA-PMA fairings and plastic replacement parts for single engine Cessnas and Pipers.

Tubular Dynamics

1207 NE Cookingham Drive
Kansas City MO 64155
Ph: 816-734-8211 • Fax: 816-734-8211
Literature/Catalog: Yes
Manufacturers of the original Jig-A-Joint and Jig-A-Joint2 tubing notchers.

Turner Aircraft Co., Inc.

Route 4, Box 115AB3 • Luscombe Acres Airport
Grandview TX 76050
Ph: 817-783-5350 • Fax: 817-783-5350
E-Mail: turnert40@aol.com
Literature/Catalog: $6.00
Aviation consultant - FAA DER; structures, and systems equipment. Homebuilt Aircraft plans.

Tuthill Corp.,Cablecraft Div.

P.O. Box 11372 • Tacoma WA 98411
Ph: 253-475-1080 • Fax: 253-474-1623
Literature/Catalog: Yes
Manufacturer of mechanical push-pull control cables.

The Ultimate Door, Inc.

N8778 Sky Lane • East Troy WI 53120
Ph: 414-642-7788 • Fax: 414-642-5163
Hardware kit containing ready to assemble parts for constructing a hangar door. Other material purchased locally saving shipping costs. Completed doors in 49 states and 3 foreign countries.

Venture Cinchers

PO Box 2163 • Ft. Smith AZ 72902
Ph: 501-648-4987
Literature/Catalog: $3.00
Custom built shoulder harnesses for ultralights and experimentals. Materials are all military grade with quick release latches.

VORTECH, Inc.

PO Box 511-KP • Fallston MD 21047
Ph: 410-692-2692 • Fax: 410-692-5902
Web: http://www.webcom.com/neatstuf/helio
Literature/Catalog: $10.00
Rotor blades, all-aluminum for helicopters and gyroplanes.

See Display Ad On Page 21

Vortecs Aircraft

10201 N.E. 51st Ave. • Vancouver WA 98686
Ph: 360-576-1245 • Fax: 360-576-8130
E-Mail: vortecs@teleport.com
Design and fabrication of experimental aircraft components. Builders assistance.

WAG-AERO

The Wag-Aero Group

1216 North Road, Box 181 • Lyons WI 53148
Ph: 800-558-6868 • Fax: 414-763-7595
E-Mail: wagaero-sales@wagaero.com
Literature/Catalog: Yes
Apparel including leather, flight, and satin jackets, gifts, pilot supplies, airport equipment, windsocks and frames, and runway lights. Seat belts and shoulder harnesses, runway lights, weather stations, windsocks and frame.

See Display Ad On Page 13

Zeftronics

1622 E. Whaley Street
Longview TX 75601-6830
Ph: 903-758-6661 • Fax: 903-236-9766
Literature/Catalog: Yes
Voltage regulators, alternator control units & electrical parts.

See Display Ad On Page 61

Services: Associations

ASC

Aero Sports Connection

PO Box 589 • Marshall MI 49068
Ph: 616-781-4021 • Fax: 616-781-7400
E-Mail: kimojim@aol.com
Literature/Catalog: Yes
ASC is an organization designed to support the ultralight community. Non-profit clubs available within organization. Holds two place training exemption.

See Display Ad On Page 49

Aircraft Owner & Pilot Assoc.

Oedenkovenstraat 39 • B-2140 Borgerhout
Antwarp BELGIUM
Ph: 32-3-236-2954 • Fax: 32-3-236-2954

Aircraft Owner & Pilot Assoc.

38 Fyshwick Plaza • 59 Wollongong Street
Fyshwick, A.C.T. 2609 AUSTRALIA
Ph: 61-6-280-4221 • Fax: 61-6-280-7341

Aircraft Owner & Pilot Assoc.

European Regional Office • 50a Cambridge Street
London SWIV 4QQ UNITED KINGDOM
Ph: 44-71-834-5631 • Fax: 44-483-35839

Aircraft Owners & Pilots Assn.

International Headquarters • 421 Aviation Way
Frederick MD 21701
Ph: 301-695-2221 • Fax: 301-695-2375

American Hatz Assn.

16225 143rd Avenue SE • Yelm WA 98597
Ph: 360-894-2862 • Fax: 360-894-2862
Literature/Catalog: Yes
International Association dedicated to building and flying the Hatz CB-1 Biplane. Quarterly newsletter. Forums at Oshkosh and Biplane Expo. Free sample issue.

American Waco Club

3546 Newhouse Place • Greenwood IN 46143
Ph: 317-736-5991 • Fax: 317-736-9020
Literature/Catalog: Yes
350 member type club, 6 newsletters annually. Annual membership dues $25.00. Factory drawings and construction data.

Antique Airplane Association

Antique Airfield • 22001 Bluegrass Road
Ottumwa IA 52501
Ph: 515-938-2773 • Fax: 515-938-2773
Literature/Catalog: Yes
Aircraft History Association for people interested in restoring and flying vintage aircraft. 24 active chapters and 11 active type clubs. Annual dues in AAA $36.00 includes Antique Airplane Association News and Antique Airplane Digest. Affiliated with Air Power Museum Inc.

The BD-5 Network

34311 Bodkin Terrace • Fremont CA 94555
Ph: 510-797-2804, E-Mail: BD5NET@AOL.COM
Literature/Catalog: Yes
The definitive BD-5 info source. Quarterly newsletter plus member/supplier list. $25/Year($29 US Overseas).

Buker Club

6438 West Millbrook Road
Remus MI 49340

Business Aircraft Users Assoc.

50 Rochester Row
London SW10 UNITED KINGDOM

Cassutt Sport Racer

National Aeronautics Co. • 5611 Kendall Court, No. 4 • Arvada CO 80002

Cherokee Pilots Association

PO Box 7927 • Tampa FL 33673

Club Newsletters, Various
PO Box 127 • Blakesburg IA 52536
Ph: 515-938-2773 • Fax: 515-938-2773
Literature/Catalog: $12.00
Newsletters-three issues of one title: $15.00. Choose from Corben, Culver, Fairchild, Great Lakes, Hatz, International Pietenpol, Parakeet, Interstate, Rearwin, Travel Air, Caribbean Breeze (6th Air Force). Future newsletters: Airmaster and Fly Baby.

Confederate Air Force
PO Box 62000 • Midland TX 79711

Corben Club
P.O. Box 127 • Blakesburg IA 52536
Ph: 515-938-2773

Dragonfly Builders
1112 Layton Drive • Olathe KS 66061
Ph: 913 764-5118
E-Mail: DBFNSpud @AOL.COM
Newsletter for Dragonfly builders and pilots. Newsletters is published 6 times a year for $18.00 in the U.S., $20.00 Alaska, Mexico and Canada and $29.00 outside of North America. Currently 360 subscribers in 19 different countries.

Europa Club
4 Saville Close • Cheltenham
Gloucestershire GL50 4NE UNITED KINGDOM • Ph: 44-1242-525-674
E-Mail: rowil@gn.apc.org
Web: http://www.avnet.co.uk/europa/
The Europa Club is for all Europa builders, future builders, operators, and enthusiasts. It exists to help us all construct and operate our Europas enjoyably, efficiently, and safely.

Experimental Aircraft Assoc.
EAA Aviation Center • PO Box 3086
Oshkosh WI 54903-3086
Ph: 920-426-4858 • Fax: 920-426-6560
E-Mail: dnocenti@eaa.org
Web: http://www.eaa.org
Literature/Catalog: Yes
The EAA is a membership organization geared towards sport flying. A wide variety of books and videos are available. Membership is $35 per year and includes 12 issues of SportAviation.
See Display Ad On Page 33

Flying Flea Squadron England
18 Bredon • VW Redditch Worc B97 5EA
UNITED KINGDOM
Ph: (44) 1527 543110 • Fax: (44) 1527 545439
Flying Flea/Pou Ou Ciel Squadron for builders of Henri Mignet designs and types. Members quarterly newsletter: "Le Sport de L'Air". Free copy sent on request. Info. 18, Bredon Vw, Redditch. Worcs. England B97 5EA.

Formula Air Racing Assoc.
2 Nowells Farm Cottages • Lower Beeding
Horsham RH13 6LN UNITED KINGDOM

Historic Aircraft Association
26A Brookhill • Woodstock Oxfordshire
OX7 1JE UNITED KINGDOM

Long-EZ
Central States Association • 9283 Lindberg Blvd.
Olmsted Falls OH 44138 • Ph: 216-826-3055
Central States Assn. is dedicated to helping canard (Rutan type) builders through activities and a quarterly technical newsletter. Cost is $20.00 per calendar year. Send for information.

National Bucker Club
Route 1, Box 419 • Moody TX 76557

PFA Pietenpol Squadron Eng.
1 Humbmer Road • Blackheath, London
SE3 UNITED KINGDOM

Porterfield Airplane Club
91 Hickory Loop • Ocala FL 34472
Ph: 352-687-4859

Pulsar Builders Association
PO Box 13941 • Scottdale AZ 85267
Fax: 602-482-7882
E-Mail: flypulsar@aol.com
Literature/Catalog: Yes
Support group for Pulsar builders and pilots. Quarterly newsletter $12.50. Complete back issue package (1-40) is $25.00.

Quicksilver Pilots Association
328 Luscombe Drive • Los Lunas NM 87031
Ph: 505-865-3466 • Fax: 505-866-7308
E-Mail: PFPA@aol.com
Publishers of Quicksilver Magazine.

Replica Fighters Association
2409 Cosmic Drive • Joliet IL 60435

Rotax Owners Association
S22, C39, RR6 Stn Main • 6235 Ohanagan Landing Road • Vernon, B.C. V1T 6V5
CANADA • Fax: 604-549-7111
Member newsletter providing service and technical information on ROTAX UL engines. Technical commentary on recent service and related information. $50.00 (US) per year in North America, $75.00 (US) per year international. Back issues available.

Sailplane Homebuilders Assoc.
21100 Angel Street • Tehachapi CA 93561
Ph: 805-822-8852
The Sailplane Homebuilder's Association provides a forum for exchanging information on building and operating soaring aircraft through its periodical "Sailplane Builder" and through workshops.

Seabee Club International
6761 NW 32nd Avenue
Fort Lauderdale FL 33309

Seaplane Pilots Association
421 Aviation Way • Frederick MD 21701

Silver Wings Fraternity
PO Box 44208 • Cincinnati OH 45244

Starduster Magazine
5725 SW Mc Ewan Road
Lakek Oswego OR 97035

T-18
Mutual Aid Society • Rt. 3, Box 295
Clinton IL 61727

The 99's, Inc.
Will Rogers Airport • PO box 59965
Oklahoma City OK 73159

UNITED STATES ULTRALIGHT ASSOCIATION
PO Box 667 • Frederick MD 21705
Ph: 301-695-9100 • Fax: 301-695-0763
Web: http://www.usua.com
Organization whose members seek to promote, protect, and represent the sport of aviation through the development of powered ultralights and microlights.

UK Jodel Club
Brooklands Church Lane • Brafield-On-The-Green • Northampton NN7 1Ba
UNITED KINGDOM

Ultralights
Phantom/Flightstar Pilots Assn
328 Luscombe Drive • Los Lunas NM 87031
Ph: 505 865-3466 • Fax: 505 866-7308
E-Mail: PFPA@ Aol.com
Publishers of Phantom Driver.

United States Ultralight Assoc
P.O. Box 667 • Frederick MD 21705
Ph: 301-695-9100 • Fax: 301-695-0763
Literature/Catalog: Yes
USUA seeks to promote and protect the sport of aviation through the development of ultralights and microlights. Monthly magazine. Representation in Washington.

Vintage Ultralight Assoc.
74 Brookwood Dr. • Marietta, CA 30064
Ph: 770-973-3860
Owners club and technical support registry committed to the preservation of ultralight planes designed between 1978 and 1988. Plans, blueprints, technical and sales documentation archived for future use as well as registry of existing planes and owners.
Contact: Scott Perkins.

Weedhopper Owners Club
74 Brookwood Dr. • Marietta, CA 30064
Ph: 770-973-3860 • 770-422-2597
Owners club and technical registry committed to the maintenance and enhancement of "Weedhopper" ultralight planes. Safety alerts, spare parts sources, buy/sell info etc. Shared via newsletter and member roster. Contact: Scott Perkins.

World War I Aeroplanes, Inc.
15 Crescent Road • Poughkeepsie NY 12601
Ph: 914-473-3679 • Literature/Catalog: $5.00
A tax-exempt service organization founded in 1961 devoted to builders, historians, modellers and restorers of historical aircraft.

Services: Builders

Aberle Custom Aircraft
2141 S. Mission Road • Fallbrook CA 92028
Ph: 760-723-1731
Over 30 years assisting show quality and performance oriented builders.

Advanced Composite Tech.
R.D. #5, Box 214C • Somerset PA 15501
Ph: 814-445-3802
Personal composite kitplane construction since 1981; sub-assemblies or complete. Long-Ez, Velocity, Lancair, Q2, Cozy, etc. Formerly Custom Composite Components.

AIRCRAFT SUPER-MARKET
#10 Ellefson Dr. • PO Box 676
De Soto IA 50069
Ph: 515-834-2225 • Fax: 515-834-2152
E-Mail: bobdurly@netins.net
Web: http://www.netins.net/showcase/t-birds
Literature/Catalog: $5.00
Handles recreational aircraft of all kinds. New and used, ultralights, powered parachutes and U/L gliders. 2600' runway.
See Display Ad On Back Cover

Alturair
1405 N. Johnson • El Cajon CA 92020
Ph: 619-449-1570 • Fax: 619-442-0481
Literature/Catalog: Yes
BD-5 parts and services including 100 hp rotary engine.

Arizona Light Aircraft
14473 Fountain Hills Blvd.
Fountain Hills AZ 85268
Ph: 602-837-5578 • Fax: 602-837-5578
E-Mail: titan@interramp.com
Web: http://webmill/titan • Literature/Catalog: Yes
Builder services for the Titan Tornado series of aircraft.

Aviation Engines, Inc.
301 Airport Street • Centreville AL 35042
Ph: 800-344-7918 • Fax: 205-926-5186
E-Mail: aviationengines@msn.com

BD Micro Technologies, Inc.
1260 Wade Road • Siletz OR 97380
Ph: 541-444-1343
Literature/Catalog: $20.00
Builders service, kits & parts for the BD-5.

Cardinal Helicopter Training
4184 Taylor Road, Suite P
Batavia OH 45103-9720
Ph: 513-732-5852 • Fax: 513-732-5853
Home builders center for airplanes and helicopters. Building is 7,000 square feet with heating and air conditioning. Power supply for both air and electric tools. Space rented by day, week, or month.

Art Chard
PO Box 492 • North Plains OR 97133
Ph: 503-647-9803
Newly constructed RV-4 wings include one-piece top skins, extended range fuel tanks (38 gallons) and foam-reinforced wing tips. Sub-assemblies for other RV models available.

Ray Collier
25 Arlington Avenue • Santa Barbara CA 93101
Ph: 805-963-2772
A&P, sheet metal, composite mechanic. FAA certification guaranteed.

Flexi-Dyne, Inc.
Rt. 1, Box 361 • S. Pittsburg TN 37380
Ph: 615-837-8537
A&P Mechanical Engineer. Experienced in all types of homebuilt construction.

Garage Built Technologies
Box 271 Dept. AC • Wilmot OH 44689-0271
Plans for building a hot wire styrofoam cutter for composite aircraft parts, models, signs and molds, hand held or stationary.

GBI - George & Becki Orndorff
2290 W. Hicks Rd., Unit 18
Ft. Worth TX 76131
Ph: 817-439-3280 • Fax: 817-439-3280
E-Mail: orndorffg@aol.com
Builder assistance to complete your project.

Grove Aviation
4360 Oakdale Place • Pittsburg CA 94565
Ph: 800-558-0048 • Literature/Catalog: Yes
Construction and design/modification of everything from basic components to complete aircraft. Fixed Wing - Helicopters - Jets. Specialist in the Seawind design.

Jones Custom Aircraft Services
PO Box 157 • Laurelton PA 17835
Ph: 717-922-4877 • E-Mail: sjonesfly@AOL.com
Literature/Catalog: Yes
Custom building of complete tube, wood, fabric, and composite aircraft, kitplanes, components, and parts. Rebuilding and restoration of FAA certified aircraft. Heliarc and machine shop services. Design and flight testing services. Prototype aircraft construction and testing.

Kits Made 4 You
2847 Tanglewood Blvd. • Orange Park FL 32065
Ph: 904-272-7253
Over seventeen years experience in structural aviation. Complete or partial kit building. All metal kits a specialty.

LesAir Inc.
HCR 31 Box 441 • 817 Lindbergh Avenue
Sandy Valley NV 89019-8909
Ph: 702-723-1950 • Fax: 702-723-1950
Gyroplane parts and construction service.

NCMD
RD 1, Box 92A • Somerset County Airport
Friedens PA 15541
Ph: 814-443-3376 • Fax: 814-443-1847
Literature/Catalog: Yes
Specializing in high performance, single engine aircraft. Aeronautical engineers, A&P specialists. VHS Video available for $25.00.

P & K Enterprises
11470 W. 800N • Colfax IN 46035-9636
Ph: 765-324-2424 • Fax: 765-324-2424
E-Mail: PKENTERPRISES@tctc.com
Web: http://pkenterprises/tctc/com/html
Literature/Catalog: Yes
Builder services for Rans, Quicksilver, Zenair and Carlson kits.

Sparcraft Aero-Marine Mfg., Co.
3001 So. Steele Street • Tacoma WA 98409
Ph: 206-474-2140 • Fax: 206-627-8395
E-Mail: sindlilj@pop3.puyallup.wednet.edu
Milling and/or assembling of wood aircraft parts.

Steelox Systems, Inc.
2412 Courseview Drive • P.O. Box 8181
Mason OH 45040-8181
Ph: 513-573-5200 • Fax: 513-573-5511
For Hangars.

Vortecs Aircraft
10201 N.E. 51st Ave. • Vancouver WA 98686
Ph: 360-576-1245 • Fax: 360-576-8130
E-Mail: vortecs@teleport.com

Zenair, Ltd.
Huronia Airport
Midland, Ont. L4R 4K8 CANADA
Ph: 705-526-2871 • Fax: 705-526-8022
E-Mail: zenair@barint.on.ca
Literature/Catalog: $15.00
Assembly Service

Services: Designers & Engineers

Aberle Custom Aircraft
2141 S. Mission Road • Fallbrook CA 92028
Ph: 760-723-1731
Over 30 years assisting show quality and performance oriented builders.

Aero Marine Engineering
9727 Hagel Circle • Lorton VA 22079
Ph: 703-541-0766 • Fax: 703-550-1236
Design, development, and engineering of low speed aircraft. Propeller design. CAD drafting. Licensed professional engineer.

Aero-K Aviation Products Co.
Redlands Municipal Airport • 1717 Sessums Dr.
Redlands CA 92374
Ph: 909-794-6631 • Fax: 909-794-9611
Literature/Catalog: Yes

AeroElectric Connection
6936 Bainbridge Road • Wichita KS 67226-1008
Ph: 316-685-8617
E-Mail: nuckolls@aeroelectric.com
Web: http://www.aeroelectric.com
Literature/Catalog: Yes
The AeroElectric Connection is an electrical systems handbook for individuals who build and maintain their own airplanes. The AeroElectric Connection is adding a line of supplies and accessories to aid builders in assembling their projects. Brochure available for #10 SASE.

AirBoss Aerospace
2198 Piccardo Circle • Stockton CA 95207
Ph: 209-952-6645 • Fax: 209-952-6645
E-Mail: mbetto@aol.com
Web: http://members.aol.com/mbetto/AirBoss.html
Specializing in aircraft design, structural analysis, project management and financial fundrasing.

Aircraft Designs, Inc.
5 Harris Court, Bldg. S • Monterey CA 93940
Ph: 408-649-6212 • Fax: 408-649-5738
Literature/Catalog: $12.00

AIRCRAFT SUPER-MARKET
#10 Ellefson Dr. • PO Box 676
De Soto IA 50069
Ph: 515-834-2225 • Fax: 515-834-2152
E-Mail: bobdurly@netins.net
Web: http://www.netins.net/showcase/t-birds
Literature/Catalog: $5.00
Handles recreational aircraft of all kinds. New and used, ultralights, powered parachutes and U/L gliders. 2600' runway.
See Display Ad On Back Cover

Amax Engineering
5 Utrecht Court • Donvale, Victoria, 3111 AUSTRALIA
Ph: 61-3-8423132 • Fax: 61-3-8418177
Literature/Catalog: $5.00
Design service for engine mounts and accessory installations for aircraft, gyroplanes and airboats.

AMTECH Services, Inc.
6643 TR 49 • Mansfield OH 44904
Literature/Catalog: $1.00
Engineering and advice for aircraft builders and designers concerning building, design, structures and aerodynamics. Providing engineering service to individuals and industry since 1964.

Avion Research Corp.
1022 W. Maude Ave. #102
Sunnyvale CA 94086
Ph: 408-738-1690 • Fax: 408-738-1678
Literature/Catalog: Yes
Replacements for all-metal, instrument panels, and lighted glare shields for Cessna, Piper, and Beechcraft.

THRUST = DRAG
Bates Engineering
2742 Swansboro Road • Placerville CA 95667
Ph: 530-622-1886 • Fax: 530-622-1886
E-Mail: bateseng@innercite.com
Web: http://www.innercite.com/~bateseng
Literature/Catalog: Yes
Airplane and propeller design services. Programs for aircraft design, performance, stability and control. Propeller design optimization software.
See Display Ad On Page 82

Bradley Aerospace
950 Muir Ave. • Chico CA 95926
Ph: 916-899-7918 • Fax: 916-899-2942
Web: http://www.kgmsoftworks.com/bradley
See Display Ad On Page 40

Campbell Aircraft
P.O. Box 610518-A • San Jose CA 95161-0518
Ph: 408-251-3109 • Literature/Catalog: Yes

Delta-Therm Corp.
PO Box 345 • 398 W. Liberty Street
Wauconda IL 60084-0345
Ph: 800-526-7887 • Fax: 847-526-4456
Literature/Catalog: Yes
Hangar door rail de-icing systems. Manufacturers and Designers of electric and hydronic heat transfer systems.

Eklund Engineering, Inc.
19960 Elliott Road • PO Box 1510
Lockeford CA 95237-1510
Ph: 209-727-0318 • Fax: 209-727-0873
E-Mail: ThorpT18@compuserve.com
Aviation engineering consulting. T-18 Thorp plans sales and builder support. T-18 tooling and fabrication facilities available for builder rental.

Richard Finch Engineering
P.O. Box 3946 • Santa Barbara CA 93130-3946
Ph: 805-683-5346 • Fax: 805-683-8868
CONSULTING SERVICES related to aircraft, welding & auto engine conversions. Books on Turbocharging, Fire Safety, Welding, Auto Engine Conversions for aircraft. Finchbird V-6 airplane plans.
See Display Ad On Page 32

Flexi-Dyne, Inc.
Rt. 1, Box 361 • S. Pittsburg TN 37380
Ph: 615-837-8537
Special project design and construction. Layouts, CAD, full-size templates.

Flight Technology Int'l., Inc.
1571 Airport Road • Charlottesville VA 22911
Ph: 804-978-4359 • Fax: 804-978-7598
Literature/Catalog: Yes
VOCALIST custom programmed audible checklist at your finger tips. 12 regular checklist positions and 12 emergency positions.

GA Airfoils
416 Riblett Lane • Wilmington DE 19808
Ph: 302-994-0479 • Literature/Catalog: $1.00
Custom airfoil design and analysis. "GA Airfoils" available from EAA bookshop, 800-843-3612.

Peter Garrison
1613 Altivo Way • Los Angeles CA 90026
Ph: 213-665-1397 • Fax: 213-953-8378
E-Mail: 73113.2165@compuserve.com
Web: http://www.iac.net/~aerol.
Aircraft design software.

Hoffmann Propeller GmbH & Co. KG
Küpferlingstrasse 9
Rosenheim D-83022 GERMANY
Ph: 49-8031-1878-O • Fax: 49-8031-1878-78
Literature/Catalog: Yes
Manufacture & overhaul of propellers for aviation and hovercraft. Wind tunnel fan blades.

Kits Made 4 You
2847 Tanglewood Blvd. • Orange Park FL 32065
Ph: 904-272-7253
Over seventeen years experience in structural aviation. Complete or partial kit building. All metal kits a specialty.

Kollsman, Inc.
220 Daniel Webster Hwy., S.
Merrimack NH 03054
Ph: 603-889-2500 • Fax: 603-889-7966
Manufacturer of avionics products including flight and engine instruments, air data test equipment, and cabin pressurization systems.

Kosola & Associates, Inc.
5601 Newton Road • P.O. Box 3529
Albany GA 31707
Ph: 800-456-7652 • Fax: 912-888-5766
Literature/Catalog: Yes

Marshall Consulting
720 Appaloosa Drive • Walnut Creek CA 94596
Ph: 510-945-6051 • Fax: 510-945-1461
Consulting on core materials, sandwich structures and composites. Author of "Composite Basics" covering all aspects of composite materials, design and manufacturing.
See Display Ad On Page 38

McCutchen Sky Wheels
555 Alter Street #15 • Broomfield CO 80020
Ph: 303-465-5794 • Fax: 303-469-2801
Literature/Catalog: Yes
Composite rotor systems for autogyros up to 1500 lbs gross weight and composite helicopter blades up to 13 feet long - 8" chord-symetrical airfoils.

McGettrick Structural Engr.
12513 E. Castlewood • Wichita KS 67206
Ph: 316-688-5668 • Fax: 316-688-5668
Literature/Catalog: Yes
Loads per FAR23, structural analysis, preliminary sizing and optimization. Structural layout and composite design.

Micro-AG Systems
Box 1932, Hwy 120 S. • Lake Geneva WI 53147
Ph: 414-248-2288 • Literature/Catalog: Yes
Design and prototype of ultra & hyper low volume crop spraying equipment.

National Products, Inc.
1017 S. Elmgrove St., Seattle WA 98108
Ph: 206-763-8361 • Fax: 206-763-9615
E-Mail: stuff@ram-mount.com
Web: http://www.ram-mount.com
Literature/Catalog: Yes

Pagnotta Engineering, Inc.
261 Beacon Drive, Phoenixville PA 19460
Ph: 610-933-2023 • Fax: 610-933-2023
E-Mail: pageng@fast.net
Literature/Catalog: Yes
Complete aircraft, CAD, and finite element analysis software. Composites, metals, hand calculations, loads, fatigue. FAA Structures Designated Engineering Representative.

Steelox Systems, Inc.
2412 Courseview Drive • P.O. Box 8181
Mason OH 45040-8181 • Ph: 513-573-5200
Fax: 513-573-5511
For Buildings.

Synergium
PO Box 21481 • Tampa FL 33622
Literature/Catalog: $10.00
NASA-style aero design CFD system. Info Pack includes demo disk.

TERF Inc.
11411 Culver • Brighton MI 48114
Ph: 810-229-2082 • Fax: 810-229-2082
E-Mail: Mtait@livgonline.com
Web: http://www.terf.com
High tech consulting on sensors and general subsystem integration. Producer of th RAF CD ROM Project.
See Display Ad On Page 52

Thinc
PO Box 16611 • Greenville SC 29606
Ph: 864-292-5821 • Literature/Catalog: $5.00
Helicar-car to helicopter conversion drawings. Car with V6 engine, 3.OL or car with V8 engine 7.4L, either $160.00.

Total Aircraft Services, Inc.
21363 Lassen, Suite 105 • Chatsworth CA 91311
Ph: 818-718-5962 • Fax: 818-718-5963
E-Mail: TASFISH@IX.netcom.com.
Literature/Catalog: Yes
Engineering design, analysis, and certification for aircraft structures, sytems, and power plants.

Turner Aircraft Co., Inc.
Route 4, Box 115AB3 • Luscombe Acres Airport
Grandview TX 76050
Ph: 817-783-5350 • Fax: 817-783-5350
E-Mail: turnert40@aol.com
Aviation consultants - FAA DER; structures, mechnical systems equipment, and administrative.

Tuthill Corp.,Cablecraft Div.
P.O. Box 11372 • Tacoma WA 98411
Ph: 253-475-1080 • Fax: 253-474-1623
Literature/Catalog: Yes
Designer and manufacturer of mechanical "push-pull" control cables.

Vortecs Aircraft
10201 N.E. 51st Ave. • Vancouver WA 98686
Ph: 360-576-1245 • Fax: 360-576-8130
E-Mail: vortecs@teleport.com

Washington Aeroprogress
8535 Perimeter Rd. S. • Seattle WA 98108
Ph: 206-763-4915 • Fax: 206-768-0661
Literature/Catalog: Yes
Develop airplanes of all types, fundamental research of flight aerodynamics, wind tunnel tests, etc.

University of Washington
Aeronautical Laboratory • Kirsten Wind Tunnel Box 352400 • Seattle WA 98195-2400
Ph: 206-543-0439 • Fax: 206-543-0217
E-Mail: uwal@aa.washington.edu
Web: http://www.aa.washington.edu/uwal/uwal.html • Literature/Catalog: Yes
Wind tunnel testing of aircraft models to determine the aerodynamic characteristics. Test results allow the designer to know how the aircraft performs and what needs improvement.

Stanley Widmer Associates, Inc
1206 Prairie Avenue W. • Staples MN 56479
Ph: 218-894-3466 • Fax: 218-894-2018
Literature/Catalog: Yes
Construction and design consulting: CAD lay-outs, airbrush renderings, pattern and mold development, glass prototypes. Finite element analysis. Over 30 years experience in thermoplastics and themo-set.

Zeftronics
1622 E. Whaley Street
Longview TX 75601-6830
Ph: 903-758-6661 • Fax: 903-236-9766
Literature/Catalog: Yes
Consulting on charging systems.
See Display Ad On Page 61

Services: Engine Overhaul & Repair

A&G Aviation Ltd.
Bldg 198, Bournemouth Intl, Christchurch, Dorset BH23 6NE UNITED KINGDOM
Ph: 0202 578558 • Fax: 0202 570203

Aircraft Specialties Services
2860 North Sheridan Road • Tulsa International Airport • Tulsa OK 74115
Ph: 800-826-9252 • Fax: 918-836-4419
Specializes in reconditioning and remachining key components of piston-driven aircraft engines. FAA repair station DD2R764K.
See Display Ad On Page 74

Aircraft Super-Market
#10 Ellefson Dr. • PO Box 676
De Soto IA 50069
Ph: 515-834-2225 • Fax: 515-834-2152
E-Mail: bobdurly@netins.net
Web: http://www.netins.net/showcase/t-birds
Literature/Catalog: $5.00
Handles recreational aircraft of all kinds. New and used, ultralights, powered parachutes and U/L gliders. 2600' runway.
See Display Ad On Back Cover

Airflow Performance, Inc.
P.O. Box 3364 • 101 Belton Dr.
Spartanburg SC 29304-3364
Ph: 864-576-4512 • Fax: 864-576-0201
Literature/Catalog: Yes
Manufacturer of fuel injection systems for experimental aircraft. Overhaul Bendix fuel injection system for experimental aircraft.

Airscrew Performance
7308 W. Griffin Avenue • Glendale AZ 85303
Ph: 602-931-6667
New, used and rebuilt engines, redrives, exhausts. Crankshaft rebuilding. CDI conversions.

Ajax Aviation, Inc.
319 Wolf Road • San Antonio TX 78216-4673
Ph: 800-531-7212 • Fax: 210-342-4361
Crankshaft repair - Lycoming and Continental. FAA certified repair station # BT2R728K.

Amax Engineering
5 Utrecht Court • Donvale, Victoria, 3111 AUSTRALIA
Ph: 61-3-8423132 • Fax: 61-3-8418177
Literature/Catalog: $5.00
Cam grinding; complete zero time rebuilt engines.

Ameratech Aviation
One Omni Dr. Ct. • Hangar 1
Cahokia IL 62206
Ph: 618-332-0082 • Fax: 618-332-0084
Overhaul and repair on Limbach engines.

Approved Aircraft Accessories
29300 Goddard Road
Romulus MI 48174
Ph: 313-946-9000 • Fax: 313-946-5547
Literature/Catalog: Yes
Overhaul of various aircraft engines, landing gear, and accessories. Manufacture test equipment.

Atlas Motors Inc.
685 North Loudoun Street
Winchester VA 22601
Ph: 800-336-9878 • Fax: 540-665-2025
See Display Ad On Page 73

Avia-Hamilton Std. Aviation
PO Box 22 • Hlavenecl 29
Stara Boleslav 250 02 CZECH REPUBLIC
Ph: 42-202-811114 • Fax: 42 202-811818
Literature/Catalog: Yes
Production, repairs, and overhauls of all-metal constant speed, automatic and gound adjustable propellers in the range of 160-220 hp for Lycoming 0-320 and LOM M 332, 337 engines. 600-777 hp for Walter M 601 engines.

Aviation Engines, Inc.
301 Airport Street • Centreville AL 35042
Ph: 800-344-7918 • Fax: 205-926-5186
E-Mail: aviationengines@msn.com
Repair station; engine/engine accessories overhaul and horsepower increases for experimentals / homebuilts.

Aviation North
1641 Airport Road • Alpena MI 49707
Ph: 517-356-6361 • Fax: 517-356-0952
Literature/Catalog: Yes
FBO, performing minor and major repairs to aircraft; hangar service; flight school; line service.

Barrett Performance Aircraft, Inc.
2870-B N. Sheridan Road • Tulsa OK 74115
Ph: 918-835-1089 • Fax: 918-835-1754

Bradley Aerospace
950 Muir Ave. • Chico CA 95926
Ph: 916-899-7918 • Fax: 916-899-2942
Web: http://www.kgmsoftworks.com/bradley
See Display Ad On Page 40

Nick Carter Engines, Inc.
2116 West G Street • Elizabethton TN 37643
Ph: 423-542-2943 • Fax: 423-542-2633
E-Mail: ncarter@usit.net • Web: http://www.carterengines.com
Overhaul and repair Continental and Lycoming engines. FAA repair station. JAR-145 approved engine rebuilding.

Crystal Shamrock, Inc.
6000 Douglas Drive N • Crystal MN 55429
Ph: 612-533-2214 • Fax: 612-533-5519
35 year old FBO. Certified repair station for upholstery, engines and service.

Custom Airmotive, Inc.
2702-B North Sheridan Road • Tulsa OK 74115
Piston engine rebuild and accessories. All models - Lycoming and Continental.

Drake Air, Inc.
4085 Southwest Blvd. • Tulsa OK 74107
Ph: 800-542-6899 • Fax: 918-445-4155
E-Mail: 102117.171@compuserve.com
Oil related equipment including coolers, pumps, and filter kits.

Dumont Aircraft Engines, Inc.
1605 West State Road 64 • Avon Park FL 33825
Ph: 941-453-2600 • Fax: 941-453-5864

EC-Elmore, Inc.
411 Hwy 301 Blvd. East • Bradenton FL 34208
Ph: 800-596-5566 • Fax: 941-737-0395
FAA repair station with complete aircraft cylinder overhaul services available.

G&N Aircraft, Inc.
Master Engine Builders • Precision Airmotive Service Center
1701 E. Main Street • Griffith IN 46319
Ph: 800-348-6504 • Fax: 219-924-1059
E-Mail: gnair@jorsm.com
Web: http://www.gnaircraft.com
Literature/Catalog: Yes
Custom overhaul, repairs and inspections for Continental and Lycoming piston engines. Overhaul and repair of all accessories: alternators, magnetos, carburetors, etc.
See Display Ad On Page 68

Don George, Inc.
1339 W. Washington St., A1 • Orlando FL 32805
Ph: 407-422-0188 • Fax: 407-422-0836
Custom building and overhaul of aircraft engines.

Great Plains Aircraft
P.O. Box 545 • Boys Town NE 68010
Ph: 402-493-6507 • Fax: 402-333-7750
E-Mail: gpasc@earthlink.net
Web: http://www.greatplainsas.com
Literature/Catalog: $3.00
See Display Ad On Page 80

Green Sky Adventures, Inc.
2395 Cream Ridge Rd. • Orwell OH 44076
Ph: 216-293-6624 • Fax: 216-293-6321
Web: http://www.greensky-adventures.com
Literature/Catalog: $2.00
America's oldest authorized Rotax service center.

Grove Aviation
4360 Oakdale Place • Pittsburg CA 94565
Ph: 800-558-0048 • Literature/Catalog: Yes
Construction and design/modification of everything from basic components to complete aircraft. Fixed Wing - Helicopters - Jets. Specialist in the Seawind design.

J.A. Air Center/Joliet Avionic
3N060 Powis Road • DuPage Airport
West Chicago IL 60185
Ph: 630-584-3200 • Fax: 630-443-0169
Web: http:////aso.solid.com/seller/650/main.htm • Literature/Catalog: Yes
J.A. is the midwest's most complete FBO, offering avionics sales, service, installations, piston and turbine aircraft maintenance.

Kosola & Associates, Inc.
5601 Newton Road • P.O. Box 3529
Albany GA 31707
Ph: 800-456-7652 • Fax: 912-888-5766
Literature/Catalog: Yes
Aeronautical engineering services.

Leading Edge Air Foils
Meadowlake Airport • 8242 Cessna Drive
Peyton CO 80831
Ph: 719-683-5323 • Fax: 719-683-5333
E-Mail: leafkp@kktv.com
Web: http://www.leadingedgeairfoils.com
Literature/Catalog: $6.00
Distributor for ultralight and experimental aircraft, parts, and engines.
See Display Ad On Page 34

Martin Induction Systems
Rt. 6, Box 611-B • Enid OK 73701
Ph: 800-848-6486 • Fax: 405-446-5411
Literature/Catalog: Yes
FAA repair station for overhauling and testing fuel systems, fuel pumps, turbochargers, waste gates, relief valves and controllers. Exchanges available. Also in-house machine shop. Worldwide service.

N-A Welding
540 California Avenue • Grants Pass OR 97526 • Ph: 503-479-0241
FAA approved repair station. Welding, machining, wet magnaflux.

P & K Enterprises
11470 W. 800N, Colfax IN 46035-9636
Ph: 765-324-2424 • Fax: 765-324-2424
E-Mail: PKENTERPRISES@tctc.com
Web: http://pkenterprises/tctc/com/html
Literature/Catalog: Yes
Rotax engine repair station.

Pacific Continental Engines
16041 Arminta Street • Van Nuys CA 91406
Ph: 818-781-4947 • Fax: 818-781-0860
Web: http://www.pcenet.com
Literature/Catalog: Yes
Overhaul and sale of Textron, Lycoming, and Continental engines, components, and parts. Avionics shop sells new, repairs, and maintains all major brands. FBO offers aircraft maintenance.

Penn Yan Aero Service, Inc.
2499 Bath Road • Penn Yan NY 14527
Ph: 8007277230 • Fax: 315-536-2335
E-Mail: PnnYanAero@Aol.Com
Web: http://www.airtourist.com/Penn Yan
Literature/Catalog: Yes
Distributors for Lycoming and Continental. Cylinder, accessories, overhaul, and repair.

Recreational Power
10888 Station Rd. • Columbia Station OH 44028
Ph: 800-583-3306 • Fax: 419-585-6004
Literature/Catalog: Yes
U.S. Hirth engine distributor.

Savage Magneto Service
1069 Langley St., PO Box 2401 • Airport Station
Oakland CA 94614 • Ph: 510-562-1593
Repair of ignition systems on small aircraft. Specialize in vintage planes.

Seneca Light Aircraft Systems, Inc.
5479 E. County Road 38 • Tiffin OH 44883
Ph: 419-585-7002 • Fax: 419-585-6004
Literature/Catalog: Yes
US distributor of Hirth two cycle aircraft engines. Sales, service, parts and engine overhauls for Hirth engines. Supplier of engine instrument systems and propellers. Catalog: $5.00.

Shannon Aircraft Motor Works
Bay 142 Industrial Estate • Shannon Co. Clare IRELAND • Ph: 353-61-472466
Fax: 353-61-472381 • E-Mail: PSN@samw.ie • Web: http://www.iol.ie/~samw
Rewinding of aircraft electrical and generator parts.

T.W. Smith Engine Co., Inc.
4510 Airport Rd. • Hangar #1 Lunken Airport
Cincinnati OH 45226
Ph: 513-871-3500 • Fax: 513-533-244
Web: http://www/twsmith
New and overhauled Continental and Lycoming engines and accessories. Certified Repair Station.

Sun-Tech Innovations
P.O. Box 9154 • Canoga Park CA 91309
Ph: 800-786-8324 • Fax: 818-882-7859
E-Mail: Sun-tech@worldnet.att.net
Web: http://www.balancemasters.com/ultraschem1.html
Balance-Master add-on for propellers and Rotax engines. Reduces vibration on ultralights up to 50%.
See Display Ad On Page 55

2 Stroke International
Beaufort Industrial Park • 8 Schein Loop
Beaufort SC 29902
Ph: 803-846-2167 • Fax: 803-846-2169
E-Mail: engines@2si.com
Web: http://www.2si.com
Literature/Catalog: Yes
Lightweight high-power fan cooled and water cooled engines for aircraft. 20 to 120 hp.

Venture Aerodome
Rt. 1 Box 11631 • Roland OK 74954
Ph: 918-427-6490 • Literature/Catalog: $4.00
Distributor of Hirth engines.

Witte, Aero Doktor
3730 Carmel Lane • Paducah KY 42003
Ph: 502-442-6279 • Fax: 502-442-6279
Restore older fabric airplanes. Also general aircraft repairs and motor overhauls.

Services: Restorations

Aircraft Placards
6 Elm Road • N. Hampton NH 03862
Ph: 603-964-8905 • Fax: 603-926-7855
Literature/Catalog: Yes

Aircraft Remanufacturing
1531 Avohill Dr. • Vista CA 92080-1325
Ph: 760-749-0239 • Fax: 760-749-6384
E-Mail: BJNash@connectnet.com
Literature/Catalog: Yes
Curved metal specialists. Repairs using hammer-mill and english wheeling machines.

Aircraft Super-Market
#10 Ellefson Dr. • PO Box 676
De Soto IA 50069
Ph: 515-834-2225 • Fax: 515-834-2152
E-Mail: bobdurly@netins.net
Web: http://www.netins.net/showcase/t-birds
Literature/Catalog: $5.00
Handles recreational aircraft of all kinds. New and used, ultralights, powered parachutes and U/L gliders. 2600' runway.
See Display Ad On Back Cover

Antique Airplane Association
Antique Airfield • 22001 Bluegrass Road
Ottumwa IA 52501
Ph: 515-938-2773 • Fax: 515-938-2773
Literature/Catalog: Yes
Aircraft History Association for people interested in restoring and flying vintage aircraft. 24 active chapters and 11 active type clubs. Annual dues in AAA $36.00 includes Antique Airplane Association News and Antique Airplane Digest. Affilliated with Air Power Museum Inc.

Approved Aircraft Accessories
29300 Goddard Road • Romulus MI 48174
Ph: 313-946-9000 • Fax: 313-946-5547
Literature/Catalog: Yes
Overhaul of various aircraft engines, landing gear, and accessories. Manufacture test equipment.

C&R Aircraft
PO Box 281 • Quincy IL 62306-0281
Ph: 217-292-5967
Wicker seat restoration and construction.

Micro Surface Finishing Prod.
1217 West Third Street • PO Box 818
Wilton IA 52778
Ph: 800-225-3006 • Fax: 319-732-3390
E-Mail: msfpi@netins.net
Web: http://www.micro-surface.com
Literature/Catalog: Yes
FAA-Certified repairs of pressurized cabin transparencies in place. Repair of crazing, chips and scratches both in-field and at our facility. FAA Repair Station #KY2R969K.

Perrone Leather Apparel, Inc.
48-50 South Main Street • Gloversville NY 12078
Ph: 518-725-9144 • Fax: 518-725-7044
E-Mail: interior@perroneleather
Web: http://www.perroneleather.com
Literature/Catalog: Yes
Fire retardant upholstery leather. "Townsend" interior leathers. FAR Part 25. Burn Certified.

Richfield Aero Partners
820 S. 300 E. • Monroe UT 84754
Ph: 801-527-4294 • Vintage Aircraft restoration. Custom made aircraft parts.

Witte, Aero Doktor
3730 Carmel Lane • Paducah KY 42003
Ph: 502-442-6279 • Fax: 502-442-6279
Restore older fabric airplanes. Also general aircraft repairs and motor overhauls.

Services: Fabric Applicators

Aircraft Super-Market
#10 Ellefson Dr., PO Box 676, De Soto IA 50069
Ph: 515-834-2225 • Fax: 515-834-2152 • E-Mail: bobdurly@net ins.net
Web: http://www.netins.net/showcase/t-birds
Literature/Catalog: $5.00
Handles recreational aircraft of all kinds. New and used, ultralights, powered parachutes and U/L gliders. 2600' runway.
See Display Ad On Back Cover

Grove Aviation
4360 Oakdale Place • Pittsburg CA 94565
Ph: 800-558-0048 • Literature/Catalog: Yes
Construction and design/modification of everything from basic components to complete aircraft. Fixed Wing - Helicopters - Jets. Specialist in the Seawind design.

Jones Custom Aircraft Services
PO Box 157 • Laurelton PA 17835
Ph: 717-922-4877 • E-Mail: sjonesfly@AOL.com
Literature/Catalog: Yes
Custom building of complete tube, wood, fabric, and composite aircraft, kitplanes, components and parts. Rebuilding and restoration of FAA certified aircraft. Prototype aircraft design and flight testing.

The Ultralight Store
36120 Pourroy Rd. • Winchester CA 92596
Ph: 909-926-3196 • Fax: 909-926-3196
Literature/Catalog: Yes
Manufacturer of ultralight aircraft replacement sail sets and storage covers. Over 45 types of aircraft models available, many in stock. In business since 1983.

Services: Financing

Green Tree Financial
345 St. Peter St. Ste 1400
St. Paul MN 55102-1639
Ph: 800-851-1367 • Fax: 800-831-8366
Web: http://www.greentree.rotor.com
See Display Ad On Page 6

NAFCO
3907 A Aero Place • Lakeland FL 33811
Ph: 800-999-3712 • Fax: 941-646-1671
E-Mail: nafco@airloans.com
Literature/Catalog: Yes
National Aircraft Finance Company. Nationwide financing for a wide range of aircraft from ultralights to experimental to jets. Financing available for restorations, engine overhauls and avionics updates.

Services: Insurance

Alexander Aviation Associates
PO Box 608067 • Orlando FL 32860-8067
Ph: 800-432-8519 • Fax: 407-889-0101
Direct insurance facilities for aircraft owners, builders, and manufacturers. Also, assistance to local insurance agents with aviation insurance placements.

Antique Airplane Association
Antique Airfield • 22001 Bluegrass Road
Ottumwa IA 52501
Ph: 515-938-2773 • Fax: 515-938-2773
Literature/Catalog: Yes
Aircraft History Association for people interested in restoring and flying vintage aircraft. 24 active chapters and 11 active type clubs. Annual dues in AAA $36.00 includes Antique Airplane Association News and Antique Airplane Digest. Affilliated with Air Power Museum Inc.

Assoc. Aviation Underwriters
51 John Kennedy Parkway • Short Hills NJ 07078 • Ph: 201-379-0806

AUA, Inc.
P.O. Box 35289 • Greensboro NC 27425
Ph: 800-727-3823 • Fax: 910-668-3697
Literature/Catalog: Yes
EAA antique & classic insurance program, experimental & homebuilt aircraft insurance, airport liability insurance, fire insurance on your hangar, & pilots universal life insurance.

Avemco Insurance Co.
411 Aviation Way • Frederick MD 21701
Ph: 800-638-8440 • Fax: 301-694-4232
E-Mail: mkt.@ave.com
Web: http://www.avemco.com
Literature/Catalog: Yes
Liability insurance and physical damage coverage for loss or damage to your homebuilt, kitbuilt or restoration aircraft. Based on where you are in the construction process. The official insurer of EAA aircraft insurance plan.

Commercial Aviation Insurance
14555 N. Scottsdale Rd. #210
Scottsdale AZ 85254
Ph: 602-922-7117 • Fax: 602-922-9779
Literature/Catalog: Yes
Aircraft physical damage and liability coverage provider through independent aviation insurance agencies.

Commercial Aviation Insurance
3675 Crestwood Pkwy, Ste. 330
Duluth GA 30136 • Ph: 404-806-1115

Cooper Aircraft Insurance
PO Box 66 • Addison TX 75001
Ph: 800-969-1902

Costell Insurance Assoc., Inc.
PO Box 28280 • Tempe AZ 85285-8280
Ph: 800-528-6483
Aviation insurance for experimental and homebuilts.

Dempsey Blanton Insurance
1401 General Aviation, #200 • PO Box 362025
Melbourne FL 32936-2025
Ph: 800-327-2222

Falcon Insurance Agency of FL
KB & Associates • P.O. Drawer 6220
Lakeland FL 33807-6220 • Ph: 813-646-9688

H R Jennings & Co. Ltd.
61 Brighton Road • So Croydon
Surrey CR2 6ED UNITED KINGDOM
Ph: 081-680-0688

Lance Toland Associates
1035 South Hill Street • Griffin GA 30223
Ph: 800-282-1219

Mesirow Insurance
350 North Clark Street
Chicago IL 60610-4796
Ph: 800-453-0600

Nat'l. Assurance Underwriters
16501 Sherman Way, Van Nuys CA 91406
Ph: 800-628-4636

PIK West Insurance Agency
PO Box 3474 • Simi Valley CA 93093
Ph: 805-522-3428 • Fax: 805-522-1739
Web: http://aviationinsurnetwork.com
All sizes of accounts in all classes of aviation insurance since 1979. Serving AL, AZ, CA, CO, FL, GA, ID, OR, NV, NM, TX, UT and WA. Hull and liability coverage.

Sedgwick Limited
PO Box 1150, Station B, Montreal, Quebec
H3B 3K9 CANADA • Ph: 800-361-1625
Fax: 514-842-1540 • Literature/Catalog: Yes
Aviation insurance. Insurance administrators to Canadian Owners and Pilots Association.

Shaw Aviation Insurance Services
11344 Coloma Rd. • Suite 190
Gold River CA 95670-4458
Ph: 916-853-8080 • Fax: 916-853-8090
Variety of aviation insurance coverages.

Skysmith International, Ltd.
2525 NW 71st Place • Ankeny IA 50021-9048
Ph: 800-743-1439 • Fax: 515-289-1235
E-Mail: SkySmith@aol.com • Skymaster, Pitts, and Smith Miniplane group insurance.
See Display Ad On Page 60

Westwind Insurance Services
PO Box 2770 • Victorville CA 92393
Ph: 619-955-4800

Services: Painters

Aerial Arts
6091-B East Young St. • Suite 110
Elma WA 98541
E-Mail: 102361.617@Compuserve.com
Literature/Catalog: $1.00
Custom paint schemes designed for all types of aircraft using the latest computer technology. Paint type and color recommendations included. For information please send 1.00

Aircraft Super-Market
#10 Ellefson Dr. • PO Box 676
De Soto IA 50069 • Ph: 515-834-2225
Fax: 515-834-2152 • E-Mail: bobdurly@netins.net
Web: http://www.netins.net/showcase/t-birds
Literature/Catalog: $5.00
Handles recreational aircraft of all kinds. New and used, ultralights, powered parachutes and U/L gliders. 2600' runway.
See Display Ad On Back Cover

Burbank Sign Company
454 N. Moss Street • Burbank CA 91502
Award winning graphics, lettering, instrument panels and aircraft markings. Special and unusual projects using state-of-the-art computer technology.

Grove Aviation
4360 Oakdale Place • Pittsburg CA 94565
Ph: 800-558-0048 • Literature/Catalog: Yes
Construction and design/modification of everything from basic components to complete aircraft. Fixed Wing - Helicopters - Jets. Specialist in the Seawind design.

KC Graphics
1899 Orchard Lake Road
Ste.Sylvan Lake MI 48320
Ph: 810-810-9119 • Fax: 810-332-9157
Literature/Catalog: Yes • Stencils

NCMD
RD 1, Box 92A • Somerset County Airport
Friedens PA 15541 • Ph: 814-443-3376
Fax: 814-443-1847 • Literature/Catalog: Yes
Award winning epoxy primer, base coat, clear coat plus system, P.P.G. VHS Video available for $25.00.

Prestige Aircraft, Inc.
Franklin County Airport (FS)
Swanton VT 05488 • Ph: 802-868-3443
E-Mail: prestige@together.net
Web: http://www.paintaircraft.com
Aircraft painters specializing in smaller aircraft, including experimental and home-built aircraft.

Services: Test Pilots

BD Micro Technologies, Inc.
1260 Wade Road • Siletz OR 97380
Ph: 541-444-1343 • Literature/Catalog: $20.00
Pilot Check out for BD-5.

Grove Aviation
4360 Oakdale Place • Pittsburg CA 94565
Ph: 800-558-0048 • Literature/Catalog: Yes
Construction and design/modification of everything from basic components to complete aircraft. Fixed Wing - Helicopters - Jets. Specialist in the Seawind design.

Jones Custom Aircraft Services
PO Box 157 • Laurelton PA 17835
Ph: 717-922-4877 • E-Mail: sjonesfly@AOL.com
Literature/Catalog: Yes
Flight test program development and experimental flight testing by professional FAR Part 23/25 experimental test pilot.

NCMD
RD 1, Box 92A • Somerset County Airport
Friedens PA 15541
Ph: 814-443-3376 • Fax: 814-443-1847
Literature/Catalog: Yes
Standard and check out programs acceptable for insurance. VHS Videos available for $25.00.

Services: Schools

Abaris Training Resources
5401 Longley Lane, Ste. 49, Reno NV 89511
Ph: 800-638-8441 • Fax: 702-827-6599
E-Mail: cn@training.abaris.com
Web: http://www.abaris.com
Literature/Catalog: Yes
Advanced composite training. Classes in design, fabrication, and repair. Since 1983.

Alabama Aviation and Tech. College
P.O.Box 1209 • U.S. Hwy. 231 South
Ozark AL 36361-1209
Ph: 800-624-3468 • Fax: 334-774-6639
E-Mail: aatc@snowhill.com
Web: http://www.snohill.com/aatc
Literature/Catalog: Yes
Public two-year certificate and AA programs in airframe and powerplant mechanics, aviation electronics and flight. VA approved. FAA certified. Financial aid available.

ALEXANDER SPORTAIR WORKSHOPS
219A Barry Whatley Way, Griffin GA 30224
Ph: 770-467-9490 • Fax: 770 467-9413
E-Mail: workshops@sportair.com
Web: http://sportair.com • Literature/Catalog: Yes
Airplane builders workshops, hands-on experience in fabric sheet metal, composites and welding.
See Display Ad On Page 22

Allied Helicopter Intl., Inc.
P.O. Box 6216 • Tulsa OK 74148
Ph: 918-425-7558
F.A.A. 141 helicopter school with over 45 years of service.

Aviation Education Consultants
P.O. Box 833 • 144 Lakeview Road
Ozark AL 36360 • Ph: 334-774-5980
Fax: 334-774-1373 • Literature/Catalog: Yes
Prep courses for all FAA exams. Specialize in the A&P Mechanic's Cource with on-site testing and oral and practical examiners.

Aviation North
1641 Airport Road • Alpena MI 49707
Ph: 517-356-6361 • Fax: 517-356-0952
Literature/Catalog: Yes
Flight school.

Aviation West
3205 Lakewood Blvd. • Long Beach CA 90808 • Ph: 310- 425-6774

BAKER'S School of Aeronautics
1645 Murfreesboro Road • Suite K
Nashville TN 37217 • Ph: 615-361-6787
Fax: 615-361-0010 • E-Mail: KINGSSCH@Aol.com
Literature/Catalog: Yes
Ground school preparation for all FAA exams, including A&P, IA, private through ATP. FAA exams administered on site after course.

Canadian Powered Parachutes Ltd.
Box 1769 • Vegreville, Alberta T9C 1S8 CANADA
Ph: 403-658-2044 • Fax: 403-658-2387
E-Mail: lindam@agt.net
Web: http://www.agt.net/public/lindam cpp2.htm • Literature/Catalog: Yes

Coastal Aviation
900 Aviation Drive • New Bern NC 28560
Ph: 919-637-9942

Colorado Unlimited Adventure
PO Box 100 • Tumacacori AZ 85640-0100
Ph: 602-961-3580 • Fax: 602-961-3581
Two and Four cycle engine training.

Dakota Aero Tech, Inc.
1801 23rd Avenue N., Room 111 • Hector Field
Fargo ND 58102
Ph: 701-237-5305 • Fax: 701-239-9073

East Coast Aero Tech. School
150 Hanscom Dr. #RM 124 • Bedford MA 01730-2630 • Ph: 508-371-9977
Fax: 508-371-3159
Certificate program in aircraft maintenance technology. FAA certifications and lifetime placement service.

Federal Exams
5602 N. Rockwell, Bethany OK 73008
Ph: 800-787-2340, Literature/Catalog: Yes
A&P writtens, oral, and practical. Housing and transportation available in Oklahoma City. Job assistance/mobile shcool. I.A. 5 day course, library, and microfiche readers.

Flightcraft, Inc.
7505 NE Airport Way • Portland OR 97218
Ph: 503-331-4200 • Fax: 503-331-4273
E-Mail: AAH07@AOL.COM
Custom panels for Glasair, Lancair. Factory authorized Lancair flight training.

Fly For Fun
71 South Shore Dr. • Cuba NY 14727
Ph: 716-968-3758 • Fax: 716-968-3758
Literature/Catalog: Yes
Aerobatic instruction from beginning to advanced.

Free Flite Enterprises
R.D. 2, Box 2121 • East Stroudsburg PA 18301
Ph: 717-421-8900

GBI - George & Becki Orndorff
2290 W. Hicks Rd., Unit 18 • Ft. Worth TX 76131
Ph: 817-439-3280 • Fax: 817-439-3280
E-Mail: orndorffg@aol.com
RV construction videos. Sheet metal working classes. Line of custom interior products.

Hemet-Ryan Aerobatic Flt. Ctr.
36890 Waldon Weaver Road, Hgr.
Hemet CA 92545
Ph: 909-658-1116 • Fax: 909-927-8415
Literature/Catalog: Yes
Aerobatic training. Tailwheel training. Aircraft include Citabria, PT-17, L-19, AT-6, BT-13 and T-28.

Kemmeries Aviation
8710 W. Carefree Hwy. • Peoria AZ 85382
Ph: 602-566-8068 • Fax: 602-561-2287
Web: http://www.sportplanes.com/ kemries.htm
Ultralight flight instruction and rentals.

King Schools
3840 Calle Fortunada • San Diego CA 92123
Ph: 619-541-2200 • Fax: 619-541-2201
E-Mail: Acharney@kingschools.com
Web: http://www.kingschools.com
Literature/Catalog: Yes
Written exam video courses for a variety of licenses.

King's School of Aeronautics
9109 Tara Blvd. • Jonesboro GA 30236
Ph: 770-477-1282
Literature/Catalog: Yes
Two day courses held around the U.S. for mechanics inspection authorization certificate.

LesAir, Inc.
HCR 31 Box 441 • 817 Lindbergh Avenue
Sandy Valley NV 89019-8909
Ph: 702-723-1950 • Fax: 702-723-1950
Gyroplane flight Instruction from ultralight to private pilot license. All instruction by Gyro CFI.

University of North Dakota
Center for Aerospace Sciences • P.O. Box 9007
Grand Forks ND 58202-9007
Ph: 701-777-2791 • Fax: 701-777-3016
E-Mail: flyund@aero.und.nodak.edu
Web: http://www.aero.und.nodak.edu
Literature/Catalog: Yes
UND Aerospace at the University of North Dakota is one of the leading collegiate aviation programs in the world offering degrees in professional flight, aviation management, airport management, air traffic control, and space studies. The program operates a training fleet of 90 aircraft.

P & K Enterprises
11470 W. 800N • Colfax IN 46035-9636
Ph: 765-324-2424 • Fax: 765-324-2424
E-Mail: PKENTERPRISES@tctc.com
Web: http://pkenterprises/tctc/com/html
Literature/Catalog: Yes
Intro flights; flight training in ultralights and light aircraft.

Stanford Flying Club
1886 Embarcadero Road • Palo Alto CA 94309
Ph: 415-858-2200
Economical pilot training. Free use of educational aviation video library for members. Single engine, high performance, aerobatic and multi-engine aircraft.

Thomason Aircraft Corporation
16700 Roscoe Blvd. • Van Nuys CA 91406
Ph: 818-908-5800
Aerobatics school.

Toccoa Aviation, Inc.
Route 3 Airport • Toccoa GA 30577
Complete flight school and FBO.

Ultralight Safaris
23450 Vista Way • Quail Valley CA 92587
Ph: 800-RANSFLY • Fax: 909-765-5022
Literature/Catalog: $20.00
Dealer for RANS aircraft AIR CREATION TRIKE, SWIFT MOTORGLIDER and flight school. Brochure available at no cost.

United States Ultralight Assoc
P.O. Box 667 • Frederick MD 21705
Ph: 301-695-9100 • Fax: 301-695-0763
Literature/Catalog: Yes
Ultralight information and instruction. Builder assistance and flight training through national network of flight instructors and Clubs.

Vista Aviation, Inc.
12653 Osborne Street • Pacoima CA 91331
Ph: 818-896-6442 • Fax: 818-896-9541
Avionics shop. Maintenance, parts and equipment. Flight school.

Daniel Webster College
20 University Drive • Nashua NH 03063-1300

Western Ultralight
4700 188th N.E. • Arlington WA 98223
Ph: 360-435-3737 • Fax: 360-435-3737
Total Ultralight facility. Including flight instruction, sales, parts, and maintenance.

Wings Aviation
T-406 Kelly Avenue - Airport
Manchester NH 03103
Ph: 603-627-8663 • Fax: 603-627-3063
Literature/Catalog: Yes
Flight training and aircraft rentals. Aircraft management program.

Zenair, Ltd.
Huronia Airport
Midland, Ont. L4R 4K8 CANADA
Ph: 705-526-2871 • Fax: 705-526-8022
E-Mail: zenair@barint.on.ca
Literature/Catalog: $15.00
ZODIAC CH 601 and STOL CH 701 weekend sheet-metal workshops at the factory. Workshops designed to introduce first-time builders to all-metal construction techniques.

Services: Miscellaneous

Abbot Brown & Sons Ltd.
Barnfleet Works • Beaminster, Dorset DT8 3EJ UNITED KINGDOM
Ph: 01308 862404 • Fax: 01308 863402
Nameplates and labels, engraving and printing, rules and protractors.

Aircraft Belts Inc.
2020 Anders Lane • PO Box 845
Kemah TX 77565-0845
Ph: 713-538-1284 • Fax: 713-538-2225
Literature/Catalog: Yes
Rewebbing services, plating, engraving.

Aircraft Fuel Injection Serv.
2731 Brookfield Avenue • Dallas TX 75235
Ph: 800 846-2515 • Fax: 214-358-5896
Literature/Catalog: Yes
Fuel injection and carburetor overhauls up to 1340 engines. Continental, Bendix, Precision, Thompson, Lear, Romec, Beechcraft, Pesco and Titan.

Aircraft Insurance Agency
P.O. Box 987 • Lancaster TX 75146
Ph: 214-227-8688 • Fax: 214-218-6114
Aircraft insurance for experimental homebuilt aircraft.

Aircraft Lighting & Components
249 Brockway Road • Chehalis WA 98532
Ph: 360-748-0234 • Fax: 360-748-6455
Literature/Catalog: Yes
Cable fabrication.

Aircraft Locators & Supply
Rt.1, Box 138A • Peculiar MO 64078
Ph: 816-779-6075 • Fax: 816-758-6054
Aviation export services for engines, propeller and complete aircraft. Now for domestic engines and propellers.

Aircraft Paint Schemes
518 Nimrod Street • Nevada City CA 95959
Ph: 800 599-5499 • Fax: 916 265-3145
E-Mail: laura@gv.net
Web: http://www.shawsantiques.com/aps.htm
Literature/Catalog: Yes
Aircraft exterior design, computer assisted. Over 6 years experience.

Aircraft Placards
6 Elm Road • N. Hampton NH 03862
Ph: 603-964-8905 • Fax: 603-926-7855
Literature/Catalog: Yes
Placards, instrument panel layout.

Aircraft Projects
2269 Maywood Ave. • San Jose CA 95128
Ph: 408-277-0645 • Fax: 408-279-6535
E-Mail: BAFRANK@att.net
Literature/Catalog: $6.00
Welding fabrication supporting V-6 STOL, Honer type wing tips, Clark 'Y' airfoil, and other Ford 3.8L V-6 engine users.

AIRCRAFT SPECIALTIES SERVICES

Aircraft Specialties Services
2860 North Sheridan Road • Tulsa International Airport • Tulsa OK 74115
Ph: 800-826-9252 • Fax: 918-836-4419
Literature/Catalog: Yes
FAA certified repair station D2R764K. Inspection, remachining of aircraft engine parts. Full service, new and used parts.
See Display Ad On Page 74

Anderson Enterprises, Inc.
405 Osage • Derby KS 67037
Ph: 800-732-6875 • Fax: 316-788-8904
Literature/Catalog: Yes
Customized mahogany replicas of any aircraft, helicopter, or ship. Aviation art and accessories.

Applied Technology
1801 Central Road • Glenview IL 60025-4228
Ph: 847-998-1892
Technical documention services (eg-builder's manuals) for the homebuilt industry.

Robert Platt Bell, P.C.
PO Box 19668 • Alexandria VA 22320
Ph: 703-683-8822 • Fax: 703-683-8823
E-Mail: ROBERTBELL@BELLPAT.COM
Literature/Catalog: Yes
Patent, Trademark, & Copyright services. First consultation free.

Burbank Sign Company
454 N. Moss Street • Burbank CA 91502
Award winning graphics, lettering, instrument panels and aircraft markings. Special and unusual projects using state-of-the-art computer technology.

Butler Parachute Systems, Inc.
1820 Loudon Avenue NW • PO Box 6098
Roanoke VA 24017-0098
Ph: 540-342-2501 • Fax: 540-342-4037
E-Mail: butlerchut@aol.com
Parachute repacks and servicing.

C.A.A. Services
6-C Herman Drive • East Granby CT 06026
Ph: 203-658-8000 • Fax: 203-651-0885
Literature/Catalog: Yes
Radio repair, custom avionics harnesses.

Continental Fuel Cell Repair Co., In
N. 2224 Locust Rd., #2 • Spokane WA 99206
Ph: 509-927-7764
Remanufacture and repair of bladder type aircraft fuel cells. Certified FAA repair station.

Delta-Therm Corp.
PO Box 345 • 398 W. Liberty Street
Wauconda IL 60084-0345
Ph: 800-526-7887 • Fax: 847-526-4456
Literature/Catalog: Yes
Hangar door rail de-icing systems. Manufacturers and Designers of electric and hydronic heat transfer systems.

Engine Oil Analysis
7820 So. 70th East Avenue • Tulsa OK 74133
Ph: 918-492-5844 • Fax: 918-492-5844
E-Mail: hfentontul@aol.com
Literature/Catalog: Yes
Wear metal analysis of used aircraft engine oils and other fluids.

Flight Instrument Service, Inc
1619 Desoto Road • Sarasota FL 34234-3063
Ph: 941-355-2221 • Fax: 941-355-2221
FAA certified instrument service shop.

Floats & Fuel Cells
4010 Pilot Drive, Ste. 3 • Memphis TN 38118
Ph: 800-674-6148 • Fax: 901-794-7845
E-Mail: ffesales@ffcfuelcelk.com
Web: http://ffcfuelcells.com
Literature/Catalog: Yes
Custom fuel bladder manufacturing and repair. Helicopter float repairs.

Free Flite Enterprises
R.D. 2, Box 2121 • East Stroudsburg PA 18301
Ph: 717-421-8900
Kit Planes and Ultra lite Sales & Services

William J. Grohs Aviation Inc.
Waterbury-Oxford Airport • Main Terminal
Oxford CT 06478-1096
Ph: 203-262-1552 • Fax: 203-262-1556
Aviation insurance brokerage.

KelPak East
840 35th Ct. SW • Vero Beach FL 32968
Ph: 800-626-0367 • Fax: 561-569-3947
We overhaul turbochargers, controllers, wastegates and pressure relief valves. We offer exchanges on parts mentioned above.

Mena Aircraft Interiors Inc.
P.O. Box 777 • Mena Municipal
Mena AR 71953 • Ph: 501-394-7717
Fax: 501-394-7717 • Literature/Catalog: Yes
Custom aircraft interior shop. Call for free quote at 1-800-354-7717.

Rhino Aviation
607 Garber Street • Caldwell ID 83605
Ph: 800-701-9128 • Fax: 208-459-0819
CNC cutting of sheet metal and metal plate to customer specifications. Items cut include instruments panels, stainless steel firewalls, metal brackets, tabs and plates. Custom prototype and production. For brochure, send self addressed stamped envelope.

Sailplane Homebuilders Assoc.
21100 Angel Street • Tehachapi CA 93561
Ph: 805-822-8852
Annual workshops in Tehachapi, CA and on east coast. May also have workshop in central US.

Sky Struck
167 North Oakwood Rd. • Oshkosh WI 54904
Ph: 800-759-7875 • Fax: 414-231-8297
E-Mail: sonerai@vbe.com
Web: http://www.vbe.com~sonerai
Aircraft construction seminars and workshops. Offering "design specific" weekend workshops and one day seminars to take builders directly from anticipation through completion. With hands-on experience, practical knowledge, and personal contact with designers and builders of your chosen model.

Southwest Aero, Inc.
659 Strander Blvd., Bldg. F • Tukwila WA 98188
Ph: 206-575-8732 • Fax: 206-575-1201
Propeller governor overhaul and repairs.

Ultralight Safaris
23450 Vista Way • Quail Valley CA 92587
Ph: 800-RANSFLY • Fax: 909-765-5022
Literature/Catalog: $20.00
Dealer for RANS aircraft AIR CREATION TRIKE, SWIFT MOTORGLIDER and flight school. Brochure available at no cost.

Wells Cargo, Inc.
P.O. Box 728 • Elkhart IN 46514
Ph: 219-264-9661 • Fax: 219-264-5938
Web: http://www.wellscargo.com
Enclosed Industrial & Commercial Trailers.

Zeftronics
1622 E. Whaley Street
Longview TX 75601-6830
Ph: 903-758-6661 • Fax: 903-236-9766
Literature/Catalog: Yes
Overhaul of voltage regulators.
See Display Ad On Page 61

Miscellaneous

Aircraft Windshield Co.
10871 Kyle Street, Los Alamitos CA 90720
Ph: 560-430-8108 • Fax: 560-598-0716
Wing tip and landing light lenses.

Aviation Engines, Inc.
301 Airport Street, Centreville AL 35042
Ph: 800-344-7918 • Fax: 205-926-5186
E-Mail: aviationengines@msn.com
Magnuflux

Aviation Products, Inc.
P.O. Box 8303, Cedar Rapids IA 52408
Ph: 319-396-6283
Literature/Catalog: Yes
A line of replica pedal toys for 3 to 7 year-olds. Replicas include the Pitts, Eagle, Grumman F3F & Gulfhawk, and others. All plans include color photo, full size patterns with materials list. Send S.A.S.E for information.

Best Sign Systems
1202 N. Park Avenue • Montrose CO 81401-3170
Ph: 970-249-2378 • Fax: 970-249-0223
Literature/Catalog: Yes
Control panel faces. Product and logo ID. Warning signs.

Vern E. Clements
308 Palo Alto Drive, Caldwell ID 83605
Ph: 208-459-7608
Literature/Catalog: $4.00
Super-scale replica model construction plans for a variety of aircraft. Used for Delmar Benjamin's GEE BEE R-2, Eicher/Kimball GEE BEE "Z", Jim Jenkins' "BULLDOG", Bill Turner "BULLDOG".

Eastwood Co.
580 Lancaster Ave. , PO Box 296
Malvern PA 19355-0714
Ph: 610-640-1450X2211 • Fax: 610-644-1028
Literature/Catalog: Yes
Transportation memorabilia.

The Fab Shop
400 N. Walnut • PO Box 422
McPherson KS 67460
Ph: 800-966-1358 • Fax: 316-241-1570
Literature/Catalog: Yes
Towbars and Electric Service Carts for small aircraft. Telescoping universal, lightweight universal and lightweight collapsable hand towbar. The electric service cart is totally electric, built-in APU system.

Falcon Watch Co.
2510 Brentwood Blvd., St. Louis MO 63144
Ph: 314-962-8222 • Fax: 314-962-8222
Web: http://www.quikpage.com/f/falconeatch/ • Literature/Catalog: Yes
Pilot supplies, watches, 24 hour dual time pilot watches. Free color brochure and information.

Fisher Aero Corp.
7118 S.R. 335, Portsmouth OH 45662
Ph: 614-820-2219 • Fax: 614-820-2219
Aviation art, custom aircraft jewelry, etc.

Forest View Enterprises
PO Box 550, Yucaipa CA 92399-0550
Ph: 909-797-7644 • Fax: 909-797-7264
Literature/Catalog: Yes
Custom made pins, medallions, key chains, emblems & embroidered patches. Design services available on request.

Hidalgo, Inc.
45 La Buena Vista, Wimberley TX 78676
Ph: 512-847-5571 • Fax: 512-847-2393
Literature/Catalog: Yes
Pilot's sunglasses, prescription glasses, model airplanes, watches, etc.

KelPak Industries, Inc.
1545 E. Acequia • Visalia CA 93292
Ph: 800-523-4809 • Fax: 209-627-2873
Literature/Catalog: Yes
Exchange or overhaul on Garrett turbochargers, wastegates, and Rajay turbochargers.

Phlogiston Products, Inc.
P.O. Box 9 • Gaston OR 97119
Ph: 503-985-3212 • Fax: 503-985-1927
Literature/Catalog: Yes
Prefabrication of main wing spars for Van's RV-4and RV-6 Kits. Service saves builder 50 to 100 hours of build time and eliminates risk of problems in spar buldup.

Precision Dial Co.
7240 West KL Avenue • Kalamazoo MI 49009
Ph: 616-375-5601 • Fax: 616-375-1370
Dial and panel printing. Manufacture and refurbish instrument dials, instrument panels, radio/autopilot faces, backlit dials and panels. For homebuilt, warbird, and antique aircraft.

RJS Accessories, Inc.
18966 G Drive North • Marshall MI 49068
Ph: 800-226-7571 • Fax: 616-781-2911
Literature/Catalog: Yes
Mail order sunglasses. Ray-Ban, Serengeti, Vuarnet and Randolph Engineering. Discount prices.

Rudy Aircraft Instruments, Inc
4711 Old Bowman Road • Rudy AR 72952
Ph: 501-474-8759 • Fax: 501-474-3306
Exchange pool available. 1 day overhaul of most aircraft instruments and vacuum pumps.

San Diego Leather Jacket Factory
1121 Bay Blvd. • Chula Vista CA 91911
Ph: 800-232-6626 • Fax: 619-575-3373
E-Mail: nancy@leather.com
Literature/Catalog: Yes
Air force, navy and commercial pilot flight jackets.

Scholarships, Inc.
PO Box 1383-AC • Goleta CA 93116
Ph: 800-333-5608 • Fax: 805-967-5781
E-Mail: g71115@aol.com
Literature/Catalog: Yes
College scholarships (including aviation).

Larry Smith
939 Office Park Rd., Suite 104
West Desmoines IA 50265
Ph: 515-222-0242
Fax: 515-247-9245
E-mail: bizopps@raccoon.com
Business brokerage services for aviation related businesses.
See Display Ad On Page 432

WAG-AERO

The Wag-Aero Group
1216 North Road, Box 181 • Lyons WI 53148
Ph: 800-558-6868 • Fax: 414-763-7595
E-Mail: wagaero-sales@wagaero.com
Literature/Catalog: Yes
Exhaust systems, engine mount and seat belt remanufacturers.
See Display Ad On Page 13

Wilco, Inc.
3502 W. Harry • Witchita KS 67213
Ph: 800-767-7593 • Fax: 316-943-9664
E-Mail: wilco@feist.com
Web: http://www.globalair.com/wilco/
De-ice timers, amp meters, de-ice boots, prop de-ice, and slip rings. L.P. Aero Products.

Kenneth Woodard Co.
103 Van Keeck Road • Mills MA 02054
Ph: 508-376-8667 • Fax: 508-376-8667
E-Mail: Kennwood@AOL.com
Literature/Catalog: Yes
"Ortho" is an easy-to-use 2- and 3-dimensional CAD program for preliminary design and advanced projection solutions.

EMC SOLUTIONS CONTINUED FROM PAGE 66

glideslope. GPS operates up in the multi-gigahertz range; it would be furthest removed in spectrum from most sources of noise.

We might conclude that Loran is the most susceptible to noise. If operating frequency were the only consideration, this would be true. The manner in which a received signal carries intelligence has a lot to do with noise immunity, irrespective of operating frequency. For example, if FM transmissions were permitted on the same frequencies as AM broadcast, the FM receiver would show little irritation by the same noise source that would make an AM receiver useless. Loran uses mathematically encoded signals which are digitally processed. Digital filters easily sort the total received spectrum to separate out orderly data from noise which is random in nature. GPS is also digitally encoded. It has the further advantage of being many more decades removed in frequency from most noise sources.

Whoa! If Loran does such a whippy job of working through the noise, why is Loran degradation the biggest single noise problem experienced by homebuilders? I'll have to seek refuge in my earlier statements about signal to noise ratios. Few homebuilders are installing both Loran and ADF receivers so it's difficult to find real life examples of relative differences. If both receivers could be compared side by side in the same noise environment, we could demonstrate a Loran installation that performs well right next to an ADF rendered useless by the same noise. I don't mean to imply that the Loran would be working at maximum performance . . . signal to noise ratio indications on the Loran may show the presence of noise. The major difference in performance is due to digital signal processing which may do an entirely adequate job in spite of whatever noise exists.

Antagonists and Their Propagation Paths

Alternators are another common noise source. They are weak emitters of RF energy (harmonics of the AC voltage frequency in the stator windings created by the rectifier diodes) and don't generally become a nuisance by radiation of noise. However, as we'll discuss later, there's a strong AUDIO rate component of noise that rides on the DC output voltage that's nearly impossible to filter directly at the output. Victims pick this kind of noise up by conduction through the power input leads or by ground loops, a special conducted interference mode. The list goes on and we'll go over the major players in detail later on. What you need to understand at the outset is that there are no hard and fast rules about who does what to whom or how it's done. There are common scenarios but they don't account for every set of similar symptoms.

Chasing down an interference problem is like the game of Clue. You get to ask a lot of questions and use the answers to formulate the next question. We'll develop techniques and recommendations for initial installation that will certainly reduce the likelihood of interference problems but there are still no guarantees. Failure to comply with all recommendations herein does not necessarily condemn the transgressor to suffer the effects of the noise in question. For a noise to make a nuisance of itself, it must enjoy the right combination of victim, propagation mode and an operating scenario where the observer judges the noise "unacceptable."

It's impossible to predict how all of the variables will add up. I recall renting a locally based training airplane for a cross country trip. Nobody had ever complained about magneto noise in the radios. Most renter-pilots talked to facilities close by and kept the squelch controls clamped down so tight that the noise wasn't annoying. Over western Texas at 8,500 feet, Flight Watch stations 50 miles away were difficult to hear through the noise.

Another stress on your decision making process will come in the form of free advice. Some readers have confessed to skepticism about my recommendations, "my hangar mate says his airplane has been wired the 'other' way and it works just fine," and it probably does.

The number of potential noise sources in an airplane may surprise you. Just about everyone is familiar with the unneighborly qualities of magnetos and alternators. But did you know that fuel tank senders have been known to emit radio noise? How about rotating beacons? Aileron and elevator hinges? Admittedly, I'm digging pretty deep into the list - these components are rarely a problem.

Sometimes, a source of interference isn't even on board the airplane enduring the problem. I recall an instance years ago where a pilot complained of poor localizer receiver performance that got worse as he approached the decision point in his approach. The problem turned out to be a spurious output from a commercial communications repeater about 1/4 mile from the runway centerline and a mile or so out. The ground based source wasn't a consideration until he discovered that approaches to other airports didn't have the problem.

FM broadcast and ground based radar stations have been known to victimize poorly designed or installed airborne equipment. So, if any broad brush statement can be made about working a noise or interference problem, it is, "Proceed with an open mind. No combination of source or propagation mode should be discounted until analysis dictates it." If the preceding gives rise to feelings of trepidation please understand that dominant noise sources and their interference propagation modes are well known. Most problems are avoided by good design and fabrication practices. In thirty-five years of hammering on airplanes, I've been involved in only a few cases where a noise problem persisted to the "hair-pulling" stage. Ultimately, all noise problems are solvable.

In summary, the chorus antagonists may include alternators, strobe light systems, motors, radio transmitters, oscillators within radio receivers, hand held transceivers, and of course, ignition systems. Less common noise sources include poor connections

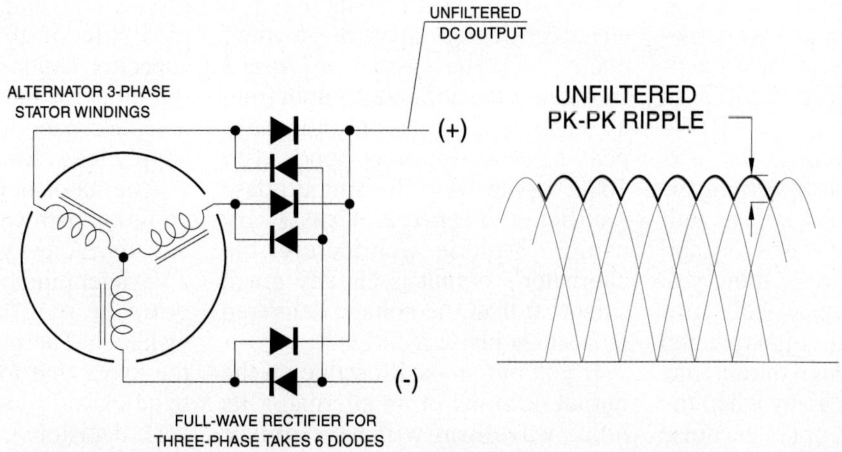

Three phase AC outputs overlap each other by 1/3 their periods. The resulting full wave rectification has a very low PK-PK ripple amplitude without filtering. Compare this with the unfiltered output in Figure 1.

Figure 3

across moving joints such as hinges depended upon for grounding a rudder mounted navigation light. And of course, large relays and contactors have the ability to dump out little, high voltage bursts of energy when current through the coil is dropped to de-energize the contactor. Okay, okay, call 'em "spikes" if you want to. I dislike the term because common perceptions consider a "spike" to be synonymous with "radio killer." Never met a spike I couldn't kill.

Potential Victims

Navigation receivers, communications receivers, audio systems, instrumentation systems, storm scopes, electronic fuel injection systems, electronic ignition systems are all potential victims. Without a modicum of protection from relatively minor excursions of bus voltage, some solid state devices can be damaged. Autopilots can get fussy when other appliances are turned on. Fortunately, protection and packaging techniques are tried and proven. I've written reams of articles and e-mail explaining performance specifications like DO-160 and Mil-Std-704 which trade off ability to limit responses to interference with ability to limit interference at the source. There's simply no excuse for a piece of modern aircraft equipment to be threatened with damage by anything the airplane can throw at it. All that leaves us is an occasional interference problem to identify, analyze and fix.

Propagation Modes

Just because potential victims are robust doesn't mean an antagonist can't be a nuisance! The noises emitted by motors, alternators and other devices may not represent a threat of damage but they can certainly exert some influence ranging from mildly annoying to outright unacceptable. The ways in which a tormentors output gets around the airplane include:

(1) Radiation through the air to be picked up on interconnecting cables, antennas, etc.

(2) Conduction though power and/ or control wiring. Special cases of conducted propagation include . . .

(a) Magnetic coupling between adjacent, parallel wires, and

(b) Electrostatic coupling between adjacent, parallel wires and,

(c) Coupling through shared pathways (ground loops).

I think it's interesting that interference problems can cut both ways. An ignition system can render a communications transceiver useless for listening while the transmitter in the same transceiver can scramble the brains of a poorly shielded electronic ignition system.

Problem Identification

For most pilots, identifying an interference problem is pretty easy. That whine in the headset or a nav receiver dropping a flag is ample notification that a problem exists. Some pieces of equipment, like Loran and GPS receivers have front panel displays that show signal to noise ratio or some other figure of merit with respect to signals. It is fortunate that most of the avionics we buy have been tested for their ability to withstand a certain amount of interference while making sure they don't become antagonists as well.

Common Causes & Their Cures

Alternators: It's a toss-up as to whether the alternator or magnetos are the worst antagonists but we'll take up alternators first. The alternator's noise spectrum covers the frequency ranges from audio (peak-to-peak ripple of rectified AC) with harmonics that extend to the low megahertz range. The higher frequency components of alternator noise are generally beaten down to size with a high quality electrolytic or plastic and film capacitor. Many automotive conversions already feature a good quality capacitor connected from the alternator's b-lead to case ground. The higher frequency emissions from an alternator are a threat to ADF and Loran receivers. There's a small possibility that radio frequency interference from alternators can be heard on AM broadcast and Citizens Band radios.

Alternator interference is characterized by its unique, almost musical whine that changes pitch in response to changes in engine RPM. First you want to confirm that it is indeed alternator noise by turning off the alternator field (switch or pull breaker). If the noise goes away and the alternator is the only thing you shut off, you've pretty well confirmed it. You can tell if it's radiated noise coming in through the antenna if (1) the noise goes away when the antenna is unhooked or (2) the noise is adjustable by the radio's volume control. In the case of Loran receivers, you can try running the receiver on a separate, 12-volt lantern battery and see if signal to noise improves when the radio is powered by a "clean" source. If the noise is still there, it's getting in by way of the antenna as opposed to power leads.

Effective RF filtering for an alternator is often possible with a simple capacitor installation. Plastic and film capacitors on the order of 4-10 microfarads may be tried. Also, computer grade electrolytics on the order of 1,000 to 20,000 microfarads have cured some problems. In either case, the capacitor needs to be mounted right on the rear of the alternator and connected with the shortest practical leads from b-terminal and case ground. Caution - electrolytic capacitors are polarity sensitive. They will feature (+) and/ or (-) symbols or have a red dot of paint on the (+) terminal. Hooking these fellers up backwards is very messy when the case splits open and the gooey stuff comes out!

If a simple capacitor doesn't get it, some amount of series inductance is called for too. In this case, the inductor has to be rated for the alternator's output current which makes the parts get BIG in a hurry. It's worth your trouble to avoid having to resort to series inductance in the alternator b-lead. Reserve it as a last resort. Audio frequency interference from an alternator is quite another matter. EVERY alternator has some amount of audio rate ripple voltage on it. It's a pesky law-of-physics thing. An alternator is a 3-phase device where separate windings 120 electrical degrees apart are feeding energy into the system. *Figure 3* shows how the unfiltered output from a 3-phase rectifier has only about 5% peak-to-peak ripple as opposed to 100% ripple from the single phase rectifier in *Figure 2, View A*. By using a 3-phase architecture, the alternator's output is already much smoother than the voltage delivered by a single phase rectifier.

If you put an oscilloscope on the output terminal of an alternator, its noise waveform will have only a very basic resemblance to the waveform in *Figure 3*. Modern automotive alternators are very compact and quite capable for their weight and size. Design compromises to achieve this performance cause the waveforms to be distorted and loaded with little bits of "trash" . . . short duration, fast moving transitions on the waveform that have HARMONIC content. Remember our discussion on harmonically related power line noise? Alternators can have it too. Our friend the capacitor can help here. If the noise from an alternator shows up where it shouldn't, a high quality capacitor from the B-terminal to ground will "smooth" those little spikes and reduce their ability to annoy.

Here's where our friends Mil-Std-704 and DO-160 come to the rescue. Mil-Std-704 says that audio rate ripple on a "properly designed bus" will not exceed 1.5 volts peak-to-peak on a 14-volt system and for the most part, a stock alternator will comply with this requirement. On the other hand, DO-160 says that we should design airplane goodies so that noises up to and including 1.5 volts peak-to-peak are not a problem. So much for specs. If you've installed a victim that wasn't designed with DO-160 in mind, there's still hope.

Getting audio rate noises filtered off the power feed from bus to a victim appliance isn't too big a deal. Fortunately, the current requirements for most panel mounted gizmos is pretty small. Radio Shack sells a kit consisting of an inductor and capacitor that are quite effective in cleaning up the noises for devices drawing up to an ampere or so. The general rule of thumb for a single capacitor, single inductor filter is that the inductor faces the noise source and the capacitor is across the power supply leads for the victim.

Alternator noise has a variety of common pathways into unwelcome territory. A few years ago we bought a Voyager minivan while at OSH (ya gotta be careful when you visit Oshkosh. There's a certain air about the event that makes us vulnerable to quick and sometimes out-of-character decisions). On the way home I noted that the front seat speakers seemed to be putting out a bit of alternator whine, even when the radio was off. "Aha," says I, "must be a ground loop. I'll bet the front speakers are on the bottom side of the front-rear fader stack and one side of the speaker is locally grounded at its mounting." Combine this with a grounding of the radio chassis at the console and conditions are right for the speakers to tell us what they "hear" in the way of noise flowing in the car's chassis ground.

I must confess that I didn't really take the problem as a serious issue (Dee's note: HE didn't drive it every day!) until Dee decided to get a CD player and a better sounding radio. Then the noise became more than just a nuisance. One Saturday morning, I got out the multimeter and screwdrivers and resolved to find the ground loop and break it. Turns out the speakers were NOT grounded locally to the car chassis. All grounds for the entertainment system were made at the back of the radio chassis. Hmmmmm . . . after some poking around I noted a "fat" wire running across the aft side of the air conditioning ducts. What's more, the noise could not be heard when the radio was hanging out of the panel on its harness. "Aha!!" says I again. I finally experienced a condition that I knew was physically possible but I'd never personally run across it. I moved the "fat" wire up to the top of the duct so that the radio harness didn't press against it while the radio was installed. The noise went away.

This was a classic case of magnetic coupling between parallel conductors. That "fat" wire was some kind of main feedwire between the dashboard loads and the alternator. The higher the loads in the wire (lights on, a/c fan running), the stronger the magnetic field and the greater the noise. Now, it's important to note that this noise could NOT have been cleaned up with shielded wire. A shielded wire simply prevents the conductor within from acting as one plate of a capacitor. A shield over a wire will not prevent magnetic coupling into or out of the wire. I've built tachometers that sense spark plug current and they work just fine on shielded aircraft plug harnesses. You can read the magnetic field around a wire right through a shield.

The easiest way to avoid this scenario in airplanes is simple: keep fat wires away from skinny wires. Fat wires tend to be bus feeds, battery leads, alternator b-lead feeds, flap or pump motor feeds, etc. They are, by their very application, conductors of the airplane's worst noises. The skinny wires associated with instruments, radios, intercom systems, and engine gauges tend to be the most vulnerable. For years I've preached the doctrine of "twisted pairs" in routing ground and power wires down the fuselage of a canard-pusher, get that b-lead breaker off the panel, and ground EVERYTHING behind the panel with a single ground bus. All of these things were recommended in the spirit of reducing the possibility of noise problems later on.

I mentioned the common grounding block idea. Using a single ground point for everything behind the panel is the best way to avoid ground loop induced noises. Our friends at B&C Specialty Products produce a "forest of spade terminals" ground blocks that are very useful in fabricating a single point ground system behind your panel. Ironically, most ground loop problems present themselves in airplanes that seem to have the most favorable grounding situation, all metal ones. It's the assumption that hooking some piece of electrical equipment to the airframe is just as likely to work well irrespective of WHERE on the airframe it is grounded. Problems are most likely in ships with aft mounted batteries grounded locally. Large, noisy alternator currents flow in the ship's bones on their way from crankcase to battery (-). Delicate circuitry with ground in TWO or more places along that pathway may sense and complain about the voltage difference between those two points.

Microphone and headset jacks are common victims of ground loops. A very few millivolts of noise can be easily detected either in the headphones or impressed upon your outgoing signal while speaking on the radio. In metal airplanes, use insulating washers to prevent the mounting bushings of headset and microphone jacks from making contact with the airframe. All audio system grounds should terminate at the audio system's distribution amplifier, usually the intercom system. Then, a single ground wire from the audio system should join all other avionics grounds at the panel equipment ground block.

In summary, alternators have strong noise components in the audio range and cause the most trouble with avionics systems. Audio frequency noise requires a larger filter than for the higher radio frequencies so it's not practical to filter anything but the weaker radio frequency components at the back of an alternator with a high quality capacitor. All other instances of alternators interference are best contained with filters in the DC, power leads to the affected device and careful attention to ground

loop avoidance, and magnetic coupling due to proximity of "fat wires" to "skinny wires."

Magnetos and Other Ignition Systems: Obviously, a system designed to generate little bolts of lightning inside a cylinder has the potential for creating a lot of noise. Magnetos don't generally make direct connection to other systems, so by-in-large magneto noise tends to find its way around an airplane by radiation. The strongest potential for launching noises into the ether is enjoyed by the spark plug wires. Connecting an oscilloscope to a spark plug wire would allow you to observe a fast rising, high amplitude waveform which occurs repetitively on a spark-by-spark basis: ideal waveforms for harmonic content. When the task is to control radiated emissions from a conductor carrying a noise source, total containment is the technique of choice. Given the close proximity between magnetos and plugs, keeping plug wire noise corralled inside a shielded environment is relatively easy: just ground the shield at both ends. This is a special case where the shield is always grounded at both ends.

Spark plug wires on cars are not shielded. Why no noise? There is SOME noise but you're right, cars don't put out much radio noise from their ignition systems. In order for the plug wire to be an effective radiator of RF energy, its resistance needs to be pretty low (strands of copper are pretty good conductors). Let's look what happens to a spark plug wire that's hanging out in the open. From a radio frequency standpoint, that piece of wire could be a fair antenna at some frequency. If you excite the wire with a little lightning strike on one end, the wire can resonate at some basic fundamental and at integer harmonics of that fundamental. Given that plug wires on aircraft engines are generally 15 to 20 inches in length, resonances in the VHF spectrum may be anticipated. Indeed, if you looked at the noise spectrum coming off of an unshielded magneto system and given that all the wires are different lengths, you would expect to observe a broad spectrum of noise in the lower to mid VHF frequencies.

Now, let us suppose the copper wires are replaced with a resistance material. The wire's ability to function as an antenna is severely curtailed. Antennas at resonance have high current nodes at some point along their length. If we're trying to develop a current through a resistive material, the material simply warms up. Energy that might have been launched toward your comm antenna is instead converted to heat. Another way of looking at resistance wire is that it lowers the "Q" (quality) of what used to be a resonant and reasonably efficient radiator. Resistance plugs figure in this effect as well. Automotive ignition designers have worked hard at this and modern gasoline ignition systems can be relatively free of radiated noise in spite of their unshielded plug harnesses.

Magnetos are such sorry sources of high voltage energy that we are reluctant to waste ANY of it in resistance wiring so the brute force technique of total containment was adopted and, for the most part, works fairly well.

If you do have ignition noise in a radio, there are some ways to deduce the propagation mode. First, turn off the mags one at a time. Does the noise go away when only one of the mags is turned off? If so, the leakage is occurring in the one system. Disconnect the p-leads from both mags right at the rear of the housings and see if the noise goes away. If so, it's getting out through the p-lead wiring. If not, then the spark plug harnesses are suspect.

Plug harnesses may be electrically tested (continuity between end of center conductor and between ends of shield braid). Unhook individual plug wires and do a continuity check between the spark plug connector shell (attaches to shield) and engine crankcase. If the noise goes away when p-leads are disconnected, check the routing of p-lead wires with respect to wires of the victim system. Does the ignition noise change in intensity as the receiver's volume control is adjusted? If so, the noise is probably radiated and coming in through the antenna system along with desired signals. If the noise is constant with respect to volume control setting, the interference is probably getting into the system via system interconnect wiring. P-leads of magnetos are routinely shielded but many builders ground the shield at both ends... not a big potential problem in all-metal airplanes but we generally have to go out of our way to FIND a ground behind the panel on a composite ship. Quite often, this added ground wire from panel end of p-lead shields CAUSES more noise problems than it fixes. I prefer to ground p-leads on one end only, preferably to the magneto housing (case screw or coaxial wiring connector for the p-leads). I use the p-lead shield to PROVIDE a ground for the center GND terminal of the classic key-type magneto switches. This one-end-only grounding philosophy insures that equal-opposite currents flow in the p-lead shields thereby insuring the effectiveness of the shields as noise control measures.

"Grounding" p-lead shields at both ends gives rise to the possibility of ground loop conducted noise and increased levels of radiated noise from an otherwise quiet conductor. Finally, "grounding" p-lead shields at both ends makes them vulnerable to generating a lot of smoke when you attempt to crank an engine with a failed or inadvertently disconnected crankcase-to-air-frame ground braid.

Electronic ignition systems require DC power from the ship's distribution system; there's the added risk of conducting noises into the power system. Fortunately, electronic ignition systems don't require a lot of energy and draw typically two amps or less on a 14-volt system. Filtering of the DC power lead as it emerges from the ignition system housing is fairly easy to accomplish. Avoid installing ignition system components on the cabin side of the firewall. A short, flexible braid jumper should complete the ground between crankcase and the forward end of the stud on a firewall ground block. This short, low impedance

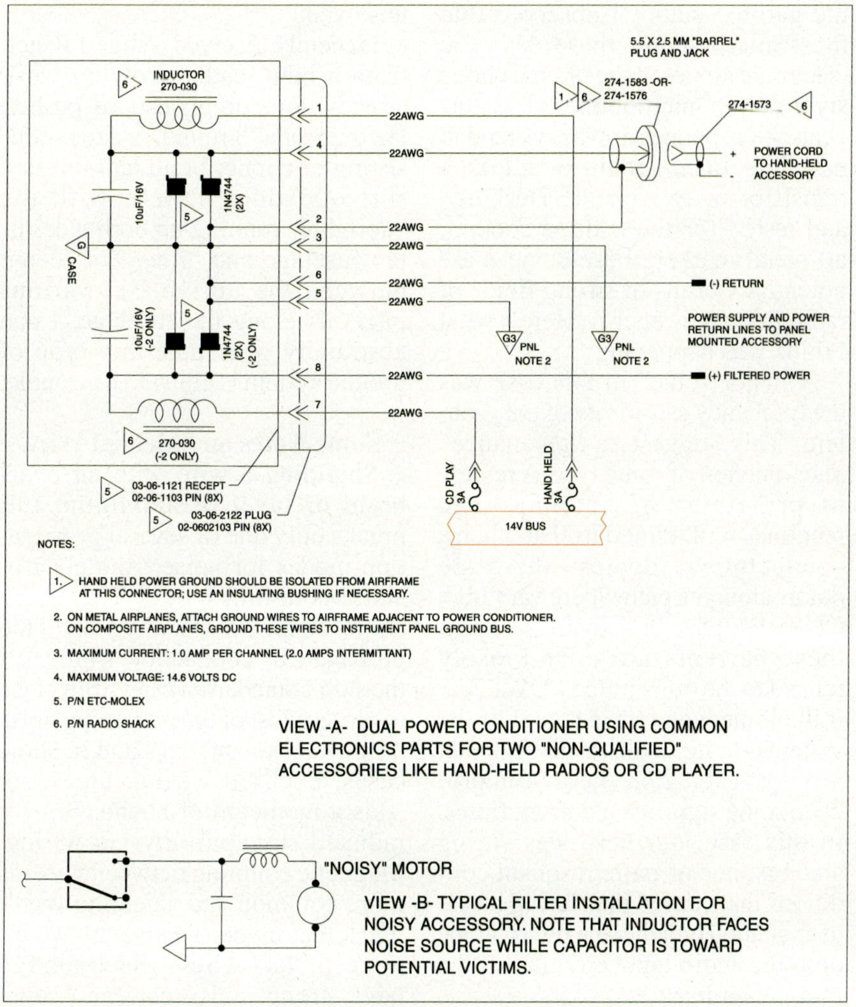

Figure 4

connection should be sufficient to keep the electronic ignition's electronics enclosure firmly "grounded" to the crankcase .

Many electronic ignition manufacturers suggest the use of stock automotive ignition wires and automotive plugs. This permits the user to take advantage of noise reduction techniques we've enjoyed in our cars for decades. The catalog of minor players expands out rapidly from here. Alternator and ignition noises have the distinction of being the biggest gorillas in the forest. Further, they run ALL the time we're in flight. The catalog of antagonists is more varied from here on out. They are less energetic noise sources. Many of them are low duty cycle (flap and gear motors), and a few require special conditions to manifest themselves. So in no special order, let's explore some additional antagonist/ propagation/victim combinations aboard airplanes.

DC motors generally have brushes running on slotted commutators that carry power into windings on the armature. They arc, they spark, and they'd sure like to make better contact with the copper surface of the commutator, but it won't hold still. Most motors on light planes are intermittent duty. They run just a few seconds per flight. I fly a number of rental Cessnas that tear up the radio during flap extension. Too small of a nuisance for the owners to mess with. Cabin vent or avionics cooling blowers might be another matter. Standard noise reduction techniques apply here. An inductor/ capacitor filter from Radio Shack was described earlier. Try a temporary installation as shown in *Figure 4*. Note the inductor faces the noise source, in this case, the motor. Make sure the motor in question does not exceed the filter's current ratings. Radio Shack and many merchants specializing in automotive audio systems have an arsenal of filters rated for various current levels. All will feature capacitors adequately rated for 14-volt operation. If you have a 28-volt airplane, be sure the filter's capacitors are rated for 30 volts or more.

Strobe light systems invariably contain transistorized power supplies designed to change low voltage DC into AC so that it can be rectified back to DC and used to power the xenon flash tubes. This power supply is a potential noise source and may be heard in the headphone on many airplanes as a sweeping, musical note that keeps time with the firing of the strobes. In metal airplanes, the strobe power supply should be grounded to airframe right where it mounts. Of course, I'd avoid mounting it on the avionics shelf with your intercom system! In a composite airplane, run power and ground wires for the strobe system parallel to each other as far as possible before splitting them apart to attach to control switch and panel ground bus. If you're hearing the sweeping audio tone in your radios, try a filter like I described for the motors.

Strobes often radiate a "pop" into the radios, especially nav radios where the omni antenna is co-located at the tip with a strobe light fixture. This interference is noticeable only while listening to the nav receiver's audio output. It generally doesn't affect the accuracy of the navigation receiver so most people don't fiddle with it. If the "pop" is finding its way into your comm radio, it may be that the comm antenna needs to be relocated. It's also possible that the shielded wire provided with most strobe installations is improperly grounded. Ground the shield only at the power supply end. On composite airplanes or metal airplanes with plastic wingtips, the shield is also connected to the strobe lamp housing. If the strobe lamp housing is mounted on a metal surface, leave the shield ground loose at the fixture end.

From here the list of potential antagonist gets pretty light weight.

Almost every device carrying an electrical current, intentional or otherwise, can become a source of interference but they're pretty weak sources. Fortunately, good wire routing and grounding practices can keep most sources from antagonizing another system.

Wiring and Installation Induced Susceptibilities

I got a call from a pilot who has been flying his composite airplane for about ten years and has put up with an interesting idiosyncrasy. Seems that when he transmits on some frequencies, other folk complain about garbled audio. Changing out the radio and other things he's tried will alter the frequencies involved but the problem is still basically there. I spoke with him at length and the following picture emerged. Like ALL composite airplanes, energy radiated from the comm antenna is pretty strong IN the cockpit. Composite airplanes have a design challenge in common, how to develop a good ground system. This particular airplane had a ground "system" that, for lack of a better term, was pretty elaborate. Lots of branches combined with an attempt to comply with common wisdom of the time, "everything that can be grounded should be grounded to every other thing that can be grounded." This statement paints a mental image of some amazing installations. I've seen airplanes where "ground" wires ran like spider webs throughout the structure without regard to the physics of what the ground wires could be expected to do. Some builders have run little wires out into the wings to attach their aileron hinges to "system ground." Generally speaking, the ONLY possible benefit this type of ground system might offer is to make sure that no static electricity potentials can exist between the various metallic components throughout the structure. Static discharge is ANOTHER topic for another article. It is sufficient to say here that this type of ground system at BEST creates few problems but at its worst, can produce some pretty bizarre effects.

Getting back to our builder with the garbled audio; I observed that most microphones these days are electronic clones of the old telephone style carbon microphones of yesteryear. Microphones nowadays tend to have integrated circuits or at least a transistor or two inside. Designed and tested for the padded cockpit, all-metal market, these devices are potential victims of strong fields of radio frequency energy. Here's what I think was happening:

A major "clue" in this case was the frequency sensitivity of the problem. This suggests a "resonance" phenomenon of some type. A review of principles of antennas and feedlines will remind us that when a conductor "resonates" there are points along its path where very high CURRENTS can prevail. Further, these currents are only loosely related to the transmitted POWER... if the impedance (resistance) of the system is quite small, relatively small levels of excitation energy can produce some significant current flows. In this case, our hero was sitting amongst one or more resonant conditions that caused interfering levels of excitation either to the microphone or to the audio input circuitry of the transceiver itself.

The offending resonator is probably part of the victimized system (VHF comm transceiver) - but not necessarily. A high quality resonating component as part of another system could be increasing general levels of energy within the cockpit. Recall that our discussion on antennas talked about passive radiators, pieces of an antenna that were not directly wired to the system. Nonetheless, these conductors could have a lot of radio frequency current flowing in them and they may exert a strong influence on the patterns of the energy's behavior. This is how antennas focus or "beam" energy in a particular direction. In this case, I believe wiring associated with a somewhat spider-webbed ground system was resonating at specific frequencies and giving his audio system fits. We discussed several experiments to see if the cause and/or a cure could be deduced but at the time of this writing, the case is unsolved.

I recently received a pencil sketch from a new reader showing work already done on his canard-pusher. He depicts a "ground" system consisting of copper braid laid out in a sort of ladder pattern with the "uprights" running up both sides of the fuselage and "rungs" soldered between the uprights at various intervals along the fuselage. I can absolutely guarantee this type of ground system is going to be trouble.

Some Notes on Shielded Wire

Shielding a wire with an overbraid or layer of aluminum foil breaks only one of several propagation modes for noise - the electrostatic coupling mode. The lion's share of all noises in an airplane propagate by conduction, riding ON the ship's standard wiring from place to place. Most of the remaining noise propagates to antennas and in some cases, exposed wiring. Shielding adds a modicum of attenuation for radiated susceptibility via wiring. Magnetic coupling between wires is more common and shielding won't break that mode. That's why we try to keep "fat" wires (which tend to have strong, noisy currents) separated from the "skinny" ones (which tend to be avionics and instrumentation). Unless the manufacturer's instructions specifically call out shielded wire for the installation of their product, you're safe in leaving the wires unshielded. Only magneto p-leads and wiring between strobe fixtures and their respective power supplies are ALWAYS shielded. Beyond that, it's generally limited to a few wires in the audio and navigation systems per installation instructions.

Really Off-the-Wall Problems

There's a lot of testing that goes on to improve on the probably of compatible operations when multiple systems coexist on an airplane. We've spoken often of DO-160 and Mil-Std-704 as common benchmarks for compatibility testing.

A new problem surfaced a few months back when it was discovered that the local oscillator in some nav receivers had strong enough 10th and 11th harmonics to interfere with GPS receivers in the same stack. GPS was just a dream when the nav transceivers were built - who would have thought this would be an issue ten years later? Try as we might, we cannot predict all the possibilities.

Rules of Thumb for Cleaner Living, Electrically!

General notes for wiring to reduce the possibility of wiring induced compatibility problems:

(1) Keep the ground system LINEAR and simple. By linear, I'll suggest that a single conductor, rated for cranking currents, run the length of the fuselage with a maximum of two extensions for under-the-cowl and behind-the-panel grounding.

(2) Common ground points should be provided for instrument panel hardware and under the cowl hardware. I'm going to relax my suggestion that the battery minus lead should run directly to the crankcase. This may be electrically ideal but it does generate another firewall penetration that's a hassle. So, a brass stud through the firewall may be used to carry cranking current grounds through the sheetmetal. Other than this, the grounding path recommendations in Chapter 5 are useful noise control techniques.

(3) Avionics and audio systems should not share wire bundles with DC power distribution. However, they can and should share common grounds behind the panel. With my favorite circuit protection (fuse blocks), I'd run DC power and control around the left side of the cockpit to accommodate switch panels on the pilot's side and keep audio/avionics wiring more or less amidships. Most avionics/instrument wiring is local to black boxes on the panel, except for antenna coaxes and extensions of audio wiring, the stuff tends to be compacted on center, to a little left of center on the panel.

(4) Engine instrumentation is pretty immune from cable conducted interference. However, it could become part of the propagation mode for bringing under the cowl noises into close proximity with audio/avionics, so plan separate bundle routing for these wires away from the radios.

(5) In metal airplanes, antennas get grounded to airframe, local to their mounting locations. In composite airplanes, artificial grounds are needed for 1/4 wave antennas while 1/2 wave antennas make their own. In any case, don't add extra ground wires from these antennas to the DC power ground system. The two are entirely separate systems and do not benefit from any form of interconnections.

(6) Avoid the "bad hair day" look in wire bundles. From a noise reduction standpoint, it's better to keep wires as close together as possible for as much length as possible. Many installations I've observed look like a tree, a fat wire bundle

CONTINUED ON PAGE 421

INTERNET CONTINUED FROM PAGE 45

Internet Glossary

Note - This glossary is designed to explain the common, everyday meanings of Internet jargon, and not its purely technical usage. Apologies in advance to all the Nerds and Geeks who may stumble upon this humble offering. About spelling, capitalize "Web" and "Internet" if you consider them to be proper names. It is no longer strictly necessary to capitalize these words or many other Web related words since they are quickly evolving into plain old nouns.

Browser - a client program that runs on a client computer like a Mac or PC in someone's home or office. A browser allows the user to view Web pages. The two most popular browsers at this time are Netscape Navigator and Microsoft Internet Explorer. Many Web sites are written to take advantage of the features of the latest versions of these two highly popular programs.Older browsers with limited features include Lynx, Cello and Mosaic.

Client - your computer and/or the related software that runs on your computer when it is connected to a network like the Internet.

Cookie - a piece of information put in a file on your computer that a particular Web site's server uses to identify you. The cookie repository is usually called the "cookie.txt" file or "cookie" folder in your computer. If you look at a cookie entry on your computer, (they are written in plain text), the URL of the server that put the cookie on your computer will identify the cookie's source. Cookies are not necessarily, or even usually, used for invasive purposes; but they may be used in negative ways — hence the concern. Turning the "cookie alert" on in your browser will notify you when you are getting a cookie and allow you to refuse a cookie, but the interruption can be truly annoying if you are trying to get anything done.

DNS - stands for domain name server. Most often seen in the phrase "DNS not found." The problem may be your ISP's name server, a very new domain name, a traffic jam, or some other temporary problem. Try the link several times before you give up. Try searching for a newer or different URL for the site you are trying to reach. If your modem seemed to connect normally, but you can't get any links to work at all, try contacting your ISP for help.

Domain name - the surname of a Web site. It is part of the URL. In the URL "http://www.aerokit.com/index.html" the "http" means it is a Web page, the "www.aerokit.com" is the domain name and "index.html" is the filename (at the time of this writing "aerokit.com" was not an active domain name.) Most newer browsers will accept the domain name without typing the "http://" as the URL and they will display the "index.html" page unless you specify a different filename.

Download - to receive information (data) that comes "down" into your computer (client) from the server. All Web pages that you see in your browser were downloaded to your computer.

Hit - a desired search result.

Home page - used loosely to mean "Web page" or "Web site." "Home" without the "page" usually means the "index.html" or main file of a Web site, but sometimes that too is called the "home page." "Home page" or "homepage" may also describe an individual's personal Web site housed on his ISP's server. An example of the URL for this kind of homepage is "http://ww.superpilot.net/~ace" (not a real URL.) In this example "superpilot.net" is the domain and "/~ace" is needed to get to the directory on the Superpilot Net server that houses Ace's Home Page. Don't worry about the "~" — it's a UNIX thing.

HTTP - stands for Hypertext Transfer Protocol. The "http" at the beginning of a URL says, "this is a Web page" to your browser and anyone else who cares.

Internet - a name for the WWW plus more. The Internet is one big, huge network that encompasses smaller networks and stand-alone servers alike. No one owns it, no one runs it. People only own and run their own parts of it. It changes from day to day. We can connect to it and use it. Don't try to understand it. Do try to make it a useful tool.

ISP - stands for "Internet service provider." An ISP is an entity, often commercial, that a user contracts with for Internet services like e-mail, a personal or business Web page, internet access and newsgroups. An ISP may be a local small business, a regional company, or it may be a large national computer service like CompuServe or America Online that offers Internet access as one of its services. In increasing numbers, traditional telephone companies are getting into the ISP business. Cable television companies are also jumping on the bandwagon. An ISP is the way most individuals access the Internet. Some other ways to connect are through an employer, government, or a school's direct Internet access.

Java - a Web page programming language. A browser can use a Java "applet" (program) to do strange and interesting things. Loosely related but different in functionality and form are Active X and JavaScript. Don't worry about the technical distinctions or how they work. If your browser can utilize Java, etc., just sit back and enjoy the show.

Jump page - a Web page or site that primarily consists of links to other pages or sites. Often seen under the heading of "cool links" or "hot links." The links on jump pages are usually arranged topically and/or alphabetically.

Link - short for "Hypertext link," "image link" or "image map" (a picture where each part section links to a different location). In today's browsers, the pointer cursor changes to a hand cursor when it is moved over a link. Clicking on a link will take the user to another Web page or file or to a particular place on a Web page. Links may lead to Web sites, multimedia files, and downloadable files. Many Web users think of links as the "click-on" words or pictures. Links can also be used by the browser with a mail client to send e-mail.

Modem - a piece of computer hardware that allows your computer to telephone and "talk" to other computers. Common modem speeds are 14.4 kbps (kilobytes per second), 28.8 kbps, and 33.6 kbps. For the Internet, faster is better. However you may not be able to connect a speeds as fast as your modem can if your phone line is noisy or if your ISP's modem is not as fast as your modem.

Network - a group of computers that are connected to each other. There are many interesting and mysterious ways to hook computers up. There are many different forms that a network may take. The Internet is just a great big, humongous network. Most individual computer users connect to the Internet via ordinary telephone line and modem.

Pathname - in this context, a long URL leading to a particular file or page. For example "http://www.aerokit.com/catalog/model/slickster.html" would lead the Web page file about the "Slickster" airplane kit which is in the "model" directory, which is in the "catalog" directory of the aerokit.com Web site. A useful trick is to truncate a URL. Say you were given the URL for the Slickster and wanted to see all the Aerokit designs you might try "http://www.aerokit.com/catalog/model/." That might get you to the jump page for all the Aerokit designs. If you wanted to see if Aerokit had a video of the Slickster for sale you could try "http://www.aerokit.com/catalog/" for the entire catalog. You can usually leave out the "http://" and the ending "/" when you type — the browsers will put them in for you. Don't be afraid to experiment with truncating URL's to find the information you want.

Search engine - a Web site that allows you to type words (keywords) into a form. The search engine server computer returns the URLs and often short descriptions of Web sites and Web pages related to that word. The best, but not the only words, for finding information about homebuilt airplanes seem to be "kitplane," and "homebuilt." The names of individual products and services often bring quick and rewarding results too. If you don't get results you like the first time, try a different search engine or slightly different words. Since each search engine works differently, consult the help section at the search engine Web site to learn how to maximize your return. Some search engines are specialized and catalog information related to one area of interest. Some search engines try to catalog the entire Internet. Current versions of MS Internet Explorer and Netscape Navigator have "search" somewhere on the menu or toolbar to get you started.

Server - a computer that is up and running and connected to the network all the time (unless it "crashes" or "goes down") to "serve" information to the client computers. Many Internet servers are run by businesses, schools and Internet service providers.

Suffix - the part of the domain name that tells you what kind of a domain it is. More suffixes have been proposed but at this time the common ones are .com (commercial), .org (organization), .gov (government), .edu (educational), and .mil (military). There are variations on this theme. There are also country suffixes like .uk (United Kingdom), se (Sweden), .us (United States), .kr (Korea), and so on.

Surfing (or Web surfing) - following links one to another, often pointlessly, for the sheer fun of seeing where they will lead. Occasionally surfing is used as a means of following leads to gather useful information and get real work done. Getting useful work done is not necessarily the best use of your personal computer, but is a good argument to use when begging a spouse, parent or business partner to agree to the purchase of a new machine.

Tag - short for "HTML tag." A coded instruction that tells a browser program how to display graphic, text and other files.

TCP/IP - the set of transfer rules (protocol) that allows computers to talk to each other over the Internet. Most of us don't need to know (or really care) how it works.

Title - the title of the Web page that appears across the top, or title bar, of most browsers. This is the name that the author or Webmaster has given the page and the name that will appear along with the URL in the results from most search engine queries.

Traffic jams - information moving along the Information Superhighway, a.k.a. the Internet, is subject to traffic jams. These jams occur especially during the rush hour which, in each respective time zone, begins roughly when school lets out in the afternoon and ends at bedtime. Downloads may be slower than the ground speed of a Piper Cub with a 40 kt headwind.

UNIX - a computer operating system that many businesses and ISP's use to run their servers. The very sound of the word strikes fear into the hearts of Windows and Mac users alike. However old DOS fans find the word vaguely comforting.

Upload - to send information (data) from your computer (client) "up" to the server. When you type a URL into your browser and hit "enter" you are uploading that information to the server which then downloads the requested information to your computer.

Web page - is a text (ASCII) file with special codes (tags) in it that the browser program uses to format text and files. Any plain text editor, in the hands of someone who knows the code, can be used to write a Web site. No mystery and no magic to it, but a little smoke and a few mirrors never hurt anyone.

Web ready - means a computer that comes with all software and hardware needed to access the Internet. Except for signing on with an ISP, entering the correct phone number, and plugging into power and phone lines, the computer really is Web ready. In this way, the user can connect to the Internet with a minimum time, energy, and computer know-how. You don't need to understand computers in order to use them.

Web site - is a Web page or pages at one address (URL) under the control of a Webmaster. The file for the main page of a Web site is often named "index.html". A Web site may be as small as one file or may consist of may files and subdirectories.

WWW (or World Wide Web or Web) - is a part of the Internet made up of many separate networks and servers. It was originally seen as the "commercial" side of the Internet without the academic or government use restriction. Domain names for Web sites on the World Wide Web start with www, www1, www2, etc. (a.k.a. the "World Wide Wait").

Gail Allinson is a WebSurfer of long standing, a commercial pilot, flight instructor, and an Earhart scholarship winner. She is half-owner of RST Engineering and a prolific Web Page author. Among her best Web pages are the RST catalog at www.rst-engr.com and her personal page at "www.gv.net/~gailla." The personal page has dozens of links that we didn't have room to list in this article and the on-line catalog is a very good example of compact, efficient use of the 'net.

Jim Weir is a regular contributor to the newsgroups "rec.aviation.*", a commercial pilot, flight instructor, and aviation safety counselor. He is half-owner of RST Engineering and the author of several hundred published aviation electronics articles. He is currently the chief designer for RST Engineering, a bimonthly contributor to KITPLANES magazine, and a teacher at the local community college.

See AeroCrafter's Web site at "http://www.baicorp.com/aerocrafter."

PROPS CONTINUED FROM PAGE 57

Around 1930, monoplanes began to replace biplanes for serious commercial and military flying. Their lower drag and more powerful and reliable engines allowed them to fly faster. Fixed pitch propellers then became a steadily increasing handicap. If given low pitch to allow engines to turn fast and deliver full power during takeoffs, cruising speed on long flights was held down. If given higher pitch for the sake of good cruising speed, takeoff performance with heavy loads aboard and from high altitude airfields was impaired.

Happily, work that went into the early Reed fixed pitch and Hamilton Standard ground adjustable propellers created a firm foundation from which to develop assorted hub types

The last two decades have seen tremendous growth in the field of homebuilt and ultralight aircraft. For simpler types of modest power and performance, the traditional two bladed, fixed pitch propeller still serves adequately and has the advantage of modest price.

Newcomers to sport aviation readily notice and are often bewildered by the number of blades the propellers of different planes have. They see two, three, four, five and six-bladers. If more is better, they assume, then six blades must be better than, for example, three blades.

But it's not as simple as that. As we remarked earlier, each airplane, engine and propeller combination addresses a particular engineering problem and must be analyzed accordingly.

There is a down side to the multiblade propeller, of course. It uses more of the engine's power to drive four blades through the air than two. Cost and weight go up. An old aviation saying is that the science of airplane design amounts to a great extent in knowing how to make acceptable compromises.

As you stroll about at a big fly-in, you will see many three bladed propellers. One pusher type ultralight might be fitted with one because the tips of a longer two blader might have inadequate clearance between its tips and the tail boom. Another might have three blades to absorb

Note glue lines indicating the location of small filler blocks in the hub area of this Cub's prop. To make economical use of expensive wood, it is common practice to this. If done expertly, strength is not affected.

the increased prop shaft torque developed by a reduction gear drive. Yet another might have three blades to provide more flywheel effect for the engine. And still another might have three blades to break up an objectionable vibration created by the interaction between the engine's power strokes and a two bladed prop.

Now, remember our admonition that pat answers are risky things. But very basically, diameter is chosen to suit engine power, with ground or aircraft structural clearances the usual limiting factors. When a fast turning propeller is fitted to a very fast airplane, blade tips will move through the air at velocities approaching the speed of sound. Then airflow begins to do strange things and efficiency can drop off. That is why many high performance, propeller driven, military planes had a reduction gear installed between their crankshafts and propellers, and why jet propulsion replaced propeller propulsion as the quest for ever higher speeds moved forward.

On the other hand, for the sake of combining very light weight with high power, many modern ultralight engines run at speeds in the 5000 to 6000 rpm range. Direct drive propellers fitted to them would be of almost comically small diameter. So to drive propellers of larger diameter, able to move more air, reduction drives are used.

Blade pitch is a function of propeller shaft rpm and aircraft forward speed. As speed varies considerably with throttle setting and angle of climb, the blades of fixed pitch props are given pitches that represent a workable and acceptable compromise. Usually good cruising speed is favored, but props of lower pitch are often available to install on planes flying out of short fields where quick takeoff and good climb are valued characteristics.

The subject of propellers is such a vast one that a very thick book could be written about it, and this is the important thing for the average sport flying enthusiast to realize. Where propellers are concerned, guesswork is out.

In the early days of homebuilt planes, it was common for people to make their own propellers. But today, most prefer to buy a ready-made prop from one of the several firms now in the business of making them to order. Most are rather small outfits but this isn't necessarily a bad thing. Such position to give customers individual attention. It makes sense to advantage of their knowledge, skill and equipment.

Our impression after meeting and talking to some of them and reading the literature of others is that they are sincere and responsible people. While attaching the propeller may be one of the last and quickest operations in assembling a homebuilt plane, it is in truth the key to good performance.

Bob Whittier, EAA 1235, has been writing for magazines for over 50 years. Since 1992 he has done the monthly "Light Plane Heritage" articles in the Experimental Aircraft association's EXPERIMENTER magazine.

Aircraft Name Index

Aircraft Name Index

EMC SOLUTIONS CONTINUED FROM PAGE 415

"trunk" with lots of fuzzy branches going everywhere. This installation technique is particularly vulnerable to radiated intrusion. When planning wire routing, envision a highway that routes close by every component in the system with short breakouts from the bundle to service each piece of equipment.

(7) Consider antenna locations with respect to their tasks. Receive antennas need to avoid proximity with noise sources, while transmit antennas should be located to minimize their coupling to potential victims. This is pretty easy in an all metal airplane: antennas on the outside are effectively shielded from stuff inside the airplane. Even graphite structures offer a high order of attenuation. The pure glass and plastic machines need careful attention to locations. Check with other builders and the editor of your type-newsletter for information on where/where not to install antennas.

The materials I've offered are necessarily limited to my own analysis and experience and cannot be all inclusive. Some of you are bound to discover some new combination with unacceptable effects. Identify the source and propagation mode. First, consider ways to control noise at the source. If the interference can be contained at the source, the possibility of multiple victims is eliminated. If the product producing the noise cannot be modified or filtered, then consider the ways to break or shield the propagation path and finally, consider ways to make the victim more tolerant or immune from the noise.

Observing the design and fabrication philosophies described here will go a long way toward elimination of noise possibilities from the outset. But use caution before you add any extraordinary features to your noise reduction campaign. A builder called me about seven years ago and described all the shielded wire and filters he'd added to his electrical system before he asked, "What else do I need?" "Gee, I dunno, have you got a noise problem?" "Oh no," says he, "I've not flown the airplane yet!" I didn't have the heart to tell him that he was probably carrying five to ten pounds too much weight and has spent hours of time installing materials and equipment that he probably didn't need.

Bob Nuckolls, ***The AeroElectric Connection***, 6936 Bainbridge Rd, Wichita, KS 67226-1008; (316) 685-8617, E-mail: nuckolls@aeroelectric.com.

Symbols

A

B

C

D

E

Q

R

S

T

You volunteer fellows do a lovely job here.
Now, to the flightline if you please, Tommy.
EXPERIMENTAL
RAY
RED 99

Hornet LV

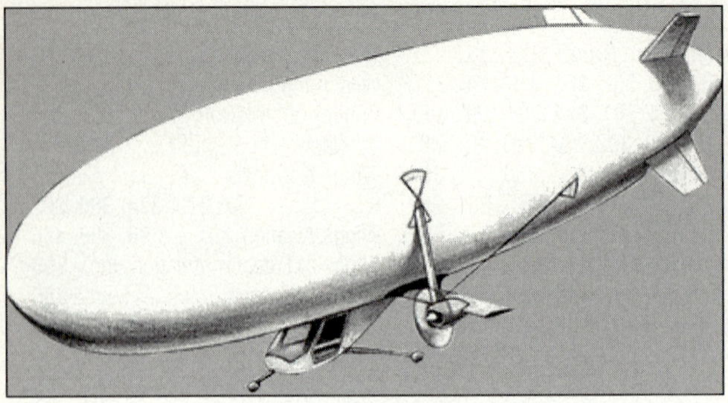

Wing Span	22.50 ft.	**Disposable Load**	204 lbs.
Length	50.00 ft.	**Range**	N/A
Diameter	12.5 ft.	**Vmax**	51 kts
Volume	4,300 cu. ft.	**Service Ceiling**	3,500 ft.
Max Gross Wt.	540 lbs.	**Seats**	1

Std. Engine/HP	Usable HP Range	Construction Mtrl.	Est. Bldg. Time
2 x Konig SC 430/24 hp	N/A	N/A	N/A.
Plans	**Kit**	**Info Packet**	**Video**
N/A	$25,000.00	Yes	No

Non-rigid, hybrid, buoyant aircraft. Available as a kit, to be assembled by the amateur builder so that he or she may operate under the FAA's Experimental category. Operates in a partially buoyant condition. One half of the total lift is provided aerostatically by the buoyancy of the helium filled envelope, while the other half is generated aerodynamically by the winglets attached to the sides of the envelope. With the tricycle landing gear it is feasible to operate the Hornet LV as if it were a conventional, winged aircraft. There is no need for a ground crew, except at the mooring mast. Flight control system is fly-by-wire.

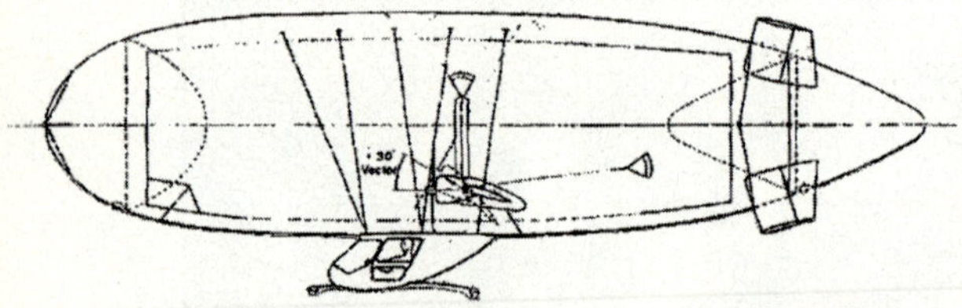

It is often common practice for builders of Experimental aircraft to carry out development modifications to their aircraft in order to improve safety, performance, and other characteristics.

AHA encourages owners of the Hornet LV to take the same approach. AHA has a policy to provide suitable kits for the completion of the aircraft in accordance with its original design. However, design variations are quite permissible, and these may yield an aircraft with special attributes, more suited to individual needs.

ADVANCED HYBRID AIRCRAFT
PO Box 144
Eugene OR 97440
Tel. 541-344-5323 • Fax 541-344-5323

Boland Balloons

Cube Balloon

A-3 Pocket Blimp

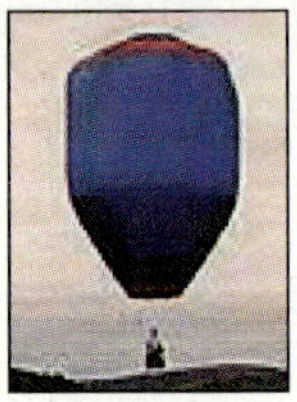

Jug Balloon

Plans are available for all of the Boland hot air balloons. In addition, they are available as Materials Kits, which include all necessary parts and materials plus a set of plans. Quick Build Kits include precut fabric panels. Balloon Building Camp available at the Post Mills Airport in Vermont.

Boland built his first Hot Air Airship in the mid seventies. It was rather large and heavy and carried four people. Since then, he has improved and simplified his designs. In order to foster the development of homebuilt, experimental, lighter-than-air craft, Brian Boland founded the Experimental Balloon and Airship Association (EBAA). The Association has an annual gathering at the Post Mills Airport.

BRIAN BOLAND
Post Mills Airport
PO Box 51 • Post Mills VT 05058
Tel: 802-333-9254

NOTES

Index to Advertisers

Company Name .... Page Number
.... Phone Number

Parting Shots

The Do-It-Yourself 747

"You Gotta Believe"

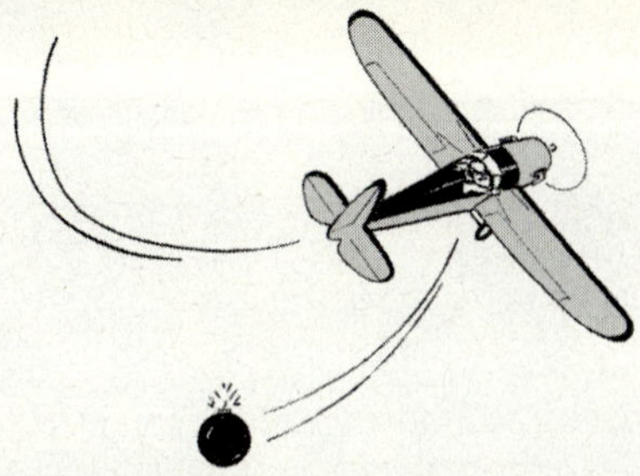

by Ron Wanttaja

The following was posted to the Internet Newsgroup ***rec.aviation.homebuilt*** *in reponse to a query about Boeing 747 plans. While Ron Wanttaja is a well respected writer of books and articles on homebuilt aircraft, this work indicates that lately he may have too much time on his hands. Ed.*

4hlk12@mda.com

"Anyone know if the construction plans for the Boeing 747 are available?"

wanttaja@halcyon.com

Boy, are YOU in luck!

A while back, I toured the 747 assembly plant in Everett, Washington. The Boeing 747 is a fantastic airplane, surely a marvel of our age. I was so impressed that I thought, "Hey, I wonder if ***I*** could build one of these?" So I went up to the tour guide and asked her if plans of the 747 were available.

"They sure are," she said.

"Great! How much?"

"Five hundred big ones."

Heck, Falco plans cost almost five hundred bucks, and here I'd be able to carry a lot more passengers a lot faster. So I handed over my AMEX gold card and went home with a complete set of "gen-u-wine" construction drawings for a Boeing 747-421.

Two days latter, I was sweeping my shop, getting ready for the big project. Not that it "really" needed cleaning, of course, but I figured the 747 would take a few clecos, and sweeping the floor is a good way to gather up the loose ones.

A FedEx truck drove up, with a statement from American Express. I nearly flipped. The plans for the 747 weren't five "hundred" dollars... Boeing had charged me "$500,000!" The plans did come in a nice set of mahogany file cabinets, but that doesn't make them worth half a million dollars.

I immediately called Boeing, of course. I offered to return them. "Uh-uh," they said. "How do we know you haven't been running Xerox™ copies for the last couple of days? If we take the plans back, that doesn't mean you won't be taxiing a brand new Wanttaja 747 out of your garage two years from now, does it?"

I could'a died! I settled down on the couch, stunned. It was just like the time I'd bought the Mrs. Field's recipe, and the Red Velvet Cake one before that. *How* can these companies keep taking advantage of the hard working American consumer?

But then it hit me: Heh, heh, I'll get back at Boeing, the same way I got my revenge on the cookie people: I'll POST the entire construction plans to the Internet!

So here is Part 1 of 17,321 on "Building Your Own 747":

For Part 1, you will need:

- 100,000 sq. ft of .040 Aluminum, alloy 7075T-6
- 2,000 gallons of BMS-142 Alodyne
- 40,000 Scotch-Brite™ pads
- 1 American LaFrance Pumper Truck, 500-gal. capacity
- 1 GMC Trent Division Street Sweeper, Model ST-401
- 1 pair tin snips
- 55 Gallons rubber cement

You'll also need a large, flat area (preferably indoors) about 400 feet square. If outdoors, you want *really* short grass.

Note: When buying the aluminum, make sure your rolls are "at least 12-feet wide." *Do not* let the vendor supply you with 12-inch-wide rolls! These are for gutters.

Step 1: Lay the Aluminum flat in the working area.

Step 2: Attach the Scotch-Brite™ pads to the buffers on the GMC Sweeper.

Step 3: Drive the sweeper over the alu minum, buffing to a dull sheen. Replace the Pads as necessary

Step 4: Load the alodyning solution into the Fire Truck and spray over the aluminum.

Step 5: Let stand 25 minutes

Step 6: Flush the alodyne off with tap water, let dry.

Step 7: Flip the aluminum over (you may want to have a friend help) and repeat.

Step 8: Print out the part layout diagram included at the end of this posting.

Step 9: Paste the layout diagram to the aluminum using the rubber cement.

Step 10: Cut the parts out using the tin snips.

Next Posting: Part 2 covers corrosion proofing. Keep your eyes open for used cattle troughs, etc., to help apply the zinc chromate.

PARTS LAYOUT PATTERN: The following pages will print out the outline of all the .040" aluminum parts for the 747 on the standard output. Numbers are printed in the corners of each page... tape the pages together, overlapping the appropriate numbers. ACCURACY IN THIS PHASE IS VITALLY IMPORTANT! Part numbers are also included.

```
        ==CUT TO THIS LINE==
|----------------------------------|
|----------/            \----------|
|Part D545-700211 Port Outer Rib #1|
```

Note: Due to space constraints, we were unable to include the remainder of the Port Outer Rib #1 Pattern. Readers who have started this project will find the pattern continued in the AeroCrafter 6th Edition. Ed.

AeroCrafter
HOMEBUILT AIRCRAFT SOURCEBOOK
Best Wishes and Safe Flying!
LUCIO
SHARON
SALINA
KIRK
REBECCA
K9
RANGER
TLM
RAY
AL
PT
YDK
JIM
SUSAN
SHERRY
DFW
DON
DALLAS